ELEMENTS OF STRING COSMOLOGY

The standard cosmological picture of our Universe emerging from a "big bang" leaves open many fundamental questions: Is the big bang a true physical singularity? What happens to the Universe at ultra-high energy densities when even gravity should be quantized? Has our cosmological history a finite or infinite past extension? Do we live in more than four space-time dimensions? String theory, a unified theory of all forces of nature, should be able to answer these questions.

This book contains a pedagogical introduction to the basic notions of string theory and cosmology. It describes the new possible scenarios suggested by string theory for the primordial evolution of our Universe. It discusses the main phenomenological consequences of these scenarios, stresses their differences from each other, and compares them with the more conventional models of inflation.

The first book dedicated to string cosmology, it summarizes over 15 years of research in this field and introduces current advances. The book is self-contained so it can be read by astrophysicists with no knowledge of string theory, and high-energy physicists with little understanding of cosmology. Detailed and explicit derivations of all the results presented provide a deeper appreciation of the subject.

MAURIZIO GASPERINI is Professor of Theoretical Physics in the Physics Department of the University of Bari, Italy. He has authored several publications on gravitational theory, high-energy physics and cosmology, and has twice received one of the Awards for Essays on Gravitation from the Gravity Research Foundation (1996 and 1998).

T0211225

ELEMENTS OF
STRING COSMOLOGY

MAURIZIO GASPERINI

University of Bari, Italy

CAMBRIDGE
UNIVERSITY PRESS

CAMBRIDGE UNIVERSITY PRESS
Cambridge, New York, Melbourne, Madrid, Cape Town,
Singapore, São Paulo, Delhi, Tokyo, Mexico City

Cambridge University Press
The Edinburgh Building, Cambridge CB2 8RU, UK

Published in the United States of America by Cambridge University Press, New York

www.cambridge.org
Information on this title: www.cambridge.org/9780521187985

First published 2007
First paperback edition 2011

A catalogue record for this publication is available from the British Library

ISBN 978-0-521-86875-4 Hardback
ISBN 978-0-521-18798-5 Paperback

To Patty
with gratitude

Contents

vii

Preface

The aim of this book is to provide an elementary, but detailed, introduction to the possible impact of string theory on the basic aspects of primordial cosmology. The content of the book includes a discussion of the new models of the Universe obtained by solving the string theory equations, as well as a systematic analysis of their phenomenological consequences, for a close comparison with more conventional inflationary scenarios based on the Einstein equations.

The book is primarily intended for graduate students, not necessarily equipped with a background knowledge of cosmology and string theory; but any reader in possession of the basic notions of general relativity and quantum field theory should be able to benefit from the use of this book (or, at least, of a great part of it). Some chapters (in particular, Chapters 1, 7 and 8) could also be used as a "soft" introduction to modern cosmology for string theorists, while other chapters (in particular, Chapters 2 and 3) as a soft introduction to string theory for astrophysicists; however, all readers are strongly advised to refer to other, more specialized books for a rigorous (independent) study of cosmology and string theory. It should be stressed, also, that this book *is not* aimed as a comprehensive and up-to-date review of all research work available in a string cosmology context: it only provides a pedagogic introduction to the basic ideas and theoretical tools, hopefully useful to the interested reader as a starting point towards more advanced research topics currently in progress in this field.

This book grew out of lectures given in May 2001 at the *First International Ph.D. Course on "Gravitational Physics and Astrophysics"*, jointly organized by the Universities of Berlin, Portsmouth, Potsdam, Salerno and Zurich. The style is that of class lectures: I have tried to be self-contained as much as possible, and I have not hesitated to insert many computational details and explanations, which may even appear to be trivial to the expert reader, but which may result in being of crucial importance for many students, as I have personally verified during the lectures. Besides organizing known material in a form appropriate to a pedagogic presentation, the book also presents explicit calculations never seen in the literature; in addition, it contains new results obtained through simple generalizations of previous studies. In particular, all topics are discussed (whenever possible) in the general context of a $(d+1)$-dimensional space-time manifold: known results in $d = 3$ are thus extended (some of them for the first time) to a generic number d of spatial dimensions.

A possible objection concerning the explicit absence of exercises and problems can be preempted by noting that the main text of the various chapters is literally "filled" with *solved* exercises, in the sense that all computations are displayed in full details, including all the explicit passages required for a reader's easy understanding. In view of such a large "equation density" in all sections and appendices, the inclusion of additional exercises seemed to be inappropriate.

Another warning concerns the appendices. In contrast with the common use of presenting technical details and computations (and with the exception of Appendix 2A), here the appendices are devoted to a self-contained discussion of specific topics which are closely related to the subject of the chapter, but which are not essential for the understanding of other chapters, and can be skipped in a first reading.

It should be explained, finally, why some chapters are characterized by a list of references much longer than others. The reason is that in some cases (for instance in Chapters 2 and 3) one can conveniently refer to existing books, which provide an excellent discussion of the subject; in other cases (for instance in Chapters 7, 8 and 10), no such book is presently available, and one has to resort to a more detailed bibliography with explicit references to the original papers on the subject.

Acknowledgements

It is a pleasure to thank all my collaborators from almost fifteen years of fun with string cosmology. In alphabetical order, they are: Luca Amendola, Nicola Bonasia, Valerio Bozza, Ram Brustein, Alessandra Buonanno, Cyril Cartier, Marco Cavaglià, Eugenio Coccia, Edmund Copeland, Giuseppe De Risi, Ruth Durrer, Massimo Giovannini, Michele Maggiore, Jnan Maharana, Kris Meissner, Slava Mukhanov, Stefano Nicotri, Federico Piazza, Roberto Ricci, Mairi Sakellariadou, Norma Sanchez, Carlo Ungarelli and Gabriele Veneziano. I would like to thank, in particular, Massimo Giovannini and Gabriele Veneziano for the numberless hours of stimulating discussions during which some of the topics discussed in this book literally grew "out of nothing".

Special thanks are due to Augusto Sagnotti, for important clarifying discussions; to Carlo Angelantonj, Carlo Ungarelli and my student Carla Coppola for their careful reading of some parts of the manuscript; to Giuseppe De Risi for his help in the preparation of the chapter on the branes; to my colleagues of the Physics Department of the University of Bari, and in particular to Leonardo Angelini, Antonio Marrone and Egidio Scrimieri who helped me improve some plots and figures. I also wish to thank all the staff of the Theory Unit at CERN for support and warm hospitality during the last stages of preparation of the final manuscript.

But, above all, I am greatly indebted to Gabriele Veneziano for his friendship and invaluable help and advice over all the years, for the generous sharing of his mastery of string theory, and for his crucial role in encouraging me to concentrate on the study of string cosmology scenarios. It is fair to say that this book would not exist without his basic input.

Finally, I would like to thank Simon Capelin (Publishing Director at Cambridge) for his kind advice and encouragement, Lindsay Barnes (Assistant Editor) for her prompt assistance, Margaret Patterson (Copy Editor) for her very careful correction of the original manuscript, and Jeanette Alfoldi (Production Editor) for her kind and efficient help during the final stages of the publishing process.

Notation, units and conventions

Unless otherwise stated, we adopt the following conventions:
- spatial indices: $i, j, k, \ldots = 1, \ldots, d$;
- space-time indices: $\mu, \nu, \alpha, \ldots = 0, 1, \ldots, d$;
- metric signature: $g_{\mu\nu} = \text{diag}(+, -, -, -, \cdots)$;
- Riemann tensor: $R_{\mu\nu\alpha}{}^{\beta} = \partial_\mu \Gamma_{\nu\alpha}{}^{\beta} + \Gamma_{\mu\rho}{}^{\beta} \Gamma_{\nu\alpha}{}^{\rho} - (\mu \leftrightarrow \nu)$;
- Ricci tensor: $R_{\nu\alpha} = R_{\mu\nu\alpha}{}^{\mu}$;
- covariant derivatives: $\nabla_\mu V^\alpha = \partial_\mu V^\alpha + \Gamma_{\mu\beta}{}^{\alpha} V^\beta$; $\quad \nabla_\mu V_\alpha = \partial_\mu V_\alpha - \Gamma_{\mu\alpha}{}^{\beta} V_\beta$.

Covariant objects are referred to the symmetric, metric-compatible Christoffel connection,

$$\Gamma_{\alpha\beta}{}^{\mu} = \frac{1}{2} g^{\mu\nu} \left(\partial_\alpha g_{\beta\nu} + \partial_\beta g_{\alpha\nu} - \partial_\nu g_{\alpha\beta} \right), \tag{1}$$

satisfying $\nabla_\alpha g_{\mu\nu} = 0$. We use natural units $\hbar = c = k_B = 1$, where k_B is the Boltzmann constant. The fundamental string mass, M_s, and string length, λ_s, are thus related to the string tension $T = (2\pi\alpha')^{-1}$ by

$$M_s^2 = \lambda_s^{-2} = (2\pi\alpha')^{-1}. \tag{2}$$

The four-dimensional (reduced) Planck mass M_P, and the Planck length λ_P, are related to the Newton constant G (in $d = 3$ spatial dimensions) by

$$M_P^2 = \lambda_P^{-2} = (8\pi G)^{-1}. \tag{3}$$

The current experimental value $G \simeq 6.709 \times 10^{-39} \, \text{GeV}^{-2}$ [1] then leads to

$$M_P = (8\pi G)^{-1/2} \simeq 2.43 \times 10^{18} \, \text{GeV} \tag{4}$$

(note the difference from an alternative – often used – definition, $M_P = G^{-1/2} \simeq 1.22 \times 10^{19} \, \text{GeV}$). In a manifold with $D = d + 1$ space-time dimensions Eq. (3) becomes

$$M_P^{d-1} = \lambda_P^{1-d} = (8\pi G_D)^{-1}, \tag{5}$$

where G_D is the D-dimensional gravitational coupling constant, and M_P, λ_P are gravitational scales, possibly different (in principle) from the numerical value (4) determined by four-dimensional phenomenology. If the geometry of the higher-dimensional manifold has a factorized, Kaluza–Klein structure, then G_D is related to the four-dimensional Newton constant G through the proper volume of the internal space $\mathcal{M}_{d-3}$ as follows:

$$(8\pi G_D)^{-1} V_{d-3} = (8\pi G)^{-1}, \qquad V_{d-3} = \int_{\mathcal{M}_{d-3}} \sqrt{|g|}\, d^{d-3}x. \qquad (6)$$

The relative strength of M_s and M_P is controlled by the scalar dilaton field ϕ, defined in such a way that, at the tree-level, and in d spatial dimensions,

$$\left(\frac{M_s}{M_P}\right)^{d-1} = e^{\phi}. \qquad (7)$$

Masses, energies and temperatures are usually expressed in eV (or multiples of eV), and distances in cm (or eV^{-1}), using the equivalence relations:

$$(1\,eV)^{-1} \simeq 1.97 \times 10^{-5}\, cm \simeq 6.59 \times 10^{-16}\, s \simeq 8.6 \times 10^{-5}\, kelvin^{-1}. \qquad (8)$$

The Planck length, defined as in Eq. (3), corresponds to

$$\lambda_P = (8\pi G)^{-1/2} \simeq 8.1 \times 10^{-33}\, cm. \qquad (9)$$

The curvature scale of the cosmological manifolds, parametrized by the Hubble parameter H, is often expressed in Planck units, and the energy densities in units of critical density $\rho_c = 3H^2/8\pi G$. For the present Universe, in particular,

$$H_0 = 3.2h \times 10^{-18}\, s^{-1} \simeq 8.7h \times 10^{-61} M_P, \qquad (10)$$

where $h = H_0/(100\ km\ s^{-1}\ Mpc^{-1})$. Recent observations suggest

$$h = 0.73^{+0.04}_{-0.03} \qquad (11)$$

as the current standard [1]. The corresponding critical density is

$$\rho_c(t_0) = \frac{3H_0^2}{8\pi G} = 3H_0^2 M_P^2 \simeq 1.88h^2 \times 10^{-29}\, g\ cm^{-3} \simeq 2.25h^2 \times 10^{-120} M_P^4. \qquad (12)$$

Reference

[1] Particle Data Group webpage at pdg.lbl.gov/

1

A short review of standard and inflationary cosmology

In this chapter we will recall some basic notions of standard and inflationary cosmology that will be used later, in a string cosmology context. We will assume that the reader is already familiar with the geometric formalism of the theory of general relativity, and with the main observational aspects of large-scale astronomy and astrophysics. We will discuss, in particular, the various assumptions of the so-called standard cosmological model, the problems associated with its initial conditions, and the basic aspects of its "inflationary" completion driven by the potential energy of a cosmic scalar field (further details on the inflationary scenario will be supplied in Chapter 8). This presentation aims at a self-contained study of the early cosmological dynamics: for a more detailed introduction, and a deeper analysis of the topics discussed in this chapter, we refer the interested reader to [1, 2, 3] for the standard cosmological model, and to [4, 5, 6] for the inflationary scenario.

1.1 The standard cosmological model

The standard cosmological model, developed during the second half of the last century, was inspired by two fundamental observational results: the recession of galaxies, discovered by Hubble [7], and the presence of the Cosmic Microwave Background (CMB), discovered by Penzias and Wilson [8]. The model relies upon a number of hypotheses – also motivated by direct and indirect observations – that we now list, with some illustrative discussion.

1.1.1 Einstein equations

The first assumption is that the gravitational interaction, on cosmological scales of distance, is well described by the classical theory of general relativity, and in particular by the equations derived from the effective four-dimensional action

$$S = -\frac{1}{16\pi\,G}\int d^4x\sqrt{-g}R + S_\Sigma + \int d^4x\sqrt{-g}\,\mathcal{L}_m. \tag{1.1}$$

Here S_Σ is the Gibbons–Hawking boundary term [9], required in order to reproduce the standard Einstein equations, and $\mathcal{L}_m$ is the Lagrangian density of the matter fields, acting as gravitational sources. The variation of the action (1.1) with respect to the metric $g_{\mu\nu}$ yields (see Chapter 2 for an explicit derivation)

$$G_{\mu\nu} \equiv R_{\mu\nu} - \frac{1}{2}g_{\mu\nu}R = 8\pi G\,T_{\mu\nu}, \tag{1.2}$$

where $G_{\mu\nu}$ is the so-called Einstein tensor, and $T_{\mu\nu}$ is the (dynamical) energy-momentum tensor of the matter sources, defined by the variation (or functional differentiation) of the matter action as

$$\delta_g\left(\sqrt{-g}\,\mathcal{L}_m\right) = \frac{1}{2}\sqrt{-g}\,T_{\mu\nu}\,\delta g^{\mu\nu}. \tag{1.3}$$

The right-hand side of Eq. (1.2) represents all the sources gravitationally coupled to the metric, and therefore includes the possible contribution of the vacuum energy density associated with a cosmological constant Λ, and described by the effective energy-momentum tensor $T_{\mu\nu} = \Lambda g_{\mu\nu}$.

1.1.2 Homogeneity and isotropy

A second assumption is that the spatial sections of the Universe, on large enough scales of distance, can be described as homogeneous and isotropic (three-dimensional) Riemann manifolds, geometrically represented by maximally symmetric spaces where rotations and translations form a six-parameter isometry group.

It may be noted that, on scales much smaller than the Hubble radius $H_0^{-1} \simeq 0.9h^{-1} \times 10^{28}$ cm, the distribution of visible matter seems to follow a "fractal" distribution (see for instance [10]), and that it is not very clear, at present, at which scale the (averaged) matter distribution becomes really homogeneous and isotropic. The hypothesis of homogeneity and isotropy refers, however, to the full set of cosmic gravitational sources (including, as we shall see, radiation, dark matter, dark energy, ...), and is quite powerful, since it allows a simplified cosmological description in which the space-time geometry can be parametrized by the so-called "comoving" chart (or set of coordinates). In that case, the fundamental space-time interval reduces to

$$ds^2 = b^2(t)dt^2 - a^2(t)d\sigma^2(\vec{r}), \tag{1.4}$$

where $a(t)$, $b(t)$ are generic functions of the time coordinate, and $d\sigma^2$ is the line-element of a three-dimensional space with constant (positive, negative or zero) curvature K. Using a set of stereographic coordinates $\{x_1, x_2, x_3\}$, the metric of such a maximally symmetric space can be parametrized as [1]

$$d\sigma^2 = dx_i \, dx^i + K \frac{(x_i \, dx^i)^2}{1 - K x_i x^i}, \tag{1.5}$$

where scalar products are performed with the Euclidean metric δ_{ij}.

An important property of the comoving chart is the fact that *static observers*, with four-velocity $u^\mu = (u^0, \vec{0})$, are also *geodesic observers*. The normalization condition $g_{\mu\nu} u^\mu u^\nu = 1$, with the metric (1.4), gives indeed $u^0 = b^{-1}(t)$ and

$$\frac{du^0}{d\tau} = -\frac{\dot{b}}{b^3}, \qquad \Gamma_{00}{}^0 (u^0)^2 = \frac{\dot{b}}{b^3}, \tag{1.6}$$

which implies that the field u^0 satisfies the geodesic equation

$$\frac{du^0}{d\tau} + \Gamma_{00}{}^0 (u^0)^2 = 0. \tag{1.7}$$

Here τ is the proper time (related to the coordinate time t by $d\tau = \sqrt{g_{00}} dt = b(t)dt$), and the dot denotes differentiation with respect to t. In addition, if $u^i = 0$, then

$$\frac{du^i}{d\tau} = -\Gamma_{00}^i (u^0)^2 = -\frac{1}{2b^2} g^{ij} \left(2 \partial_0 g_{j0} - \partial_j g_{00} \right) \equiv 0. \tag{1.8}$$

Thus, in the absence of non-gravitational forces, static observers are always at rest with respect to comoving coordinates, even if the geometry is time dependent.

The existence of such observers provides a natural reference frame for synchronizing clocks, and suggests the use of a convenient time coordinate, the so-called *cosmic time*, which corresponds to the proper time of the static observers. The choice of this time coordinate leads to the *synchronous gauge*, defined by the condition $g_{00} = 1$. It is also convenient to parametrize the maximally symmetric space of Eq. (1.5) with spherical coordinates $\{r, \theta, \varphi\}$. By setting $x_1 = r \sin\theta \cos\varphi$, $x_2 = r \sin\theta \sin\varphi$, $x_3 = r \cos\theta$, and differentiating to compute $d\sigma^2$, in the synchronous gauge of the comoving chart, one finally arrives at the well-known Robertson–Walker metric, defined by

$$ds^2 = dt^2 - a^2(t) \left[\frac{dr^2}{1 - K r^2} + r^2 (d\theta^2 + \sin^2\theta \, d\varphi^2) \right]. \tag{1.9}$$

Here t is the cosmic-time coordinate, and the constant K (with dimensions L^{-2}) controls the intrinsic curvature of the space-like $t = \text{const}$ hypersurfaces, representing three-dimensional sections of the space-time manifold. With our conventions the function $a(t)$, called the "scale factor", is dimensionless, while the comoving radial coordinate r has conventional dimensions of length.

Another choice of time coordinate (often used in this book) is the so-called *conformal gauge*, defined by the condition $g_{00} = a^2$. The time parameter of this gauge, usually denoted by η, is thus related to the cosmic time t by $dt = a \, d\eta$. The choice of the conformal gauge is particularly convenient for spatially flat manifolds ($K = 0$), whose metric can then be written in conformally flat form, using cartesian coordinates, as

$$ds^2 = a^2(\eta) \left(d\eta^2 - dx_i \, dx^i \right). \tag{1.10}$$

A space-time described by the Robertson–Walker metric is characterized by a number of interesting kinematical properties concerning the motion of test bodies and the propagation of signals (see for instance [1]). For the purposes of this book it will be enough to recall two effects.

The first effect concerns the spectral shift of a periodic signal, a shift originating from the well-known temporal slow-down produced by gravity. Indeed, at any given time t, all points of the three-dimensional spatial sections at constant curvature will be affected by exactly the same gravitational field, so that any local process will be equally slowed-down with respect to the same process occurring in the flat Minkowski space, quite independently of its spatial position. However, if the scale factor $a(t)$ varies with time, then the curvature radius of the spatial sections (and the associated intensity of the local effective gravitational field) will also vary with time. This will produce a difference in the local gravitational field (and in the local "slow-down") between the time $t_{\rm em}$ of emission of a periodic signal of pulsation $\omega_{\rm em}$, and the time $t_{\rm obs} > t_{\rm em}$ when the same signal is observed with pulsation $\omega_{\rm obs}$. The ratio of the two pulsations will be clearly proportional to the ratio of the local gravitational intensities at $t_{\rm em}$ and $t_{\rm obs}$, and thus inversely proportional to the spatial curvature radius.

For a more precise computation of the spectral shift $\omega_{\rm em}/\omega_{\rm obs}$ we may consider a photon of four-momentum p^μ, traveling along a null geodesic of a spatially flat Robertson–Walker metric. In the cosmic-time gauge such a null path has differential equation $dt = a \widehat{n}_i \, dx^i$, where $\widehat{n}$ is a unit vector ($|\widehat{n}| = 1$) specifying the photon direction; the null photon momentum is, in this gauge, $p^\mu = p^0(1, \widehat{n}^i/a)$, with $g_{\mu\nu} p^\mu p^\nu = 0$. The momentum is parallelly transported along the geodesic, and for the energy p^0 we have, in particular,

$$dp^0 = -\Gamma_{\alpha\beta}{}^0 \, dx^\alpha p^\beta = \Gamma_{ij}^0 \, dx^i p^j$$

$$= -\dot{a} p^0 \widehat{n}_i \, dx^i = -\frac{\dot{a}}{a} p^0 \, dt. \tag{1.11}$$

The integration gives $p^0 = \overline{\omega}/a(t)$, where the integration constant $\overline{\omega}$ represents the proper frequency of the photon in the Minkowski space locally tangent to the given cosmological manifold.

The local frequency measured by a static, comoving observer u^μ is thus time dependent, being determined by the projection $p^\mu u_\mu = \overline{\omega}/a(t)$. A photon emitted at $t = t_{em}$ and received at $t = t_{obs}$, even in the absence of a (possible) Doppler effect due to the relative motion of source and emitter, will be characterized by the spectral shift

$$\frac{\omega_{em}}{\omega_{obs}} = \frac{(p^\mu u_\mu)_{em}}{(p^\mu u_\mu)_{obs}} = \frac{a_{obs}}{a_{em}} \tag{1.12}$$

(see also Eqs. (8.172)–(8.173), and the discussion of Section 8.2). If the Universe is expanding, then $a_{obs} > a_{em}$ for $t_{obs} > t_{em}$, and the Robertson–Walker metric produces an effective redshift of the signals received from distant sources, i.e. $\omega_{obs} < \omega_{em}$. In particular, since observations are carried out at the present time, $t_{obs} = t_0$, it may be useful to introduce a redshift parameter $z(t)$ defined as

$$1 + z(t) = \frac{a(t_0)}{a(t)} \equiv \frac{a_0}{a(t)}, \tag{1.13}$$

which controls the relative "stretching" of the wavelengths of the received radiation,

$$z = \frac{\Delta\lambda}{\lambda} = \frac{\lambda_{obs} - \lambda_{em}}{\lambda_{em}}. \tag{1.14}$$

A second important feature of the Robertson–Walker kinematics, which we recall here for later applications, is the possible existence of "horizons", i.e. of surfaces with relevant causal properties. For any given observer we may consider, in particular, the *particle horizon*, which divides the portion of space-time already observed from the one yet to be observed, and the *event horizon*, which divides the observable portion of space-time from the one causally disconnected [11]. For their precise definition we must refer to the limiting times t_m and t_M corresponding, respectively, to the maximum *past* extension and *future* extension of the time coordinate on the given cosmological manifold.

Let us consider a signal propagating towards the origin along a null radial geodesic of the metric (1.9) ($ds^2 = 0$, $d\theta = 0 = d\varphi$), satisfying the equation $dt/a = dr/\sqrt{1 - Kr^2}$, and received by a comoving observer at rest at the origin of the polar coordinate system. A signal emitted from a radial position $r = r_1$, at a time $t = t_1$, will be received at $r = 0$ at a time $t = t_0 > t_1$, such that

$$\int_0^{r_1} \frac{dr}{\sqrt{1 - Kr^2}} = \int_{t_1}^{t_0} \frac{dt}{a(t)}. \tag{1.15}$$

The considered signal was emitted at a *proper* distance $d(t)$ from the origin which, at time t_0, is determined by

$$d(t_0) = a(t_0) \int_0^{r_1} \frac{dr}{\sqrt{1 - Kr^2}} = a(t_0) \int_{t_1}^{t_0} \frac{dt}{a(t)}. \tag{1.16}$$

In the limit $t_1 \to t_m$ we then define the "**particle horizon**", for the given observer at time t_0, as the spherical surface centered at the origin $r = 0$ with proper radius

$$d_p(t_0) = a(t_0) \int_{t_m}^{t_0} \frac{dt}{a(t)}. \tag{1.17}$$

This surface encloses the maximal portion of space physically accessible to direct observation from the origin of the coordinate system at the time t_0. Points located at a proper spatial distance $d > d_p(t_0)$ *cannot* be causally connected with the given observer at the given time t_0 (they may become causally connected at later times, at least in principle).

Consider now a radial signal emitted towards the origin at time t_0, from a point located at a comoving position r_2, and received at the origin at a time $t_2 > t_0$. The proper distance of the emitter from the origin, at time t_0, is then

$$d(t_0) = a(t_0) \int_0^{r_2} \frac{dr}{\sqrt{1 - Kr^2}} = a(t_0) \int_{t_0}^{t_2} \frac{dt}{a(t)}. \tag{1.18}$$

In the limit $t_2 \to t_M$ we can then define the "**event horizon**", at the time t_0, as the spherical surface centered at the origin with proper radius

$$d_e(t_0) = a(t_0) \int_{t_0}^{t_M} \frac{dt}{a(t)}. \tag{1.19}$$

Signals emitted from points located at a proper distance $d > d_e(t_0)$ will *never* be able to reach the origin. In other words, points with spatial separations $d > d_e$ will never become causally connected, even extending the time coordinate to the extremal future limit allowed by the given cosmological manifold.

The above horizons exist if the integrals of Eqs. (1.17) and (1.19) are convergent, of course. Consider, for instance, a cosmological solution describing a Universe expanding for ever from an initial singularity, and parametrized in cosmic time by the power-law scale factor $a(t) = t^\alpha$, with $\alpha > 0$, and $0 \le t \le \infty$: it can be easily checked that the particle horizon exists if $0 < \alpha < 1$, while the event horizon exists if $\alpha > 1$. For $\alpha = 1$ neither the particle horizon nor the event horizon exists. The definitions of horizon given here will be used in the following chapters, and will be applied in particular in Section 5.3 to illustrate some important differences between standard and string cosmology models of inflation.

1.1.3 Perfect fluid sources

A third assumption (or set of assumptions) of the standard cosmological model refers to the gravitational sources that we need to specify in order to solve the Einsten equations. According to the standard model the sources of the cosmological gravitational field on large scales, after averaging over possible spatial fluctuations, can be represented as a barotropic, perfect fluid with energy-momentum tensor

$$T_\mu^\nu = (\rho + p) u_\mu u^\nu - p \delta_\mu^\nu, \tag{1.20}$$

where the energy density ρ and pressure p depend only on time, and are related by the equation of state

$$\frac{p}{\rho} = \gamma = \text{const.} \tag{1.21}$$

In addition, the fluid is assumed to be at rest in the comoving frame. Thus, in the synchronous gauge, $u^\mu = (1, \vec{0})$ and T_μ^ν becomes diagonal,

$$T_0^0 = \rho(t), \qquad T_i^j = -p(t) \delta_i^j. \tag{1.22}$$

With the given sources we are now able to write explicitly the Einstein equations (1.2), using the following (more convenient, but equivalent) form:

$$R_\mu^\nu = 8\pi G \left(T_\mu^\nu - \frac{1}{2} T \delta_\mu^\nu \right). \tag{1.23}$$

For the Robertson–Walker metric (1.9) the non-zero components of the Ricci tensor, in mixed form, depend only on time, and are given by

$$R_1^1 = R_2^2 = R_3^3 = -\frac{\ddot{a}}{a} - 2 \left(H^2 + \frac{K}{a^2} \right),$$

$$R_0^0 = -3 \frac{\ddot{a}}{a}, \tag{1.24}$$

where $H = \dot{a}/a$ (the dot indicates the derivative with respect to cosmic time). The time and spatial components of Eqs. (1.23) then provide, respectively, the following independent equations:

$$\frac{\ddot{a}}{a} = -\frac{4\pi G}{3} (\rho + 3p),$$

$$\frac{\ddot{a}}{a} + 2 \left(H^2 + \frac{K}{a^2} \right) = 4\pi G (\rho - p). \tag{1.25}$$

Combining them in order to eliminate $\ddot{a}/a$, and differentiating the energy density ρ with respect to time, leads to the system of first-order differential equations:

$$H^2 + \frac{K}{a^2} = \frac{8\pi G}{3}\rho, \tag{1.26}$$

$$\dot{\rho} + 3H(\rho + p) = 0. \tag{1.27}$$

The last equation can also be directly obtained from the covariant conservation of the energy-momentum tensor, $\nabla_\nu T^\nu_\mu = 0$, which is a consequence of the contracted Bianchi identity $\nabla_\nu G^\nu_\mu = 0$ (see Eq. (1.2)).

In order to solve the above system of equations for the three unknown functions $a(t)$, $\rho(t)$, $p(t)$, it is necessary to use the equation of state $p = p(\rho)$, which in our case corresponds to the barotropic condition (1.21). In general, the gravitational sources of the standard cosmological model can be represented as a mixture of barotropic perfect fluids,

$$\rho = \sum_n \rho_n, \qquad p = \sum_n p_n, \qquad p_n = \gamma_n \rho_n, \tag{1.28}$$

with no energy transfer between the different fluid components, so that the energy-momentum tensor of each fluid is separately conserved. Equation (1.27) then yields, for each component,

$$\rho_n(t) = \rho_n(t_0)\left(\frac{a}{a_0}\right)^{-3(1+\gamma_n)}, \tag{1.29}$$

where $\rho_n(t_0)$ is an integration constant. Since the energy density of the different components has a different time behavior, the evolution of the Universe will then be characterized by different phases, each of them dominated by different fluid components.

In each cosmological phase the time evolution of the scale factor can be obtained by substituting Eq. (1.29) into (1.26), and solving the corresponding differential equation for $a(t)$. If, in particular, we are interested in the very early time evolution we can neglect the spatial curvature term (see below), and we obtain the scale factor

$$a_n(t) = \left(\frac{t}{t_0}\right)^{2/3(1+\gamma_n)}, \qquad \gamma_n \neq -1, \tag{1.30}$$

where t_0 is an integration constant. The case $\gamma_n = -1$ corresponds to the energy-momentum tensor of a cosmological constant

$$T^\nu_\mu = \Lambda \delta^\nu_\mu, \tag{1.31}$$

which describes an effective fluid with equation of state $p_n = -\rho_n = -\Lambda = \text{const}$ (see Eq. (1.22)). In this case Eq. (1.29) is still valid, and the integration of Eq. (1.26) (with $K = 0$) gives the exponential solution

$$a_n(t) = \exp[H(t - t_0)], \qquad H = \left(\frac{8\pi G\Lambda}{3}\right)^{1/2} = \text{const.} \qquad (1.32)$$

The standard cosmological model, in its original formulation [1], assumes that the cosmic fluid consists of two fundamental components: incoherent matter (ρ_m) with zero pressure $p_m = 0$, and radiation (ρ_r) with pressure $p_r = \rho_r/3$. The radiation component of the cosmic fluid represents the contribution of all massless (or very light) relativistic particles (photons, gravitons, neutrinos, ...), while the pressureless matter component takes into account the large-scale contribution of the macroscopic gravitational sources (galaxies, clusters, interstellar gas, ...), and the contribution of cosmic backgrounds of heavy, non-relativistic particles (baryons, as well as other, more exotic, possible dark-matter components). As we shall see later in more detail (see Eq. (1.39)), the present energy density of incoherent matter is roughly of the same order of magnitude as the critical density, $\rho_m(t_0) \sim \rho_c(t_0)$, where [12]

$$\rho_c(t_0) = \frac{3 H_0^2}{8\pi G} = 3 H_0^2 M_P^2 \simeq 2.25 h^2 \times 10^{-120} M_P^4, \qquad (1.33)$$

and is thus much greater than the radiation energy density today, since [12]

$$\rho_r(t_0) \simeq 4.15 h^{-2} \times 10^{-5} \rho_c(t_0). \qquad (1.34)$$

Therefore, according to the standard cosmological model, the present scale factor (assuming negligible spatial curvature) should evolve in time as $a(t) \sim t^{2/3}$.

As the Universe expands, however, the energy density of the matter component decreases in time as the inverse of the proper volume, $\rho_m \sim a^{-3}$, i.e. more slowly than the radiation component, $\rho_r \sim a^{-4}$ (see Eq. (1.29)). Going backwards in time one thus necessarily reaches the so-called *equality* time, $t = t_{eq}$, characterized by the same amount of matter and radiation energy density, $\rho_m(t_{eq}) = \rho_r(t_{eq})$. At earlier times, $t < t_{eq}$, the standard model then predicts the existence of a primordial phase where the radiation is the dominant component of the total energy density, and the scale factor evolves with different kinematics, $a(t) \sim t^{1/2}$, according to Eq. (1.30).

It is worth stressing that both the matter-dominated and the radiation-dominated regimes, according to the standard model, correspond to a phase of expansion which is *decelerated* and has *decreasing curvature*, i.e. satisfies

$$\dot{a} > 0, \qquad \ddot{a} < 0, \qquad \dot{H} < 0, \qquad (1.35)$$

as one can easily verify by differentiating Eq. (1.30) for $\gamma = 0$ and $\gamma = 1/3$ (with a power-law scale factor, we can take H as a good indicator of the time behavior of the space-time curvature scale). However, the recent large-scale observations concerning both the Hubble diagram of Type Ia Supernovae [13, 14] and the harmonic analysis of the CMB anisotropies [15, 16, 17] seem to indicate, with a growing level of precision and confidence [18, 19, 20], that the present Universe is undergoing a phase of accelerated expansion, $\ddot{a} > 0$.

Such observations are thus compatible with the first of Eqs. (1.25) only if the sources of cosmic gravity are presently dominated by a component with negative enough pressure (i.e. $\rho + 3p < 0$), so as to produce a kind of "cosmic repulsion" on large scales. Adding explicitly this new source ρ_q (dubbed "*quintessence*", or "*dark energy*") to the usual dust matter sources ρ_m, Eq. (1.26) becomes

$$H^2 + \frac{K}{a^2} = \frac{8\pi G}{3}(\rho_m + \rho_q),$$
(1.36)

where $\rho_q > \rho_m$, and $p_q/\rho_q \equiv \gamma_q < -1/3$. Dividing by H^2 we can then obtain a relation between the various components of the cosmic fluid in critical units, i.e.

$$1 = \Omega_m + \Omega_q + \Omega_K,$$
(1.37)

where

$$\Omega_m = \frac{\rho_m}{\rho_c}, \qquad \Omega_q = \frac{\rho_q}{\rho_c}, \qquad \Omega_K = -\frac{K}{a^2 H^2}.$$
(1.38)

The simplest model of dark energy is a cosmological constant, $\rho_q = \Lambda = \text{const}$ (which corresponds to $\gamma_q = -1$). In this case, replacing Ω_q with $\Omega_\Lambda = \Lambda/\rho_c$, the results of present observations can be summarized as follows [12]:

$$\Omega_m = 0.24^{+0.03}_{-0.04}, \qquad \Omega_\Lambda = 0.76^{+0.04}_{-0.06}.$$
(1.39)

These results refer to the particular case $K = 0$, but can be consistently applied to the present cosmological state where the allowed deviations of $\Omega_m + \Omega_\Lambda$ from 1 are very small: indeed,

$$\Omega_K = -0.015^{+0.020}_{-0.016}$$
(1.40)

according to a recent combination of supernovae and CMB data [20].

The experimental results are not very different from those of Eq. (1.39) even if ρ_q does not correspond to a cosmological constant, but represents the contribution of some weakly coupled, time-dependent field, as will be discussed in Section 9.3. In such a case, the effective equation of state $\gamma_q = p_q/\rho_q$ of the dark-energy component is presently constrained by the limits

$$\gamma_q = -0.97^{+0.07}_{-0.09},$$
(1.41)

obtained by combining supernovae and CMB data [20] (and assuming $K = 0$). In any case, it may be noted that the dilution of the dark-energy density due to the expansion of the Universe, $\rho_q \sim a^{-3(1+\gamma_q)}$, is much slower than the corresponding dilution of the matter component, $\rho_m \sim a^{-3}$. Therefore, going backward in time, the dominance of ρ_q and the associated cosmic acceleration tend to disappear quickly. In a decelerated Universe, on the other hand, the contribution of the spatial curvature decreases as $\Omega_K(t) \sim (aH)^{-2}$, going backward in time. Considering the present limits (1.40) we are thus fully entitled to neglect the spatial curvature during the early stages of the standard cosmological evolution.

It is also worth mentioning that the addition of ρ_q (with negative pressure) to the Einstein equation (1.36) may drastically change the conventional, well-known picture (see for instance [1]) where a Universe with positive spatial curvature $\Omega_K < 0$ (also called a "closed" Universe) will collapse in a finite time with a future "big crunch", while a Universe with $\Omega_K > 0$ (also called an "open" Universe) will expand forever. If $\Omega_q \neq 0$ there are indeed closed models with $\Omega_m + \Omega_q > 1$ which are of the "hyperbolic" type, and evergrowing, and open models with $\Omega_m + \Omega_q < 1$ which are of the "elliptic" type, and recollapsing. This possibility can be easily explored by assuming for instance $\rho_q = \Lambda$, and performing the numerical integration of Eq. (1.36) for various different initial values of ρ_m and ρ_q (see for instance [21]). If, in addition, $\gamma_q \neq -1$, and/or γ_q is time dependent, we can find different types of singularities eventually characterizing the future configuration of our Universe: *"big rip"* singularities [22] and *"sudden"* singularities [23].

In order to obtain experimental information on the parameters characterizing our present cosmological state, such as $\Omega_m(t_0)$, $\Omega_q(t_0)$, γ_q, H_0, we can use two important quantities which can be directly confronted with observations: (1) the so-called **"age of the Universe"**, t_0, and (2) the **luminosity distance**, $d_L(t_0)$.

(1) The first parameter t_0 simply (and more properly) represents the time scale of our present cosmological state, and can be defined starting from Eq. (1.36). Expressing the scale factor in terms of the redshift parameter (1.13), i.e. $a(t) = a_0(1+z)^{-1}$, and using the explicit time evolution (1.29) of the ρ_m, ρ_q components, Eq. (1.36) can then be recast in the form

$$(1+z)^{-2} \left(\frac{dz}{dt} \right)^2 = H^2(z), \tag{1.42}$$

where

$$H(z) = H_0 \left[\Omega_m(t_0)(1+z)^3 + \Omega_q(t_0)(1+z)^{3+3\gamma_q} + \Omega_K(t_0)(1+z)^2 \right]^{1/2}$$

$$= H_0(1+z) \left\{ 1 + z\Omega_m(t_0) + \Omega_q(t_0) \left[(1+z)^{1+3\gamma_q} - 1 \right] \right\}^{1/2}. \tag{1.43}$$

(We have used the definitions (1.38) and, in the second line of the equation, we have eliminated Ω_K through Eq. (1.37).) Let us now integrate Eq. (1.42) from $t = 0$ to $t = t_0$, assuming that $z \to \infty$ for $t \to 0$ (in other words, we are extrapolating the standard model up to the so-called "big bang" singularity $a = 0$ at $t = 0$). We obtain

$$t_0 = \int_0^\infty \frac{dz}{(1+z)H(z)}, \qquad (1.44)$$

which defines t_0 as a function of the four parameters H_0, $\Omega_m(t_0)$, $\Omega_q(t_0)$ and γ_q.

The precision of this definition can be improved by inserting into Eq. (1.43) the contribution of the radiation energy density, which scales as $\rho_r \sim (1+z)^4$, and becomes important at earlier epochs than equality (i.e. for $z \gtrsim 10^4$, see below). In any case, because of our ignorance about the very early cosmological evolution, one cannot determine t_0 from any direct observation; however, given the age of some component of our present Universe, one can put lower limits on t_0, and then derive indirect constraints on the dark-matter and dark-energy parameters.

(2) What can be directly measured is the correlation between the luminosity and the redshift of signals received from very distant sources. To obtain such a correlation we can consider a signal propagating towards the origin along a null radial geodesic, and satisfying the differential condition

$$\frac{dr}{\sqrt{1-Kr^2}} = \frac{dr}{[1+a_0^2 H_0^2 \Omega_K(t_0)r^2]^{1/2}} = \frac{dt}{a} = \frac{(1+z)}{a_0}dt = \frac{dz}{a_0 H(z)} \qquad (1.45)$$

(we have used the definitions of Ω_K and of z, and Eq. (1.42) for dt/dz). Suppose that the signal was emitted at a distance r from the origin: by integrating between 0 and r the first term of the above equation, taking into account the intrinsic sign of Ω_K, and using the elementary results

$$\int \frac{dx}{\sqrt{1+\alpha x^2}} = \begin{cases} x, & \alpha = 0, \\ \alpha^{-1/2} \sinh^{-1}(\sqrt{\alpha}\,x), & \alpha > 0, \\ \alpha^{-1/2} \sin^{-1}(\sqrt{\alpha}\,x), & \alpha < 0, \end{cases} \qquad (1.46)$$

we can then obtain from Eq. (1.45) the comoving distance of the source as a function of the redshift of the received signal, $r(z)$, as follows:

$$a_0 r(z) = \begin{cases} \int_0^z \frac{dz'}{H(z')}, & K = 0, \\ H_0^{-1} |\Omega_K|^{-1/2} \sinh\left[H_0 |\Omega_K|^{1/2} \int_0^z \frac{dz'}{H(z')}\right], & K < 0, \\ H_0^{-1} |\Omega_K|^{-1/2} \sin\left[H_0 |\Omega_K|^{1/2} \int_0^z \frac{dz'}{H(z')}\right], & K > 0, \end{cases} \qquad (1.47)$$

where $H(z)$ is defined by Eq. (1.43).

The above relation for $r(z)$ cannot be directly applied to observations, however, because we do not know the comoving radial distance of the various astrophysical sources. For pratical applications we must use, instead of the radial distance, the notion of apparent and absolute magnitude, commonly used to set the distance scales of astronomical observations. Let us consider, for this purpose, a source of massless radiation located at a distance r_{em} from the origin, with absolute emitting power (or luminosity) $L_{em} = (dE/dt)_{em}$. The energy flux received at $r = 0$, per unit of time and surface, at time $t = t_0$, is then given by

$$L_0 = \frac{1}{4\pi d_0^2}\left(\frac{dE}{dt}\right)_0, \tag{1.48}$$

where

$$d_0 \equiv d(t_0) = a(t_0)\int_0^{r_{em}}\frac{dr}{\sqrt{1 - Kr^2}} \equiv a_0\, r_{em}(z) \tag{1.49}$$

is the proper distance of the source at time t_0, expressed as in Eq. (1.47). Because of the frequency shift (1.12), the received energy will be shifted by the factor $(dE)_0 = (dE)_{em}(a_{em}/a_0)$. The time intervals will also be shifted, for the same reason, as $(dt)_0 = (dt)_{em}(a_0/a_{em})$. Taking into account the total shift of the received power, we can thus express the apparent luminosity L_0, for a source at a distance r, at the time t_0, as

$$L_0 = \frac{L_{em}}{4\pi d_0^2}\left(\frac{a_{em}}{a_0}\right)^2 = \frac{L_{em}}{4\pi a_0^2 r_{em}^2(1 + z_{em})^2}. \tag{1.50}$$

We can now introduce the so-called "luminosity distance" of the source, defined as the proper distance $d_L(z)$ such that $L_{em}/L_0 = 4\pi d_L^2(z)$. For a source located at a distance $r(z)$ we obtain, using Eq. (1.47),

$$d_L(z) = a_0\, r(z)(1 + z) = \frac{(1 + z)}{H_0 |\Omega_K|^{1/2}}\,\mathcal{F}\left[H_0 |\Omega_K|^{1/2}\int_0^z\frac{dz'}{H(z')}\right], \tag{1.51}$$

where the function $\mathcal{F}$ is defined as $\mathcal{F}(x) = \sinh x$ if $\Omega_K > 0$, $\mathcal{F}(x) = \sin x$ if $\Omega_K < 0$, and $\mathcal{F}(x) = x$ if $\Omega_K = 0$. The conventional astronomical unit of luminosity is the *apparent magnitude m*, defined by

$$m = -2.5\log_{10} L_0 + \text{const}, \tag{1.52}$$

where the constant is conventionally fixed by defining the apparent magnitude of the pole star to be $m = 2.15$. Comparing Eqs. (1.50) and (1.52) we finally obtain

$$m(z) = 5\log_{10} d_L(z) + c_M, \tag{1.53}$$

where c_M is a z-independent quantity related to the absolute magnitude M (i.e. to the absolute luminosity) of the source (see also [1]).

Fitting the experimental data for $m(z)$ in terms of the curves generated by the theoretical predictions (1.51) and (1.53) it becomes possible, in principle, to determine the parameters H_0, $\Omega_m(t_0)$, $\Omega_q(t_0)$ and γ_q contained in $H(z)$. This analysis is usually performed using the luminosity distance of the various models, computing the so-called "distance modulus", i.e. the difference between the apparent and absolute magnitude $m - M$, and finally plotting the difference $\Delta(m - M)$ between the distance modulus of a given model and the distance modulus of the hyperbolic empty model with $\Omega_m = 0 = \Omega_q$.

This last special model is characterized by $\Omega_K = 1$, and corresponds to the well-known Milne parametrization of the globally flat Minkowski space, represented in Robertson–Walker form by a linear (cosmic time) evolution, $a = t/t_0$, and by spatial sections with constant negative curvature $K = -1/t_0^2$. The luminosity distance for such a model, according to Eqs. (1.51) and (1.43), is given by

$$d_L^0(z) = \frac{(1+z)}{H_0} \sinh \ln(1+z) = \frac{z(2+z)}{2H_0}. \tag{1.54}$$

A convenient phenomenological representation of the distance–redshift relation is then obtained through the variable

$$\Delta(m - M) = 5\log_{10} d_L(z) - 5\log_{10} d_L^0(z)$$

$$= 5\log_{10} \left\{ \frac{2(1+z)}{z(2+z)|\Omega_K|^{1/2}} \mathcal{F}\left[H_0 |\Omega_K|^{1/2} \int_0^z \frac{dz'}{H(z')} \right] \right\}, \tag{1.55}$$

where $H(z)$ is given by Eq. (1.43). Note that such a relation can be easily extended to a generic model containing an arbitrary number of sources $\rho_n(t)$, evolving independently according to Eq. (1.29), provided we replace $H(z)$ with the more general expression

$$H(z) = H_0 \left[\Omega_K(t_0)(1+z)^2 + \sum_n \Omega_n(t_0)(1+z)^{3+3\gamma_n} \right]^{1/2}. \tag{1.56}$$

1.1.4 Thermal equilibrium

Another important assumption of the standard cosmological model concerns the spectral distribution of the radiation fluid. Following present observational evidence, the radiation is assumed to be in a state of thermodynamic equilibrium at a proper temperature T, with a Planck or Fermi–Dirac distribution according to the bosonic or fermionic character of the various radiation components. The energy distribution, per unit volume and per unit logarithmic frequency, can then be written in the form

$$\frac{d\rho(\omega, t)}{d\log\omega} \equiv \omega \frac{d\rho}{d\omega} = \frac{N}{2\pi^2}\frac{\omega^4}{e^{\omega/T}\pm 1} \tag{1.57}$$

(see [1, 5]), where the + and − signs correspond to the fermionic and bosonic cases, respectively, and N is the number of independent polarization states (for instance, $N = 2$ for photons, ultrarelativistic electrons and positrons, $N = 1$ for any relativistic neutrino/antineutrino species). Integrating over all modes we obtain the total energy density, which is given by

$$\rho_b(t) = \frac{\pi^2}{30}N_b T_b^4 \tag{1.58}$$

for pure bosonic radiation, and by

$$\rho_f(t) = \frac{7}{8}\frac{\pi^2}{30}N_f T_f^4 \tag{1.59}$$

for pure fermionic radiation. For a thermal mixture of N_b bosonic and N_f fermionic states the total energy density can then be written as

$$\rho_r = \frac{\pi^2}{30}N_\star T^4, \tag{1.60}$$

where

$$N_\star = \sum_b N_b \left(\frac{T_b}{T}\right)^4 + \frac{7}{8}\sum_f N_f \left(\frac{T_f}{T}\right)^4 \tag{1.61}$$

is the total effective number of degrees of freedom in thermal equilibrium at temperature T.

It may be useful, for later applications, to recall that the entropy S of a system in thermal equilibrium at temperature T, characterized by proper volume $V \sim a^3$, pressure p, and energy density ρ, must satisfy the differential thermodynamic condition

$$dS = \frac{1}{T}[d(\rho V) + p\,dV]. \tag{1.62}$$

It follows that the thermal entropy S is exactly conserved during the standard cosmological evolution, thanks to the conservation equation (1.27), which (multiplied by V) can be rewritten in differential form as

$$V\,d\rho = -3\frac{da}{a}(\rho + p)V = -(\rho + p)dV. \tag{1.63}$$

Substituting into Eq. (1.62) this condition leads in fact to $dS = 0$, which implies an exact adiabatic evolution for each decoupled component of the cosmic fluid in thermal equilibrium.

Using T and V as independent variables, differentiating Eq. (1.62) twice, and imposing the integrability condition $\partial^2 S/\partial V\,\partial T = \partial^2 S/\partial T\,\partial V$, one also obtains [1]

$$T\,dp = (\rho + p)dT\,, \tag{1.64}$$

which allows one to rewrite Eq. (1.62) as

$$dS = \frac{1}{T}[d(\rho V) + d(pV) - V\,dp] = d\left[\frac{V}{T}(\rho + p)\right]. \tag{1.65}$$

The integration provides the entropy density σ, for a generic equation of state $p = \gamma\rho$,

$$\sigma = \frac{S(T, V)}{V} = \frac{1+\gamma}{T}\rho. \tag{1.66}$$

For a radiation fluid, in particular, $\gamma = 1/3$ and $\sigma = 4\rho/3\,T$, a result valid for bosons as well as for fermions. For a thermal mixture, with N_b bosonic and N_f fermionic states, we can use Eqs. (1.58) and (1.59) to obtain

$$\sigma_r(t) = \frac{2\pi^2}{45}g_\star T^3\,, \tag{1.67}$$

where

$$g_\star = \sum_b N_b\left(\frac{T_b}{T}\right)^3 + \frac{7}{8}\sum_f N_f\left(\frac{T_f}{T}\right)^3 \tag{1.68}$$

is the effective number of degrees of freedom contributing to the thermal entropy density at a given time t. It is important to stress that this number, as well as the number $N_\star$ of Eq. (1.61), is in general time dependent in a cosmological context, and that a change in $g_\star$ (due for instance to the disappearance of some degrees of freedom from the thermal mixture) must necessarily be accompanied by a corresponding variation of the temperature, in order for the total entropy to be conserved.

The direct integration of Eq. (1.64) provides another important relation between ρ and T for a fluid in thermal equilibrium, namely

$$T \sim \rho^{\gamma/(1+\gamma)}. \tag{1.69}$$

The time evolution of ρ, on the other hand, is determined by the solution (1.29) of the conservation equation. The combination of these two results implies that the proper temperature of the thermal mixture is not a constant in the Robertson–Walker geometry, but evolves in time as

$$T(t) \sim a^{-3\gamma}. \tag{1.70}$$

A radiation fluid, in particular, has $\gamma = 1/3$ and $T(t) \sim a^{-1}(t)$. It follows that the radiation temperature is redshifted by the cosmological expansion exactly as the proper frequency $\omega(t)$ (see Eq. (1.12)). The same conclusion can be reached by combining Eqs. (1.60) and (1.29).

Thus, although the radiation becomes colder because of the expansion, the ratio ω/T is constant, and *the shape* (1.57) of the (bosonic and fermionic) spectral distributions does not change in time (even if the overall height of the peak of the distributions decreases, being controlled by T^4). This means that the condition of thermal equilibrium is preserved in the course of the standard cosmological expansion: as a consequence, one can take the CMB temperature T_γ as a useful evolution parameter – like the cosmic time, or the space-time curvature radius – to which to refer the various phases of the history of our Universe. Using as a reference the present value of the CMB temperature [12],

$$T_0 \equiv T_\gamma(t_0) = 2.725 \pm 0.001 \, \text{K} \sim 2.3 \times 10^{-4} \, \text{eV}, \tag{1.71}$$

one can compute, for instance, the temperature at the epoch of matter–radiation equality, $t = t_{eq}$, when $\rho_m = \rho_r$. From the Einstein equations we have $\rho_r/\rho_m \sim a^{-1}(t)$, hence,

$$\frac{\rho_r(t_0)/\rho_m(t_0)}{\rho_r(t_{eq})/\rho_m(t_{eq})} = \frac{a_{eq}}{a_0} = \frac{1}{1+z_{eq}}. \tag{1.72}$$

From the condition of thermodynamic equilibrium, on the other hand, $(1+z_{eq})^{-1} = T_0/T_{eq}$. We can therefore write

$$T_{eq} = T_0 \, (1+z_{eq}) = T_0 \, \frac{\rho_m(t_0)}{\rho_r(t_0)}$$

$$\simeq 0.7 \times 10^4 T_0 h^2 \left(\frac{\Omega_m}{0.3}\right) \simeq 1.6 \, h^2 \left(\frac{\Omega_m}{0.3}\right) \text{eV}$$

$$\simeq 2 \times 10^4 h^2 \left(\frac{\Omega_m}{0.3}\right) \text{K}, \tag{1.73}$$

where we have used Eqs. (1.34) and (1.71), and a typical value of the matter density suggested by the present data (see Eq. (1.39)).

During the radiation-dominated epoch the temperature is directly related to another important evolution parameter, the curvature scale $H(t)$. Indeed, using Eqs. (1.26) and (1.60), and neglecting the contribution of the spatial curvature, one obtains

$$H(t) = \left(\frac{\pi^2 N_\star}{90}\right)^{1/2} \frac{T^2(t)}{M_P}. \tag{1.74}$$

As a simple application, also useful for future discussions, this equation can be used to estimate the curvature scale at the epoch of matter–radiation equality, when the radiation temperature is given by Eq. (1.73).

For a precise computation of N_* we need to take into account that the cosmic radiation fluid at $t = t_{eq}$, according to the standard model of particle interactions, should contain two bosonic degrees of freedom, associated with the polarization states of the photon, and six fermionic degrees of freedom, associated with the three neutrino flavors and the corresponding antineutrinos (we are neglecting other, possibly sub-leading, contributions such as that of gravitons, see Chapter 7). However, neutrinos are slightly colder than photons, since the photon gas has been heated up by the annihilation of the electron/positron pairs taking place well before t_{eq}, at a temperature of about 0.43 MeV [1]. Indeed, at the epoch of electron annihilation, the conservation of the entropy associated with the thermal mixture of photons (γ) and electron/positron ($e^\pm$) pairs has caused a jump in the electromagnetic temperature, from an initial value identical to the neutrino temperature, $T = T_\nu$, to a new value $T = T_\gamma$ such that $\sigma(\gamma) = \sigma(\gamma, e^\pm)$. Taking into account that the $e^\pm$ pairs contribute to the fermionic degrees of freedom before annihilation, and using Eqs. (1.67) and (1.68), one obtains from entropy conservation

$$g_*(\gamma) T_\gamma^3 = 2 T_\gamma^3 = g_*(\gamma, e^\pm) T_\nu^3 = \left(2 + \frac{7}{8} \times 4 \right) T_\nu^3, \tag{1.75}$$

from which

$$T_\gamma = \left(\frac{11}{4} \right)^{1/3} T_\nu. \tag{1.76}$$

Let us now consider the total radiation energy density (1.60), where we take T_γ as the reference temperature. After the $e^\pm$ annihilation epoch we have $N_b = 2$, $N_f = 6$, so that

$$N_* = \sum_b N_b + \frac{7}{8} \sum_f N_f \left(\frac{T_\nu}{T_\gamma} \right)^4 = 2 + \frac{42}{8} \left(\frac{4}{11} \right)^{4/3} \simeq 3.36. \tag{1.77}$$

At the time of matter–radiation equality, neglecting the spatial curvature and a possible dark-energy contribution, the total energy density is $\rho(t_{eq}) = \rho_m + \rho_r = 2\rho_r(t_{eq})$. Using Eqs. (1.60), (1.73) and (1.77) one finally obtains

$$H_{eq} = \left(\frac{3.36\pi^2}{45} \right)^{1/2} \frac{T_{eq}^2}{M_P} \simeq 3.7 \times 10^{-55} h^4 \left(\frac{\Omega_m}{0.3} \right)^2 M_P. \tag{1.78}$$

This value is still a very tiny fraction of the Planck mass, but is nevertheless much greater than the present curvature scale [12]:

$$H_0 \simeq 8.7 h \times 10^{-61} M_P \simeq 2.35 \times 10^{-6} H_{eq} \left(\frac{0.3}{\Omega_m} \right)^2 h^{-3}. \tag{1.79}$$

The two curvatures H_0 and H_{eq} will be used as convenient reference scales in the following chapters.

The standard cosmological model provides a detailed thermal history of the Universe [1, 5], and suggests an evolution scenario where an initially hot, dense and highly curved configuration expands, becoming cooler and flatter. This scenario is in excellent agreement with important observational data referring to our present cosmological state, such as those concerning the galactic recession velocities and the relic background of microwave radiation. It is also consistent with the primordial mechanisms of nucleosynthesis and baryogenesis, which can only take place in the presence of a sufficiently high temperature.

The cosmological solutions of the standard model, however, cannot be extended indefinitely backward in time. In the radiation-dominated solution, for instance, the energy density and the temperature diverge at a fixed instant of time, conventionally chosen to coincide with $t = 0$:

$$t \to 0 \quad \Rightarrow \quad \rho \sim T^4 \sim a^{-4} \sim t^{-2} \to \infty. \tag{1.80}$$

At the same instant of time the curvature invariants also diverge:

$$t \to 0 \quad \Rightarrow \quad \left(R_{\mu\nu}R^{\mu\nu}\right)^{1/2} \sim \left(R_{\mu\nu\alpha\beta}R^{\mu\nu\alpha\beta}\right)^{1/2} \sim H^2 \sim t^{-2} \to \infty. \tag{1.81}$$

This singularity is a consequence of rigorous theorems formulated within the theory of general relativity (see for instance [24]), and cannot be removed even by abandoning the symmetry hypotheses underlying the Robertson–Walker metric (see for instance the discussion of the Kasner solution in [2]). It can be shown, in particular, that a geodesic time-like curve of the standard model, evolved backward in time from any given finite epoch t_0, reaches the singularity at $t = 0$ in a *finite* value of its affine parameter (i.e. in a finite proper time interval). At the classical level, the space-time cannot be extended beyond a singuarity, and this implies that the time $t = 0$ (the so-called "big-bang" singularity) should be interpreted, within the standard cosmological model, as the *beginning* of space-time, and as the *birth* of the Universe itself.

This conclusion could be avoided if some drastic modification of the standard scenario were to take place before reaching the initial singularity. After all, the standard cosmological model is based on general relativity, i.e. on a classical theory which is not guaranteed to be valid when the space-time curvature becomes large in Planck units, and the Universe enters the quantum gravity regime.

On the other hand, as already stressed in this section, recent observations indicate that the standard cosmological model has to be modified even at the present low-curvature scales, in order to account for the "cosmic repulsion" producing an accelerated space-time expansion. In addition, as we shall see in the next section, there are many valid reasons (other than the existence of a singularity) why the standard cosmological model should be modified when the curvature reaches high enough scales (i.e. at early enough cosmological epochs). These primordial modifications lead to the introduction of the so-called inflationary scenario, which will be illustrated in the next section.

1.2 The inflationary cosmological model

The present observed values of the cosmological parameters, if taken as initial conditions to evolve our Universe backward in time according to the standard picture, lead us to a primordial state characterized by somewhat "unnatural" properties, even without reaching the initial singularity. A dynamical explanation of such peculiar properties of the primordial Universe is possible, provided the epoch of standard (decelerated) expansion is preceded by an appropriate phase of accelerated evolution, dubbed "inflation" [25, 26, 27]. We start this section by presenting simple arguments that motivate the introduction of such a phase.

1.2.1 Standard kinematic problems

A first argument is based on the so-called "flatness problem". As already pointed out in Section 1.1, the spatial curvature today provides only a small, sub-dominant contribution to the total space-time curvature (see Eq. (1.40)). This contribution, however, is not a constant,

$$|\Omega_K| \sim (aH)^{-2} \equiv r^2(t), \tag{1.82}$$

being controlled by the parameter $r(t)$ which is a monotonically increasing function of time in the standard cosmological model. For an expanding, power-law scale factor, $a(t) \sim t^\alpha$, with $0 < \alpha < 1$ and $t \to \infty$, one obtains

$$t \to \infty \quad \Rightarrow \quad r(t) = (aH)^{-1} = \dot{a}^{-1} \sim t^{1-\alpha} \to \infty, \tag{1.83}$$

so that $r(t)$ is increasing in both the matter-dominated ($\alpha = 2/3$) and radiation-dominated ($\alpha = 1/2$) eras. This implies that the contribution of the spatial curvature becomes less and less significant as we go back in time, according to the standard-model equations.

Let us consider, for instance, the Planck epoch $t = t_P$, defined as the time when $H = M_P$ and $\rho = \rho_c = 3M_P^4$. Using the kinematic properties of the standard model ($a \sim H^{-2/3}$, $a \sim H^{-1/2}$ for the matter- and radiation-dominated eras, respectively), we can rescale the parameter $r(t)$ to the time t_P as follows:

$$\frac{r_P}{r_0} \equiv \frac{r(t_P)}{r(t_0)} = \frac{(aH)_0}{(aH)_{eq}} \frac{(aH)_{eq}}{(aH)_P} = \left(\frac{H_0}{H_{eq}} \right)^{1/3} \left(\frac{H_{eq}}{M_P} \right)^{1/2}. \tag{1.84}$$

Using Eqs. (1.78) and (1.79) for the values of H_0 and H_{eq}, and adopting the conservative constraint $|\Omega_K(t_0)| < 0.1$, we obtain

$$|\Omega_K(t_P)| = |\Omega_K(t_0)| \left(\frac{r_P}{r_0}\right)^2 \lesssim 10^{-60}. \tag{1.85}$$

Such a large suppression of the spatial curvature with respect to the space-time curvature represents a rather unnatural initial condition, and requires a significant amount of fine-tuning. Also, what makes the problem even more serious is the fact that a violation of the above upper limit by only a few orders of magnitude would be enough to forbid the formation of our present cosmological configuration. In that case the Universe would enter (much before the present epoch) a curvature-dominated phase ($|\Omega_K| \sim 1$), which would lead to a subsequent collapse if $K > 0$, and which would not be appropriate to sustain the formation of large-scale structures if $K < 0$.

The initial condition (1.85) can be dynamically explained, however, if the phase of standard evolution is preceded by a primordial phase during which the function $r(t)$ *decreases* in time (instead of growing, as in Eq. (1.83)). It then becomes possible to start from a "natural" set of initial conditions for this primordial epoch, characterized by $r \sim 1$, provided that the decrease of r during such an epoch is large enough to compensate the subsequent growth produced by the standard evolution.

As an example of this non-standard phase let us consider again a power-law scale factor, $a(t) \sim t^\beta$, with $t \to \infty$. If β is large enough (in particular, $\beta > 1$), then the function $r(t)$ decreases in time:

$$t \to \infty \quad \Rightarrow \quad r(t) = \dot{a}^{-1} \sim t^{1-\beta} \to 0. \tag{1.86}$$

It is straightforward to check that such a scale factor describes accelerated expansion,

$$\frac{\dot{a}}{a} = \frac{\beta}{t} > 0, \qquad \frac{\ddot{a}}{a} = \frac{\beta(\beta-1)}{t^2} > 0, \tag{1.87}$$

hence the term *inflation* used to denote this phase, complementary to the decelerated evolution of the standard cosmological model (see Chapter 5 for a general classification of the various classes of inflationary kinematics). As will be shown later, this kind of accelerated expansion can be obtained from the Einstein equation using, as gravitational source, a scalar field with an appropriate exponential potential (see Eqs. (1.120) and (1.121)).

The presence of such an inflationary epoch, besides solving the flatness problem, also provides a solution to the so-called *horizon problem* of the standard cosmological model. The standard cosmological evolution ($a \sim t^\alpha$, $\alpha < 1$) is characterized, in fact, by the existence of particle horizons (see Section 1.1), which control, at any given instant of time, the maximum size of the spatial regions within which causal interactions take place. The proper size of the particle horizon is of the order of the Hubble radius H^{-1}, and grows linearly with the cosmic time, according to Eq. (1.17):

$$d_p(t) = a(t) \int_0^t \frac{dt'}{a(t')} = \frac{t}{1-\alpha} = \frac{\alpha}{1-\alpha} H^{-1}(t). \qquad (1.88)$$

Let us consider the spatial section of the Universe included within our current particle horizon, namely the portion of space of typical size H_0^{-1}, currently accessible to our direct observation. Going backward in time, the proper volume of this spatial region decreases as a^3, and therefore its proper radius decreases as $a \sim t^\alpha$. The radius of the particle horizon, i.e. of the causally connected portion of space, also decreases going backward in time, but goes linearly with the cosmic time (according to Eq. (1.88)), and thus faster than the scale factor (recall that $\alpha < 1$). As a consequence, the portion of space that we are currently observing was, in the past, much bigger than the corresponding extension of the particle horizon: in other words, many parts of the currently observed Universe were not causally connected. If we rescale, for instance, the proper size of the present observable Universe, H_0^{-1}, down to the Planck epoch $t = t_P$, when the horizon size was $H_P^{-1} = M_P^{-1}$, we obtain a proper radius much larger than the horizon:

$$\frac{a(t_P)}{a(t_0)} H_0^{-1} = \frac{r(t_0)}{r(t_P)} M_P^{-1} \sim 10^{29} M_P^{-1} \gg M_P^{-1}. \qquad (1.89)$$

Given such initial conditions, we are led to the questions: why is the current Universe so homogeneous and isotropic, or why is the average CMB temperature everywhere the same, as if all the portions of space we are now observing were in the past in causal contact, and had time to interact and thermalize?

An interesting solution of this problem arises from noticing that the ratio between the horizon size ($\sim H^{-1}$) and the proper size of a spatial region ($\sim a$) is governed by the same function $r(t) = (aH)^{-1}$ as controls the ratio between the spatial curvature and the space-time curvature. A sufficiently long inflationary phase, which makes $r(t)$ decreasing and which is able to solve the flatness problem, can thus simultaneously also solve the horizon problem. Indeed, if $r(t)$ decreases as time goes on, the causally connected regions expand faster than the Hubble horizon: at the end of inflation one then precisely obtains a configuration which corresponds to the "unnatural" initial conditions of the standard cosmological scenario (see Fig. 1.1).

How long does the inflationary phase have to be in order to solve the flatness and horizon problems? The answer depends on both the expansion rate and the

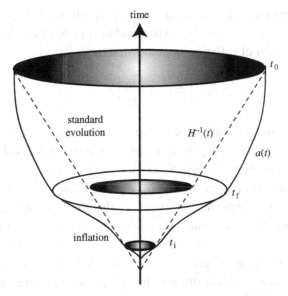

Figure 1.1 Qualitative evolution of the Hubble horizon (dashed line) and of the scale factor (solid curve). The time coordinate is on the vertical axis, while the horizontal axes are space coordinates spanning a two-dimensional spatial section of the cosmological manifold. The inflationary phase extends from t_i to t_f, the standard cosmological phase from t_f to the present time t_0. The shaded areas represent causally connected regions at different epochs. At the beginning of the standard evolution the size of the currently observed Universe (bounded by $a(t)$) is larger than the corresponding Hubble radius (bounded by H^{-1}); all its parts, however, emerge from a spatial region that is causally connected at the beginning of inflation.

beginning of the inflationary epoch. In any case, the decrease of the function $r(t)$, from the beginning t_i to the end t_f of inflation, has to be large enough to compensate for its subsequent increase from t_f to the present time t_0. This defines the following necessary condition to be satisfied by a successful inflationary epoch:

$$\left(\frac{r_f}{r_i}\right) \lesssim \left(\frac{r_f}{r_0}\right), \tag{1.90}$$

where $r_f \equiv r(t_f)$, and so on. Assuming for the inflationary phase the power-law evolution (1.86), one then obtains the condition

$$\left(\frac{t_f}{t_i}\right)^{1-\beta} = \left(\frac{H_i}{H_f}\right)^{1-\beta} \lesssim \left(\frac{r_f}{r_{eq}}\right)\left(\frac{r_{eq}}{r_0}\right) = \left(\frac{H_{eq}}{H_f}\right)^{1/2}\left(\frac{H_0}{H_{eq}}\right)^{1/3}, \tag{1.91}$$

which determines t_f, i.e. the scale H_f at which inflation ends, as a function of β and t_i. It may be useful to notice that, for scale factors following a power-law evolution in time, the function r is proportional to the conformal time coordinate: $r \sim \dot{a}^{-1} \sim \eta = \int a^{-1} \, dt$. The condition (1.90) then directly provides the minimum duration of inflation in conformal time, i.e.

$$\left| \frac{\eta_f}{\eta_i} \right| \gtrsim \left| \frac{\eta_f}{\eta_0} \right| . \tag{1.92}$$

We have inserted the absolute value because, as we shall see later, inflationary (accelerated) scale factors are parametrized by a power-law evolution within a range of negative values of the conformal time coordinate.

The presence of a primordial inflationary phase, characterized by accelerated kinematics, is today universally accepted as the most natural complement of the subsequent decelerated expansion, driven by the standard radiation/matter fluids. The presence of such an inflationary phase allows us to explain the peculiar initial conditions of the standard cosmological model; in addition, it provides a dynamical mechanism for the origin of the large-scale structures and of the small CMB anisotropies (as will be discussed in Chapter 8). It is therefore natural to try to address in the inflationary context the crucial problem of the standard cosmological model, i.e. the presence of the initial singularity.

The premises are encouraging. One of the *necessary* (although not sufficient) conditions for avoiding the singularity is the violation of the so-called condition of "geodesic convergence" [24], for any time-like or null geodesic u^μ. This condition reads, in our notations,

$$R_{\mu\nu} u^\mu u^\nu \geq 0 , \qquad u_\mu u^\mu \geq 0 , \tag{1.93}$$

and is also equivalent, using the Einstein equations, to the so-called "strong energy condition" imposed on the gravitational sources,

$$T_{\mu\nu} u^\mu u^\nu \geq \frac{1}{2} T u_\mu u^\mu . \tag{1.94}$$

For a comoving geodesic $u^\mu = (1, \vec{0})$ of the Robertson–Walker metric, and for a perfect fluid source, the above conditions are violated when $p < -\rho/3$, which implies $R_0^0 = -3\ddot{a}/a < 0$, i.e. just when the expansion is accelerated ($\ddot{a} > 0$), and hence inflationary. It is worth noticing, at this point, that a typical example of an inflationary solution (which is also, historically, the first example [25, 28]) is the de Sitter solution, which describes a maximally symmetric, four-dimensional manifold with constant positive curvature, and which is indeed a regular solution of the Einstein equations (all curvature invariants are constant and finite everywhere).

1.2.2 de Sitter inflation

The matter source for the de Sitter solution corresponds to an effective energy-momentum tensor of the type (1.31), where the cosmological constant Λ may be interpreted as the "vacuum" energy density associated with a scalar field (the "inflaton"), frozen at the minimum $\phi = \phi_0$ of an appropriate potential. Consider the action for a self-interacting scalar field ϕ, minimally coupled to gravity,

$$S_m = \frac{1}{2} \int d^4x \sqrt{-g} \left[g^{\mu\nu} \partial_\mu \phi \, \partial_\nu \phi - 2V(\phi) \right]. \tag{1.95}$$

The variation of S_m with respect to ϕ yields the equation of motion

$$\nabla_\mu \nabla^\mu \phi + \frac{\partial V}{\partial \phi} = 0, \tag{1.96}$$

which admits the constant solution $\phi = \phi_0$, $\partial_\mu \phi_0 = 0$, provided ϕ_0 extremizes the scalar potential $(\partial V / \partial \phi)_{\phi = \phi_0} = 0$. The energy-momentum tensor of the scalar field,

$$T_{\mu\nu} = \partial_\mu \phi \, \partial_\nu \phi - \frac{1}{2} g_{\mu\nu} \left[(\nabla \phi)^2 - 2V(\phi) \right], \tag{1.97}$$

for $\phi = \phi_0$ assumes the form (1.31), with $\Lambda = V(\phi_0) = $ const, and the corresponding Einstein equations (1.26) and (1.27) are identically satisfied by the regular de Sitter solution with constant scalar curvature $R = -8\pi GT = -32\pi GV(\phi_0)$.

Using an appropriate, spatially flat ($K = 0$) chart, the solution can be represented in exponential form (see Eq. (1.32)) as

$$ds^2 = dt^2 - a^2(t)|d\vec{x}|^2, \qquad a(t) = e^{Ht},$$

$$H = \left[\frac{V(\phi_0)}{3 M_P^2} \right]^{1/2} = \text{const}, \qquad -\infty \leq t \leq \infty. \tag{1.98}$$

This solution describes accelerated expansion at constant curvature, $\dot{a} > 0$, $\ddot{a} > 0$, $\dot{H} = 0$. Introducing the conformal time coordinate,

$$\eta = \int^t \frac{dt'}{a(t')} = -\frac{e^{-Ht}}{H} = -\frac{1}{aH}, \qquad -\infty \leq \eta \leq 0, \tag{1.99}$$

it can be written in a conformally flat form,

$$ds^2 = a^2(\eta) \left(d\eta^2 - |d\vec{x}|^2 \right), \qquad a(\eta) = (-H\eta)^{-1}, \tag{1.100}$$

and the condition (1.90) of sufficient inflation becomes

$$\frac{r_i}{r_f} = \left| \frac{\eta_i}{\eta_f} \right| = \frac{a_f}{a_i} = e^{H(t_f - t_i)} \gtrsim \frac{r_0}{r_f}. \tag{1.101}$$

It can be easily checked, in this form, that a very short duration (in units of H^{-1}) of the accelerated phase may be enough to compensate for large variations of the function r, even if inflation occurs at very primordial epochs.

Suppose, for instance, that the inflationary phase ends when $H_f = 10^{-5} M_P$ (in many models, higher values of the curvature scale are inconsistent with CMB observations, as we shall see in Chapter 7), and that the Universe immediately enters the radiation-dominated regime. The radiation temperature associated with t_f is then of the order of the grand unification theory (GUT) scale, $T_f \sim 10^{15} - 10^{16}$ GeV, according to Eq. (1.74). For exponential (or quasi-exponential) inflation, on the other hand, the time duration of the accelerated phase $\Delta t = t_f - t_i$ can be conveniently expressed in terms of the "e-folding factor", $N = \ln(a_f/a_i)$. In terms of N, the condition (1.101) then becomes

$$N = \ln \left(\frac{a_f}{a_i} \right) = H \Delta t$$

$$\gtrsim \ln \left(\frac{r_0}{r_f} \right) = \frac{1}{2} \ln \left(\frac{H_f}{H_{eq}} \right) + \frac{1}{3} \ln \left(\frac{H_{eq}}{H_0} \right) \qquad (1.102)$$

(we have used Eq. (1.91)). For $H_f = 10^{-5} M_P$ one obtains $N \gtrsim \ln 10^{27} \simeq 62$, i.e. $\Delta t \gtrsim 62 H^{-1}$, where H is the curvature scale of the de Sitter manifold.

The de Sitter solution may give an appropriate description of the primordial inflationary phase; however, it cannot be extended forward in time towards "too late" epochs, since the Universe must enter into the standard decelerated phase that allows nucleosynthesis and the formation of large-scale structures, and that eventually converges to our present cosmological configuration. The transition (also called "graceful exit") between the inflationary and the standard regime is usually implemented, in conventional models of inflation, by assuming that the scalar field is not exactly constant, frozen at the minimum of its potential; instead, it is initially displaced from this minimum, and "slow-rolls" towards it.

1.2.3 Slow-roll inflation

To illustrate this possibility we start by considering the cosmological equations with the energy-momentum tensor of the scalar field as the only gravitational source. Also, we assume that we are given an initial spatial domain of size smaller than (or comparable to) the initial horizon radius H^{-1}, in which the spatial inhomogeneities of the scalar field are negligible, $|\partial_i \phi| \ll |\dot{\phi}|$. Restricting to this spatial domain we can then neglect the spatial dependence of our variables, and we can treat the scalar field as a perfect fluid *at rest* in the comoving frame, with the following effective energy density and pressure,

$$\rho = \frac{\dot{\phi}^2}{2} + V(\phi), \qquad p = \frac{\dot{\phi}^2}{2} - V(\phi) \tag{1.103}$$

(see Eq. (1.97)), acting as the source of a homogeneous and isotropic Robertson–Walker geometry. Combining Eqs. (1.25) and (1.26), neglecting K, and using the identity $\ddot{a}/a = \dot{H} + H^2$, we are then led to the following independent Einstein equations:

$$3H^2 = 8\pi G\rho = 8\pi G\left(\frac{\dot{\phi}^2}{2} + V\right), \tag{1.104}$$

$$2\dot{H} = -8\pi G(\rho + p) = -8\pi G\dot{\phi}^2. \tag{1.105}$$

We may add to this system the scalar field equation (1.96), which reads

$$\ddot{\phi} + 3H\dot{\phi} + \frac{\partial V}{\partial \phi} = 0. \tag{1.106}$$

For $\dot{\phi} \neq 0$ this equation is not independent, however, since it can be obtained by the differentiation of Eq. (1.104) and its combination with Eq. (1.105).

The dynamics of the slow-roll regime can be conveniently illustrated by using the scalar field as the independent variable (replacing cosmic time) of our differential equations. Denoting with a prime the differentiation with respect to ϕ, and dividing Eq. (1.105) by $\dot{\phi}$ (assuming a monotonic evolution with $\dot{\phi} \neq 0$), we obtain

$$2H' = -\lambda_P^2 \dot{\phi} \tag{1.107}$$

(recall that $\lambda_P^2 = 8\pi G$). Inserting $\dot{\phi}$ from this equation into Eq. (1.104) we are led to the first-order equation

$$H'^2 - \frac{3}{2}\lambda_P^2 H^2 = -\frac{1}{2}\lambda_P^4 V, \tag{1.108}$$

which is equivalent to the Hamilton–Jacobi equation for the gravity-scalar field system [29]. Let us also define, for later applications, the following useful parameters:

$$\epsilon_H = -\frac{\dot{H}}{H^2} = -\frac{d\ln H}{d\ln a} = \frac{2}{\lambda_P^2}\frac{H'^2}{H^2}, \tag{1.109}$$

$$\eta_H = -\frac{\ddot{\phi}}{H\dot{\phi}} = -\frac{d\ln\dot{\phi}}{d\ln a} = \frac{2}{\lambda_P^2}\frac{H''}{H} \tag{1.110}$$

(we have used Eq. (1.107) to obtain the last equalities of both definitions). The subscript H makes explicit reference to the Hamilton–Jacobi formalism.

The so-called slow-roll regime corresponds to a sufficiently slow evolution of the scalar field, initially dominated by the "geometrical friction" term $H\dot{\phi}$, and characterized by a kinetic energy which is negligible with respect to the scalar potential. More precisely, the scalar field is slow-rolling if the following conditions are valid:

$$\ddot{\phi} \ll H\dot{\phi}, \qquad \dot{\phi}^2 \ll V, \qquad \dot{H} \ll H^2. \tag{1.111}$$

The slow-roll regime is thus implemented in the limit in which the parameters (1.109) and (1.110) are very small ($\epsilon_H \ll 1$, $\eta_H \ll 1$), and very slowly varying ($\dot{\epsilon}_H \simeq 0$, $\dot{\eta}_H \simeq 0$, to first order). In this limit we can immediately integrate Eq. (1.109) to obtain $H = (\epsilon_H t)^{-1}$. A second integration leads to the scale factor $a(t) \sim t^{1/\epsilon_H}$. Using conformal time,

$$a(\eta) \sim (-\eta)^{-(1+\epsilon_H)}. \tag{1.112}$$

One thus obtains an inflationary (i.e. accelerated) scale factor which approximates the de Sitter metric in the limit $\epsilon_H \to 0$ (see Eq. (1.100)).

The cosmological dynamics during the slow-roll regime is well described by the two independent equations

$$3H^2 = \lambda_P^2 V, \qquad 3H\dot{\phi} = -V', \tag{1.113}$$

obtained from Eqs. (1.108) and (1.106), respectively, using the conditions $\epsilon_H \ll 1$, $\eta_H \ll 1$. Differentiating with respect to ϕ the first equation, we obtain

$$\frac{H'}{H} = \frac{V'}{2V}. \tag{1.114}$$

Inserting this condition into the exact definitions (1.109) and (1.110) we are led to approximate relations defining two new parameters, ϵ and η [30], satisfying

$$\epsilon_H \simeq \frac{1}{2\lambda_P^2}\left(\frac{V'}{V}\right)^2 \equiv \epsilon, \qquad \eta_H \simeq -\epsilon + \eta, \qquad \eta \equiv \frac{V''}{\lambda_P^2 V}, \tag{1.115}$$

and often used for the computation of the spectra of the metric perturbations amplified by inflation (see Section 8.2). The smallness of these two parameters guarantees the "flatness" of the potential $V(\phi)$, and the consequent slowness of the motion of ϕ towards the minimum.

The slow-roll equations (1.113) can be formally integrated, for any given $V(\phi)$, using the exact differential relations $da/a = Hdt$, $dt = d\phi/\dot{\phi}$, and writing the scale factor in the form

$$a(t) = a_i \exp\left(\int_{t_i}^{t} H \, dt\right) = a_i \exp\left(\int_{\phi_i}^{\phi(t)} \frac{H \, d\phi}{\dot\phi}\right)$$

$$= a_i \exp\left(-\lambda_P^2 \int_{\phi_i}^{\phi(t)} \frac{V}{V'} d\phi\right), \tag{1.116}$$

while $\phi(t)$ is obtained by integrating the equation

$$\dot\phi = -\frac{V'}{3H} = -\frac{1}{\sqrt{3}\lambda_P} \frac{V'}{V^{1/2}}. \tag{1.117}$$

This solution is consistent provided the evolution of ϕ given by Eq. (1.117) is sufficiently slow ($\ddot\phi \to 0$), and the scale factor (1.116) approximates the exponential de Sitter solution ($H \to$ const). A useful parameter, in such a context, is the number of e-folds $N(t)$ between a given time t and the end of inflation t_f,

$$N(t) = \ln \frac{a_f}{a(t)} = \int_t^{t_f} H \, dt. \tag{1.118}$$

Using Eq. (1.116) we can relate $N(t)$ to the corresponding value of the inflaton field at the same time t, namely,

$$N(t) = N(\phi(t)) = \lambda_P^2 \int_{\phi_f}^{\phi} \frac{V}{V'} d\phi. \tag{1.119}$$

This relation will be applied in Chapter 8 to parametrize the primordial spectrum of metric perturbations obtained in the context of slow-roll inflation.

A simple example of an inflationary solution of the slow-roll type can be implemented, in practice, using an appropriate exponential potential [31],

$$V(\phi) = V_0 e^{-\lambda_P \phi \sqrt{2/p}}, \tag{1.120}$$

where p and V_0 are positive parameters. In this case the Einstein equations (1.104) and (1.105) are solved by the particular exact solution

$$a = a_0 t^p,$$

$$\lambda_P \phi = \sqrt{2p} \ln\left[\lambda_P t \sqrt{\frac{V_0}{p(3p-1)}}\right], \tag{1.121}$$

which for $p > 1$ satisfies the kinematic conditions of power-law inflation (see Eq. (1.87)). The computation of H, $\dot H$, $\dot\phi$ and $\ddot\phi$ for this solution, together with the use of the exact definitions (1.109) and (1.110), leads to

$$\epsilon_H = \eta_H = p^{-1}. \tag{1.122}$$

For $p \gg 1$ the above solution (1.121) thus describes a phase of slow-roll inflation, which approaches de Sitter inflation in the limit $p \to \infty$.

Another efficient mechanism for generating slow-roll solutions is based on simple polynomial potentials of the type $V \sim \phi^n$, provided they are flat enough to satisfy the conditions $\epsilon \ll 1$ and $\eta \ll 1$ (a typical example is the so-called "chaotic" inflationary scenario [32], which includes the simplest case $n = 2$). For such potentials $V'/V = n/\phi$, and the total e-folding factor (computed from Eqs. (1.118) and (1.119)) takes the form

$$N = \ln \left(\frac{a_f}{a_i} \right) = \frac{\lambda_P^2}{2n} (\phi_i^2 - \phi_f^2). \tag{1.123}$$

Moreover, $V''/V \sim \phi^{-2}$, and the condition $\eta \ll 1$ requires very large values of the initial inflaton field in Planck units, $(\lambda_P \phi_i)^2 \gg 1$, for the slow-roll regime to be valid. But this automatically guarantees an efficient inflationary expansion, $N \gg 1$, according to Eq. (1.123).

It must be noted that the slow-roll parameters associated with a polynomial potential are (slowly) evolving in time during inflation, in contrast with the case of the exponential potential where the parameters are constant (see Eq. (1.122)), and are in principle associated with an "eternal" duration of the phase of inflation. For models based on polynomial potentials the end of the inflationary phase may automatically occur as soon as the rolling velocity of the inflation increases, near to the minimum of the potential. In particular, when the effective mass V'' becomes of order H, the inflaton enters a regime of rapid oscillations characterized by the approximate equality of kinetic and potential energy, $\langle \dot{\phi}^2 \rangle \simeq 2 \langle V \rangle$. This regime preludes the inflaton decay and the consequent production of a cosmic background of relativistic particles, eventually becoming the dominant source of the standard, radiation-dominated era [5].

1.2.4 Initial singularity

A phase of slow-roll evolution, of the type illustrated by the above examples, seems to provide a more realistic (and probably even more natural) model of inflation than the one based upon the de Sitter solution, which requires instead a scalar field rigidly trapped at the minimum of its potential. Slow-roll solutions, however, do not describe a regular geometry like the de Sitter manifold, and therefore do not provide a solution to the singularity problem of the standard cosmological model. Indeed, the curvature decreases (even if slowly) during the slow-roll phase and this implies that, going backward in time, the Universe emerges from a singular state. The suppression of $\dot{H}$ during the slow-roll evolution moves the singularity backward in time towards much earlier epochs than in the standard scenario, but it does not remove it.

It should be noted, on the other hand, that even the exact inflationary solution (1.98), describing exponential expansion at constant curvature, does not completely remove the initial singularity. This solution, in fact, represents the de Sitter manifold in a chart that is *not* geodesically complete: a geodesic observer of such a coordinate system, starting from the origin, can reach a point at infinite spatial distance during a finite proper-time interval.

The geodesic incompleteness of the solution (1.98) is shown by recalling that the four-dimensional de Sitter manifold can be represented [33] as a pseudo-hypersphere (or hyperboloid) of radius $R_0 = \left(3M_P^2/\Lambda\right)^{1/2} = H^{-1} =$ const, embedded into a five-dimensional pseudo-Euclidean space with metric $\eta_{AB} = (+, -, -, -, -)$, spanned by the cartesian coordinates $z^A = (z^0, z^1, \ldots z^4)$. The hyperboloid has equation

$$-\eta_{AB} z^A z^B = (z^i)^2 + (z^4)^2 - (z^0)^2 = H^{-2}, \tag{1.124}$$

where $A, B = 0, \ldots 4$, and $i = 1, 2, 3$. The metric (1.98) can then be obtained by defining on the hyperboloid the intrinsic, four-dimensional cartesian chart $x^\mu = (t, x^i)$, and embedding the hypersurface into the higher-dimensional manifold through the following parametric equations,

$$z^i = e^{Ht} x^i,$$
$$z^0 = \frac{1}{H} \sinh(Ht) + \frac{H}{2} e^{Ht} x_i^2, \tag{1.125}$$
$$z^4 = \frac{1}{H} \cosh(Ht) - \frac{H}{2} e^{Ht} x_i^2,$$

satisfying Eq. (1.124). Differentiating, and substituting into the five-dimensional form $ds^2 = \eta_{AB} dz^A dz^B$, one obtains the line-element (1.98) with exponential scale factor. However, even for x_i and t ranging from $-\infty$ to $+\infty$, the given parametrization does not cover the full de Sitter manifold, but only a portion of it, defined by the condition $z^0 \geq -z^4$ (for $t \to -\infty$ one reaches the border of the parametrized region, marked by the null ray $z^0 = -z_4$).

A geodesically complete chart, covering the whole de Sitter hyperboloid, is obtained by considering a solution of Eq. (1.26) with $\rho = -p = \Lambda$ *and* with non-vanishing (constant, positive) spatial curvature $K = H^2 = (\Lambda/3M_P^2)$. The corresponding four-dimensional metric can then be written in the form

$$ds^2 = dt^2 - \cosh^2(Ht) \left[\frac{dr^2}{1 - H^2 r^2} + r^2 d\Omega^2 \right], \tag{1.126}$$

and is related to the five-dimensional hyperboloid through the parametric equations

$$z^0 = H^{-1}\sinh(Ht),$$

$$z^1 = H^{-1}\cosh(Ht)\cos\chi,$$

$$z^2 = H^{-1}\cosh(Ht)\sin\chi\cos\theta, \qquad(1.127)$$

$$z^3 = H^{-1}\cosh(Ht)\sin\chi\sin\theta\cos\phi,$$

$$z^4 = H^{-1}\cosh(Ht)\sin\chi\sin\theta\sin\phi.$$

Their differentiation, and substitution into the five-dimensional Minkowski form, leads to

$$ds^2 = dt^2 - H^{-2}\cosh^2(Ht)\left[d\chi^2 + \sin^2\chi d\Omega^2\right], \qquad(1.128)$$

which reduces to Eq. (1.126) after setting $H^{-1}\sin\chi = r$. It is straightforward to check that the intrinsic chart $x^\mu = (t, \chi, \theta, \phi)$, with $-\infty \le t \le \infty$, $0 \le \chi \le \pi$, $0 \le \theta \le \pi$, $0 \le \phi \le 2\pi$, provides a full coverage of the hypersurface (1.124) (see for instance [33]).

By using the regular, complete de Sitter solution to eliminate the initial singularity we are led to a picture in which the primordial Universe enters a phase that can be extended (in a geodesically complete way) towards past infinity, according to the metric (1.126), keeping a constant, finite curvature controlled by Λ. However, the kinematic properties of such a phase are determined by the scale factor $a(t) = \cosh(Ht)$, describing a Universe which is initially contracting (at $t \to -\infty$), starting from an infinitely large spatial extension, and which becomes inflationary expanding only at large enough positive times, $t \to +\infty$.

Unfortunately, in models where the complete de Sitter solution is due to the potential energy of a scalar field satisfying standard causality and weak energy ($\rho \ge 0$) conditions, it seems impossible (using the Einstein equations) to include a smooth transition from the contracting to the expanding phase [34, 35]: starting from the exponentially contracting state, the Universe is doomed to collapse towards the singularity $a \to 0$, without "bouncing" to reach eventually the phase of accelerated expansion. In other words, known models of standard, potential-dominated inflation cannot be "past-eternal" [36].

Thus, for a successful model of de Sitter (or quasi de Sitter) potential-dominated inflation, the Universe has to enter the exponential regime already in the state of expansion. Such a state, as we have seen, cannot be arbitrarily extended backward in time without singularities, even in the case of the exact solution (1.98). We can say, therefore, that an inflationary phase driven by the potential energy of a scalar field mitigates the rapid growth of the curvature typical of the standard cosmological model, and shifts back in time the position of the initial singularity, without completely removing it, however (see Fig. 1.2)).

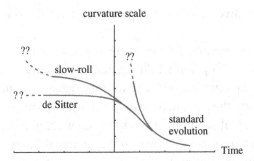

Figure 1.2 Qualitative evolution of the curvature scale in the standard cosmological model, and in models of de Sitter inflation and slow-roll inflation.

The problem of the initial configuration of the standard cosmological model, solved by inflation, then reappears (even if in a more relaxed form) for the inflationary phase, whose effectiveness still depends on the choice of an appropriate initial state. The question that arises is, in particular, the following: does there exist a dynamical mechanism able to "prepare" the appropriate initial inflationary state, producing (for instance) a homogeneous space-time domain that is already characterized by an exponential expansion (the "second half" of the de Sitter solution), and that can smoothly evolve towards the standard cosmological configuration?

One possible approach to this issue is provided by the methods of quantum cosmology (see Chapter 6). Using the Wheeler–De Witt equation [37, 38] it is possible to compute, for instance, the probability that our Universe emerges in the appropriate inflationary state directly from the vacuum (through a process conventionally called "tunneling from nothing" [39, 40, 41]). Such a probability, unfortunately, is strongly dependent on the initial quantum state representing the Universe before the transition, and this state is unknown, as it should be determined in correspondence with the initial singularity. There are various possible prescriptions for choosing the appropriate boundary conditions [39–43]: they are however "ad hoc", and lead to different (and strongly contrasting) results, leaving the debate still open.

Another possible, semiclassical approach is the one based on the "chaotic" inflationary scenario [32, 44]. In this approach the initial values of the scalar field are randomly distributed over different space-time regions, and those regions are characterized by different degrees of homogeneity with respect to the horizon scale. If, in some region, the scalar field happens to be sufficiently homogeneous, sufficiently large (in Planck units) and displaced from the minimum, then a phase of slow-roll inflation is triggered, and that initial region can evolve towards a configuration similar to the Universe in which we are living. In other space-time regions, where such conditions are not satisfied, inflation does not occur, and the subsequent evolution diverges from the path leading to the present cosmological state.

We should mention, finally, that even after a satisfactory explanation of the initial conditions, the scalar potential-dominated inflationary scenario suffers from other conceptual difficulties (see [45] for a recent discussion), such as the cosmological constant problem, the so-called "trans-Planckian" problem (see Section 5.3). String theory, as we shall see in the following chapters, may support inflationary mechanisms different from those based on the potential energy of a scalar field. As a consequence, different primordial scenarios are also possible, based on initial configurations other than the highly curved, hot and dense state approaching the initial singularity, typical of the standard model and of the inflationary models considered in this section.

References

[1] S. Weinberg, *Gravitation and Cosmology* (New York: John Wiley & Sons, 1972).

[2] J. B. Zeldovic and I. D. Novikov, *The Structure and Evolution of the Universe* (Chicago: University of Chicago Press, 1983).

[3] P. Coles and F. Lucchin, *Cosmology* (Chichester, UK: John Wiley & Sons, 2000).

[4] A. D. Linde, *Particle Physics and Inflationary Cosmology* (New York: Harwood, 1990).

[5] E. W. Kolb and M. S. Turner, *The Early Universe* (Redwood City, CA: Addison-Wesley, 1990).

[6] A. R. Liddle and D. H. Lyth, *Cosmological Inflation and Large-Scale Structure* (Cambridge, UK: Cambridge University Press, 2000).

[7] E. P. Hubble, *Proc. Nat. Acad. Sci.* **15** (1927) 168.

[8] A. A. Penzias and R. W. Wilson, *Ap. J.* **142** (1965) 419.

[9] G. W. Gibbons and S. Hawking, *Phys. Rev.* **D15** (1977) 2752.

[10] F. Sylos Labini, M. Montuori and L. Pietronero, *Phys. Rep.* **293** (1998) 61.

[11] W. Rindler, *Mon. Not. R. Astron. Soc.* **116** (1956) 6.

[12] Particle Data Group web pages at pdg.lbl.gov/.

[13] S. Pelmutter *et al.*, *Nature* **391** (1998) 51.

[14] A. G. Riess *et al.*, *Astronom. J.* **116** (1998) 1009.

[15] P. de Bernardis *et al.*, *Nature* **404** (2000) 955.

[16] S. Hanay *et al.*, *Ap. J. Lett.* **545** (2000) L5.

[17] N. W. Alverson *et al.*, *Ap. J.* **568** (2002) 38.

[18] D. N. Spergel *et al.*, *Ap. J. Suppl.* **148** (2003) 175.

[19] A. G. Riess *et al.*, *Ap. J.* **607** (2004) 665.

[20] D. N. Spergel *et al.*, astro-ph/0603449.

[21] P. B. Pal, *Pramana* **54** (2000) 79.

[22] R. R. Caldwell, M. Kamionkowski and N. N. Weinberg, *Phys. Rev. Lett.* **91** (2003) 07130.

[23] J. D. Barrow, *Class. Quantum Grav.* **21** (2004) 5619.

[24] S. W. Hawking and G. R. F. Ellis, *The Large Scale Structure of Space-Time* (Cambridge: Cambridge University Press, 1973).

[25] A. Guth, *Phys. Rev.* **D23** (1981) 347.

[26] A. D. Linde, *Phys. Lett.* **B108** (1982) 389.

[27] A. Albrecht and P. J. Steinhardt, *Phys. Rev. Lett.* **48** (1982) 1220.

[28] A. A. Starobinski, *Phys. Lett.* **B91** (1980) 99.

[29] D. S. Salopek and J. R. Bond, *Phys. Rev.* **D42** (1990) 3936.

[30] A. D. Liddle and D. Lyth, *Phys. Lett.* **B291** (1992) 391.

[31] F. Lucchin and S. Matarrese, *Phys. Rev.* **D32** (1985) 1316.
[32] A. D. Linde, *Phys. Lett.* **B129** (1983) 177.
[33] W. Rindler, *Essential Relativity* (New York: Van Nostrand Company, 1969).
[34] A. Vilenkin, *Phys. Rev.* **D46** (1992) 2355.
[35] A. Borde and A. Vilenkin, *Phys. Rev. Lett.* **72** (1994) 3305.
[36] A. Borde, A. Guth and A. Vilenkin, *Phys. Rev. Lett.* **90** (2003) 151301.
[37] J. A. Wheeler, in *Battelle Rencontres*, eds. C. De Witt and J. A. Wheeler (New York: Benjamin, 1968).
[38] B. S. De Witt, *Phys. Rev.* **D160** (1967) 1113.
[39] A. Vilenkin, *Phys. Rev.* **D30** (1984) 509.
[40] A. D. Linde, *Sov. Phys. JETP* **60** (1984) 211.
[41] V. A. Rubakov, *Phys. Lett.* **B148** (1984) 280.
[42] J. B. Hartle and S. W. Hawking, *Phys. Rev.* **D28** (1983) 2960.
[43] S. W. Hawking, *Nucl. Phys.* **B239** (1984) 257.
[44] A. D. Linde, *Phys. Lett.* **B351** (1995) 99.
[45] R. Brandenberger, in *Proc. YKIS 2005* (Yukawa Institute for Theoretical Physics, Kyoto, Japan, July 2005), hep-th/0509159.

2

The basic string cosmology equations

The aim of this chapter is to present the effective string theory equations governing the low-energy dynamics of the gravitational field and of its sources. Such equations are not postulated ad hoc but, as we shall see in the next chapter, they are required for the consistency of a quantum theory of strings propagating in a curved manifold, and interacting with other fields possibly present in the background. For a more systematic approach to these equations, the analysis of this chapter should probably follow the discussion of string quantization and the computation of the spectrum of the bosonic string states, which will be presented in Chapter 3. However, in the context of a cosmologically oriented book, we have preferred to postpone the string theory motivations in favor of a more immediate presentation of the basic string gravity equations, lying at the foundations of string cosmology just like the Einstein equations are at the foundations of standard cosmology.

For our purposes we only need to recall that the exact string theory equations, for all fields (including gravity) present in the string spectrum, can be approximated by a perturbative expansion, in general in two ways [1]: (*i*) as a higher-derivatives expansion (namely, as an expansion in powers of the "curvatures", or field strengths), and (*ii*) as an expansion in powers of the coupling parameter g_s^2, controlling the intensity of the string interactions. This second expansion is similar to the "loop" expansion of conventional quantum field theory, while the first one is peculiar to strings, since it is controlled by the fundamental length λ_s appearing in the (two-dimensional) string action integral (see Chapter 3); such an expansion disappears in the point-particle limit $\lambda_s \to 0$.

The discussion of this chapter will concentrate on the tree-level equations for the fundamental massless (boson) fields present in all models of strings and superstrings [1, 2] (here "tree-level" means that the equations are truncated to lowest order in both the curvature expansion, controlled by the parameter $2\pi\alpha' = \lambda_s^2$, and the loop expansion, controlled by g_s^2). Such equations can be derived from an effective action which is valid for manifolds with low enough curvature, $\alpha' R \ll 1$, and for fields with weak enough interactions, $g_s^2 \ll 1$ (the action is valid, therefore, in the low-energy, perturbative regime). Nevertheless, they can be basic equations even in a primordial cosmological context where there are scenarios – possibly suggested by string-duality symmetries – with perturbative initial configurations, well described by the low-energy equations [3] (see Chapter 4). Also, such equations are used in the context of the so-called "string gas cosmology" [4] that will be discussed in Chapter 6.

These low-energy equations will be explicitly derived from the action in the string frame (Section 2.1) and in the Einstein frame (Section 2.2). In the last section (and in the appendix) we will discuss the possible corrections induced by the addition of quadratic curvature terms to the effective gravitational equations, to first order in the α' expansion.

2.1 Tree-level equations

The gravitational (massless, bosonic) sector of the string effective action contains not only the metric but also (and even to lowest order) at least one more fundamental field: a scalar field ϕ, called the "dilaton". The corresponding tree-level action can be written as follows:

$$S = -\frac{1}{2\lambda_s^{d-1}} \int d^{d+1}x \sqrt{|g|}\, e^{-\phi} \left[R + (\nabla\phi)^2 \right] + S_\Sigma + S_m. \qquad (2.1)$$

Here S_Σ is the boundary term required to reproduce the standard Einstein equations in the general-relativistic limit, and S_m is the action of all other fields, possibly coupled to ϕ and to $g_{\mu\nu}$ as prescribed by the conformal invariance of fundamental string interactions (see the discussion of the next chapter). Note that we have used (and we shall often use) the compact notation $(\nabla\phi)^2 = \nabla_\mu \phi \nabla^\mu \phi$.

The above equation is written adopting the so-called "string frame" (S-frame) parametrization of the action, where ϕ is dimensionless, and where the metric $g_{\mu\nu}$ is the same metric to which a fundamental string is minimally coupled, and with respect to which a free "test" string evolves geodesically. Otherwise stated, the action (2.1) is parametrized by the same metric field present in the two-dimensional action integral governing the motion of a fundamental string in a curved background (as illustrated in Chapter 3).

It should be noted, also, that we have generically considered the action for a $D = (d+1)$-dimensional space-time manifold. As we shall see later, the quantum theory of an extended object like a string can be consistently formulated only if the number of dimensions is fixed at a critical value $D = D_{\text{crit}}$ (for instance, $D_{\text{crit}} = 26$ for the bosonic string, $D_{\text{crit}} = 10$ for a superstring [1, 2]). We will often also consider a number of dimensions less than critical – in particular,

$D = 4$ – assuming, in that case, that the background fields have a factorizable structure, that the integral over the remaining $D_{\text{crit}} - D$ spatial dimensions gives only a trivial (finite) volume factor, and that such an extra factor has been absorbed by an appropriate rescaling of the dilaton.

The constant length λ_s appearing in the action (2.1) represents the characteristic proper extension of a quantized one-dimensional object like a fundamental string, and provides the natural units of length (λ_s) and energy $(\lambda_s^{-1} = M_s)$ for a physical model based on the S-frame action [5]. The comparison with the $(d+1)$-dimensional Einstein–Hilbert action,

$$S = -\frac{1}{2\lambda_P^{d-1}} \int d^{d+1}x \sqrt{|g|} \, R, \tag{2.2}$$

immediately provides the (tree-level) relation between the string length and the Planck length,

$$\frac{\lambda_P}{\lambda_s} = \frac{M_s}{M_P} = e^{\phi/(d-1)}, \tag{2.3}$$

which clearly shows how the effective gravitational coupling, $8\pi G_D \equiv \lambda_P^{d-1}$, is controlled by the dilaton, in string units, as $8\pi G_D = \lambda_s^{d-1} \exp \phi$.

For a ϕ-independent matter action S_m, the action (2.1) would seem to describe a scalar-tensor model of gravity of the Brans–Dicke (BD) type, with BD parameter $\omega = -1$. In fact, if we set

$$\frac{e^{-\phi}}{\lambda_s^{d-1}} = \frac{\Phi}{8\pi G_D}, \tag{2.4}$$

the gravi-dilaton part of the action can be rewritten in the "canonical" BD form (see for instance [6]),

$$S_{\text{BD}} = \frac{1}{8\pi G_D} \int d^{d+1}x \sqrt{|g|} \left[-\Phi R + \omega \Phi^{-1} (\nabla \Phi)^2 \right], \tag{2.5}$$

provided ω is fixed to the value -1.

Even for the gravi-dilaton sector, however, the analogy with a "pure" BD model is possibly valid only at the tree-level: in fact, after including the higher-loop corrections required by string theory in the strong coupling limit, the effective action may be rewritten in the form (2.5), only at the cost of defining a BD parameter which is dilaton dependent, $\omega = \omega(\Phi)$ (see for instance [7, 8]). In addition, the tree-level analogy with BD models only holds for a particular class of fields, whose S-frame action S_m is decoupled from the dilaton (for example, for the bosonic forms present in the Ramond–Ramond sector of type IIA and type IIB superstrings, see e.g. [1, 2] and Appendix 3B). String theory, in general, predicts a *non-minimal* and *non-universal* coupling of the various fields to the dilaton (see Chapter 9): it is thus impossible, in principle (even at the tree-level), to

introduce an appropriate "Jordan frame" where all fields (except the dilaton) are
geodesically coupled to the metric, and satisfy the principle of equivalence.

The variation of the gravi-dilaton action (2.1) with respect to $g_{\mu\nu}$ can be
performed as in general relativity, using the standard definition (1.3) of the
dynamical energy-momentum tensor $T_{\mu\nu}$, and exploiting the standard result for
the variation of the scalar curvature density:

$$\delta_g\left(\sqrt{|g|}R\right) = \sqrt{|g|}\left(G_{\mu\nu}\delta g^{\mu\nu} + g_{\mu\nu}\nabla^2\delta g^{\mu\nu} - \nabla_\mu\nabla_\nu\delta g^{\mu\nu}\right) \qquad (2.6)$$

(here $G_{\mu\nu} = R_{\mu\nu} - g_{\mu\nu}R/2$ is the Einstein tensor, and $\nabla^2 \equiv \nabla_\alpha\nabla^\alpha$ denotes the
covariant d'Alembert operator). In our context, however, the integral over the
second derivatives of $\delta g^{\mu\nu}$ is no longer equivalent to a surface integral, because
of the dilaton pre-factor multiplying the scalar curvature. After a first integration
by parts, and the application of the metricity condition $\nabla_\alpha g_{\mu\nu} = 0$, the variation
of the action (2.1) gives

$$\delta_g S = \frac{1}{2}\int_\Omega d^{d+1}x\sqrt{|g|}\,T_{\mu\nu}\,\delta g^{\mu\nu} + \delta_g S_\Sigma - \frac{1}{2\lambda_s^{d-1}}\int_\Omega d^{d+1}x\sqrt{|g|}\,e^{-\phi}$$

$$\times\left[G_{\mu\nu} + \nabla_\alpha\phi\,g_{\mu\nu}\nabla^\alpha - \nabla_\mu\phi\nabla_\nu + \nabla_\mu\phi\nabla_\nu\phi - \frac{1}{2}g_{\mu\nu}\left(\nabla\phi\right)^2\right]\delta g^{\mu\nu}$$

$$-\frac{1}{2\lambda_s^{d-1}}\int_\Omega d^{d+1}x\sqrt{|g|}\,\nabla_\alpha\left[e^{-\phi}g_{\mu\nu}\nabla^\alpha\delta g^{\mu\nu} - e^{-\phi}\nabla_\nu\delta g^{\nu\alpha}\right], \qquad (2.7)$$

where Ω is the $(d+1)$-dimensional portion of the manifold where we impose
that the action be stationary ($\delta_g S = 0$), with the condition of no variation on
the boundary, $(\delta g)_{\partial\Omega} = 0$. A second integration by parts of the gradients of δg
cancels the bilinear term $\nabla_\mu\phi\nabla_\nu\phi$, and introduces the second derivatives of the
dilaton field. Collecting all similar terms we obtain

$$\delta_g S = \frac{1}{2}\int_\Omega d^{d+1}x\sqrt{|g|}\,T_{\mu\nu}\,\delta g^{\mu\nu} + \delta_g S_\Sigma$$

$$-\frac{1}{2\lambda_s^{d-1}}\int_\Omega d^{d+1}x\sqrt{|g|}\,e^{-\phi}\left[G_{\mu\nu} + \frac{1}{2}g_{\mu\nu}\left(\nabla\phi\right)^2 - g_{\mu\nu}\nabla^2\phi + \nabla_\mu\nabla_\nu\phi\right]\delta g^{\mu\nu}$$

$$-\frac{1}{2\lambda_s^{d-1}}\int_{\partial\Omega}\sqrt{|g|}\,e^{-\phi}\left[g_{\mu\nu}\nabla^\alpha\delta g^{\mu\nu} - \nabla_\nu\delta g^{\nu\alpha}\right]d\Sigma_\alpha$$

$$-\frac{1}{2\lambda_s^{d-1}}\int_{\partial\Omega}\sqrt{|g|}\,e^{-\phi}\left[g_{\mu\nu}\nabla^\alpha\phi\delta g^{\mu\nu} - \nabla_\nu\phi\,\delta g^{\nu\alpha}\right]d\Sigma_\alpha, \qquad (2.8)$$

where we have used the Gauss theorem to transform the integrals of a divergence
over the hypervolume Ω into hypersurface integrals over the boundary $\partial\Omega$.

Imposing that the action be stationary, we have no contribution from the last surface integral, because $\delta g = 0$ on the boundary. However, the gradients of δg are *not* required to be vanishing by the variational procedure: in order to obtain the desired field equations we have thus to cancel the other surface integral through an appropriate boundary term S_Σ, as in general relativity. By defining

$$S_\Sigma = \frac{1}{2\lambda_s^{d-1}} \int_{\partial\Omega} \sqrt{|g|}\, e^{-\phi} K^\alpha d\Sigma_\alpha, \tag{2.9}$$

we must require, in particular, that

$$\delta_g \left(\int_{\partial\Omega} \sqrt{|g|}\, e^{-\phi} K^\alpha d\Sigma_\alpha \right) = \int_{\partial\Omega} \sqrt{|g|}\, e^{-\phi} \left[g_{\mu\nu} \nabla^\alpha \delta g^{\mu\nu} - \nabla_\nu \delta g^{\nu\alpha} \right] d\Sigma_\alpha. \tag{2.10}$$

It follows that S_Σ differs from the Gibbons–Hawking boundary term [9], used in general relativity to reproduce the Einstein equations, only by the presence of the dilaton pre-factor $\exp(-\phi)$. Hence, as in general relativity, we can geometrically identify K^α as $K^\alpha = Kn^\alpha$, where K and n^α are, respectively, the trace of the extrinsic curvature and the normal vector of the d-dimensional hypersurface $\partial\Omega$, bounding the portion of space-time Ω over which we are varying the action.

Taking into account the cancelation of all surface integrals, the condition $\delta_g S = 0$ then leads to the equations

$$G_{\mu\nu} + \nabla_\mu \nabla_\nu \phi + \frac{1}{2} g_{\mu\nu} (\nabla\phi)^2 - g_{\mu\nu} \nabla^2 \phi = \lambda_s^{d-1} e^\phi\, T_{\mu\nu}. \tag{2.11}$$

In string units $\lambda_s^{d-1} = 1$ (which are the natural units of this frame), the exponential $\exp(\phi)$ thus parametrizes the effective gravitational coupling $8\pi G_D$.

The independent equation governing the dynamics of the dilaton field is obtained by varying the action (2.1) with respect to ϕ. In analogy with the definition (1.3) of the energy-momentum tensor $T_{\mu\nu}$ – which represents the tensor "current density" of the matter sources – we can define the scalar charge density σ of the sources by the variation with respect to ϕ as

$$\delta_\phi S_m = -\frac{1}{2} \int d^{d+1} \sqrt{|g|}\, \sigma\, \delta\phi. \tag{2.12}$$

As already noted, this scalar charge is zero for a pure BD model in the Jordan frame, but is non-zero, in general in all frames, for the string effective action. The variation of the full action with respect to ϕ leads to the Euler–Lagrange equation for the dilaton,

$$\partial_\mu \left[-2\sqrt{|g|} e^{-\phi} \partial^\mu \phi \right] = e^{-\phi} \sqrt{|g|} \left[R + (\nabla\phi)^2 \right] - \lambda_s^{d-1} \sqrt{|g|}\, \sigma, \tag{2.13}$$

which can be written in explicit covariant form as

$$R + 2\nabla^2 \phi - (\nabla\phi)^2 = \lambda_s^{d-1} e^\phi \sigma. \tag{2.14}$$

Using this equation to eliminate the scalar curvature present in the Einstein tensor, we can eventually recast Eq. (2.11) in the simplified form

$$R_\mu{}^\nu + \nabla_\mu \nabla^\nu \phi = \lambda_s^{d-1} e^\phi \left(T_\mu{}^\nu + \frac{1}{2} \delta_\mu^\nu \sigma \right). \tag{2.15}$$

The system of equations (2.14) and (2.15) (or, equivalently, (2.11)) replaces the Einstein equations in the description of the low-energy gravitational dynamics for a given distribution of sources, and can be applied to discuss possible modifications of the standard cosmological scenario. It is important to note, even at this stage, that with the above system of equations the contracted Bianchi identity, $\nabla_\nu G_\mu{}^\nu = 0$, no longer implies the covariant conservation of the stress tensor $T_{\mu\nu}$. Computing the covariant divergence of Eq. (2.11), and using the Bianchi identity, we obtain

$$\nabla_\nu(\nabla_\mu \nabla^\nu \phi) + (\nabla_\mu \nabla_\nu \phi) \nabla^\nu \phi - \nabla_\mu(\nabla^2 \phi) = \lambda_s^{d-1} e^\phi \left(\nabla_\nu T_\mu{}^\nu + T_\mu{}^\nu \nabla_\nu \phi \right). \tag{2.16}$$

The commutator of two covariant derivatives, on the other hand, is determined by the properties of the Riemann tensor, which imply

$$(\nabla_\mu \nabla_\nu - \nabla_\nu \nabla_\mu) A^\alpha = R_{\mu\nu\beta}{}^\alpha A^\beta, \tag{2.17}$$

for any vector A^α. The application of this commutation relation to the first term of Eq. (2.16) then gives

$$\nabla_\nu(\nabla_\mu \nabla^\nu \phi) = \nabla_\mu(\nabla^2 \phi) + R_{\mu\nu} \nabla^\nu \phi. \tag{2.18}$$

Inserting this result into Eq. (2.16), and using Eq. (2.15) for the Ricci tensor, we are finally led to the generalized conservation law

$$\nabla_\nu T_\mu{}^\nu = \frac{1}{2} \sigma \nabla_\mu \phi. \tag{2.19}$$

This result represents the crucial difference between the string gravity equations and the equations of a conventional BD model of scalar-tensor gravity, where $\sigma = 0$ (in the Jordan frame), so that $T_{\mu\nu}$ is separately conserved. The above equation implies, as we shall see in Chapter 9, that the motion of a free test body is no longer geodesic when the body has an intrinsic scalar charge and the gravitational background contains a non-trivial dilaton component.

Let us now come back to the action (2.1), to note that the gravitational sector can be generalized by the possible addition of a dilaton potential, $V(\phi)$, and by the presence of a third, fundamental string theory field belonging to the massless multiplet of bosonic string states (see Chapter 3). Such a field is represented by the antisymmetric tensor potential $B_{\mu\nu} = -B_{\nu\mu}$, usually called "Neveu-Schwarz–Neveu-Schwarz" (NS–NS) two-form, with totally antisymmetric field strength $H_{\mu\nu\alpha}$:

$$H_{\mu\nu\alpha} = \partial_\mu B_{\nu\alpha} + \partial_\nu B_{\alpha\mu} + \partial_\alpha B_{\mu\nu}. \tag{2.20}$$

For historical reasons the field B is also called "torsion" since, to lowest order, it can be identified with the antisymmetric part $\Gamma_{[\mu\nu]}{}^{\alpha}$ of the affine connection, in the context of a non-Riemannian geometric structure. An alternative, often used, name is "Kalb–Ramond axion", in reference to the pseudo-scalar (axion) field φ related to H, in four dimensions, by the space-time "duality" transformation $H_{\mu\nu\alpha} = \epsilon_{\mu\nu\alpha\beta}\nabla^{\beta}\varphi$.

Including the additional contributions of V and B, the gravi-dilaton part of the action (2.1) turns out to be generalized as follows:

$$S = -\frac{1}{2\lambda_s^{d-1}} \int d^{d+1}x \sqrt{|g|}\, e^{-\phi} \left[R + (\nabla\phi)^2 + 2\lambda_s^{d-1}V(\phi) - \frac{1}{12}H^2 \right], \quad (2.21)$$

where $H^2 = H_{\mu\nu\alpha}H^{\mu\nu\alpha}$. Note that we have normalized the dilaton potential in such a way that V has the canonical dimensions of an energy density, $[V] = M^{d+1}$. With these new terms, there are new contributions to the variation of the action, with respect to g:

$$\delta_g S = -\frac{1}{2\lambda_s^{d-1}} \int d^{d+1}x \sqrt{|g|}\, e^{-\phi} \left[-g_{\mu\nu}\lambda_s^{d-1}V(\phi) + \frac{1}{24}H^2 g_{\mu\nu} \right.$$

$$\left. -\frac{1}{4}H_{\mu\alpha\beta}H_\nu{}^{\alpha\beta} \right] \delta g^{\mu\nu}, \quad (2.22)$$

and with respect to ϕ:

$$\delta_\phi S = \frac{1}{2\lambda_s^{d-1}} \int d^{d+1}x \sqrt{|g|}\, e^{-\phi} \left[2\lambda_s^{d-1}\left(V - \frac{\partial V}{\partial\phi} \right) - \frac{1}{12}H^2 \right] \delta\phi. \quad (2.23)$$

Adding up these contributions to Eqs. (2.8) and (2.13), we obtain

$$G_\mu{}^\nu + \nabla_\mu\nabla^\nu\phi - \frac{1}{4}H_{\mu\alpha\beta}H^{\nu\alpha\beta}$$

$$+ \frac{1}{2}\delta_\mu^\nu \left[(\nabla\phi)^2 - 2\nabla^2\phi - 2\lambda_s^{d-1}V(\phi) + \frac{1}{12}H^2 \right] = \lambda_s^{d-1}e^{\phi}T_\mu{}^\nu, \quad (2.24)$$

$$R + 2\nabla^2\phi - (\nabla\phi)^2 + 2\lambda_s^{d-1}\left(V - \frac{\partial V}{\partial\phi} \right) - \frac{1}{12}H^2 = \lambda_s^{d-1}e^{\phi}\sigma, \quad (2.25)$$

$$R_\mu{}^\nu + \nabla_\mu\nabla^\nu\phi - \lambda_s^{d-1}\delta_\mu^\nu \frac{\partial V}{\partial\phi} - \frac{1}{4}H_{\mu\alpha\beta}H^{\nu\alpha\beta} = \lambda_s^{d-1}e^{\phi}\left(T_\mu{}^\nu + \frac{1}{2}\delta_\mu^\nu\sigma \right), \quad (2.26)$$

replacing Eqs. (2.11), (2.14) and (2.15), respectively. As we shall see in Chapter 4, the presence of the antisymmetric field is crucial in order to implement a generalized form of duality symmetry of the cosmological equations in the case of homogeneous backgrounds. The dilaton potential tends to break such a symmetry, but its presence is important for the formulation of a realistic cosmological scenario (as we shall see in many phenomenological applications).

The above system has to be completed by the equation of motion of the two-form field. By varying the action (2.21) with respect to $B_{\mu\nu}$, and defining the "axion" current density $J_{\mu\nu} = -J_{\nu\mu}$ of the matter sources as

$$\delta_B S_m = \frac{1}{2} \int d^{d+1} \sqrt{|g|} \, J_{\mu\nu} \, \delta B^{\mu\nu}, \qquad (2.27)$$

the corresponding Euler–Lagrange equations can be written in the form

$$\frac{1}{\sqrt{|g|}} \partial_\mu \left(\sqrt{|g|} \, e^{-\phi} H^{\mu\alpha\beta} \right) \equiv \nabla_\mu \left(e^{-\phi} H^{\mu\alpha\beta} \right) = 2\lambda_s^{d-1} J^{\alpha\beta}. \qquad (2.28)$$

As an example of matter with non-zero $J_{\mu\nu}$ we quote here the case of a gas of free fundamental strings [10], and we note that the role of the antisymmetric current $J_{\mu\nu}$ is essential for extending to the matter sector the duality symmetries of the low-energy string effective action.

Let us conclude this section by providing the explicit form of the above equations for a homogeneous, isotropic and spatially flat background (representing a typical cosmological configuration). We assume that $B_{\mu\nu}$ is vanishing and that the matter sources can be represented in the form of a fluid, with energy density $\rho(t)$, pressure $p(t)$ and dilaton charge $\sigma(t)$. In the synchronous gauge of the comoving frame we can set, therefore,

$$
\begin{aligned}
g_{\mu\nu} &= \mathrm{diag}(1, -a^2 \delta_{ij}), & a &= a(t), & \phi &= \phi(t), \\
T_\mu{}^\nu &= \mathrm{diag}(\rho, -p\delta_i^j), & \rho &= \rho(t), & p &= p(t), & \sigma &= \sigma(t).
\end{aligned}
\qquad (2.29)
$$

An explicit computation for a $(d+1)$-dimensional manifold gives the following non-zero components of the connection:

$$\Gamma_{0i}{}^j = H\delta_i^j, \qquad \Gamma_{ij}{}^0 = a\dot{a}\,\delta_{ij} \qquad (2.30)$$

(where $H = \dot{a}/a$) and of the Ricci tensor:

$$
\begin{aligned}
R_0{}^0 &= -d\dot{H} - dH^2, \\
R_i{}^j &= -\delta_i^j(\dot{H} + dH^2).
\end{aligned}
\qquad (2.31)
$$

The scalar curvature is

$$R = -2d\dot{H} - d(d+1)H^2. \qquad (2.32)$$

For the dilaton field we have

$$
\begin{aligned}
(\nabla\phi)^2 &= \dot{\phi}^2, & \nabla^2\phi &= \ddot{\phi} + dH\dot{\phi}, \\
\nabla_0\nabla^0\phi &= \ddot{\phi}, & \nabla_i\nabla^j\phi &= H\dot{\phi}\,\delta_i^j.
\end{aligned}
\qquad (2.33)
$$

The (00) component of Eq. (2.24) then gives

$$\dot{\phi}^2 - 2dH\dot{\phi} + d(d-1)H^2 = 2\lambda_s^{d-1}\left(e^{\phi}\rho + V\right),\qquad (2.34)$$

the space component (ij) of Eq. (2.26) gives

$$\dot{H} - H\dot{\phi} + dH^2 = \lambda_s^{d-1}\left[e^{\phi}\left(p - \frac{\sigma}{2}\right) - \frac{\partial V}{\partial\phi}\right],\qquad (2.35)$$

and the dilaton equation (2.25) gives

$$2\ddot{\phi} - \dot{\phi}^2 + 2dH\dot{\phi} - 2d\dot{H} - d(d+1)H^2 = 2\lambda_s^{d-1}\left(e^{\phi}\frac{\sigma}{2} + \frac{\partial V}{\partial\phi} - V\right).\qquad (2.36)$$

By differentiating Eq. (2.34) with respect to time, eliminating $\dot{H}$ through Eq. (2.35), and $\ddot{\phi}$ through Eq. (2.36), we are led to the generalized conservation equation,

$$\dot{\rho} + dH(\rho + p) = \frac{1}{2}\sigma\dot{\phi},\qquad (2.37)$$

which also directly follows from Eq. (2.19). We note, finally, that by eliminating $\dot{H}$ and H^2 through Eqs. (2.34) and (2.35) the dilaton equation can be rewritten in a useful form which explicitly contains the scalar coupling to the trace of the matter stress tensor:

$$\ddot{\phi} + dH\dot{\phi} - \dot{\phi}^2 + 2\lambda_s^{d-1}\left[V + \left(\frac{d-1}{2}\right)\frac{\partial V}{\partial\phi}\right]$$
$$+ \lambda_s^{d-1}e^{\phi}(\rho - dp) + \frac{d-1}{2}\lambda_s^{d-1}e^{\phi}\,\sigma = 0.\qquad (2.38)$$

This set of low-energy cosmological equations will be used repeatedly in many parts of this book.

2.2 The Einstein-frame representation

The discussion of the previous section is based on the S-frame representation of the string effective action, i.e. on the frame in which the coupling to a constant dilaton is unambiguosly fixed (at the tree-level) for all fields, and to all orders in the higher-derivative α' expansion [1, 2, 3]. The S-frame is, in a sense, the "preferred" string theory frame, where physical intuition is often more direct and easily applicable. However, in various practical applications it may be convenient to work in other frames, using a representation of the action different from that of Eq. (2.1). Of course – as will be again emphasized in this book – all observable results must be "frame independent", namely independent of the particular set of fields (generically called "frame") chosen to parametrize the action.

For a direct comparison of different cosmological scenarios, or for the direct application of known general relativistic results, it may be useful to rewrite the effective string equations in the so-called "Einstein frame" (E-frame), where the dilaton enters as a canonically normalized scalar field, minimally coupled to gravity, and where the Planck scale provides the natural units of length (λ_P) and energy ($\lambda_P^{-1} = M_P$). This frame is preferred, in particular, for performing canonical quantization, for identifying the particle content of a theory, and for defining the effective low-energy masses and couplings (as will be discussed in Chapter 9).

The E-frame parametrization is obtained from the S-frame action (2.1) by performing the conformal transformation (or "Weyl rescaling") that diagonalizes the gravi-dilaton kinetic term $\exp(-\phi)R$. To this purpose we introduce the rescaled metric $\widetilde{g}$, related to the S-frame metric g by the field redefinition

$$g_{\mu\nu} = \widetilde{g}_{\mu\nu} \left(\frac{\lambda_s}{\lambda_P}\right)^2 e^{\psi(x)}, \tag{2.39}$$

where ψ is an arbitrary (scalar) space-time function. A computation of the transformed scalar curvature then gives [11, 12]

$$R = \left(\frac{\lambda_s}{\lambda_P}\right)^2 e^{-\psi(x)} \left[\widetilde{R} - d\widetilde{\nabla}^2\psi - \frac{d}{4}(d-1)(\widetilde{\nabla}\psi)^2\right], \tag{2.40}$$

where $R = R(g)$, $\widetilde{R} = \widetilde{R}(\widetilde{g})$ and $\widetilde{\nabla}$ is the covariant derivative associated with $\widetilde{g}$. By using

$$\sqrt{|g|} = \sqrt{|\widetilde{g}|} \left(\frac{\lambda_s}{\lambda_P}\right)^{d+1} e^{\psi(d+1)/2}, \tag{2.41}$$

the action becomes

$$S = S_m(\phi, \widetilde{g}) - \frac{1}{2\lambda_P^{d-1}} \int d^{d+1}x \sqrt{|\widetilde{g}|}\, e^{-\phi+\psi(d-1)/2}$$

$$\times \left[\widetilde{R} - d\widetilde{\nabla}^2\psi - \frac{d}{4}(d-1)(\widetilde{\nabla}\psi)^2 + \left(\widetilde{\nabla}\phi\right)^2 + 2\frac{\lambda_s^{d+1}}{\lambda_P^2}e^{\psi}V(\phi)\right] \tag{2.42}$$

(we have included the potential, normalized as in Eq. (2.21)). We now fix the conformal factor as

$$\psi = \frac{2\phi}{d-1}, \tag{2.43}$$

in such a way that the exponential pre-factor disappears from the action, and $\widetilde{\nabla}^2\psi$ becomes a total divergence which can be neglected. The action parametrized by $\widetilde{g}$ then assumes the standard Einstein form, with the dilaton field minimally coupled to the metric, and the gravitational coupling strength determined (as usual) by the Planck length λ_P:

$$S = \frac{1}{2\lambda_P^{d-1}} \int d^{d+1}x \sqrt{|\tilde{g}|} \left[-\tilde{R} + \frac{1}{d-1} \left(\tilde{\nabla}\phi \right)^2 - 2\frac{\lambda_s^{d+1}}{\lambda_P^2} V(\phi) \, e^{2\phi/(d-1)} \right]$$

$$+ S_m(\phi, \tilde{g}). \tag{2.44}$$

We can finally introduce the rescaled dilaton $\tilde{\phi}$, such that

$$\tilde{\phi} = \mu\phi, \qquad \mu = \left(\frac{M_P^{d-1}}{d-1} \right)^{1/2}, \tag{2.45}$$

where $\lambda_P^{d-1} = M_P^{1-d} = 8\pi G_D$. This new field has canonical dimensions, $[\tilde{\phi}] = M^{(d-1)/2}$, and its kinetic term is canonically normalized. The action becomes

$$S = \int d^{d+1}x \sqrt{|\tilde{g}|} \left[-\frac{\tilde{R}}{16\pi G_D} + \frac{1}{2} \left(\tilde{\nabla}\tilde{\phi} \right)^2 - \tilde{V}(\tilde{\phi}) \right] + S_m(\tilde{\phi}, \tilde{g}), \tag{2.46}$$

where

$$\tilde{V} = \left(\frac{\lambda_s}{\lambda_P} \right)^{d+1} V(\tilde{\phi}) \, e^{2\tilde{\phi}/\mu(d-1)}. \tag{2.47}$$

It is important to note the flipped sign of the dilaton kinetic term with respect to the action (2.1). Note, also, that the factor $(\lambda_s/\lambda_P)^{d+1}$ rescales in Planck units the potential V, originally expressed in string units in the S-frame action.

The E-frame equations for the gravi-dilaton sector of the action (2.46) are identical to the equations one obtains in general relativity for a canonical scalar field $\tilde{\phi}$, possibly self-interacting through the potential $\tilde{V}$, and minimally coupled to the metric $\tilde{g}$ with strength fixed by the Newton constant $G_D = \lambda_P^{d-1}/8\pi$. All known general-relativistic results for the system $\{\tilde{g}, \tilde{\phi}\}$ can thus be safely applied to this case. The inclusion of other matter fields, however, leads to a set of equations in principle different from the corresponding equations of general relativity, because of the different dilaton couplings possibly generated by the transformation from g to $\tilde{g}$ for the matter fields present in S_m.

The full set of E-frame equations can be derived in two ways: either by varying the action (2.46), or by directly transforming the S-frame equations, using the relations (2.39), (2.43) and (2.45) between the two sets of variables. Here we follow the first procedure, starting from the definition of the E-frame sources of tensor and scalar interactions:

$$\delta_{\tilde{g}} S_m = \frac{1}{2} \int d^{d+1}x \sqrt{|\tilde{g}|} \, \tilde{T}_{\mu\nu} \, \delta\tilde{g}^{\mu\nu},$$

$$\delta_{\tilde{\phi}} S_m = -\frac{1}{2} \int d^{d+1}x \sqrt{|\tilde{g}|} \, \tilde{\sigma}\delta\tilde{\phi}. \tag{2.48}$$

It is instructive, before proceeding with the variational computation, to derive the explicit relation between $\widetilde{T}$, $\widetilde{\sigma}$ and the S-frame sources T, σ, following from the transformation laws (2.39) and (2.43).

For the stress tensor we have

$$\int d^{d+1}x\sqrt{|g|}\,T_{\mu\nu}\delta g^{\mu\nu} = \int d^{d+1}x\sqrt{|\widetilde{g}|}\left(\frac{\lambda_s}{\lambda_P}\right)^{d-1}e^{\frac{d+1}{d-1}\phi}T_{\mu\nu}e^{-\frac{2\phi}{d-1}}\delta\widetilde{g}^{\mu\nu}, \quad (2.49)$$

from which

$$\widetilde{T}_{\mu\nu} = \left(\frac{\lambda_s}{\lambda_P}\right)^{d-1}e^{\phi}T_{\mu\nu},$$

$$\widetilde{T}_{\mu}{}^{\nu} = \widetilde{g}^{\nu\alpha}\widetilde{T}_{\mu\alpha} = \left(\frac{\lambda_s}{\lambda_P}\right)^{d+1}e^{\frac{d+1}{d-1}\phi}T_{\mu}{}^{\nu}.$$
$$(2.50)$$

Reintroducing $\widetilde{\phi}$, and considering the diagonal stress tensor of a perfect fluid, one obtains, in particular,

$$\widetilde{\rho} = \left(\frac{\lambda_s}{\lambda_P}\right)^{d+1}e^{\frac{d+1}{d-1}\frac{\widetilde{\phi}}{\mu}}\rho,$$

$$\widetilde{p} = \left(\frac{\lambda_s}{\lambda_P}\right)^{d+1}e^{\frac{d+1}{d-1}\frac{\widetilde{\phi}}{\mu}}p,$$
$$(2.51)$$

where the constant ratio $(\lambda_s/\lambda_P)^{d+1}$ is needed, as before, to rescale in Planck units the physical variables originally expressed in string units in the S-frame.

For the density of scalar charge we have, similarly,

$$\int d^{d+1}x\sqrt{|g|}\,\sigma\delta\phi = \int d^{d+1}x\sqrt{|\widetilde{g}|}\left(\frac{\lambda_s}{\lambda_P}\right)^{d+1}e^{\frac{d+1}{d-1}\phi}\left(\frac{\sigma}{\mu}\right)\delta\widetilde{\phi}, \quad (2.52)$$

from which

$$\widetilde{\sigma} = \left(\frac{\lambda_s}{\lambda_P}\right)^{d+1}e^{\frac{d+1}{d-1}\frac{\widetilde{\phi}}{\mu}}\left(\frac{\sigma}{\mu}\right). \quad (2.53)$$

Note that σ has the dimension of an energy density, while $\widetilde{\sigma}$ has different dimensions, being defined with respect to $\widetilde{\phi}$, which is not dimensionless like ϕ.

Let us now consider the variation of the action (2.46). By varying with respect to $\widetilde{g}$, and eliminating the variational contribution of $\widetilde{\nabla}\delta\widetilde{g}$ through an appropriate boundary term (see Section 2.1), we obtain

$$\int d^{d+1}x\sqrt{|\widetilde{g}|}\left[-\frac{\widetilde{G}_{\mu\nu}}{2\lambda_P^{d-1}} + \frac{1}{2}\left(\partial_{\mu}\widetilde{\phi}\partial_{\nu}\widetilde{\phi} - \frac{1}{2}\widetilde{g}_{\mu\nu}(\widetilde{\nabla}\widetilde{\phi})^2\right) + \frac{1}{2}\widetilde{g}_{\mu\nu}\widetilde{V}\right]\delta\widetilde{g}^{\mu\nu}$$

$$+ \delta_{\widetilde{g}}S_m = 0. \quad (2.54)$$

A second variation with respect to $\widetilde{\phi}$, taking explicitly into account the dependence of S_m on $\widetilde{\phi}$ induced by the conformal transformation (2.39), gives

$$\int d^{d+1}x \sqrt{|\widetilde{g}|} \left[-\frac{1}{\sqrt{|\widetilde{g}|}} \partial_\mu \left(\sqrt{|\widetilde{g}|} \, \partial^\mu \widetilde{\phi} \right) - \frac{\partial \widetilde{V}}{\partial \widetilde{\phi}} \right] \delta\widetilde{\phi}$$

$$+ \int d^{d+1}x \left[\frac{\partial \sqrt{|\widetilde{g}|} \mathcal{L}_m}{\partial \widetilde{\phi}} + \frac{\delta \sqrt{|g|} \mathcal{L}_m}{\delta g^{\mu\nu}} \frac{\partial g^{\mu\nu}}{\partial \widetilde{\phi}} \right] \delta\widetilde{\phi} = 0. \tag{2.55}$$

We can also use the transformation (2.39), and the relation (2.50) between T and $\widetilde{T}$, to obtain

$$\frac{\delta S_m}{\delta g^{\mu\nu}} \frac{\partial g^{\mu\nu}}{\partial \widetilde{\phi}} \delta\widetilde{\phi} = \frac{1}{2} \int d^{d+1}x \sqrt{|g|} \, T_{\mu\nu} \frac{\partial g^{\mu\nu}}{\partial \widetilde{\phi}} \delta\widetilde{\phi}$$

$$= -\frac{1}{\mu(d-1)} \int d^{d+1}x \sqrt{|\widetilde{g}|} \, \widetilde{T} \, \delta\widetilde{\phi}. \tag{2.56}$$

By inserting the explicit sources into the variational equations we are thus led to the system of equations

$$\widetilde{G}_\mu{}^\nu = \lambda_{\rm P}^{d-1} \left[\widetilde{T}_\mu{}^\nu + \partial_\mu \widetilde{\phi} \, \partial^\nu \widetilde{\phi} - \frac{1}{2} \delta_\mu^\nu (\widetilde{\nabla}\widetilde{\phi})^2 + \delta_\mu^\nu \, \widetilde{V} \right], \tag{2.57}$$

$$\widetilde{\nabla}^2 \widetilde{\phi} + \frac{\partial \widetilde{V}}{\partial \widetilde{\phi}} + \frac{\widetilde{\sigma}}{2} + \frac{\widetilde{T}}{\mu(d-1)} = 0, \tag{2.58}$$

describing the low-energy gravitational dynamics according to the E-frame representation of the string effective action ($\widetilde{\sigma}$ is defined by the first term in the second line of Eq. (2.55)). The application of the contracted Bianchi identity, $\widetilde{\nabla}_\nu \widetilde{G}_\mu^\nu = 0$, leads to the associated conservation equation

$$\widetilde{\nabla}_\nu \widetilde{T}_\mu^\nu = \left[\frac{\widetilde{\sigma}}{2} + \frac{\widetilde{T}}{\mu(d-1)} \right] \widetilde{\nabla}_\mu \widetilde{\phi}. \tag{2.59}$$

Finally, we note that the standard equations of general relativity are recovered in the limit in which we neglect the non-minimal coupling of the dilaton to the trace of the stress tensor $\widetilde{T}$ (induced by the conformal transformation) and to $\widetilde{\sigma}$, induced by the intrinsic (S-frame) scalar charge σ.

For a conformally flat metric background, and a perfect-fluid representation of the matter sources, we may now directly transfer to the E-frame the results of Eqs. (2.29)–(2.33) (simply by adding a tilde over all variables, $\widetilde{a}, \widetilde{p} \dots$), and write the independent components of the system of equations (2.57) and (2.58). We use the explicit form (2.47) of $\widetilde{V}$ in terms of V (which is convenient for further applications), and we denote $\widetilde{H} = \dot{\widetilde{a}}/\widetilde{a}$ the Hubble parameter for the E-frame metric $\tilde{g}$, where the dot denotes differentiation with respect to the E-frame cosmic time. The (00) component of Eq. (2.57) then gives

$$d(d-1)\tilde{H}^2 = 2\lambda_P^{d-1}\left[\tilde{\rho}+\frac{1}{2}\dot{\tilde{\phi}}^2+\left(\frac{\lambda_s}{\lambda_P}\right)^{d+1} e^{\frac{2}{d-1}\frac{\tilde{\phi}}{\mu}} V(\tilde{\phi})\right], \tag{2.60}$$

the (ij) component gives

$$2(d-1)\dot{\tilde{H}}+d(d-1)\tilde{H}^2 = 2\lambda_P^{d-1}\left[-\tilde{p}-\frac{1}{2}\dot{\tilde{\phi}}^2+\left(\frac{\lambda_s}{\lambda_P}\right)^{d+1} e^{\frac{2}{d-1}\frac{\tilde{\phi}}{\mu}} V(\tilde{\phi})\right], \tag{2.61}$$

and the dilaton equation (2.58) gives

$$\ddot{\tilde{\phi}}+d\tilde{H}\dot{\tilde{\phi}}+\frac{\tilde{\sigma}}{2}+\frac{1}{\mu(d-1)}(\tilde{\rho}-d\tilde{p})+\left(\frac{\lambda_s}{\lambda_P}\right)^{d+1} e^{\frac{2}{d-1}\frac{\tilde{\phi}}{\mu}}\left[\frac{2V(\tilde{\phi})}{\mu(d-1)}+\frac{\partial V}{\partial\tilde{\phi}}\right]=0. \tag{2.62}$$

Differentiating Eq. (2.60), and eliminating $\dot{\tilde{H}}$ and $\ddot{\tilde{\phi}}$ through Eqs. (2.61) and (2.62), respectively, we obtain the conservation equation

$$\dot{\tilde{\rho}}+d\tilde{H}(\tilde{\rho}+\tilde{p}) = \frac{1}{2}\tilde{\sigma}\dot{\tilde{\phi}}+\frac{1}{\mu(d-1)}(\tilde{\rho}-d\tilde{p})\dot{\tilde{\phi}}, \tag{2.63}$$

which obviously corresponds to the homogeneous and isotropic limit of the general equation (2.59).

We conclude the section with an instructive exercise showing that the set of cosmological equations (2.60)–(2.62), obtained by varying the E-frame action, can also be obtained by directly transforming to the E-frame the S-frame cosmological equations (2.34)–(2.38).

We start by defining $k = (\lambda_s/\lambda_P)$, and representing the conformal rescaling in compact form as follows:

$$\begin{aligned} \tilde{a} &= k^{-1}a e^{-\frac{\phi}{d-1}}, & d\tilde{t} &= k^{-1}dt\, e^{-\frac{\phi}{d-1}}, & \tilde{\phi} &= \mu\phi, \\ \tilde{\rho} &= k^{d+1}e^{\frac{d+1}{d-1}\phi}\rho, & \tilde{p} &= k^{d+1}e^{\frac{d+1}{d-1}\phi}p, & \tilde{\sigma} &= k^{d+1}e^{\frac{d+1}{d-1}\phi}\mu^{-1}\sigma. \end{aligned} \tag{2.64}$$

Here $\tilde{t}$ is the cosmic-time coordinate in the E-frame, and μ is the factor (2.45) required for the canonical rescaling of the dilaton. Using the above transformations we obtain, in particular,

$$\dot{\phi} = \frac{d\phi}{dt} = \frac{1}{\mu}\frac{d\tilde{\phi}}{d\tilde{t}}\frac{d\tilde{t}}{dt} = \frac{1}{k\mu}\dot{\tilde{\phi}}e^{-\phi/(d-1)}, \tag{2.65}$$

where $\dot{\tilde{\phi}} \equiv d\tilde{\phi}/d\tilde{t}$, and

$$\ddot{\phi} = \frac{d\tilde{t}}{dt}\frac{d\dot{\phi}}{d\tilde{t}} = \frac{e^{-\frac{2\phi}{d-1}}}{k^2}\left[\frac{\ddot{\tilde{\phi}}}{\mu} - \frac{\dot{\tilde{\phi}}^2}{\mu^2(d-1)}\right], \qquad (2.66)$$

where $\ddot{\tilde{\phi}} = d^2\tilde{\phi}/d\tilde{t}^2$. In the same way

$$H = \frac{1}{a}\frac{da}{dt} = \frac{e^{-\frac{\phi}{d-1}}}{k}\left[\tilde{H} + \frac{\dot{\tilde{\phi}}}{\mu(d-1)}\right], \qquad (2.67)$$

where $\tilde{H} = \dot{\tilde{a}}/\tilde{a}$ with $\dot{\tilde{a}} = d\tilde{a}/d\tilde{t}$, and

$$\dot{H} = \frac{e^{-\frac{2\phi}{d-1}}}{k^2}\left[\dot{\tilde{H}} + \frac{\ddot{\tilde{\phi}}}{\mu(d-1)} - \frac{\tilde{H}\dot{\tilde{\phi}}}{\mu(d-1)} - \frac{\dot{\tilde{\phi}}^2}{\mu^2(d-1)^2}\right], \qquad (2.68)$$

where $\dot{\tilde{H}} = d\tilde{H}/d\tilde{t}$, and so on.

The S-frame dilaton equation (2.38), using the above relations, can then be written in terms of the E-frame variables as

$$\ddot{\tilde{\phi}} + d\tilde{H}\dot{\tilde{\phi}} + k^{d+1}e^{\frac{d+1}{d-1}\phi}\left[\frac{\rho - dp}{\mu(d-1)} + \frac{\sigma}{2\mu}\right] + \frac{2k^{d+1}}{d-1}e^{\frac{2\phi}{d-1}}\left[\frac{V}{\mu} + \frac{(d-1)}{2}\frac{\partial V}{\partial\tilde{\phi}}\right] = 0, \qquad (2.69)$$

so that we exactly recover Eq. (2.62) after use of the definitions (2.64) for ρ, p and σ. In a similar way, starting from the S-frame equation (2.34), we obtain

$$d(d-1)\tilde{H}^2 = 2\lambda_{\mathrm{P}}^{d-1}\left[\frac{\dot{\tilde{\phi}}^2}{2} + k^{d+1}\left(e^{\frac{d+1}{d-1}\phi}\rho + e^{\frac{2\phi}{d-1}}V\right)\right], \qquad (2.70)$$

which reduces to Eq. (2.60) after use of the definitions (2.64). Finally, starting from Eq. (2.35), we obtain

$$\dot{\tilde{H}} + d\tilde{H}^2 + \frac{\ddot{\tilde{\phi}}}{\mu(d-1)} + \frac{d\tilde{H}\dot{\tilde{\phi}}}{\mu(d-1)} + \mu\lambda_{\mathrm{P}}^{d-1}k^{d+1}e^{\frac{2\phi}{d-1}}\frac{\partial V}{\partial\tilde{\phi}}$$

$$= \lambda_{\mathrm{P}}^{d-1}k^{d+1}e^{\frac{d+1}{d-1}\phi}\left(p - \frac{\sigma}{2}\right). \qquad (2.71)$$

The last three terms on the left-hand side, as well as the last term on the right-hand side of this equation, can be eliminated through Eq. (2.69), which implies

$$\frac{\ddot{\tilde{\phi}}}{\mu(d-1)} + \frac{d\widetilde{H}\dot{\tilde{\phi}}}{\mu(d-1)} + \mu\lambda_P^{d-1}k^{d+1}\,e^{\frac{2\phi}{d-1}}\frac{\partial V}{\partial\tilde{\phi}} + \frac{\sigma}{2}\lambda_P^{d-1}k^{d+1}\,e^{\frac{d+1}{d-1}\phi}$$

$$= -\frac{k^{d+1}}{\mu^2(d-1)}\left[2e^{\frac{2\phi}{d-1}}V + e^{\frac{d+1}{d-1}\phi}(\rho - dp)\right]$$

$$\equiv -\frac{2\lambda_P^{d-1}}{d-1}\left[\widetilde{V} + \frac{1}{2}(\widetilde{\rho} - d\widetilde{p})\right]. \qquad (2.72)$$

Inserting this result into Eq. (2.71), and eliminating $\widetilde{\rho}$ through Eq. (2.60), we exactly recover the spatial equation (2.61) which completes the system of cosmological equations in the E-frame representation.

2.3 First-order α' corrections

The effective action introduced in the previous sections is compatible with the conformal invariance of a quantized string in a curved background only to zeroth order of the expansion in powers of $\alpha' = \lambda_s^2/2\pi$ (as we shall see in Chapter 3). To first order in α', the condition of quantum conformal invariance introduces higher-derivative terms in the equations for the background fields, in such a way that their equations can be derived from an effective action containing quadratic curvature corrections, of the type $\sim \alpha' R^2$.

The α' corrections to the effective action become more and more important as the curvature grows: in principle, all higher-order contributions should be included when the curvature radius of the space-time – or, more generally, the inverse of the gradients of the background fields – becomes comparable with (or smaller than) the fundamental string length (namely when $\lambda_s^2 R \gtrsim 1$, $\lambda_s^2(\nabla\phi)^2 \gtrsim 1$, and so on). In that regime, the perturbative expansion of the effective action fails to give a consistent description of the background dynamics: one should instead adopt an exact conformal field-theory model, which automatically takes into account (in a non-perturbative way) the α' corrections to all orders (see [13] for possible examples in a cosmological context).

The α' corrections to the classical field equations are a peculiar string theory effect, due to the finite extension of the fundamental components of the theory. They may be expected to play an important role in the possible regularization of the singularities appearing in the gravitational Einstein theory (and, more generally, in any field theory based on the notion of point-like particles). Here we limit our discussion to the first-order α' corrections, but we stress that even to this order the higher-derivative terms seem to have promising applications to the problem of removing the curvature singularities, not only in a cosmological context [14] but also in the case of static and spherically symmetric gravitational fields [15].

It must be stressed, also, that the whole series of α' corrections is rigidly prescribed by the condition of conformal invariance applied to the scattering amplitudes determined by the string S-matrix [1] (at the tree-level in the string coupling g_s). At each approximation level – corresponding to the truncation of the action at a given order of the α' expansion – there is, however, an intrinsic (and unavoidable) ambiguity, due to field redefinitions which preserve the general covariance and the gauge invariance of the action [16]. Performing such redefinitions one can obtain a large class of different effective actions which are all of the same order in α', and all acceptable, in the sense that they are all equivalent to the same S-matrix, and thus perfectly compatible with the condition of conformal invariance. Such an ambiguity remains even when imposing on the field redefinitions the restriction of preserving a given frame representation.

The discussion of this section concerns the gravi-dilaton sector of the fundamental multiplet of massless string states, and is referred to the S-frame, where the coupling to a constant dilaton is unambiguously fixed. The most general action quadratic in the curvature then contains nine different additional terms: it can be written as $S = S_0 + S_1$, where S_0 is the gravi-dilaton action (2.1), and

$$
S_1 = \frac{\alpha'}{2\lambda_s^{d-1}} \int d^{d+1}x \sqrt{|g|}\, e^{-\phi} a_0 \left[R_{\mu\nu\alpha\beta}^2 + a_1 R_{\mu\nu}^2 + a_2 R^2 + \frac{a_3}{4} R^{\mu\nu} \nabla_\mu \phi \nabla_\nu \phi \right.
$$
$$
\left. + \frac{a_4}{4} R(\nabla\phi)^2 + \frac{a_5}{4} R\nabla^2\phi + \frac{a_6}{4}(\nabla^2\phi)^2 + \frac{a_7}{8}\nabla^2\phi(\nabla\phi)^2 + \frac{a_8}{16}(\nabla\phi)^4 \right].
$$

$$(2.73)$$

Any other quadratic invariant built up with the metric and the dilaton can be reduced to one of these nine terms, up to a total divergence.

Given a set of parameters $\{a_0, a_1, \ldots, a_8\}$, determined in such a way that the condition of conformal invariance of the quantized string model is satisfied, a new, equivalent set of parameters can be obtained through an appropriate field redefinition (truncated to first order in α'), which preserves the general covariance of the action (and also the gauge invariance under tranformations $\delta B_{\mu\nu} = \partial_\mu \lambda_\nu - \partial_\nu \lambda_\mu$, were the NS–NS two-form included in S_1). Choosing to preserve the S-frame representation, and working with the gravi-dilaton sector, one can still perform the following general transformations [16]:

$$
g'_{\mu\nu} = g_{\mu\nu} + \alpha' \left\{ b_1 R_{\mu\nu} + \frac{b_2}{4} \nabla_\mu \phi \nabla_\nu \phi + g_{\mu\nu} \left[b_3 R + \frac{b_4}{4}(\nabla\phi)^2 + \frac{b_5}{2} \nabla^2\phi \right] \right\},
$$
$$
\frac{\phi'}{2} = \frac{\phi}{2} + \alpha' \left\{ c_1 R + \frac{c_2}{4}(\nabla\phi)^2 + \frac{c_3}{2} \nabla^2\phi \right\},
$$

$$(2.74)$$

with arbitrary coefficients b_i, c_i. The computation of the transformed action, truncated to first order in α', leads again to the form (2.73), but with a new set of coefficients $\{a_0', a_1', \ldots, a_8'\}$, given in terms of a_i, b_i, c_i as follows [16]:

$$a_0' = a_0,$$

$$a_1' = a_1 - b_1,$$

$$a_2' = a_2 - 2c_1 + \frac{b_1}{2} + \frac{b_3}{2}(d-1),$$

$$a_3' = a_3 - 4b_1 - b_2,$$

$$a_4' = a_4 + 8c_1 - 2(d+1)b_3 - 2c_2 + \frac{b_2}{2} + \frac{b_4}{2}(d-1), \qquad (2.75)$$

$$a_5' = a_5 - 8c_1 + b_1 + 2db_3 - 2c_3 + \frac{b_5}{2}(d-1),$$

$$a_6' = a_6 - 8c_3 + 2db_5,$$

$$a_7' = a_7 + 3b_2 + 2db_4 - 8c_2 + 8c_3 - 2(d+1)b_5,$$

$$a_8' = a_8 - 4b_2 + 8c_2 - 2(d+1)b_4.$$

It can be easily checked that, besides a_0, the following combination of coefficients is also invariant:

$$a_8 + 2a_7 + 4a_6 - 8a_5 - 4a_4 + 16a_2 = \text{const.} \qquad (2.76)$$

It follows that one can eliminate at most seven of the nine parameters present in the action (2.73), through an appropriate transformation of the type (2.74). A convenient choice is then represented by the field redefinition which fixes $a_1 = a_2 = \cdots = a_7 = 0$, in view of the fact that the condition of conformal invariance requires (in the S-frame) that $a_8 = 0$, and that $a_0 = k/4$, where k is a model-dependent numerical coefficient depending on the considered model of string [16]. One obtains in this way the simplest form of the gravi-dilaton action compatible with a consistent string quantization, up to first order in α'. For the bosonic string, in particular, $k = 1$, so that

$$S = -\frac{1}{2\lambda_s^{d-1}} \int d^{d+1}x \sqrt{|g|} \, e^{-\phi} \left[R + (\nabla\phi)^2 - \frac{\alpha'}{4} R_{\mu\nu\alpha\beta}^2 \right]. \qquad (2.77)$$

The application of this action as an effective model of string gravity – for instance in a cosmological context – is complicated by the fact that the equations following from the variation of the Riemann-squared term contain, in general, higher than second derivatives of the metric tensor. Such a formal complication can be avoided, however, by performing an appropriate field redefinition shifting the action from the set of parameters $a_1 = a_2 = \cdots = a_8 = 0$ to a new, equivalent set in which $a_1 = -4$ and $a_2 = 1$, in such a way that the Riemann-squared term turns out to be replaced by the so-called Euler–Gauss–Bonnet invariant,

$$R_{\text{GB}}^2 \equiv R_{\mu\nu\alpha\beta}^2 - 4R_{\mu\nu}^2 + R^2. \tag{2.78}$$

In that case it is known that the gravitational field equations contain at most second-order derivatives of the metric tensor (see for instance [17]).

Considering the general transformation rules (2.75) we can see that it is always possible to obtain the desired result, provided we introduce α' corrections also to the dilaton kinetic term. We may consider, for instance, the following field redefinitions:

$$g'_{\mu\nu} = g_{\mu\nu} + 4\alpha' \left[R_{\mu\nu} - \nabla_\mu \phi \nabla_\nu \phi + g_{\mu\nu} (\nabla\phi)^2 \right],$$

$$\phi' = \phi + \alpha' \left[R + (2d - 3)(\nabla\phi)^2 \right], \tag{2.79}$$

corresponding to the transformation (2.74) with $b_1 = 4$, $b_2 = -16 = -b_4$, $c_1 = 1/2$, $c_2 = 2(2d - 3)$. When applied to the action (2.77) one finds that the new coefficients $a'_3, a'_4, \ldots, a'_7$ are still vanishing, but $a'_8 = -16 \neq 0$, and one is led to the following effective action (truncated to first order in α'):

$$S = -\frac{1}{2\lambda_{\text{s}}^{d-1}} \int d^{d+1}x \sqrt{|g|} \, e^{-\phi} \left[R + (\nabla\phi)^2 - \frac{\alpha'}{4} R_{\text{GB}}^2 + \frac{\alpha'}{4} (\nabla\phi)^4 \right]. \tag{2.80}$$

It will be shown, in the following chapters, that the cosmological equations obtained from this action (as well as from the action (2.77), and from other, equivalent forms of the first-order action) admit particular solutions describing a phase of constant curvature and linearly evolving dilaton – which represents a possible exact solution of the string theory equations even *to all orders* of the α' expansion [10]. However, we will concentrate our attention on the particular parametrization associated with the action (2.80) because, in that case, there are solutions in which such a (typically "stringy") phase of high, constant curvature may act as an asymptotic attractor of the primordial cosmological evolution: in particular, for those solutions, the constant curvature phase is smoothly connected to the string perturbative vacuum, which may represent a natural initial configuration in the context of "self-dual" inflationary scenarios (see Chapter 6).

The property of smooth cosmological evolution, unfortunately, is *not* invariant under field redefinitions when the action is truncated at any given finite order of the perturbative expansion. There are, for instance, other choices of the coefficients a_i corresponding to actions which are still free from higher derivatives in the field equations, compatible with the required condition of conformal invariance [18], and also compatible with a higher-order extension of the tree-level T-duality symmetry [19]. In those cases the solutions at constant curvature exist, and are fixed points of the cosmological evolution, but they are classically disconnected from the perturbative regions of phase space.

2.3.1 Higher-order gravi-dilaton equations

We conclude this section with a detailed derivation of the set of gravi-dilaton equations following from the higher-order action (2.80). Using the method presented here, we believe that the reader will be able to apply the same procedure to derive the field equations for any other model of higher-derivative action.

In view of the presence of quadratic curvature terms it is convenient to adopt the language of the external (differential) forms, for a more compact notation (a few technical details and definitions, to introduce the reader to this formalism, will be provided in Appendix 2A). This formalism is based on the projection of the gravitational dynamics on the Minkowski space locally tangent to the Riemannian manifold and uses, as fundamental variables, the vielbein (V^a) and the Lorentz connection (ω^{ab}) one-forms,

$$V^a = V^a_\mu \, dx^\mu, \qquad \omega^{ab} = \omega_\mu{}^{ab} \, dx^\mu \qquad (2.81)$$

(see for instance [11]). The notations that will be adopted from now on to the end of this chapter will be as follows: Greek indices $\mu, \nu, \cdots = 0, 1, \ldots, d$ (usually called *holonomic* indices, or world indices) will denote tensor components transforming covariantly under general coordinate reparametrizations of the curved Riemannian manifold, with metric $g_{\mu\nu}$; Roman indices $a, b, \cdots = 0, 1, \ldots, d$ (called *anholonomic* indices, or flat indices) will denote components of the tensor representation of the local Lorentz group acting in the flat Minkowski space-time, with metric η_{ab}, locally tangent to the given world manifold.

In such a context, any world tensor $A^{\mu\nu\cdots}$ can be locally projected into a corresponding flat-space (Lorentz) tensor $A^{ab\cdots} = A^{\mu\nu\cdots} V^a_\mu V^b_\nu \cdots$ (and vice versa) through the vielbein fields V^a_μ (and their inverse V^μ_a, such that $V^a_\mu V^\mu_b = \delta^a_b$), which represent an orthonormal base in the locally tangent Minkowski space, and which satisfy the orthonormality conditions

$$g^{\mu\nu} V^a_\mu V^b_\nu = \eta^{ab}, \qquad \eta_{ab} V^a_\mu V^b_\nu = g_{\mu\nu}. \qquad (2.82)$$

The reparametrization invariance of the Riemannian manifold is thus translated into the *local* Lorentz invariance of the tangent-space formulation of the gravitational equations. In the presence of a local symmetry, on the other hand, we need a "connection", which in this case is represented by the Lorentz connection ω_μ (also called the "spin connection" [6]), representing the "gauge potential" which compensates the non-homogeneous transformations of the gradient with respect to local Lorentz rotations $\Lambda(x)$, and which transforms as

$$\omega_\mu \to \Lambda \omega_\mu \Lambda^{-1} - (\partial_\mu \Lambda) \Lambda^{-1}. \qquad (2.83)$$

This connection defines a Lorentz-covariant derivative, which will be denoted by $D_\mu = \partial_\mu + \omega_\mu$, and which transforms homogeneously even locally:

$$A \to \Lambda(x)A \quad \Longrightarrow \quad D_\mu A \to \Lambda(x)\,(D_\mu A). \tag{2.84}$$

In particular, for variables which are tensor-valued on the local Lorentz group, the linear action of the connection is represented by the $(d+1) \times (d+1)$ anti-symmetric matrices $\omega_\mu{}^{ab} = -\omega_\mu{}^{ba}$. For a generic (world-scalar) Lorentz tensor $A^{a\cdots}{}_{b\cdots}$ we have, for instance,

$$D_\mu A^{a\cdots}{}_{b\cdots} = \partial_\mu A^{a\cdots}{}_{b\cdots} + \omega_\mu{}^a{}_c A^{c\cdots}{}_{b\cdots} - \omega_\mu{}^c{}_b A^{a\cdots}{}_{c\cdots} + \cdots. \tag{2.85}$$

For an object carrying both curved and flat space indices the full covariant derivative will be defined, of course, by using both the Lorentz and the Christoffel connection. An important example of this type is represented by the covariant derivative of the vielbein field,

$$\nabla_\mu V_\nu^a = \partial_\mu V_\nu^a + \omega_\mu{}^a{}_b V_\nu^b - \Gamma_{\mu\nu}{}^\alpha V_\alpha^a \equiv D_\mu V_\nu^a - \Gamma_{\mu\nu}{}^a. \tag{2.86}$$

It is important to note that, for consistency with the relations (2.82), the previous covariant derivative has to be vanishing for a gravitational theory which is of the metric type (i.e. which satisfies $\nabla_\alpha g_{\mu\nu} = 0$), and for a local symmetry group which (like the Lorentz group) contains $d(d+1)/2$ parameters, and is thus associated with an antisymmetric connection (which implies $D_\mu \eta_{ab} = 0$).

The metricity condition $\nabla_\mu V_\nu^a = 0$ provides a relation between the Lorentz and the Christoffel connection which enables us to express, by direct substitution, the Riemann tensor in terms of the local Lorentz connection ω. Such an expression can also (and more easily) be obtained by considering the commutator of two covariant derivatives applied to a generic Lorentz vector A^a:

$$\begin{aligned}
[\nabla_\mu, \nabla_\nu] A^a &= [D_\mu, D_\nu] A^a \\
&= \left(\partial_\mu \omega_\nu{}^a{}_b - \partial_\nu \omega_\mu{}^a{}_b + \omega_\mu{}^a{}_c \omega_\nu{}^c{}_b - \omega_\nu{}^a{}_c \omega_\mu{}^c{}_b \right) A^b
\end{aligned} \tag{2.87}$$

(the terms containing the Christoffel connection cancel because of its symmetry in the two lower indices, $\Gamma_{\mu\nu}{}^\alpha = \Gamma_{\nu\mu}{}^\alpha$). Using the projection $A^a = A^\alpha V_\alpha^a$, the metricity condition $\nabla V = 0$, and the property (2.17) of the Riemann tensor, we have also

$$[\nabla_\mu, \nabla_\nu] A^\alpha V_\alpha^a = V_\alpha^a R_{\mu\nu\beta}{}^\alpha A^\beta = V_\alpha^a V_\beta^b R_{\mu\nu}{}^{\beta\alpha} A_b, \tag{2.88}$$

from which

$$\begin{aligned}
R_{\mu\nu}{}^{\beta\alpha}(\Gamma) &= V_b^a V_\beta^\beta R_{\mu\nu}{}^{ab}(\omega), \\
R_{\mu\nu}{}^{ab}(\omega) &= \partial_\mu \omega_\nu{}^{ab} + \omega_\mu{}^a{}_c \omega_\nu{}^{cb} - (\mu \leftrightarrow \nu).
\end{aligned} \tag{2.89}$$

Note the different position of the second pair of (antisymmetric) indices ($\beta\alpha$ and ab) of the curvature tensor, which leads to a difference of sign between $R(\Gamma)$ and $R(\omega)$, due to the fact that we are following the standard conventions for $R(\omega)$, while we have adopted different conventions for $R(\Gamma)$ (see the definition of the Riemann tensor in the preliminary section specifying the adopted conventions).

The gravitational equations, usually formulated in the curved space-time manifold in terms of the Riemannian variables $\{g, \Gamma(g), R(\Gamma)\}$, can thus be equivalently expressed in the local tangent space in terms of the set of related variables $\{V, \omega(V), R(\omega)\}$. In this second case one can make the formalism explicitly independent of the choice of the coordinate system, using as basic "gauge" variables the one-forms (2.81). The associated "field strengths" are, respectively, the torsion (R^a) and curvature (R^{ab}) two-forms, whose definitions fully specify the geometric and algebraic structure of the gravitational theory under consideration. In our case, from the antisymmetric part of Eq. (2.86), and the definition (2.89) of $R(\omega)$, we find, respectively,

$$R^a \equiv \Gamma_{[\mu\nu]}{}^a \, dx^\mu \wedge dx^\nu = DV^a \equiv dV^a + \omega^a{}_b \wedge V^b, \tag{2.90}$$

$$R^{ab} \equiv \frac{1}{2} R_{\mu\nu}{}^{ab} \, dx^\mu \wedge dx^\nu = d\omega^{ab} + \omega^a{}_c \wedge \omega^{cb}, \tag{2.91}$$

where d denotes the external derivative, $D = d + \omega$ the (Lorentz) covariant external derivative, and the wedge the (antisymmetric) external product of forms (see Appendix 2A). By taking the external covariant derivative of the previous two equations (also called "structure equations") one easily finds the identities

$$DR^a = R^a{}_b \wedge V^b, \qquad DR^{ab} = 0. \tag{2.92}$$

These identities, when rewritten in the usual tensor language for a torsionless connection ($R^a = 0$), exactly correspond to the first and second Bianchi identities, respectively, characterizing the standard Riemann geometry. We shall restrict our subsequent computations always to the case of vanishing torsion, $DV^a = 0$.

Using the above variables, we can now rewrite the higher-order action (2.80) in differential form as

$$S = -\frac{1}{2\lambda_s^{d-1}} \int e^{-\phi} \left[R^{ab} \wedge {}^\star(V_a \wedge V_b) + D\phi \wedge {}^\star D\phi \right.$$

$$\left. -\frac{\alpha'}{4} R^{ab} \wedge R^{cd} \wedge {}^\star(V_a \wedge V_b \wedge V_c \wedge V_d) + \frac{\alpha'}{4} D\phi \wedge {}^\star D\phi \wedge {}^\star(D\phi \wedge {}^\star D\phi) \right],$$

$$\tag{2.93}$$

where $D\phi = d\phi = \nabla_\mu \phi \, dx^\mu$ is the one-form corresponding to the gradient of the dilaton field, and the star denotes the "Hodge dual" map, defined in Appendix 2A.

The absence of an explicit measure in the action integral is due to the fact that the infinitesimal volume element is already implicitly contained in the definition of the integrated D-forms. Consider, for instance, the quadratic dilaton kinetic term: it can be written explicitly as

$$D\phi \wedge {}^*D\phi = \frac{1}{d!}\nabla_\nu \phi \nabla^\mu \phi\, \eta_{\mu\nu_1\cdots\nu_d} dx^\nu \wedge dx^{\nu_1} \wedge \cdots dx^{\nu_d}$$

$$= \nabla_\nu \phi \nabla^\mu \phi\, \delta^\nu_\mu \sqrt{|g|}\, d^{d+1}x, \tag{2.94}$$

where we have used Eqs. (2A.15), (2A.18) and (2A.20). In the same way, the other three terms of the above action correspond, respectively, to the Einstein action and to the α' corrections appearing in the action (2.80).

We now vary the action (2.93) with respect to ϕ and V^a, to obtain the corresponding field equations written in the language of differential forms. In order to simplify notations, we adopt the convenient definition $V_{a_1} \wedge V_{a_2} \wedge \cdots \equiv V_{a_1 a_2 \cdots}$. The variation with respect to ϕ of the action, integrated over the space-time volume Ω, then gives

$$2\lambda_s^{d-1} \delta_\phi S = \int_\Omega \delta\phi\, e^{-\phi} \left[R^{ab} \wedge {}^*V_{ab} + D\phi \wedge {}^*D\phi - \frac{\alpha'}{4} R^{ab} \wedge R^{cd} \wedge {}^*V_{abcd} \right.$$

$$\left. + \frac{\alpha'}{4} D\phi \wedge {}^*D\phi \wedge {}^*(D\phi \wedge {}^*D\phi) \right] - 2\int_\Omega e^{-\phi}(\delta D\phi) \wedge {}^*D\phi$$

$$- \alpha' \int_\Omega e^{-\phi}(\delta D\phi) \wedge {}^*D\phi \wedge {}^*(D\phi \wedge {}^*D\phi)$$

$$= \int_\Omega \delta\phi\, e^{-\phi}\left[R^{ab} \wedge {}^*V_{ab} - D\phi \wedge {}^*D\phi + 2D^*D\phi \right]$$

$$- \frac{\alpha'}{4}\int_\Omega \delta\phi\, e^{-\phi} \left\{ R^{ab} \wedge R^{cd} \wedge {}^*V_{abcd} \right.$$

$$+ \left[3D\phi \wedge {}^*D\phi - 4D^*D\phi \right] \wedge {}^*(D\phi \wedge {}^*D\phi)$$

$$\left. - 4D^*(D\phi \wedge {}^*D\phi) \wedge {}^*D\phi \right\}. \tag{2.95}$$

To obtain the second equality, the variational contribution of $\delta D\phi$ has been computed by using the properties of the Hodge dual and of the external product (see Appendix 2A), and by exploiting the two identities

$$e^{-\phi}(\delta D\phi) \wedge {}^*D\phi$$

$$= d(e^{-\phi}\delta\phi {}^*D\phi) - e^{-\phi}\delta\phi D^*D\phi + e^{-\phi}\delta\phi D\phi \wedge {}^*D\phi, \quad (2.96)$$

$$e^{-\phi}(\delta D\phi) \wedge {}^*D\phi \wedge {}^*(D\phi \wedge {}^*D\phi)$$

$$= d(e^{-\phi}\delta\phi {}^*D\phi) \wedge {}^*(D\phi \wedge {}^*D\phi)$$

$$- e^{-\phi}\delta\phi \,(D^*D\phi - D\phi \wedge {}^*D\phi) \wedge {}^*(D\phi \wedge {}^*D\phi)$$

$$- e^{-\phi}\delta\phi \, D^*(D\phi \wedge {}^*D\phi) \wedge {}^*D\phi. \quad (2.97)$$

The integral over the external-derivative terms, appearing on the right-hand sides of the above identities, can be transformed into a surface integral over the boundary $\partial\Omega$ of the integration volume, with no contribution to the field equations because of the variational condition $(\delta\phi)_{\partial\Omega} = 0$. Imposing that the action be stationary, $\delta S/\delta\phi = 0$, and summing up all contributions proportional to $\delta\phi$, one then obtains from Eq. (2.95) the dilaton equation, including to first order the α' corrections. Shifting eventually to the usual tensor language – using the correspondence $D^*D\phi \to \nabla^2\phi$, $D\phi \wedge {}^*D\phi \to (\nabla\phi)^2$, and so on (see Eqs. (2A.27)–(2A.29)) – such an equation can be finally written as follows:

$$2\nabla^2\phi - (\nabla\phi)^2 + R - \frac{\alpha'}{4}\left[R_{\rm GB}^2 + 3(\nabla\phi)^4 - 4(\nabla^2\phi)(\nabla\phi)^2\right.$$

$$\left. - 4(\nabla^\alpha\phi)\nabla_\alpha(\nabla\phi)^2\right] = 0. \quad (2.98)$$

In the limit $\alpha' \to 0$ one exactly recovers the tree-level result (2.14).

In order to vary the action with respect to the base field V^a we first observe that, by projecting the dual forms in the local tangent space according to Eq. (2A.2), we can explicitly rewrite the dual of the product of p base vectors as follows:

$$^*V_{a_1 \dots a_p} = \frac{1}{(D-p)!} \, \epsilon_{a_1 \dots a_D} \, V^{a_{p+1+}} \wedge \cdots V^{a_D}. \quad (2.99)$$

Its variation gives

$$\delta_V({}^*V_{a_1 \dots a_p}) = \frac{D-p}{(D-p)!} \, \delta V^{a_{p+1}} \wedge \cdots V^{a_D} \epsilon_{a_1 \dots a_D} \equiv \delta V^{a_{p+1}} \wedge {}^*V_{a_1 \dots a_{p+1}}. \quad (2.100)$$

From the first and third terms of the action (2.93) we then obtain a first variational contribution:

$$2\lambda_s^{d-1}\delta_V S_1 = -\int e^{-\phi}\delta V^i \wedge \left(R^{ab} \wedge {}^*V_{abi} - \frac{\alpha'}{4}R^{ab} \wedge R^{cd} \wedge {}^*V_{abcdi}\right). \quad (2.101)$$

A second variational contribution is obtained from the dilaton terms. Since $\delta_V D\phi = 0$, their contribution arises (as in the case of the tensor formalism) from the scalar products, represented by the Hodge dual operator. Using the tangent space representation,

$$^\star D\phi = \frac{1}{d!}\nabla^a\phi\ \epsilon_{ab_1...b_d}\ V^{b_1}\wedge\cdots V^{b_d} = \frac{1}{d!}V_c^\mu\nabla_\mu\phi\ \epsilon^c{}_{b_1...b_d}\ V^{b_1}\wedge\cdots V^{b_d}, \quad (2.102)$$

we obtain

$$\delta_V(^\star D\phi) = \frac{1}{(d-1)!}\nabla^a\phi\ \epsilon_{ab_1...b_d}\ \delta V^{b_1}\wedge\cdots V^{b_d}$$

$$-\frac{1}{d!}\delta V_\mu^a\ \nabla_a\phi\ V_c^\mu\ \epsilon^c{}_{b_1...b_d} V^{b_1}\wedge\cdots V^{b_d} \quad (2.103)$$

(in the second term we have used the identity $V_\mu^c(\delta V_a^\mu) = -(\delta V_\mu^c)V_a^\mu$, following from the definition (2.82) of the inverse vielbein field). Thus, using again the definition of dual,

$$\delta_V(^\star D\phi) = \nabla^a\phi\ \delta V^b\wedge{}^\star V_{ab} - \nabla_a\phi\ ^\star\delta V^a. \quad (2.104)$$

In the same way we obtain

$$\delta_V\left[^\star(D\phi\wedge{}^\star D\phi)\right] = \delta_V(\nabla_a\phi\nabla^a\phi) = 2(\delta V_c^\mu)\nabla_\mu\phi\nabla^c\phi = -2(\delta V_\mu^a)\nabla_a\phi\nabla^\mu\phi$$

$$\equiv -2\nabla_a\phi\nabla_b\phi\ ^\star(\delta V^a\wedge{}^\star V^b). \quad (2.105)$$

Exploiting these results, the variational contribution of the two dilaton terms of the action (2.93) can be written as follows:

$$2\lambda_s^{d-1}\delta_V S_2 = \int e^{-\phi}\delta V^i\wedge\left[\theta_i + \frac{\alpha'}{4}(\theta_i + 2\nabla_i\phi\nabla_b\phi\ ^\star V^b)(\nabla\phi)^2\right], \quad (2.106)$$

where

$$\theta_i = \nabla^b\phi D\phi\wedge{}^\star V_{bi} + \nabla_i\phi{}^\star D\phi \quad (2.107)$$

is the d-form representing the usual stress tensor of a scalar field.

At this point, we still need to compute the variational contribution of the curvature two-forms appearing in the action. To this purpose we first express the variation of the curvature in terms of the variation of the Lorentz connection, using the definition (2.91):

$$\delta R^{ab}(\omega) = d\delta\omega^{ab} + \delta\omega^a{}_c\wedge\omega^{cb} + \omega^a{}_c\wedge\delta\omega^{cb} \equiv D\delta\omega^{ab}. \quad (2.108)$$

This expression has to be used in the first and third terms of the action (2.93). Integrating by parts, using the torsionless property of the connection, $D^\star V_{ab...} = 0$, and the Bianchi identity, $DR^{ab} = 0$, the variational contribution of the curvature terms can be written as

$$2\lambda_s^{d-1}\delta_V S_3 = -\int e^{-\phi}D\phi\wedge\delta_V\omega^{ab}\wedge\left(^\star V_{ab} - \frac{\alpha'}{2}R^{cd}\wedge{}^\star V_{abcd}\right), \quad (2.109)$$

modulo a surface integral to be canceled by an appropriate generalization of the standard Gibbons–Hawking boundary term (as discussed in Section 2.1).

In order to express $\delta\omega$ as a function of δV we can now use the torsionless condition $DV^a = 0$, whose variation implies

$$\delta_V\omega^a{}_b \wedge V^b = -d\delta V^a - \omega^a{}_b \wedge \delta V^b \equiv -D\delta V^a. \tag{2.110}$$

This equation can be solved exactly by cyclic permutation of its components, and by repeated use of the antisymmetry property of the Lorentz connection, $\omega_{\mu ab} = -\omega_{\mu ba}$. The full result for $\delta\omega$ is the one-form

$$\delta_V\omega^{ab} = V^c\left(F^{ba}{}_c - F^a{}_c{}^b - F_c{}^{ba}\right), \qquad F_{ab}{}^c = V^\mu_a V^\nu_b \nabla_{[\mu}\delta V_{\nu]}{}^a. \tag{2.111}$$

Such a variational contribution has to be saturated by external multiplication with ${}^\star V_{ab}, {}^\star V_{abcd}$, inserted into Eq. (2.109), and integrated by parts. Collecting all terms providing equivalent contributions, and neglecting the boundary terms because of the variational constraint $(\delta V)_{\partial\Omega} = 0$, it turns out that the net variational contribution of $\delta\omega$ can be conveniently represented by the simpler effective expression

$$\delta_V\omega^{ab} = \nabla^b\delta V^a - \nabla^a\delta V^b. \tag{2.112}$$

Inserting this result into Eq. (2.109), and integrating by parts, we can then express the variation of the curvature terms as

$$2\lambda_s^{d-1}\delta_V S_3 = -\int e^{-\phi}\delta V^i \wedge (\nabla^a\phi D\phi - \nabla^a D\phi) \wedge \left(2{}^\star V_{ai} - \alpha' R^{cd} \wedge {}^\star V_{aicd}\right). \tag{2.113}$$

Summing up the contributions (2.101), (2.106) and (2.113), we can finally write the string gravity equation, to first order in α', as follows:

$$R^{ab} \wedge {}^\star V_{abi} - \nabla_i\phi{}^\star D\phi + (\nabla^a\phi D\phi - 2\nabla^a D\phi) \wedge {}^\star V_{ai}$$

$$-\frac{\alpha'}{4}\left[R^{ab} \wedge R^{cd} \wedge {}^\star V_{abcdi} + \left(\nabla^b\phi D\phi \wedge {}^\star V_{bi} + \nabla_i\phi{}^\star D\phi + 2\nabla_i\phi\nabla_b\phi{}^\star V^b\right)(\nabla\phi)^2\right.$$

$$\left.+4(\nabla^a\phi D\phi - \nabla^a D\phi) \wedge R^{cd} \wedge {}^\star V_{aicd}\right] = 0. \tag{2.114}$$

The quadratic curvature term, which would represent the sole contribution of the Gauss–Bonnet term in the absence of non-minimal coupling to the dilaton, is non-zero only in $D > 4$ dimensions. The above equation can be easily rewritten in the conventional tensor language by exploiting the correspondence between forms and tensor components, according to the equations (2A.31)–(2A.38) reported in the appendix. The gravitational equation then takes the form

$$G_\mu{}^\nu + A_\mu{}^\nu + \alpha' B_\mu{}^\nu = 0, \tag{2.115}$$

where $G_\mu{}^\nu$ is the Einstein tensor, $A_\mu{}^\nu$ represents the tree-level dilaton contributions

$$A_\mu{}^\nu = \nabla_\mu \nabla^\nu \phi - \delta_\mu^\nu \nabla^2 \phi + \frac{1}{2}(\nabla\phi)^2 \delta_\mu^\nu \qquad (2.116)$$

(see Eq. (2.11)), while $B_\mu{}^\nu$ contains all α' corrections:

$$B_\mu{}^\nu = \frac{1}{2}(\phi_\alpha \phi^\alpha)\phi^\nu \phi_\mu - \frac{1}{8}\delta_\mu^\nu (\phi_\alpha \phi^\alpha)^2 + \frac{R}{2}(\phi^\nu \phi_\mu - \phi_\mu{}^\nu) + G_\mu{}^\nu(\phi_\alpha \phi^\alpha - \phi_\alpha{}^\alpha)$$
$$- R_\mu{}^\alpha(\phi^\nu \phi_\alpha - \phi_\alpha{}^\nu) - R_\alpha{}^\nu(\phi_\mu \phi^\alpha - \phi^\alpha{}_\mu)$$
$$+ (\delta_\mu^\nu R_\alpha{}^\beta + R_{\alpha\mu}{}^{\beta\nu})(\phi^\alpha \phi_\beta - \phi_\beta{}^\alpha) + L_\mu{}^\nu \qquad (2.117)$$

(for simplicity, we have denoted with an index the covariant derivative acting on the dilaton field, namely $\phi_\alpha \equiv \nabla_\alpha \phi$, $\phi_\alpha{}^\alpha \equiv \nabla_\alpha \nabla^\alpha \phi$, ...). The last term of the above expression denotes the so-called "Lanczos tensor",

$$L_\mu{}^\nu = \frac{1}{8}\delta_\mu^\nu R_{\rm GB}^2 - \frac{1}{2}RR_\mu^\nu + R_\mu{}^\alpha R_\alpha{}^\nu + R_{\mu\alpha}{}^{\beta\nu}R_\beta{}^\alpha + \frac{1}{2}R_{\alpha\beta}{}^{\gamma\nu}R_{\mu\gamma}{}^{\alpha\beta}, \qquad (2.118)$$

which is absent in $D = 4$ and which, in $D > 4$, represents the sole correction to the Einstein equations induced by the Gauss–Bonnet invariant in the absence of the dilaton.

Let us conclude this chapter with a remark – useful in view of subsequent cosmological applications – concerning the possibility of solving Eq. (2.114) through "de Sitter-like" configurations at constant curvature, characterized by $R^{ab} = \Lambda V^a \wedge V^b$, with $\Lambda = {\rm const}$. Consider, for instance, the case $D > 4$, and the simple situation in which the dilaton is frozen, $\nabla_a \phi = 0 = D\phi$. Inserting this ansatz into Eq. (2.114) we obtain an algebraic equation for Λ,

$$d!\left[\frac{\Lambda}{(d-2)!} - \frac{\alpha'\Lambda^2}{4(d-4)!}\right] {}^\star V_i = 0, \qquad (2.119)$$

with non-trivial solution

$$\Lambda = \frac{8\pi}{\lambda_s^2(d-2)(d-3)}. \qquad (2.120)$$

It will be shown in Chapter 6 that non-trivial cosmological solutions at constant curvature can be obtained from Eq. (2.114) even in $D = 4$, provided $D\phi = 0$.

Appendix 2A
Differential forms in a Riemannian manifold

A differential form A of degree p, or p-form, is an element of the linear vector space Λ^p spanned by the (totally antisymmetric) external composition of p differentials, which can be represented as follows:

$$A \in \Lambda^p \implies A = A_{\mu_1 \cdots \mu_p} dx^{\mu_1} \wedge \cdots dx^{\mu_p}, \tag{2A.1}$$

where $dx^\mu \wedge dx^\nu = -dx^\nu \wedge dx^\mu$ for any pairs of indices, and where $A_{\mu_1 \cdots \mu_p}$, called "components" of the p-form, correspond to the components of a totally antisymmetric world tensor of rank p. Using the basic one-forms on the local tangent space, $V^a = V^a_\mu dx^\mu$, where V^a_μ is the vielbein field, and considering the local projection $A_{\mu_1 \cdots \mu_p} = A_{a_1 \cdots a_p} V^{a_1}_{\mu_1} \cdots V^{a_p}_{\mu_p}$, it follows that any p-form also admits the coordinate-independent representation

$$A = A_{a_1 \cdots a_p} V^{a_1} \wedge \cdots V^{a_p} \in \Lambda^p, \tag{2A.2}$$

equivalent to (2A.1).

In a $D = (d+1)$-dimensional manifold, the direct sum of the vector spaces Λ^p defines the so-called "Cartan's algebra" $\Lambda = \bigoplus_{p=0}^{D} \Lambda^p$, i.e. the linear vector space spanned by the composition of $0, 1, 2, \ldots, p$ differentials (or basic one-forms). This space is equipped with a map $\Lambda \times \Lambda \to \Lambda$ denoted by the wedge symbol $\wedge$, called "external product", and satisfying the properties of bilinearity, associativity and skewness. Such properties, when referred to the elements of the (coordinate) differential base $dx^{\mu_1} \wedge dx^{\mu_2} \cdots$, can be expressed, respectively, as follows:

(1) $(\alpha\, dx^{\mu_1} \wedge \cdots dx^{\mu_p} + \beta\, dx^{\mu_1} \wedge \cdots dx^{\mu_p}) \wedge dx^{\nu_1} \wedge \cdots dx^{\nu_q}$

$\qquad = \alpha\, dx^{\mu_1} \wedge \cdots dx^{\mu_p} \wedge dx^{\nu_1} \wedge \cdots dx^{\nu_q}$

$$\qquad + \beta\, dx^{\mu_1} \wedge \cdots dx^{\mu_p} \wedge dx^{\nu_1} \wedge \cdots dx^{\nu_q}, \tag{2A.3}$$

(2) $(dx^{\mu_1} \wedge \cdots dx^{\mu_p}) \wedge (dx^{\mu_{p+1}} \wedge \cdots dx^{\mu_{p+q}})$

$$\qquad = dx^{\mu_1} \wedge \cdots dx^{\mu_{p+q}}, \tag{2A.4}$$

(3) $$dx^{\mu_1} \wedge \cdots dx^{\mu_p} = dx^{[\mu_1} \wedge \cdots dx^{\mu_p]}, \tag{2A.5}$$

where α and β are real numbers, the indices in square brackets are totally antisymmetrized, and $p + q \leq D$. The exterior multiplication of a number of differentials larger than the dimensions of the space-time manifold is identically vanishing, due to the third property. It follows, in particular, that the external product of a p-form $A \in \Lambda^p$ and a q-form

$B \in \Lambda^q$ is a mapping $\wedge : \Lambda^p \times \Lambda^q \to \Lambda^{p+q}$, bilinear, associative and antisymmetric, which defines the $(p+q)$-form C:

$$C = A \wedge B = A_{\mu_1 \ldots \mu_p} B_{\mu_{p+1} \ldots \mu_{p+q}} dx^{\mu_1} \wedge \cdots dx^{\mu_{p+q}} \quad \in \quad \Lambda^{p+q}, \tag{2A.6}$$

and which satisfies the commutation property

$$A \wedge B = (-1)^{pq} B \wedge A. \tag{2A.7}$$

The external derivative of a form $A \in \Lambda^p$ can be interpreted as the external product of the one-form gradient and of the p-form A, and is thus represented by the mapping $d : \Lambda^p \to \Lambda^{p+1}$ which defines the $(p+1)$-form dA:

$$dA = \partial_{\mu_1} A_{\mu_2 \ldots \mu_{p+1}} dx^{\mu_1} \wedge \ldots dx^{\mu_{p+1}}$$

$$= \nabla_{\mu_1} A_{\mu_2 \ldots \mu_{p+1}} dx^{\mu_1} \wedge \cdots dx^{\mu_{p+1}} \quad \in \quad \Lambda^{p+1}. \tag{2A.8}$$

The second equality, i.e. the replacement

$$\partial_\mu A_{\nu\rho\ldots} \to \partial_\mu A_{\nu\rho\ldots} - \Gamma_{\mu\nu}{}^\alpha A_{\alpha\rho\ldots} - \Gamma_{\mu\rho}{}^\alpha A_{\nu\alpha\ldots} \cdots, \tag{2A.9}$$

is justified by the symmetry of the two lower indices of the Christoffel connection, which has vanishing contraction with the (antisymmetric) exterior product of differentials. An obvious consequence of the definition (2A.8) is the property

$$d^2 A = d \wedge dA \equiv 0, \tag{2A.10}$$

valid for all forms. Another consequence of the definition is the (generalized) Leibniz rule for the external derivative of a product: given $A \in \Lambda^p$ and $B \in \Lambda^q$ one obtains, from Eq. (2A.8),

$$d(A \wedge B) = dA \wedge B + (-1)^p A \wedge dB,$$

$$d(B \wedge A) = dB \wedge A + (-1)^q B \wedge dA. \tag{2A.11}$$

The Lorentz-covariant external derivative is represented by the one-form $D = d + \omega$, where $\omega = \omega_\mu^{ab} S_{ab} dx^\mu$ is given by the contraction of the matrix-valued spin connection one-form ω_μ^{ab} and of the (tangent space) Lorentz generators S_{ab}, appropriate to the fields we are differentiating. For a p-form A which is scalar-valued with respect to local Lorentz rotations, for instance, $DA = dA$, while for a tensor-valued p-form $A^{a\ldots}{}_{b\ldots} \in \Lambda^p$ the connection ω acts linearly on all indices of the local Lorentz representation:

$$DA^{a\ldots}{}_{b\ldots} = dA^{a\ldots}{}_{b\ldots} + \omega^a{}_c \wedge A^{c\ldots}{}_{b\ldots} - \omega^c{}_b \wedge A^{a\ldots}{}_{c\ldots} + \cdots$$

$$\equiv \left[\partial_{\mu_1} (A_{\mu_2 \ldots \mu_{p+1}})^{a\ldots}{}_{b\ldots} + \omega_{\mu_1}{}^a{}_c (A_{\mu_2 \ldots \mu_{p+1}})^{c\ldots}{}_{b\ldots} - \cdots \right] dx^{\mu_1} \wedge \cdots dx^{\mu_{p+1}}. \tag{2A.12}$$

The above expression also applies to zero-forms, namely to world-scalar objects ($p = 0$) which are tensor-valued in the representation of the local Lorentz group. Two examples are in order, also in view of further applications: the Lorentz-covariant derivative of the local Minkowski metric, η^{ab}, and of the local Levi-Civita tensor density, $\epsilon^{a_1 \ldots a_D}$. Both geometrical objects are covariantly constant with respect to D: by applying Eq. (2A.12), and using the antisymmetry of the Lorentz connection, $\omega^{ab} = \omega^{[ab]}$, one obtains

$$D\eta^{ab} = \omega^a{}_b \eta^{cb} + \omega^b{}_c \eta^{ac} \equiv 0, \tag{2A.13}$$

$$D\epsilon^{a_1 \ldots a_D} = \omega^{a_1}{}_c \epsilon^{ca_2 \ldots} + \omega^{a_2}{}_c \epsilon^{a_1 c \ldots} + \cdots = \omega^c{}_c \epsilon^{a_1 \ldots a_D} \equiv 0. \tag{2A.14}$$

It should be noticed that the Leibniz rule (2A.11) can also be applied to the external covariant derivative D, which, like d, is a one-form. The property (2A.10), however, is no longer valid when d is replaced by D, because the commutator of two covariant derivatives is non-zero, in general, and proportional to the space-time curvature. One finds, in fact, $D^2 \equiv D \wedge D = d\omega + \omega \wedge \omega$, which, after explicit insertion of the flat-space indices, exactly reproduces the curvature two-form of Eq. (2.91).

For a consistent formulation of the gravitational equations in an arbitrary number D of dimensions we need, finally, the so-called "Hodge duality" operation, a mapping $\star : \Lambda^p \to \Lambda^{D-p}$ which associates, to any p-form A, its $(D-p)$-dimensional "complement" $^\star A$:

$$^\star A = \frac{1}{(D-p)!} A^{\mu_1 \cdots \mu_p} \eta_{\mu_1 \cdots \mu_D} dx^{\mu_{p+1}} \wedge \cdots dx^{\mu_D} \quad \in \quad \Lambda^{D-p}. \tag{2A.15}$$

Here η is the totally antisymmmmetric tensor, related to the Levi-Civita tensor density ϵ by

$$\eta_{\mu_1 \cdots \mu_D} = \sqrt{|g|}\, \epsilon_{\mu_1 \cdots \mu_D}. \tag{2A.16}$$

The dual of the identity thus corresponds, according to the definition (2A.15), to the scalar measure representing the covariant world-volume element:

$$^\star 1 = \frac{1}{D!}\, \eta_{\mu_1 \cdots \mu_D} dx^{\mu_1} \wedge \cdots dx^{\mu_D} = \sqrt{|g|}\, dx^1 \wedge dx^2 \ldots dx^D = \sqrt{|g|}\, d^D x. \tag{2A.17}$$

Using the well-known multiplication rule $\eta_{\mu_1 \cdots \mu_D} \eta^{\mu_1 \cdots \mu_D} = D!$, we obtain the useful result

$$dx^{\mu_1} \wedge \cdots dx^{\mu_D} = \eta^{\mu_1 \cdots \mu_D} \sqrt{|g|}\, d^D x. \tag{2A.18}$$

It follows that the integral of the external product of a p-form with the dual of a form of the same degree automatically reproduces a volume integral, which is invariant under general coordinate reparametrizations of the D-dimensional world manifold. This, in particular, is what allows us to rewrite the action in terms of p-form variables. Considering, for instance, $A \in \Lambda^p$ and $B \in \Lambda^p$ one obtains

$$\int A \wedge {}^\star B = \frac{1}{(D-p)!} \int A_{\mu_1 \cdots \mu_p} B^{\nu_1 \cdots \nu_p} \eta_{\nu_1 \cdots \nu_p \mu_{p+1} \cdots \mu_D}\, dx^{\mu_1} \wedge \cdots dx^{\mu_D}$$

$$= \int d^D x \sqrt{|g|} A_{\mu_1 \cdots \mu_p} B^{\nu_1 \cdots \nu_p} \delta^{\mu_1 \cdots \mu_p}_{\nu_1 \cdots \nu_p}$$

$$= p! \int d^D x \sqrt{|g|} A_{\mu_1 \cdots \mu_p} B^{\mu_1 \cdots \mu_p}. \tag{2A.19}$$

We have used the multiplication rules of two totally antisymmetric tensors with a number $D - p$ of contracted indices, namely

$$\eta_{\nu_1 \cdots \nu_p \mu_{p+1} \cdots \mu_D} \eta^{\mu_1 \cdots \mu_p \mu_{p+1} \cdots \mu_D} = (D-p)! \, \delta^{\mu_1 \cdots \mu_p}_{\nu_1 \cdots \nu_p}, \tag{2A.20}$$

where the so-called generalized Kronecker symbol is defined by the following determinant

$$\delta^{\mu_1 \cdots \mu_p}_{\nu_1 \cdots \nu_p} = \begin{vmatrix} \delta^{\mu_1}_{\nu_1} & \delta^{\mu_2}_{\nu_1} & \cdots & \delta^{\mu_p}_{\nu_1} \\ \delta^{\mu_1}_{\nu_2} & \delta^{\mu_2}_{\nu_2} & \cdots & \delta^{\mu_p}_{\nu_2} \\ \cdots & \cdots & \cdots & \cdots \\ \delta^{\mu_1}_{\nu_p} & \delta^{\mu_2}_{\nu_p} & \cdots & \delta^{\mu_p}_{\nu_p} \end{vmatrix}. \tag{2A.21}$$

We can also rewrite the result of Eq. (2A.19) as

$$A \wedge {}^*B = B \wedge {}^*A = {}^*1 \, p! \, A_{\mu_1 \dots \mu_p} B^{\mu_1 \dots \mu_p}, \qquad (2A.22)$$

where A and B are forms of the same degree.

After introducing all the required tools we can now check that the various terms of the action (2.93), written in the language of external forms, exactly reproduce all the terms of the action (2.80), written in the conventional tensor language.

The correspondence of the dilaton kinetic terms has already been illustrated in Eq. (2.94). For the Einstein term we have

$$R^{ab} \wedge {}^*(V_a \wedge V_b) = \frac{1}{2(D-2)!} R_{\mu\nu}{}^{ab} V_a^\alpha V_b^\beta \, \eta_{\alpha\beta\nu_1 \dots \nu_{D-2}} dx^\mu \wedge dx^\nu \wedge dx^{\nu_1} \wedge \cdots dx^{\nu_{D-2}}$$

$$= \frac{1}{2} R_{\mu\nu}{}^{\beta\alpha}(\Gamma) \delta^{\mu\nu}_{\alpha\beta} \sqrt{|g|} \, d^D x = \frac{1}{2} \left(R_{\mu\nu}{}^{\nu\mu} - R_{\mu\nu}{}^{\mu\nu} \right) \sqrt{|g|} \, d^D x$$

$$= R \sqrt{|g|} \, d^D x, \qquad (2A.23)$$

where we have used the relation (2.89) between the Lorentz and the Christoffel connection. In the same way we obtain, for the Gauss–Bonnet term,

$$R^{a_1 a_2} \wedge R^{a_3 a_4} \wedge {}^*(V_{a_1} \wedge \cdots V_{a_4})$$

$$= \frac{1}{4(D-4)!} R_{\mu_1 \mu_2}{}^{a_1 a_2} R_{\mu_3 \mu_4}{}^{a_3 a_4} V_{a_1}^{\nu_1} \cdots V_{a_4}^{\nu_4} \, \eta_{\nu_1 \dots \nu_4 \nu_5 \dots \nu_D} dx^{\mu_1} \wedge \cdots dx^{\mu_4} \wedge \cdots dx^{\nu_D}$$

$$= \frac{1}{4} R_{\mu_1 \mu_2}{}^{\nu_2 \nu_1}(\Gamma) R_{\mu_3 \mu_4}{}^{\nu_4 \nu_3}(\Gamma) \delta^{\mu_1 \dots \mu_4}_{\nu_1 \dots \nu_4} \sqrt{|g|} \, d^D x$$

$$= \frac{1}{4} \sqrt{|g|} \, d^D x \left\{ R_{\mu\nu}{}^{\nu\mu} R_{\alpha\beta}{}^{\beta\alpha} + R_{\mu\nu}{}^{\nu\beta} R_{\alpha\beta}{}^{\alpha\mu} + R_{\mu\nu}{}^{\nu\alpha} R_{\alpha\beta}{}^{\mu\beta} - R_{\mu\nu}{}^{\nu\mu} R_{\alpha\beta}{}^{\alpha\beta} \right.$$

$$\left. - R_{\mu\nu}{}^{\nu\beta} R_{\alpha\beta}{}^{\mu\alpha} - R_{\mu\nu}{}^{\nu\alpha} R_{\alpha\beta}{}^{\beta\mu} - R_{\mu\nu}{}^{\mu\beta} R_{\alpha\beta}{}^{\alpha\nu} - \cdots \right\}$$

$$= \sqrt{|g|} \, d^D x \left(R^2 - 4 R_\mu{}^\nu R_\nu{}^\mu + R_{\mu\nu}{}^{\alpha\beta} R_{\alpha\beta}{}^{\mu\nu} \right) \equiv \sqrt{|g|} \, d^D x \, R_{\mathrm{GB}}^2. \qquad (2A.24)$$

The curly brackets contain the $4! = 24$ terms arising from the possible permutations of the four contravariant indices ν, μ, β, α (at fixed covariant indices) of the products $R_{\mu\nu}{}^{\nu\mu} R_{\alpha\beta}{}^{\beta\alpha}$, with the $(+)$ sign for even permutations and the $(-)$ sign for odd permutations.

Finally, for the quartic dilaton term present in the action (2.93), we can note that $D\phi \wedge {}^*D\phi$ is a $(d+1)$-form, with components

$$\frac{1}{d!} \nabla_\mu \phi \nabla^\nu \phi \, \eta_{\nu\mu_1 \dots \mu_d}. \qquad (2A.25)$$

The associated dual, according to the definition (2A.15), is the zero-form $(\nabla\phi)^2$. Thus, using Eqs. (2A.18) and (2A.20),

$$D\phi \wedge {}^*D\phi \wedge {}^*(D\phi \wedge {}^*D\phi) = (\nabla\phi)^2 \nabla_\mu \phi \nabla^\nu \phi \, \delta^\mu_\nu \sqrt{|g|} \, d^D x = (\nabla\phi)^4 \sqrt{|g|} \, d^D x. \qquad (2A.26)$$

Consider now the dilaton equation (2.95), written in terms of differential forms. For its translation to the tensor language we still need to compute the three terms containing the second derivatives of the dilaton. Using Eqs. (2A.15), (2A.18) and (2A.20), and replacing D_μ with ∇_μ in the form components (because of the total antisymmetrization of the tensor indices), we obtain

$$D^\star D\phi = \frac{1}{d!}\nabla_\mu\nabla^\nu\phi\,\eta_{\nu\mu_1\ldots\mu_d}\,dx^\mu \wedge dx^{\mu_1} \wedge \cdots dx^{\mu_d} = \nabla^2\phi\,\sqrt{|g|}\,d^D x, \quad (2\text{A}.27)$$

$$D^\star D\phi \wedge {}^\star(D\phi \wedge {}^\star D\phi) = (\nabla\phi)^2 \nabla^2\phi\,\sqrt{|g|}\,d^D x, \quad\quad\quad (2\text{A}.28)$$

$$D^\star(D\phi \wedge {}^\star D\phi) \wedge {}^\star D\phi = \frac{\nabla_\mu}{d!}(\nabla\phi)^2\nabla^\nu\phi\,\eta_{\nu\mu_1\ldots\mu_d}\,dx^\mu \wedge dx^{\mu_1} \wedge \cdots dx^{\mu_d}$$

$$= \nabla^\mu\phi\nabla_\mu(\nabla\phi)^2\,\sqrt{|g|}\,d^D x. \quad (2\text{A}.29)$$

Inserting these results into Eq. (2.95) one then recovers the dilaton equation (2.98).

We close the appendix by computing the tensor counterpart of the various terms appearing in the gravitational equation (2.114). The first term can be written explicitly as

$$R^{ab} \wedge {}^\star V_{abi} = \frac{1}{2(D-3)!}R_{\mu_1\mu_2}{}^{ab}V_a^{\nu_1}V_b^{\nu_2}V_i^{\nu_3}\,\eta_{\nu_1\nu_2\nu_3\mu_3\ldots\mu_d}\,dx^{\mu_1} \wedge \cdots dx^{\mu_d}. \quad (2\text{A}.30)$$

Let us antisymmetrize the components of this d-form by multiplying them by $\eta^{\rho\mu_1\ldots\mu_d}$, and use the multiplication rule (2A.20). We are thus led to

$$\frac{1}{2}R_{\mu_1\mu_2}{}^{\nu_2\nu_1}(\Gamma)V_i^{\nu_3}\,\delta^{\rho\mu_1\mu_2}_{\nu_1\nu_2\nu_3}$$

$$= \frac{1}{2}\Bigg\{R_{\mu_1\mu_2}{}^{\mu_1\rho}V_i^{\mu_2} + R_{\mu_1\mu_2}{}^{\rho\mu_2}V_i^{\mu_1} + R_{\mu_1\mu_2}{}^{\mu_2\mu_1}V_i^{\rho}$$

$$-R_{\mu_1\mu_2}{}^{\mu_1\mu_2}V_i^{\rho} - R_{\mu_1\mu_2}{}^{\rho\mu_1}V_i^{\mu_2} - R_{\mu_1\mu_2}{}^{\mu_2\rho}V_i^{\mu_1}\Bigg\}$$

$$= \frac{1}{2}\left(2RV_i^\rho - 4R^\rho{}_i\right) = -2G^\rho{}_\beta V_i^\beta, \quad (2\text{A}.31)$$

where $G^\rho{}_\beta$ is the Einstein tensor. We proceed in the same way for the other d-forms, following the order of Eq. (2.114). The second term is

$$\nabla_i\phi^\star D\phi = \frac{1}{d!}\nabla_i\phi\nabla^\nu\phi\,\eta_{\nu\mu_1\ldots\mu_d}\,dx^{\mu_1} \wedge \cdots dx^{\mu_d}. \quad (2\text{A}.32)$$

Antisymmetrizing its components, and exploiting the multiplication of the antisymmetric tensors, we are led to

$$\frac{1}{d!}\nabla_i\phi\nabla^\nu\phi\,\eta_{\nu\mu_1\ldots\mu_d}\,\eta^{\rho\mu_1\ldots\mu_d} = \nabla_i\phi\nabla^\rho\phi. \quad (2\text{A}.33)$$

Using the same procedure we can associate a tensor term to each d-form, according to the following correspondence:

$$\nabla^a\phi D\phi \wedge {}^\star V_{ai} \Rightarrow \nabla^\rho\phi\nabla_i\phi - (\nabla\phi)^2 V_i^\rho, \quad (2\text{A}.34)$$

$$\nabla^a D\phi \wedge {}^\star V_{ai} \Rightarrow \nabla^\rho\nabla_i\phi - \nabla^2\phi V_i^\rho, \quad (2\text{A}.35)$$

$$\nabla_i \phi \nabla_b \phi^* V^b \Rightarrow \nabla^\rho \phi \nabla_i \phi \tag{2A.36}$$

$$
\begin{aligned}
(\nabla^a \phi D\phi - \nabla^a D\phi) \wedge R^{cd} \wedge {}^* V_{aicd} \Rightarrow & R(\nabla^\rho \phi \nabla_i \phi - \nabla^\rho \nabla_i \phi) \\
& + (2R^\rho_{\ i} - RV^\rho_i)(\nabla \phi^2 - \nabla^2 \phi) \\
& - R_i^{\ a}(\nabla^\rho \phi \nabla_a \phi - \nabla^\rho \nabla_a \phi) \\
& + 2(R_a^{\ b} V_i^\rho + R_{ai}^{\ \ b\rho})(\nabla^a \phi \nabla_b \phi - \nabla^a \nabla_b \phi) \\
& - 2R_a^{\ \rho}(\nabla_i \phi \nabla^a \phi - \nabla_i \nabla^a \phi). \tag{2A.37}
\end{aligned}
$$

The last term we must consider is the quadratic curvature term of Eq. (2.114), which does not contribute to the gravitational equations in $D = 4$. In $D > 4$ its contribution is represented by the so-called Lanczos tensor, according to the correspondence

$$
\begin{aligned}
R^{ab} \wedge R^{cd} \wedge {}^* V_{abcdi} \Rightarrow & R^2_{\mathrm{GB}} V_i^\rho - 4RR_i^{\ \rho} + 8R_i^{\ \alpha} R_\alpha^{\ \rho} \\
& + 8R_{ia}^{\ \ \beta\rho} R_\beta^{\ \alpha} + 4R_{i\gamma}^{\ \ \alpha\beta} R_{\alpha\beta}^{\ \ \gamma\rho}, \tag{2A.38}
\end{aligned}
$$

where R^2_{GB} is the Gauss–Bonnet invariant.

References

[1] M. B. Green, J. Schwartz and E. Witten, *Superstring Theory* (Cambridge: Cambridge University Press, 1987).

[2] J. Polchinski, *String Theory* (Cambridge: Cambridge University Press, 1998).

[3] M. Gasperini and G. Veneziano, *Phys. Rep.* **373** (2003) 1.

[4] T. Battefeld and S. Watson, *Rev. Mod. Phys.* **78** (2006) 435.

[5] G. Veneziano, *Europhys. Lett.* **2** (1986) 133.

[6] S. Weinberg, *Gravitation and Cosmology* (New York: John Wiley & Sons, 1972).

[7] I. Antoniadis, J. Rizos and K. Tamvakis, *Nucl. Phys.* **B415** (1994) 497.

[8] S. Foffa, M. Maggiore and R. Sturani, *Nucl. Phys.* **B552** (1999) 395.

[9] G. W. Gibbons and S. W. Hawking, *Phys. Rev.* **D15** (1977) 2752.

[10] M. Gasperini and G. Veneziano, *Phys. Lett.* **B277** (1992) 256.

[11] R. M. Wald, *General Relativity* (Chicago: University of Chicago Press, 1984).

[12] J. L. Synge, *Relativity: The General Theory* (Amsterdam: North-Holland, 1960).

[13] E. Kiritsis and C. Kounnas, *Phys. Lett.* **B331** (1994) 51.

[14] M. Gasperini, M. Maggiore and G. Veneziano, *Nucl. Phys.* **B494** (1997) 315.

[15] A. Buonanno, M. Gasperini and C. Ungarelli, *Mod. Phys. Lett.* **A12** (1997) 1883.

[16] R. R. Metsaev and A. A. Tseytlin, *Nucl. Phys.* **B293** (1987) 385.

[17] B. Zwiebach, *Phys. Lett.* **B156** (1985) 315.

[18] N. E. Mavromatos and J. L. Miramontes, *Phys. Lett.* **B201** (1988) 473.

[19] K. A. Meissner, *Phys. Lett.* **B392** (1997) 298.

3

Conformal invariance and string effective actions

In this chapter we will illustrate the quantum/stringy origin of the effective field equations introduced in the previous chapter. In particular, we will show that such equations must be satisfied by the background fields through which a bosonic string is propagating, in order to implement a consistent quantization of the string motion without anomalies (i.e. without quantum breakdown of the symmetries already present at the classical level).

The main content of this chapter has no direct application in cosmology, and will not often be referred to in the rest of this book. The simple introduction presented here, even if incomplete and approximate in many respects, is nonetheless compulsory for a reader with no previous knowledge of string theory, in order to understand how the cosmological equations used throughout the book are rigidly prescribed by the theory and cannot undergo ad hoc modifications, unlike the equations of other, more conventional models of gravity based on the notions of fields and point-like sources.

The motion of a point-like particle, in fact, can be consistently quantized without imposing any constraints on the background fields with which the point is interacting. In particular, the geometry of the space-time manifold in which the particle trajectory is embedded may be generated by arbitrary sources, may be characterized by an arbitrary number of dimensions, and may be governed by dynamical equations arbitrarily prescribed (or arbitrarily modified with respect to the Einstein equations), without dramatically affecting the quantization of the point motion.

In the string case the situation is different. Assuming that the string represents a unified model of all fundamental interactions, then every field with which the string can interact should be contained in the spectrum of states associated with the quantization of its free oscillations. The dynamics of all background fields (even at the *classical* level) must thus be determined so as to be consistent with the results of string quantization: in particular, their interactions must not destroy the (two-dimensional) conformal invariance which is present in all string models, and which is crucial to obtain the quantum spectrum in which the background fields themselves are included.

This constraint, as we will see, imposes a set of differential conditions on the background fields interacting with the string: such conditions fully determine the dynamical evolution of these fields, in principle to all orders of a perturbative expansion. The gravitational dynamics (in particular) and the possible cosmological scenarios (as a consequence) are thus rigidly fixed by the chosen model of string, and the only possible freedom left is in the choice of the initial configurations of the various fields.

In order to illustrate such a revolutionary aspect of string theory we will discuss in Section 3.1 the conformal invariance of the so-called "sigma model" action, which describes the dynamical evolution of a string in the presence of "condensates" of its massless modes (including, in particular, the gravitational field), and we will explicitly determine the background field equations to the lowest perturbative order. In Section 3.2 we will briefly comment on the expected (perturbative and non-perturbative) corrections introduced by the high-curvature and strong-coupling regimes. In Appendix 3A we will present a detailed discussion (and computation) of the spectrum of the physical states associated with the quantization of open and closed bosonic strings in Minkowski space. In Appendix 3B we will finally sketch the various models, and the corresponding effective actions, obtained in the context of the supersymmetric string. For a complete introduction to, and a deeper insight of, the string theory aspects discussed in this chapter interested readers are referred to the excellent books already existing on this subject [1, 2].

3.1 Strings in curved backgrounds and conformal anomalies

The action of a string freely evolving in a background gravitational field – classically described as a curved, Riemannian manifold with metric $g_{\mu\nu}$ – can be easily constructed following the analogy with the case of a free point-like particle.

Let us recall that the motion of a point describes a "world-line" in the space-time manifold in which the point is embedded, and that the corresponding action is proportional to the length of the trajectory, computed as the *line integral* over an appropriate time parameter τ. If the trajectory is represented by the parametric equation $x^\mu = x^\mu(\tau)$, we then have the action

$$S = m \int_a^b \mathrm{d}s = m \int_a^b \sqrt{\mathrm{d}x^\mu \, \mathrm{d}x^\nu \, g_{\mu\nu}} = m \int_{\tau_1}^{\tau_2} \mathrm{d}\tau \sqrt{\dot{x}^\mu \dot{x}^\nu g_{\mu\nu}}, \qquad (3.1)$$

where $\dot{x} = dx/d\tau$. The square root appearing in the above equation can be eliminated by introducing an appropriate auxiliary field (or Lagrange multiplier) $V(\tau)$, and reformulating the action as follows:

$$S = \frac{1}{2} \int d\tau \left(V^{-1} \dot{x}^\mu \dot{x}^\nu g_{\mu\nu} + m^2 V \right). \tag{3.2}$$

It can be easily checked that the variation with respect to V provides the constraint

$$\dot{x}^\mu \dot{x}^\nu g_{\mu\nu} = V^2 m^2, \tag{3.3}$$

which, solved for V, and inserted into Eq. (3.2), exactly reproduces the action (3.1). The action (3.2), however, is also defined for massless particles, unlike the action (3.1). Its variation with respect to x^μ leads to the Euler–Lagrange equation describing the relativistic motion of the particle. Choosing, in particular, an appropriate temporal "gauge" in which the field V (determined by the condition (3.3)) is fixed to a constant, one obtains the equation

$$\frac{d}{d\tau} \frac{\partial L}{\partial \dot{x}^\mu} = \frac{d}{d\tau} (\dot{x}^\nu g_{\mu\nu}) = \ddot{x}_\mu + \frac{1}{2} \dot{x}^\nu \dot{x}^\alpha \left(\partial_\alpha g_{\mu\nu} + \partial_\nu g_{\mu\alpha} \right)$$

$$= \frac{\partial L}{\partial x^\mu} = \frac{1}{2} \dot{x}^\nu \dot{x}^\alpha \partial_\mu g_{\nu\alpha}. \tag{3.4}$$

Rewriting the equation in explicitly covariant form one then recovers the standard geodesic motion of the particle in the given background metric:

$$\ddot{x}^\mu + \Gamma_{\nu\alpha}{}^\mu \dot{x}^\nu \dot{x}^\alpha = 0. \tag{3.5}$$

The action for a free string in a gravitational background can be written following the analogy with the point particle, and noting that the time evolution of a one-dimensional object describes, instead of a curve, a two-dimensional surface Σ, called a "world-sheet", in the space-time manifold (also called the "target space") in which it is embedded. The corresponding action can thus be expressed as a two-dimensional *surface integral* and, as discussed in more detail in Appendix 3A, is proportional to the proper area of the world-sheet spanned by the string motion (one obtains, in this way, the so-called Nambu–Goto action [3]).

As in the particle case, it is convenient to eliminate the square root present in the action and to work with the (equivalent) Polyakov form [4, 5], which extends the action (3.2) to objects with one-dimensional spatial extension. Let us call $X^\mu = X^\mu(\xi^a)$, with $\mu = 0, 1, \ldots, D-1$, the parametric equations governing the embedding of the world-sheet Σ into the D-dimensional target space (with a prescribed number of dimensions, as will be explained in Appendix 3A); also, let us call ξ^a, with $a, b = 0, 1$, the intrinsic world-sheet coordinates. The Polyakov action then takes the form

$$S \equiv \frac{1}{\lambda_s^2} \int_\Sigma d^2 \xi \, L(X, \partial X) = \frac{1}{2\lambda_s^2} \int_\Sigma d^2 \xi \sqrt{-\gamma} \, \gamma^{ab} \partial_a X^\mu(\xi) \partial_b X^\nu(\xi) g_{\mu\nu}(X), \quad (3.6)$$

where $\partial_a \equiv \partial/\partial\xi^a$, and where the symmetric tensor $\gamma_{ab}(\xi)$, with Lorentzian signature $(+, -)$ and determinant $\gamma \equiv \det \gamma_{ab}$, represents the *intrinsic* metric of the world-sheet surface, here playing the role of an auxiliary field analogous to the field $V(\tau)$ of the action (3.2). Finally, $\lambda_s^{-2} = M_s^2$ is the so-called string tension (or mass per unit length), determining the natural units to which to refer the world-sheet area, and representing the *only* arbitrary fundamental parameter of the considered model. In the rest of this chapter we will put $\lambda_s^2 = 2\pi\alpha'$, to follow the conventional notations usually adopted in the string theory literature. Note that, for a flat background metric $g_{\mu\nu} = \eta_{\mu\nu}$, the above action is quadratic in the fields X^μ; for a curved background metric, $g_{\mu\nu}(X)$, we have instead a *non-linear* action, called the sigma model, for the fields $X^\mu(\xi)$ defined on the given world-sheet surface.

The string equations of motion are now obtained by varying the action (3.6) with respect to the coordinates X^μ, as in the particle case. By imposing appropriate boundary conditions (see Appendix 3A for a discussion of the various possibilities), we obtain the standard Euler–Lagrange equations,

$$\partial_a \frac{\partial L}{\partial(\partial_a X^\mu)} = \partial_a \left(\sqrt{-\gamma} \, \gamma^{ab} \partial_b X^\nu \right) g_{\mu\nu} + \sqrt{-\gamma} \, \gamma^{ab} \partial_b X^\nu \partial_a X^\alpha \partial_\alpha g_{\mu\nu}$$

$$= \frac{\partial L}{\partial X^\mu} = \frac{1}{2} \sqrt{-\gamma} \, \gamma^{ab} \partial_a X^\alpha \partial_b X^\nu \partial_\mu g_{\alpha\nu}. \quad (3.7)$$

Dividing by $\sqrt{-\gamma}$, and introducing the target space connection Γ, we can rewrite the equations in a form which closely resembles the point-like geodesic equation:

$$\gamma^{ab} \nabla_a \nabla_b X^\rho + \gamma^{ab} \partial_a X^\alpha \partial_b X^\beta \Gamma_{\alpha\beta}{}^\rho = 0, \quad (3.8)$$

where $\gamma^{ab} \nabla_a \nabla_b = (\sqrt{-\gamma})^{-1} \partial_a (\sqrt{-\gamma} \gamma^{ab} \partial_b)$ is the covariant d'Alembert operator for the curved world-sheet metric. By varying the action with respect to the auxiliary field γ^{ab} we also obtain the constraint

$$T_{ab} \equiv \partial_a X^\mu \partial_b X^\nu g_{\mu\nu} - \frac{1}{2} \gamma_{ab} \partial_c X^\mu \partial^c X^\nu g_{\mu\nu} = 0, \quad (3.9)$$

which generalizes the "mass-shell" condition (3.3), previously obtained for the point-like particle. We note that T_{ab}, defined by the variation of the intrinsic metric γ_{ab}, corresponds to the geometric stress tensor of the two-dimensional world-sheet theory. The conformal invariance of this theory (see below) is then reflected in the fact that the trace of such a tensor is identically vanishing, as can be checked by its definition, quite independently of the equation of motion $T_{ab} = 0$.

The above equations can be applied to study the free motion of a string (or of a perfect gas of non-interacting strings) in a given gravitational field and, in particular, in a cosmological background. One then finds, even at the classical level, interesting results which will be reported later [6–9]. For the purpose of this chapter, the point we must focus on is the invariance of the two-dimensional model (3.6) under conformal transformations (i.e. local scale transformations, or Weyl transformations) of the world-sheet metric, defined by

$$\gamma_{ab} \to e^{2\omega(\xi)} \gamma_{ab}, \qquad \gamma^{ab} \to e^{-2\omega(\xi)} \gamma^{ab}. \tag{3.10}$$

In a generic, D-dimensional manifold we have

$$\sqrt{-\gamma}\gamma^{ab} \to e^{(D-2)\omega}\sqrt{-\gamma}\gamma^{ab}, \tag{3.11}$$

so that, for $D = 2$, the action (3.6) turns out to be exactly invariant.

Thanks to this invariance, it is always possible to represent the string dynamics in the so-called "conformal gauge" in which the world-sheet has a flat intrinsic metric, $\gamma_{ab} = \eta_{ab}$. Using the reparametrization invariance of the world-sheet surface, we can first perform a suitable coordinate transformation $\xi^a \to \tilde{\xi}^a$, imposing two conditions on the three independent components of γ_{ab}, in such a way that the transformed metric is diagonal and conformally flat, $\gamma_{ab} \to a^2(\xi)\eta_{ab}$. Then, by exploiting the conformal invariance (3.11), we can eventually obtain the Minkowski metric through a last rescaling with conformal factor $\exp(2\omega) = a^{-2}$. In the conformal gauge (which will be repeatedly used in Appendix 3A) the string equations of motion are considerably simplified: Eq. (3.8) reduces to

$$\ddot{X}^\mu - X''^\mu + \Gamma_{\alpha\beta}{}^\mu \left(\dot{X}^\alpha + X'^\alpha\right)\left(\dot{X}^\beta - X'^\beta\right) = 0, \tag{3.12}$$

where the dot denotes differentiation with respect to the time-like world-sheet coordinate, $\xi^0 = \tau$, and the prime with respect to the space-like coordinate $\xi^1 = \sigma$. Also, the constraints (3.9) can be written in the form

$$g_{\mu\nu}\left(\dot{X}^\mu \dot{X}^\nu + X'^\mu X'^\nu\right) = 0, \qquad g_{\mu\nu}\dot{X}^\mu X'^\nu = 0, \tag{3.13}$$

where the first condition is obtained from the components T_{00}, T_{11}, while the second one is from T_{10}.

It is important to stress that the conformal invariance of the action (3.6) holds, at a classical level, *quite independently* of the given target space geometry (namely, it holds for *any* given metric $g_{\mu\nu}(X)$). At the quantum level, however, such an invariance tends to break because of the required loop corrections, being preserved only for particular metric backgrounds satisfying appropriate differential conditions imposed to avoid conformal anomalies, order by order in the perturbative expansion of the sigma model quantization [1].

The property of conformal invariance is expected to be valid also in a quantum context for the formulation of a consistent theory in which quantum corrections can be expanded around a compatible classical solution; after all, conformal transformations are part of the two-dimensional reparametrization group [10], and the background fields present in the sigma model action – in particular, the gravitational field represented by the metric $g_{\mu\nu}$ – are part of the quantum spectrum of physical states obtained by using the conformal invariance of the action. It follows that all background field configurations admissible in a string theory context – not only for the metric of the space-time manifold, but also for any (boson and fermion) matter field on it – are not arbitrary: the *only* allowed configurations are those satisfying the differential conditions imposed by the requirement of conformal invariance of the quantized sigma model expansion. The effective action introduced in Chapter 2 is indeed the action which reproduces the appropriate differential conditions for the basic gravitational multiplet, to lowest order in the loop expansion of the sigma model quantization.

To provide an explicit example of this important string theory result we now consider the case which is the most relevant one in the context of this book, considering the geodesic evolution of a string in a pure gravitational background, described by the action (3.6), and interpreted as an effective action for the quantum field X^μ.

For the computation of the one-loop corrections we consider the fluctuations $\widehat{x}^\mu$ of the field around a classical expectation value, performing the shift $X^\mu \to X^\mu + \widehat{x}^\mu$, and expanding the action in powers of the quantum fluctuations. The shift represents a coordinate transformation of the target space-time manifold, and thus induces a transformation of the metric tensor, $g(X) \to g(X + \widehat{x})$, which also has to be expanded in powers of the quantum fluctuations. Such an expansion is in general complicated, but it can be expressed in a rather simple form by exploiting the general covariance of the theory: in particular, if we introduce the so-called Riemann "normal" coordinates $z^\mu(\widehat{x})$ (see e.g. [11]), then all explicitly non-covariant terms disappear from the expansion, and the shifted fields, g and X, can be expressed in powers of these new variables as follows:

$$g_{\mu\nu}(X) \to g_{\mu\nu}(X) - \frac{1}{3}R_{\mu\alpha\nu\beta}(X)z^\alpha z^\beta + \cdots,$$

$$\partial_a X^\mu \to \partial_a X^\mu + \partial_a z^\mu - \frac{1}{3}\partial_a X^\nu R^\mu{}_{\alpha\nu\beta}(X)z^\alpha z^\beta \cdots \tag{3.14}$$

By inserting this expansion in the sigma model action (3.6), up to terms of order z^2, we obtain for the fluctuations the following quadratic action:

$$S_2 = \frac{1}{4\pi\alpha'} \int d^2\xi \sqrt{-\gamma} \gamma^{ab} \left[\partial_a z^\mu \partial_b z^\nu g_{\mu\nu}(X) - \partial_a X^\mu \partial_b X^\nu R_{\mu\alpha\nu\beta}(X) z^\alpha z^\beta \right].$$
(3.15)

This action includes two terms which we can interpret, respectively, as the kinetic term and the effective mass term for the quantum field $z(\xi)$ in a classical (curved) background.

A quantum model of this type contains divergences, which are to be renormalized by introducing appropriate counterterms; such counterterms can be taken as proportional to the effective mass, and in this case are of the form [1]

$$S_\infty = -\frac{1}{4\pi\alpha'} \int d^2\xi \sqrt{-\gamma} \gamma^{ab} \partial_a X^\mu \partial_b X^\nu R_{\mu\alpha\nu\beta} \langle z^\alpha z^\beta \rangle_{\xi=\xi'}.$$
(3.16)

Here

$$\langle z^\alpha z^\beta \rangle_{\xi=\xi'} = \lim_{\xi \to \xi'} \left[-i 2\pi\alpha' g^{\alpha\beta} \int \frac{d^2 k}{(2\pi)^2} \frac{e^{-ik\cdot(\xi-\xi')}}{k^2} \right]$$
(3.17)

is the one-loop term obtained from the two-point correlation function $\langle z^\alpha(\xi) z^\beta(\xi') \rangle$ (namely from the Fourier transform of the propagator of the free field z^μ), in the limit $\xi \to \xi'$. Notice that this contribution is of order α' with respect to the classical terms of the action, in agreement with the fact that the quantum effects must become negligible when the action integral becomes larger than one in natural units, namely in the limit $\alpha' \to 0$. In other words, the expansion in quantum loops of the sigma model must correspond to a perturbative expansion in powers of α'.

Adding such a term we obtain a model which is regular but no longer conformally invariant, in general. For an explicit check of this important point we can consider an infinitesimal deformation of the number of dimensions of the world-sheet manifold, from $D = 2$ to $D = 2 + \epsilon$, in such a way as to make the integral (3.17) computable in finite form. In $D = 2 + \epsilon$ dimensions the factor $\sqrt{-\gamma} \gamma^{ab}$ is not conformally invariant, and the same is true for the action S_∞. Considering an infinitesimal conformal transformation with parameter $\delta\omega$, and expanding the result (3.11) to first order in $\delta\omega$, we can easily obtain the corresponding infinitesimal variation of the one-loop action (3.16), in $2 + \epsilon$ dimensions, as

$$\delta S_\infty = -\frac{\epsilon}{4\pi\alpha'} \int d^{2+\epsilon}\xi \sqrt{-\gamma} \gamma^{ab} \partial_a X^\mu \partial_b X^\nu R_{\mu\alpha\nu\beta} \langle z^\alpha z^\beta \rangle_{\xi=\xi'} \delta\omega.$$
(3.18)

The conformal invariance would seem to be restored in the limit $\epsilon \to 0$, where one recovers a two-dimensional world-sheet integral. For $\epsilon \neq 0$, however, the one-loop contribution can be computed explicitly, and it can be shown to contain a pole which diverges as ϵ^{-1} for $\epsilon \to 0$.

For the explicit computation of the integral (3.17) we can conveniently absorb the imaginary factor $-i$ inside the space-like integration variable, so as to obtain an integral with an effective Euclidean metric. We can then evaluate the one-loop contribution as follows:

$$
\begin{aligned}
\int \frac{d^2k}{(2\pi)^2} \frac{1}{k^2} &= \lim_{\epsilon,m\to 0} \int_{-\infty}^{+\infty} \frac{d^{2+\epsilon}k}{(2\pi)^{2+\epsilon}} \frac{1}{k^2+m^2} \\
&= \lim_{\epsilon,m\to 0} \int_{-\infty}^{+\infty} \frac{d^{2+\epsilon}k}{(2\pi)^{2+\epsilon}} \int_0^\infty ds\, e^{-s(m^2+k^2)} \\
&= \lim_{\epsilon,m\to 0} \int_0^\infty \frac{ds}{(2\pi)^{2+\epsilon}} \left(\frac{\pi}{s}\right)^{\frac{2+\epsilon}{2}} e^{-sm^2} \\
&= \lim_{\epsilon,m\to 0} \frac{m^\epsilon}{(4\pi)^{\frac{2+\epsilon}{2}}} \Gamma\left(-\frac{\epsilon}{2}\right) \\
&= -\lim_{\epsilon\to 0} \frac{1}{2\pi\epsilon},
\end{aligned} \tag{3.19}
$$

where we have used the Gauss result for the integral over k, and the definition of the Euler Gamma function for the integral over s. Therefore, according to the definition (3.17),

$$
\langle z^\alpha z^\beta \rangle_{\xi=\xi'} = -\lim_{\epsilon\to 0} \frac{\alpha'}{\epsilon} g^{\alpha\beta}. \tag{3.20}
$$

When inserted into Eq. (3.18), this one-loop contribution exactly cancels the ϵ-dependence introduced by the infinitesimal conformal transformation, so that the variation of the action remains non-zero ($\delta S_\infty / \delta\omega \neq 0$) even in the limit $\epsilon \to 0$:

$$
\delta S_\infty = -\frac{1}{4\pi} \int d^2\xi \sqrt{-\gamma}\, \gamma^{ab} \partial_a X^\mu \partial_b X^\nu R_{\mu\nu}\, \delta\omega \tag{3.21}
$$

(we have used $R_{\mu\nu} = R_{\alpha\mu\nu}{}^\alpha$). As a consequence, this regularized quantum model turns out to be conformally invariant, if and only if the classical background geometry satisfies the differential conditions

$$
R_{\mu\nu}(X) = 0, \tag{3.22}
$$

namely, if and only if we restrict the gravitational background to field configurations satisfying the vacuum Einstein equations!

Similar arguments, applied to a sigma model which includes the interaction with a background gauge field, lead us to conclude that the conformal invariance is preserved at the quantum level provided the gauge field satisfies the appropriate Yang–Mills equations (see [10, 12, 13] for an explicit computation in the case of the heterotic superstring model). In the same way one finds that the requirement of "superconformal" invariance (i.e. world-sheet conformal symmetry) supplies the dynamics for the space-time fermion fields [10].

The conceptual importance of these results, showing that the consistency of the quantum string motion uniquely fixes the dynamics of the fields with which the string is interacting, can hardly be underestimated. Thanks to this effect, on one hand, we can recover – through a purely theoretical argument, and as a first, low-energy approximation – the field equations of the classical forces, originally postulated on the grounds of phenomenological motivations only. On the other hand, we can compute – through an unambiguous, automatic procedure – the quantum corrections to such classical equations, pushing to higher orders the α' expansion generated by the loop corrections (and, in addition, considering world-sheet manifolds with non-trivial topologies, as we will see in Section 3.2).

However, the example we have just discussed cannot be regarded as a complete description of the string motion in a gravitational background, since the condition of conformal invariance cannot be separately imposed on the metric, on the gauge fields, and so on: it should be simultaneously imposed on all background fields with which a string can interact. Such fields are contained in the spectrum of states (with growing level of masses) that one obtains in the first quantization of the given (open or closed, bosonic or supersymmetric) model of string, and that associates with the metric other important gravitational partners. For the cosmological applications of this book we here focus our attention on the multiplet of three massless states present in the spectrum of the closed bosonic string (see Appendix 3A), as well as in the various models of superstrings (see Appendix 3B): the graviton, the dilaton and the Kalb–Ramond axion, respectively represented by a tensor field (the metric $g_{\mu\nu}$), a scalar field ϕ and a second-rank antisymmetric tensor (the NS–NS two-form $B_{\mu\nu}$).

The interaction of a string with these background fields is described by the following sigma model action:

$$S = \frac{1}{4\pi\alpha'} \int_\Sigma d^2\xi \left\{ \partial_a X^\mu \partial_b X^\nu \left[\sqrt{-\gamma}\gamma^{ab} g_{\mu\nu}(X) + \epsilon^{ab} B_{\mu\nu}(X) \right] \right.$$

$$\left. + \frac{\alpha'}{2}\sqrt{-\gamma}\,R^{(2)}\phi(X) \right\}. \tag{3.23}$$

We have added to the Polyakov action the so-called Wess–Zumino term [14, 15] describing the interaction with the field $B_{\mu\nu} = -B_{\nu\mu}$ ($\epsilon^{ab} = -\epsilon^{ba}$ is the two-dimensional Levi-Civita tensor density on the world-sheet surface). We have also included the explicit coupling between the dilaton and $R^{(2)}$, the intrinsic scalar curvature associated with the world-sheet metric [16]. If ϕ is a constant such a term does not contribute to the sigma model equations, as $\int d^2\xi R^{(2)}$ is a pure Euler two-form in two dimensions, just like the Gauss–Bonnet four-form in four dimensions, see Section 2.3.

It is important to notice that the first two terms of the above action are classically invariant under conformal transformations of the metric γ_{ab}, while the dilaton term breaks conformal invariance at the classical level. For dimensional reasons, however, the dilaton term is of order α' with respect to the others, namely it is of the same order as the quantum one-loop corrections: the (classical) contribution of the dilaton term is thus to be included into the condition of one-loop conformal invariance of the total action. In order to obtain the differential conditions governing the motion of g, B and ϕ, we thus proceed by computing the variation of the action (3.23) under infinitesimal conformal transformations to the one-loop order for g and B, and at the classical level for ϕ.

We need, first of all, a generalization of the string equations of motion (3.8), so as to include the interaction with the new background field $B_{\mu\nu}$. Without repeating the computation already performed for the gravitational background, it will be enough to add to the left-hand side of Eq. (3.8) the contribution of the Wess–Zumino Lagrangian, given by

$$\partial_a \frac{\partial L_{WZ}}{\partial(\partial_a X^\mu)} - \frac{\partial L_{WZ}}{\partial X^\mu}$$

$$= \epsilon^{ab}\partial_a\partial_b X^\nu B_{\mu\nu} + \epsilon^{ab}\partial_b X^\nu \partial_a X^\alpha \partial_\alpha B_{\mu\nu} - \frac{1}{2}\epsilon^{ab}\partial_a X^\alpha \partial_b X^\nu \partial_\mu B_{\alpha\nu}. \qquad (3.24)$$

The first term is vanishing for the antisymmetry of ϵ_{ab}. Antisymmetrizing the second term with respect to α and ν, and adding the new terms to Eq. (3.8) (remembering that we have divided by $\sqrt{-\gamma}$), we obtain the equation of motion

$$\nabla^a\nabla_a X^\mu + \gamma^{ab}\partial_a X^\alpha \partial_b X^\nu \Gamma_{\alpha\nu}{}^\mu - \frac{1}{2}\frac{\epsilon^{ab}}{\sqrt{-\gamma}}\partial_a X^\alpha \partial_b X^\nu H^\mu{}_{\alpha\nu} = 0, \qquad (3.25)$$

where $H_{\mu\alpha\nu} = \partial_\mu B_{\alpha\nu} + \partial_\alpha B_{\nu\mu} + \partial_\nu B_{\mu\alpha}$.

We can then compute the conformal transformation of the classical dilaton term appearing in the action (3.23). Considering the transformation (3.10), and applying the general result (2.40), (2.41) we obtain, in a total number of $D = 2$ dimensions,

$$\sqrt{-\gamma}R^{(2)} \rightarrow \sqrt{-\gamma}\left(R^{(2)} - 2\nabla^2\omega\right). \qquad (3.26)$$

For an infinitesimal transformation with parameter $\delta\omega$ we have, to first order, $\delta(\sqrt{-\gamma}R^{(2)}) = -2\sqrt{-\gamma}\,\nabla^2\delta\omega$, and we find that the variation of the dilaton part of the action (3.23) is

$$\delta S_\phi = -\frac{1}{4\pi}\int d^2\xi\sqrt{-\gamma}(\nabla_a\partial^a\delta\omega)\,\phi(X). \qquad (3.27)$$

Integrating by parts two times, and setting $\partial_a \phi = \partial_a X^\mu \partial_\mu \phi$, we obtain

$$\delta S_\phi = \frac{1}{4\pi} \int d^2\xi \sqrt{-\gamma} \gamma^{ab} \partial_a X^\mu \partial_b \delta\omega \partial_\mu \phi$$

$$= -\frac{1}{4\pi} \int d^2\xi \sqrt{-\gamma}\, \delta\omega \left[\frac{1}{\sqrt{-\gamma}} \partial_b (\sqrt{-\gamma} \gamma^{ab} \partial_a X^\mu) \partial_\mu \phi + \gamma^{ab} \partial_a X^\mu \partial_b X^\nu \partial_\nu \partial_\mu \phi \right].$$

$$(3.28)$$

Let us finally write the result in explicit covariant form, using the definition $\nabla_\mu \nabla_\nu \phi = \partial_\mu \partial_\nu \phi - \Gamma_{\mu\nu}{}^\rho \partial_\rho \phi$, and applying the string equation of motion (3.25). The infinitesimal variation of the dilaton action then takes the form

$$\delta S_\phi = -\frac{1}{4\pi} \int d^2\xi\, \delta\omega \left[\sqrt{-\gamma} \gamma^{ab} \partial_a X^\mu \partial_b X^\nu \nabla_\mu \nabla_\nu \phi \right.$$

$$\left. + \frac{1}{2} \epsilon^{ab} \partial_a X^\mu \partial_b X^\nu H_{\mu\nu\alpha} \nabla^\alpha \phi \right], \qquad (3.29)$$

which represents the classical conformal anomaly of the sigma model (3.23) due to the dilaton, to be added to the one-loop contributions due to $g_{\mu\nu}$ and $B_{\mu\nu}$.

For the computation of the B contribution we may follow the procedure used for the gravitational field, by shifting the coordinates around a classical solution, and expanding the Wess–Zumino term in Riemann normal coordinates. We must then complement Eq. (3.14) with the two-form expansion,

$$B_{\mu\nu}(X) \to B_{\mu\nu}(X) + \nabla_\alpha B_{\mu\nu} z^\alpha + \frac{1}{2} \left(\nabla_\alpha \nabla_\beta B_{\mu\nu} - \frac{1}{3} R^\rho{}_{\alpha\mu\beta} B_{\rho\nu} \right.$$

$$\left. - \frac{1}{3} R^\rho{}_{\alpha\nu\beta} B_{\rho\mu} \right) z^\alpha z^\beta + \cdots \qquad (3.30)$$

to be inserted into the Wess–Zumino term of the action (3.23). Stopping the expansion to order z^2, we are led to the following quadratic action for the quantum fluctuations z^μ:

$$S_2^{WZ} = \frac{1}{4\pi\alpha'} \int d^2\xi\, \epsilon^{ab} \left[\partial_a z^\mu \partial_b z^\nu B_{\mu\nu}(X) + 2\partial_a X^\mu \nabla_\alpha B_{\mu\nu}(X) \partial_b z^\nu z^\alpha \right.$$

$$\left. + \partial_a X^\mu \partial_b X^\nu \left(\frac{1}{2} \nabla_\alpha \nabla_\beta B_{\mu\nu}(X) - R^\rho{}_{\alpha\mu\beta}(X) B_{\rho\nu}(X) \right) z^\alpha z^\beta \right]. \qquad (3.31)$$

For the simplification of the final equations it is convenient to rewrite this action in terms of the field strength $H_{\mu\nu\alpha}$, eliminating the antisymmetric potential $B_{\mu\nu}$. By using the definition of H, the properties of the Riemann tensor, integrating by parts, and using the definitions $\partial_a B_{\mu\nu} = \partial_a X^\alpha \partial_\alpha B_{\mu\nu}$ and so on, we obtain the equivalent form

$$S_2^{WZ} = \frac{1}{4\pi\alpha'} \int d^2\xi \, \epsilon^{ab} \left[z^\alpha \partial_a z^\beta \partial_b X^\mu H_{\mu\alpha\beta}(X) - \frac{1}{2}\partial_a X^\mu \partial_b X^\nu \nabla_\alpha H_{\beta\mu\nu}(X) z^\alpha z^\beta \right],$$

(3.32)

to be added to the gravitational action (3.15).

With the addition of S_2^{WZ}, the effective action for the quantum field z acquires a new contribution to the mass term $z^\alpha z^\beta$, plus an additional "soft" mass term proportional to $z^a \partial z^b$. In order to regularize the new divergences we have thus to introduce another counterterm, to be added to Eq. (3.16) used for the gravitational background. In the one-loop approximation, the new term is given by

$$S_\infty^{WZ} = -\frac{1}{4\pi\alpha'} \int d^2\xi \, \partial_a X^\mu \partial_b X^\nu \left(\frac{1}{2}\epsilon^{ab}\nabla_\alpha H_{\beta\mu\nu} + \frac{1}{4}\sqrt{-\gamma}\gamma^{ab} H_{\mu\alpha}{}^\rho H_{\nu\beta\rho} \right)$$

$$\times \langle z^\alpha z^\beta \rangle_{\xi=\xi'}$$

(3.33)

(in fact by rewriting in canonical form the full quadratic action for z^μ, it can be shown that the effective mass term is given by the sum of the initial mass term, minus the coefficient of the soft term divided by two and squared (see e.g. A. Sagnotti, *Lectures on String Theory* (Università di Roma "Tor Vergata" and "Scuola Normale Superiore", Pisa, 2004), unpublished.)).

Again, this counterterm induces a breaking of the conformal symmetry, as can be easily checked by shifting to $d = 2 + \epsilon$ dimensions, and performing an infinitesimal conformal transformation which introduces into S_∞^{WZ} the factor $\epsilon\delta\omega$. The ϵ factor is canceled by the one-loop contribution (3.20), so that the infinitesimal conformal anomaly survives even in the limit $\epsilon \to 0$:

$$\delta S_\infty^{WZ} = \frac{1}{4\pi} \int d^2\xi \, \partial_a X^\mu \partial_b X^\nu \left(\frac{1}{2}\epsilon^{ab}\nabla^\alpha H_{\alpha\mu\nu} + \frac{1}{4}\sqrt{-\gamma}\gamma^{ab} H_{\mu\alpha\beta}H_\nu{}^{\alpha\beta} \right)\delta\omega.$$

(3.34)

We are now able to determine the full set of differential conditions to be satisfied by the background fields for a consistent cancelation of the conformal anomalies, to leading order in the α' (one-loop) corrections. Using the results (3.21), (3.29) and (3.34), imposing $\delta S_\infty + \delta S_\infty^{WZ} + \delta S_\phi = 0$, and separating the symmetric and antisymmetric factors of $\partial_a X^\mu \partial_b X^\nu$ we obtain, respectively,

$$R_{\mu\nu} + \nabla_\mu\nabla_\nu\phi - \frac{1}{4}H_{\mu\alpha\beta}H_\nu{}^{\alpha\beta} = 0,$$

(3.35)

and

$$\nabla^\alpha H_{\alpha\mu\nu} - \nabla^\alpha\phi \, H_{\alpha\mu\nu} = 0 = \nabla^\alpha \left(e^{-\phi} H_{\alpha\mu\nu} \right).$$

(3.36)

These equations exactly coincide with the effective equations (2.26) and (2.28) (without sources) discussed in the previous chapter.

The above set of equations has to be completed by the addition of the dilaton equation of motion, which can be obtained through a combination of the above two equations. In fact, if we compute the covariant divergence of Eq. (3.35), using the contracted Bianchi identity $\nabla^\nu R_{\mu\nu} = \nabla_\mu R/2$, the commutation relation (2.18), and the torsion equation (3.36), we obtain

$$\nabla_\mu \left(\frac{R}{2} + \nabla^2\phi - \frac{1}{2}\nabla\phi^2 \right) - \frac{1}{4} H^{\nu\alpha\beta}\nabla_\nu H_{\alpha\beta\mu} = 0. \tag{3.37}$$

The last term, using the explicit definition of the field strength H and of the covariant derivative, can be rewritten as

$$-\frac{1}{4} H^{\nu\alpha\beta} \frac{1}{3} \left(\nabla_\nu H_{\alpha\beta\mu} + \nabla_\alpha H_{\beta\nu\mu} + \nabla_\beta H_{\nu\alpha\mu} \right)$$

$$= -\frac{1}{12} H^{\nu\alpha\beta}\nabla_\mu H_{\alpha\beta\nu} = -\frac{1}{24}\nabla_\mu \left(H_{\alpha\beta\nu} H^{\alpha\beta\nu} \right). \tag{3.38}$$

Thus

$$\nabla_\mu \left(R + 2\nabla^2\phi - \nabla\phi^2 - \frac{1}{12}H^2 \right) = 0, \tag{3.39}$$

which coincides with the gradient of the dilaton equation (2.25) (without sources and potential).

The above equation actually implies

$$R + 2\nabla^2\phi - \nabla\phi^2 - \frac{1}{12}H^2 = c, \tag{3.40}$$

where c is any constant number. Such a constant, however, is found to be zero in a perturbative approach to string quantization, where the quantum theory is consistently formulated only in a critical number $D = D_c$ of space-time dimensions (see Appendix 3A). For the bosonic string, in particular, one finds [5, 10] that $c = (D-26)/3\alpha'$ (for the superstring $c \sim (D-10)$). Taking into account this result, the set of equations (3.35), (3.36) and (3.40), for the multiplet of fundamental fields g, B and ϕ, can be deduced from the effective action

$$S = -\frac{1}{2\lambda_s^{d-1}} \int d^{d+1}x \sqrt{|g|} e^{-\phi} \left(R + \nabla\phi^2 - \frac{1}{12}H^2 + \frac{26-D}{3\alpha'} \right), \tag{3.41}$$

as discussed in the previous chapter (see Eq. (2.21)). In a non-critical number of dimensions ($D \neq 26$) the action should contain a dilaton potential $V \propto (D_c - D)$, which is not in general constant at higher levels of the perturbative approximation, as we will see in the next section.

3.2 Higher-curvature and higher-genus expansion

As noted in the previous section, the quantum loop expansion of the sigma model quantization corresponds to an expansion in powers of the string length parameter $\alpha' = \lambda_s^2/2\pi$. This parameter represents the natural (i.e. model-provided) unit of distance to which to refer the geometry described by the sigma model metric $g_{\mu\nu}$: in particular, curvature radii which are large with respect to $\sqrt{\alpha'}$ will correspond to small curvatures in string units (i.e. to small gradients of the gravitational background). The quantum sigma model expansion in powers of α' is thus equivalent to a geometric expansion in powers of the curvature and of its derivatives, namely to a higher-derivative expansion of the metric (and of all the other background fields).

This fact is also evident from the computations of the previous section, where the expansion (3.14) around the classical solution can be continued to higher orders, in Riemann normal coordinates, as follows:

$$g_{\mu\nu}(X) \to g_{\mu\nu}(X) - \frac{1}{3} R_{\mu\alpha\nu\beta}(X) z^\alpha z^\beta + c_1 R_{\mu\alpha\nu\beta}(X) R^\mu{}_\rho{}^\nu{}_\sigma(X) z^\alpha z^\beta z^\rho z^\sigma + \cdots ,$$
(3.42)

where c_1 is an appropriate numerical coefficient. Extending the computation of the quantum corrections to higher orders one then finds that the condition of conformal invariance (3.22), at the two-loop level, acquires quadratic curvature corrections [1],

$$R_{\mu\nu} + \frac{\alpha'}{2} R_{\mu\alpha\beta\rho} R_\nu{}^{\alpha\beta\rho} = 0,$$
(3.43)

with associated quadratic corrections (of order α') to the effective action (3.41). The same is true for the other background fields B, ϕ included in the equations of motion.

Such corrections, as already noted in Section 2.3, suffer from an intrinsic ambiguity: by performing appropriate field redefinitions it turns out to be possible to satisfy the condition of conformal invariance even when the field equations have different higher-derivative corrections – and thus admit solutions with different geometrical properties – at any given order of the (truncated) α' expansion [17].

This ambiguity may be resolved only in the context of an exact conformal theory, where the background fields satisfy the equations of the non-linear sigma model (to all orders of the α' expansion, not only to a given truncated order). For the purposes of this book, i.e. for a simple illustration of the possibile higher-derivative effects in a cosmological context, we will adopt in the following chapters the "minimal" gravi-dilaton model with quadratic curvature corrections, parametrized by the action (2.80), and discussed in detail in the previous chapter.

We refer the reader to the literature for perturbative actions including α' corrections also in the $B_{\mu\nu}$ sector [17] – possibly satisfying particular duality properties [18, 19] – and for a study of exact conformal models with possible applications in a string cosmology context [20, 21].

The higher-derivative expansion, controlled by the parameter α', is a peculiar string theory property, closely related to the finite one-dimensional extension $\lambda_s \sim \sqrt{\alpha'}$ of the fundamental objects of the theory. However, it is not the only possible expansion that we can apply in the context of the sigma model action for a perturbative approach to the effective background equations.

Another important (and useful) sigma model approximation is the expansion in the growing level of complexity of the topology of the world-sheet surface, the so-called "higher-genus" expansion [1]. This is also a typical aspect of string theory which has, however, a well-known counterpart in a quantum field theory context, represented by the conventional loop expansion in powers of the coupling constant. The various topological levels of the world-sheet surface, in fact, can be closely correlated to the various levels of Feynman graphs of a field theory with point-like sources. Higher levels of topological complexity correspond to higher powers of the coupling constant g_s^2 which controls the strength of the interactions among strings. It is thus of crucial importance to stress that such a topological expansion can be expressed as an expansion in powers of $\exp\langle\phi\rangle$, a result which clearly identifies the expectation value of the dilaton, $\langle\phi\rangle$, as the parameter controlling the effective coupling constant of perturbative string theory.

Consider, in fact, the usual loop expansion of a scattering process in a quantum field theory context, graphically represented by a sum of Feynman diagrams as in Fig. 3.1. String interactions can be described by a similar sum of elementary processes: the time evolution of a string, however, describes a world-surface, not a world-line, and the interactions correspond to modifications of the world-sheet topology, as illustrated in Fig. 3.1. For a closed string, for instance, the interaction region (represented by the shaded areas of the figure) has the topology of a sphere in the tree-level graphic, of a torus in the one-loop graphic, and so on. The interaction region of a general, n-loop graph thus corresponds to a two-dimensional Riemannian surface Σ_n of genus n, i.e. to a manifold with n "handles".

For a two-dimensional closed orientable manifold, on the other hand, the genus n is completely determined by the so-called Euler characteristic χ, which, by virtue of the Gauss–Bonnet theorem, is given by the topological invariant

$$\chi = \frac{1}{4\pi} \int_{\Sigma_n} d^2\xi \sqrt{-\gamma}\, R^{(2)}(\gamma) \equiv 2 - 2n. \tag{3.44}$$

If we take the dilaton part of the sigmal model action (3.23), and we set $\phi = \phi_0 + \Phi$ (extracting from the dilaton field a classical, constant part $\phi_0 = \langle\phi\rangle$,

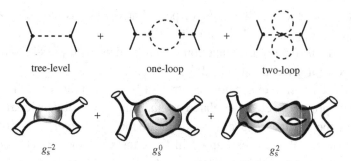

Figure 3.1 Loop expansion for particle interactions (top diagrams) and string interactions (bottom diagrams). The string-loop counting parameter is the string coupling g_s^2, and the n-loop amplitude of the process is proportional to $g_s^{2(n-1)}$.

averaged with respect to the sigma model partition function), then the action is simply shifted by a constant term which is genus dependent:

$$S_\chi(\phi) = S_\chi(\Phi) + \frac{1}{2}\chi\phi_0. \tag{3.45}$$

The index χ of the action identifies a sigma model defined on a world-sheet with Euler characteristic χ, according to Eq. (3.44). The partition function describing the string scattering of Fig. 3.1, written as a topological expansion of growing genus,

$$\mathcal{Z} = \sum_\chi \mathcal{Z}_\chi, \qquad \mathcal{Z}_\chi = \int \mathcal{D}X \, e^{-S_\chi(\phi)} = e^{-\frac{\chi}{2}\phi_0} \int \mathcal{D}X \, e^{-S_\chi(\Phi)}, \tag{3.46}$$

can then be rewritten, using Eq. (3.44), as an expansion in growing powers of $\exp(\phi_0)$:

$$\mathcal{Z} = \sum_\chi \mathcal{Z}_\chi = \sum_{n=0}^{\infty} e^{(n-1)\phi_0} \int \mathcal{D}X \, e^{-S_n(\Phi)}. \tag{3.47}$$

A comparison with the conventional loop expansion clearly identifies the exponential of the dilaton expectation value with the parameter controlling the tree-level string coupling (i.e. with the string-loop counting parameter g_s^2 of Fig. 3.1), namely

$$g_s^2 = \exp\langle\phi\rangle. \tag{3.48}$$

The two approximations defined in the context of the sigma model action – the higher-derivative expansion, controlled by α', and the higher-genus expansion, controlled by g_s^2 – are in principle independent, and can be applied simultaneously. Taking into account both possibilities, the most general perturbative form of the string effective action can be schematized as follows (for simplicity, we include here only the gravi-dilaton sector):

$$S = -\frac{1}{2\lambda_s^{d-1}} \int d^{d+1}x\sqrt{|g|} \left\{ e^{-\phi}\left[R + (\nabla\phi)^2 - \frac{\alpha'}{4}R^2 + \cdots\right]\right.$$

$$+ \left[c_R^1 R + c_\phi^1(\nabla\phi)^2 + \alpha' c_{\alpha'}^1 R^2 + \cdots\right]$$

$$+ e^{\phi}\left[c_R^2 R + c_\phi^2(\nabla\phi)^2 + \alpha' c_{\alpha'}^2 R^2 + \cdots\right]$$

$$\left. + e^{2\phi}[\cdots] + \cdots\right\}, \tag{3.49}$$

where the first line contains the tree-level terms in g_s^2, the second line contains the one-loop terms in g_s^2, and so on. At any given order of the topological expansion there is a full expansion in α', for all fields, and to all orders. Conversely, at any given order in α', the contribution of the topological loops introduces dilatonic corrections, $c^n e^{n\phi}$, which are in general different *for different fields* and *at different orders*. Therefore, the growth of the coupling strength tends to break the universality of the dilaton coupling to the various matter fields: this effect, as we shall discuss in Chapter 9, may induce an effective violation of the equivalence principle, with possibly interesting phenomenological implications (unless the dilaton-mediated force is too weak, or too short-range).

In particular, the lowest-order action discussed in Chapter 2 is valid at low enough energy scales (when the field gradients are small enough in string units) *and* at weak enough coupling. In the following chapters we will often consider background configurations (possibly implemented in a primordial cosmological context) in which the coupling is weak ($g_s^2 \ll 1$, $\phi \rightarrow -\infty$), while the curvature is non-negligible in string units, and for which we can limit the expansion to the lowest topological order, including only the α' corrections. In the present cosmological configuration the curvatures are low ($\alpha' R \ll 1$), but the couplings are probably inside (or very near to) the strong coupling regime ($g_s \sim 1$): in that case, we can use the tree-level approximation in α', but we should include the loop contributions of the topological expansion, possibly to all orders. In that regime, a realistic form of the (string-frame) gravi-dilaton effective action is then the following:

$$S = -\frac{1}{2\lambda_s^{d-1}} \int d^{d+1}x\sqrt{|g|} \left[Z_R(\phi)R + Z_\phi(\phi)(\nabla\phi)^2 + V(\phi)\right], \tag{3.50}$$

where the loop contributions are included into the dilaton "form factors" Z_R, Z_ϕ (see Chapter 9).

We have also included in the above action a possible dilaton potential, which we have seen to be present, at the perturbative level, for a model formulated in non-critical dimensions ($D \neq D_c$). At the n-loop order, such a potential is the source of an exponential contribution of the form

$$V_{\text{pert}}(\phi) \sim \frac{D_c - D}{\lambda_s^2} e^{(n-1)\phi}. \tag{3.51}$$

When the coupling is large we should also take into account the possible presence of a non-perturbative potential, which is non-zero even in critical dimensions, and which is required, for instance, by superstring models of supersymmetry breaking [22], producing an effective mass term for the dilaton [23].

In order to discuss the possible form of the functions $Z(\phi)$ and $V(\phi)$ we first note that in the weak coupling regime ($g_s^2 \to 0$) one must recover the tree-level action (3.41), so that $Z_R \to Z_\phi \to \exp(-\phi)$ for $\phi \to -\infty$. In the same regime we know that the non-perturbative potential has to be extremely flat, typically characterized by an instantonic suppression of the type

$$V(\phi) \sim e^{-\alpha/g_s^2} \sim e^{-\alpha e^{-\phi}}, \qquad \phi \to -\infty, \tag{3.52}$$

where α is a model-dependent coefficient of order one. As ϕ is growing, the loop contributions lead the potential to develop a structure with local maxima and minima (see for instance [24]), which could trap the dilaton, and freeze out the string coupling g_s^2 to a realistic value compatible with our present cosmological scenario (see Fig. 3.2). However, there are at present no firm and unambiguous theoretical predictions for the behavior of the dilaton potential in the limit of extremely strong coupling, $\phi \to +\infty$: the perturbative component of the potential is exponentially divergent (see Eq. (3.51)), but such a divergence could be suppressed by large non-perturbative effects. The two possible limiting cases, $V \to 0$ and $V \to \infty$, are illustrated, respectively, by the dashed curves (a) and (b) of Fig. 3.2.

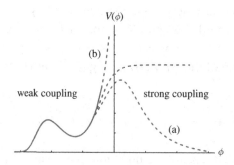

Figure 3.2 Possible qualitative behavior of the non-perturbative dilaton potential.

An important remark is in order at this point. In the context of a theory in which the fundamental coupling parameter is fully controlled by a field, the dilaton – and thus is free to run, following the dilaton evolution – the stabilization (or quasi-stabilization) of the coupling, needed for a realistic description of present interactions, requires an effective dilaton potential. Quite independently of the possible origin of such a potential (see [25] for recent progress towards the stabilization of all moduli fields in a cosmological context), there are in principle two alternative scenarios for the phenomenology of the stabilization mechanism.

A first scenario is based on a potential which is rapidly increasing at large values of ϕ, so as to make the strong-coupling regime hardly accessible, and which has a local minimum $V'(\phi_0) = 0$ at a value ϕ_0 located in the semi-perturbative region in which $\phi_0 < 0$, $|\phi_0| = \mathcal{O}(1)$, so as to be phenomenologically consistent with the tree-level relation (2.3) between string mass and Planck mass. For models in which $D-4$ spatial dimensions are compactified at the string scale, and in which $M_s \simeq 0.1 M_P$ (as required by a consistent string unification of gravitational and gauge interactions [26]), one can estimate the value of ϕ_0 as

$$\phi_0 = 2 \ln \left(\frac{M_s}{M_P} \right) \simeq -4.6. \tag{3.53}$$

A possible "minimal" example of such a potential, corresponding to the curve (b) of Fig. 3.2, and with an amplitude controlled by the single dimensional parameter m^2, is the following [24]:

$$V(\phi) = m^2 \left[e^{k_1(\phi-\phi_1)} + \beta e^{-k_2(\phi-\phi_1)} \right] e^{-\alpha \exp[-\gamma(\phi-\phi_1)]}, \tag{3.54}$$

where $k_1, k_2, \alpha, \beta, \gamma, \phi_1$ are dimensionless numbers of order one. Choosing appropriate values of such parameters one can easily obtain a local minimum at $\phi_0 \simeq \phi_1$, satisfying the requirement (3.53). However, in this case the dilaton may tend to be shifted away from the equilibrium position in the course of the cosmological evolution because of its non-minimal coupling to the trace of the matter stress tensor, unless the effective dilaton mass m^2 is restricted to an appropriate range of values [27].

A second possibility for the stabilization scenario is based on a model in which both the potential and the dilaton form factors, $Z(\phi)$, admit an asymptotic expansion in inverse power of the "bare" (i.e. tree-level) coupling $g_s^2 = \exp \phi$, for $\phi \to +\infty$ [28]:

$$V(\phi) = V_0 e^{-\phi} + V_1 e^{-2\phi} + \cdots,$$

$$Z_\phi(\phi) = -c_2^2 + b_2^2 e^{-\phi} + a_2^2 e^{-2\phi} + \cdots, \tag{3.55}$$

$$Z_R(\phi) = c_1^2 + b_1^2 e^{-\phi} + a_1^2 e^{-2\phi} + \cdots, \qquad \phi \to +\infty.$$

The leading-order (dimensionless) coefficients c_i of this expansion are determined by the number N of gauge fields present in the given model of superstring unification, and contributing to the loop corrections. In realistic grand-unification models such a number is large: one has, typically, $c_1^2 \sim c_2^2 \sim N \sim 10^2$ (the coefficient c_2^2 enters with a minus sign, in order to guarantee the correct normalization of the dilaton kinetic term in the limit $\phi \to +\infty$). As a consequence, the effective coupling constant $\bar{g}_s^2$, which in this context asymptotically controls the ratio between string mass and Planck mass, turns out to be saturated to a (realistic) moderately perturbative value in the limit in which the bare coupling parameter $g_s^2(\phi) = \exp\phi$ becomes sufficiently large. We obtain, from Eqs. (3.55) and (3.50),

$$\left(\frac{M_s}{M_P}\right)^{d-1} = Z_R^{-1}(\phi) \to \left(c_1^2 + b_1^2 e^{-\phi} + \cdots\right)^{-1}$$

$$\to \frac{g_s^2(\phi)}{c_1^2 g_s^2(\phi) + b_1} \to \frac{1}{c_1^2} \equiv \bar{g}_s^2, \qquad \phi \to +\infty. \tag{3.56}$$

A possible example of non-perturbative potential which agrees with the asymptotic behavior of Eqs. (3.52) and (3.55) is the following [29]:

$$V(\phi) = m^2 \left[e^{-\frac{1}{\beta_1}\exp(-\phi)} - e^{-\frac{1}{\beta_2}\exp(-\phi)}\right], \qquad \beta_1 > \beta_2 > 0. \tag{3.57}$$

Here β_1, β_2 are dimensionless parameters of order one, and the typical scale of the potential is determined, as before, by the dimensional parameter m^2 (a comparison with Eq. (3.55) gives $V_0 = m^2(\beta_1 - \beta_2)/\beta_1\beta_2$). The above potential, which has a "bell-like" structure like the curve labelled (a) of Fig. 3.2, obviously requires a dilaton field running towards $+\infty$, in order to implement the saturation mechanism of Eq. (3.56).

The two examples of non-perturbative potential presented in this section will be applied in Chapter 9 to discuss possible scenarios of "late time" cosmology, in which the Universe is characterized by a phase of low curvatures and (moderately) strong couplings, as seems to be typical of our present cosmological configuration.

Appendix 3A
The massless multiplet of the bosonic string in Minkowski space

The action of a one-dimensional object freely evolving in a D-dimensional Minkowski space, in close analogy to the action of a free particle, is proportional to the proper area of the world-sheet Σ spanned by its motion. Let us call ξ^a, $a = 0, 1$, the coordinates on the world-sheet surface, and $X^\mu = X^\mu(\xi)$, $\mu = 0, 1, \ldots, D-1$, the parametric equations governing the embedding of Σ in the Minkowski space-time $\mathcal{M}$: the action can then be written as follows:

$$S = \frac{1}{2\pi\alpha'} \int_\Sigma d^2\xi \sqrt{|h|}. \tag{3A.1}$$

Here $(2\pi\alpha')^{-1} = M_s^2 = \lambda_s^{-2}$ is the string tension (i.e. the proper mass per unit proper length of the string), and $h = \det h_{ab}$, where

$$h_{ab} = \frac{\partial X^\mu}{\partial \xi^a} \frac{\partial X^\nu}{\partial \xi^b} \eta_{\mu\nu} \tag{3A.2}$$

is the so-called "induced metric", derived from the mapping $X : \Sigma \to \mathcal{M}$ parametrized by $X^\mu(\xi)$. Following the usual convention in which $\xi^0 = \tau$ and $\xi^1 = \sigma$ are, respectively, time-like and space-like (dimensionless) coordinates, we obtain the metric

$$h_{00} = \dot{X}^\mu \dot{X}_\mu, \qquad h_{01} = h_{10} = \dot{X}^\mu X'_\mu, \qquad h_{11} = X'^\mu X'_\mu, \tag{3A.3}$$

where $\dot{X} = \partial X / \partial \tau$, $X' = \partial X / \partial \sigma$; the action (3A.1) can then be written in explicit **Nambu–Goto** form [3] as

$$S = \frac{1}{2\pi\alpha'} \int_0^\pi d\sigma \int_{\tau_1}^{\tau_2} d\tau \left| \dot{X}^\mu \dot{X}_\mu X'^\nu X'_\nu - (\dot{X}^\mu X'_\mu)^2 \right|^{1/2}. \tag{3A.4}$$

We have assumed, as usual, that for an open string the spatial ends correspond to the values $\sigma = 0$ and $\sigma = \pi$, and that for a closed string the periodicity condition $X(\sigma) = X(\sigma + \pi)$ is satisfied.

As in the case of a point-like particle, the formal problems associated with the presence of a square root in the action can be avoided by introducing a new auxiliary field γ_{ab}, which in this case represents the "intrinsic" metric of the world-sheet surface, and which leads to defining the so-called **Polyakov action** [4, 5],

$$S = \frac{1}{4\pi\alpha'} \int_\Sigma d^2\xi \sqrt{-\gamma} \gamma^{ab} \partial_a X^\mu \partial_b X^\nu \eta_{\mu\nu}, \tag{3A.5}$$

where $\partial_a X \equiv \partial X / \partial \xi^a$. This action can be interpreted either geometrically, as the action for a relativistic string propagating in the physical space-time $\mathcal{M}$, or as the action for a two-dimensional conformal field theory living on the world-sheet Σ, with fields X^μ taking values in the target space $\mathcal{M}$. The equivalence with the Nambu–Goto action can be easily checked by varying Eq. (3A.5) with respect to γ^{ab}, which gives the constraint

$$T_{ab} \equiv \partial_a X^\mu \partial_b X^\nu \eta_{\mu\nu} - \frac{1}{2}\gamma_{ab}\gamma^{ij}\partial_i X^\mu \partial_j X^\nu \eta_{\mu\nu} = 0. \qquad (3A.6)$$

Solving the constraint with respect to γ_{ab} (using the definition (3A.2) of the induced metric), one obtains $\gamma_{ab} = h_{ab}$, and inserting back into the Polyakov action (using the property $\gamma^{ac}\gamma_{bc} = \delta^a_b$) one immediately recovers the action (3A.1).

The string equations of motion can be deduced by varying the Polyakov action with respect to X^μ – at fixed metric and world-sheet coordinates – and by imposing the standard boundary conditions of zero variation at the initial and final times of the motion, $\delta X^\mu(\tau_1) = 0 = \delta X^\mu(\tau_2)$. Defining

$$S = \int_{\tau_1}^{\tau_2} d\tau \int_0^\pi d\sigma L(\dot{X}, X'),$$

$$L(\dot{X}, X') = \frac{1}{4\pi\alpha'}\sqrt{-\gamma}\gamma^{ab}\partial_a X^\mu \partial_b X_\mu, \qquad (3A.7)$$

and integrating by parts, we obtain

$$\delta S = \int_{\tau_1}^{\tau_2} d\tau \int_0^\pi d\sigma \frac{\partial L}{\partial(\partial_a X^\mu)} \delta\partial_a X^\mu$$

$$= -\int_{\tau_1}^{\tau_2} d\tau \int_0^\pi d\sigma \left[\partial_a \frac{\partial L}{\partial(\partial_a X^\mu)}\right] \delta X^\mu$$

$$+ \int_0^\pi d\sigma \left[\frac{\partial L}{\partial \dot{X}^\mu} \delta X^\mu\right]_{\tau_1}^{\tau_2} + \int_{\tau_1}^{\tau_2} d\tau \left[\frac{\partial L}{\partial X'^\mu} \delta X^\mu\right]_0^\pi. \qquad (3A.8)$$

The first term after the equality corresponds to the usual Euler–Lagrange equations, the second term does not contribute because $\delta X = 0$ at the time boundaries, and the third term represents the variational contribution at the spatial ends of the action integral, to be fixed by appropriate boundary conditions. Using the Lagrangian (3A.7) one then finds that the action is stationary provided the string satisfies the equations of motion

$$\partial_a \left(\sqrt{-\gamma}\gamma^{ab}\partial_b X_\mu\right) = 0, \qquad (3A.9)$$

and the following boundary conditions:

$$\left[\frac{\partial L}{\partial X'^\mu}\delta X^\mu\right]_{\sigma=0}^{\sigma=\pi} = 0. \qquad (3A.10)$$

The boundary conditions will be discussed later, when presenting the classical solution for open and closed strings separately. It is first appropriate to note that the equations of motion can be simplified using the symmetries of the Polyakov action (3A.5), which are of three types: (*i*) global invariance under Poincaré transformations of the target space coordinates, $X^\mu(\xi) \to \Lambda^\mu{}_\nu X^\nu(\xi) + a^\mu$, where Λ, a are constants, and ξ is fixed; (*ii*) local invariance under reparametrization of the world-sheet manifold, $\xi^a \to \widetilde{\xi}^a(\xi)$; (*iii*) conformal invariance under local scale transformations of the world-sheet metric, $\gamma_{ab} \to \gamma_{ab}\exp[2\omega(\xi)]$. Thanks to the last two symmetries, in particular, it is always possible to choose the "conformal gauge" in which the world-sheet manifold is characterized by a globally flat (Minkowskian) geometry, $\gamma_{ab} = \eta_{ab}$, as already discussed in Section 3.1. In this gauge, the equations of motion (3A.9) simplify to

$$\ddot{X}_\mu - X_\mu'' = 0, \tag{3A.11}$$

and the constraints $T_{ab} = 0$ of Eq. (3A.6) can be combined to give the conditions

$$\frac{1}{2}(T_{00} + T_{10}) = \frac{1}{4}\eta_{\mu\nu}\left(\dot{X}^\mu + X'^\mu\right)\left(\dot{X}^\nu + X'^\nu\right) = 0,$$
$$\frac{1}{2}(T_{00} - T_{10}) = \frac{1}{4}\eta_{\mu\nu}\left(\dot{X}^\mu + X'^\mu\right)\left(\dot{X}^\nu - X'^\nu\right) = 0, \tag{3A.12}$$

also known as "Virasoro constraints" [30], written in a form which is useful for later applications. One can easily check that the world-sheet "stress tensor" T_{ab} is covariantly conserved, $\nabla^a T_{ab} = 0$: thus, if the constraints $T_{ab} = 0$ are satisfied at a given time $\tau = \tau_0$, they will be satisfied at all times along the string trajectory.

For solving the equations of motion explicitly it is finally convenient to introduce in the flat world-sheet manifold a non-cartesian base, the so-called "light-cone" coordinates $\xi^\pm$, defined in such a way that

$$\xi^\pm = \tau \pm \sigma, \qquad \partial_\pm = \frac{1}{2}\left(\partial_\tau \pm \partial_\sigma\right),$$
$$\tau = \frac{1}{2}\left(\xi^+ + \xi^-\right), \qquad \sigma = \frac{1}{2}\left(\xi^+ - \xi^-\right). \tag{3A.13}$$

The two-dimensional wave equation (3A.11) then becomes

$$\partial_+\partial_- X^\mu = 0, \tag{3A.14}$$

and can be solved, in general, by a linear combination of left- and right-moving waves,

$$X^\mu(\xi) = X_L^\mu(\xi^+) + X_R^\mu(\xi^-). \tag{3A.15}$$

In these coordinates the Virasoro constraints take the form

$$T_{++} \equiv \frac{1}{2}(T_{00} + T_{10}) = \partial_+ X^\mu \partial_+ X_\mu = 0,$$
$$T_{--} \equiv \frac{1}{2}(T_{00} - T_{10}) = \partial_- X^\mu \partial_- X_\mu = 0. \tag{3A.16}$$

We are now able to present the full explicit solutions governing the classical motion of a bosonic string in Mikowski space. In the following, we will analyze separately the cases of closed and open strings.

3A.1 Classical closed string

If the string is a closed loop, without free ends, topologically equivalent to a circle, the associated world-sheet is a "tube" whose spatial section is the surface enclosed inside the string loop. The boundary conditions (3A.10) can be identically satisfied in this case by having a spatial coordinate σ which varies between 0 and π, and by imposing the periodicity condition

$$X^\mu(\tau, \sigma) = X^\mu(\tau, \sigma + \pi). \tag{3A.17}$$

The solutions of the wave equation satisfying this condition, expanded in Fourier series, and separated into right- and left-moving modes, can be parametrized, respectively, as follows:

$$X_R^\mu(\xi^-) = \frac{1}{2}x_0^\mu + \alpha' p^\mu(\tau - \sigma) + i\sqrt{\frac{\alpha'}{2}} \sum_{n\neq 0} \frac{\alpha_n^\mu}{n} e^{-2in(\tau-\sigma)},$$

$$X_L^\mu(\xi^+) = \frac{1}{2}x_0^\mu + \alpha' p^\mu(\tau + \sigma) + i\sqrt{\frac{\alpha'}{2}} \sum_{n\neq 0} \frac{\widetilde{\alpha}_n^\mu}{n} e^{-2in(\tau+\sigma)}$$

(3A.18)

(we are following the conventions of [1]). Here x_0^μ and p^μ are integration constants representing, respectively, the initial position and the total (constant) momentum of the center of mass of the string (see below, Eq. (3A.44)). The sum is extended to all non-zero (negative and positive) integers, and the Fourier coefficients α_n, $\widetilde{\alpha}_n$ must satisfy the reality condition $(\alpha_n^\mu)^* = \alpha_{-n}^\mu$, $(\widetilde{\alpha}_n^\mu)^* = \widetilde{\alpha}_{-n}^\mu$, for the reality of the coordinates X^μ. Summing up the left and right components we obtain the full general solution

$$X^\mu(\sigma, \tau) = x_0^\mu + 2\alpha' p^\mu \tau + i\sqrt{\frac{\alpha'}{2}} \sum_{n\neq 0} \frac{1}{n} \left(\alpha_n^\mu e^{2in\sigma} + \widetilde{\alpha}_n^\mu e^{-2in\sigma}\right) e^{-2in\tau}. \tag{3A.19}$$

In order to impose on this solution the constraints (3A.16) we first compute the light-cone gradients $\partial_\pm X^\mu$: defining $\alpha_0^\mu = p^\mu \sqrt{\alpha'/2}$ we can include the case $n = 0$ in the sum of Fourier modes, and we obtain

$$\partial_+ X^\mu = \sqrt{2\alpha'} \sum_{n=-\infty}^{\infty} \widetilde{\alpha}_n^\mu e^{-2in(\tau+\sigma)},$$

$$\partial_- X^\mu = \sqrt{2\alpha'} \sum_{n=-\infty}^{\infty} \alpha_n^\mu e^{-2in(\tau-\sigma)}. \tag{3A.20}$$

It is also convenient to introduce the so-called Virasoro functionals [30], defined (at any given fixed time τ) by

$$L_m = \frac{1}{4\pi\alpha'} \int_0^\pi d\sigma \, T_{--} e^{2im(\tau-\sigma)},$$

$$\widetilde{L}_m = \frac{1}{4\pi\alpha'} \int_0^\pi d\sigma \, T_{++} e^{2im(\tau+\sigma)}, \tag{3A.21}$$

and satisfying the complex conjugation relations $L_m^* = L_{-m}$, $\widetilde{L}_m^* = \widetilde{L}_{-m}$. The contraints (3A.16) are then implemented by the conditions $L_m = 0 = \widetilde{L}_m$. By using the explicit definition of T_{--} the first constraint can then be written in the (explicitly time independent) form

$$L_m = \frac{1}{2\pi} \int_0^\pi d\sigma \sum_n \sum_k \alpha_n^\mu \alpha_{k\mu} \, e^{-2i(\tau-\sigma)(n+k-m)}$$

$$= \frac{1}{2} \sum_n \sum_k \alpha_n^\mu \alpha_{k\mu} \delta_{k,m-n}$$

$$= \frac{1}{2} \sum_{n=-\infty}^{+\infty} \alpha_{m-n}^\mu \alpha_{n\mu} = 0 \tag{3A.22}$$

(we have used the orthonormality of the base $e^{2in\sigma}/\sqrt{\pi}$ in the space $L^2[0, \pi]$). In the same way, the computation of T_{++} gives, for the left-moving modes,

$$\tilde{L}_m = \frac{1}{2} \sum_{n=-\infty}^{+\infty} \tilde{\alpha}^\mu_{m-n} \tilde{\alpha}_{n\mu} = 0. \qquad (3A.23)$$

The solution (3A.19), complemented by the constraints (3A.22) and (3A.23), provides the full general description of the motion of a closed string in Minkowski space.

3A.2 Classical open string

When the string has two non-coincident ends, corresponding to the values 0 and π of the spatial coordinate σ, the boundary conditions (3A.10) must be imposed at each end of the string, and can be satisfied in two ways. A first possibility is to impose $\partial L/\partial X'^\mu = 0$ at the two spatial boundaries. In the conformal gauge, where

$$L = \frac{1}{4\pi\alpha'} \left(\dot{X}^\mu \dot{X}_\mu - X'^\mu X'_\mu \right), \qquad (3A.24)$$

one then directly obtains the so-called **Neumann** boundary conditions,

$$X'^\mu|_{\sigma=0} = 0, \qquad X'^\mu|_{\sigma=\pi} = 0, \qquad (3A.25)$$

which guarantee that no momentum is flowing off the ends of the string. The condition (3A.10), however, can be rewritten as

$$\left[\frac{\partial L}{\partial X'^\mu} \delta X^\mu \right]_{\sigma=0}^{\sigma=\pi} = \left[\frac{\partial L}{\partial X'^\mu} \left(\dot{X}^\mu \delta\tau + X'^\mu \delta\sigma \right) \right]_{\sigma=0}^{\sigma=\pi} = \left[\frac{\partial L}{\partial X'^\mu} \dot{X}^\mu \delta\tau \right]_{\sigma=0}^{\sigma=\pi} = 0, \qquad (3A.26)$$

and we see that it can be satisfied also by imposing the so-called **Dirichlet** boundary conditions,

$$\dot{X}^\mu|_{\sigma=0} = 0, \qquad \dot{X}^\mu|_{\sigma=\pi} = 0, \qquad (3A.27)$$

corresponding to the case in which the ends of the strings are kept fixed.

Generally, one can impose Neumann conditions on the time coordinate and on p spatial directions, $\{X^0, X^1, \ldots, X^p\}$, and the Dirichlet conditions on the other $D - p - 1$ directions, $\{X^{p+1}, \ldots, X^{D-1}\}$ assuming, in the simplest configuration, that all open strings begin and end on a p-dimensional plane located at a fixed position X^i, with $p + 1 \le i \le D - 1$. Such a hyperplane is called a Dirichlet membrane, or D_p-brane. The ends of the strings are fixed in the Dirichlet directions, but can still move freely along the $p + 1$ Neumann directions, spanning the world-volume of the brane. Neumann and Dirichlet directions are also called "parallel" and "transverse" directions, with reference to the space-time orientation of the brane.

In the rest of this appendix we will mainly concentrate our discussion on the case of Neumann boundary conditions (see [2] for a systematic discussion of D_p-branes in a string theory context). The solutions for the open string motion can be separated, again, into left- and right-moving modes, and expanded in Fourier series as in Eq. (3A.18), with the only difference being that the expansion is now referred to the base $\exp[-in(\tau \pm \sigma)]$ (without the factor 2 required for closed strings by the boundary condition (3A.17)). One then finds that the Neumann condition (3A.25) is satisfied provided n is an integer, and provided $p^\mu = \tilde{p}^\mu$, $\alpha^\mu_n = \tilde{\alpha}^\mu_n$. With this condition, when summing up the two solutions

X_R^μ and X_L^μ, it turns out that the left and right components combine to give a stationary wave, and one obtains

$$X^\mu(\sigma, \tau) = x_0^\mu + 2\alpha' p^\mu \tau + i\sqrt{2\alpha'} \sum_{n \neq 0} \frac{\alpha_n^\mu}{n} e^{-in\tau} \cos(n\sigma). \tag{3A.28}$$

The open string solution satisfying the Dirichlet boundary conditions can be written, instead, in the form

$$X^\mu(\sigma, \tau) = x_1^\mu + (x_2^\mu - x_1^\mu)\frac{\sigma}{\pi} + i\sqrt{2\alpha'} \sum_{n \neq 0} \frac{\alpha_n^\mu}{n} e^{-in\tau} \sin(n\sigma), \tag{3A.29}$$

where x_1 and x_2 are the positions of the ends of the strings, corresponding to $\sigma = 0$ and $\sigma = \pi$, respectively.

In order to impose the Virasoro constraints we can then apply the same procedure as in the closed string case, by defining $\alpha_0^\mu = p^\mu \sqrt{2\alpha'}$ so as to include the mode $n = 0$ in the Fourier series. The computation of the light-cone gradients then gives

$$\partial_\pm X^\mu = \sqrt{\frac{\alpha'}{2}} \sum_{n=-\infty}^{\infty} \alpha_n^\mu e^{-in(\tau \pm \sigma)}. \tag{3A.30}$$

There is, however, a subtle difference from the previous case, due to the fact that the functions $e^{-in\sigma}$ do not provide a complete orthonormal base in the considered $[0, \pi]$ interval. Such a problem can be solved analytically by extending the solution over the interval $[-\pi, \pi]$, imposing the conditions [1]

$$X_R^\mu(\sigma + \pi) = X_L^\mu(\sigma), \qquad X_L^\mu(\sigma + \pi) = X_R^\mu(\sigma), \tag{3A.31}$$

from which

$$X'^\mu(-\sigma) = -X'^\mu(\sigma), \qquad \dot{X}^\mu(-\sigma) = \dot{X}^\mu(\sigma). \tag{3A.32}$$

With the given boundary conditions the open string solution becomes periodic, with period 2π, over the extended interval $-\pi \leq \sigma \leq \pi$. The Virasoro constraints in such a case can be expressed by a unique condition, since they emerge as the even and odd part (with respect to spatial reflections $\sigma \rightarrow -\sigma$) of the condition $T_{++} = 0$. The Virasoro functional can be defined on the extended interval $[-\pi, \pi]$, and gives the constraint

$$L_m = \frac{1}{2\pi\alpha'} \int_{-\pi}^{\pi} d\sigma\, T_{++} e^{im(\tau+\sigma)} = \frac{1}{2\pi\alpha'} \int_{-\pi}^{\pi} d\sigma\, \partial_+ X^\mu \partial_+ X_\mu e^{im(\tau+\sigma)}$$

$$= \frac{1}{4\pi} \int_{-\pi}^{\pi} d\sigma \sum_n \sum_k \alpha_n^\mu \alpha_{k\mu} e^{-i(\tau+\sigma)(n+k-m)}$$

$$= \frac{1}{2} \sum_{n=-\infty}^{+\infty} \alpha_{m-n}^\mu \alpha_{n\mu} = 0. \tag{3A.33}$$

Let us conclude the classical part of this discussion by noting that there is a close connection between L_0, $\tilde{L}_0$ and the canonical Hamiltonian associated with the Polyakov action, for both closed and open strings.

In the conformal gauge, described by the Lagrangian density (3A.24), we have the following canonical expressions for the momentum density, Π_μ, and for the Hamiltonian density, $\mathcal{H}$:

$$\Pi_\mu = \frac{\dot{X}_\mu}{2\pi\alpha'},$$

$$\mathcal{H} = \dot{X}^\mu \Pi_\mu - L = \frac{1}{4\pi\alpha'}\left(\dot{X}^\mu \dot{X}_\mu + X'^\mu X'_\mu\right). \tag{3A.34}$$

Integrating over the spatial dimension, and using Eqs. (3A.6) and (3A.16), we can write the Hamiltonian as

$$
\begin{aligned}
H &= \frac{1}{4\pi\alpha'}\int_0^\pi d\sigma \left(\dot{X}^\mu \dot{X}_\mu + X'^\mu X'_\mu\right) \\
&= \frac{1}{2\pi\alpha'}\int_0^\pi d\sigma\, T_{00} = \frac{1}{2\pi\alpha'}\int_0^\pi d\sigma\,(T_{++} + T_{--}).
\end{aligned} \tag{3A.35}
$$

For a closed string, using the definition (3A.21) of L_m, one immediately obtains

$$H = 2(L_0 + \tilde{L}_0). \tag{3A.36}$$

For the open string we have to recall that L_m is defined over the extended interval $[-\pi, \pi]$, so that

$$L_m = \frac{1}{2\pi\alpha'}\left(\int_0^\pi d\sigma + \int_{-\pi}^0 d\sigma\right) T_{++}\, e^{im(\tau+\sigma)}. \tag{3A.37}$$

The extended open string solution, however, satisfies the periodicity conditions (3A.31) and (3A.32), so that the reflection $\sigma \to -\sigma$ transforms T_{++} into T_{--}, i.e.

$$\sigma \to -\sigma, \qquad \partial_+ X_L^\mu \partial_+ X_{L\mu} \to \partial_- X_R^\mu \partial_- X_{R\mu}. \tag{3A.38}$$

We can then rewrite

$$L_m = \frac{1}{2\pi\alpha'}\int_0^\pi d\sigma\left(T_{++}\, e^{im(\tau+\sigma)} + T_{--}\, e^{im(\tau-\sigma)}\right), \tag{3A.39}$$

and the comparison with (3A.35) immediately gives

$$H = L_0 \tag{3A.40}$$

for the canonical Hamiltonian of the open string. It may be anticipated that the above difference between the Hamiltonians of closed and open strings will produce a factor of 4 difference between the two mass spectra, after quantization.

3A.3 Quantization

The first quantization of the bosonic string model can now be performed according to the standard canonical procedure, in which the classical variables are promoted to operators defined in an appropriate Hilbert space, and the classical Poisson brackets are replaced by commutators according to the well-known prescription $\{A, B\} \to -i[A, B]$. We thus impose on the (canonically conjugate) position-momentum variables the fundamental (equal time) commutation relations

$$[\Pi^\mu(\tau, \sigma), X^\nu(\tau, \sigma')] = \left[\frac{\dot{X}^\mu}{2\pi\alpha'}(\tau, \sigma), X^\nu(\tau, \sigma')\right] = i\eta^{\mu\nu}\delta(\sigma - \sigma'),$$

$$[X^\mu(\tau, \sigma), X^\nu(\tau, \sigma')] = 0 = [\dot{X}^\mu(\tau, \sigma), \dot{X}^\nu(\tau, \sigma')]$$

(3A.41)

(recall that $\eta^{ij} = -\delta^{ij}$ in our notations). Inserting the solutions for X^μ previously determined we can then obtain the corresponding commutators for the Fourier coefficients $\alpha^\mu, \tilde{\alpha}^\mu$, which now become operators satisfying the hermitian conjugation conditions

$$(\alpha_n^\mu)^\dagger = \alpha_{-n}^\mu, \qquad (\tilde{\alpha}_n^\mu)^\dagger = \tilde{\alpha}_{-n}^\mu,$$

(3A.42)

replacing the classical complex conjugation.

Consider for instance a closed string, described by the solution (3A.19). Differentiating with respect to τ, and imposing Eq. (3A.41), we obtain for the zero modes the canonical commutation relation for the position and the center of mass of the string,

$$[P_{CM}^\mu, X_{CM}^\nu] = i\eta^{\mu\nu},$$

(3A.43)

where

$$X_{CM}^\mu = \frac{1}{\pi}\int_0^\pi d\sigma X^\mu(\sigma, \tau) = x_0^\mu + 2\alpha' p^\mu \tau,$$

$$P_{CM}^\mu = \frac{1}{\pi}\int_0^\pi d\sigma \Pi^\mu(\sigma, \tau) = \frac{p^\mu}{\pi}.$$

(3A.44)

For the other Fourier coefficients, imposing null commutation brackets between left and right modes,

$$[\alpha_n^\mu, \tilde{\alpha}_m^\nu] = 0,$$

(3A.45)

we find the condition

$$\frac{i}{2\pi}\sum_{n\neq 0}\sum_{m\neq 0}\frac{1}{m}\left([\alpha_n^\mu, \alpha_m^\nu]e^{2i(n\sigma+m\sigma')} + [\tilde{\alpha}_n^\mu, \tilde{\alpha}_m^\nu]e^{-2i(n\sigma+m\sigma')}\right)e^{-2i\tau(n+m)}$$

$$= i\eta^{\mu\nu}\delta(\sigma - \sigma').$$

(3A.46)

This condition is satisfied provided

$$[\alpha_n^\mu, \alpha_m^\nu] = m\eta^{\mu\nu}\delta_{n+m,0} = [\tilde{\alpha}_n^\mu, \tilde{\alpha}_m^\nu],$$

(3A.47)

as can be easily checked by including the case $n = 0$, $m = 0$ in the sums of Eq. (3A.46), and by exploiting the distributional convergence of the functions $e^{2in\sigma}/\sqrt{\pi}$ to the Dirac delta function:

$$\sum_{n=-\infty}^{+\infty} e^{2in(\sigma-\sigma')} = \pi\delta(\sigma - \sigma').$$

(3A.48)

The above commutators are very similar to those encountered in the quantization of the harmonic oscillator problem. We can obtain an even closer analogy by changing n to $-n$ in Eq. (3A.47), and introducing new operators a_m^μ, defined by

$$\alpha_m^\mu = \sqrt{m}a_m^\mu, \qquad \alpha_{-m}^\mu = (\alpha_m^\mu)^\dagger = \sqrt{m}(a_m^\mu)^\dagger,$$

(3A.49)

through which we can rewrite the previous commutators with the conventional oscillator normalization as follows:

$$[(a_n^\mu)^\dagger, a_m^\mu] = \eta^{\mu\nu}\delta_{mn}. \tag{3A.50}$$

The operators $a_n, a_n^\dagger$ (or their equivalent version α_n, α_{-n}) thus represent creation and annihilation operators for the various energy levels of the quantum string. Through their action we can build up a spectrum characterized by a ground state $|p, 0\rangle$, which is an eigenstate of the momentum p^μ of the center of mass of the string, and which is annihilated by all annihilation operators,

$$\alpha_m^\mu |p, 0\rangle = 0 = \tilde{\alpha}_m^\mu |p, 0\rangle, \qquad \forall m > 0. \tag{3A.51}$$

The corresponding Fock space of the system, spanned by the states obtained by applying to the vacuum an arbitrary number of creation operators,

$$|p, n_1, m_2, \ldots\rangle = (\alpha_{-n}^\mu)^{n_1} (\alpha_{-m}^\nu)^{m_2} \ldots |p, 0\rangle, \tag{3A.52}$$

is not positive-definite, however. If we consider, for instance, the state $(a_m^0)^\dagger |p, 0\rangle$, we can easily check, by using the commutation relations (3A.50), that its norm is negative:

$$\langle 0, p| a_m^0 (a_m^0)^\dagger |p, 0\rangle = \langle 0, p| \left((a_m^0)^\dagger a_m^0 + [a_m^0, (a_m^0)^\dagger] \right) |p, 0\rangle$$
$$= -\langle 0, p|p, 0\rangle = -1 \tag{3A.53}$$

(we are assuming that the ground state is normalized to one). In order to obtain the physical states associated with the quantum string spectrum we must thus consider a subset of this Fock space, by imposing appropriate restrictions in order to eliminate all "ghost" (i.e. negative norm) states.

We should recall that even at the classical level there are restrictions on the solutions of the string equations of motion, due to the Virasoro constraints, which impose the conditions $L_m = 0 = \tilde{L}_m$. We may thus expect that the elements $|\psi\rangle$ of the Fock subspace containing the physical states must satisfy the conditions $L_m |\psi\rangle = 0 = \tilde{L}_m |\psi\rangle$. In a quantum context, however, the Virasoro functionals L_m are promoted to operators, and their definition is in general affected by ordering ambiguities because of the presence of products of non-commuting operators, like $\alpha_m \alpha_n$. We can then adopt the usual "normal ordering" prescription, defining $2L_m = \sum_n : \alpha_{m-n}^\mu \alpha_{n\mu} :$ (see e.g. Eq. (3A.22)), in which all annihilation operators are moved to the right of the creation operators. This is unambiguous, except for L_0 and $\tilde{L}_0$, since according to the rules (3A.47) it is just for $m = 0$ that the commutators of $\alpha_{m-n}^\mu \alpha_{n\mu}$ are non-zero, and the definition of the operator is ordering dependent. Taking into account this effect, we can impose the *quantum* Virasoro constraints on the physical states as follows:

$$(L_0 - \delta)|\psi\rangle = 0 = (\tilde{L}_0 - \delta)|\psi\rangle, \qquad L_m |\psi\rangle = 0 = \tilde{L}_m |\psi\rangle, \qquad m > 0, \tag{3A.54}$$

where L and $\tilde{L}$ are the ordered operators, and δ is a *finite* constant depending on the chosen ordering prescription. Note that the number of imposed conditions (one for each value of $m \geq 0$) is equal to the number of temporal oscillators α_m^0 associated with the negative norm states, and is thus sufficient, in principle, to eliminate the ghosts from the physical subspace provided the value of δ is chosen appropriately.

In order to fix δ we follow a non-covariant procedure, in which we rewrite the classical solutions of the string equations of motion by adopting the light-cone gauge not only for the world-sheet but also for the target space manifold, introducing the coordinates $X^\mu = \{X^+, X^-, X^i\}$, where $i = 1, \ldots, D-2$, and $X^\pm = (X^0 \pm X^{D-1})/\sqrt{2}$. We are able, in this way, to linearize the Virasoro constraints. Also, we fix the residual degrees of freedom left by the conformal gauge by assuming that the motion of the string along the X^+ direction is a pure translation,

$$X^+ = x^+ + 2\alpha' p^+ \tau, \qquad (3A.55)$$

without any residual oscillation, i.e. that $\alpha_n^+ = 0 = \tilde{\alpha}_n^+$, $\forall n \neq 0$ (see Eqs. (3A.19) and (3A.28)). Using the Virasoro constraints we can then eliminate the α_n^- oscillators in terms of α_n^i ones, and we are eventually left with the oscillations along the $D-2$ transverse directions, parametrized by α_n^i, $n \neq 0$, as the only independent degrees of freedom to be quantized. It is convenient, at this point, to discuss separately the open and closed string spectrum.

3A.4 Open string spectrum

In the light-cone gauge, the Virasoro functional L_0 can be rewritten as follows:

$$L_0 = \frac{1}{2} \sum_{n=-\infty}^{+\infty} \alpha_{-n}^\mu \alpha_{n\mu} = \frac{1}{2} \alpha_0^\mu \alpha_{0\mu} + \frac{1}{2} \sum_{n\neq 0} \alpha_{-n}^\mu \alpha_{n\mu}$$

$$= \alpha' p^\mu p_\mu + \frac{1}{2} \sum_{n\neq 0} \left(2\alpha_{-n}^+ \alpha_n^- - \alpha_{-n}^i \alpha_n^i \right)$$

$$= \alpha' p^\mu p_\mu - \frac{1}{2} \sum_{n=1}^{\infty} \left(\alpha_{-n}^i \alpha_n^i + \alpha_n^i \alpha_{-n}^i \right), \qquad (3A.56)$$

where we have used the definition $\alpha_0^\mu = \sqrt{2\alpha'} p^\mu$ (valid in the open string case), and the condition $\alpha_n^+ = 0$. By applying the commutation rule $[\alpha_n^i, \alpha_{-n}^j] = n\delta^{ij}$ we then obtain

$$L_0 = \alpha' p^\mu p_\mu - \frac{1}{2} \sum_{n=1}^{\infty} \left[2\alpha_{-n}^i \alpha_n^i + (D-2)n \right]. \qquad (3A.57)$$

The term quadratic in the α^i corresponds, in a quantum context, to a normal ordered operator, while the divergent sum over n represents the infinite contribution of the vacuum energy of the oscillators, and has to be appropriately regularized for extracting meaningful physical predictions. We can use, in particular, an exponential regularization scheme, by rewriting the divergent sum as follows:

$$\lim_{\epsilon \to 0} \sum_{n=0}^{\infty} n e^{-\epsilon n} = \lim_{\epsilon \to 0} \left(-\frac{d}{d\epsilon} \sum_{n=0}^{\infty} e^{-\epsilon n} \right) = -\lim_{\epsilon \to 0} \frac{d}{d\epsilon} (1 - e^{-\epsilon})^{-1}$$

$$= -\lim_{\epsilon \to 0} \frac{d}{d\epsilon} \left(\epsilon - \frac{\epsilon^2}{2} + \frac{\epsilon^3}{6} + \cdots \right)^{-1}$$

$$= -\lim_{\epsilon \to 0} \frac{d}{d\epsilon} \frac{1}{\epsilon} \left(1 - \frac{\epsilon}{2} + \frac{\epsilon^2}{6} + \cdots \right)^{-1}$$

$$= -\lim_{\epsilon \to 0} \frac{d}{d\epsilon} \frac{1}{\epsilon} \left(1 + \frac{\epsilon}{2} + \frac{\epsilon^2}{12} + \cdots \right)$$

$$= \lim_{\epsilon \to 0} \left(\frac{1}{\epsilon^2} - \frac{1}{12} \right). \qquad (3A.58)$$

By subtracting the infinite part we thus obtain the following regularized expression

$$L_0 = \alpha' p^2 - \sum_{n=1}^{\infty} N_n + \frac{D-2}{24}, \qquad (3A.59)$$

where we have introduced the number operator $N_n = \alpha^i_{-n} \alpha^i_n$.

We now recall that the classical constraint $L_0 = 0$, for an open string, corresponds to the canonical mass-shell condition $H = 0$, according to Eq. (3A.40). The comparison with the quantum constraint (3A.54) defines the parameter δ for the normal ordered version of L_0, i.e. $\delta = (D-2)/24$, and determines the mass spectrum of the open string states as

$$\alpha' M^2 = N - \frac{D-2}{24}, \qquad (3A.60)$$

where we have denoted with $N = \sum_n N_n$ the sum of the number operators of all oscillator modes, and with $M^2 = p^\mu p_\mu$ the square of the proper energy of the Nth excited level in D-dimensional Minkowski space. This result is consistently defined only in a fixed number $D = D_c$ of dimensions, as we discuss in the following.

First of all we note that the spectrum of the number operators N_n ranges over all non-negative integers, and that the ground state $|p, 0\rangle$, corresponding to the case in which all number eigenvalues are zero, is associated with a mass

$$M^2 = -\frac{D-2}{24\alpha'}. \qquad (3A.61)$$

For $D > 2$ one obtains $M^2 < 0$, so that this level describes a scalar "tachyonic" configuration which seems to signal an instability of the quantum theory (such a negative eigenvalue of M^2 disappears, as we shall see, in a superstring context).

The first excited level corresponds to the eigenvalue 1 of N, and describes a vector-like configuration $\alpha^\mu_{-1}|p, 0\rangle$, associated with a mass

$$M^2 = \frac{1}{\alpha'}\left(1 - \frac{D-2}{24}\right). \qquad (3A.62)$$

Contracting with the polarization tensor ϵ_μ of the produced vector state, and imposing the constraint generated by the Virasoro operator L_1, it can be easily checked that the polarization of this state has to be transverse, i.e. that $p^\mu \epsilon_\mu = 0$. Indeed, using the normal ordering of L_1 and the commutator (3A.47), we have

$$L_1 \epsilon_\mu \alpha^\mu_{-1}|p, 0\rangle = \frac{1}{2}\epsilon_\mu \sum_n : \alpha^\nu_{1-n}\alpha_{n\nu} : \alpha^\mu_{-1}|p, 0\rangle = \frac{1}{2}\epsilon_\mu \, a^\nu_0 \alpha_{1\nu} \alpha^\mu_{-1}|p, 0\rangle$$

$$= \frac{1}{2}\epsilon_\mu \, a^\nu_0 \left[\alpha_{1\nu}, \alpha^\mu_{-1}\right]|p, 0\rangle = \frac{1}{2}\epsilon_\mu \alpha^\mu_0 |p, 0\rangle = 0, \qquad (3A.63)$$

from which, using $\alpha^\mu_0 \sim p^\mu$, one finally obtains $p^\mu \epsilon_\mu = 0$.

It must be recalled now that, for any oscillator α^μ, there are only $D - 2$ physical degrees of freedom that can always be associated with the string oscillations along the transverse directions X^i, as discussed in the light-cone gauge. On the other hand, a transverse vector ϵ_μ with only $D - 2$ independent space-like components corresponds to the irreducible vector representation of the so-called "little group" $SO(D-2)$, and is associated with a light-like momentum p^μ (a transverse, massive vector field has, in fact, $D - 1$ independent components). The spectrum of the open bosonic string is thus compatible with a Lorentz-invariant description of the physical states only if the vector level $N = 1$ is characterized by the condition $M^2 = 0$: this implies, according to Eq. (3A.62), that the theory must be formulated in a space-time with critical number of dimensions

$$D = D_c = 26.$$ (3A.64)

3A.5 Closed string spectrum

As in the previous case we introduce light-cone coordinates, and impose the conditions $\alpha_n^+ = 0 = \tilde{\alpha}_n^+$, $\forall n \neq 0$. For a closed string we have $\alpha_0^\mu = p^\mu \sqrt{\alpha'/2} = \tilde{\alpha}_0^\mu$, and we can rewrite the Virasoro functionals, after the regularization, as follows:

$$L_0 = \frac{\alpha'}{4}p^2 - \sum_{n=1}^{\infty} \alpha_{-n}^i \alpha_n^i + \frac{D-2}{24},$$

$$\tilde{L}_0 = \frac{\alpha'}{4}p^2 - \sum_{n=1}^{\infty} \tilde{\alpha}_{-n}^i \tilde{\alpha}_n^i + \frac{D-2}{24}.$$ (3A.65)

In a classical context the two constraints $L_0 = 0$, $\tilde{L}_0 = 0$ must be separately satisfied by both left- and right-moving modes. In a quantum context we may thus impose that the physical states be annihilated by the sum and by the difference of the normal ordered version of the above operators. The difference provides the so-called "level-matching" condition, $(L_0 - \tilde{L}_0)|\psi\rangle = 0$, which guarantees the same eigenvalue of the number operator for the α and $\tilde{\alpha}$ oscillators, namely $N|\psi\rangle = \tilde{N}|\psi\rangle$, where $N = \sum_n \alpha_{-n}^i \alpha_n^i$ and $\tilde{N} = \sum_n \tilde{\alpha}_{-n}^i \tilde{\alpha}_n^i$. The sum, which is proportional to the Hamiltonian, provides instead the canonical mass-shell condition, and defines the allowed energy levels of the quantum closed string:

$$\frac{\alpha'}{2}M^2 = N + \tilde{N} - \frac{D-2}{12}.$$ (3A.66)

The ground state $|p, \widetilde{00}\rangle$ is obtained as the tensor product of the eigenstates of N and $\tilde{N}$ with zero eigenvalue, and describes again a tachyonic configuration, with mass

$$M^2 = -\frac{D-2}{6\alpha'}.$$ (3A.67)

The first excited level allowed by the level-matching condition corresponds to the eigenvalue 1 of N and $\tilde{N}$, and describes the tensor configuration $\alpha_{-1}^\mu \tilde{\alpha}_{-1}^\nu |p, \widetilde{00}\rangle$ associated with the mass

$$M^2 = \frac{2}{\alpha'}\left(2 - \frac{D-2}{12}\right).$$ (3A.68)

Again, by multiplying by the polarization tensor $\epsilon_{\mu\nu}$, and imposing the constraints generated by the Virasoro operators L_1 and $\tilde{L}_1$, one finds a state of transverse polarization, $p^\nu \epsilon_{\mu\nu} = 0$. On the other hand, as already discussed in the open string case, the number of independent degrees of freedom is $D - 2$ for any vector index, which is only consistent for massless field configurations, $M^2 = 0$, and which again implies $D = D_c = 26$, using Eq. (3A.68), as in the open string case.

To extract the particle content of this massless level we note that the transverse polarization tensor ϵ_{ij} can be decomposed into components transforming as irreducible representations of $SO(D-2)$, namely,

$$\epsilon_{ij} = h_{ij} + A_{ij} + \frac{\phi}{D-2}\delta_{ij},$$

$$h_{ij} = \epsilon_{(ij)} - \frac{\phi}{D-2}\delta_{ij}, \qquad A_{ij} = \epsilon_{[ij]}, \qquad \phi = \delta^{ij}\epsilon_{ij}, \qquad (3A.69)$$

where h_{ij} is the symmetric, trace-free part, A_{ij} the antisymmetric part, and ϕ the scalar trace. This level thus contains a transverse, traceless symmetric tensor, an antisymmetric tensor, and a scalar field: they are all massless fields, representing the graviton, torsion and dilaton multiplet introduced in Chapter 2, and studied in this chapter in Section 3.1.

We can observe, as a useful check, that the traceless symmetric tensor h_{ij} has a total number of $(D^2 - 3D)/2$ components, exactly the same number of polarization states as a spin-two gravitational wave $h_{\mu\nu}$ in D space-time dimensions, as we shall discuss in Chapter 7. Also, in $D-2$ transverse directions, the antisymmetric tensor A_{ij} has $[(D-2)^2 - (D-2)]/2$ independent components, which is exactly the number of $(D-3)(D-2)/2$ independent degrees of freedom of a two-form "gauge potential" $B_{\mu\nu}$ living in D space-time dimensions. The sum of these components plus one (the scalar component) obviously reproduces the $(D-2)(D-2)$ components of the rank-two tensor ϵ_{ij}.

We can finally note that, after removing the negative-norm states, a physical state $|\psi\rangle$, belonging to a positive-definite subset of the Hilbert space, is only defined up to the addition of null states, i.e. $|\psi\rangle \sim |\psi\rangle + |\chi\rangle$, where $\langle\chi|\chi\rangle = 0$, and $|\chi\rangle$ is orthogonal to all physical vectors $|\psi\rangle$. Adding such null states, for the massless levels of the closed string, corresponds to adding new polarization components as follows:

$$\epsilon_{(\mu\nu)} \sim \epsilon_{(\mu\nu)} + p_\mu\xi_\nu + p_\nu\xi_\mu, \qquad \epsilon_{[\mu\nu]} \sim \epsilon_{[\mu\nu]} + p_\mu\lambda_\nu - p_\nu\lambda_\mu, \qquad (3A.70)$$

where ξ_μ and λ_μ are arbitrary vectors orthogonal to p_μ. By Fourier transforming we see that adding null states is equivalent to performing local transformations generated by ξ_μ and λ_μ, exploiting the residual gauge freedom left by the transversality condition, in agreement with the properties of gauge invariance typical of a massless spin-two field and of a massless antisymmetric tensor field in Minkowski space.

Appendix 3B
Superstring models and effective actions

We will now generalize the bosonic string model presented in Appendix 3A by including fermions (required for a realistic description of all fundamental interactions), and eliminating the tachyons present in the ground level of the spectrum (see Eqs. (3A.61, 3A.67)). An appropriate generalization, which satisfies the above requirements and is consistent with an anomaly-free quantization, can be achieved by making the world-sheet action invariant under supersymmetry transformations generated by a number $N = 1$ of "supercharges" [1].

In that case, as we shall see in this appendix, a consistent string quantization requires a number $D_c = 10$ of critical dimensions, and needs a truncation of the spectrum which makes the theory supersymmetric not only on the world-sheet, but also on the 10-dimensional space-time in which the string is embedded. Such an "induced" space-time supersymmetry may be characterized by $N = 1$ or $N = 2$ supercharges, depending on the choice of the boundary conditions imposed on the fermionic fields present in the action: one then obtains, respectively, the so-called type I (open and closed) or type II (closed) superstring model. Another possibility of implementing $N = 1$ space-time supersymmetry is provided by the so-called "heterotic" model, in which only the right-moving modes of the closed bosonic string are supersymmetrized, while the left-moving sector is independently quantized following a different scheme.

The supersymmetric generalization of the bosonic string action is based on the introduction of new fermion fields on the world-sheet, by adding to the bosonic coordinates $X^\mu(\sigma, \tau)$ an equal number of fermionic "partners", represented by the fields $\psi_A^\mu(\sigma, \tau)$. These new fields transform as a two-component Majorana spinor (with index $A = 1, 2$) with respect to world-sheet transformations, and as a vector (with index $\mu = 0, 1, \ldots, D-1$) with respect to Lorentz transformations in the target space manifold. We recall, also in view of later applications, that a spinor in an even number D of space-time dimensions has, in general, $2^{D/2}$ components, which may be chosen to be real if the spinor satisfies the Majorana condition, i.e. if it is invariant (modulo a phase) under the action of the charge conjugation operator (see below).

In a curved world-sheet geometry, however, supersymmetry can be consistently implemented only as a *local* invariance of the action (the commutator of two infinitesimal supersymmetry transformations generates in fact a translation [31], and only local translations make sense in a curved manifold). A local supersymmetric action, on the other hand, requires the presence of an additional Rarita–Schwinger field, the vector-spinor (or "gravitino") field, represented by the variable $\chi_A^a(\sigma, \tau)$ which transforms as a two-component Majorana world-sheet spinor in the index A, and as a world-sheet vector (with respect to local reparametrizations) in the index $a = 0, 1$. The components of this field provide the fermionic partners of the "zweibein" field $V_a^i(\sigma, \tau)$ associated with the two-dimensional world-sheet metric, $\gamma_{ab} = V_a^i V_b^j \eta_{ij}$, where $i, j = 0, 1$ are Lorentz indices in the flat Minkowski space locally tangent to the curved world-sheet manifold (see Section 2.3).

Using these new variables we are now able to write a locally supersymmetric world-sheet action, generalizing the bosonic Polyakov action (3A.5) as follows:

$$S = \frac{1}{4\pi\alpha'} \int d^2\xi \sqrt{-\gamma} \left[\gamma^{ab} \partial_a X^\mu \partial_b X^\nu + i \overline{\psi}^\mu \gamma^a \nabla_a \psi^\nu \right.$$

$$\left. - 2\overline{\chi}_a \gamma^b \gamma^a \psi^\mu \partial_b X^\nu - \frac{1}{2} \overline{\psi}^\mu \psi^\nu \overline{\chi}_a \gamma^b \gamma^a \chi_b \right] \eta_{\mu\nu} \qquad (3B.1)$$

(a sum over the spin indices is to be understood, as usual). Here $\overline{\psi} = \psi^\dagger \gamma^0$, $\overline{\chi} = \chi^\dagger \gamma^0$ and $\gamma^a(\sigma, \tau) = V_i^a \gamma^i$ are two-dimensional matrices defined on the curved world-sheet in terms of the constant (flat space) Dirac matrices γ^i, $i = 0, 1$. We recall that the Dirac matrices obey the standard anticommutation relations

$$\{\gamma^i, \gamma^j\} \equiv \gamma^i \gamma^j + \gamma^j \gamma^i = 2\eta^{ij}, \qquad (3B.2)$$

and satisfy $\gamma^0 = (\gamma^0)^\dagger$, $\gamma^1 = -(\gamma^1)^\dagger$. We adopt a convenient basis in which all the γ^i have imaginary components,

$$\gamma^0 = \begin{pmatrix} 0 & -i \\ i & 0 \end{pmatrix}, \qquad \gamma^1 = \begin{pmatrix} 0 & i \\ i & 0 \end{pmatrix}, \qquad (3B.3)$$

and in which the Majorana spinors are real. Finally, $\nabla_a \psi$ is the covariant spinor derivative, computed in terms of the spin connection ω_a^{ij} associated with local Lorentz transformations on the world-sheet manifold. Notice that, in the two-dimensional action (3B.1), there are no kinetic terms for the V_a^i and χ_a fields (which are instead present in all supergravity models formulated in $D > 2$).

The action (3B.1) is invariant under global Poincaré transformations in the flat target space, and under local Lorentz transformations and general coordinate transformations in the world-sheet manifold, like the bosonic string action. The new property is the invariance under local supersymmetry transformations, mixing the bosonic and fermionic degrees of freedom. In infinitesimal form, such transformations can be represented as follows:

$$\delta X^\mu = \overline{\epsilon}\psi^\mu, \qquad \delta\psi^\mu = -i\gamma^a\epsilon\left(\partial_a X^\mu - \overline{\psi}^\mu \chi_a\right),$$

$$\delta V_a^i = -2i\overline{\epsilon}\gamma^i\chi_a, \qquad \delta\chi_a = \nabla_a\epsilon, \qquad (3B.4)$$

where $\epsilon(\sigma, \tau)$ is an anticommuting Majorana spinor. The presence of only one spinor ϵ in the above transformations is associated with the presence of only one conserved "supercurrent", and reflects the $N = 1$ character of the considered supersymmetry transformations. The invariance of the action (modulo a total divergence) can be checked by using the properties of the Dirac matrices and of the two-dimensional Majorana spinors, which are defined by

$$\epsilon = \epsilon^c \equiv C\overline{\epsilon}^T. \qquad (3B.5)$$

Here C is the charge conjugation operator, which satisfies

$$C^T = -C, \qquad C^{-1}\gamma^i C = -(\gamma^i)^T \qquad (3B.6)$$

$(C = -\gamma^0$ in the representation of Eq. (3B.3)). One then obtains, in particular, that $\bar{\epsilon}\psi_\mu = \bar{\psi}_\mu \epsilon$. It is also useful to note that $\gamma^a \nabla_a \psi \equiv \gamma^a \partial_a \psi$ for a two-dimensional Majorana spinor.

The action (3B.1) has two further local symmetries. One represents the extension of the conformal (Weyl) invariance associated with the rescaling of the world-sheet metric (see Eq. (3.10)) and thus of the vielbein field, according to the transformation $V_a^i \to V_a^i \exp(\omega)$. Indeed, for a local rescaling induced by the infinitesimal parameter $\omega(\sigma, \tau)$, each term of the action (3B.1) is left invariant by the following infinitesimal transformation:

$$\delta X^\mu = 0, \qquad\qquad \delta V_a^i = \omega V_a^i,$$

$$\delta \psi^\mu = -\frac{1}{2}\omega \psi^\mu, \qquad \delta \chi_a = \frac{1}{2}\omega \chi_a. \tag{3B.7}$$

The other local symmetry is represented by the infinitesimal gravitino transformation,

$$\delta \chi_a = i\gamma_a \eta, \qquad\qquad \delta \bar{\chi}_a = -i\bar{\eta}\gamma_a,$$

$$\delta X^\mu = \delta \psi^\mu = \delta V_a^i = 0, \tag{3B.8}$$

where $\eta(\sigma, \tau)$ is an arbitrary two-component Majorana spinor (the invariance of the action can be easily checked using the identity $\gamma_a \gamma^b \gamma^a = 0$, valid in two dimensions). The combination of these two local symmetries is also called "superconformal" symmetry.

For the bosonic string we have seen, in Section 3.1, that the conformal symmetry can be used to simplify the description of the string dynamics by choosing an appropriate "conformal gauge". In the same way, for the superstring, we can impose the "superconformal gauge" in which the metric of the world-sheet is flat ($\gamma_{ab} = \eta_{ab}$), and the gravitino field is vanishing ($\chi_a = 0$).

On the world-sheet, in fact, there are four local bosonic symmetries (two general coordinate transformations, one local Lorentz transformation and one Weyl rescaling), which can be used to set the four components of the zweibein in the trivial form $V_a^i = \delta_a^i$. Also, there are four fermionic symmetries (two supersymmetry transformations with spinor parameter ϵ_A, and two superconformal transformations with spinor parameter η_A), which can be used to set to zero the four-component of χ_A^a. In this superconformal gauge the action (3B.1) reduces to the so-called Ramond–Neveu–Schwarz (RNS) superstring action [32, 33],

$$S = \frac{1}{4\pi\alpha'} \int d\sigma\, d\tau\, \eta^{ab} \left(\partial_a X^\mu \partial_b X^\nu + i\bar{\psi}^\mu \gamma_a \partial_b \psi^\nu \right) \eta_{\mu\nu}, \tag{3B.9}$$

which is invariant under the *global* infinitesimal supersymmetry transformations

$$\delta X^\mu = \bar{\epsilon}\psi^\mu, \qquad \delta \psi^\mu = -i\gamma^a \partial_a X^\mu \epsilon, \tag{3B.10}$$

where ϵ is a constant, anticommuting Majorana spinor.

The equations of motion and the constraints, for the model of superstring that we are considering, can be obtained by varying the action (3B.1) with respect to the variables X, ψ, χ and V. Imposing (after the variation) the superconformal gauge, we can considerably simplify the dynamics, and we are led, respectively, to the following equations of motion for the bosonic variables,

$$\eta^{ab} \partial_a \partial_b X^\mu = 0, \tag{3B.11}$$

for the fermionic variables,

$$\eta^{ab} \gamma_a \partial_b \psi^\mu = 0, \tag{3B.12}$$

and to the set of constraints,

$$J^a \equiv \gamma^b \gamma^a \psi^\mu \partial_b X_\mu = 0, \qquad (3B.13)$$

$$T_{ab} \equiv \partial_a X^\mu \partial_b X_\mu + \frac{i}{2} \overline{\psi}^\mu \gamma_{(a} \partial_{b)} \psi_\mu - \frac{1}{2} \gamma_{ab} \left(\partial_i X^\mu \partial^i X_\mu + \frac{i}{2} \overline{\psi}^\mu \gamma_i \partial^i \psi_\mu \right) = 0. \qquad (3B.14)$$

The last two conditions are associated with the vanishing of the so-called world-sheet supercurrent, J^a, and energy-momentum tensor, T_{ab}, defined (in units $4\pi\alpha' = 1$) by

$$2VJ^a = \frac{\delta S}{\delta \overline{\chi}_a},$$

$$T_{ab} = V_{i(b} T_{a)}{}^i, \qquad 2VT_a{}^i = \frac{\delta S}{\delta V_i^a}, \qquad (3B.15)$$

where $V \equiv \det V_a^i$. The conditions (3B.13) and (3B.14) represent the supersymmetric generalizations of the bosonic Virasoro constraints (3A.6).

The solution of the equations of motion is simplified by introducing the light-cone coordinates $\xi^\pm$ defined on the world-sheet as in the bosonic case, according to Eqs. (3A.13). For the fermionic variables ψ^μ it is also convenient to introduce the (one-component) Majorana–Weyl spinors $\psi_\pm^\mu$, defined by the chiral projections

$$\frac{1}{2}(1+\gamma^3)\psi^\mu = \begin{pmatrix} \psi_-^\mu \\ 0 \end{pmatrix}, \qquad \frac{1}{2}(1-\gamma^3)\psi^\mu = \begin{pmatrix} 0 \\ \psi_+^\mu \end{pmatrix}, \qquad (3B.16)$$

and satisfying

$$\gamma^3 \psi_\pm^\mu = \mp \psi_\pm^\mu, \qquad (3B.17)$$

where $\gamma^3 = \gamma^0 \gamma^1$ is the chirality operator (equivalent to the γ^5 operator in four dimensions). The action (3B.9) can then be rewritten as

$$S = \frac{1}{\pi\alpha'} \int d\sigma \, d\tau \left(\partial_+ X^\mu \partial_- X_\mu + \frac{i}{2} \psi_+^\mu \partial_- \psi_{+\mu} + \frac{i}{2} \psi_-^\mu \partial_+ \psi_{-\mu} \right), \qquad (3B.18)$$

and provides the boson equations of motion in the form (3A.14), plus the fermion equations of motion

$$\partial_- \psi_+^\mu = 0, \qquad \partial_+ \psi_-^\mu = 0. \qquad (3B.19)$$

Their general solutions, $\psi_+ = \psi_+(\xi^+)$, $\psi_- = \psi_-(\xi^-)$, clearly show that the chirality states $\psi_- \equiv \psi_R$ and $\psi_+ \equiv \psi_L$ describe right- and left-moving modes, respectively. Also, from the light-cone components of the stress tensor (3B.14) we obtain the super-Virasoro constraints in the form

$$T_{++} = \frac{1}{2}(T_{00} + T_{10}) = \partial_+ X^\mu \partial_+ X_\mu + \frac{i}{2} \psi_+^\mu \partial_+ \psi_{+\mu} = 0,$$

$$T_{--} = \frac{1}{2}(T_{00} - T_{10}) = \partial_- X^\mu \partial_- X_\mu + \frac{i}{2} \psi_-^\mu \partial_- \psi_{-\mu} = 0. \qquad (3B.20)$$

Finally, we have to include the condition of vanishing supercurrent, according to Eq. (3B.13). The two conditions $J^0 = 0$ and $J^1 = 0$ are equivalent, and both provide two independent constraints which, in the light-cone gauge, can be written as follows:

$$J_+ = \psi_+^\mu \partial_+ X_\mu = 0, \qquad J_- = \psi_-^\mu \partial_- X_\mu = 0. \qquad (3B.21)$$

All the above constraints are associated with quantities which are covariantly conserved on the world-sheet, like the bosonic stress tensor appearing in the constraint (3A.12): thus, if the constraints are imposed at a given time τ, they will also be valid at all later times along the string trajectory.

3B.1 Fermionic boundary conditions

Before presenting explicit solutions to the equations of motion we need to choose from the possible boundary conditions. For the bosonic coordinates we can apply the discussion of Appendix 3A, which is still valid. For the fermionic coordinates we have a similar situation, which we discuss in the following. The variation of the fermionic part of the action (3B.18) gives two types of boundary terms,

$$
\frac{i}{4\pi\alpha'} \int_{\tau_1}^{\tau_2} d\tau \int_0^\pi d\sigma \left[\partial_\tau \left(\psi_+ \cdot \delta\psi_+ + \psi_- \cdot \delta\psi_- \right) + \partial_\sigma \left(\psi_- \cdot \delta\psi_- - \psi_+ \cdot \delta\psi_+ \right) \right]
$$

$$
= \frac{i}{4\pi\alpha'} \int_0^\pi d\sigma \left[\psi_+ \cdot \delta\psi_+ + \psi_- \cdot \delta\psi_- \right]_{\tau_1}^{\tau_2} + \frac{i}{4\pi\alpha'} \int_{\tau_1}^{\tau_2} d\tau \left[\psi_- \cdot \delta\psi_- - \psi_+ \cdot \delta\psi_+ \right]_0^\pi ,
$$

$$
\tag{3B.22}
$$

where we have denoted with a dot the contraction of the target space indices, $\psi \cdot \delta\psi \equiv \psi^\mu \delta\psi_\mu$. The first term, integrated over the σ variable, is identically vanishing because the variational principle requires $\delta\psi_\pm = 0$ at the time boundaries τ_1 and τ_2. We are thus left with the condition

$$
\left[\psi_- \cdot \delta\psi_- - \psi_+ \cdot \delta\psi_+ \right]_{\sigma=0}^{\sigma=\pi} = 0, \tag{3B.23}
$$

which can be satisfied in various ways.

For an **open superstring** the ends of the strings are independent, and we must require $\psi_- \cdot \delta\psi_- = \psi_+ \cdot \delta\psi_+$ at each end of the string, $\sigma = 0$ and $\sigma = \pi$. This can be satisfied by imposing either *periodic* (Ramond) or *antiperiodic* (Neveu–Schwarz) boundary conditions,

$$
\psi_+^\mu(\tau, 0) = \psi_-^\mu(\tau, 0), \qquad \psi_+^\mu(\tau, \pi) = \pm\psi_-^\mu(\tau, \pi) \tag{3B.24}
$$

(and the same for $\delta\psi_\pm^\mu$). Here the plus sign corresponds to Ramond (R) boundary conditions, and the minus sign to Neveu–Schwarz (NS) boundary conditions. For both choices of boundary conditions we can then express the solutions of Eqs. (3B.19) in explicit form, expanding in Fourier series and separating left- and right-moving modes. In particular, in the case of R boundary conditions we have the solution

$$
\text{(R)}: \qquad \psi_\pm^\mu = \sqrt{\frac{\alpha'}{2}} \sum_{n=-\infty}^\infty d_n^\mu\, e^{-in(\tau\pm\sigma)}, \tag{3B.25}
$$

where the sum runs over all integers n (we have used the same normalization as in the bosonic case, see Eq. (3A.30)). In the case of NS boundary conditions, on the contrary, we have the solution

$$(\text{NS}): \qquad \psi_\pm^\mu = \sqrt{\frac{\alpha'}{2}} \sum_{n=-\infty}^{\infty} b_{n+1/2}^\mu \, e^{-i(n+1/2)(\tau\pm\sigma)}, \qquad (3\text{B}.26)$$

where the sum runs over all half-integers $r = n + 1/2$.

For a **closed superstring** the two chirality components $\psi_\pm^\mu$ are independent, and the boundary conditions (3B.23) can be satisfied by imposing periodicity (R) or antiperiodicity (NS) on each component of ψ^μ separately, i.e.

$$\psi_-^\mu(\sigma) = \pm\psi_-^\mu(\sigma+\pi), \qquad \psi_+^\mu(\sigma) = \pm\psi_+^\mu(\sigma+\pi) \qquad (3\text{B}.27)$$

(and the same for $\delta\psi_\pm$). Depending on the behavior of ψ_+ and ψ_- we thus have four possible choices of boundary conditions: R–R, R–NS, NS–R and NS–NS, which correspond to different sectors of the closed string spectrum. As in the bosonic case, we denote with a tilde the Fourier coefficients of the left-moving modes. For R boundary conditions the solutions of Eq. (3B.19) can then be expanded as follows:

$$(\text{R}): \qquad \psi_-^\mu = \sqrt{\alpha'} \sum_{n=-\infty}^{\infty} d_n^\mu \, e^{-2in(\tau-\sigma)},$$

$$(\widetilde{\text{R}}): \qquad \psi_+^\mu = \sqrt{\alpha'} \sum_{n=-\infty}^{\infty} \widetilde{d}_n^\mu \, e^{-2in(\tau+\sigma)}. \qquad (3\text{B}.28)$$

In the case of NS boundary conditions we have instead the expansion

$$(\text{NS}): \qquad \psi_-^\mu = \sqrt{\alpha'} \sum_{n=-\infty}^{\infty} b_{n+1/2}^\mu \, e^{-2i(n+1/2)(\tau-\sigma)},$$

$$(\text{NS}): \qquad \psi_+^\mu = \sqrt{\alpha'} \sum_{n=-\infty}^{\infty} \widetilde{b}_{n+1/2}^\mu \, e^{-2i(n+1/2)(\tau+\sigma)}. \qquad (3\text{B}.29)$$

3B.2 *Classical constraints*

For both open and closed strings the classical solutions are completed by imposing the constraints (3B.20) and (3B.21). At the quantum level these constraints will remove the states of negative norm, and will fix the levels of the energy spectrum, just as in the case of the bosonic string.

To impose the constraints it is convenient to introduce the generalized Virasoro operators L_m and $\widetilde{L}_m$ which are associated with the conditions (3B.20), and which can be separated into bosonic and fermionic parts as follows:

$$L_m = L_m^X + L_m^\psi, \qquad \widetilde{L}_m = \widetilde{L}_m^X + \widetilde{L}_m^\psi. \qquad (3\text{B}.30)$$

The bosonic part refers to the solutions $\partial_\pm X$, and has already been computed in Appendix 3A (see Eqs. (3A.21)–(3A.23) for the closed string, and Eq. (3A.33) for the open string case). For the fermionic part of the constraint we can follow the same procedure, recalling that the open string solution has to be analytically extended over the whole interval $-\pi \leq \sigma \leq \pi$, using the prescription

$$
\psi^\mu = \begin{cases} \psi_+^\mu(\sigma), & 0 \leq \sigma \leq \pi, \\ \psi_-^\mu(-\sigma), & -\pi \leq \sigma \leq 0. \end{cases} \tag{3B.31}
$$

We are then led to the definitions (valid at any given fixed value of τ, as in the bosonic case)

$$
\begin{aligned}
L_m^\psi &= \frac{1}{\pi\alpha'} \int_0^\pi d\sigma \left(\frac{i}{2} \psi_-^\mu \partial_- \psi_{-\mu} e^{im(\tau-\sigma)} + \frac{i}{2} \psi_+^\mu \partial_+ \psi_{+\mu} e^{im(\tau+\sigma)} \right) \\
&= \frac{1}{\pi\alpha'} \int_{-\pi}^\pi d\sigma \frac{i}{2} \psi_+^\mu \partial_+ \psi_{+\mu} e^{im(\tau+\sigma)},
\end{aligned} \tag{3B.32}
$$

for the open superstring (see Eq. (3A.39)), and

$$
\begin{aligned}
L_m^\psi &= \frac{1}{2\pi\alpha'} \int_0^\pi d\sigma \frac{i}{2} \psi_-^\mu \partial_- \psi_{-\mu} e^{2im(\tau-\sigma)}, \\
\widetilde{L}_m^\psi &= \frac{1}{2\pi\alpha'} \int_0^\pi d\sigma \frac{i}{2} \psi_+^\mu \partial_+ \psi_{+\mu} e^{2im(\tau+\sigma)},
\end{aligned} \tag{3B.33}
$$

for the closed superstring (see Eq. (3A.21)). In both cases, the fermionic operator L_m^ψ has to be separately evaluated for R and NS boundary conditions.

Inserting the explicit solutions for $\psi_\pm^\mu$ it is possible to express the above Virasoro functionals in terms of the Fourier coefficients d_n^μ, b_r^μ. For the open string we can exploit the orthonormality of the base $e^{in\sigma}/2\pi$ on the interval $[-\pi, \pi]$, while for the closed string the orthonormality of $e^{2in\sigma}/\pi$ on the interval $[0, \pi]$: in both cases we obtain the same result in terms of the Fourier coefficients. We are interested, in particular, in the zero-frequency part of the Virasoro operators, which is the part generating the mass-shell condition. For the R boundary conditions, using Eqs. (3B.25) and (3B.28), we obtain the time-independent result

$$
\text{(R)}: \qquad L_0^R = \frac{1}{2} \sum_{n=-\infty}^\infty n\, d_{-n}^\mu d_{n\mu}, \tag{3B.34}
$$

while in the case of NS boundary conditions, using Eqs. (3B.26) and (3B.29), we obtain

$$
\begin{aligned}
\text{(NS)}: \qquad L_0^{NS} &= \frac{1}{2} \sum_{n=-\infty}^\infty \left(n + \frac{1}{2} \right) b_{-n-1/2}^\mu b_{n+1/2,\mu} \\
&= \frac{1}{2} \sum_r r\, b_{-r}^\mu b_{r\mu},
\end{aligned} \tag{3B.35}
$$

where the sum over $r = n + 1/2$ denotes a sum over all half-integer numbers. In the closed superstring case we have similar expansions for $\widetilde{L}_0^R, \widetilde{L}_0^{NS}$ in terms of $\widetilde{d}, \widetilde{b}$.

For the sake of completeness we also report here the fermionic generators of the supercurrent constraints (3B.21). In the open string case one finds that the generator associated with the R boundary conditions, using Eq. (3B.25), is given by

$$(R): \qquad F_m = \frac{1}{\pi\alpha'} \int_0^\pi d\sigma \left(J_+ e^{im(\tau+\sigma)} + J_- e^{im(\tau-\sigma)} \right)$$

$$\equiv \frac{1}{\pi\alpha'} \int_0^\pi d\sigma \, \psi_+^\mu \partial_+ X_\mu \, e^{im(\tau+\sigma)} = \sum_{n=-\infty}^\infty \alpha_{-n}^\mu d_{n+m,\mu}, \qquad (3B.36)$$

while the generator associated with the NS boundary conditions, using Eq. (3B.26), is

$$(NS): \qquad G_r = \frac{1}{\pi\alpha'} \int_0^\pi d\sigma \left(J_+ e^{ir(\tau+\sigma)} + J_- e^{ir(\tau-\sigma)} \right)$$

$$= \sum_{n=-\infty}^\infty \alpha_{-n}^\mu b_{n+r,\mu}. \qquad (3B.37)$$

Similarly, in the closed string case, we have from Eq. (3B.28) the R supercurrent generators,

$$(R): \qquad F_m = \frac{1}{\sqrt{2}\pi\alpha'} \int_0^\pi d\sigma J_- e^{2im(\tau-\sigma)} = \sum_{n=-\infty}^\infty \alpha_{-n}^\mu d_{n+m,\mu},$$

$$\tilde{F}_m = \sum_{n=-\infty}^\infty \tilde{\alpha}_{-n}^\mu \tilde{d}_{n+m,\mu}, \qquad (3B.38)$$

and, from Eq. (3B.29), the NS supercurrent generators,

$$(NS): \qquad G_r = \frac{1}{\sqrt{2}\pi\alpha'} \int_0^\pi d\sigma J_- e^{2ir(\tau-\sigma)} = \sum_{n=-\infty}^\infty \alpha_{-n}^\mu b_{n+r,\mu},$$

$$\tilde{G}_r = \sum_{n=-\infty}^\infty \tilde{\alpha}_{-n}^\mu \tilde{b}_{n+r,\mu}. \qquad (3B.39)$$

3B.3 Quantization

We are now in the position of computing the energy spectrum of the quantized superstring models, imposing on the physical states to be annihilated by the application of the (normal-ordered) operators associated with the classical constraints:

$$(L_0 - \delta)|\psi\rangle = 0 = (\tilde{L}_0 - \tilde{\delta})|\psi\rangle, \qquad L_m|\psi\rangle = 0 = \tilde{L}_m|\psi\rangle, \quad m > 0,$$

$$F_m|\psi\rangle = 0 = \tilde{F}_m|\psi\rangle, \quad m \geq 0, \qquad G_r|\psi\rangle = 0 = \tilde{G}_r|\psi\rangle, \quad r > 0 \qquad (3B.40)$$

(compare with Eq. (3A.54)). The first constraint, in particular, will provide the quantum mass-shell condition after determining the parameters $\delta, \tilde{\delta}$ through the normal ordering of the operators $L_0, \tilde{L}_0$.

In the bosonic case the computation was performed by introducing light-cone coordinates in the target manifold, and fixing the residual gauge degrees of freedom by eliminating the oscillations along the X^+ direction (see Eq. (3A.55)). We have used, in particular, the condition $\alpha_m^+ = 0 = \tilde{\alpha}_m^+$ for all modes m, where $\alpha^+ = (\alpha^0 + \alpha^{D-1})/\sqrt{2}$. In the superstring case we can still use this convenient gauge choice, extending it to the fermionic sector of the action through the definition of transverse (ψ^i) and longitudinal ($\psi^{\pm}$) fermionic coordinates, such that

$$\psi^{\pm} = \frac{1}{\sqrt{2}}\left(\psi^0 \pm \psi^{D-1}\right), \qquad \psi^i, \ i = 1, \ldots, D-2,$$

$$\psi^{\mu}\psi_{\mu} = 2\psi^+\psi_- - \psi^i\psi^i. \tag{3B.41}$$

Thanks to the superconformal symmetry, in fact, we have the freedom of applying a local supersymmetry transformation to gauge away the ψ^+ component of the fermionic coordinates. We can thus complete the gauge specification by imposing the conditions

$$d_m^+ = 0 = \tilde{d}_m^+, \qquad b_r^+ = 0 = \tilde{b}_r^+, \tag{3B.42}$$

which greatly simplifies the Virasoro operators, leaving only the transverse fermionic modes $d_n^i, b_r^i, i = 1, \ldots, D-2$.

To obtain the superstring spectrum we have now to compute the regularized version of $: L_0 :$, promoting to operators the coefficients of the classical Fourier expansion, and imposing canonical commutation relations for the bosonic modes and anticommutation relations for the fermionic ones. Concerning the bosonic part of the operators we can safely apply all results presented in the previous appendix. For the fermionic part we shall impose the anticommutation brackets,

$$\{d_m^{\mu}, d_n^{\nu}\} = \{\tilde{d}_m^{\mu}, \tilde{d}_n^{\nu}\} = -\eta^{\mu\nu}\delta_{m+n,0},$$

$$\{b_r^{\mu}, b_s^{\nu}\} = \{\tilde{b}_r^{\mu}, \tilde{b}_s^{\nu}\} = -\eta^{\mu\nu}\delta_{r+s,0}, \tag{3B.43}$$

which define (for $m > 0$, $r > 0$) d_m, b_r as annihilation and $d_{-m} = d_m^{\dagger}$, $b_{-m} = b_r^{\dagger}$ as creation fermionic operators. Notice that, according to their statistical properties, d and b are nilpotent operators, $(d_m^{\mu})^2 = 0$, $(b_m^{\mu})^2 = 0$, and the associated number operators are projectors, satisfying $(d_{-m}^{\mu}d_{m\mu})^2 = d_{-m}^{\mu}d_{m\mu}$, $(b_{-r}^{\mu}b_{r\mu})^2 = b_{-r}^{\mu}b_{r\mu}$, with discrete eigenvalues $0, 1$.

Let us now separately consider the cases of R and NS boundary conditions. In the first case we start from Eq. (3B.34) and we obtain, in the light-cone gauge,

$$L_0^R = \frac{1}{2}\sum_{n=-\infty}^{\infty} n\, d_{-n}^{\mu} d_{n\mu} = -\frac{1}{2}\sum_{n=-\infty}^{\infty} n\, d_{-n}^i d_n^i$$

$$= -\frac{1}{2}\sum_{n=1}^{\infty}\left(n\, d_{-n}^i d_n^i - n\, d_n^i d_{-n}^i\right)$$

$$= -\sum_{n=1}^{\infty} n\, d_{-n}^i d_n^i + \frac{D-2}{2}\sum_{n=1}^{\infty} n$$

$$= -N_R - \frac{D-2}{24}, \tag{3B.44}$$

where we have defined the operator $N_R = \sum_n n\, d_{-n}^i d_n^i$, with a spectrum of non-negative integer eigenvalues, $N_R = 0, 1, 2, \ldots$ We notice, for the sake of clarity, that the second line of the above equalities follows from the transformation $n \to -n$, the third line from the anticommutation relations (3B.43), and the fourth line from the exponential regularization (3A.58) of the infinite sum over n.

In the case of NS boundary conditions we obtain, from Eq. (3B.35),

$$L_0^{NS} = \frac{1}{2}\sum_r r\, b_{-r}^\mu b_{r\mu} = -\frac{1}{2}\sum_r r\, b_{-r}^i b_r^i$$

$$= -\frac{1}{2}\sum_{r=1/2}^\infty \left(r b_{-r}^i b_r^i - r b_r^i b_{-r}^i\right)$$

$$= -\sum_{r=1/2}^\infty r\, b_{-r}^i b_r^i + \frac{D-2}{2}\sum_{r=1/2}^\infty r. \tag{3B.45}$$

Applying again the exponential regularization procedure one finds

$$\sum_{r=1/2}^\infty r = \sum_{n=0}^\infty \left(n+\frac{1}{2}\right) = \lim_{\epsilon\to 0}\sum_{n=0}^\infty \left(n+\frac{1}{2}\right)e^{-\epsilon(n+1/2)}$$

$$= -\lim_{\epsilon\to 0}\frac{d}{d\epsilon}e^{-\epsilon/2}\sum_n e^{-\epsilon n}$$

$$= \lim_{\epsilon\to 0}\frac{e^{-\epsilon/2}}{1-e^{-\epsilon}}\left(\frac{1}{2}+\frac{e^{-\epsilon}}{1-e^{-\epsilon}}\right) = \lim_{\epsilon\to 0}\left(\frac{1}{\epsilon^2}+\frac{1}{24}\right). \tag{3B.46}$$

Thus, after subtracting the infinite contribution of the vacuum,

$$L_0^{NS} = -N_{NS} + \frac{D-2}{48}, \tag{3B.47}$$

where we have denoted by N_{NS} the operator $\sum_r r\, b_{-r}^i b_r^i$, with a spectrum including the zero and all positive half-integer eigenvalues, $N_{NS} = 0, 1/2, 3/2, \ldots$

Summing up the bosonic and fermionic parts of the Virasoro operator we can then write down the mass-shell condition determining the superstring energy spectrum, for both R and NS boundary conditions. In the **open string** case, using the bosonic result (3A.59), and imposing $L_0^X + L_0^\psi = 0$, we obtain two possible spectra:

- the R sector, with $L_0^\psi = L_0^R$, and

$$\text{(R)}: \qquad \alpha' M^2 = N + N_R; \tag{3B.48}$$

- the NS sector, with $L_0^\psi = L_0^{NS}$, and

$$\text{(NS)}: \qquad \alpha' M^2 = N + N_{NS} - \frac{D-2}{16}. \tag{3B.49}$$

In the **closed string** case, right- and left-moving modes are independent, and we must separately impose that the sum and the difference of $L_0 = L_0^X + L_0^\psi$ and $\widetilde{L}_0 = \widetilde{L}_0^X + \widetilde{L}_0^\psi$ are vanishing on the physical states, thus obtaining the mass-shell condition and a generalized version of the "level-matching" condition, respectively. Using the result (3A.65) for the bosonic operator we then obtain four different sectors of the spectrum:

- the R–R sector, with L_0^R and $\widetilde{L}_0^R$, characterized by

$$\text{(R–R)}: \qquad \alpha' M^2 = 2(N + \widetilde{N} + N_R + \widetilde{N}_R),$$

$$N + N_R = \widetilde{N} + \widetilde{N}_R; \tag{3B.50}$$

- the R–NS sector, with L_0^R and $\widetilde{L}_0^{NS}$, characterized by

$$(\text{R–NS}): \qquad \alpha' M^2 = 2\left(N + \widetilde{N} + N_R + \widetilde{N}_{NS} - \frac{D-2}{16}\right),$$

$$N + N_R = \widetilde{N} + \widetilde{N}_{NS} - \frac{D-2}{16};$$

$$(3B.51)$$

- the NS–R sector, with L_0^{NS} and $\widetilde{L}_0^R$, characterized by

$$(\text{NS–R}): \qquad \alpha' M^2 = 2\left(N + \widetilde{N} + N_{NS} + \widetilde{N}_R - \frac{D-2}{16}\right),$$

$$N + N_{NS} = \widetilde{N} + \widetilde{N}_R + \frac{D-2}{16};$$

$$(3B.52)$$

- the NS–NS sector, with L_0^{NS} and $\widetilde{L}_0^{NS}$, characterized by

$$(\text{NS–NS}): \qquad \alpha' M^2 = 2\left(N + \widetilde{N} + N_{NS} + \widetilde{N}_{NS} - \frac{D-2}{8}\right),$$

$$N + N_{NS} = \widetilde{N} + \widetilde{N}_{NS}.$$

$$(3B.53)$$

It should be noted that, for a generic value of D, only the R–R and NS–NS sectors are characterized by an equal number of left- and right-moving modes (and then satisfy a true level-matching condition). Also, only in the closed R–R and in the open R sectors is the ground state massless: in the other spectra the lowest-energy level is tachyonic, $M^2 < 0$, for a generic number D of the space-time dimensions.

We are now able to determine the number of critical dimensions for a consistent quantization of the considered model of superstring. We can start, for instance, from the open NS spectrum of Eq. (3B.49), and consider the lowest-mass, non-tachyonic state, corresponding to the eigenvalues $N = 0$ and $N_{NS} = 1/2$: this state can be obtained from the vacuum by applying the creation operator $b^i_{-1/2}$, and thus describes the transverse, vector-like configuration $b^i_{-1/2}|0_{NS}\rangle$, with $i = 1, \ldots, D-2$, and with mass eigenvalue

$$M^2 = \frac{1}{\alpha'}\left(\frac{1}{2} - \frac{D-2}{16}\right). \qquad (3B.54)$$

As discussed in the bosonic string case, a transverse vector with only $D-2$ independent degrees of freedom is compatible with the D-dimensional Lorentz symmetry only if such a vector is massless. The condition $M^2 = 0$ then immediately implies that the model of superstring is consistent in a critical number

$$D = D_c = 10 \qquad (3B.55)$$

of space-time dimensions (differently from the bosonic string which requires $D_c = 26$, see Eq. (3A.64)).

The same result holds for the closed superstring model, as can be checked starting from the NS–NS spectrum of Eq. (3B.53) and considering the eigenvalues $N = \widetilde{N} = 0$, $N_{NS} = \widetilde{N}_{NS} = 1/2$. One obtains in this way the lowest excited level, represented by the (transverse) tensor configuration $b^i_{-1/2}\widetilde{b}^i_{-1/2}|0_{NS}\widetilde{0}_{NS}\rangle$: this configuration can be decomposed into a graviton, a dilaton and an antisymmetric tensor as in the case of the bosonic string, and is consistent with the D-dimensional Lorentz group only if these fields are massless, i.e. if $D - 2 = 8$ according to Eq. (3B.53).

It may be noticed that, once we have fixed $D_c = 10$, we have also fixed the value of the ordering parameters δ, $\widetilde{\delta}$ appearing in Eq. (3B.40). Remembering that $\delta = (D-2)/24$ for the bosonic case (Eq. (3A.59)), and summing the fermionic contribution, separately for R and NS boundary conditions (Eqs. (3B.44) and (3B.47), respectively), we obtain

$$\delta_R = \widetilde{\delta}_R = \frac{D_c - 2}{24} - \frac{D_c - 2}{24} = 0,$$
$$\delta_{NS} = \widetilde{\delta}_{NS} = \frac{D_c - 2}{24} + \frac{D_c - 2}{48} = \frac{D_c - 2}{16} = \frac{1}{2}.$$

(3B.56)

It is only for these values that we may have a consistent model of quantum superstring.

In order to determine the statistical properties of the different sectors of the spectrum we now observe that the operators d_0^μ, associated with R boundary conditions, satisfy the Clifford algebra

$$\{d_0^\mu, d_0^\nu\} = -\eta^{\mu\nu} \qquad (3B.57)$$

(see Eq. (3B.43)), where $\eta_{\mu\nu}$ is the metric of the 10-dimensional Minkowski space-time: thus, they can be represented by the $D = 10$ canonical Dirac matrices Γ^μ, by setting $d_0^\mu = i\Gamma^\mu/\sqrt{2}$ in order to agree with the standard Dirac algebra, Eq. (3B.2).

The operators d_0^μ, on the other hand, do not introduce transitions between different mass-levels, but generate maps of the given state onto itself. Each mass-level of the open R spectrum must then provide a representation of the Clifford algebra (3B.57): the ground state $|0_R\rangle$, in particular, must correspond to an irreducible representation, since there is no additional degeneracy of this level produced by other zero-mode operators. Since the unique irreducible representation of this algebra is the spinor representation of the Lorentz group $SO(1,9)$, it follows that the R ground state has to be a $D = 10$ spinor, with $2^{D/2} = 2^5 = 32$ independent components. All the other states of the R spectrum, obtained by applying to $|0_R\rangle$ the vector-like creation operators α_{-m}^μ, d_{-m}^μ, $m > 0$, are thus space-time spinors, and this is how space-time fermions appear in the quantized superstring model in spite of their absence in the initial sigma model action (3B.9).

The ground state of the NS spectrum, on the contrary, is non-degenerate and transforms as a scalar representation of the $SO(1,9)$ group. All the levels of the open NS spectrum, obtained by applying to $|0_{NS}\rangle$ an arbitrary number of vector creation operators α_{-m}^μ, b_{-r}^μ, are thus space-time tensors, and represent the bosonic content of the open superstring spectrum.

This analysis can be easily extended to the closed superstring spectrum, considering the ground states of the four different sectors (3B.50)–(3B.53) and noting that the product of two spinor representations of the Lorentz group, $|0_R \widetilde{0}_R\rangle \equiv |0_R\rangle \otimes |\widetilde{0}_R\rangle$, corresponds to a vector representation with bosonic statistical properties, while the product of a tensor and a spinor, like $|0_{NS} \widetilde{0}_R\rangle \equiv |0_{NS}\rangle \otimes |\widetilde{0}_R\rangle$, is again a spinor. Thus, the R–R and NS–NS sectors of the closed superstring spectrum describe bosons, while the R–NS and NS–R sectors describe space-time fermions.

A consistent theory of interacting string can now be constructed by putting together the Hilbert subspaces of the physical states of the R and NS sectors: the model obtained in this way contains bosons and fermions, is consistent in a critical number of $D = 10$ space-time dimensions, but is still unsatisfactory for various reasons.

First of all, there is still a tachyon in the ground state of the NS spectrum (see Eqs. (3B.49) and (3B.53)). Furthermore, there are states which correspond to space-time bosons, and which are connected by transformations generated by an *odd* number of anticommuting (fermion-like) operators. Consider, for instance, the second-excited level of the open NS spectrum (3B.49), with $\alpha' M^2 = 1/2$: it contains the state $\alpha_{-1}^{\mu}|0_{NS}\rangle$, with eigenvalues $N = 1$, $N_{NS} = 0$, but contains also the state $b_{-1/2}^{\mu} b_{-1/2}^{\nu}|0_{NS}\rangle$, with eigenvalues $N = 0$, $N_{NS} = 1$. This second configuration can be obtained by applying the fermionic operator $b_{-1/2}^{\mu}$ to the massless level $b_{-1/2}^{\nu}|0_{NS}\rangle$ of the open NS spectrum, so that we are in a situation in which two bosonic states (the first and the second excited level of the NS spectrum) are mapped onto each other by an operator with anticommuting statistical properties.

A third difficulty is that, if we compare the bosonic states of the NS spectrum with the corresponding fermionic ones of the R spectrum, we find in general a different number of bosonic and fermionic degrees of freedom, even at the level of the same mass eigenvalue. This prevents a possible implementation of supersymmetry in the context of the 10-dimensional physical space-time, in contrast with the two-dimensional supersymmetry already present in the world-sheet action.

A concrete example of this problem can be illustrated by considering the massless levels of the open string spectrum. In the NS sector of the spectrum we find a transverse vector state, represented by $b_{-1/2}^{i}|0_{NS}\rangle$, associated with a total number of $D - 2 = 8$ independent degrees of freedom. In the R sector we find a spinor, $|0_R\rangle$, which has in general $2^{D/2} = 32$ complex components. For a Majorana spinor we can always choose a representation in which the 10-dimensional Dirac matrices Γ^{μ} have imaginary components and the operators $d_0^{\mu} = i\Gamma^{\mu}/\sqrt{2}$ are real, so that we are left with a spinor with 32 *real* components. In addition, this state has to be annihilated by the constraints generated by the supercurrent operator (see Eq. (3B.40), and which implies $F_m|\psi\rangle = 0$, $m \geq 0$, for all states of the open R spectrum. The vacuum $|0_R\rangle$, in particular, is the lowest-mass solution of the condition $F_0|\psi\rangle = 0$, which is equivalent to the Fourier transform of the massless Dirac equation $\Gamma^{\mu}\partial_{\mu}\psi = 0$. Such a condition reduces the number of independent components by a factor $1/2$, since it relates half of the spinor components to the other ones. Thus, one is left with $32/2 = 16$ real components, which is still twice, however, the number of independent components of a massless vector.

All the difficulties mentioned above can be automatically solved by performing the so-called GSO projection [34, 35], i.e. by imposing on the physical states a condition which eliminates (or "projects out") in the NS sector all states obtained from the vacuum by applying an *even* number of anticommuting operators b_{-r}^{μ}, and in the R sector eliminates one of the two chirality components of the spinor. This procedure, in particular, removes the NS tachyon, as well as the fermionic mapping among states with bosonic statistics. Also, and most importantly, the chirality condition imposed on the fermionic states $|\psi\rangle$ of the R sector implies that this sector describes Majorana–Weyl spinors, i.e. spinors satisfying the reality condition *and* the chirality condition

$$\Gamma^{11}|\psi\rangle = \pm|\psi\rangle, \qquad \Gamma^{11} = \Gamma^0 \Gamma^1 \dots \Gamma^9, \qquad \{\Gamma^{11}, \Gamma^{\mu}\} = 0 \qquad (3B.58)$$

(remarkably, the two conditions are only compatible in $D = 2$ (modulo 8) space-time dimensions). The chirality condition eliminates half of the spinor components and thus, in particular, reduces from 16 to 8 the number of real components of the R vacuum $|0_R\rangle$. Hence, after the GSO projection, one obtains that the massless states of the open string spectrum represent a 10-dimensional "supermultiplet"

$$\left\{ b_{-1/2}^{i}|0_{NS}\rangle, \, |0_R\rangle \right\}, \qquad (3B.59)$$

with an equal number (8) of boson and fermion components, as appropriate to a 10-dimensional supersymmetric gauge theory [1].

The GSO projection produces the same number of bosonic and fermionic components not only in the case of the massless level, but also for all levels of the superstring spectrum. In a generic level the projection is represented by the condition

$$P_{\text{GSO}}^{\text{NS}}|\phi\rangle = |\phi\rangle, \qquad P_{\text{GSO}}^{\text{NS}} = -(-1)^{\sum_{r=1/2}^{\infty} b_{-r}^{\mu} b_{r\mu}}, \qquad (3\text{B}.60)$$

to be satisfied by all physical (bosonic) states of the NS spectrum, and by the condition

$$P_{\text{GSO}}^{\text{R}}|\psi\rangle = |\psi\rangle, \qquad P_{\text{GSO}}^{\text{R}} = \Gamma^{11}(-1)^{\sum_{n=1}^{\infty} d_{-n}^{\mu} d_{n\mu}}, \qquad (3\text{B}.61)$$

to be satisfied by all physical (fermionic) states of the R spectrum. Without discussing the technical details of these conditions we should explain, however, why the elimination of one of the two chirality components is consistent also at the level of the massive fermion states. It is well known, indeed, that massive fermions cannot be represented by chiral (or Weyl) spinors, since the chirality operator (Γ^{11}, in our case) does not anticommute with the massive Dirac operator ($i\Gamma^{\mu}\partial_{\mu} + m$). The crucial point is that, after imposing the condition (3B.61), one is left with R states having either (*i*) *positive* chirality and an *even* number of d_{-m}^{μ} creation operators, or (*ii*) *negative* chirality and an *odd* number of d_{-m}^{μ} creation operators. At each mass-level these two states of opposite handedness can be combined to give a Majorana spinor that corresponds to a full (non-chiral) massive representation of the Lorentz group [1].

It should be mentioned, finally, that the equality of the number of boson and fermion degrees of freedom is a necessary condition for the construction of a model with space-time supersymmetry, but does not constitute a proof of the existence of such a symmetry, of course. The supersymmetry becomes manifest only if the considered string model is reformulated starting from a world-sheet action which already contains space-time supersymmetry, unlike the actions (3B.1) or (3B.9), which are supersymmetric only on the two-dimensional world-sheet manifold. A presentation of this formalism is outside the scope of this appendix (see for instance [1]); it is important to mention, however, that quantizing the space-time supersymmetric action one finds that the GSO conditions are automatically built in from the outset, without having to make any truncation of the spectrum (which might appear a contrived sort of procedure in the model of superstring so far considered).

3B.4 Type IIA and type IIB superstrings

In the closed string case, the GSO conditions must be separately applied both to the left-moving and to the right-moving sectors of the spectra. For the fermionic R sector, however, we have two possibilities, depending on which chirality component we want to keep in the ground state: there are two inequivalent GSO projections, in which we take either *opposite* chiralities, $|0_{\text{R}}^{+} \widetilde{0}_{\text{R}}^{-}\rangle$, or *the same* (for instance positive) chiralities, $|0_{\text{R}}^{+} \widetilde{0}_{\text{R}}^{+}\rangle$ (the plus and minus superscripts refer to the eigenvalues of the chirality equations (3B.58)). The resulting theories are called, respectively, type IIA and type IIB superstrings.

Let us look at the particle content of their fundamental massless level, $M^2 = 0$, after imposing the GSO conditions, considering separately the four possible sectors of the closed superstring spectrum.

(1) The NS–NS sector, Eq. (3B.53), is not affected by the choice of chirality, so that the particle content is the same for both types of strings. The level $M^2 = 0$ corresponds to the eigenvalues $N = \widetilde{N} = 0$, $N_{\text{NS}} = \widetilde{N}_{\text{NS}} = 1/2$, and is represented by the tensor (bosonic)

configuration $b^i_{-1/2} \tilde{b}^j_{-1/2} |0_{\rm NS} \tilde{0}_{\rm NS}\rangle$ (which survives the GSO projections as it contains an odd number of b and $\tilde{b}$ operators). Such a state transforms as the product of two irreducible vector representations of the group $SO(8)$, i.e. $8_V \otimes \tilde{8}_V$. The total number of on-shell components is thus $8 \times 8 = 64$, which can be decomposed into the $D(D-3)/2 = 35$ components of the symmetric trace-free tensor h_{ij}, the $(D-3)(D-2)/2 = 28$ components of the antisymmetric tensor A_{ij}, and the scalar trace: they correspond, respectively, to the graviton, the antisymmetric tensor (or NS–NS two-form) and the dilaton. The situation is exactly the same as that we have already encountered in the bosonic string case, see Eq. (3A.69). Summarizing,

$$ b^i_{-1/2} \tilde{b}^j_{-1/2} |0_{\rm NS} \tilde{0}_{\rm NS}\rangle \longrightarrow \phi, \; g_{\mu\nu}, \; B_{\mu\nu}. \tag{3B.62} $$

(2) In the R–R sector, Eq. (3B.50), the massless level corresponds to zero eigenvalue of all number operators, and after the GSO projection is associated with the tensor product of the two eight-dimensional spinors $|0_R\rangle$ and $|\tilde{0}_R\rangle$: we obtain thus a representation of the Lorentz group of bosonic type, again with $8 \times 8 = 64$ independent components. Such a representation can be decomposed into antisymmetric tensor components, whose explicit form depends on the chirality of the two initial spinors.

To discuss this point let us introduce two Majorana–Weyl spinors with components θ_a and $\theta_{\dot{a}}$, forming an eight-dimensional representation of the transverse $SO(8)$ group in the light-cone gauge, and corresponding to opposite eigenvalues of the chirality operator: we may assume, for instance, that θ_a belongs to the irreducible spinor representation 8_+ with positive chirality, and that $\theta_{\dot{a}}$ belongs to the irreducible spinor representation 8_- with negative chirality. We then obtain terms like $\theta_a \tilde{\theta}_{\dot{b}}$, associated with the representation $8_+ \otimes 8_-$ and typical of the state $|0_R^+ \tilde{0}_R^-\rangle$ of type IIA strings, and terms like $\theta_a \tilde{\theta}_b$, associated with the representation $8_+ \otimes \tilde{8}_+$ and typical of the state $|0_R^+ \tilde{0}_R^+\rangle$ of type IIB strings.

On the other hand, given the eight-dimensional spinors $\theta_a, \theta_{\dot{a}}$, and the 16×16 Dirac matrices γ^i of $SO(8)$, satisfying

$$ \gamma^i = \begin{pmatrix} 0 & \gamma^i_{a\dot{a}} \\ \gamma^i_{\dot{b}b} & 0 \end{pmatrix}, \qquad \gamma^i_{a\dot{a}} \gamma^j_{\dot{a}b} + \gamma^j_{a\dot{a}} \gamma^i_{\dot{a}b} = 2\delta^{ij} \delta_{ab}, \tag{3B.63} $$

we can construct two different classes of antisymmetric tensor objects through the following contractions:

$$ \gamma^{ijk\cdots} = \theta_a \left(\gamma^{[i} \gamma^j \cdots \gamma^{k]} \right)^{ab} \tilde{\theta}_b, \qquad \gamma^{ij\cdots} = \theta_a \left(\gamma^{[i} \cdots \gamma^{j]} \right)^{ab} \tilde{\theta}_b, \tag{3B.64} $$

where the first contraction contains an *odd* number, and the second an *even* number, of gamma matrices. This clearly shows how odd-rank antisymmetric tensors (like γ^i, γ^{ijk}) are associated with terms of the type $\theta_a \tilde{\theta}_{\dot{b}}$, i.e. to the decomposition of the representation $8_+ \otimes 8_-$, while even-rank antisymmetric tensors (like 1, γ^{ij}, γ^{ijkl}) are associated with terms of the type $\theta_a \tilde{\theta}_b$, i.e. to the decomposition of the representation $8_+ \otimes \tilde{8}_+$. Counting the 64 on-shell degrees of freedom, available in $D = 10$, we are then led to the following decomposition of the product of two spinor representations of opposite chirality (type IIA superstrings):

$$ \text{IIA}: \qquad 8_+ \otimes \tilde{8}_- = 8_V \oplus 56_V \tag{3B.65} $$

(the subscript V denotes the vector representation). For the product of spinor representations of the same chirality (type IIB superstrings), on the contrary, we have the decomposition:

$$\text{IIB}: \qquad 8_+ \otimes \tilde{8}_+ = 1 \oplus 28_V \oplus 35_V. \qquad (3B.66)$$

In other terms, the massless R–R sector of type IIA superstrings describes a vector (or one-form) and a third-rank antisymmetric tensor (or three-form):

$$\text{IIA}: \qquad |0_R^+ \tilde{0}_R^-\rangle \longrightarrow A_\mu, \; A_{\mu\nu\alpha}. \qquad (3B.67)$$

They contain, respectively, $D - 2 = 8$ and $(D-4)(D-3)(D-2)/3! = 56$ physical degrees of freedom. The massless R–R sector of type IIB superstrings describes, instead, a scalar (or zero-form), a two-form, and a four-form with self-dual field strength:

$$\text{IIB}: \qquad |0_R^+ \tilde{0}_R^+\rangle \longrightarrow A, \; A_{\mu\nu}, \; A_{\mu\nu\alpha\beta}, \qquad (3B.68)$$

containing, respectively, 1, 28 and 35 degrees of freedom. It can be checked, as a useful exercise, that the supercurrent constraints $F_0|\psi\rangle = 0$ and $\tilde{F}_0|\psi\rangle = 0$ provide the equations of motion and the Bianchi identities for the antisymmetric tensors associated with these states. In terms of the field strengths, $H_{\mu\nu\alpha...} = \partial_{[\mu}A_{\nu\alpha...]}$, such conditions are represented, respectively, by the first-order differential equations

$$\partial^\mu H_{\mu\nu\alpha...} = 0, \qquad \partial_{[\mu}H_{\nu\alpha\beta...]} = 0. \qquad (3B.69)$$

Using the language of differential forms, exterior derivatives and Hodge dual (see Appendix 2A), and denoting with a subscript the rank of a form, the previous equations can be rewritten, respectively, as

$$H_p = dA_{p-1}, \qquad d^*H_p = 0, \qquad dH_p = 0. \qquad (3B.70)$$

It should be noted, finally, that the five-form field strength $H_5 = dA_4$ must satisfy the self-duality condition $H_5 = {}^*H_5$, which is needed in order to halve the number of components of A_4 and to match the representation 35_V of the decomposition (3B.66). A totally antisymmetric potential of rank p, in D dimensions, has indeed $(D-p-1)(D-p)(D-p+1)\cdots(D-2)/p!$ independent components, which become 70 in $D = 10$ for $p = 4$, and which are reduced to 35 only if H_5 satisfies the self-dual condition.

(3) In the NS–R sector, Eq. (3B.52), the massless level corresponds to the eigenvalues $N = \tilde{N} = \tilde{N}_R = 0$, $N_{NS} = 1/2$, and is represented by the state $b^i_{-1/2}|0_{NS}\tilde{0}_R^\pm\rangle$, which transforms according to the product of the vector and spinor representations, $8_V \otimes \tilde{8}_\pm$. The 64 components of this representation can be decomposed into a spinor χ (the so-called "dilatino"), and a vector-spinor χ^i (or gravitino) carrying both vector and spinor indices, with a total of $8 + 56 = 64$ physical degrees of freedom (the gravitino has 7×8 components, instead of 8×8, because of the additional conditions $\gamma^i\chi_i = 0$ following from the supercurrent constraints). Depending on the chirality imposed by the GSO projection on the R sector we have two possibilities:

$$\text{IIA}: \qquad b^i_{-1/2}|0_{NS}\tilde{0}_R^-\rangle \longrightarrow \chi_-, \chi^\mu_-; \qquad (3B.71)$$

$$\text{IIB}: \qquad b^i_{-1/2}|0_{NS}\tilde{0}_R^+\rangle \longrightarrow \chi_+, \chi^\mu_+. \qquad (3B.72)$$

(4) In the R–NS sector, Eq. (3B.51), the massless level corresponds to the eigenvalues $N = \tilde{N} = N_R = 0$, $\tilde{N}_{NS} = 1/2$, and is represented by the state $\tilde{b}^i_{-1/2}|0_R^+\tilde{0}_{NS}\rangle$, which transforms according to the representation, $8_+ \otimes \tilde{8}_V$. As in the previous case we can decompose

this level into the sum of two spinor states: one dilatino ξ and one gravitino ξ^i, with a total of 64 independent components. In this case, however, there is no difference due to the chirality choice, and we obtain the same fields in both types of strings, IIA and IIB:

$$\widetilde{b}^i_{-1/2}|0^+_R \widetilde{0}_{NS}\rangle \longrightarrow \xi_+, \ \xi^\mu_+. \tag{3B.73}$$

The fermion sector of the spectrum, for the two superstring models, contains two gravitinos and two dilatinos, with opposite chirality in type IIA, and with the same chirality in type IIB. We have thus a non-chiral theory (type IIA superstring) and a chiral theory (type IIB superstring). In both cases, the presence of two gravitinos, and the equality of the number ($64 \times 2 = 128$) of bosonic and fermionic degrees of freedom, suggests that both superstring models are compatible with the invariance under space-time supersymmetry transformations associated with $N = 2$ conserved supercharges – as indeed confirmed by a reformulation of the superstring action in a form which is explicitly supersymmetric in the target space-time manifold [1].

Putting together the results of the various sectors, we can summarize the content of the massless (bosonic and fermionic) particles for type II superstrings as follows:

$$\text{type IIA} \longrightarrow \left\{\phi, g_{\mu\nu}, B_{\mu\nu}\right\}_{NS-NS} + \left\{A_\mu, A_{\mu\nu\rho}\right\}_{R-R} + \left\{\chi_-, \chi^\mu_-, \xi_+, \xi^\mu_+\right\}_{NS\leftrightarrow R}, \tag{3B.74}$$

$$\text{type IIB} \longrightarrow \left\{\phi, g_{\mu\nu}, B_{\mu\nu}\right\}_{NS-NS} + \left\{A, A_{\mu\nu}, A_{\mu\nu\rho\sigma}\right\}_{R-R}$$

$$+ \left\{\chi_+, \chi^\mu_+, \xi_+, \xi^\mu_+\right\}_{NS\leftrightarrow R}. \tag{3B.75}$$

We conclude the discussion by reporting the tree-level effective action for the pure bosonic part of these two superstring models, in $D_c = 10$ space-time dimensions and in the S-frame (see for instance [2]). For the type IIA superstring we have the action

$$S_{IIA} = -\frac{1}{2\lambda_s^8} \int d^{10}x\sqrt{-g}\left[e^{-\phi}\left(R + \nabla\phi^2 - \frac{1}{12}H_3^2\right)\right.$$

$$\left. + \frac{1}{4}H_2^2 + \frac{1}{48}\left(H_4 - A_1 \wedge H_3\right)^2\right]$$

$$+ \frac{1}{4\lambda_s^8} \int B_2 \wedge H_4 \wedge H_4, \tag{3B.76}$$

while for the type IIB superstring we have the action

$$S_{IIB} = -\frac{1}{2\lambda_s^8} \int d^{10}x\sqrt{-g}\left[e^{-\phi}\left(R + \nabla\phi^2 - \frac{1}{12}H_3^2\right) - \frac{1}{2}H_1^2 - \frac{1}{12}\left(\overline{H}_3 - A_0 H_3\right)^2\right.$$

$$\tag{3B.77}$$

$$\left. - \frac{1}{240}\left(H_5 - \frac{1}{2}A_2 \wedge H_3 + \frac{1}{2}B_2 \wedge \overline{H}_3\right)^2\right] + \frac{1}{4\lambda_s^8} \int A_4 \wedge H_3 \wedge \overline{H}_3.$$

We have defined the field strengths of various ranks as $H_p = dA_{p-1}$, with the only exception being the case $p = 3$, where $H_3 = dB_2$ is the field strength of the NS–NS two-form, while $\overline{H}_3 = dA_2$ is the field strength of the R–R two-form. Finally, the square of a form simply denotes the tensor scalar product, i.e. $H_p^2 = H_{\mu_1\mu_2\cdots\mu_p}H^{\mu_1\mu_2\cdots\mu_p}$.

A few comments are in order. The action for the NS–NS fields (the first three terms in the square brackets) is the same for both models, and also coincides with the tree-level action obtained from the closed bosonic string, Eq. (3.41), modulo a different number of critical dimensions. The R–R fields (the remaining terms in the square brackets) are associated with antisymmetric tensor potentials A_p of odd rank $(p = 1, 3)$ in type IIA theory, and of even rank $(p = 0, 2, 4)$ in type IIB theory; in both cases, it should be stressed that the R–R fields are *not* directly coupled to the dilaton, in the string frame. The last term of the above actions is a Chern–Simons term, required by space-time supersymmetry; such a term, however, is identically vanishing for the homogeneous and isotropic backgrounds that will be the main object of our cosmological applications in the following chapters. Finally, it is important to note that the field equations obtained from the type IIB action are to be supplemented by the additional constraint $H_5 = {}^\star H_5$, which *does not* follow from the action (3B.77), but which is compatible with it, and required by the self-duality (or chirality) of the R–R gauge field A_4, as stressed before.

3B.5 Type I superstring

The previous models, based on the consistent quantization of closed superstrings, provide an elegant (and very interesting) supersymmetric description of gravitational interactions. It seems hard, however, to connect these models to the phenomenology of the other low-energy fundamental interactions, and to the standard model of elementary particle physics. The main difficulty is probably the explicit absence of Yang–Mills fields and of non-Abelian symmetry groups, related to the standard description of nuclear and subnuclear interactions. A possible solution to this problem, in the context of type II superstrings, is that the gauge groups may arise as isometries of the internal geometry upon reduction from ten to four dimensions, through some appropriate compactification mechanism.

A more direct solution of this phenomenological problem is provided by models based, at least in part, on open superstrings. These strings, in fact, may have "charges" on their ends, associated with a non-Abelian symmetry group which coincides with $U(n)$ for open oriented strings, and with $SO(n)$ or $USp(2n)$ for open non-oriented strings.

We should recall that non-oriented strings are strings with a quantum spectrum of states which is invariant under the so-called "world-sheet parity" transformation P_σ, defined by

$$P_\sigma: \qquad \sigma \to \pi - \sigma = -\sigma \quad (\text{modulo } \pi). \tag{3B.78}$$

Let us consider for simplicity a bosonic string, whose mode expansion for the open and closed solution is given, respectively, in Eqs. (3A.28) and (3A.19). Acting on those solutions with the transformation (3B.78), one can easily derive the transformation rules of α and $\tilde{\alpha}$, so as to determine also the transformation of a bosonic string state with total occupation number given by N. One then finds

$$P_\sigma |N\rangle = (-1)^N |N\rangle \tag{3B.79}$$

for open string states, and

$$P_\sigma |N, \tilde{N}\rangle = |\tilde{N}, N\rangle \tag{3B.80}$$

for closed string states. Non-oriented string models are defined by keeping only those states which are left invariant by P_σ (namely, which are "insensitive" to the world-sheet orientation). So, for the bosonic string, the only states which survive this projection are the open string states with *even* occupation number, and the closed string states which are *symmetric* in the exchange of left- and right-moving modes, $N \leftrightarrow \tilde{N}$.

Non-Abelian gauge fields can be included in the open string model by introducing additional degrees of freedom, or "charges" (called Chan–Paton factors [36]) at the ends of the string, transforming according to the fundamental representation $[n]$ of $U(n)$ at one end, and according to the complex-conjugated representation $[\bar{n}]$ at the other end of the string. The massless vector state of the open string spectrum thus acquires two additional labels, $a, \overline{b}$, and can be written in the form

$$|A^\mu\rangle_{a\overline{b}} = \alpha^\mu_{-1}|0\,a\,\overline{b}\rangle, \tag{3B.81}$$

where a is an index of $[n]$, and $\overline{b}$ of $[\overline{n}]$ (again, for simplicity, we are considering the bosonic string case). The tensor product of the two representations, $[n] \otimes [\overline{n}]$, corresponds to the so-called adjoint representation, so that the massless states A_μ can be interpreted as the gauge boson of the non-Abelian gauge group $U(n)$.

In the case of open non-oriented strings, however, the two ends of the string must be equivalent, so that the allowed states of changed non-oriented strings are those left invariant by the generalized parity transformation $\overline{P}_\sigma$ which inverts σ and, simultaneously, exchanges the indices a and b. Such a transformation can be defined as follows:

$$\overline{P}_\sigma|N\,a\,b\rangle = \epsilon\,(-1)^N|N\,b\,a\rangle, \tag{3B.82}$$

where $\epsilon = \pm 1$. Imposing that the spectrum is invariant under the action of $\overline{P}_\sigma$ then leaves us with two possibilities, depending on the value of ϵ. If $\epsilon = 1$ we find that the vector state (3B.81) (corresponding to $N = 1$) transforms as the gauge boson of the group $SO(n)$, whose adjoint representation is antisymmetric with respect to the exchange of the indices. If $\epsilon = -1$ we find, instead, that the vector state transforms as the gauge boson of the symplectic group $USp(2n)$, whose adjoint representation is symmetric. The same discussion and results apply to the superstring case, with the only difference that the sign of ϵ has to be inverted with respect to the bosonic case [1].

Considering open strings we can thus include non-Abelian gauge fields at the level of the 10-dimensional model, without resorting to the dimensional reduction mechanism. A consistent quantum model, however, cannot contain only open strings (after all, open strings must be allowed to join ends, giving rise to closed strings), but may contain either closed strings only, or both open and closed strings. Open and closed *oriented* strings, however, cannot be combined to form a supersymmetric theory, because closed oriented strings contain two gravitinos in their spectrum and, as already stressed, provide an explicit realization of $N = 2$ space-time supersymmetry, while open strings are only compatible with $N = 1$ supersymmetry.

In order to combine open and closed strings in a consistent supersymmetric way we must "project out" one of the two gravitinos of the closed superstring spectrum, so as to reduce down to $N = 1$ the degree of supersymmetry of the system. The required projection turns out to be equivalent to imposing on the states the property of world-sheet parity, thus producing a model of *closed non-oriented* superstrings. We can use the type IIB model of superstring, but not the type IIA superstring in which the left and right modes of the spectrum have opposite chirality, and there are no states symmetric in their exchange. On the other hand, the model of closed non-oriented type IIB superstring at the quantum level is affected by divergences and anomalies which can only be canceled by adding to the model open strings with gauge group $SO(32)$ [37, 38] (and then open non-oriented superstrings, according to our previous discussion of the Chan–Paton factors). We are eventually led, in this way, to a quantum-mechanically consistent theory of non-oriented open and closed strings, with $N = 1$ supersymmetry and $SO(32)$ gauge invariance, which is known as the type I superstring.

To extract the particle content of the theory let us first consider the type IIB spectrum, and note that the projection over non-oriented states imposes the symmetry in the exchange of left- and right-moving quantities, $b_r^\mu \leftrightarrow \tilde{b}_r^\mu$, $|0_R\rangle \leftrightarrow |\tilde{0}_R\rangle$, according to Eq. (3B.80). In the NS–NS sector, the symmetrization of the massless level (3B.62) leads to the state

$$\text{NS–NS}: \qquad \frac{1}{2}\left(b_{-1/2}^i \tilde{b}_{-1/2}^j + \tilde{b}_{-1/2}^i b_{-1/2}^j\right)|0_{\text{NS}}\,\tilde{0}_{\text{NS}}\rangle. \qquad (3B.83)$$

This clearly shows that the antisymmetric tensor component disappears from the spectrum, and that we are left with the graviton and the scalar dilaton only, with a total of $35 + 1 = 36$ bosonic degrees of freedom.

In the R–R sector, taking into account the change of sign due to the exchange of two fermionic states, we are led to the following massless combination:

$$\text{R–R}: \qquad \frac{1}{2}\left(|0_R^+ \tilde{0}_R^+\rangle - |\tilde{0}_R^+ 0_R^+\rangle\right). \qquad (3B.84)$$

Among the different terms associated with the decomposition of the spinor representation $8_+ \otimes \tilde{8}_+$ (see Eq. (3B.64)), we have to select those with an even number of tensor indices (because of the type IIB model), and those which are antisymmetric in the exchange $\theta \leftrightarrow \tilde{\theta}$ (because of the world-sheet parity). We are only left with a two-index antisymmetric tensor, which means that only the two-form $A_{\mu\nu}$ survives the projection, for a total of 28 independent bosonic components.

Finally, the symmetrization of the NS–R sector, Eq. (3B.72), and R–NS sector, Eq. (3B.73), leads to the same (non-oriented) massless level, associated with the state

$$\text{NS}\leftrightarrow\text{R}: \qquad \frac{1}{2}\left(b_{-1/2}^i|0_{\text{NS}}\,\tilde{0}_R^+\rangle + \tilde{b}_{-1/2}^i|0_R^+ 0_{\text{NS}}\rangle\right). \qquad (3B.85)$$

We are left with one gravitino χ_+^μ and one dilatino χ_+ only, for a total of 64 fermionic degrees of freedom, equal to the $36 + 28 = 64$ bosonic degrees of freedom of the other sectors (as required by space-time supersymmetry).

In addition to these states, belonging to the closed superstring spectrum, the model also contains the massless levels of the open non-oriented superstring, with gauge group $SO(32)$. In the NS sector, Eq. (3B.49), the massless level with $N = 0$, $N_{\text{NS}} = 1/2$ corresponds to the state $b_{-1/2}^i|0_{\text{NS}}\,ab\rangle$, associated with the gauge vector A_μ^a of the group $SO(32)$. In the R sector, Eq. (3B.48), the massless level $|0_R\,a\rangle$ corresponds to the eigenvalues $N = 0 = N_R$, and is associated with the massless spinor multiplet ψ^a (also called "gaugino"), whose combination with A_μ^a forms the vector supermultiplet of the gauge group $SO(32)$.

In conclusion, we can summarize the particle content of the massless levels of type I superstring as follows:

$$\text{type I} \longrightarrow \ \{\phi, g_{\mu\nu}\}_{\text{NS–NS}} + \{A_{\mu\nu}\}_{\text{R–R}} + \{\chi_+, \chi_+^\mu\}_{\text{NS}\leftrightarrow\text{R}} + \{A_\mu^a, \psi^a\}_{SO(32)},$$

$$(3B.86)$$

where the first three contributions arise from the closed, non-oriented spectrum, while the last contribution is from the open, non-oriented superstring spectrum. The corresponding tree-level, $N = 1$ supersymmetric action for the bosonic fields, in the S-frame, can be written as [2]

$$S_I = -\frac{1}{2\lambda_s^8} \int d^{10}x \sqrt{-g} \left[e^{-\phi}\left(R + \nabla\phi^2\right) - \frac{1}{12}\overline{H}_3^2 + \frac{1}{4}e^{-\phi/2}\,\text{Tr}\,F_2^2 \right], \qquad (3B.87)$$

where F_2 is the matrix valued two-form representing the Yang–Mills field for the gauge group $SO(32)$, and the trace refers to the vector representation of $SO(32)$. Finally,

$$\overline{H}_3 = \mathrm{d}A_2 - \mathrm{Tr}\left(A_1 \wedge \mathrm{d}A_1 + \frac{2}{3}A_1 \wedge A_1 \wedge A_1\right) \tag{3B.88}$$

represents the mixed contribution of the R–R two-form A_2 and of the matrix valued one-form A_1, associated with the gauge potential of the $SO(32)$ group.

3B.6 Heterotic superstrings

A different possibility for including non-Abelian gauge interactions in string theory is provided by the so-called "heterotic" model [39, 40], based on closed oriented strings in which only one half of the physical degrees of freedom (for instance, those associated with right-moving modes) is supersymmetrized, while the other half keeps its bosonic properties and is quantized without fermionic partners. This procedure is allowed because, in closed string theory, left- and right-moving modes are decoupled and can be treated independently.

The bosonic sector, on the other hand, can be consistently quantized only in the presence of 26 space-time dimensions (as discussed in the previous appendix), while the quantization of the supersymmetric sector has to be performed in a 10-dimensional subspace (with the appropriate GSO projection, and all the required constraints). The extra $26 - 10 = 16$ spatial dimensions are then to be compactified, or identified periodically, by setting $X^\alpha = X^\alpha + w_{(i)}^\alpha$, $\alpha = 11, 12, \dots, 26$, where the 16 vectors $w_{(i)}$ generate a 16-dimensional lattice. We obtain, in this way, a 10-dimensional theory with $N = 1$ supersymmetry and a non-Abelian gauge group, associated with the symmetries of the 16 extra spatial dimensions. The consistent quantization of the left-moving modes, as we shall see, is only compatible with two groups, $SO(32)$ and $E_8 \times E_8$: as a consequence, we end up with only two possible models of superstring, heterotic $SO(32)$ and heterotic $E_8 \times E_8$.

For an explicit formulation of the model through a world-sheet action we can separate the 10-dimensional supersymmetric part from the part referring to the extra bosonic coordinates, containing only left-moving modes. Instead of using a different number of dimensions for the different sectors of the model, however, it is convenient to notice that 16 free left-moving bosons (the coordinates X_+^α, $\alpha = 11, 12, \dots, 26$) are equivalent, for what concerns quantum anomalies in a two-dimensional world-sheet [1], to 32 left-moving Majorana–Weyl spinors λ_+^A, $A = 1, 2, \dots, 32$. We can thus write the action in the form

$$S = \frac{1}{4\pi\alpha'} \int \mathrm{d}\sigma\,\mathrm{d}\tau \left[\sum_{\mu=0}^{9}\left(\partial_a X^\mu \partial^a X_\mu + 2\mathrm{i}\psi_-^\mu \partial_+ \psi_{-\mu}\right) + 2\mathrm{i}\sum_{A=1}^{32} \lambda_+^A \partial_- \lambda_+^A\right], \tag{3B.89}$$

where the $(-)$ and $(+)$ subscripts denote right- and left-moving modes, respectively. This action has a global world-sheet supersymmetry mixing the 10 right-moving fermions ψ_-^μ and the right-moving part X_-^μ of the 10 bosonic coordinates, like the action (3B.9).

The left-moving (non-supersymmetric) part of the model corresponds to the 10 bosonic coordinates X_+^μ and to the 32 spinors λ_+^A. If such spinors are all quantized with the same boundary conditions we have a model with an evident $SO(32)$ symmetry. The only other consistent quantization, as we shall see, is the case in which the spinors are separated into two independent groups of 16, and we consider periodic and antiperiodic boundary conditions for each group: in that case one is led to the symmetry group $E_8 \times E_8$.

The discussion of this appendix is concentrated on the case of $SO(32)$ symmetry, and on the corresponding particle content of the massless sector of the spectrum. We start by noting that the spinors λ_+^A can be quantized using periodic (P) or antiperiodic (A) boundary conditions, by applying, respectively, the procedure of the R and NS sectors of the superstring spectrum. In the periodic case we shall expand the (closed string) solutions as

$$\text{P}: \qquad \lambda_+^A = \sum_{n=-\infty}^{\infty} \lambda_n^A e^{-2in(\tau+\sigma)}, \qquad (3\text{B}.90)$$

where the sum is over all integers (see Eq. (3B.28)); in the antiperiodic case we shall expand the solutions as

$$\text{A}: \qquad \lambda_+^A = \sum_{r=-\infty}^{\infty} \lambda_r^A e^{-2ir(\tau+\sigma)}, \qquad (3\text{B}.91)$$

where the sum is over all half-integers $r = n + 1/2$ (see Eq. (3B.29)). In the quantum version of the model the Fourier coefficients λ_n^A, λ_r^A become operators satisfying the canonical anticommutation relations

$$\{\lambda_m^A, \lambda_n^B\} = \delta^{AB}\delta_{m+n,0}, \qquad \{\lambda_r^A, \lambda_s^B\} = \delta^{AB}\delta_{r+s,0}. \qquad (3\text{B}.92)$$

The heterotic spectrum can then be easily determined by exploiting the previous computations of the ordered and regularized version of the right-moving and left-moving operators, L_0, $\widetilde{L}_0$, for both the bosonic and the fermionic part of the spectrum.

Let us first consider the right-mode sector, supersymmetrized by the GSO projection. Summing up the bosonic part (relative to X_-^μ) and the fermionic part (relative to ψ_-^μ), the regularized Virasoro operator can be written as $L_0 = (\alpha'/4)p^2 - N$, where N is a number operator with integer eigenvalues which takes into account all the states left in the spectrum after the projection. Then, we have to consider the left-mode sector: we decompose the operator as $\widetilde{L}_0 = \widetilde{L}_0^X + \widetilde{L}_0^\lambda$, where $\widetilde{L}_0^X$ is the bosonic part (relative to X_+^μ) and $\widetilde{L}_0^\lambda$ is the fermionic part (relative to λ_+^A). For the bosonic part we can always apply the regularized result (3A.65), which can be written as

$$\widetilde{L}_0^X = \frac{\alpha'}{4}p^2 - \widetilde{N}_X + \frac{8}{24} \qquad (3\text{B}.93)$$

(we have set $D - 2 = 8$ because the X_+^μ coordinates are defined on the 10-dimensional space-time of the superstring model). For the fermionic part we must separately consider the case in which we impose periodic or antiperiodic boundary conditions.

In the periodic case, using the anticommutation relations of λ_m, λ_n, and repeating exactly the same computations as those leading to the result (3B.44) for the R superstring spectrum, we obtain

$$\widetilde{L}_0^P = -\widetilde{N}_P - \frac{32}{24}, \qquad (3\text{B}.94)$$

where $\widetilde{N}_P = \sum_{n=1}^{\infty} n\lambda_{-n}^A \lambda_n^A$, and where the index P denotes the periodic sector of the spectrum. The different numerical factor, 32 instead of $D - 2$, is due to the sum of the indices A, B of the 32 spinor variables λ, which appears by taking the trace of their anticommutation relation. In the same way, for antiperiodic boundary conditions, we can repeat the computations applied to the NS superstring sector (see Eq. (3B.47)) to obtain

$$\tilde{L}_0^A = -\tilde{N}_A + \frac{32}{48}, \tag{3B.95}$$

where the index A denotes the antiperiodic sector, and $\tilde{N}_A = \sum_{r=1/2}^{\infty} r\lambda_{-r}^A \lambda_r^A$.

The sum of the bosonic and fermionic parts of the left-moving spectrum now gives $\tilde{L}_0$, where we define $\tilde{N}$ as the total occupation number of all fermionic and bosonic left-moving modes. Using also the right-moving operator L_0, and imposing the conditions $L_0 + \tilde{L}_0 = 0$, $L_0 - \tilde{L}_0 = 0$, we then obtain, respectively, the mass-shell condition and the generalized level-matching conditions. For the periodic sector this leads to the spectrum

$$P: \qquad \alpha' M^2 = 2(N + \tilde{N} + 1),$$

$$N = \tilde{N} + 1; \tag{3B.96}$$

for the antiperiodic sector we have, instead,

$$A: \qquad \alpha' M^2 = 2(N + \tilde{N} - 1),$$

$$N = \tilde{N} - 1. \tag{3B.97}$$

Two comments are now in order. The first is that, since the minimal eigenvalue of the number operators is zero, the massless states will belong to the antiperiodic sector of the spectrum. The second is that the GSO projection, which is imposed within the supersymmetric (right-moving) sector of the spectrum, and which leaves the operator N with a set of integer eigenvalues, implies the necessity of a GSO-like projection also for the P and A left-moving sectors. One has indeed to eliminate all states with half-integer values of $\tilde{N}$, which would be in contrast with the conditions $N = \tilde{N} \pm 1$.

Before moving to the analysis of the massless levels of the spectrum, it seems appropriate to discuss the possible existence of symmetries other than $SO(32)$ in the 32-dimensional space spanned by the fermionic coordinates λ_+^A. Let us suppose that we break the $SO(32)$ symmetry of the previous example by separating the fermions into two groups of n and $32 - n$ components, respectively. Imposing periodic and antiperiodic boundary conditions separately, and independently, on the two groups, we may then obtain, for the left-moving modes, four possible different sectors that we denote as $P_n P_{32-n}$, $P_n A_{32-n}$, $A_n P_{32-n}$, $A_n A_{32-n}$, where P_n means that the group of n fermions is assigned periodic boundary conditions, A_{32-n} that the group of $32 - n$ fermions is assigned antiperiodic boundary conditions, and so on. Repeating the previous computations, and summing the bosonic and fermionic parts of the (normal-ordered, regularized) operator $\tilde{L}_0$, for the different sectors, we obtain the following possible results:

$$P_n P_{32-n}: \qquad \tilde{L}_0 = \frac{\alpha'}{4} p^2 - \tilde{N} - 1, \tag{3B.98}$$

$$P_n A_{32-n}: \qquad \tilde{L}_0 = \frac{\alpha'}{4} p^2 - \tilde{N} + 1 - \frac{n}{16}, \tag{3B.99}$$

$$A_n P_{32-n}: \qquad \tilde{L}_0 = \frac{\alpha'}{4} p^2 - \tilde{N} - 1 + \frac{n}{16}, \tag{3B.100}$$

$$A_n A_{32-n}: \qquad \tilde{L}_0 = \frac{\alpha'}{4} p^2 - \tilde{N} + 1, \tag{3B.101}$$

We have to use, at this point, the Virasoro operator of the right-moving (supersymmetric) sector, $L_0 = (\alpha'/4)p^2 - N$, and to impose the level-matching condition $L_0 = \widetilde{L}_0$. We obtain a relation between N and $\widetilde{N}$ which depends on the considered sector, and which can be written as

$$N = \widetilde{N} \pm 1, \qquad N = \widetilde{N} \pm \left(1 - \frac{n}{16}\right), \tag{3B.102}$$

where the first condition refers to the PP and AA sectors, and the second one to the AP and PA sectors. We have to take into account, also, that N has only integer eigenvalues (because of the GSO supersymmetrization), while $\widetilde{N}$ takes integer values in the P sector, integer and half-integer values in the A sector (remember the definitions of $\widetilde{N}_P$ and $\widetilde{N}_A$, Eqs. (3B.94) and (3B.95)). It follows that there are two possibilities to satisfy the conditions (3B.102).

(1) If n is not divisible by eight, then there are no physical states in the AP and PA sectors (the unphysical ones are to be eliminated by a suitable projection). In this case we are back to the situation in which all 32 fermions obey the same boundary conditions, and the symmetry group is $SO(32)$, as before (actually, $SO(32)/Z_2$ because of the two classes of periodic and antiperiodic conditions).

(2) If n is divisible by eight, then we have the possible values $n = 0, 8, 16, 24, 32$. The cases $n = 0, 32$ correspond again to the full $SO(32)$ symmetry. The cases $n = 8, 24$ lead to models affected by quantum anomalies [1], and will not be considered further. We are left with $n = 16$, which corresponds to the subdivision of the 32 spinor variables λ^A into two equal groups, and which implies an integer shift between the left- and right-moving number operators, i.e. $N - \widetilde{N} = 0, \pm 1$. By performing on this spectrum the projections required by a consistent quantization, one then finds that the associated symmetry group is $E_8 \times E_8$ [1].

Coming back to the heterotic model with symmetry group $SO(32)$, let us finally analyze the field content of the massless states which, as already stressed, belong to the antiperiodic sector of the spectrum (Eq. (3B.97)), and are characterized by the eigenvalues $N = 0$, $\widetilde{N} = 1$. Let us first consider, separately, the left- and right-moving contributions.

In the left-moving sector we must recall the definition of the number operator $\widetilde{N}$ for antiperiodic boundary conditions,

$$\widetilde{N} = \widetilde{N}_X + \sum_r r\lambda^A_{-r}\lambda^A_{-r} = \sum_{n=1}^{\infty} \alpha^i_{-n}\alpha^i_n + \sum_{r=1/2}^{\infty} r\lambda^A_{-r}\lambda^A_{-r} \tag{3B.103}$$

(see Eq. (3B.95)). One then immediately finds that there are two left-moving states with $\widetilde{N} = 1$:

$$\left\{ \alpha^i_{-1}|0_+\rangle, \ \lambda^A_{-1/2}\lambda^B_{-1/2}|0_+\rangle \right\}. \tag{3B.104}$$

The first is an $SO(32)$ scalar, which transforms as a transverse vector under the action of the Lorentz group in 10-dimensional Minkowski space; the second is a Lorentz singlet which transforms as an antisymmetric, second-rank tensor (i.e. according to the adjoint representation) of the gauge group $SO(32)$.

In the right-moving (supersymmetric) sector we find that there are two states with $N = 0$ (see e.g. the superstring spectrum (3B.48), (3B.49)), and which contribute to the massless heterotic spectrum:

$$\left\{ b^i_{-1/2}|0_-\rangle_{NS}, \ |0_-\rangle_R \right\}. \tag{3B.105}$$

The first is a transverse vector of the Lorentz group, with eight independent components, the second is an eight-component, right-moving spinor: their combination $8_V \oplus 8_-$ represents the vector supermultiplet, already encountered in the open superstring spectrum.

The complete set of states of the massless level of the heterotic string can now be obtained as the tensor product of the left- and right-moving massless sectors (3B.104) and (3B.105), as typical of all closed string models. The product of the two Lorentz-vector states gives a tensor state with $8 \times 8 = 64$ components which can be decomposed, as usual, into a symmetric traceless part (the graviton), an antisymmetric part (the torsion) and a scalar part (the dilaton):

$$\alpha^i_{-1}|0_+\rangle \otimes b^j_{-1/2}|0_-\rangle_{\text{NS}} \longrightarrow \{\phi, g_{\mu\nu}, B_{\mu\nu}\}. \qquad (3\text{B}.106)$$

Similarly, the product of the left-moving vector with the right-moving fermion reproduces the vector-spinor representation with 64 independent components, which can be decomposed into a 56-component gravitino field χ^i (satisfying the condition $\gamma_i \chi^i = 0$) and the 8-component spinor χ:

$$\alpha^i_{-1}|0_+\rangle \otimes |0_-\rangle_{\text{R}} \longrightarrow \{\chi, \chi^\mu\}. \qquad (3\text{B}.107)$$

Finally, the tensor product of the 16 components of the vector supermultiplet (3B.105), with the $32 \times 31/2 = 496$ components associated with the antisymmetric tensor

$$\lambda^A_{-1/2} \lambda^B_{-1/2}|0_+\rangle, \qquad (3\text{B}.108)$$

gives the super Yang–Mills multiplet for the gauge group $SO(32)$:

$$\lambda^A_{-1/2} \lambda^B_{-1/2}|0_+\rangle \otimes \left(b^i_{-1/2}|0_-\rangle_{\text{NS}} \oplus |0_-\rangle_{\text{R}}\right) \longrightarrow \{A_\mu, \psi\} \qquad (3\text{B}.109)$$

(we have suppressed, for simplicity, the group indices). The total number of components is 16×496, equally distributed between bosonic and fermionic degrees of freedom.

Comparing these results with Eq. (3B.86) we can notice, at this point, that the field content of the massless levels of the heterotic string is exactly the same as that of the type I superstring: a graviton, a dilaton, a two-form, a gravitino, a dilatino, a gauge boson and a gaugino. The massive levels, however, have a different particle content. But even at the massless level there are important differences, due to the coupling of the massless fields to the dilaton. In particular, the two-form $A_{\mu\nu}$ of type I superstrings arises from the R–R sector of the spectrum, and is uncoupled from the dilaton in the string frame (see the action (3B.87)). In the heterotic model, on the contrary, the two-form $B_{\mu\nu}$ belongs to the gravitational multiplet (3B.106), and couples to the dilaton exactly as in the type II string (or bosonic string) case. The coupling to the dilaton of the gauge field strength is also different, in the two cases.

Indeed, the tree-level, S-frame heterotic effective action can be written as [2]

$$S_{\text{het}} = -\frac{1}{2\lambda_s^8} \int d^{10}x \sqrt{-g}\, e^{-\phi} \left(R + \nabla\phi^2 - \frac{1}{12}\overline{H}_3^2 + \frac{1}{4}\text{Tr}\,F_2^2\right), \qquad (3\text{B}.110)$$

where F_2 is the matrix valued two-form for the gauge field strength ($SO(32)$ or $E_8 \times E_8$), and where

$$\overline{H}_3 = dB_2 - \text{Tr}\left(A_1 \wedge dA_1 + \frac{2}{3}A_1 \wedge A_1 \wedge A_1\right). \qquad (3\text{B}.111)$$

This action is different from the type I action (3B.87). However, it is interesting to point out that the heterotic action can be obtained from Eq. (3B.87) by performing the rescaling

$$g_{\mu\nu} = \widetilde{g}_{\mu\nu} e^{4\phi/(D-2)} \equiv \widetilde{g}_{\mu\nu} e^{\phi/2}, \qquad \phi = -\widetilde{\phi}. \qquad (3B.112)$$

In fact, applying the general transformation rules (2.40) and (2.41) presented in Chapter 2, one can easily check that, in terms of the "tilded" variables $\widetilde{g}$, $\widetilde{\phi}$, the type I action exactly reproduces the heterotic action (3B.110). The transformation $\phi \rightarrow -\phi$, in particular, inverts the coupling parameter $g_s = \exp(\phi/2)$, and suggests that the strong coupling limit of one theory should be appropriately described by the other theory, in the perturbative regime.

The superstring effective actions presented in this appendix, type I, type IIA, type IIB, heterotic $SO(32)$ and heterotic $E_8 \times E_8$, will be used in the following chapters for various cosmological applications. In particular, the heterotic model with $E_8 \times E_8$ gauge symmetry, with six "internal" dimensions appropriately compactified on a complex (three-dimensional) Calabi–Yau manifold (see for instance [41])), seems to represent a promising model for a realistic description of the standard low-energy phenomenology.

These five superstring theories, however, are not independent, being connected by the so-called "duality" transformations: in particular, target space duality, or T-duality (see Chapter 4), acting on the background geometry, and S-duality, acting on the string coupling constant parametrized by the dilaton (see for instance [2]). In addition, all these five models seem to represent the weak coupling limit of a more fundamental, 11-dimensional theory, called M-theory [42, 43].

The low-energy limit of this fundamental theory is believed to be represented by the 11-dimensional, $N = 1$ supergravity theory [44], which has the largest space-time dimensionality allowed in a supersymmetric theory of gravity. Its particle content includes, in 11 dimensions, the graviton $g_{\mu\nu}$ (with $(D^2 - 3D)/2 = 44$ degrees of freedom), an antisymmetric three-form potential $A_{\mu\nu\rho}$ (with $(D-4)(D-3)(D-2)/3! = 84$ degrees of freedom) and a gravitino ψ^μ (a Majorana spinor with 128 degrees of freedom). The number 128 is obtained by multiplying the vector degrees of freedom, $D - 2 = 9$, by the number of components of a Dirac–Majorana spinor, which in an odd number D of dimensions is given by $2^{(D-1)/2}/2 = 16$ (the overall $1/2$ factor is given to the Majorana condition). From the product $9 \times 16 = 144$ we must subtract, however, the 16 conditions $\gamma_\mu \psi^\mu = 0$ (following from the local gauge invariance of the gravitino action), and we are left with $144 - 16 = 128$ independent degrees of freedom. The bosonic sector of this supergravity theory is described by the action

$$S_{11} = \frac{1}{16\pi G_{11}} \left[\int d^{11}x \sqrt{|g|} \left(R - \frac{1}{48} F_4^2 \right) + \frac{1}{6} \int A_3 \wedge F_4 \wedge F_4 \right], \qquad (3B.113)$$

where $F_4 = dA_3$, and where, as in the previous models, the Chern–Simons term arises as a direct consequence of the space-time supersymmetry.

One can show, in particular, that the compactification of this theory on a circle exactly leads to the type IIA superstring action, and that the 10-dimensional dilaton field arises as the dynamical scale factor of the eleventh dimension, $\phi \propto \ln a_{11}$. It is thus clear that, in such a context, the weak coupling limit $\phi \rightarrow -\infty$ corresponds to a_{11} shrinking to zero, and is thus associated with the dynamical dimensional reduction from 11 to 10 dimensions, and to the appearance of the 10-dimensional superstring models playing the role of perturbative approximations of a more fundamental theory. We refer the reader to the existing literature for a detailed discussion of the interesting relations connecting the five 10-dimensional superstring models and the 11-dimensional supergravity theory.

References

[1] M. B. Green, J. Schwartz and E. Witten, *Superstring Theory* (Cambridge: Cambridge University Press, 1987).

[2] J. Polchinski, *String Theory* (Cambridge: Cambridge University Press, 1998).

[3] Y. Nambu, in *Proceedings of the International Conference on Symmetries and Quark Models* (New York: Gordon and Breach, 1970).

[4] L. Brink, P. Di Vecchia and P. Howe, *Phys. Lett.* **B65** (1976) 471.

[5] A. M. Polyakov, *Phys. Lett.* **B103** (1981) 207.

[6] R. H. Brandenberger and C. Vafa, *Nucl. Phys.* **B316** (1989) 391.

[7] A. A. Tseytlin and C. Vafa, *Nucl. Phys.* **B372** (1992) 443.

[8] M. Gasperini, N. Sanchez and G. Veneziano, *Int. J. Mod. Phys.* **A6** (1991) 3853.

[9] M. Gasperini, N. Sanchez and G. Veneziano, *Nucl. Phys.* **B364** (1991) 365.

[10] C. G. Callan, D. Friedan, E. J. Martinec and M. J. Perry, *Nucl. Phys.* **B262** (1985) 593.

[11] C. W. Misner, K. S. Thorne and J. A. Wheeler, *Gravitation* (San Francisco: W. H. Freeman and Co., 1973).

[12] A. Sen, *Phys. Rev. Lett.* **55** (1985) 1846.

[13] A. Sen, *Phys. Rev.* **D32** (1985) 2102.

[14] J. Wess and B. Zumino, *Phys. Lett.* **B37** (1971) 95.

[15] E. Witten, *Commun. Math. Phys.* **92** (1984) 455.

[16] E. S. Fradkin and A. A. Tseytlin, *Phys. Lett.* **B160** (1985) 69.

[17] R. R. Metsaev and A. A. Tseytlin, *Nucl. Phys.* **B293** (1987) 385.

[18] K. A. Meissner, *Phys. Lett.* **B392** (1997) 298.

[19] N. Kaloper and K. A. Meissner, *Phys. Rev.* **D56** (1997) 7940.

[20] E. Kiritsis and C. Kounnas, *Phys. Lett.* **B331** (1994) 51.

[21] H. J. de Vega, A. L. Larsen and N. Sanchez, *Phys. Rev.* **D61** (2000) 066003.

[22] P. Binétruy and M. K. Gaillard, *Phys. Rev.* **D34** (1986) 3069.

[23] T. R. Taylor and G. Veneziano, *Phys. Lett.* **B213** (1988) 459.

[24] N. Kaloper and K. A. Olive, *Astropart. Phys.* **1** (1993) 185.

[25] S. Kachru, R. Kallosh, A. Linde and S. P. Trivedi, *Phys. Rev.* **D68** (2003) 046005.

[26] V. Kaplunovsky, *Phys. Rev. Lett.* **55** (1985) 1036.

[27] M. Gasperini, *Phys. Rev.* **D64** (2001) 043510.

[28] G. Veneziano, *JHEP* **0206** (2002) 051.

[29] M. Gasperini, F. Piazza and G. Veneziano, *Phys. Rev.* **D65** (2002) 023508.

[30] M. A. Virasoro, *Phys. Rev.* **D1** (1970) 2933.

[31] P. van Nieuwenhuizen, *Phys. Rep.* **68** (1981) 189.

[32] P. Ramond, *Phys. Rev.* **D3** (1971) 2415.

[33] A. Neveu and J. H. Schwarz, *Nucl. Phys.* **B31** (1971) 86.

[34] F. Gliozzi, J. Scherk and D. Olive, *Phys. Lett.* **B65** (1976) 282.

[35] F. Gliozzi, J. Scherk and D. Olive, *Nucl. Phys.* **B122** (1977) 253.

[36] H. M. Chan and J. E. Paton, *Nucl. Phys.* **B10** (1969) 516.

[37] M. B. Green and J. H. Schwarz, *Phys. Lett.* **B149** (1984) 117.

[38] M. B. Green and J. H. Schwarz, *Nucl. Phys.* **B255** (1985) 93.

[39] D. J. Gross, J. Harvey, E. J. Martinec and R. Rohm, *Phys. Rev. Lett.* **54** (1985) 502.

[40] D. J. Gross, J. Harvey, E. J. Martinec and R. Rohm, *Nucl. Phys.* **B256** (1985) 253.

[41] B. R. Greene, in *Boulder 1996, TASI Lectures on Fields, String and Dualities*, eds. C. Efthimiou and B. Greene (Singapore: World Scientific, 1997), p. 543.

[42] E. Witten, *Nucl. Phys.* **B443** (1995) 85.

[43] P. Townsend, *Phys. Lett.* **B350** (1995) 184.

[44] E. Cremmer, B. Julia and J. Scherk, *Phys. Lett.* **B76** (1978) 409.

4

Duality symmetries and cosmological solutions

The lowest-order string-gravity equations introduced in the previous chapters will be applied in this chapter to the case of homogeneous cosmological backgrounds, and will be shown to be invariant under an important class of transformations associated with the so-called "duality" symmetries of the string effective action. These symmetries cannot be implemented in the context of the standard general-relativity equations, as they require the presence of the full massless multiplet of states of the closed bosonic string spectrum (the metric, the dilaton and the antisymmetric tensor field), coupled exactly as predicted by the string effective action (see Chapter 3). It will be shown that, by exploiting these symmetries, it is possible to obtain new cosmological solutions starting from known configurations, typical of the standard scenario. These new solutions may suggest possible scenarios for the primordial evolution of our Universe.

It should be recalled that the above-mentioned cosmological symmetries represent the extension to time-dependent backgrounds of the so-called "target space duality" (or T-duality) symmetry, present in the spectrum of a closed bosonic string propagating in a manifold in which some spatial dimensions are compact [1, 2]. It is well known, in fact, that the spatial periodicity along such directions (topologically equivalent to a circle or, more generally, to a higher-dimensional torus), implies the quantization of the conjugate momenta, $p \rightarrow p_n = n/R$, where n is an integer and R is the (constant) radius of the compact dimensions. A closed string, however, can also "wrap" an arbitrary number m of times around the compact dimensions (in that case, the integer m is called the "winding number"). Taking into account both momentum quantization and winding, the solution of the equations of motion for a closed string along a compact direction Y can then be written as

$$Y = y_0 + 2\alpha' \frac{n}{R} \tau + 2mR\sigma + i\sqrt{\frac{\alpha'}{2}} \sum_{n \neq 0} \frac{1}{n} \left(\alpha_n^\mu e^{2in\sigma} + \widetilde{\alpha}_n^\mu e^{-2in\sigma} \right) e^{-2in\tau}. \quad (4.1)$$

This is still a solution of the flat-space wave equation (3A.14), but differs from the solution (3A.19), valid in Minkowski space, because of the term with linear dependence on sigma, describing the wrapping of the string around Y. Separating left- and right-moving modes, and defining two new variables,

$$p_L = \frac{1}{2}\left(\frac{n}{R} + m\frac{R}{\alpha'}\right),$$

$$p_R = \frac{1}{2}\left(\frac{n}{R} - m\frac{R}{\alpha'}\right), \tag{4.2}$$

the above solution can also be rewritten as

$$Y = y_0 + 2\alpha' p_L(\tau + \sigma) + 2\alpha' p_R(\tau - \sigma) + \text{oscillations}. \tag{4.3}$$

If we interpret the momenta p_L, p_R as "internal" degrees of freedom, and refer the mass-shell condition $M^2 = p^\mu p_\mu$ to the translational motion of the center of mass of the string along the "external" non-compact directions (see Appendix 3A), we can then rewrite the regularized Virasoro constraints (in $D = 26$ dimensions) by separating p_L, p_R from the external momenta p_μ. It follows that Eq. (3A.65) is generalized as

$$L_0 = \frac{\alpha'}{4}M^2 - \alpha' p_R^2 - N + 1 = 0,$$

$$\tilde{L}_0 = \frac{\alpha'}{4}M^2 - \alpha' p_L^2 - \tilde{N} + 1 = 0. \tag{4.4}$$

From the sum and the difference of these conditions we are led, respectively, to a generalized form of mass-shell condition,

$$\frac{\alpha'}{2}M^2 = \frac{\alpha'}{2}\left(\frac{n^2}{R^2} + m^2\frac{R^2}{\alpha'^2}\right) + N + \tilde{N} - 2, \tag{4.5}$$

and level-matching condition,

$$N - \tilde{N} + nm = 0. \tag{4.6}$$

These new energy levels differ from the Minkowski spectrum of Eq. (3A.66) as they depend on the "internal" quantum numbers n, m, and are clearly invariant with respect to the so-called T-duality transformation

$$R \leftrightarrow \frac{\alpha'}{R}, \qquad n \leftrightarrow m \tag{4.7}$$

(according to the definitions (4.2), such a transformation can also be represented as a reflection of the right-moving momenta, $p_R \leftrightarrow -p_R$, $p_L \leftrightarrow p_L$). Interestingly enough, the invariance under the transformation (4.7) suggests the existence of an intrinsic "indistinguishability" (for the string) between geometric configurations of "small radius" and "large radius", in units α': in other words, it suggests the actual existence of a "minimal" length scale $R = \sqrt{\alpha'}$, below which the same physics as that of scales above $\sqrt{\alpha'}$ is reproduced.

In a cosmological context, where the target space geometry becomes time dependent, the above invariance property of the closed string spectrum is reflected by the invariance property of the background field equations, which in the simplest case corresponds to the so-called "scale-factor duality" symmetry [3, 4] that will be discussed in Section 4.1. The possible extension of such a symmetry to include the presence of hydrodynamical matter, antisymmetric tensor backgrounds and an appropriate (non-local) dilaton potential will be illustrated in Sections 4.2, 4.3 and Appendix 4A, respectively. We will also present and discuss various examples of regular and non-regular cosmological solutions, together with their duality properties, in Appendix 4B.

Throughout this chapter we will work in the context of the low curvature, weak coupling regime described by the lowest order, S-frame effective action (2.21), possibly supplemented by the contribution of fluid-matter sources. The corresponding equations for the metric, the dilaton and the antisymmetric field are given by Eqs. (2.24), (2.25) and (2.28), respectively.

4.1 Scale-factor duality and the pre-big bang scenario

Let us consider a $(d + 1)$-dimensional space-time manifold, homogeneous but anisotropic, spatially flat, described by a diagonal metric of the Bianchi-I type. We start with the study of the gravi-dilaton system, setting $B_{\mu\nu} = 0$, but including the possible contribution of (anisotropic) perfect fluid sources: they have no viscosity and friction terms in their stress tensor, but may have different pressure components p_i along the different spatial directions. In the synchronous gauge of the comoving system of coordinates (see Section 1.1) we can set, therefore,

$$
\begin{aligned}
g_{\mu\nu} &= \text{diag}(1, -a_i^2 \delta_{ij}), & a_i &= a_i(t), & \phi &= \phi(t), \\
T_\mu{}^\nu &= \text{diag}(\rho, -p_i \delta_i^j), & \rho &= \rho(t), & p_i &= p_i(t), & \sigma &= \sigma(t).
\end{aligned}
\tag{4.8}
$$

An explicit computation then gives the following non-zero components of the connection:

$$
\Gamma_{0i}{}^j = H_i \delta_i^j, \qquad \Gamma_{ij}{}^0 = a_i \dot{a}_i \delta_{ij}
\tag{4.9}
$$

($H_i = \dot{a}_i/a_i$, no sum over i), and of the Ricci tensor:

$$R_0{}^0 = -\sum_i \left(\dot{H}_i + H_i^2\right),$$

$$R_i{}^j = -\delta_i^j \left(\dot{H}_i + H_i \sum_k H_k\right). \tag{4.10}$$

The associated scalar curvature is

$$R = -\sum_i \left(2\dot{H}_i + H_i^2\right) - \left(\sum_i H_i\right)^2. \tag{4.11}$$

For the dilaton field we have:

$$(\nabla\phi)^2 = \dot{\phi}^2, \qquad \nabla^2\phi = \ddot{\phi} + \dot{\phi}\sum_i H_i,$$

$$\nabla_0\nabla^0\phi = \ddot{\phi}, \qquad \nabla_i\nabla^j\phi = \dot{\phi}H_i\,\delta_i^j. \tag{4.12}$$

We are now in the position of writing the anisotropic version of the string cosmology equations presented in Section 2.1. We work for simplicity in string units, setting everywhere $2\lambda_s^{d-1} = 1$. The dilaton equation (2.25) then gives

$$2\ddot{\phi} - \dot{\phi}^2 + 2\dot{\phi}\sum_i H_i - \sum_i \left(2\dot{H}_i + H_i^2\right) - \left(\sum_i H_i\right)^2 + V - V' = \frac{1}{2}e^{\phi}\sigma, \tag{4.13}$$

the (00) component of Eq. (2.24) gives

$$\dot{\phi}^2 - 2\dot{\phi}\sum_i H_i + \left(\sum_i H_i\right)^2 - \sum_i H_i^2 - V = e^{\phi}\rho, \tag{4.14}$$

while the diagonal part (ii) of the space components, after some simplifications performed using Eq. (4.13), finally gives

$$\dot{H}_i - H_i\left(\dot{\phi} - \sum_k H_k\right) + \frac{1}{2}V' = \frac{1}{2}e^{\phi}\left(p_i - \frac{\sigma}{2}\right), \tag{4.15}$$

where $V' = \partial V/\partial\phi$.

The above set of $d+2$ equations contains $2d+3$ variables, $\{a_i, p_i, \rho, \phi, \sigma\}$. Their solution obviously requires additional information concerning the sources, in the form of $d+1$ "equations of state", $p_i = p_i(\rho)$, $\sigma = \sigma(\rho)$, able to eliminate the pressure and the dilaton charge density from the set of unknown variables. However, before studying the solutions, and independently of the particular types of sources, we are first interested here in the invariance properties of the given equations.

The first property, which also characterizes the Einstein equations, is the classical invariance under the "time-reversal" transformation $t \to -t$, for which

$$\dot{\phi} \to -\dot{\phi}, \qquad H \to -H, \qquad \ddot{\phi} \to \ddot{\phi}, \qquad \dot{H} \to \dot{H}. \qquad (4.16)$$

According to this invariance property, given any set of variables $\mathcal{S} = \{a_i(t), \phi(t), \rho(t)\}$ representing an exact solution of the above equations, it follows that the time-reversed set $\tilde{\mathcal{S}} = \{a_i(-t), \phi(-t), \rho(-t)\}$ is also another acceptable solution of the same cosmological equations. In addition, the string cosmology equations are invariant under other transformations which have no analogue in the corresponding Einstein equations.

For a simple illustration of the new invariance properties let us first consider the equations in vacuum, $T_{\mu\nu} = 0 = \sigma$, and in the absence of the dilaton potential, $V = 0$. It is convenient to introduce the so-called "shifted dilaton" variable $\overline{\phi}$, defined by

$$\overline{\phi} = \phi - \ln (\Pi_i a_i) = \phi - \sum_i \ln a_i,$$

$$\dot{\overline{\phi}} = \dot{\phi} - \sum_i H_i, \qquad \ddot{\overline{\phi}} = \ddot{\phi} - \sum_i \dot{H}_i. \qquad (4.17)$$

Equations (4.14), (4.15) and the dilaton equation (4.13) can then be rewritten, respectively, as follows:

$$\dot{\overline{\phi}}^2 - \sum_i H_i^2 = 0,$$

$$\dot{H}_i - H_i \dot{\overline{\phi}} = 0, \qquad (4.18)$$

$$2\ddot{\overline{\phi}} - \dot{\overline{\phi}}^2 - \sum_i H_i^2 = 0.$$

Consider then the transformation $a_i \to \tilde{a}_i = a_i^{-1}$ (at fixed t), for which

$$H \to \tilde{H} = \frac{\dot{\tilde{a}}}{\tilde{a}} = a \left(\frac{da^{-1}}{dt} \right) = -H. \qquad (4.19)$$

It can be immediately checked that all the equations (4.18) are invariant under the transformations

$$a_i \to a_i^{-1}, \qquad \overline{\phi} \to \overline{\phi}, \qquad (4.20)$$

called "scale-factor duality" transformations [3, 4].

The above transformations extend the T-duality transformations (4.7) to the case of a time-dependent gravi-dilaton background, and are only a particular case of a more general class of transformations that will be discussed in Section 4.3.

Unlike in the case of the T-duality transformations (4.7) there is no need for the inverted scale factor to be associated with compact dimensions, in this case. Furthermore, the invariance property does not concern the energy levels of the quantum string spectrum, being a property of the "classical" fields present in the background in which the string is moving. Finally, the invariance under the duality transformations (4.20) cannot be implemented for the scale factor alone without an associated transformation of the dilaton field [5, 6], which is required to guarantee the invariance of the shifted variable $\overline{\phi}$ of Eq. (4.17). Because of this last point, there is no analogue of this symmetry in the gravitational Einstein equations.

Thanks to the invariance property (4.20), given a set of variables representing an exact solution of Eqs. (4.18),

$$S = \{a_1, a_2, \ldots, a_d, \phi\}, \tag{4.21}$$

we can then automatically generate another set of variables $\widetilde{S}$ representing a new (and physically different) solution of the same equations simply by inverting k scale factors ($1 \le k \le d$),

$$\widetilde{S} = \{a_1^{-1}, a_2^{-1}, \ldots, a_k^{-1}, a_{k+1}, \ldots, a_d, \widetilde{\phi}\}. \tag{4.22}$$

The new dilaton $\widetilde{\phi}$ is determined by the condition $\overline{\phi} = \widetilde{\overline{\phi}}$, namely,

$$\overline{\phi} \equiv \phi - \sum_{i=1}^{d} \ln a_i = \widetilde{\overline{\phi}} \equiv \widetilde{\phi} - \sum_{i=1}^{k} \ln a_i^{-1} - \sum_{i=k+1}^{d} \ln a_i, \tag{4.23}$$

which implies

$$\widetilde{\phi} = \phi - 2 \sum_{i=1}^{k} \ln a_i. \tag{4.24}$$

In particular, by inverting all scale factors, we obtain the full duality transformation

$$\{a_i, \phi\} \rightarrow \{a_i^{-1}, \phi - 2 \sum_{i=1}^{d} \ln a_i\}. \tag{4.25}$$

At this point, an important remark is in order. The combination of time-reversal and duality transformations leads to defining, for any given solution associated with a scale factor $a_i(t)$, four different (and, in principle, physically distinct) branches:

$$a_i(t), \qquad a_i(-t), \qquad a_i^{-1}(t), \qquad a_i^{-1}(-t). \tag{4.26}$$

Consider, for instance, an isotropic d-dimensional background, in which the vacuum equations (4.18) are reduced to

$$\dot{\overline{\phi}}^2 = dH^2, \qquad \dot{H} = H\dot{\overline{\phi}}, \qquad 2\ddot{\overline{\phi}} - \dot{\overline{\phi}}^2 = dH^2 \tag{4.27}$$

Table 4.1 Kinematic classification of the four branches (4.29)

1	$\dot{a} > 0$, expansion	$\ddot{a} < 0$, decelerated	$\dot{H} < 0$, decreasing curvature
2	$\dot{a} < 0$, contraction	$\ddot{a} < 0$, accelerated	$\dot{H} < 0$, increasing curvature
3	$\dot{a} < 0$, contraction	$\ddot{a} > 0$, decelerated	$\dot{H} > 0$, decreasing curvature
4	$\dot{a} > 0$, expansion	$\ddot{a} > 0$, accelerated	$\dot{H} > 0$, increasing curvature

(in this case, for $\dot{\overline{\phi}} \neq 0$, only two equations are independent). Such equations are satisfied by the particular exact solution

$$a(t) = \left(\frac{t}{t_0}\right)^{1/\sqrt{d}}, \qquad \overline{\phi} = -\ln\left(\frac{t}{t_0}\right), \tag{4.28}$$

for any value of $t_0 = \text{const}$. Using duality and time-reversal transformations we can associate with this solution four different cosmological configurations:

$$
\boxed{1} \qquad \qquad \text{time reversal} \qquad \qquad \boxed{2}
$$
$$
\{a = t^{1/\sqrt{d}}, \quad \overline{\phi} = -\ln t\} \quad \Longleftrightarrow \quad \{a = (-t)^{1/\sqrt{d}}, \quad \overline{\phi} = -\ln(-t)\}
$$
$$
\updownarrow \text{ duality} \qquad \qquad \qquad \text{duality } \updownarrow \qquad \qquad (4.29)
$$
$$
\{a = t^{-1/\sqrt{d}}, \quad \overline{\phi} = -\ln t\} \quad \Longleftrightarrow \quad \{a = (-t)^{-1/\sqrt{d}}, \quad \overline{\phi} = -\ln(-t)\}.
$$
$$
\boxed{3} \qquad \qquad \text{time reversal} \qquad \qquad \boxed{4}
$$

The solutions $\boxed{1}$ and $\boxed{3}$ are defined for $t > 0$, the solutions $\boxed{2}$ and $\boxed{4}$ for $t < 0$. In fact, all four solutions are affected by a curvature singularity at $|t| \to 0$, and are thus classically defined only on the real (positive or negative) half-line. They have different (and complementary) kinematic properties, as summarized in Table 4.1. Two branches describe expansion ($H > 0$), two branches contraction. Two branches (those defined for $t < 0$) are characterized by growing curvature (H^2, or $|H|$, is growing in time), the other two branches ($t > 0$) by decreasing curvature, as clearly illustrated in Fig. 4.1. Finally, two branches ($t < 0$) are accelerated (i.e. sign $\dot{a} = $ sign $\ddot{a}$), two branches ($t > 0$) are decelerated (sign $\dot{a} = -$ sign $\ddot{a}$).

It is worth noticing, in particular, that any given solution $H(t)$ of type $\boxed{1}$, with the typical properties of the standard cosmological evolution, is always associated – through duality and time-reversal transformations – with a "*dual partner*" $\tilde{H}(-t)$, namely with a solution of type $\boxed{4}$, describing accelerated (i.e.

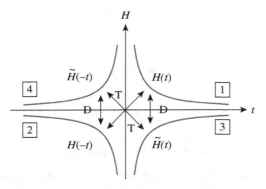

Figure 4.1 Time evolution of the Hubble parameter in the four branches of the isotropic vacuum solution (4.29). The vertical arrows, labeled by "D", represent duality transformations. The diagonal arrows, labeled by "T", represent time reflections.

inflationary, see Chapter 5) expansion and growing curvature. This possibility of symmetry-correlated solutions, which are absent in the context of the standard equations of general relativity, suggests a model of "*self-dual*" cosmological evolution characterized by a solution satisfying $a(t) = a^{-1}(-t)$, and associated with a "bell-like" shape of the curvature scale, as illustrated in Fig. 4.2 (see Appendix 4B for explicit examples). The curvature, in such a context, would avoid the standard cosmological singularity: moving back in time from the present epoch, it would reach a maximum – possibly controlled by the fundamental string scale λ_s^2 – after which it would become decreasing again, asymptotically approaching the flat space configuration. The phase of high (but finite) string-scale curvature would replace the big bang singularity of the standard cosmological scenario, so that it becomes natural in this context to call "*pre-big bang*" [7, 8] the initial phase ($t < 0$), accelerated and at growing curvature, in contrast to the subsequent, "*post-big bang*" phase ($t > 0$) of standard evolution at decreasing curvature.

The vacuum solutions that we are considering, on the other hand, must satisfy Eqs. (4.27): the first of these equations implies $\dot{\overline{\phi}} = \pm\sqrt{d}H$, which means that the solutions (4.29) actually represent the bisecting lines of the plane $\{\dot{\overline{\phi}}, \sqrt{d}H\}$. Looking at the signs of H and $\dot{\overline{\phi}}$, the four branches can then be plotted as in Fig. 4.3, where each branch ranges from the origin to infinity (or vice versa). The origin corresponds to the trivial solution $H = 0$, $\phi = $ const, and the singularity is at infinity. The flow of the various branches, following the increasing-time direction, is illustrated by the big (black and white) arrows. The growing-curvature solutions $\boxed{2}$ and $\boxed{4}$ (of pre-big bang type) are characterized by $\dot{\overline{\phi}} > 0$, while the decreasing-curvature solutions $\boxed{1}$ and $\boxed{3}$ (of post-big bang type) are characterized by $\dot{\overline{\phi}} < 0$.

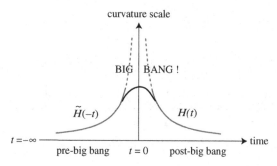

Figure 4.2 Qualitative example of self-dual solution which smoothly interpolates between an initial (pre-big bang) phase with growing curvature and accelerated expansion, and a final (post-big bang) phase with decreasing curvature and decelerated expansion.

A transition from the pre- to the post-big bang regime must thus correspond to a transition from $\ddot{\phi} > 0$ to $\ddot{\phi} < 0$.

It is also appropriate, in this context, to consider the sign of $\dot{\phi}$, in order to have information on the behavior of the dilaton (and thus on the tree-level coupling $\exp \phi$) in the various branches. Rewritten in terms of the ϕ variable, the four branches (4.29) are given by

$$a_\pm(\pm t) = (\pm t)^{\pm 1/\sqrt{d}},$$

$$\phi_\pm(\pm t) = \overline{\phi}(\pm t) + d \ln a_\pm(\pm t) = (\pm\sqrt{d} - 1) \ln(\pm t). \tag{4.30}$$

It can be easily checked that among the branches $\{a_\pm(-t), \phi_\pm(-t)\}$ at growing curvature, the solution $a_-(-t)$ – corresponding to the curve $\boxed{4}$ and describing an *expanding* pre-big bang configuration – is associated with a *growing* dilaton, while the solution $a_+(-t)$ – corresponding to the curve $\boxed{2}$ and describing a *contracting* pre-big bang configuration – is associated with a *decreasing* dilaton.

On the other hand, as we will see in the next section, a genuinely self-dual cosmological scenario seems to require a growth of the coupling constant (and thus of the dilaton) during the initial pre-big bang phase: this requirement thus selects an expanding metric, according to the above solutions. In such a case, we should expect that the transition from the pre- to the post-big bang phase (the dashed curve of Fig. 4.3) may possibly occur in the strong coupling regime, as we will discuss in Chapter 6. However, one could also construct a scenario which is not self-dual, in which the pre-big bang phase corresponds to a contraction of the (S-frame) metric, and is associated with a decreasing dilaton coupling [9] (see Chapter 10). In that case the transition to the expanding, post-big bang phase (the

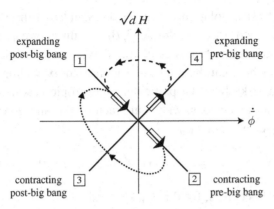

Figure 4.3 The four branches of Fig. 4.1 in the "phase space" plane spanned by $\dot{\bar{\phi}}$ and $\sqrt{d}\,H$. Growing curvature (pre-big bang) phases correspond to $\dot{\bar{\phi}} > 0$, decreasing curvature (post-big bang) phases to $\dot{\bar{\phi}} < 0$.

dotted curve of Fig. 4.3) is represented by a "bounce" of the scale factor, and is expected to occur in the weak coupling regime.

It should be stressed, finally, that the close relation that we have considered between the time behavior of a and ϕ (sign $\dot{a} = $ sign $\dot{\phi}$) is referred to a simple isotropic solution of the string cosmology equations. In more general cases of anisotropic, higher-dimensional solutions it is possible to find that both the curvature and the dilaton are growing even in the presence of contracting dimensions, provided there is a sufficient number of expanding dimensions.

To elucidate this point we may consider the generalization of the solution (4.28) to the case of an anisotropic D-dimensional metric, by setting

$$a_i = \left(\frac{\pm t}{t_0}\right)^{\beta_i}, \qquad \bar{\phi} = -\ln\left(\frac{\pm t}{t_0}\right). \tag{4.31}$$

We obtain

$$H_i = \frac{\beta_i}{t}, \qquad \dot{H}_i = -\frac{\beta_i}{t^2}, \qquad \dot{\bar{\phi}} = -\frac{1}{t}, \qquad \ddot{\bar{\phi}} = \frac{1}{t^2}, \tag{4.32}$$

and all the vacuum equations (4.18) are satisfied provided

$$\sum_i \beta_i^2 = 1. \tag{4.33}$$

The corresponding solution for the dilaton is

$$\phi = \bar{\phi} + \sum_i \ln a_i = \left(\sum_i \beta_i - 1\right)\ln\left(\frac{\pm t}{t_0}\right) + \text{const} \tag{4.34}$$

(the well-known Kasner solution, valid for the standard Einstein equations and characterized by the additional condition $\sum_i \beta_i = 1$, thus corresponds to a constant dilaton). The negative-time branches of the above solution describe a growing-curvature regime, which can be associated with both expanding and contracting dimensions. Let us take here the particular, anisotropic case in which there are d expanding spatial dimensions, with scale factor $a(t)$, and n contracting spatial dimensions, with scale factor $b(t)$:

$$a(t) = (-t)^{-1/\sqrt{d+n}}, \qquad b(t) = (-t)^{1/\sqrt{d+n}}, \qquad t < 0, \qquad t \to 0_-. \quad (4.35)$$

The solution (4.34) becomes, for this background,

$$\phi(t) = \frac{n - d - \sqrt{d+n}}{\sqrt{d+n}} \ln(-t) + \text{const}, \qquad (4.36)$$

so that the dilaton (together with the coupling parameter $g_s^2 = \exp \phi$) is growing for $t \to 0_-$, provided

$$d + \sqrt{d+n} > n, \qquad (4.37)$$

i.e. provided the number of expanding dimensions is sufficiently large with respect to the number of contracting ones.

It may be noted that a frozen dilaton ($\dot\phi = 0$) with $d = 3$ expanding dimensions is only allowed, in such a context, for $n = 6$ contracting dimensions, i.e. for a total number of dimensions ($D = d + n + 1 = 10$) which exactly corresponds to the critical number $D_c = 10$ required by superstring theory (see Chapter 3). It should be mentioned, however, that the above solution is not expected to provide a realistic description of the final cosmological configuration with stabilized dilaton, as it describes a vacuum scenario which does not include the contribution of other fields and, in particular, of the ordinary matter sources. The presence of matter, as we shall see in the following section, modifies the conditions required for obtaining solutions at constant dilaton.

4.2 Duality with matter sources

In the previous section we have illustrated the simplest example of duality symmetry for the free gravi-dilaton system in vacuum. The invariance under the transformations (4.20) also applies in the presence of a dilaton potential V, provided V depends on the dilaton only through the shifted variable $\bar\phi$ (see Appendix 4A). Such an invariance is preserved in the presence of matter sources, provided they satisfy appropriate transformation laws.

To discuss this second possibility let us start with sources which can be approximated as perfect fluids, and consider the full set of equations (4.13)–(4.15), with $V = 0$. Introducing for the fluid the "shifted" variables,

$$\bar{\rho} = \rho\, \Pi_i a_i, \qquad \bar{p} = p\, \Pi_i a_i, \qquad \bar{\sigma} = \sigma\, \Pi_i a_i, \qquad (4.38)$$

we can complete the vacuum equations (4.18) as follows:

$$\dot{\phi}^2 - \sum_i H_i^2 = \bar{\rho}\, e^{\bar{\phi}}, \qquad (4.39)$$

$$\dot{H}_i - H_i \dot{\bar{\phi}} = \frac{1}{2} e^{\bar{\phi}} \left(\bar{p}_i - \frac{\bar{\sigma}}{2} \right), \qquad (4.40)$$

$$2\ddot{\phi} - \dot{\phi}^2 - \sum_i H_i^2 = \frac{1}{2} \bar{\sigma} e^{\bar{\phi}}. \qquad (4.41)$$

These $d+2$ equations are all independent, and their combination leads to the covariant conservation equation. Differentiating Eq. (4.39), eliminating $\ddot{\phi}$ with Eq. (4.41), $\dot{H}_i$ with Eq. (4.40), and $\dot{\bar{\phi}}$ with Eq. (4.39), we obtain,

$$\dot{\bar{p}} + \sum_i H_i \bar{p}_i = \frac{1}{2} \bar{\sigma} \left(\dot{\bar{\phi}} + \sum_i H_i \right), \qquad (4.42)$$

which can be rewritten in terms of the non-shifted variables as

$$\dot{\rho} + \sum_i H_i (\rho + p_i) = \frac{1}{2} \sigma \dot{\phi}, \qquad (4.43)$$

in agreement with the general equation (2.19). When $\sigma = 0$ one recovers the usual covariant conservation of the matter stress tensor.

It is evident, from the explicit form of the above equations, that they are invariant under the time-reversal transformation (4.16). In addition, they are invariant under the duality transformation (4.20) if the dilaton charge is vanishing, $\bar{\sigma} = 0$, and if the inversion of the scale factor is accompanied by an appropriate transformation of ρ and p which reads, in terms of the shifted variables [4],

$$a_i \to a_i^{-1}, \qquad \bar{\phi} \to \bar{\phi}, \qquad \bar{\rho} \to \bar{\rho}, \qquad \bar{p}_i \to -\bar{p}_i, \qquad \bar{\sigma} = 0. \qquad (4.44)$$

In the perfect fluid approximation, scale-factor duality thus requires a "reflection" of the equation of state for its implementation. Note that the transformation of $\bar{\rho}$ is trivial, but the transformation of the physical variable ρ is non-trivial: in fact, from the definitions (4.38), we obtain

$$\rho \to \tilde{\rho} = \rho\, \Pi_i a_i^2, \qquad (4.45)$$

if all scale factors are inverted.

It should be mentioned that the above symmetry can be extended to include a non-vanishing dilaton charge, provided the sources are coupled to the dilaton only through the shifted variable $\bar{\phi}$: in that case it can be shown that the contribution of $\bar{\sigma}$ disappears from the spatial equation (4.40), and that the duality symmetry of the dilaton equation requires the invariance of the shifted charge density, $\bar{\sigma}$, to be added to the transformation laws (4.44) (see e.g. [10]). In such a case, however, the general covariance of the effective action requires a non-local interaction of the dilaton with the matter sources, as in the case of the dilaton potential discussed in Appendix 4A. We will not discuss such a possibility here so that, in our context, the presence of the dilaton charge will always signal a breakdown of duality invariance.

In the next section we will present an explicit example of physical sources transforming in agreement with Eq. (4.44). In this section we first discuss some important aspects of the duality transformations (4.44), in view of possibly realistic cosmological applications.

Let us first observe that, even in the presence of sources, the string cosmology solutions are in general characterized by four different branches. A simple example can be given by considering an isotropic background with d spatial dimensions, sourced by a barotropic perfect fluid, with equation of state $p/\rho = \gamma = \text{const}$. In the limit $\sigma = 0$ we are thus left with a system of three equations for the three variables a, ϕ, ρ, and we can conveniently choose Eqs. (4.39), (4.41) and (4.42) as our independent equations.

Looking for a particular exact solution we set

$$a = \left(\frac{t}{t_0}\right)^{\alpha}, \qquad \bar{\phi} = -\beta \ln\left(\frac{t}{t_0}\right), \qquad p = \gamma\rho, \qquad (4.46)$$

from which

$$\dot{\bar{\phi}} = -\frac{\beta}{t}, \qquad \ddot{\bar{\phi}} = \frac{\beta}{t^2},$$

$$H = \frac{\alpha}{t}, \qquad \dot{H} = -\frac{\alpha}{t^2}, \qquad e^{\bar{\phi}} = \left(\frac{t}{t_0}\right)^{-\beta}. \qquad (4.47)$$

The integration of the conservation equation (4.42) then gives

$$\bar{\rho} = \rho_0 a^{-d\gamma}, \qquad (4.48)$$

where ρ_0 is an integration constant. Equation (4.39) is satisfied provided

$$d\gamma\alpha + \beta = 2, \qquad (4.49)$$

and, once satisfied, fixes ρ_0 as a function of t_0. Equation (4.41) finally provides the additional constraint

$$2\beta - \beta^2 - d\alpha^2 = 0. \qquad (4.50)$$

The elimination of α leads to the equation

$$\beta^2(1+d\gamma^2) - \beta(4+2d\gamma^2) + 4 = 0, \qquad (4.51)$$

with two possible solutions for our set of parameters:

$$(1) \quad \beta = 2, \qquad\qquad \alpha = 0, \qquad\qquad \gamma = 0, \qquad (4.52)$$

$$(2) \quad \beta = \frac{2}{1+d\gamma^2}, \qquad \alpha = \frac{2\gamma}{1+d\gamma^2}, \qquad \gamma \neq 0. \qquad (4.53)$$

The first case corresponds to a particular (S-frame) solution describing a globally flat space-time and a non-trivial evolution of the dilaton, $\overline{\phi} = \phi = -2\ln t$, sustained by a perfect fluid with vanishing pressure and constant energy density, $\rho = \overline{\rho} = \text{const}$. Such a peculiar string cosmology configuration provides a possible example of initial pre-big bang evolution, in a regime where the space-time curvature is negligible with respect to the dilaton kinetic energy [11]. In such a case the two duality-related branches of the solution obviously coincide, while the branches connected by time-reversal transformations are different.

To obtain four different branches we may consider the second set of parameters (4.53), associated with $\gamma \neq 0$. The corresponding solution, expressed in terms of the conventional variables $\{a, \phi, \rho\}$, has the form [12]

$$a = \left(\frac{t}{t_0}\right)^{\frac{2\gamma}{1+d\gamma^2}},$$

$$\rho = \overline{\rho} a^{-d} = \rho_0 a^{-d(1+\gamma)}, \qquad (4.54)$$

$$\phi = \overline{\phi} + d\ln a = \frac{2(d\gamma - 1)}{1+d\gamma^2} \ln\left(\frac{t}{t_0}\right) + \text{const}.$$

In this case the general transformation (4.44) (which preserves $\overline{\phi}$ and $\overline{\rho}$) simply corresponds to the reflection

$$\gamma \leftrightarrow -\gamma, \qquad (4.55)$$

and the four different branches of the solution (connected by duality and time-reversal transformations) can be written as

$$a_{\pm}(\pm t) \sim (\pm t)^{\pm\frac{2|\gamma|}{(1+d\gamma^2)}},$$

$$\rho_{\pm}(\pm t) \sim (\pm t)^{\mp 2\frac{d|\gamma|(1\pm|\gamma|)}{(1+d\gamma^2)}},$$

$$p_{\pm} = \pm|\gamma|\rho_{\pm}, \qquad (4.56)$$

$$\phi_{\pm}(\pm t) \sim -2\frac{1\mp d|\gamma|}{1+d\gamma^2} \ln(\pm t) + \text{const}.$$

The positive-time branches are decelerated, with $a_+(t)$ which describes expansion and $a_-(t)$ which describes contraction (the curves $\boxed{1}$ and $\boxed{3}$ of Fig. 4.1). The negative-time branches are accelerated, of the pre-big bang type, with $a_+(-t)$ which describes contraction and $a_-(-t)$ which describes expansion (the curves $\boxed{2}$ and $\boxed{4}$ of Fig. 4.1). Again, as in the vacuum case, the solution $a_+(t)$, typical of the standard cosmological evolution, has a dual partner $a_-(-t)$ describing inflationary expansion (see Chapter 5 for a kinematic classification of the various types of inflation). And, again, this dual "complement" is associated with a dilaton field which (at least in the isotropic case) is always growing as $t \to 0_-$, for any given equation of state:

$$e^{\phi_-(-t)} = (-t)^{-2\frac{1+d|\gamma|}{1+d\gamma^2}} \to +\infty, \qquad t \to 0_-. \tag{4.57}$$

It is instructive to consider the realistic example of a Universe with $d = 3$ isotropic spatial dimensions, dominated by a radiation fluid with $\gamma = 1/3$. In this case, the expanding decelerated branch $\{a_+(t), \phi_+(t), \rho_+(t)\}$ of Eq. (4.56) exactly reproduces the well-known solution of the standard cosmological scenario,

$$a = t^{1/2}, \qquad \phi = \text{const}, \qquad \rho = \rho_0 a^{-4}, \qquad p = \rho/3, \tag{4.58}$$

describing decelerated expansion, decreasing curvature and frozen dilaton for $0 < t < \infty$:

$$\dot{a} > 0, \qquad \ddot{a} < 0, \qquad \dot{H} < 0, \qquad \dot{\phi} = 0. \tag{4.59}$$

Through a dual inversion and a time reflection we obtain the associated inflationary partner $\{a_-(-t), \phi_-(-t), \rho_-(-t)\}$, i.e.

$$a = (-t)^{-1/2}, \qquad \phi = -3\ln(-t), \qquad \rho = \rho_0 a^{-2}, \qquad p = -\rho/3. \tag{4.60}$$

This solution is defined for $-\infty < t < 0$, and describes a phase of accelerated expansion (driven by the dilaton and by negative-pressure sources), growing curvature and *growing dilaton*:

$$\dot{a} > 0, \qquad \ddot{a} > 0, \qquad \dot{H} > 0, \qquad \dot{\phi} > 0. \tag{4.61}$$

This simple example may give us interesting suggestions on how to extrapolate back in time the evolution of the standard cosmological scenario, which we know to be characterized by a time-decreasing curvature and, at least locally, by a nearly constant dilaton – it is the dilaton, indeed, which controls the effective gravitational coupling (see Eq. (2.3)), determining a value of G which is found by present measurements to be constant (or, at most, very slowly varying, see e.g. [13]). Starting from the present cosmological phase, and looking for a past extension of the cosmological history on the grounds of a principle of *self-duality* – i.e. assuming that the past evolution of our Universe should represent the dual "complement" of the present one (assuming also the possible smoothing out of the big bang singularity) – string theory then suggests for the Universe, at very early epochs, a phase characterized not only by *growing curvature*, but also by *growing dilaton* [4, 7, 8].

The dilaton, on the other hand, represents the exponential (tree-level) parametrization of the fundamental string coupling g_s^2 (as discussed in Chapter 3). As a consequence, it automatically controls the strength of all (gravitational and gauge) interactions. The principle of self-duality thus suggests that the Universe reached its present state after a long evolution starting from an extremely simple – almost trivial – initial configuration, characterized by a nearly flat geometry and a very small coupling parameter,

$$H^2 \to 0, \qquad e^\phi \to 0, \tag{4.62}$$

the so-called "*string perturbative vacuum*". With this assumption, which is the basis of all models of pre-big bang inflation [7, 8], the initial Universe evolves in a low-energy, extremely perturbative regime, in which the curvatures (i.e. the field gradients) are small, and the couplings are weak (in string units),

$$\frac{H^2}{M_s^2} \ll 1, \qquad \frac{\dot{\phi}^2}{M_s^2} \ll 1, \qquad \dots, \qquad g_s^2 \ll 1, \tag{4.63}$$

so that the dynamics may be appropriately described by the lowest-order string effective action, at tree-level in the α' and quantum-loop expansion (see Fig. 4.4).

This picture is in remarkable contrast with the conventional cosmological scenario in which the Universe evolves starting from a very *hot, curved* and *dense* initial configuration: in that context, the more we go back in time, the more we enter a Planckian and (possibly) trans-Planckian non-perturbative regime of ultra-high energies (see [14] for a recent discussion), requiring, for its correct description, the full inclusion of all quantum gravity effects, to all orders. The

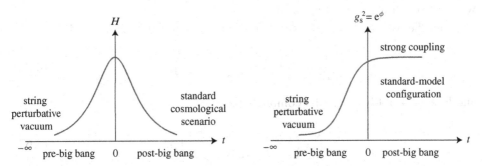

Figure 4.4 Qualitative time evolution of the curvature scale (left panel) and of the dilaton (right panel) for a typical self-dual solution of the string cosmology equations: the present standard configuration is the outcome of a smooth evolution from the string perturbative vacuum.

self-duality principle, instead, suggests a picture in which the more we go back in time (after reaching the maximum curvature scale), the more we approach a *flat*, *cold* and *vacuum* configuration, which can be perfectly described by the classical background equations. Quantum effects, in the form of both higher-derivative corrections, $\mathcal{O}(H^n/M_s^n)$, and higher-loop contributions, $\mathcal{O}(g_s^n)$, are expected eventually to become important only *towards the end* of the phase of pre-big bang inflation, when the background approaches the string scale. In that limit the higher-order string theory effects should come into play, and possibly trigger the transition from the pre- to the post-big bang regime, as discussed in Chapter 6.

4.2.1 General integration of the lowest-order equations

The last part of this section is devoted to the exact integration of the string cosmology equations (4.39)–(4.41) in the fully anisotropic case, under the assumption that the sources are represented by a barotropic fluid ($p/\rho = $ const), with a uniformly distributed dilaton charge proportional to the energy density ($\sigma/\rho = $ const). In this case we can set

$$\bar{p}_i = \gamma_i \bar{\rho}, \qquad \bar{\sigma} = \gamma_0 \bar{\rho}, \tag{4.64}$$

where γ_i, γ_0 are $d+1$ constant parameters specifying the equation of state of a given model of source. This parametrization includes, in particular, the case of a homogeneous scalar field $\psi = \psi(t)$, exponentially coupled to the dilaton according to the matter action

$$S_m \sim \int d^{d+1}x \sqrt{|g|} \left(\partial_\mu \psi \partial^\mu \psi\right) e^{k\phi}, \tag{4.65}$$

which is typical of a massless axion or of a higher-dimensional modulus field, as we shall see in the following chapters.

Using the assumption (4.64) we can rewrite the spatial equation (4.40) as

$$\frac{d}{dt} \left(e^{-\bar{\phi}} H_i\right) = \frac{1}{2} \left(\gamma_i - \frac{\gamma_0}{2}\right) \bar{\rho}, \tag{4.66}$$

and Eqs. (4.41), (4.39) can be combined to give

$$\frac{d^2}{dt^2} \left(e^{-\bar{\phi}}\right) = \frac{1}{2} \left(1 - \frac{\gamma_0}{2}\right) \bar{\rho}. \tag{4.67}$$

Defining a convenient (dimensionless) time parameter x, such that

$$\frac{dx}{dt} = \frac{L}{2} \bar{\rho} \tag{4.68}$$

(L is a constant length parameter), we can then integrate a first time the previous two equations, obtaining

$$\frac{1}{2}\bar{\rho}\left(e^{-\bar{\phi}}\right)' = -\frac{1}{2}\bar{\rho}e^{-\bar{\phi}}\bar{\phi}' = \left(1 - \frac{\gamma_0}{2}\right)\frac{x + x_0}{L^2}, \qquad (4.69)$$

$$\frac{1}{2}\bar{\rho}e^{-\bar{\phi}}\left(\frac{a_i'}{a_i}\right) = \left(\gamma_i - \frac{\gamma_0}{2}\right)\frac{x + x_i}{L^2}, \qquad (4.70)$$

where x_0 and x_i are integration constants (the derivative with respect to x has been denoted by a prime).

We now need the conservation equation (4.42), which can be rewritten in terms of the new variable x as follows:

$$\bar{\rho}' + \bar{\rho}\sum_i \left(\frac{a_i'}{a_i}\right)\left(\gamma_i - \frac{\gamma_0}{2}\right) = \frac{1}{2}\gamma_0\bar{\rho}\bar{\phi}'. \qquad (4.71)$$

Multiplying by $e^{-\bar{\phi}}$, eliminating a_i' through Eq. (4.70) and $\bar{\phi}'$ through Eq. (4.69), we obtain

$$\frac{1}{2}e^{-\bar{\phi}}\bar{\rho}' = -\sum_i \left(\gamma_i - \frac{\gamma_0}{2}\right)^2 \frac{x + x_i}{L^2} - \frac{1}{2}\gamma_0\left(1 - \frac{\gamma_0}{2}\right)\frac{x + x_0}{L^2}. \qquad (4.72)$$

The sum of this last equation and of Eq. (4.69) gives

$$\left(\frac{1}{2}e^{-\bar{\phi}}\bar{\rho}\right)' = \left(1 - \frac{\gamma_0}{2}\right)^2 \frac{x + x_0}{L^2} - \sum_i \left(\gamma_i - \frac{\gamma_0}{2}\right)^2 \frac{x + x_i}{L^2}, \qquad (4.73)$$

and its integration leads to

$$L^2 e^{-\bar{\phi}}\bar{\rho} = D(x),$$

$$D(x) = \left(1 - \frac{\gamma_0}{2}\right)^2 (x + x_0)^2 - \sum_i \left(\gamma_i - \frac{\gamma_0}{2}\right)^2 (x + x_i)^2 + \beta, \qquad (4.74)$$

where β is a (dimensionless) integration constant, still to be fixed. Inserting this result into Eqs. (4.69) and (4.70) we can then obtain two independent equations for $\bar{\phi}$ and a_i:

$$\bar{\phi}' = -2\left(1 - \frac{\gamma_0}{2}\right)\frac{x + x_0}{D(x)}, \qquad \gamma_0 \neq 2, \qquad (4.75)$$

$$\frac{a_i'}{a_i} = 2\left(\gamma_i - \frac{\gamma_0}{2}\right)\frac{x + x_i}{D(x)}, \qquad \gamma_0 \neq 2\gamma_i. \qquad (4.76)$$

Up to now we have used Eq. (4.40) and a combination of Eqs. (4.39) and (4.41). Imposing that Eq. (4.39) is also separately satisfied, and using the last three equations for a_i', $\overline{\phi}'$, D, we finally obtain the condition

$$4\left(1-\frac{\gamma_0}{2}\right)^2\left(\frac{x+x_0}{D}\right)^2 - 4\sum_i\left(\gamma_i-\frac{\gamma_0}{2}\right)^2\left(\frac{x+x_i}{D}\right)^2 = \frac{4}{D},\qquad(4.77)$$

which identically fixes the integration constant β, giving, in this case, $\beta = 0$.

We must consider now the values of γ_i and γ_0 excluded from the previous analysis. If $\gamma_0 = 2$ then Eq. (4.75) is to be replaced by

$$\overline{\phi}' = -\frac{2x_0}{D(x)},\qquad \gamma_0 = 2,\qquad(4.78)$$

and the quadratic form $D(x)$ is given by

$$D(x) = x_0^2 - \sum_i\left(\gamma_i-\frac{\gamma_0}{2}\right)^2(x+x_i)^2.\qquad(4.79)$$

If, instead, $\gamma_0 \neq 2$, but $\gamma_0 = 2\gamma_i$ for $i = 1, 2, \ldots, k$, then the corresponding k components of Eq. (4.76) are to be replaced by

$$\frac{a_i'}{a_i} = \frac{2x_i}{D(x)},\qquad \gamma_0 = 2\gamma_i,\qquad i = 1, 2, \ldots, k,\qquad(4.80)$$

and the quadratic form $D(x)$ is given by

$$D(x) = \left(1-\frac{\gamma_0}{2}\right)^2(x+x_0)^2 - \sum_{i=1}^{k}x_i^2 - \sum_{i=k+1}^{d}\left(\gamma_i-\frac{\gamma_0}{2}\right)^2(x+x_i)^2.\qquad(4.81)$$

In both cases the relation $L^2 e^{-\overline{\phi}}\overline{\rho} = D(x)$ is left unchanged.

It can be easily checked, from the equations for $\overline{\phi}'$ and a', that a necessary condition for obtaining regular solutions (without singularities in both the curvature and the dilaton kinetic energy) is the absence of real zeros of the quadratic form $D(x)$, on the whole real line $-\infty \leq x \leq +\infty$. Examples with such properties can be obtained by including in the above equations the contribution of a non-local, duality-invariant dilaton potential, as will be shown in Appendix 4B. Without such a contribution $D(x)$ always has real zeros in the isotropic case, as can be checked by setting $D = \alpha x^2 + bx + c$, and noting that $b^2 - 4\alpha c \geq 0$ when all the x_i and γ_i parameters are equal. In the anisotropic case, on the contrary, it is possible to obtain regular solutions [15], provided $\gamma_0 \neq 2$ and $\gamma_0 \neq 2\gamma_i$ (see Appendix 4B). However, for such solutions it turns out that $D(x)$ is everywhere negative, and thus $\rho < 0$ according to Eq. (4.74). Sources of this type may possibly be interpreted as a classical, phenomenological description of the backreaction of the quantum fluctuations outside the horizon [16, 17, 18].

Particular examples of regular, self-dual solutions, smoothly interpolating between the phase of pre-big bang evolution and the standard cosmological regime, will be presented in Section 4.3 and Appendix 4B. However, we may expect that the regularization of the big bang singularity in general needs the effects of the higher-order loop and α' corrections, possibly introducing violations of the weak (or strong) energy conditions, as we shall discuss in Chapter 6. The discussion of this section will be restricted to the pure low-energy string effective action, without dilaton potential, and with conventional matter sources satisfying the weak energy condition $\rho > 0$. The solutions associated with the pre- and post-big bang branches will be disconnected by a curvature singularity, and thus appropriate to describe the scenario of Fig. 4.4 only sufficiently far from the transition regime $|t| \to 0$.

Let us consider, in particular, the integration of Eqs. (4.75) and (4.76), so as to include in our discussion the case $\gamma_0 = 0$ (compatible with the duality transformations (4.44)). Computing the zeros of $D(x)$ we can set

$$D(x) = \alpha x^2 + bx + c \equiv \alpha(x - x_+)(x - x_-), \qquad (4.82)$$

where

$$\alpha = \left(1 - \frac{\gamma_0}{2}\right)^2 - \sum_i \left(\gamma_i - \frac{\gamma_0}{2}\right)^2,$$

$$x_\pm = \alpha^{-1} \left(\sum_i x_i \alpha_i^2 - x_0 \alpha_0^2 \pm \Delta\right), \qquad (4.83)$$

$$2\Delta = (b^2 - 4\alpha c)^{1/2} = \left[\left(x_0 \alpha_0^2 - \sum_i x_i \alpha_i^2\right)^2 - \alpha\left(x_0^2 \alpha_0^2 - \sum_i x_i^2 \alpha_i^2\right)\right]^{1/2},$$

and where we have introduced the convenient notation

$$\alpha_0 = 1 - \gamma_0/2, \qquad \alpha_i = \gamma_i - \gamma_0/2. \qquad (4.84)$$

We can now integrate the two equations (4.75) and (4.76), for the case $\Delta^2 \geq 0$, to obtain the following general solution (see e.g. [19]):

$$a_i = a_{i0} \left|(x - x_+)(x - x_-)\right|^{\alpha_i/\alpha} \left|\frac{x - x_+}{x - x_-}\right|^{\delta_i}, \qquad (4.85)$$

$$e^{\overline{\phi}} = e^{\phi_0} \left|(x - x_+)(x - x_-)\right|^{-\alpha_0/\alpha} \left|\frac{x - x_+}{x - x_-}\right|^{-\delta_0}, \qquad (4.86)$$

where a_{i0}, ϕ_0 are integration constants, and where

$$\delta_i = \frac{\alpha_i}{\alpha\Delta}\left(\alpha x_i - x_0\alpha_0^2 + \sum_i x_i\alpha_i^2\right),$$

$$\delta_0 = \frac{\alpha_0}{\alpha\Delta}\left(\alpha x_0 - x_0\alpha_0^2 + \sum_i x_i\alpha_i^2\right). \tag{4.87}$$

This result generalizes to the case $\gamma_0 \neq 0$ the solutions presented in [12, 11]. From Eq. (4.74) we can then obtain the time evolution of the energy density,

$$\rho = \frac{D(x)}{L^2}e^{\bar\phi} = \frac{\alpha}{L^2}e^{\phi_0}(x - x_+)(x - x_-)\left|(x - x_+)(x - x_-)\right|^{-\alpha_0/\alpha}\left|\frac{x - x_+}{x - x_-}\right|^{-\delta_0}, \tag{4.88}$$

which also fixes the evolution of $\bar{p}_i = \gamma_i\bar\rho$ and $\bar\sigma = \gamma_0\bar\rho$.

The above class of solutions is characterized by two (in general different) singular points $x = x_\pm$, where both the curvature and the kinetic energy of the dilaton diverge. In the range $x < x_-$ and $x > x_+$ one recovers the two branches describing, respectively, the pre-big bang and the post-big bang regime. These two branches are disconnected by the singularity and (if $x_+ \neq x_-$) also by an extended intermediate region where $\bar p$ becomes negative (see for instance Fig. 4.5, where we have plotted a particular isotropic solution with $d = 3$, $\gamma_0 = 0$ and $\gamma_i = 1/3$). This confirms, as anticipated, the difficulties one encounters in a naive extrapolation of the lowest-order solutions outside their validity regime, without the appropriate corrections in powers of α' and g_s^2.

It should be mentioned, finally, that the integration procedure applied here to the anisotropic, Bianchi-I-type background (4.8) can be extended and applied to other classes of homogeneous Bianchi models, characterized by a d-dimensional group of non-Abelian isometries. In that case the metric can be parametrized (in the synchronous frame) as

$$g_{00} = 1, \qquad g_{0i} = 0, \qquad g_{ij} = e_i^m(x)\gamma_{mn}(t)e_j^n(x), \tag{4.89}$$

where all dependence on the spatial coordinates is contained in the "spatial" *vielbein* e_i^m, whose Ricci rotation coefficients

$$C_{mn}{}^p = e_m^i e_n^j\left(\partial_i e_j^p - \partial_j e_i^p\right) \tag{4.90}$$

are constant. They are determined by the algebraic structure of the isometry group as

$$[\xi_m, \xi_n] = C_{mn}{}^p\xi_p, \qquad \xi_p = \xi_p^i\partial_i, \tag{4.91}$$

where ξ_p^i, with $p = 1, \dots, d$, are the d Killing vectors generating the given non-Abelian isometries (see for instance [20]). The integration method presented

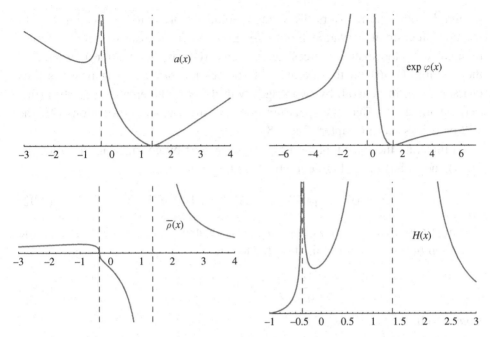

Figure 4.5 Plot of the solutions (4.85), (4.86), (4.88), for the isotropic case with $d = 3$, $\gamma_0 = 0$, $\gamma_i = 1/3$, $x_0 = 0$, $x_i = 1$, $a_{i0} = 1$, $\phi_0 = 0$ and $L^2 = 2/3$ (a similar plot can be obtained for the dual case with $\gamma = -1/3$). There are two singular points at $x_\pm = (1 \pm \sqrt{3})/2$ where the curvature diverges; the asymptotic behavior (4.56) is recovered at small enough curvature, in the limit $x \to \pm\infty$.

here has been applied to the low-energy equations in four dimensions, with homogeneous (i.e. space-independent) dilaton, without sources ($T_{\mu\nu} = 0$), torsion background ($B_{\mu\nu} = 0$) and dilaton potential ($V = 0$), to obtain the general exact solution for homogeneous metrics of Bianchi type I, II, III, V, VI_0 and VI_h [21]. In all those cases one finds that the solution exhibits, in general, two physical branches characterized by an initial or final singularity, except for the trivial case in which the dilaton is constant and the metric globally flat (up to reparametrizations). Thus, the singularity cannot be avoided in the context of homogeneous gravi-dilaton backgrounds, at least if other sources and/or higher-derivative and quantum loop corrections are not included in the effective action.

4.2.2 Asymptotic limits

The possible regularizing effect of the higher-order corrections will be discussed in Chapter 6. Here we conclude the discussion by illustrating an important property of the general low-energy solution (4.85), (4.86): as one approaches the singular points $x \to x_\pm$ the contribution of the matter sources becomes more and more negligible, and one recovers the vacuum solutions presented in Section 4.1. In the opposite, small curvature limit $x \to \pm\infty$ one finds instead a configuration dominated by the matter sources, of the same type as the solution presented in the example (4.56). In the presence of sources the pre-big bang phase is thus characterized, in general, by two stages with different kinematic properties (this may significantly affect the spectrum of the cosmological perturbations [22], as will be discussed in Chapters 7 and 8).

To illustrate these asymptotic properties we start by considering the solutions (4.85) and (4.86) in the high-curvature limit $x \to x_\pm$, where

$$a_i \sim |x - x_\pm|^{\frac{\alpha_i}{\alpha} \pm \delta_i}, \qquad \overline{\phi} \sim -\ln|x - x_\pm|^{\frac{\alpha_0}{\alpha} \pm \delta_0}. \tag{4.92}$$

The time coordinate x, on the other hand, is related to the cosmic time t by the definition (4.68), which in this limit reduces to

$$\frac{dx}{dt} \sim |x - x_\pm|^{1 - \frac{\alpha_0}{\alpha} \mp \delta_0}, \tag{4.93}$$

from which

$$|x - x_\pm| \sim |t - t_\pm|^{\alpha/(\alpha_0 \pm \alpha\delta_0)}. \tag{4.94}$$

Using this result in Eq. (4.92) we obtain

$$a_i \sim |t - t_\pm|^{\beta_i}, \qquad \overline{\phi} \sim -\ln|t - t_\pm|, \tag{4.95}$$

where

$$\beta_i^\pm = \frac{\alpha_i \pm \alpha\delta_i}{\alpha_0 \pm \alpha\delta_0}, \qquad \sum_i (\beta_i^\pm)^2 = 1, \tag{4.96}$$

and we exactly recover the Kasner-like, dilaton-dominated solution for anisotropic backgrounds, already given in Eqs. (4.31) and (4.33). It is important to stress that for this gravi-dilaton system there is no occurrence of the so-called BKL oscillations [23] of the Kasner exponents β_i near the singularity [24] (however, such oscillations generically reappear when all the massless, bosonic p-form fields present in superstring models are turned on, leading the system to a chaotic approach to the singularity [25]).

In the opposite, small-curvature limit, where $|x| \gg |x_\pm|$, the solutions (4.85) and (4.86) become

$$a_i \sim |x|^{2\alpha_i/\alpha}, \qquad \overline{\phi} \sim -\ln|x|^{2\alpha_0/\alpha}, \tag{4.97}$$

and the definition (4.68) for $x(t)$ gives

$$|x| \sim |t|^{\frac{\alpha}{2\alpha_0 - \alpha}}. \tag{4.98}$$

Using the cosmic-time coordinate, the asymptotic solution for $x \to \pm\infty$ takes the form

$$a_i \sim |t|^{2\alpha_i/(1-\gamma_0^2/4+\sum_i \alpha_i^2)},$$

$$\overline{\phi} \sim \frac{\gamma_0-2}{1-\gamma_0^2/4+\sum_i \alpha_i^2} \ln|t|,$$

$$\phi \sim 2 \frac{\sum_i \alpha_i - \alpha_0}{1-\gamma_0^2/4+\sum_i \alpha_i^2} \ln|t|, \qquad (4.99)$$

$$\overline{\rho} \sim |t|^{-(\gamma_0-\gamma_0^2/2+2\sum_i \alpha_i^2)/(1-\gamma_0^2/4+\sum_i \alpha_i^2)},$$

which generalizes to the case $\sigma = \gamma_0 \rho \neq 0$ the results already presented in [12, 11]. For $\gamma_0 = 0$, and in the isotropic limit, one recovers the particular solution of Eq. (4.54).

Let us conclude the section by reporting, in view of future applications, the isotropic form of the asymptotic solutions (in the small and large curvature limits) as a function of the conformal time η, for both the S-frame and the E-frame representations.

The large-time, matter-dominated solution (4.99), in a $(d+1)$-dimensional, isotropic background where $\gamma_i \equiv \gamma$ has the same value along all directions, takes the form

$$a(t) \sim |t|^{(2\gamma-\gamma_0)/[1+d\gamma^2+(d-1)\gamma_0^2/4-d\gamma\gamma_0]},$$

$$\phi \sim \frac{2d\gamma - 2 + (1-d)\gamma_0}{2\gamma - \gamma_0} \ln a, \qquad (4.100)$$

$$\rho \sim a^{-[2d\gamma(1+\gamma)+\gamma_0(1-d-2d\gamma)+(d-1)\gamma_0^2/2]/(2\gamma-\gamma_0)}$$

(it should be noted that for $\gamma_0 \neq 0$ the reflection $\gamma \to -\gamma$ is not a symmetry transformation generating new solutions, since the duality invariace is broken for matter with dilaton charge). Using the conformal-time coordinate η, defined by $dt = a \, d\eta$, the above expressions for ϕ and ρ in terms of a are still valid, while the scale factor becomes

$$a(\eta) \sim |\eta|^{(2\gamma-\gamma_0)/[1+d\gamma^2-2\gamma+\gamma_0(1-d\gamma)+(d-1)\gamma_0^2/4]}. \qquad (4.101)$$

In the opposite, large-curvature limit we have a dilaton-dominated solution which takes the isotropic form (4.30), and which can also be rewritten as

$$a_\pm(t) \sim |t|^{\pm 1/\sqrt{d}}, \qquad \phi_\pm \sim \sqrt{d}(\sqrt{d}\mp 1)\ln a_\pm. \qquad (4.102)$$

In conformal time

$$a_\pm(\eta) \sim |\eta|^{\pm 1/(\sqrt{d}\mp 1)}. \qquad (4.103)$$

Now we move to the E-frame, parametrized by the "tilded" variables $\tilde{a}, \tilde{\phi}, \ldots$, related to the S-frame variables by the set of transformations (2.64). We absorb, for simplicity, the powers of λ_s/λ_P into the transformed variables, with the understanding that we are switching from string to Planck units. We also recall that we are using units in which $2\lambda_s^{d-1} = 1$, so that $\mu = \sqrt{2/(d-1)}$. The matter-dominated, isotropic solution (4.100), expressed in terms of the E-frame cosmic time $\tilde{t}$, then becomes

$$\tilde{a}(\tilde{t}) \qquad\qquad \sim |\tilde{t}|^{\beta_E},$$

$$\tilde{\rho} \qquad\qquad \sim \tilde{a}^{-2/\beta_E}, \tag{4.104}$$

$$\tilde{\phi} \sim (d-1)\frac{2(d\gamma-1)-(d-1)\gamma_0}{2(1-\gamma)}\ln\tilde{a},$$

where

$$\beta_E = \frac{2(1-\gamma)}{(d-1)(1+d\gamma^2)+2(1-d\gamma)+(d-1)\gamma_0[1-d\gamma+(d-1)\gamma_0/4]}. \tag{4.105}$$

We note that for a radiation fluid ($d\gamma = 1$) with zero dilaton charge ($\gamma_0 = 0$) the dilaton is trivially constant, and the solution retains the same form in both the E-frame and the S-frame (compare Eqs. (4.105) and (4.100)). However, this triviality is broken by the direct coupling of the dilaton to the matter sources. The conformal-time version of the above solution can be easily obtained by recalling that the conformal-time parameter is the same in both frames,

$$d\eta = \frac{dt}{a} = \frac{d\tilde{t}}{\tilde{a}} = d\tilde{\eta} \tag{4.106}$$

(from the transformations (2.64)), and that a and $\tilde{a}$ are related by

$$\tilde{a} = a^{2(1-\gamma)/(d-1)(2\gamma-\gamma_0)} \tag{4.107}$$

(again, using Eq. (2.64)). We thus obtain, from Eq. (4.101),

$$\tilde{a}(\eta) \sim |\eta|^{2(1-\gamma)/(d-1)[1-2\gamma+d\gamma^2+\gamma_0(1-d\gamma)+(d-1)\gamma_0^2/4]}. \tag{4.108}$$

Let us finally report the E-frame transformed version of the vacuum, dilaton-dominated solution (4.102). By applying the transformations (2.64) we obtain

$$\tilde{a}_\pm = a_\pm^{(\pm\sqrt{d}-1)/(d-1)}, \qquad |t| = |\tilde{t}|^{(d-1)/(d\mp\sqrt{d})}. \tag{4.109}$$

Their combination, and the use of the solution (4.102), leads to

$$\tilde{a}_\pm(\tilde{t}) \sim |\tilde{t}|^{1/d}, \qquad \tilde{\phi}_\pm \sim \pm\sqrt{2d(d-1)}\ln\tilde{a}_\pm. \tag{4.110}$$

In conformal time $\eta = \tilde{\eta}$,

$$\tilde{a}_\pm(\eta) \sim |\eta|^{1/(d-1)}. \tag{4.111}$$

For this E-frame solution, a duality transformation inverting the S-frame scale factor, $a_+ \leftrightarrow a_-$, is associated with a dilaton reflection, i.e. with a transformation inverting the tree-level string coupling $(\widetilde{g}_s^\pm)^2 = \exp(\widetilde{\phi}_\pm)$:

$$-\widetilde{\phi}_- = \widetilde{\phi}_+ \leftrightarrow \widetilde{\phi}_- = -\widetilde{\phi}_+, \quad \Rightarrow \quad \widetilde{g}_s^+ \leftrightarrow (\widetilde{g}_s^+)^{-1}, \quad \widetilde{g}_s^- \leftrightarrow (\widetilde{g}_s^-)^{-1}. \quad (4.112)$$

The scale-factor duality typical of the S-frame is thus mapped into a weak coupling – strong coupling transformation, with no effect on the E-frame metric background, since $\widetilde{a}_+ = \widetilde{a}_-$.

4.3 Global $O(d, d)$ symmetry

The generalized properties of target space duality illustrated in the previous section are not limited to the case of Bianchi-I-type metric backgrounds, but are expected to be valid (with the appropriate modifications) for generic string theory backgrounds, possibly to all orders [26]. Already at the tree-level, for instance, it turns out that the symmetry associated with scale-factor duality transformations is only a particular case of a more general invariance under global transformations of the pseudo-orthogonal group $O(d, d)$, which induce a non-trivial mixing of the metric and of the NS–NS two-form B [27, 28], and which is valid in time-dependent backgrounds with d Abelian isometries (see [29] for a general discussion). Such an invariance property can also be extended to backgrounds with non-Abelian isometries [30], as will be briefly discussed at the end of this section.

In order to illustrate the $O(d, d)$ covariance of the string cosmology equations we need the full multiplet of massless states of the closed bosonic string: $\{\phi, g_{\mu\nu}, B_{\mu\nu}\}$. We assume that the $(d+1)$-dimensional background is isometric with respect to the d spatial translations, so that we can choose a "synchronous" system of coordinates where

$$g_{00} = 1, \qquad g_{0i} = 0 = B_{0\mu}, \qquad (4.113)$$

and where $\{\phi, g_{ij}, B_{ij}\}$ are functions of the cosmic time only. The symmetry properties of this multiplet can then be studied by rewriting the action (2.21) directly in the synchronous gauge (to obtain the field equations, however, one must reintroduce g_{00} in the action, imposing the gauge only after performing the variational procedure).

For the purposes of this section it is convenient to adopt a matrix notation, defining a $d \times d$ matrix G representing the spatial part of the covariant metric tensor g_{ij} (obviously, G^{-1} will represent the contravariant components g^{ij}). We can then write in matrix form the non-vanishing components of the connection,

$$\Gamma_{ij}{}^0 = -\frac{1}{2}\dot{G}_{ij}, \qquad \Gamma_{0i}{}^j = \frac{1}{2}g^{jk}\dot{g}_{ik} = \frac{1}{2}\left(G^{-1}\dot{G}\right)_i{}^j, \qquad (4.114)$$

and of the Ricci tensor,

$$R_0{}^0 = -\frac{1}{4}\mathrm{Tr}\left(G^{-1}\dot{G}\right)^2 - \frac{1}{2}\mathrm{Tr}\left(G^{-1}\ddot{G}\right) - \frac{1}{2}\mathrm{Tr}\left(\dot{G}^{-1}\dot{G}\right),$$

$$R_i{}^j = -\frac{1}{2}\left(G^{-1}\ddot{G}\right)_i{}^j - \frac{1}{4}\left(G^{-1}\dot{G}\right)_i{}^j\mathrm{Tr}\left(G^{-1}\dot{G}\right) + \frac{1}{2}\left(G^{-1}\dot{G}G^{-1}\dot{G}\right)_i{}^j,$$

$$(4.115)$$

where

$$\mathrm{Tr}\left(G^{-1}\dot{G}\right) = \left(G^{-1}\right)^{ij}\dot{G}_{ji} = g^{ij}\dot{g}_{ji}, \tag{4.116}$$

and so on, and where we have used the notation $\dot{G}^{-1} \equiv \mathrm{d}\left(G^{-1}\right)/\mathrm{d}t$. In the same way, by introducing the matrix B representing the spatial components B_{ij} of the antisymmetric field, we obtain

$$H_{0ij} = \dot{B}_{ij},$$

$$H^{0ij} = g^{ik}g^{jl}\dot{B}_{kl} = \left(G^{-1}\dot{B}G^{-1}\right)^{ij}, \tag{4.117}$$

$$H_{\mu\nu\alpha}H^{\mu\nu\alpha} = 3H_{0ij}H^{0ij} = 3\dot{B}_{ij}(G^{-1}\dot{B}G^{-1})^{ij} = -3\mathrm{Tr}\left(G^{-1}\dot{B}\right)^2.$$

Finally, we introduce the rescaled variable $\overline{\phi}$, defined by

$$e^{-\overline{\phi}} = \lambda_s^{-d} \int d^d x \sqrt{|\det g_{ij}|}\, e^{-\phi}, \tag{4.118}$$

which can also be interpreted as the homogeneous limit of the non-local (but general-covariant) scalar variable defined in Appendix 4A, as we will see later. Assuming that our background has spatial sections of finite volume, $\left(\int d^d x \sqrt{|g|}\right)_{t=\mathrm{const}} = \left(\sqrt{|g|}V_d\right)_{t=\mathrm{const}} < \infty$, and absorbing into ϕ the constant $\ln(V_d/\lambda_s^d)$, we obtain

$$\overline{\phi} = \phi - \frac{1}{2}\ln|\det G|, \qquad \dot{\overline{\phi}} = \dot{\phi} - \frac{1}{2}\mathrm{Tr}(G^{-1}\dot{G}) \tag{4.119}$$

(we have used the identity $|\det G| = \exp[\mathrm{Tr}\ln G]$).

Summing up all contributions arising from ϕ, R and $H_{\mu\nu\alpha}$, the effective action (2.21) can then be written, in the synchronous gauge, as follows:

$$S = -\frac{\lambda_s}{2}\int dt\, e^{-\overline{\phi}}\left[\dot{\overline{\phi}}^2 + V + \frac{1}{4}\mathrm{Tr}\left(G^{-1}\dot{G}\right)^2 - \frac{1}{2}\mathrm{Tr}\left(\dot{G}^{-1}\dot{G}\right) + \frac{1}{4}\mathrm{Tr}\left(G^{-1}\dot{B}\right)^2\right.$$

$$\left. - \mathrm{Tr}\left(G^{-1}\ddot{G}\right) + \dot{\overline{\phi}}\mathrm{Tr}\left(G^{-1}\dot{G}\right)\right]. \tag{4.120}$$

The last two terms of this equation can be eliminated (modulo a total derivative) by noting that

$$\frac{d}{dt}\left[e^{-\overline{\phi}}\,\mathrm{Tr}\left(G^{-1}\dot{G}\right)\right] = e^{-\overline{\phi}}\left[\mathrm{Tr}\left(G^{-1}\ddot{G}\right) + \mathrm{Tr}\left(\dot{G}^{-1}\dot{G}\right) - \dot{\overline{\phi}}\,\mathrm{Tr}\left(G^{-1}\dot{G}\right)\right].$$
(4.121)

Using the identity $G^{-1}G = I$, from which $\dot{G}^{-1}G = -G^{-1}\dot{G}$, and

$$\dot{G}^{-1}\dot{G} = -G^{-1}\dot{G}G^{-1}\dot{G},$$
(4.122)

we can finally obtain for the action a quadratic expression in the first derivatives of the fields,

$$S = -\frac{\lambda_s}{2}\int dt\, e^{-\overline{\phi}}\left[\dot{\overline{\phi}}^2 - \frac{1}{4}\mathrm{Tr}\left(G^{-1}\dot{G}\right)^2 + \frac{1}{4}\mathrm{Tr}\left(G^{-1}\dot{B}\right)^2 + V\right].$$
(4.123)

We are now in the position of discussing the properties of $O(d, d)$ invariance, introducing a $2d \times 2d$ matrix M constructed from the spatial components of the metric and of the antisymmetric field as follows:

$$M = \begin{pmatrix} G^{-1} & -G^{-1}B \\ BG^{-1} & G - BG^{-1}B \end{pmatrix}.$$
(4.124)

Using the invariant metric η of the $O(d, d)$ group in the off-diagonal representation,

$$\eta = \begin{pmatrix} 0 & I \\ I & 0 \end{pmatrix}$$
(4.125)

(I is the unit d-dimensional matrix), we have

$$M\eta = \begin{pmatrix} -G^{-1}B & G^{-1} \\ G - BG^{-1}B & BG^{-1} \end{pmatrix}.$$
(4.126)

and

$$\mathrm{Tr}\left(\dot{M}\eta\right)^2 = 2\mathrm{Tr}\left[\dot{G}^{-1}\dot{G} + \left(G^{-1}\dot{B}\right)^2\right] = -2\mathrm{Tr}\left(G^{-1}\dot{G}\right)^2 + 2\mathrm{Tr}\left(G^{-1}\dot{B}\right)^2,$$
(4.127)

so that the action (4.123) can be rewritten as

$$S = -\frac{\lambda_s}{2}\int dt\, e^{-\overline{\phi}}\left[\dot{\overline{\phi}}^2 + \frac{1}{8}\mathrm{Tr}\left(\dot{M}\eta\right)^2 + V\right].$$
(4.128)

The matrix M, on the other hand, is symmetric, and belongs itself to the $O(d, d)$ group, since

$$M^T\eta M = M\eta M = \eta,$$
(4.129)

as can be checked from the definition (4.124). Thus $M\eta = \eta M^{-1}$ and

$$\dot{M}\eta = \eta\dot{M}^{-1}, \qquad \left(\dot{M}\eta\right)^2 = \eta\dot{M}^{-1}\dot{M}\eta,$$
(4.130)

whose trace gives

$$\text{Tr}\left(\dot{M}\eta\right)^2 = \text{Tr}\left(\dot{M}\dot{M}^{-1}\right) \qquad (4.131)$$

(we have used the cyclic property of the trace). The action (4.128) can thus be recast in the form

$$S = -\frac{\lambda_s}{2}\int dt\, e^{-\overline{\phi}}\left[\dot{\overline{\phi}}^2 + \frac{1}{8}\text{Tr}\left(\dot{M}\dot{M}^{-1}\right) + V\right]. \qquad (4.132)$$

The kinetic part of this action is explicitly invariant under global transformations of the $O(d, d)$ group preserving the shifted dilaton $\overline{\phi}$, i.e. under the transformations

$$\overline{\phi} \to \overline{\phi}, \qquad M \to \widetilde{M} = \Omega^T M \Omega, \qquad (4.133)$$

where Ω is a constant matrix satisfying

$$\Omega^T \eta \Omega = \eta. \qquad (4.134)$$

Indeed,

$$M^{-1} \to \Omega^{-1} M^{-1} (\Omega^T)^{-1}, \qquad (4.135)$$

so that

$$\text{Tr}\left(\dot{M}\dot{M}^{-1}\right) \to \text{Tr}\left[\Omega^T \dot{M}\Omega\Omega^{-1}\dot{M}^{-1}(\Omega^T)^{-1}\right] = \text{Tr}\left[\dot{M}\dot{M}^{-1}(\Omega^T)^{-1}\Omega^T\right]$$

$$= \text{Tr}\left(\dot{M}\dot{M}^{-1}\right). \qquad (4.136)$$

Such an invariance is still valid in the presence of a dilaton potential only if V is a constant, or is a function of the variable $\overline{\phi}$ (see Appendix 4A), or, more generally, is a function of some $O(d, d)$ scalar formed with M and $\overline{\phi}$.

In this general context, the scale-factor duality transformations (4.20) can be retrieved as a particular $O(d, d)$ transformation. Let us consider a pure gravi-dilaton background with $B = 0$, and a global transformation generated by the particular matrix $\eta \in O(d, d)$. We have

$$M = \begin{pmatrix} G^{-1} & 0 \\ 0 & G \end{pmatrix}, \qquad \widetilde{M} = \Omega^T M \Omega = \eta M \eta = \begin{pmatrix} G & 0 \\ 0 & G^{-1} \end{pmatrix}, \qquad (4.137)$$

so that $G \to \widetilde{G} = G^{-1}$, i.e. the considered transformation produces an effective inversion of the spatial part of the metric. For a diagonal isotropic metric, in particular, $G = -a^2 I$, and we obtain the scale-factor inversion $a \to \widetilde{a} = a^1$.

4.3.1 $O(d, d)$ symmetry and matter sources

As in the case of scale-factor duality, the invariance under global $O(d, d)$ transformations can be extended to the case in which there are sources (other than the ϕ and B fields) in the cosmological equations, provided the sources satisfy appropriate transformation rules [31]. To illustrate this possibility it is convenient to introduce "shifted" variables associated with the various source terms, namely

$$\bar{\rho} = \sqrt{|g|}\, T_0^0, \qquad \bar{\theta} = \sqrt{|g|}\, T^{ij}, \qquad \bar{\sigma} = \sqrt{|g|}\, \sigma, \qquad \bar{J} = \sqrt{|g|}\, J^{ij}, \qquad (4.138)$$

where $\bar{\theta}$ and $\bar{J}$ are $d \times d$ matrices representing, respectively, the spatial parts of the symmetric stress tensor and of the antisymmetric current density (2.27). In terms of these variables we can rewrite in compact matrix form the full set of equations (2.24), (2.25) and (2.28).

Let us consider, in particular, the case of an $O(d, d)$ covariant background with $V = V(\bar{\phi})$ and $\sigma = 0$. The variation of the action (4.132) with respect to $\bar{\phi}$ gives

$$\dot{\bar{\phi}}^2 - 2\ddot{\bar{\phi}} - \frac{1}{8}\mathrm{Tr}(\dot{M}\eta)^2 + \frac{\partial V}{\partial \bar{\phi}} - V = 0, \qquad (4.139)$$

which represents the matrix version of the dilaton equation (2.25). Re-inserting the variable g_{00} in the action, and performing the variation, we obtain the constraint

$$\dot{\bar{\phi}}^2 + \frac{1}{8}\mathrm{Tr}(\dot{M}\eta)^2 - V = \bar{\rho} e^{\bar{\phi}}, \qquad (4.140)$$

which exactly corresponds to the (00) component of Eq. (2.24) (in units $2\lambda_s^{d-1} = 1$). Finally, the variation with respect to M gives [27, 28, 31]

$$\frac{\mathrm{d}}{\mathrm{d}t}(M\eta\dot{M}) - \dot{\bar{\phi}}(M\eta\dot{M}) = e^{\bar{\phi}}\bar{T}, \qquad (4.141)$$

where

$$\bar{T} = \begin{pmatrix} -\bar{J}, & -\bar{\theta}G + \bar{J}B \\ G\bar{\theta} - B\bar{J}, & G\bar{J}G + B\bar{J}B - G\bar{\theta}B - B\bar{\theta}G \end{pmatrix} \qquad (4.142)$$

is a $2d \times 2d$ matrix representing the variational contribution of the matter action. This last equation contains the combination of the spatial part of the gravitational equations (2.26), and of Eq. (2.28) for the antisymmetric tensor field. Note however that there is an important difference from the spatial components of Eq. (2.26) (which is written for $V = V(\phi)$): the difference concerns the complete absence of the dilaton potential, and is due to the intrinsic "non-locality" of the variable $\bar{\phi}$ (see Appendix 4A for a detailed computation).

The matrix equations given above are all independent, and their combination leads to a generalized energy-conservation equation. The differentiation of Eq. (4.140), and the elimination of $\ddot{\bar{\phi}}$ and $\dot{\bar{\phi}}^2 - V$, through Eqs. (4.139) and (4.140), respectively, gives

$$\dot{\bar{\rho}} = e^{\bar{\phi}}\frac{\mathrm{d}}{\mathrm{d}t}\left[\frac{1}{8}\mathrm{Tr}(\dot{M}\eta)^2 e^{-2\bar{\phi}}\right]. \qquad (4.143)$$

Let us also exploit the spatial equation (4.141), written in the form

$$\frac{\mathrm{d}}{\mathrm{d}t}\left(\mathrm{e}^{-\overline{\phi}}M\eta\dot{M}\eta\right) = \overline{T}\eta, \tag{4.144}$$

and note that

$$\frac{\mathrm{d}}{\mathrm{d}t}\left(\mathrm{e}^{-\overline{\phi}}M\eta\dot{M}\eta\mathrm{e}^{-\overline{\phi}}M\eta\dot{M}\eta\right) = \mathrm{e}^{-\overline{\phi}}\left(\overline{T}\eta M\eta\dot{M}\eta + M\eta\dot{M}\eta\overline{T}\eta\right). \tag{4.145}$$

Tracing the last equation, using the identity

$$(M\eta\dot{M}\eta)^2 = -(\dot{M}\eta)^2 \tag{4.146}$$

(following from the fact that $M \in O(d,d)$), and inserting the result into Eq. (4.143) we can finally rewrite the conservation equation in the following useful form:

$$\dot{\overline{\rho}} + \frac{1}{4}\mathrm{Tr}(\overline{T}\eta M\eta\dot{M}\eta) = 0. \tag{4.147}$$

For an explicit demonstration of the possible physical effects of the antisymmetric tensor we consider here the example of a perfect fluid source, with diagonal stress tensor $\overline{\theta}G = -\overline{p}I$, evolving in a diagonal metric background with $G = -a^2 I$, and $G^{-1}\dot{G} = 2HI$. The conservation equation (4.147) then reduces to

$$\dot{\overline{\rho}} + dH\overline{p} + \frac{1}{2}\mathrm{Tr}\left(\overline{J}\dot{B}\right) = 0, \tag{4.148}$$

that is

$$\dot{\rho} + dH(\rho + p) + \frac{1}{2}J^{ik}\dot{B}_{ik} = 0. \tag{4.149}$$

We can thus note that the antisymmetric source-density J^{ik}, in this equation, plays the same role as that of an intrinsic vorticity tensor in the context of cosmological models with spinning fluid sources (see e.g. [32, 33]). In those models, on the other hand, it is known that vorticity is a source of repulsive contributions to the gravitational equations, possibly smoothing out the initial singularity [34]. We are thus led to the speculation that the same effect could be induced in string models with $B \neq 0$ (at the end of this section we will present a model confirming this conjecture).

The point we wish to stress, before further applications, is that the entire set of cosmological equations (4.139)–(4.141), with matter sources and with potential $V(\overline{\phi})$, is globally $O(d,d)$ covariant under the generalized transformations

$$\overline{\phi} \to \overline{\phi}, \qquad \overline{p} \to \overline{p}, \qquad M \to \Omega^T M\Omega, \qquad \overline{T} \to \Omega^T \overline{T}\Omega, \tag{4.150}$$

as can be easily checked using the transformation properties of M and η previously discussed. In other words, the $O(d,d)$ symmetry is still valid in the presence of sources, provided $\overline{T}$ transforms exactly like M. It seems to be appropriate to show immediately an example of sources satisfying this property, consisting of a gas of classical, non-interacting strings, minimally coupled to the massless background B, g, ϕ.

The evolution of a single string in the given background is described by the sigma model action (3.23). Following the discussion (and using the notations) of Chapter 3, we choose the world-sheet metric in the conformal gauge. The total action S_m for the string gas is then given by the sum over the single component strings, $S_m = \sum_i S^i_{\text{string}}$, where

$$S_{\text{string}} = \frac{1}{4\pi\alpha} \int d^{d+1}x \int_\Sigma d\sigma \, d\tau \, \delta^{d+1}(x - X(\sigma, \tau)) \left[(\partial_\tau X^\mu \partial_\tau X^\nu - X'^\mu X'^\nu) g_{\mu\nu}(x) \right.$$
$$\left. + (\partial_\tau X^\mu X'^\nu - \partial_\tau X^\nu X'^\mu) B_{\mu\nu}(x) \right] \tag{4.151}$$

(we have omitted the index i, for simplicity, and we have written explicitly partial derivatives with respect to τ, as we reserve the dot for cosmic-time differentiation). Note that there is no direct dilaton coupling because $R^{(2)} = 0$ in the conformal gauge (the world-sheet metric is globally flat). We have thus a gas with zero dilaton charge, in agreement with our assumption leading to Eqs. (4.139)–(4.141).

The variations with respect to X_μ and to the world-sheet metric $\gamma_{ab} = \eta_{ab}$ lead, respectively, to the "geodesic" equations of motion and to the associated constraints (see Chapter 3). The variation with respect to $g_{\mu\nu}$ and $B_{\mu\nu}$ leads, instead, to the energy-momentum density, $T_{\mu\nu}$, and to the axion current density, $J_{\mu\nu}$, according to the standard definitions (1.3) and (2.27). For any string we can write, in particular,

$$\bar{\theta}^{\mu\nu} \equiv \sqrt{|g|} T^{\mu\nu}(t) = \frac{1}{2\pi\alpha} \int d\sigma \frac{d\tau}{dX^0} (\partial_\tau X^\mu \partial_\tau X^\nu - X'^\mu X'^\nu), \tag{4.152}$$

$$\bar{J}^{\mu\nu} \equiv \sqrt{|g|} J^{\mu\nu}(t) = \frac{1}{2\pi\alpha} \int d\sigma \frac{d\tau}{dX^0} (\partial_\tau X^\mu X'^\nu - \partial_\tau X^\nu X'^\mu), \tag{4.153}$$

where we have integrated with respect to $t = X^0$, and we have absorbed the spatial part of the delta function into the trivial volume integral of the action (for a background with the given isometries, all fields and sources are only time dependent). The total stress tensor and axion current density are then obtained by summing over the single string contributions. We recall that this model of cosmological fluid may be characterized by a *negative* effective pressure [35, 36], and is used as the dominant background source in the context of the so-called string-gas cosmology that will be discussed in Chapter 6 (see [37] for a recent review).

In order to compute the duality transformations of the matrix $\overline{T}$, and thus of $\bar{\theta}$ and $\bar{J}$, we must determine how a solution of the string equations of motion, $X^\mu(\tau, \sigma)$, is changed when the background fields B and G are changed by a duality transformation represented by $M \to \Omega^T M \Omega$. For a convenient approach to this problem we can rewrite the string equations of motion (3.25), and the associated constraints, in explicit $O(d, d)$ covariant form. Introducing the $2d$-dimensional vector Z, with components

$$Z^T = (P_i, X'^i), \qquad P_i = G_{ij}\partial_\tau X^j + B_{ij}X'^j, \qquad (4.154)$$

the equations of motion and constraints can be written in matrix form, respectively, as [31]

$$\partial_\tau^2 X^0 - X''^0 = \frac{1}{2}Z^T \dot{M}Z, \qquad \partial_\tau Z = (\eta M Z)', \qquad (4.155)$$

$$(\partial_\tau X^0)^2 + (X'^0)^2 + Z^T M Z = 0, \qquad Z^T \eta Z + 2\partial_\tau X^0 X'^0 = 0. \quad (4.156)$$

It can be easily verified, in this form, that if a given set of string variables $\{\partial_\tau X^0, X'^0, Z\}$ satisfies the above equations for a given background M, then the transformed set

$$\partial_\tau X^0 \to \partial_\tau X^0, \qquad X'^0 \to X'^0, \qquad Z \to \Omega^{-1} Z \qquad (4.157)$$

satisfies the same equations in the transformed background

$$M \to \Omega^T M \Omega \qquad (4.158)$$

(this computation requires the use of the identity $\Omega^T = \eta \Omega^{-1} \eta$, following from Eq. (4.134)). To obtain the corresponding transformations of the source terms $\bar{\rho}$, $\bar{T}$, we may notice that, from Eq. (4.152),

$$\bar{\rho} = \sqrt{|g|}T_0^0 = \frac{1}{4\pi\alpha} \int d\sigma \frac{d\tau}{dX^0} \left[(\partial_\tau X^0)^2 - (X'^0)^2 \right]. \qquad (4.159)$$

It follows that $\bar{\rho}$ is automatically invariant under the transformation (4.157). The spatial matrix $\bar{T}$, on the other hand, can be rewritten in terms of Z and M, in compact form, as [31]

$$\bar{T} = \frac{1}{4\pi\alpha} \int d\sigma \frac{d\tau}{dX^0} \left(M Z Z^T \eta - \eta Z Z^T M \right). \qquad (4.160)$$

By applying the transformations (4.157) and (4.158), and exploiting the $O(d, d)$ invariance of η, Eq. (4.134), we immediately obtain

$$\bar{T} \to \Omega^T \bar{T} \Omega. \qquad (4.161)$$

Both source terms $\bar{\rho}$ and $\bar{T}$, for the model of string gas we have considered, thus transform exactly as required by Eq. (4.150) to preserve the $O(d, d)$ symmetry of the cosmological background.

We can also check that, in the absence of torsion, the particular duality transformation (4.137) generated by $\Omega = \eta$ acts on the sources as

$$\overline{T} = \begin{pmatrix} 0 & -\overline{\theta}G \\ G\overline{\theta} & 0 \end{pmatrix} \rightarrow \eta \overline{T} \eta = \begin{pmatrix} 0 & \overline{\theta}G \\ -G\overline{\theta} & 0 \end{pmatrix}, \qquad (4.162)$$

and thus induces a "reflection" of the spatial part of the stress tensor, $\sqrt{|g|}T_i{}^j \rightarrow -\sqrt{|g|}T_i{}^j$, exactly as predicted by the scale-factor duality transformation (4.44) for diagonal sources. More general $O(d, d)$ transformations, however, do not preserve the diagonal form of the matter stress tensor, and automatically introduce shear and/or bulk-viscosity terms (even if the background is torsionless).

A simple illustration of this important cosmological effect can be obtained by considering a perfect fluid source, characterized by $\overline{\theta}^i_j = -\overline{p}\delta^i_j$, $J^{ij} = 0$, in a torsionless, $B_{ij} = 0$, isotropic metric background, $g_{ij} = -a^2\delta_{ij}$. The initial configuration is then represented by

$$M = \begin{pmatrix} G^{-1} & 0 \\ 0 & G \end{pmatrix}, \qquad G = -a^2 I, \qquad \overline{T} = \begin{pmatrix} 0 & \overline{p}I \\ -\overline{p}I & 0 \end{pmatrix}. \qquad (4.163)$$

Let us consider, for simplicity, the case of $d = 2$ spatial dimensions, and let us apply to the configuration (4.163) the one-parameter $O(2, 2)$ transformation generated by the 4×4 matrix

$$\Omega(\alpha) = \frac{1}{2} \begin{pmatrix} 1+c & s & c-1 & -s \\ -s & 1-c & -s & 1+c \\ c-1 & s & 1+c & -s \\ s & 1+c & s & 1-c \end{pmatrix}, \qquad c \equiv \cosh\alpha, \quad s \equiv \sinh\alpha, \qquad (4.164)$$

where α is a real parameter ranging from 0 to ∞. It can be easily checked that $\Omega^T \eta \Omega = \eta$, and that for $\alpha \rightarrow 0$ this transformation simply generates the discrete inversion of one of the two scale factors [38]. The transformed sources $\Omega^T \overline{T} \Omega$ still have $\overline{J} = 0$, but are associated with a non-diagonal stress tensor,

$$\widetilde{\overline{\theta}}^i_j = -\overline{p}\tau^i{}_j, \qquad \tau^i{}_j = \begin{pmatrix} c & -s \\ -s & -c \end{pmatrix}, \qquad (4.165)$$

in a non-diagonal metric background [31]

$$\widetilde{G} = -\frac{1}{2ca^2} \begin{pmatrix} c(1+a^4)+a^4-1 & -s(1+a^4) \\ -s(1+a^4) & c(1+a^4)-a^4+1 \end{pmatrix}. \qquad (4.166)$$

We should recall now that the stress tensor of a comoving fluid, including possible viscosity terms, can be parametrized in general as [39]

$$\theta^i{}_j = -(p - \xi V)\delta^i_j + 2S\sigma^i{}_j. \qquad (4.167)$$

Here ξ and S are, respectively, the bulk and shear viscosity coefficients, $V = \nabla_\mu u^\mu$ is the expansion parameter, u^μ is the geodesic velocity field of the comoving fluid and

$$\sigma^i_{\ j} = \nabla^i u_j - \frac{V}{D-1}\delta^i_j \tag{4.168}$$

is the traceless shear tensor in D space-time dimensions. For the transformed metric (4.166) we can easily find that $V = 0$, and that $\sigma^i_{\ j} = \Gamma^i_{\ 0j} = H\tau^i_{\ j}$, with $H = \dot{a}/a$. Comparing Eqs. (4.165) and (4.167) we are thus led to conclude that the transformed source can be consistently described as a pressureless fluid with no bulk viscosity, and with shear viscosity proportional to the original pressure, i.e.

$$\widetilde{\overline{p}} = 0, \qquad \widetilde{\xi} = 0, \qquad \widetilde{S} = -\frac{p}{2H}. \tag{4.169}$$

Therefore, we can say that the perfect fluid approximation is not compatible, in general, with the property of $O(d, d)$ covariance of the string cosmology equations.

4.3.2 General integration of the matrix equations

The property of $O(d, d)$ symmetry is crucial for obtaining an exact integration of the equations with antisymmetric field and sources, for a wide class of equations of state which generalize the barotropic case (4.64) discussed in the previous section.

For such an integration we have to work, indeed, with the matrix form of the equations; in particular, we need the combination of Eqs. (4.139) and (4.140), which we write in the form

$$\frac{\mathrm{d}^2}{\mathrm{d}t^2}\left(e^{-\overline{\phi}}\right) + \frac{1}{2}e^{-\overline{\phi}}\left(\frac{\partial V}{\partial\overline{\phi}} - 2V\right) = \frac{1}{2}\overline{\rho}, \tag{4.170}$$

and which we couple to Eq. (4.144), multiplied from the right by η. Let us also re-introduce the time variable x defined by Eq. (4.68), and assume that the sources satisfy the differential relation

$$\overline{\rho}\,\mathrm{d}\Gamma = \overline{T}\,\mathrm{d}x, \tag{4.171}$$

where Γ is a $2d \times 2d$ (possibly time-dependent) matrix relating $\overline{\rho}$ and $\overline{T}$. In this sense, Γ specifies a particular "equation of state" for the sources. Finally, assuming that

$$\frac{\partial V}{\partial\overline{\phi}} = 2V, \tag{4.172}$$

we find that the equations (4.170) and (4.144) can be integrated a first time to give

$$\frac{1}{2}\overline{\rho}\left(e^{-\overline{\phi}}\right)' = \frac{x + x_0}{L^2},\tag{4.173}$$

$$\overline{\rho}M\eta M' = \frac{4}{L^2}e^{\overline{\phi}}\Gamma(x).\tag{4.174}$$

Using the last equation, and the definition $\overline{T}/\overline{\rho} = \Gamma'$, we can also rewrite the conservation equation (4.147) as

$$\frac{1}{2}e^{-\overline{\phi}}\overline{\rho}' + \frac{1}{4L^2}\mathrm{Tr}\,(\Gamma\eta\Gamma\eta)' = 0.\tag{4.175}$$

Summing Eqs. (4.173) and (4.175), and integrating, we obtain

$$L^2\overline{\rho}e^{-\overline{\phi}} = \beta + (x + x_0)^2 - \frac{1}{2}\mathrm{Tr}\,(\Gamma\eta)^2 \equiv D(x),\tag{4.176}$$

where β is a dimensionless integration constant. Inserting this result into Eqs. (4.173) and (4.174) we are finally led to separate equations for $\overline{\phi}$ and M,

$$\overline{\phi}' = -\frac{2}{D(x)}(x + x_0),\tag{4.177}$$

$$M\eta M' = \frac{4}{D(x)}\Gamma(x),\tag{4.178}$$

generalizing the equations (4.75) and (4.76) which are valid for the pure gravi-dilaton system.

We have still to fix the constant β, which can be determined by imposing the separate validity of Eq. (4.140). Using the previous two equations, together with the identity (4.146), we then obtain the condition

$$(x + x_0)^2 - \frac{1}{2}\mathrm{Tr}\,(\Gamma\eta)^2 = D(x) + \frac{D^2(x)}{L^2\overline{\rho}^2}V(\overline{\phi})\tag{4.179}$$

which, together with the condition (4.172) on the potential, can be satisfied in two ways. A first possibility is the trivial case $V = 0$, which implies $\beta = 0$. In this case we are led to the same situation as that discussed in the previous section, and in which regular isotropic solutions are excluded. Now, however, there is a second, non-trivial possibility, corresponding to the potential

$$V = -V_0 e^{2\overline{\phi}}, \qquad V_0 = \mathrm{const},\tag{4.180}$$

which also satisfies Eq. (4.172) and leads, through Eq. (4.176), to

$$\beta = L^2 V_0.\tag{4.181}$$

In this case one can obtain smooth solutions even in the isotropic case, *provided* $V_0 > 0$, as will be shown in Appendix 4B.

4.3.3 Non-trivial solutions via duality transformations

New exact solutions of the cosmological equation can also be obtained without performing an explicit integration, and by applying appropriate $O(d, d)$ transformations to already known (even trivial) solutions. The procedure is conceptually the same as that we have exploited in the case of scale-factor duality transformations, and here will be applied to derive a regular class of low-energy cosmological backgrounds (with a non-vanishing two-form field, but without fluid sources).

Let us start from the trivial solution with $V = 0$, $B = 0$, constant dilaton and flat space-time metric, parametrized in Milne coordinates as

$$ds^2 = dt^2 - \left(\frac{t}{t_0}\right)^2 dx^2 - dy^2 - dz_i^2, \qquad \phi = \text{const}, \qquad (4.182)$$

where z_i are cartesian coordinates spanning a Euclidean $(d-2)$-dimensional manifold. In spite of the non-trivial scale factor $a(t)$ along the x-axes, this metric describes a globally flat space-time, as can be checked by performing the transformation

$$x' = t \sinh\left(\frac{x}{t_0}\right), \qquad t' = t \cosh\left(\frac{x}{t_0}\right), \qquad (4.183)$$

and noting that, with the new coordinates (t', x', y, z_i), the quadratic form (4.182) assumes everywhere the standard Minkowski form. This trivial solution is associated with the "dual partner"

$$ds^2 = dt^2 - \left(\frac{t}{t_0}\right)^{-2} dx^2 - dy^2 - dz_i^2, \qquad \phi = -2\ln\left|\frac{t}{t_0}\right| + \text{const}, \qquad (4.184)$$

obtained through the scale-factor duality transformation (4.25). This is also an exact solution of the low-energy string cosmology equations but, unlike the previous one, is characterized by a non-trivial geometry of the two-dimensional sections spanned by x and t. The corresponding space-time curvature is non-zero, and the two branches $t < 0$ and $t > 0$ (describing, respectively, pre-big bang accelerated expansion and post-big bang decelerated contraction) are separate by a singularity at $t = 0$.

Such a singular background can be regularized through a global $O(d, d)$ transformation (in our case, $d = 2$), in the sense that it can be transformed into a new exact solution of the same equations containing an additional dynamical dimension (along the y direction), and in which the pre- and post-big bang branches are smoothly connected near the origin, without singularities [38]. The same $O(d, d)$

transformation, applied to the trivial solution (4.182), also provides a new non-trivial solution which is smooth everywhere, and which asymptotically reproduces the regimes of linear expansion and contraction at $t \to \pm\infty$, typical of the Milne metric.

Let us write explicitly the initial matrix configuration representing the solutions in the spatial (x, y) plane, namely,

$$
B = 0, \qquad G_\pm = -\begin{pmatrix} a_\pm^2 & 0 \\ 0 & 1 \end{pmatrix}, \qquad M_\pm = \begin{pmatrix} G_\pm^{-1} & 0 \\ 0 & G_\pm \end{pmatrix},
$$

$$
a_\pm = \left| \frac{t}{t_0} \right|^{\pm 1}, \qquad \overline{\phi} = -\ln\left| \frac{t}{t_0} \right| \tag{4.185}
$$

(the $\pm$ signs correspond, respectively, to the Milne metric (4.182) and to its dual (4.184)). This two-dimensional sector of the solution has two Abelian isometries (indeed, it is invariant under spatial translations along x and y), and we can thus consider a global "boost" of the pseudo-orthogonal group $O(2, 2)$, generated by the one-parameter matrix already introduced in Eq. (4.164). By applying the transformations (4.133) to our initial background (4.185), we obtain a new set of background fields $\{\widetilde{G}, \widetilde{B}, \widetilde{\phi}\}$, defined as follows [38]:

$$
\widetilde{G}_\pm(\alpha) = -\begin{pmatrix} \frac{c\mp 1 + (c\pm 1)(t/t_0)^2}{c\pm 1 + (c\mp 1)(t/t_0)^2} & \frac{s[1+(t/t_0)^2]}{c\pm 1 + (c\mp 1)(t/t_0)^2} \\ \frac{s[1+(t/t_0)^2]}{c\pm 1 + (c\mp 1)(t/t_0)^2} & 1 \end{pmatrix},
$$

$$
\widetilde{B}_\pm(\alpha) = \begin{pmatrix} 0 & \frac{-s[1+(t/t_0)^2]}{c\pm 1 + (c\mp 1)(t/t_0)^2} \\ \frac{s[1+(t/t_0)^2]}{c\pm 1 + (c\mp 1)(t/t_0)^2} & 0 \end{pmatrix}, \tag{4.186}
$$

$$
\widetilde{\phi}_\pm = -\ln\left[c\pm 1 + (c\mp 1)(t/t_0)^2 \right].
$$

These new backgrounds exactly satisfy the string cosmology equations with $T_{\mu\nu} = 0 = V$, as can be explicitly checked by Eqs. (2.24), (2.25) and (2.28), or by their matrix versions (4.139)–(4.141). Unlike the initial backgrounds, however, the new ones are non-trivial and everywhere regular, since all curvature invariants are bounded, as well as the string couplings $(\widetilde{g}_s^2)_\pm = \exp(\widetilde{\phi}_\pm)$. The new solutions represent a $(2+1)$-dimensional background evolving smoothly from (anisotropic) contraction to expansion, or vice-versa, according to the behavior of the initial metric (the Milne solution or its dual, respectively). These kinematic properties can be displayed by computing, for instance, the rate of change $\widetilde{H}$ of the relative distance along the x direction between two comoving observers, which for the

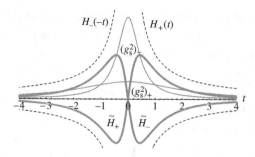

Figure 4.6 The dashed curves represent the Hubble parameters $H_\pm(\pm t)$ for the pre-boosted solutions of Eq. (4.185). The full bold curves, labeled by $\tilde{H}_\pm$, represent the corresponding variables (4.187) for the boosted solution (4.186). Also shown is the smooth behavior of the string coupling $(g_s^2)_\pm$ (thin curves).

metric $\tilde{G}_\pm$ is given by

$$\tilde{H}_\pm = V_{\mu\nu} n^\mu n^\nu = \pm \frac{4ct}{t_0^2[c+1+(c-1)(t/t_0)^2][c-1+(c+1)(t/t_0)^2]}. \quad (4.187)$$

Here $V_{\mu\nu} = \nabla_{(\mu}u_{\nu)}$ is the so-called expansion tensor for a congruence of comoving geodesics u^μ, and n^μ is a unit space-like vector along the x direction, $n_\mu n^\mu = -1$, $n_\mu u^\mu = 0$. By taking $u^\mu = \delta_0^\mu$ then, in the synchronous frame, $\tilde{H} = \Gamma_{11}{}^0 = -\dot{g}_{11}/2$. The plots of $\tilde{H}_\pm$, together with those of $(g_s^2)_\pm$, are given in Fig. 4.6 for the particular values $t_0 = 1$ and $\alpha = 1$ (for $\alpha \gg 1$ the two coupling parameters $(g_s^2)_+$ and $(g_s^2)_-$ tend to coincide).

In both cases, the boosted backgrounds $\tilde{M}_+$ and $\tilde{M}_-$ describe the evolution from a phase of growing curvature and growing dilaton, to a phase of decreasing curvature and decreasing dilaton, as appropriate to a transition from a pre-big bang to a post-big bang configuration. Being $(2+1)$-dimensional, and highly anisotropic, this class of solutions does not seem appropriate to provide a realistic description of the present cosmological state. The situation drastically changes, however, if the final configuration for $t \to \infty$ is modified by including the back-reaction of the radiation produced by the transition from the accelerated to the decelerated regime (according to the mechanism discussed in Chapter 7).

Let us consider, for instance, the class of metric backgrounds $\tilde{G}_-$, which for $t < 0$ describe a phase of pre-big bang inflationary expansion, and let us extend the solution to a third spatial dimension, which has initially a trivial dynamics. Also, let us include, for $t > 0$, a small amount of radiation as a possible (initially negligible) source of the background geometry. A numerical analysis [40] of the string cosmology equations for $t > 0$ then shows that, after a short period of post-big bang contraction, the radiation tends to become dominant with respect to the dilaton kinetic energy. Once it is dominant, it rapidly stabilizes the dilaton and isotropizes the metric background, eventually turning the initial contraction into a phase of decelerated and isotropic expansion of all the three spatial dimensions (as appropriate to a realistic configuration of the standard cosmological scenario).

Other examples of regular, homogeneous and isotropic backgrounds will be presented in Appendix 4B. We recall here that the reported example of regular solution, obtained by applying an appropriate $O(d, d)$ transformation to a singular background, is not the only example available in the literature of this interesting property of the string cosmology equations. Another example has been obtained by applying an $O(3, 3)$ transformation to the four-dimensional, non-homogeneous and singular solution of Nappi and Witten [41]: the transformed five-dimensional axion-gravi-dilaton background [42] is still non-homogeneous, but does not contain singularities, either in the curvature of the transformed metric $\tilde{G}$, or in the tree-level coupling $\tilde{g}_s^2 = \exp \tilde{\phi}$.

4.3.4 Non-Abelian duality

The presence of the $O(d, d)$ symmetry is a general property of the string backgrounds admitting a d-parameter group of Abelian isometries. When the isometries are spatial translations, in particular, the corresponding background is described by a Bianchi-I type metric; this background, however, is only a particular case of a larger class of homogeneous manifolds – the so-called Bianchi models, whose d-dimensional spatial sections are invariant under the action of a d-parameter Lie group of non-Abelian isometries [20]. A question which naturally arises, therefore, is whether a new class of duality transformations may exist in a generic homogeneous background, generalizing the $O(d, d)$ transformations only valid for Abelian isometry groups.

The answer to this question is positive [30], and, in general, the modified duality transformations associated with a non-Abelian group can be implemented through the standard prescriptions, using either a Lagrangian or a Hamiltonian approach [43] (see also [44]).

Consider, for instance, a Wess–Zumino–Witten (WZW) model represented by the two-dimensional action (3.23), and describing a string conformally coupled to a homogeneous backgroud (with non-Abelian isometries). The metric $g_{\mu\nu}$ can be parametrized as in Eq. (4.89), and the same factorization can be used for the NS–NS two-form $B_{\mu\nu}$,

$$B_{0i} = 0, \qquad B_{ij} = e_i^m(X)\beta_{mn}(t)e_j^n(X). \qquad (4.188)$$

Here $\beta_{mn} = -\beta_{nm}$, and the spatial dependence of the *vielbein* fields e_i^m is fixed (up to reparametrizations) by the isometry group according to Eqs. (4.90) and (4.91). Performing an appropriate transformation of the spatial coordinates of the string, $X_i \rightarrow \widetilde{X}_i$, of the spatial part of the metric, $G \rightarrow \widetilde{G}$, and of the two-form, $B \rightarrow \widetilde{B}$, one can obtain a "dual" conformal action representing a new WZW model parametrized by the coordinates $\widetilde{X}$ and by the transformed background fields $\widetilde{G}$, $\widetilde{B}$, where [30]

$$\widetilde{G}^{-1}(\widetilde{X}, t) = (\gamma + \beta + K)\gamma^{-1}(\gamma - \beta - K),$$

$$\widetilde{B}(\widetilde{X}, t) = -(\gamma - \beta - K)^{-1}(\beta + K)(\gamma + \beta + K)^{-1}. \qquad (4.189)$$

Here γ and β are $d \times d$ matrices (symmetric and antisymmetric, respectively), representing the background tensors γ_{mn} and β_{mn} of Eqs. (4.89) and (4.188), and K is an antisymmetric $d \times d$ matrix representing the tensor

$$K_{mn} = C_{mn}{}^p \widetilde{X}_p, \qquad (4.190)$$

directly associated with the non-Abelian part of the isometry group.

It can be easily checked that the above transformation extends to the non-Abelian case the global $O(d, d)$ transformation generated by the off-diagonal metric η of Eq. (4.125). In fact, starting with a generic non-symmetric background ($G \neq 0$ and $B \neq 0$), and applying the transformation (4.133) with $\Omega = \eta$, one finds

$$\widetilde{M} = \eta M \eta = M = \begin{pmatrix} G - BG^{-1}B & BG^{-1} \\ -G^{-1}B & G^{-1} \end{pmatrix}. \qquad (4.191)$$

Comparing $\widetilde{M}$ with the definition (4.124) of the background matrix M one then obtains

$$\widetilde{G}^{-1}(t) = G - BG^{-1}B = (G + B)G^{-1}(G - B),$$

$$\widetilde{B}(t) = B^{-1} - G^{-1}BG^{-1} = -(G - B)^{-1}B(G + B)^{-1}, \qquad (4.192)$$

reproducing the transformation (4.189) for the special case $K = 0$ of an Abelian isometry group.

An important point to be stressed for the non-Abelian case is that the initial background, $\{G, B\}$, and the duality-transformed one, $\{\widetilde{G}, \widetilde{B}\}$, do not share, in general, the same isometry group, as evident from Eq. (4.189). This is to be contrasted with the case of Abelian duality transformations. An even more important difference concerns the transformation of the dilaton field, which is required to complete the set of background transformations.

The standard prescription, dictated by the conformal invariance of the integration measure in the path-integral representation of the partition function [3, 5, 6, 45, 46, 47], is only valid for non-Abelian isometry groups which are semisimple [44]. The homogeneous Bianchi models, on the other hand, provide important examples of physically relevant background solutions with non-semisimple, non-Abelian isometry groups, whose dual counterpart fails to satisfy the expected requisite of conformal invariance, no matter what choice is made for the transformed dilaton. Explicit examples of this anomalous behavior have been presented for models of Bianchi types V [43], IV [44] and III [21].

The reason for this failure is that, for non-semisimple groups, there are generators with non-vanishing trace in the adjoint representation, which are sources of a mixed gauge and gravitational anomaly in the integration measure of the integral representation of the model [44] (see also [48]). Such a mixed anomaly generates a contribution to the trace anomaly (of the world-sheet stress tensor) which cannot be absorbed by any shift or redefinition of the dilaton. This implies a breakdown of conformal invariance, and requires an additional anomaly cancelation condition for the quantum consistency of the non-Abelian duality transformations [44].

Appendix 4A
A non-local, general-covariant dilaton potential

The shifted variable $\overline{\phi}$ defined in Eq. (4.119) is invariant under global $O(d, d)$ transformations, but is not a scalar under general coordinate transformations. If we want to add a potential energy term $V(\overline{\phi})$, without breaking the general covariance of the effective action, we must include the full $(d + 1)$-dimensional proper volume into the definition of $\overline{\phi}$, thus leading to a scalar potential which is a *non-local* function of the dilaton. We will show in this appendix that the self-dual action (4.132), and the derived duality-covariant equations (4.139)–(4.141), can be obtained as the homogeneous limit of an action in which the dilaton potential is a local function of the non-local (but scalar under general coordinate transformations) variable $\overline{\phi} = \overline{\phi}(\phi)$, defined by [49]

$$e^{-\overline{\phi}(x)} = \lambda_s^{-d} \int d^{d+1}x' \sqrt{|g(x')|} \; e^{-\phi(x')} \sqrt{\partial_\mu \phi(x') \partial^\mu \phi(x')} \; \delta(\phi(x) - \phi(x')). \qquad (4A.1)$$

This definition of $\overline{\phi}$ is obviously appropriate to backgrounds with $\partial_\mu \phi \partial^\mu \phi > 0$, but it can be easily extended to include the case of non-cosmological backgrounds, in which space-like dilaton gradients may be dominant.

We start by noting that in the limit of a homogeneous and isotropic background we can use the cosmic-time gauge, where $g_{00} = 1$, $\phi = \phi(t)$, and the previous definition reduces to

$$e^{-\overline{\phi}(x)} = \lambda_s^{-d} V_d \int dt' \frac{d\phi}{dt'} \sqrt{|g(t')|} \; e^{-\phi(t')} \; \delta(\phi(t) - \phi(t')) = \lambda_s^{-d} V_d \sqrt{|g(t)|} \; e^{-\phi(t)}, \qquad (4A.2)$$

where $V_d = \int dx'$. Assuming that the background has spatial sections of finite volume, $V_d < \infty$, we can absorb the constant volume factor inside ϕ, and we exactly recover the definition of $\overline{\phi}$ already used in Eqs. (4.17) and (4.119) of this chapter. Since $\exp \overline{\phi}$ plays the role of a "dimensionally reduced" coupling constant, we may expect (at least in the perturbative regime) the dilaton potential to go as some power of the coupling; we thus set $V = V(\exp \overline{\phi})$, and consider the following scalar-tensor action,

$$S = -\frac{1}{2\lambda_s^{d-1}} \int d^{d+1}x \sqrt{|g|} \; e^{-\phi} \left[R + (\nabla\phi)^2 + 2\lambda_s^{d-1} V(e^{-\overline{\phi}}) \right] + S_m, \qquad (4A.3)$$

invariant under general coordinate reparametrizations, as well as global $O(d, d)$ transformations of background fields with Abelian isometries. To obtain the field equations we shall impose that the action be stationary, by equating to zero the functional derivatives of S with respect to $g_{\mu\nu}$ and ϕ.

Let us first consider the case of the metric tensor. For the local part of the action (i.e. for S_m and the kinetic terms S_k) we can directly apply the results of Section 2.1 (in particular Eq. (2.8)) to obtain

$$
\frac{\delta}{\delta g_{\mu\nu}(x)}(S_k + S_m) = \int d^{d+1}x'\, \delta^{d+1}(x-x')\left[\frac{1}{2}\left(\sqrt{|g|}\, T_{\mu\nu}\right)_{x'}\right.
$$
$$
\left. -\frac{1}{2\lambda_s^{d-1}}\left(\sqrt{|g|}\, e^{-\phi}\right)_{x'}\left(G_{\mu\nu} + \nabla_\mu\nabla_\nu\phi + \frac{1}{2}g_{\mu\nu}\nabla\phi^2 - g_{\mu\nu}\nabla^2\phi\right)_{x'}\right].
\tag{4A.4}
$$

We have included in S_k the surface term required to cancel the second derivatives of the metric. Note that we are using the convenient notation in which a variable appended to round brackets, $(\ldots)_x$, means that all quantities inside the brackets are functions of the appended variable. We also use the notation $\phi_x \equiv \phi(x)$.

The functional differentiation of the potential gives

$$
A_{\mu\nu}(x) \equiv \frac{\delta}{\delta g^{\mu\nu}(x)}\int d^{d+1}x'\left(\sqrt{|g|}\, e^{-\phi}V\right)_{x'}
$$
$$
= \int d^{d+1}x'\left[-\frac{1}{2}\left(\sqrt{|g|}\, e^{-\phi}g_{\mu\nu}V\right)_{x'}\delta^{d+1}(x-x')\right.
$$
$$
\left. +\frac{1}{\lambda_s^d}\left(\sqrt{|g|}\, e^{-\phi}V'\right)_{x'}\frac{\delta}{\delta g^{\mu\nu}(x)}\int d^{d+1}y\left(\sqrt{|g|}\, e^{-\varphi}\sqrt{(\partial\varphi)^2}\right)_y \delta(\varphi_x - \varphi_y)\right],
\tag{4A.5}
$$

where V' denotes the derivative of the potential with respect to its argument,

$$
V' \equiv \frac{\partial V}{\partial e^{-\phi}} = -e^{\bar\phi}\frac{\partial V}{\partial\bar\phi}.
\tag{4A.6}
$$

The functional derivative of the last term of Eq. (4A.5) can be written explicitly as

$$
-\frac{1}{2}\int d^{d+1}y\left(\sqrt{|g|}\, e^{-\phi}\gamma_{\mu\nu}\sqrt{(\partial\phi)^2}\right)_y \delta^{d+1}(x-y)\delta(\varphi_x - \varphi_y),
\tag{4A.7}
$$

where

$$
\gamma_{\mu\nu} = g_{\mu\nu} - \frac{\partial_\mu\phi\partial_\nu\phi}{(\partial\phi)^2}.
\tag{4A.8}
$$

Inserting this result into Eq. (4A.5), and integrating the first term with respect to x', and the second with respect to y, we obtain

$$
-A_{\mu\nu} = \frac{1}{2}\left(\sqrt{|g|}\, e^{-\phi}g_{\mu\nu}V\right)_x + \frac{1}{2}\left(\sqrt{|g|}\, e^{-2\phi}\gamma_{\mu\nu}\sqrt{(\partial\phi)^2}\right)_x I_1(x),
\tag{4A.9}
$$

where

$$
I_1(x) = \lambda_s^{-d}\int d^{d+1}x'\left(\sqrt{|g|}\, V'\right)_{x'}\delta(\phi_{x'} - \phi_x).
\tag{4A.10}
$$

Finally, integrating Eq. (4A.4), and summing all terms to those of Eq. (4A.9), we are led to the integro-differential, generally covariant, gravitational equation

$$G_{\mu\nu} + \nabla_\mu \nabla_\nu \phi + \frac{1}{2} g_{\mu\nu} \left[(\nabla\phi)^2 - 2\nabla^2\phi - 2\lambda_s^{d-1} V \right] - \lambda_s^{d-1} e^{-\phi} \sqrt{(\partial\phi)^2} \; \gamma_{\mu\nu} I_1$$

$$= \lambda_s^{d-1} e^\phi T_{\mu\nu}, \tag{4A.11}$$

which generalizes Eq. (2.11) to the case of a dilaton potential depending on the non-local variable (4A.1).

The derivation of the dilaton equation proceeds along the same lines, through the computation of the functional derivative with respect to $\phi(x)$. For the local part of the action one finds, in agreement with the results of Section 2.1,

$$\frac{\delta}{\delta\phi(x)} (S_k + S_m) = \int d^{d+1}x' \, \delta^{d+1}(x - x') \left[-\frac{1}{2} \left(\sqrt{|g|} \, \sigma \right)_{x'} \right.$$

$$\left. + \frac{1}{2\lambda_s^{d-1}} \left(\sqrt{|g|} \, e^{-\phi} \right)_{x'} \left(R + 2\nabla^2\phi - \nabla\phi^2 \right)_{x'} \right]. \tag{4A.12}$$

The derivative of the non-local potential gives

$$B \equiv \frac{\delta}{\delta\phi(x)} \int d^{d+1}x' \left(\sqrt{|g|} \, e^{-\phi} V \right)_{x'}$$

$$= - \left(\sqrt{|g|} e^{-\phi} V \right)_x + \frac{1}{\lambda_s^d} \int d^{d+1}x' \left(\sqrt{|g|} e^{-\phi} V' \right)_{x'}$$

$$\times \int d^{d+1}y \left[- \left(\sqrt{|g|} e^{-\phi} \sqrt{(\partial\phi)^2} \right)_y \delta(\phi_{x'} - \phi_y) \delta^{d+1}(x - y) \right.$$

$$\left. + \left(\sqrt{|g|} e^{-\phi} \sqrt{(\partial\phi)^2} \right)_y \delta'(\phi_{x'} - \phi_y) \left[\delta^{d+1}(x - x') - \delta^{d+1}(x - y) \right] \right]$$

$$- \frac{1}{\lambda_s^d} \int d^{d+1}x' \left(\sqrt{|g|} e^{-\phi} V' \right)_{x'}$$

$$\times \int d^{d+1}y \, \partial_\mu \left[\frac{\sqrt{|g|} e^{-\phi} \partial^\mu \phi}{\sqrt{(\partial\phi)^2}} \delta(\phi_{x'} - \phi_y) \right] \delta^{d+1}(x - y), \tag{4A.13}$$

where δ' denotes the derivative of the delta distribution with respect to its argument. Integrating, and using the properties of the delta distribution, one finds that there are exact cancelations between the first and the third integral, and the terms of the last integral containing $\partial_\mu \exp(-\phi)$ and $\partial_\mu [\delta(\phi_{x'} - \phi_y)]$, respectively. Thus, we are left with

$$B = - \left(\sqrt{|g|} e^{-\phi} V \right)_x - e^{-2\phi} \partial_\mu \left(\frac{\sqrt{|g|} \partial^\mu \phi}{\sqrt{(\partial\phi)^2}} \right) I_1(x)$$

$$+ \frac{1}{\lambda_s^d} \int d^{d+1}x' \left(\sqrt{|g|} e^{-\phi} V' \right)_{x'} \int d^{d+1}y \left(\sqrt{|g|} e^{-\phi} \sqrt{(\partial\phi)^2} \right)_y \delta'(\phi_x - \phi_y). \tag{4A.14}$$

The second term on the right-hand side, which multiplies I_1, can now be conveniently rewritten as

$$-e^{-2\phi}\nabla_\mu\left(\frac{\partial^\mu\phi}{\sqrt{(\partial\phi)^2}}\right) = -e^{-2\phi}\frac{\sqrt{|g|}}{\sqrt{(\partial\phi)^2}}\gamma_{\mu\nu}\nabla^\mu\nabla^\nu\phi. \qquad (4A.15)$$

In the term containing δ' we can set $dy_0 = d\phi_y/\dot{\phi}_y$, $(d/d\phi_y) = \dot{\phi}_y^{-1}(d/dy_0)$, and we obtain

$$\lambda_s^{-d}\int d^d y\, d\phi_y\left(\frac{\sqrt{|g|}\,e^{-\phi}\sqrt{(\partial\phi)^2}}{\dot{\phi}}\right)_y \delta'(\phi_x-\phi_y)$$

$$= \lambda_s^{-d}\int d^d y\,\frac{d\phi_y}{\dot{\phi}_y}\frac{d}{dy_0}\left(\frac{\sqrt{|g|}\,e^{-\phi}\sqrt{(\partial\phi)^2}}{\dot{\phi}}\right)_y \delta(\phi_x-\phi_y)$$

$$= -e^{-\bar{\phi}(x)} + e^{-\phi(x)}\lambda_s^{-d}\int d^d y\,\frac{d\phi_y}{\dot{\phi}_y}\frac{d}{dy_0}\left(\frac{\sqrt{|g|}\,\sqrt{(\partial\phi)^2}}{\dot{\phi}}\right)_y \delta(\phi_x-\phi_y)$$

$$= -e^{-\bar{\phi}(x)} + e^{-\phi(x)}\lambda_s^{-d}\int d^d y\,\frac{d\phi_y}{\dot{\phi}_y}\left(\sqrt{|g|}\,\sqrt{(\partial\phi)^2}\right)_y \delta'(\phi_x-\phi_y)$$

$$= -e^{-\bar{\phi}(x)} + e^{-\phi(x)}I_2(x), \qquad (4A.16)$$

where

$$I_2(x) = \lambda_s^{-d}\int d^{d+1}y\left(\sqrt{|g|}\,\sqrt{(\partial\phi)^2}\right)_y \delta'(\phi_x-\phi_y). \qquad (4A.17)$$

Thus

$$-B = \left(\sqrt{|g|}\,e^{-\phi}\right)_x\left(V + \frac{e^{-\phi}}{\sqrt{(\partial\phi)^2}}\gamma_{\mu\nu}\nabla^\mu\nabla^\nu\phi\,I_1 + e^{-\bar{\phi}}V' - e^{-\phi}V'I_2\right)_x. \qquad (4A.18)$$

Summing to this equation the contribution of Eq. (4A.12) we are finally led to the dilaton equation

$$R + 2\nabla^2\phi - (\nabla\phi)^2 + 2\lambda_s^{d-1}\left[V - \frac{\partial V}{\partial\bar{\phi}} + \frac{e^{-\phi}}{\sqrt{(\partial\phi)^2}}\gamma_{\mu\nu}\nabla^\mu\nabla^\nu\phi\,I_1 - e^{-\phi}V'I_2\right]$$

$$= \lambda_s^{d-1}e^\phi\sigma, \qquad (4A.19)$$

which generalizes Eq. (2.14) to the case of the non-local potential $V(\exp(-\bar{\phi}))$. This equation can be used to eliminate the scalar curvature in Eq. (4A.11), and to obtain

$$R_\mu{}^\nu + \nabla_\mu\nabla^\nu\phi - \lambda_s^{d-1}\delta_\mu^\nu\left(\frac{\partial V}{\partial\bar{\phi}} + e^{-\phi}V'I_2\right)$$

$$+ \lambda_s^{d-1}e^{-\phi}\left(\delta_\mu^\nu\frac{\gamma_{\beta\alpha}\nabla^\alpha\nabla^\beta\phi}{\sqrt{(\partial\phi)^2}} - \gamma_\mu{}^\nu\sqrt{(\partial\phi)^2}\right)I_1 = \lambda_s^{d-1}e^\phi\left(T_\mu{}^\nu + \frac{1}{2}\delta_\mu^\nu\sigma\right), \qquad (4A.20)$$

which generalizes Eq. (2.15).

It is important to stress that the equations following from the non-local action (4A.3) are qualitatively different from an action which has the same form but a local dilaton potential (see for instance Eqs. (2.24)–(2.26), with $H_{\mu\nu\alpha} = 0$). The differences tend to disappear in a metric background of the cosmological type, but they are not completely eliminated even if the geometry is homogeneous, isotropic and spatially flat, as is shown in detail in the following.

Let us consider, in fact, the conformally flat background introduced at the end of Section 2.1, and described in the cosmic-time gauge by Eqs. (2.29)–(2.33). We find, in this background,

$$e^{-\bar{\phi}} = \Omega_d \, a^d e^{-\phi}, \qquad \gamma_{00} = 0, \qquad \gamma_i{}^j = \delta_i{}^j,$$

$$I_1 = \Omega_d \, a^d \frac{V'}{\dot{\phi}}, \qquad I_2 = d\Omega_d \, a^d \frac{H}{\dot{\phi}}, \tag{4A.21}$$

where $\Omega_d = V_d/\lambda_s^d$ is the volume of the spatial sections in string units. In this homogeneous limit we can thus establish the particular relations

$$\frac{\gamma_{\beta\alpha} \nabla^\alpha \nabla^\beta \phi}{\sqrt{(\partial\phi)^2}} I_1 = V' I_2, \qquad e^{-\phi} \sqrt{(\partial\phi)^2} \, I_1 = -\frac{\partial V}{\partial\bar{\phi}}, \tag{4A.22}$$

which lead to a considerable simplification of the corresponding equations. From the (00) component of Eq. (4A.11), in particular, we obtain

$$\dot{\phi}^2 - 2dH\dot{\phi} + d(d-1)H^2 = 2\lambda_s^{d-1} \left(e^\phi \rho + V \right). \tag{4A.23}$$

In the spatial components (ij) of Eq. (4A.20) all terms induced by the non-local potential cancel, and we obtain

$$\dot{H} - H\dot{\phi} + dH^2 = \lambda_s^{d-1} e^\phi \left(p - \frac{\sigma}{2} \right). \tag{4A.24}$$

From the dilaton equation (4A.19) we obtain, finally,

$$2\ddot{\phi} + 2dH\dot{\phi} - \dot{\phi}^2 - 2d\dot{H} - d(d+1)H^2 = 2\lambda_s^{d-1} \left(\frac{\partial V}{\partial\bar{\phi}} - V + \frac{1}{2} e^\phi \sigma \right). \tag{4A.25}$$

Their combination leads to the conservation equation

$$\dot{\rho} + dH(\rho + p) = \frac{1}{2}\sigma\dot{\phi}, \tag{4A.26}$$

while the elimination of $\dot{H}$ and H^2 through Eqs. (4A.23) and (4A.24), respectively, leads to rewriting the dilaton equation in more conventional form as

$$\ddot{\phi} + dH\dot{\phi} - \dot{\phi}^2 + 2\lambda_s^{d-1} \left(V - \frac{1}{2}\frac{\partial V}{\partial\bar{\phi}} \right) + \lambda_s^{d-1} e^\phi (\rho - dp) + \frac{1}{2}(d-1)\lambda_s^{d-1} e^\phi \sigma = 0. \tag{4A.27}$$

We are now in the position of comparing these isotropic equations (4A.23)–(4A.27) with the equations (2.34)–(2.38), derived in the presence of a local potential $V = V(\phi)$. The (00) and the conservation equation are the same, while in the dilaton equation (4A.25) the term $\partial V/\partial\phi$ is simply replaced by $\partial V/\partial\bar{\phi}$. The most important difference appears in the spatial equation (4A.24), where the contribution of the potential completely disappears: this is to be contrasted with the corresponding equation (2.35), where the contribution of the local potential survives in the term $\partial V/\partial\phi$. It is just because of this difference that the low-energy equations with some particular potential $V(\overline{\phi})$ are exactly integrable and admit regular solutions, as will be illustrated in Appendix 4B.

In order to facilitate their integration, it will be useful to rewrite the equations in terms of the shifted variables:

$$\overline{\phi} = \phi - d\ln a, \qquad \overline{\rho} = \rho a^d, \qquad \overline{p} = p a^d, \qquad \overline{\sigma} = \sigma a^d. \tag{4A.28}$$

From Eqs. (4A.23)–(4A.25) we obtain, respectively (in units $2\lambda_s^{d-1} = 1$),

$$\dot{\overline{\phi}}^2 - dH^2 - V = e^{\overline{\phi}}\overline{\rho},$$

$$\dot{H} - H\dot{\overline{\phi}} = \frac{1}{2}e^{\overline{\phi}}\left(\overline{p} - \frac{\overline{\sigma}}{2}\right), \tag{4A.29}$$

$$2\ddot{\overline{\phi}} - \dot{\overline{\phi}}^2 - dH^2 + V - \frac{\partial V}{\partial \overline{\phi}} = \frac{1}{2}e^{\overline{\phi}}\overline{\sigma},$$

and the conservation equation becomes

$$\dot{\overline{\rho}} + dH\overline{p} = \frac{1}{2}\overline{\sigma}(\dot{\overline{\phi}} + dH). \tag{4A.30}$$

These generalize to the case $V = V(\overline{\phi}) \neq 0$, Eqs. (4.39)–(4.41), and to the case $\sigma \neq 0$, the homogeneous, isotropic, torsionless limit of the matrix equations (4.139)–(4.141).

We conclude this appendix by presenting the E-frame version of the above equations, which is useful for later computations of the scalar and tensor spectrum of metric perturbations. The E-frame form of the non-local contributions can be determined either by transforming the action (4A.3), and then determining the corresponding cosmological equations [50], or by directly transforming into the E-frame the cosmological S-frame equations. Here we apply the second procedure, using the standard relations (2.64)–(2.68) connecting the geometric and matter variables of the two frames, for a homogeneous and isotropic background.

Starting from Eq. (4A.23), and denoting with the tilde the E-frame variables, we obtain

$$d(d-1)\widetilde{H}^2 = 2\lambda_P^{d-1}\left(\widetilde{\rho} + \frac{1}{2}\dot{\widetilde{\phi}}^2 + e^{\frac{2}{d-1}\frac{\widetilde{\phi}}{\mu}}k^{d+1}V\right), \tag{4A.31}$$

identical to Eq. (2.60) for the local potential (we recall that $k = \lambda_s/\lambda_P$). From the dilaton equation (4A.27) we obtain

$$\ddot{\widetilde{\phi}} + d\widetilde{H}\dot{\widetilde{\phi}} + \frac{\widetilde{\sigma}}{2} + \frac{1}{\mu(d-1)}(\widetilde{\rho} - d\widetilde{p}) + \frac{2k^{d+1}}{d-1}e^{\frac{2}{d-1}\frac{\widetilde{\phi}}{\mu}}\left(\frac{V}{\mu} - \frac{1}{2\mu}\frac{\partial V}{\partial \overline{\phi}}\right) = 0, \tag{4A.32}$$

which differs from the corresponding local equation in the terms containing $\partial V/\partial\overline{\phi}$. From the spatial equation (4A.24) we obtain

$$\dot{\widetilde{H}} + d\widetilde{H}^2 + \frac{\ddot{\widetilde{\phi}}}{\mu(d-1)} + \frac{d\widetilde{H}\dot{\widetilde{\phi}}}{\mu(d-1)} = \lambda_P^{d-1}\left(\widetilde{p} - \frac{\mu}{2}\widetilde{\sigma}\right). \tag{4A.33}$$

Eliminating $\ddot{\widetilde{\phi}}, \dot{\widetilde{\phi}}$ through the dilaton equation (4A.32), and $\widetilde{p}$ through Eq. (4A.31), we can recast the spatial equation in the form

$$2(d-1)\dot{\widetilde{H}} + d(d-1)\widetilde{H}^2 = 2\lambda_{\rm P}^{d-1}\left[-\widetilde{p} - \frac{1}{2}\dot{\widetilde{\phi}}^2 + k^{d+1}\mathrm{e}^{\frac{2}{d-1}\frac{\widetilde{\phi}}{\mu}}\left(V - \frac{\partial V}{\partial\overline{\phi}}\right)\right], \qquad (4A.34)$$

to be compared with Eq. (2.61), valid for a local potential. The difference, again, is induced by the presence of the potential derivative $\partial V/\partial\overline{\phi}$.

Appendix 4B
Examples of regular and self-dual solutions

In this appendix we apply the integration procedure presented in Sections 4.2 and 4.3 to obtain explicit examples of bouncing solutions, smoothly interpolating between an initial accelerated, growing curvature phase to a final decelerated, decreasing curvature phase. Throughout this section we limit ourselves to the case of torsionless (but possibly anisotropic) gravi-dilaton backgrounds, described by a Bianchi-I-type metric, possibly sourced by perfect fluid matter.

Let us start by recalling that, without the dilaton potential, regular isotropic solutions are impossible as already remarked in the discussion following Eq. (4.81). Regular anisotropic metric backgrounds are not forbidden [15], however, as will be shown here by considering an example in which the spatial geometry can be factorized as the direct product of two conformally flat manifolds, with d and n dimensions, respectively. Following the notations of Section 4.2 we set

$$a_i = a_1, \qquad \gamma_i = \gamma_1, \qquad x_i = x_1, \qquad i = 1, \ldots, d,$$
$$a_i = a_2, \qquad \gamma_i = \gamma_2, \qquad x_i = x_2, \qquad i = d+1, \ldots, d+n, \tag{4B.1}$$

and we choose a convenient set of integration constants, such that the linear term bx of the quadratic form (4.82) disappears. For instance,

$$x_0 = 0, \qquad x_1 = -x_2 \frac{n\alpha_2^2}{d\alpha_1^2}. \tag{4B.2}$$

In this case the discriminant is simply $4\Delta^2 = -4\alpha c$, and the constant term c turns out to be always negative,

$$c = -dx_1^2\alpha_1^2 - nx_2^2\alpha_2^2 < 0. \tag{4B.3}$$

The presence of real zeros of the quadratic form $D(x)$ can then be avoided provided

$$\alpha = \alpha_0^2 - d\alpha_1^2 - n\alpha_2^2 < 0. \tag{4B.4}$$

Assuming that this condition is satisfied, the integration of Eqs. (4.75) and (4.76) leads to the following exact solution:

$$a_i = a_{i0} E_i(x) |D(x)|^{\alpha_i/\alpha}, \qquad\qquad i = 1, 2,$$
$$e^\phi = a_{10}^d a_{20}^n e^{\phi_0} E_1^d E_2^n |D(x)|^{-(\alpha_0 - d\alpha_1 - n\alpha_2)/\alpha}, \tag{4B.5}$$
$$\rho = -L^{-2} e^{\phi_0} a_{10}^{-d} a_{20}^{-n} E_1^{-d} E_2^{-n} |D(x)|^{1-(\alpha_0 + d\alpha_1 + n\alpha_2)/\alpha},$$

where

$$E_i(x) = \exp\left[\frac{2x_i\alpha_i}{\sqrt{\alpha c}}\tan^{-1}\left(\frac{\alpha x}{\sqrt{\alpha c}}\right)\right], \tag{4B.6}$$

and a_{10}, a_{20}, ϕ_0 are integration constants.

As discussed in [15], it can be shown that in the space of the parameters γ_i, γ_0 there is a non-vanishing region where the condition (4B.4) is satisfied, together with the conditions required to guarantee that the curvature, the dilaton kinetic energy, the effective string coupling and the matter energy density are bounded everywhere. It can also be imposed, simultaneously, that the energy density goes to zero, asymptotically, and that at large positive times the solutions describe a final configuration with d expanding and n contracting dimensions (as seems appropriate for a "realistic" phase of dynamical dimensional reduction).

The condition (4B.4) implies, however, $D(x) < 0$ everywhere and thus, according to Eq. (4.74), $\rho < 0$ everywhere (as also evident from the last term of the explicit solution (4B.5)). Therefore, the matter sources present in this context cannot represent the ordinary macroscopic fluids appearing in the standard cosmological equations, but are possibly a classical, effective representation of the backreaction of quantum fluctuations outside the horizon [16, 17, 18]. Thus, the above solutions are possibly appropriate for a description of the cosmological background in close proximity to the transition regime from the pre- to the post-big bang configuration, where it is known that a violation of the classical energy conditions is required to avoid the singularity in a spatially flat background [51]. In that regime, solutions regularized by the contributions of ghost fields [52] or ghost condensation [53] have also been suggested.

Regular solutions, implementing more realistic configurations, asymptotically dominated by conventional matter sources, can be obtained in the context of duality-invariant equations by including an appropriate potential $V(\overline{\phi})$. Before introducing sources, however, it is worth stressing that with a dilaton potential depending on the non-local, duality-invariant variable $\overline{\phi}$ we can obtain regular solutions even for isotropic, vacuum metric backgrounds.

Consider, for instance, the following potential,

$$V(\overline{\phi}) = -V_0\,e^{4\overline{\phi}}, \qquad V_0 > 0, \tag{4B.7}$$

possibly interpreted as a four-loop correction in a perturbative context. With this potential, and for $T_{\mu\nu} = 0 = \sigma$, the isotropic $(d+1)$-dimensional equations (4A.29) are exactly solved by the following particular solution [54]:

$$a(t) = a_0\left[\frac{t}{t_0} + \left(1 + \frac{t^2}{t_0^2}\right)^{1/2}\right]^{1/\sqrt{d}}$$

$$\overline{\phi} = -\frac{1}{2}\ln\left[\sqrt{V_0}\,t_0\left(1 + \frac{t^2}{t_0^2}\right)\right], \tag{4B.8}$$

where a_0 and t_0 are integration constants. This regular, "bouncing" solution is exactly self-dual, in the sense that it satisfies the property $a(t)/a_0 = a_0/a(-t)$, and is characterized by a bounded, "bell-like" shape not only of the curvature but also of the dilaton kinetic energy (see Fig. 4.7, left panel). The solution smoothly interpolates between the two expanding branches of the isotropic vacuum solutions (4.29), connecting the pre-big bang, inflationary configuration

$$t \to -\infty \quad \Rightarrow \quad a \sim (-t)^{-1/\sqrt{d}}, \qquad \overline{\phi} \sim \sqrt{d}\ln a, \tag{4B.9}$$

to the final, post-big bang, decelerated configuration

$$t \to \infty \quad \Rightarrow \quad a \sim t^{1/\sqrt{d}}, \qquad \overline{\phi} \sim -\sqrt{d}\ln a. \tag{4B.10}$$

Together with its time-reversed partner such a solution describes a perfect "figure-eight-shaped" curve in the phase-space plane spanned by $\dot{\overline{\phi}}$ and $\sqrt{3}H$ (Fig. 4.7, right panel).

It should be noted that, for the above solution, the dilaton keeps monotonically growing (indeed, $\phi \sim (\sqrt{d}-1)\ln t$, for $t \to \infty$), but the curvature is bounded even in the E-frame, where the background smoothly evolves from accelerated contraction to decelerated expansion. To illustrate this point we have plotted in Fig. 4.7 also the E-frame Hubble parameter H_{E}, defined according to Eq. (2.64) (with $k = 1$) as

$$H_{\mathrm{E}}(t) = \frac{\mathrm{d}\ln a_{\mathrm{E}}}{\mathrm{d}t_{\mathrm{E}}} = \left(H - \frac{\dot{\phi}}{d-1} \right) e^{\phi/(d-1)}. \tag{4B.11}$$

(here $H = \mathrm{d}\ln a/\mathrm{d}t$ and $\dot{\phi} = \mathrm{d}\phi/\mathrm{d}t$ are both referred to the S-frame cosmic time). However, the fact that the string coupling is unbounded is not consistent, asymptotically, with the use of a tree-level effective action. Also, when the dilaton is unbounded, the curvature might be singular in some frame, different from the string and the Einstein frames.

The potential (4B.7) is only a particular case of a general class of potentials, defined by

$$V(e^{-\overline{\phi}}) = m^2 e^{2\overline{\phi}} \left[\left(\beta - e^{2n\overline{\phi}} \right)^{\frac{2n-1}{n}} - d \right], \tag{4B.12}$$

parametrized by the dimensionless coefficient β and by the "loop-counting" parameter n (for $n = 1$, $\beta = d$ and $m^2 = V_0$ one recovers the particular potential (4B.7)). For this class of potentials there are particular exact solutions of the vacuum, duality-invariant equations (4A.29) which are regular, for any given value of β and n [49, 50].

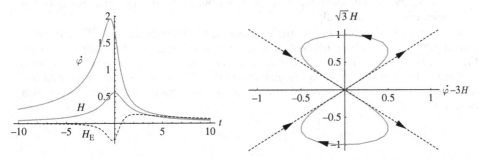

Figure 4.7 The left panel shows the plot of the curvature and of the dilaton kinetic energy for the solution (4B.8). The thin dashed curve illustrates the evolution of the E-frame Hubble parameter (4B.11), as a function of the S-frame cosmic time t. The right panel shows the trajectory of the solution and of its time-reversed partner in the phase space of Fig. 4.3 (the dashed bisecting lines represent the asymptotic vacuum solutions (4.29)). All curves are plotted for $t_0 = 1$, $V_0 = 1$ and $d = 3$.

We consider here some particular examples for the case $\beta > 0$ and $n > 0$. Assuming $\dot{\bar{\phi}} \neq 0$, Eqs. (4A.29) (in the absence of sources other than the potential) can be reduced to quadratures as follows:

$$H = H_0 e^{\bar{\phi}}, \qquad t = \pm \int d\bar{\phi} \left(V + dH_0^2 e^{2\bar{\phi}} \right)^{-1/2}, \qquad (4B.13)$$

where H_0 is an integration constant. By inserting the potential (4B.12), and considering the class of solutions with $H_0 = m$, the integral can be solved exactly,

$$mt = \int dy \left(\beta - y^{-2n} \right)^{\frac{1-2n}{2n}} = \beta^{-1} \left(\beta y^{2n} - 1 \right)^{1/2n} \qquad (4B.14)$$

(we have set $y = \exp(-\bar{\phi})$). We are thus led to the class of particular, exact solutions

$$H = m e^{\bar{\phi}} = m \left[\frac{\beta}{1 + (\beta mt)^{2n}} \right]^{1/2n}, \qquad (4B.15)$$

whose "bell-like" shape describes a bouncing evolution of the curvature scale, i.e. a smooth transition from growing to decreasing curvature, and from accelerated to decelerated expansion.

Consider the asymptotic limit $|t| \to \infty$, where $H \to \beta^{(1-2n)/n} |t|^{-1}$. In this limit we can easily integrate Eq. (4B.15) to obtain $a(t)$, and we find an initial, accelerated configuration with growing dilaton,

$$t \to -\infty \quad \Rightarrow \quad a \sim (-t)^{-\beta^{(1-2n)/2n}}, \qquad \phi \sim -\left[d\beta^{(1-2n)/2n} + 1 \right] \ln(-t), \qquad (4B.16)$$

evolving towards a final "dual" and time-reversed configuration

$$t \to \infty \quad \Rightarrow \quad a \sim t^{\beta^{(1-2n)/2n}}, \qquad \varphi \sim \left[d\beta^{(1-2n)/2n} - 1 \right] \ln t. \qquad (4B.17)$$

The "minimal", vacuum solutions (4B.9) and (4B.10), dominated by the dilaton kinetic energy, are thus recovered for

$$\beta = d^{n/(2n-1)}. \qquad (4B.18)$$

If this condition is not satisfied, then the contribution of the potential remains non-negligible, even asymptotically.

The asymptotic sign of $\dot{\phi}$ is controlled, instead, by the product $d\beta^{(1-2n)/2n}$. In particular, if $d\beta^{(1-2n)/2n} > 1$, the dilaton keeps growing also in the post-big bang branch (see Eq. (4B.17)), with the possible occurrence of curvature singularities when the solution is transformed to other frame representations. In the E-frame, for instance, we find from the definition (4B.11) that the curvature of the post-big bang branch has the following behavior:

$$H_E \sim \left[\frac{1 - \beta^{(1-2n)/2n}}{d-1} \right] \times t^{\frac{d}{d-1} \left[\beta^{(1-2n)/2n} - 1 \right]}. \qquad (4B.19)$$

Thus, the solution is regular and bouncing even in the Einstein frame, only provided

$$\beta^{(1-2n)/2n} < 1 \qquad (4B.20)$$

(this condition is always satisfied by the free vacuum solutions characterized by Eq. (4B.18), as already stressed).

Given the interesting regularizing properties of $V(\bar{\phi})$, we are led to investigate, in the same context, the possibility of more realistic solutions containing also the conventional fluid sources, and reducing asymptotically to some standard cosmological configuration (with stabilized dilaton). Examples of this type will be given here using the duality-invariant potential satisfying Eqs. (4.75) and (4.76), in a $(d+1)$-dimensional isotropic background, sourced by a perfect barotropic fluid with equation of state $p/\rho = \gamma = \text{const.}$

In such a case $G = -a^2 I$, $G\bar{\theta} = -\bar{p}I$ so that, using the definitions of Section 4.3 for $M, \bar{T}$ and Γ,

$$M\eta M' = 2\frac{a'}{a}\begin{pmatrix} 0 & I \\ -I & 0 \end{pmatrix}, \quad \bar{T} = \bar{p}\begin{pmatrix} 0 & I \\ -I & 0 \end{pmatrix},$$

$$\Gamma' = \frac{\bar{T}}{\bar{p}}\begin{pmatrix} 0 & I \\ -I & 0 \end{pmatrix}, \quad \Gamma = \gamma(x+x_1)\begin{pmatrix} 0 & I \\ -I & 0 \end{pmatrix}, \tag{4B.21}$$

where x_1 is an integration constant. Equations (4.177) and (4.178) reduce to

$$\bar{\phi}' = -\frac{2(x+x_0)}{D}, \quad \frac{a'}{a} = \frac{2\gamma(x+x_1)}{D}, \quad \gamma \neq 0, \tag{4B.22}$$

where, from Eq. (4.176),

$$D(x) = (x+x_0)^2 - d\gamma^2(x+x_1)^2 + L^2 V_0 \equiv \alpha x^2 + bx + c \tag{4B.23}$$

(the same result can be obtained by the direct integration of Eqs. (4A.29) for $p = \gamma\rho$ and $V = -V_0\exp(2\bar{\phi})$). Because of the presence of V_0, it becomes possible to choose the integration constants x_0, x_1 in such a way that $D(x)$ has no real zeros, even if the background is isotropic and the sources have a positive energy density. Indeed, for the quadratic form (4B.23),

$$\Delta^2 = \frac{1}{4}(b^2 - 4\alpha c) = d\gamma^2(x_1 - x_0)^2 + L^2 V_0(d\gamma^2 - 1); \tag{4B.24}$$

we can thus can obtain $\Delta^2 < 0$ provided the equation of state is not too "stiff", i.e. for $d\gamma^2 < 1$.

This result is valid also for the case $\gamma = 0$ ("dust" fluid source), which is not included, however, in the general integration (4B.22). For this particular case we can start directly from Eqs. (4A.29) with $p = 0 = \sigma$, and, after a first integration, we are led to

$$\bar{\phi}' = -\frac{2(x+x_0)}{D}, \quad \frac{a'}{a} = \frac{2x_1}{D}, \quad \gamma = 0, \tag{4B.25}$$

where

$$D(x) = (x+x_0)^2 - dx_1^2 + L^2 V_0, \tag{4B.26}$$

replacing Eqs. (4B.22) and (4B.23). One then finds that regular exact solutions are allowed for $\Delta^2 = dx_1^2 - L^2 V_0 < 0$. A first, very simple example can thus be obtained by choosing $x_1 = 0 = x_0$, which leads to the (almost) trivial solution describing a flat (S-frame) space-time, with the dilaton performing a time-symmetric, bell-like evolution sustained by the presence of the constant energy density of the dust sources:

$$a = a_0, \quad \rho = \rho_0 = \text{const}, \quad p = 0, \quad e^\phi = \frac{e^{\phi_0}}{1 + (t/t_0)^2}. \tag{4B.27}$$

Here a_0, ρ_0 are integration constants, satisfying

$$e^{\phi_0}\rho_0 = V_0 e^{2\phi_0} = \frac{4}{t_0^2}. \tag{4B.28}$$

Such a solution acquires a less trivial representation in the E-frame where one finds, using the standard transformations (2.64), in conformal time,

$$a_E(\eta) = a_0 \, e^{-\frac{\phi_0}{d-1}} \left(1 + \frac{\eta^2}{\eta_0^2}\right)^{\frac{1}{d-1}},$$

$$e^\phi = e^{\phi_0} \left(1 + \frac{\eta^2}{\eta_0^2}\right)^{-1},$$

$$\rho_E(\eta) = \rho_0 a_0^d \left(1 + \frac{\eta^2}{\eta_0^2}\right)^{-\frac{d}{d-1}}, \tag{4B.29}$$

with $\eta_0 = t_0/a_0$. In this frame the metric describes a non-trivial evolution from accelerated contraction to decelerated expansion, sustained by the dilaton and by a symmetric, bell-like evolution of the energy density of the pressureless fluid. The two asymptotic branches of such a "self-dual" solution are simply related by a time-reversal transformation, as appropriate to the E-frame metric [12].

Another, possibly realistic example can be obtained by considering a radiation fluid, with equation of state $\gamma = 1/d$, and its "dual" partner, with equation of state $\gamma = -1/d$ (this second type of source can be phenomenologically interpreted as a gas of strings "frozen" outside the horizon [35, 36, 55], or also as a gas of "winding" strings [56, 57], see Section 6.3). For this example let us come back to Eq. (4B.22), and assume that $\Delta^2 < 0$ (we can satisfy this condition, for instance, by setting $x_1 = x_0$ in Eq. (4B.24)). The integration of Eqs. (4B.22) then provides regular solutions which can be written in the form [7]

$$\bar\phi = \phi_0 + \ln |D|^{-1/\alpha} - \frac{2\alpha x_0 - b}{\alpha |\Delta|} T(x),$$

$$a = a_0 |D|^{\gamma/\alpha} \exp\left[\frac{\gamma(2\alpha x_1 - b)}{\alpha |\Delta|} T(x)\right], \tag{4B.30}$$

where

$$\alpha = 1 - d\gamma^2 = \frac{d-1}{d}, \qquad T(x) = \tan^{-1}\left(\frac{\alpha x + b/2}{|\Delta|}\right) \tag{4B.31}$$

(ϕ_0 and a_0 are integration constants). Their combination gives

$$e^\phi = a_0^d \, e^{\phi_0} |D|^{\frac{d\gamma-1}{\alpha}} \exp\left[\frac{2\alpha(x_1 d\gamma - x_0) + b(1 - d\gamma)}{\alpha |\Delta|} T(x)\right], \tag{4B.32}$$

while from the result (4.176) we obtain the (positive) energy density of the fluid source,

$$\bar\rho = L^{-2} \, e^{\phi_0} D^{\frac{\alpha-1}{\alpha}} \exp\left[-\frac{2\alpha x_0 - b}{\alpha |\Delta|} T(x)\right]. \tag{4B.33}$$

It may be interesting to note that for a radiation fluid we have

$$\frac{\gamma}{\alpha} = \frac{1}{d-1} = \frac{1-\alpha}{\alpha}, \tag{4B.34}$$

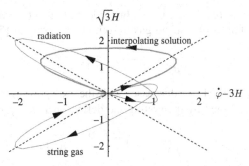

Figure 4.8 The plots of the solutions (4B.30) with constant equation of state $\gamma = 1/d$ (upper thin curve) and $\gamma = -1/d$ (lower thin curve) compared to the plot of the solution (4B.41) with time-dependent equation of state, interpolating between $\gamma = -1/d$ and $\gamma = 1/d$ (bold curve). The vacuum asymptotic solutions (4.29), represented by the dashed lines, are not approached because of the presence of sources. The solutions (4B.30) have been plotted for $d = 3$, $x_1 = 1$, $x_0 = 1/3$ and $\beta = L^2 V_0 = 8/9$. The solution (4B.41) has been plotted for $d = 3$, $x_0 = 0$ and $x_1 = \beta = L^2 V_0 = 1$.

so that, asymptotically, $\overline{\rho} \sim a^{-1}$. As $\overline{\rho} \sim dx/dt$ (from the definition (4.68)), it turns out that the time parameter x asymptotically coincides with the conformal time coordinate: $dt \sim a\,dx$, from which $x \sim t^{(d-1)/(d+1)}$. Also, for the radiation solution, the dilaton (4B.32) goes to a constant as $x \to \pm\infty$. The radiation-dominated solution with $\gamma = 1/d$ thus provides a smooth interpolation between an initial phase of accelerated contraction, $a \sim (-t)^{2/(d+1)}$, constant dilaton $\phi = \phi_-$ and growing curvature, to a final phase of decelerated expansion, $a \sim t^{2/(d+1)}$, constant dilaton $\phi = \phi_+$ and decreasing curvature. There are neither horizons nor singularities in the curvature and in the string coupling.

The dual solution with $\widetilde{\gamma} = -\gamma = -1/d$ is related to the radiation solution by a scale-factor duality transformation, i.e. by

$$\widetilde{a} = a^{-1}, \qquad \widetilde{\overline{\rho}} = \overline{\rho}, \qquad \widetilde{\overline{\phi}} = \overline{\phi}, \tag{4B.35}$$

as also evident from the solutions (4B.30) and (4B.33), written explicitly in terms of γ. One obtains, in that case, a smooth evolution from an initial phase of accelerated expansion, $\widetilde{a} \sim (-t)^{-2/(d+1)}$, and logarithmically increasing dilaton, to a final phase of decelerated contraction and logarithmically decreasing dilaton. Again, the maximal curvature regime is crossed over without singularities.

The two duality-related solutions, with $\gamma = \pm 1/d$, thus perform a smooth transition from the pre- to the post-big bang sector of Fig. 4.3. The evolution of the scale factor, however, is non-monotonic, as the gravi-dilaton system evolves from contraction to expansion, or vice-versa. This is clearly illustrated in Fig. 4.8, where we have plotted the radiation-dominated solution and its dual partner for a particular set of parameters corresponding to $\Delta^2 < 0$ (we have plotted, in particular, the case with $b = 0$ and $c = \alpha = 1 - d\gamma^2$, so that $D(x) = (1 - d\gamma^2)(1 + x^2)$).

The two examples we have given suggest the possibility of describing a monotonic evolution, from pre- to post-big bang expansion, based on a non-barotropic fluid source in which the equation of state is time dependent, and evolves smoothly from $\gamma = -1/d$ at

$-\infty$ to $\gamma = 1/d$ at $+\infty$. A particular, integrable model of such behavior (justified by the study of the equation of state of a string gas in rolling backgrounds [55]) is given by [7]:

$$\frac{p}{\rho} = \frac{x}{d(x^2 + x_1^2)^{1/2}}.$$

(4B.36)

We can still apply Eq. (4B.21) to this case, with the exception of the result for the Γ matrix, which is now obtained by integrating the definition $\Gamma' = \overline{T}/\overline{\rho}$:

$$\Gamma(x) = \frac{1}{d} \left(x^2 + x_1^2\right)^{1/2} \begin{pmatrix} 0 & I \\ -I & 0 \end{pmatrix}.$$

(4B.37)

The new equations for a and $\overline{\phi}$ are then

$$\overline{\phi}' = -\frac{2(x + x_0)}{D(x)}, \qquad \frac{a'}{a} = \frac{2}{d} \frac{(x^2 + x_1^2)^{1/2}}{D(x)},$$

(4B.38)

where

$$D(x) = (x + x_0)^2 - \frac{1}{d}(x^2 + x_1^2) + L^2 V_0.$$

(4B.39)

We integrate here the above equations for the particularly simple choice of parameters $x_0 = 0$ and $x_1^2 = \beta = L^2 V_0$, for which

$$D(x) = \frac{d-1}{d}(x^2 + x_1^2).$$

(4B.40)

More general choices of the parameters may lead to solutions characterized by a short contraction stage between the initial inflationary expansion and the final radiation-dominated epoch [58]. This may have interesting phenomenological applications, but does not modify the global properties of the cosmological background, so that the following discussion concentrates on the simplifying choice (4B.40). In such a case, the integration of Eqs. (4B.38) gives

$$a = a_0 \left(x + \sqrt{x^2 + x_1^2}\right)^{\frac{2}{d-1}}, \qquad e^{\overline{\phi}} = e^{\phi_0} \left(x^2 + x_1^2\right)^{-\frac{d}{d-1}},$$

(4B.41)

where a_0, ϕ_0 are integration constants. Their combination, together with Eqs. (4.176) and (4B.36), leads to

$$e^{\phi} = e^{\overline{\phi}} a^d = a_0^d e^{\phi_0} \left(1 + \frac{x}{\sqrt{x^2 + x_1^2}}\right)^{\frac{2d}{d-1}},$$

$$\overline{\rho} = \frac{d-1}{dL^2} e^{\phi_0} \left(x^2 + x_1^2\right)^{-\frac{1}{d-1}},$$

(4B.42)

$$\rho e^{\phi} = \frac{d-1}{dL^2} e^{2\phi_0} \left(x^2 + x_1^2\right)^{-\frac{d+1}{d-1}},$$

$$p e^{\phi} = \frac{d-1}{d^2 L^2} e^{2\phi_0} x \left(x^2 + x_1^2\right)^{-\frac{3d+1}{2(d-1)}}.$$

The smooth behavior of the solution (4B.41) in the plane $(\dot{\overline{\phi}}, \sqrt{3}H)$ is illustrated in Fig. 4.8.

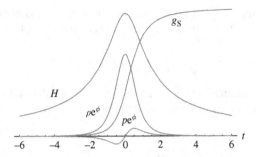

Figure 4.9 Time-evolution of the relevant geometric and matter variables for the isotropic, self-dual solution given in Eqs. (4B.41) and (4B.42). The plots correspond to the particular case $d = 3$, $x_1 = 1$, $L = 1$, $\phi_0 = 0$ and $a_0 = e^{-2/3}$.

The above solution is self-dual, in the sense that $\overline{\phi}(x) = \overline{\phi}(-x)$, and

$$\left[\frac{a}{a_0 x_1^{2/(d-1)}} \right](x) = \left[\frac{a}{a_0 x_1^{2/(d-1)}} \right]^{-1}(-x) \tag{4B.43}$$

(with an appropriate choice of the integration constant a_0 it is always possible to set to 1 the fixed point of the scale-factor inversion). We also note that, asymptotically,

$$
\begin{aligned}
x \to -\infty &\quad\Rightarrow\quad a \sim (-x)^{-2/(d-1)} \sim \overline{\rho} \sim \frac{dx}{dt}, \\
x \to +\infty &\quad\Rightarrow\quad a \sim x^{2/(d-1)} \quad\sim \frac{1}{\overline{\rho}} \sim \frac{dt}{dx}.
\end{aligned}
\tag{4B.44}
$$

In the asymptotic limit we can then easily re-express the solution in terms of the cosmic-time coordinate, to find that it describes a monotonic evolution from an initial state of accelerated expansion and growing dilaton, associated with negative-pressure matter,

$$t \to -\infty \quad\Rightarrow\quad a \sim (-t)^{-\frac{2}{d+1}}, \quad e^{\phi} \sim (-t)^{-\frac{4d}{d+1}}, \quad p = -\frac{\rho}{d}, \tag{4B.45}$$

towards a final, radiation-dominated state of decelerated expansion and asymptotically frozen dilaton,

$$t \to \infty \quad\Rightarrow\quad a \sim t^{\frac{2}{d+1}}, \quad e^{\phi} \sim \text{const}, \quad p = \frac{\rho}{d}. \tag{4B.46}$$

The curvature parameters H and $\dot{H}$, the effective energy density and pressure, and the string coupling g_s^2, are everywhere bounded as illustrated in Fig. 4.9.

The interpolating solution (4B.41) is associated with a dual partner which can be obtained by applying the transformation (4.44), and which corresponds to the same type of time-dependent equation of state as that of Eq. (4B.36), but with the opposite sign. It describes a background in monotonic contraction, with the sources evolving from positive to negative pressure. The combination of the two types of solutions may suggest a $(d+n)$-dimensional, anisotropic scenario, in which the d-dimensional "external" spatial sections of the geometry are isotropically and monotonically expanding, with scale factor $a(t)$, sourced by an effective, time-dependent pressure p such that

$$\frac{p}{\rho} = \frac{1}{d+n} \frac{x}{(x^2+x_1^2)^{1/2}},$$ (4B.47)

while n-dimensional "internal" spatial sections are isotropically and monotonically contracting, with scale factor $b(t) = a^{-1}(t)$, sourced by an effective, time-dependent pressure q such that

$$\frac{q}{\rho} = -\frac{1}{d+n} \frac{x}{(x^2+x_1^2)^{1/2}}.$$ (4B.48)

Equations (4B.38) become, in this case,

$$\overline{\phi}' = -\frac{2(x+x_0)}{D(x)}, \qquad \frac{a'}{a} = -\frac{b'}{b} = \frac{2}{d+n} \frac{(x^2+x_1^2)^{1/2}}{D(x)},$$ (4B.49)

where

$$D(x) = (x+x_0)^2 - \frac{1}{d+n}(x^2+x_1^2) + L^2 V_0.$$ (4B.50)

Choosing, as before, $x_0 = 0$ and $x_1^2 = L^2 V_0$, we can avoid the real zeros of $D(x)$, and the integration leads to the following exact solution [7]:

$$a = a_0 \left(x + \sqrt{x^2+x_1^2} \right)^{\frac{2}{d+n-1}} = b^{-1},$$ (4B.51)

$$e^{\overline{\phi}} = e^{\phi_0} \left(x^2+x_1^2 \right)^{-\frac{d+n}{d+n-1}},$$

from which

$$e^{\phi} = a_0^{d-n} e^{\phi_0} \left(x + \sqrt{x^2+x_1^2} \right)^{\frac{2(d-n)}{d+n-1}} \left(x^2+x_1^2 \right)^{-\frac{d+n}{d+n-1}},$$ (4B.52)

$$\overline{\rho} = \frac{(d+n-1)e^{\phi_0}}{(d+n)L^2} \left(x^2+x_1^2 \right)^{-\frac{1}{d+n-1}}.$$

This anisotropic solution describes a higher-dimensional scenario which includes inflation and dynamical dimensional reduction. By exploiting the asymptotic behavior of $\overline{\rho}$, and the relation between x and t, we may recover for $x \to \pm\infty$ a particular example of the general anisotropic solution (4.99). For $x \to -\infty$ we find, in particular, that the initial configuration is characterized by the accelerated expansion of d dimensions, driven by a negative exterior pressure, and by the accelerated contraction of n dimensions, driven by a positive internal pressure:

$$a \sim (-t)^{-2/(d+n+1)} = b^{-1}, \qquad p = -\frac{\rho}{(d+n)} = -q < 0.$$ (4B.53)

In the opposite limit $x \to \infty$ we find instead a final configuration characterized by the decelerated expansion of d dimensions, driven by positive-pressure radiation, and by the decelerated contraction of n dimensions, damped by a negative internal pressure:

$$a \sim t^{2/(d+n+1)} = b^{-1}, \qquad p = \frac{\rho}{(d+n)} = -q > 0.$$ (4B.54)

It is important to remark that the non-local potential $V(\bar{\phi})$ plays no role in determining the kinematics of such asymptotic configurations, but is only effective in arranging a smooth connection between the two branches when approaching the high-curvature regime.

For this last anisotropic solution the background curvature scale has the same qualitative behavior as that illustrated in Fig. 4.9. There is an important difference, however, for the dilaton field at large positive times: the dilaton is not stabilized but is decreasing asymptotically, driven by the decelerated contraction of the internal dimensions. From Eq. (4B.52) we find that, for $x \to +\infty$,

$$e^\phi \sim x^{-\frac{4n}{d+n-1}} \sim t^{-\frac{4n}{d+n+1}}. \tag{4B.55}$$

One thus obtains, in this background, an interesting link between the variation of the coupling constants (controlled by the dilaton in the context of string models of unifications), and the contraction of the internal dimensions. For a sufficiently slow contraction the scenario could be phenomenologically acceptable, and in agreement with recent observational results [13].

References

[1] K. Kikkawa and M. Y. Yamasaki, *Phys. Lett.* **B149** (1984) 357.

[2] N. Sakai and I. Senda, *Prog. Theor. Phys.* **75** (1986) 692.

[3] A. A. Tseytlin, *Mod. Phys. Lett.* **A6** (1991) 1721.

[4] G. Veneziano, *Phys. Lett.* **B265** (1991) 287.

[5] T. H. Buscher, *Phys. Lett.* **B194** (1987) 59.

[6] P. Ginsparg and C. Vafa, *Nucl. Phys.* **B289** (1989) 414.

[7] M. Gasperini and G. Veneziano, *Astropart. Phys.* **1** (1993) 317.

[8] M. Gasperini and G. Veneziano, *Phys. Rep.* **373** (2003) 1.

[9] J. Khouri, B. A. Ovrut, N. Seiberg, P. J. Steinhardt and N. Turok, *Phys. Rev.* **D65** (2002) 086007.

[10] M. Gasperini, Dilaton cosmology and phenomenology, hep-th/0702166.

[11] M. Gasperini and G. Veneziano, *Phys. Rev.* **D50** (1994) 2519.

[12] M. Gasperini and G. Veneziano, *Mod. Phys. Lett.* **A8** (1993) 3701.

[13] J. D. Barrow, *Phil. Trans. Roy. Soc. Lond.* **A363** (2005) 2139.

[14] R. H. Brandenberger and J. Martin, *Phys. Rev.* **D71** (2005) 023504.

[15] G. De Risi and M. Gasperini, *Phys. Lett.* **B521** (2001) 335.

[16] L. W. R. Abramo, R. Brandenberger and V. F. Mukhanov, *Phys. Rev.* **D56** (1997) 3248.

[17] Y. Nambu, *Phys. Rev.* **D63** (2001) 044013.

[18] A. Ghosh, R. Madden and G. Veneziano, *Nucl. Phys.* **B570** (2000) 207.

[19] H. B. Dwight, *Tables of Integrals* (New York: MacMillan Publishing Co., 1961).

[20] M. P. Ryan and L. C. Shepley, *Homogenous Relativistic Cosmologies* (Princeton: Princeton University Press, 1975).

[21] M. Gasperini and R. Ricci, *Class. Quantum Grav.* **12** (1995) 677.

[22] M. Gasperini and G. Veneziano, *Phys. Rev.* **D59** (1999) 43503.

[23] V. A. Belinskii, I. M. Khalatnikov and E. M. Lifshitz, *Adv. Phys.* **19** (1970) 525.

[24] J. D. Barrow and M. P. Dabrowski, *Phys. Rev.* **D57** (1998) 7204.

[25] T. Damour and M. Henneaux, *Phys. Rev. Lett.* **85** (2000) 920.

[26] A. Sen, *Phys. Lett.* **B271** (1991) 295.

[27] K. A. Meissner and G. Veneziano, *Mod. Phys. Lett.* **A6** (1991) 3397.

[28] K. A. Meissner and G. Veneziano, *Phys. Lett.* **B267** (1991) 33.

[29] A. Giveon, M. Porrati and E. Rabinovici, *Phys. Rep.* **244** (1994) 77.

[30] X. C. de la Ossa and F. Quevedo, *Nucl. Phys.* **B403** (1993) 377.

[31] M. Gasperini and G. Veneziano, *Phys. Lett.* **B277** (1992) 256.

[32] J. R. Ray and L. L. Smalley, *Phys. Rev.* **D27** (1983) 183.

[33] M. Gasperini, *Phys. Rev. Lett.* **56** (1986) 2873.

[34] M. Gasperini, *Gen. Rel. Grav.* **30** (1998) 1703.
[35] M. Gasperini, N. Sanchez and G. Veneziano, *Int. J. Mod. Phys.* **A6** (1991) 3853.
[36] M. Gasperini, N. Sanchez and G. Veneziano, *Nucl. Phys.* **B364** (1991) 365.
[37] T. Battenfeld and S. Watson, *Rev. Mod. Phys.* **78** (2006) 435.
[38] M. Gasperini, J. Maharana and G. Veneziano, *Phys. Lett.* **B272** (1991) 277.
[39] G. F. R. Ellis, in *Proc. Int. School of Physics "E. Fermi"*, Course XLVII (New York: Academic Press, 1965), p. 104.
[40] M. Gasperini, *Mod. Phys. Lett.* **A14** (1999) 1059.
[41] C. R. Nappi and E. Witten, *Phys. Lett.* **B293** (1992) 309.
[42] M. Gasperini, J. Maharana and G. Veneziano, *Phys. Lett.* **B296** (1992) 51.
[43] M. Gasperini, R. Ricci and G. Veneziano, *Phys. Lett.* **B319** (1993) 438.
[44] E. Alvarez, L. Alvarez-Gaumé and Y. Lozano, *Nucl. Phys.* **B424** (1994) 155.
[45] E. Kiritsis, *Mod. Phys. Lett.* **A6** (1991) 2871.
[46] M. Rocek and A. Verlinde, *Nucl. Phys.* **B373** (1992) 630.
[47] A. S. Schwarz and A. A. Tseytlin, *Nucl. Phys.* **B399** (1993) 691.
[48] A. Given and M. Rocek, *Nucl. Phys.* **B380** (1992) 128.
[49] M. Gasperini, M. Giovannini and G. Veneziano, *Phys. Lett.* **B569** (2003) 113.
[50] M. Gasperini, M. Giovannini and G. Veneziano, *Nucl. Phys.* **B694** (2004) 206.
[51] R. Brustein and R. Madden, *Phys. Lett.* **B410** (1997) 110.
[52] L. E. Allen and D. Wands, *Phys. Rev.* **D70** (2004) 063515.
[53] P. Creminelli, M. A. Luty, A. Nicolis and L. Senatore, *JHEP* **0612** (2006) 080.
[54] M. Gasperini, J. Maharana and G. Veneziano, *Nucl. Phys.* **B472** (1996) 349.
[55] M. Gasperini, M. Giovannini, K. A. Meissner and G. Veneziano, *Nucl. Phys. (Proc. Suppl.)* **B49** (1996) 70.
[56] R. Brandenberger and C. Vafa, *Nucl. Phys.* **B316** (1989) 391.
[57] A. A. Tseytlin and C. Vafa, *Nucl. Phys.* **B372** (1992) 443.
[58] C. Angelantonj, L. Amendola, M. Litterio and F. Occhionero, *Phys. Rev.* **D51** (1995) 1607.

5

Inflationary kinematics

We have shown in Chapter 4 that the symmetries of the low-energy string cosmology equations can be used to obtain new solutions, characterized by kinematic properties which are of "dual" type with respect to the known solutions of the standard cosmological scenario. In particular, we have stressed that such new solutions are characterized by an accelerated evolution of the background geometry, and thus describe a phase of cosmological inflation.

For a better illustration of this important point it should be recalled that the "naturalness" problem of the initial conditions of the standard cosmological scenario can be solved (as discussed in Chapter 1) by introducing a suitable period of "inflationary" evolution, during which the function

$$r(t) = (aH)^{-1} \tag{5.1}$$

decreases in time – instead of growing as in the epoch of standard decelerated evolution. In particular, if the inflationary kinematics is parametrized by a power-law scale factor, $a(t) \sim t^\beta$, we obtain the explicit condition

$$r(t) = (\dot{a})^{-1} \sim t^{1-\beta} \to 0, \tag{5.2}$$

to be satisfied as the cosmic-time parameter increases from the beginning to the end of inflation.

In this chapter we will illustrate the different ways in which this condition can be implemented in general, and in the context of models based on the string cosmology equations in particular. The present discussion will purely concentrate on the kinematic aspects of the various classes of metric backgrounds, while the important phenomenological consequences of the different kinematics (concerning amplification of fluctuations, anisotropy production, structure formation, ...) will be discussed in the following chapters (see, in particular, Chapters 7 and 8).

5.1 Four different types of inflation

The condition (5.2) can be satisfied, in general, by two different classes of (homogeneous and isotropic) metric backgrounds. The first class of inflationary solutions is defined over a positive range of values of the cosmic-time coordinate, and can be parametrized by the following scale factor,

$$\boxed{\text{Class I}} \qquad a \sim t^\beta, \qquad \beta > 1, \qquad t > 0, \qquad t \to +\infty. \qquad (5.3)$$

This parametrization, which satisfies Eq. (5.2), represents a phase of "**power-inflation**" [1], characterized by accelerated expansion

$$\dot{a} \sim \beta t^{\beta-1} > 0, \qquad \ddot{a} \sim \beta(\beta-1)t^{\beta-2} > 0, \qquad (5.4)$$

and decreasing curvature,

$$H = \frac{\dot{a}}{a} = \frac{\beta}{t} > 0, \qquad \dot{H} = -\frac{\beta}{t^2} < 0. \qquad (5.5)$$

It is probably useful to point out explicitly that we call "accelerated" the case in which sign $\{\dot{a}\}$ = sign $\{\ddot{a}\}$, quite independently of the sign of $\dot{a}$. Also, taking H^2 as a generic indicator of the time behavior of the space-time curvature scale, the curvature will be growing for a phase in which $d(H^2)/dt > 0$, namely sign $\{H\}$ = sign $\{\dot{H}\}$, and decreasing in the opposite case, quite independently from the sign of H.

This first class of backgrounds includes the limiting case $\beta \to \infty$, which we may call "**de Sitter inflation**" [2],

$$a \sim e^{kt}, \qquad k = \text{const} > 0, \qquad t > 0, \qquad t \to +\infty, \qquad (5.6)$$

since the scale factor grows exponentially in cosmic time, as in the case of the spatially flat chart of the de Sitter solution (see Eq. (1.98)). This scale factor also satisfies Eq. (5.2), and represents a phase of accelerated expansion,

$$\dot{a} \sim k e^{kt} > 0, \qquad \ddot{a} \sim k^2 e^{kt} > 0, \qquad (5.7)$$

at constant curvature,

$$H = k = \text{const}, \qquad \dot{H} = 0. \qquad (5.8)$$

The second class of inflationary backgrounds is defined over a negative range of values of the cosmic-time coordinate, and is parametrized by the following scale factor,

$$\boxed{\text{Class II}} \qquad a \sim (-t)^\beta, \qquad \beta < 1, \qquad t < 0, \qquad t \to 0_-. \qquad (5.9)$$

This case also satisfies Eq. (5.2), and includes two possible subclasses. One is

$$\boxed{\text{Class IIa}} \qquad \beta < 0, \qquad (5.10)$$

representing a phase of "**super-inflation**" [3, 4, 5] (or "pole-inflation"), characterized by accelerated expansion,

$$\dot{a} \sim -\beta(-t)^{\beta-1} > 0, \qquad \ddot{a} \sim \beta(\beta-1)(-t)^{\beta-2} > 0, \tag{5.11}$$

and growing curvature scale,

$$H = -\frac{\beta}{(-t)} > 0, \qquad \dot{H} = -\frac{\beta}{t^2} > 0. \tag{5.12}$$

The other is

$$\boxed{\text{Class IIb}} \qquad 0 < \beta < 1, \tag{5.13}$$

representing a phase of "**accelerated contraction**" [6]

$$\dot{a} \sim -\beta(-t)^{\beta-1} < 0, \qquad \ddot{a} \sim \beta(\beta-1)(-t)^{\beta-2} < 0, \tag{5.14}$$

and growing curvature scale,

$$H = -\frac{\beta}{(-t)} < 0, \qquad \dot{H} = -\frac{\beta}{t^2} < 0. \tag{5.15}$$

An important remark, at this point, is in order. All inflationary backgrounds of Class I are defined in a range of *positive* values of the cosmic-time coordinate, and describe accelerated evolution from the higher-curvature towards the lower-curvature regime, as occurs in typical models of slow-roll inflation [7] (with the exception of the limiting case of de Sitter inflation, where the curvature stays constant; see however the discussion of Section 1.2 on the problems concerning an infinite past extension of such an expanding solution). If extrapolated up to their maximum (classical) limit of validity, these geometries are characterized by an *initial* curvature singularity, located at some finite time in the past, and are thus appropriate to describe a phase of inflation occurring *after the big bang*, much in agreement with the standard cosmological picture in which the Universe is expected to emerge from the initial singularity.

The evolution of the Class II backgrounds, defined in a *negative* time range, goes instead from the lower- to the higher-curvature regime and, if extrapolated to its maximum validity limit, is characterized by a *final* singularity, occurring at a given future value of the time coordinate. This class of backgrounds is thus appropriate to describe a phase of inflation occurring *before the big bang* (assuming the singularity smoothed out by string effects), and is typical of the scenario suggested by the self-duality principle discussed in the previous chapter, as well as of models characterized by a BPS initial configuration and by a shrinking scale factor, formulated in the context of the ekpyrotic [8, 9] or cyclic [10, 11] scenarios. For the duality-generated inflationary solutions in particular, the two subclasses IIa and IIb turn out to be dynamically equivalent, in the sense that they may be interpreted as two possible kinematic descriptions of the same inflationary model, represented in two different (but conformally related) frames.

This possible equivalence of Classes IIa and IIb will be illustrated in the next section. Here we will conclude the discussion of the different inflationary classes by reporting their convenient representation in terms of the conformal-time coordinate η, such that $dt = a\, d\eta$. Indeed, *all* types of inflation that we have introduced can be parametrized in the *negative* range of values of the conformal time coordinate, with an appropriate power α ranging from $-\infty$ to ∞. This parametrization will be very useful to emphasize the phenomeno-logical differences among the various classes, as we shall see in Chapters 7 and 8.

Let us define

$$a(\eta) = (-\eta)^{\alpha}, \qquad -\infty \leq \eta < 0, \qquad -\infty \leq \alpha \leq +\infty. \tag{5.16}$$

The particular case $\alpha = -1$ gives $t \sim \ln(-\eta)^{-1}$, and we recover for $a(t)$ the exponential de Sitter parametrization, Eq. (5.6). For $\alpha \neq -1$, we obtain, in general,

$$-(1+\alpha)t = (-\eta)^{1+\alpha}, \tag{5.17}$$

so that

$$a = (-\eta)^{\alpha} = [-(1+\alpha)t]^{\beta}, \qquad \beta = \frac{\alpha}{1+\alpha}. \tag{5.18}$$

We may thus consider three cases.

(1) The case $\alpha < -1$. In this case t in Eq. (5.18) varies over a range of positive values, and the power β is always larger than one. The scale factor (5.16) thus describes a phase of power-inflation, according to the definition (5.3).
(2) The case $-1 < \alpha < 0$. In this case the range of values of t is negative, and the power β also turns out to be negative. The scale factor describes super-inflation, according to Eqs. (5.9) and (5.10).
(3) The case $\alpha > 0$. The range of values of t is negative, again, and we find that $0 < \beta < 1$. The scale factor describes accelerated contraction, according to Eqs. (5.9) and (5.13).

These three different cases, plus the de Sitter one, are graphically summarized in Fig. 5.1. Considering the behavior of the geometry at the two temporal boundaries, it may be noted that the scale factor (5.16) goes to zero for $\eta \to -\infty$ if $\alpha < 0$, and for $\eta \to 0_-$ if $\alpha > 0$. The curvature, on the other hand, diverges for $\eta \to -\infty$ only if $\alpha < -1$, while it diverges for $\eta \to 0_-$, if $\alpha > -1$. This clearly shows that the scale factor may safely go to zero without being associated with a curvature singularity. Conversely, the curvature may diverge even if the scale factor goes

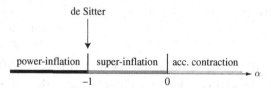

Figure 5.1 The four types of inflationary kinematics as a function of the power α, according to the conformal-time parametrization of Eq. (5.16).

to infinity. An example of this last possibility, the so-called "big rip" singularity, has been recently discussed as the possible fate of our Universe in the context of "phantom" models of the dark energy field [12], describing the present stage of accelerated cosmic evolution (see Chapter 9).

5.2 Dynamical equivalence of super-inflation and accelerated contraction

In the context of the duality-generated solutions of the string cosmology equations, the background geometries of Class IIa and Class IIb may correspond to different kinematic representations of the same inflationary model, in two different frames. A metric background representing a phase of super-inflationary expansion in the S-frame, in fact, tends to become a metric describing a phase of accelerated contraction when it is transformed to the E-frame [6, 13].

This effect can be conveniently illustrated by recalling here the S-frame, anisotropic and vacuum solution already presented in Eqs. (4.35) and (4.36), in which the spatial sections of the background geometry are factorizable as the direct product of two conformally flat manifolds, with d and n dimensions, respectively expanding with scale factor a and contracting with scale factor $b = a^{-1}$. In the cosmic-time gauge we have

$$a(t) = (-t)^{-1/\sqrt{d+n}}, \qquad b(t) = (-t)^{1/\sqrt{d+n}}, \qquad t < 0. \qquad (5.19)$$

The associated dilaton field,

$$\phi(t) = \left(d - n + \sqrt{d+n}\right) \ln a, \qquad t < 0, \qquad (5.20)$$

is growing as $t \to 0_-$, provided $d + \sqrt{d+n} > n$.

This solution of the low-energy string effective action describes an accelerated evolution of all dimensions, as can be easily checked in the conformal-time gauge, where

$$a = (-\eta)^{\alpha_1}, \qquad b = (-\eta)^{\alpha_2}, \qquad \alpha_1 = -\alpha_2 = -\frac{1}{1+\sqrt{d+n}}. \qquad (5.21)$$

The first power satisfies $-1 < \alpha_1 < 0$, so that the d "external" dimensions expand in a super-inflationary way; the second power satisfies $\alpha_2 > 0$, implying that the n

"internal" dimensions undergo accelerated contraction (see Fig. 5.1). Let us now ask if (and how) the kinematic is different when this solution is transformed to the E-frame.

The transformation of an anisotropic solution like this could be performed by directly applying the general prescription presented in Chapter 2, see Eqs. (2.39) and (2.43). It is instructive, however, to rederive the transformation by working with the explicit form of the free gravi-dilaton action (2.1), written for a Bianchi-I-type metric without fixing the temporal gauge, i.e. by setting

$$\phi = \phi(t), \qquad g_{00} = N^2(t), \qquad g_{ij} = -a_i^2(t)\delta_{ij}. \tag{5.22}$$

Defining $H_i = \dot{a}_i/a_i$, $F = \dot{N}/N$, and inserting explicitly (whenever needed) the sum symbol, we obtain

$$(\nabla\phi)^2 = \frac{\dot{\phi}^2}{N^2}, \qquad \sqrt{-g} = N\prod_i a_i,$$

$$\Gamma_{0i}{}^j = H_i\delta_i^j, \qquad \Gamma_{ij}{}^0 = \frac{a_i\dot{a}_i}{N^2}\delta_{ij}, \qquad \Gamma_{00}{}^0 = \frac{\dot{N}}{N} \equiv F, \tag{5.23}$$

$$R = \frac{1}{N^2}\left[2F\sum_i H_i - 2\sum_i \dot{H}_i - \sum_i H_i^2 - \left(\sum_i H_i\right)^2\right].$$

By noting that

$$\frac{\mathrm{d}}{\mathrm{d}t}\left[2\frac{e^{-\phi}}{N}\prod_k a_k \sum_i H_i\right]$$

$$= \frac{e^{-\phi}}{N}\left(\prod_k a_k\right)\left[2\sum_i \dot{H}_i - 2F\sum_i H_i - 2\dot{\phi}\sum_i H_i + 2\left(\sum_i H_i\right)^2\right], \tag{5.24}$$

we can rewrite the S-frame action (2.1) in the standard quadratic form, by eliminating the terms linear in the first derivative of H and N. We obtain, modulo a total derivative (in units $2\lambda_s^{d-1} = 1$),

$$S = -\int \mathrm{d}^{d+1}x\sqrt{|g|}\,e^{-\phi}(R + \nabla\phi^2)$$

$$= -\int \mathrm{d}^d x\,\mathrm{d}t\left(\prod_{i=1}^d a_i\right)\frac{e^{-\phi}}{N}\left[\dot{\phi}^2 - \sum_i H_i^2 + \left(\sum_i H_i\right)^2 - 2\dot{\phi}\sum_i H_i\right]. \tag{5.25}$$

In the same anisotropic background, on the other hand, the Einstein action takes the form (again, modulo a total derivative, and in units $2\lambda_P^{d-1} = 1$)

$$\widetilde{S}(\widetilde{g}, \widetilde{\phi}) = -\int d^{d+1}x\sqrt{|\widetilde{g}|}\left(\widetilde{R} - \frac{1}{2}\widetilde{\nabla}\widetilde{\phi}^2\right)$$

$$= -\int d^d x \frac{dt}{\widetilde{N}}\left(\prod_{i=1}^{d}\widetilde{a}_i\right)\left[-\frac{1}{2}\dot{\widetilde{\phi}}^2 - \sum_i\widetilde{H}_i^2 + \left(\sum_i\widetilde{H}_i\right)^2\right]. \quad (5.26)$$

It can be immediately checked that the field redefinition (at *fixed* coordinates!)

$$\widetilde{a}_i = a_i e^{-\phi/(d-1)}, \qquad \widetilde{N} = N e^{-\phi/(d-1)}, \qquad \widetilde{\phi} = \phi\sqrt{\frac{2}{d-1}}, \quad (5.27)$$

gives

$$\widetilde{H}_i = H_i - \frac{\dot{\phi}}{d-1}, \quad (5.28)$$

and directly transforms the E-frame action (5.26) into the S-frame action (5.25). The above transformation generalizes Eq. (2.64) to the anisotropic case, for an arbitrary choice of the temporal gauge.

Let us now apply this result to the particular solution (5.19). Using the explicit form of the dilaton solution, Eq. (5.20), we obtain

$$\widetilde{a} = a^{(2n-1-\sqrt{d+n})/(d+n-1)}, \qquad \widetilde{b} = b\, a^{(n-d-\sqrt{d+n})/(d+n-1)}. \quad (5.29)$$

(notice that $\widetilde{a} \neq \widetilde{b}^{-1}$). For the study of the kinematic properties of $\widetilde{a}$ and $\widetilde{b}$, it is then convenient to use the conformal-time gauge, by imposing $\widetilde{N}\,dt = \widetilde{a}\,d\widetilde{\eta}$. Since the conformal-time coordinate is the same in the two frames (in fact, if we choose $N = a$, then $\widetilde{N} = \widetilde{a}$ from the transformation (5.27)), we can directly use for $a(\eta)$ and $b(\eta)$ the expressions (5.21). The result is

$$\widetilde{a}(\eta) = (-\eta)^{\widetilde{\alpha}_1}, \qquad \widetilde{\alpha}_1 = -\frac{2n-1-\sqrt{d+n}}{(1+\sqrt{d+n})(d+n-1)},$$

$$\widetilde{b}(\eta) = (-\eta)^{\widetilde{\alpha}_2}, \qquad \widetilde{\alpha}_2 = \frac{2d-1+\sqrt{d+n}}{(1+\sqrt{d+n})(d+n-1)}. \quad (5.30)$$

Using the kinematic definitions illustrated in Fig. 5.1 we can now identify the type of inflationary behavior of the transformed solution.

In the isotropic case, corresponding to $n = 0$, it can be easily checked that the S-frame super-inflationary expansion, $a \sim (-\eta)^{-1/(1+\sqrt{d})}$, becomes in the E-frame accelerated contraction, $\widetilde{a} \sim (-\eta)^{1/(d-1)}$. For the anisotropic configuration with

$n \neq 0$, on the contrary, the E-frame evolution of the external dimensions described by $\tilde{a}$ may, in principle, remain of the super-inflationary type, provided

$$2n > 1 + \sqrt{d+n}. \tag{5.31}$$

If we require, however, that the S-frame solution describe a growing dilaton (as suggested, for instance, by the self-dual scenario introduced in the previous chapter), then the condition (5.31) turns out to be in competition with the growing-dilaton condition $d + \sqrt{d+n} > n$ (see Eq. (5.20)), which forbids too large values of n. By choosing, for instance, $d = 3$, and imposing both conditions, we find that the E-frame scale factor $\tilde{a}$ is (super-inflationary) expanding only for $n = 2$. In all other cases the power $\tilde{\alpha}_1$ turns out to be positive, so that the d-dimensional part of the metric is transformed from super-inflation to accelerated contraction.

It should be stressed, in any case, that the inflationary properties of the accelerated metric – namely, the solution of the kinematic problems through the condition (5.2), the possible amplification of the vacuum fluctuations (discussed in the following chapters), and so on – keep their validity, quite apart from the particular choice of S-frame or E-frame representation, and from the expanding or contracting behavior of the scale factor. The condition which fixes the minimum duration of inflation as a necessary condition of any successful scenario, in particular, has the same form in both frames (see Eq. (1.92)): if satisfied in a frame it is thus automatically satisfied also in the other, conformally related representation [6, 13]. This means that the inflationary properties of a background are frame independent, at least for frames corresponding to consistent representations of the same underlying physical model.

Let us finally notice that the kinematical properties of the transformed solutions can be deduced also through a direct computation in the cosmic-time gauge, by setting $\tilde{N} \, dt = d\tilde{t}$, and using the relation (5.27) connecting N and $\tilde{N}$. Starting from the S-frame solution (5.20), in the gauge $N = 1$, we obtain

$$d\tilde{t} = dt \, e^{-\phi/(d+n-1)}, \qquad (-\tilde{t}) \sim (-t)^{[\sqrt{d+n}(d+n)+d-n]/[\sqrt{d+n}(d+n-1)]}. \tag{5.32}$$

Thus, from Eq. (5.29),

$$\tilde{a}(\tilde{t}) = (-\tilde{t})^{\tilde{\beta}_1}, \qquad \tilde{\beta}_1 = -\frac{2n - 1 - \sqrt{d+n}}{\sqrt{d+n}(d+n) + d - n},$$

$$\tilde{b}(\tilde{t}) = (-\tilde{t})^{\tilde{\beta}_2}, \qquad \tilde{\beta}_2 = \frac{2d - 1 + \sqrt{d+n}}{\sqrt{d+n}(d+n) + d - n}. \tag{5.33}$$

The computation of $\dot{\tilde{a}}$, $\ddot{\tilde{a}}$ and $\tilde{H}$ (where the dots denote differentiation with respect to $\tilde{t}$) then confirms the previous results concerning the kinematical properties of the transformed background. A similar analysis, with similar results, can be repeated for other examples of accelerated backgrounds, possibly including matter sources and other fields in the effective action.

5.3 Horizons and kinematics

As stressed in Section 5.1, the accelerated backgrounds of Class I are typical of the conventional (i.e. post-big bang) inflationary scenario, while those of Class II are peculiar to string cosmology models of (pre-big bang) inflation. The difference between the two classes is not only of kinematical type (as for Class IIa and Class IIb), but is also dynamical, and may be conveniently illustrated using the notion of "event horizon" already introduced in Chapter 1.

We should recall that the proper distance $d_e(t)$ of the event horizon from a comoving observer, at rest in a homogeneous and isotropic background, is given by

$$d_e(t) = a(t) \int_t^{t_M} dt' a^{-1}(t').$$ (5.34)

Here t_M is the maximal allowed extension, towards the future, of the cosmic-time coordinate in the given space-time manifold. This integral converges for all types of accelerated (expanding or contracting) scale factors. For Class I metrics and, in particular, for power-inflation ($\beta > 1$, $t > 0$), we have

$$d_e(t) = t^\beta \int_t^\infty dt' t'^{-\beta} = \frac{t}{\beta - 1} = \frac{\beta}{\beta - 1} H^{-1}(t),$$ (5.35)

while, for de Sitter inflation,

$$d_e(t) = e^{Ht} \int_t^\infty dt' e^{-Ht'} = H^{-1} = \text{const.}$$ (5.36)

For Class II metrics ($\beta < 1$, $t < 0$), we have

$$d_e(t) = (-t)^\beta \int_t^0 dt' (-t')^{-\beta} = \frac{(-t)}{1 - \beta} = \frac{\beta}{\beta - 1} H^{-1}(t)$$ (5.37)

(note that d_e is always positive, even for accelerated contraction where $H < 0$, but $\beta > 0$ and $\beta - 1 < 0$). In all cases the proper distance d_e evolves in time as the so-called "Hubble horizon" (also called the "Hubble radius", i.e. the inverse of the modulus of the Hubble parameter), and then as the inverse of the curvature scale. Thus, the proper size of the horizon will be *constant* or *growing* in models of post-big bang inflation (Class I), *decreasing* in pre-big bang inflation (Class II), in both the S-frame and E-frame representations.

Such an important difference is illustrated in Figs. 5.2 and 5.3, where the dashed lines represent the time evolution of the horizon, and the solid curves the evolution of the scale factor. The shaded area at the time t_0 represents the

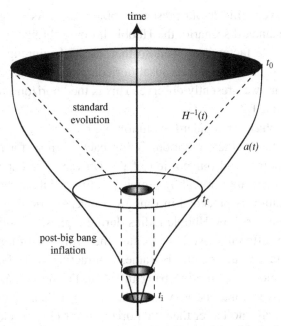

Figure 5.2 Qualitative evolution of the Hubble horizon (dashed lines) and of the scale factor (solid curves) in a model of standard, post-big bang inflation at constant curvature.

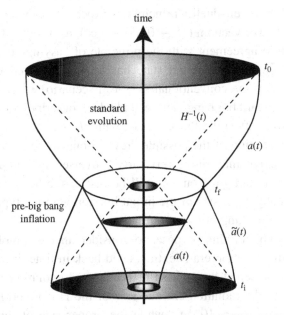

Figure 5.3 Qualitative evolution of the Hubble horizon (dashed lines) and of the scale factor (solid curves) in a model of pre-big bang inflation, represented by the S-frame expansion $a(t)$ or by the E-frame contraction $\widetilde{a}(t)$.

portion of the Universe inside our present Hubble radius. As we go back in time, according to the standard scenario, the Hubble horizon shrinks linearly while the decrease of the scale factor is slower, so that, at the beginning of the standard evolution $t = t_f$, we are left with a causal horizon much smaller than the portion of the Universe that we presently observe. This is the "horizon problem" already presented in Section 1.2.

In Fig. 5.2 the phase of standard evolution is preceded by a phase of conventional (in particular de Sitter) inflation. Going back in time, for $t < t_f$, the scale factor keeps shrinking, and our portion of the Universe "re-enters" the horizon during a phase of constant (or slightly growing in time) Hubble radius. In Fig. 5.3 the standard evolution is preceded in time by a phase of pre-big bang inflation. The Universe "re-enters" the Hubble radius during a phase of shrinking horizon. To emphasize the difference, we have plotted the evolution of both the S-frame, expanding scale factor, a, and the E-frame, contracting scale factor, $\tilde{a}$. It may be noted that, in models of pre-big bang inflation, the proper size of the initial portion of our Universe may be very large in string (or Planck) units at the time of re-entry, but always no larger than the horizon itself [14], as clearly illustrated in the picture.

The initial horizon H_i^{-1}, on the other hand, is large because the initial curvature scale is small (in string models, in particular, the initial configuration satisfies $H_i \ll \lambda_s^{-1}$). This is a basic consequence of the choice of the initial state which, for models inspired by the self-duality principle, is expected to approach the flat, cold and empty perturbative vacuum ($H \to 0$, $\phi \to -\infty$), as discussed in the previous chapter (and also in agreement with the principle of "asymptotic past triviality" introduced in [15]). Such an initial state has to be contrasted with the highly curved initial state of the conventional slow-roll scenario, which is more deeply anchored to the standard big bang picture, and which describes a Universe starting to inflate at (or soon after) the Planck scale, with $H_i \sim \lambda_P^{-1}$.

For a better illustration of the possible "revolutionary" approach suggested by string theory to the dynamics of the very early Universe – and, in particular, to the initial cosmological configuration – it will be instructive to provide an explicit, quantitative estimate of the size of the initial horizon within a typical example of pre-big bang (or "shrinking-horizon") inflation.

It will be enough, for this purpose, to consider an oversimplified model in which the standard radiation era is extrapolated back in time down to the Planck scale, and is just preceded by a phase of accelerated evolution. One finds that, at the beginning of the radiation era, the size of the Hubble horizon is of order of the Planck length, $\lambda_P \sim 10^{-1} \lambda_s$, while the proper size of the homogeneous and causally connected region inside our present Hubble radius (rescaled down according to the standard scenario) is unnaturally larger than the horizon by the factor $\ell_f/\lambda_P \sim 10^{30}$ (or $\ell_f/\lambda_s \sim 10^{29}$), as already remarked in Chapter 1. Going back in time during the inflationary epoch, the ratio $r = (aH)^{-1}$ must thus increase at least by the factor 10^{30}, so as to push the homogeneous region inside the horizon by the amount required to compensate for the effects of the subsequent decelerated evolution. Using the relation $r \sim \eta$ we can express the condition of efficient inflation (1.92), in this case, as

$$\frac{r_f}{r_i} = \frac{|\eta_f|}{|\eta_i|} \lesssim 10^{-30}, \tag{5.38}$$

where η_f and η_i are, respectively, the epochs at which inflation ends and begins.

Let us choose, as our explicit example, a phase of d-dimensional isotropic super-inflation, described by the S-frame scale factor

$$a \sim (-t)^{-1/\sqrt{d}} \sim (-\eta)^{-1/(1+\sqrt{d})} \tag{5.39}$$

(according to Eqs. (5.19), (5.21) with $n = 0$). By applying the condition (5.38) it follows that, going back in time during inflation, the scale factor decreases by the factor $a_i/a_f \lesssim 10^{-30/(1+\sqrt{d})}$. Therefore, the proper size of the homogeneous region at the beginning of a phase of "minimal" efficient inflation is still very large in string units (which are the natural units for the S-frame variables that we are considering):

$$\ell_i = \ell_f \left(\frac{a_i}{a_f}\right) \sim 10^{-\frac{30}{1+\sqrt{d}}} 10^{29} \lambda_s \sim 10^{\frac{30\sqrt{d}}{1+\sqrt{d}}} 10^{-1}\lambda_s. \tag{5.40}$$

For instance, $\ell_i \sim 10^{18}\lambda_s$ if $d = 3$. This result has to be contrasted with the case of standard slow-roll inflation which, for the purpose of the present discussion, we can approximate as a phase of de Sitter inflation with $a \sim (-\eta)^{-1}$. In this case, going back in time during inflation, the scale factor turns out to be reduced by the factor 10^{-30}, so that the size of the initial homogeneous region is just of order $10^{-1}\lambda_s$, like the horizon which stays (approximately) frozen (see Fig. 5.2).

The contrast is even more striking if we express the initial size ℓ_i in Planck units, and we take into account that, in the S-frame, the effective Planck length varies (according to Eq. (2.3)) as $\lambda_P = \lambda_s \exp[\phi/(d-1)]$. Using the isotropic limit ($n = 0$) of the solutions (5.20) and (5.21) we can then estimate that the initial value of the Planck length, $\lambda_P(\eta_i)$, is smaller than the (standard) final value $\lambda_P(\eta_f) \sim 10^{-1}\lambda_s$ by the factor

$$\frac{\lambda_P(\eta_i)}{\lambda_P(\eta_f)} = \left(\frac{\eta_f}{\eta_i}\right)^{\frac{\sqrt{d}}{d-1}} \lesssim 10^{-30\frac{\sqrt{d}}{d-1}}. \tag{5.41}$$

As a consequence, the lower bound (5.40) is enhanced (in Planck units) by the factor $\lambda_P(\eta_f)/\lambda_P(\eta_i)$, and we obtain

$$\ell_i = \ell_f \left(\frac{a_i}{a_f}\right) \frac{\lambda_P(\eta_f)}{\lambda_P(\eta_i)} \sim 10^{\frac{30d}{d-1}} \lambda_P, \tag{5.42}$$

corresponding to $\ell_i \gtrsim 10^{45} \lambda_P$ for $d = 3$. We may note that the left-hand side of Eq. (5.42) exactly represents the initial size of the homogeneous region evaluated in the E-frame, where $\ell_i = \ell_f(\tilde{a}_i/\tilde{a}_f)$. In this frame, where λ_P is constant, the inflation is represented as a contraction: following the evolution of the scale factor back in time we are led to a much larger size of the initial homogeneous region, as clearly illustrated also in Fig. 5.3.

However, it should be stressed that a large size of the initial homogeneous region (in Planck or string units) is in contrast neither with the spirit of the classical inflationary paradigm, nor with the solution of the kinematic problems of the standard cosmological scenario. Without the presence of the phase characterized by accelerated evolution, in fact, the initial homogeneous region would be much larger than the horizon itself (see the spatial sections of Figs. 5.2, 5.3 at the time $t = t_f$). For those models in which inflation occurs at growing curvature (or shrinking horizon) the initial region is large in string units, but not larger than the corresponding initial horizon, as we now verify explicitly.

Consider again the super-inflationary, S-frame solution (5.39). As we go back in time, the size of the horizon H^{-1} grows linearly in cosmic time. Starting with the value $H_f^{-1} = \lambda_P = 10^{-1}\lambda_s$, we can compute, for the given example, the ratio of the initial to final horizon size as

$$\frac{H_i^{-1}}{H_f^{-1}} = \frac{t_i}{t_f} = \left(\frac{\eta_i}{\eta_f}\right)^{\frac{\sqrt{d}}{1+\sqrt{d}}}. \tag{5.43}$$

Taking into account the condition (5.38), we find that the initial horizon is at least as large as

$$H_i^{-1} \sim 10^{\frac{30\sqrt{d}}{1+\sqrt{d}}} 10^{-1}\lambda_s, \tag{5.44}$$

i.e. exactly as large as the proper size (5.40) of the initial homogeneous region. The same result is obtained if we repeat the same analysis in Planck units for the E-frame metric (as illustrated in Fig. 5.3).

We should mention, in addition, that a large horizon in Planck units at the beginning of inflation makes the model free from the so-called "trans-Planckian problem" [16], which affects all models of inflation where the size of the initial horizon is Planckian. It is clear, in fact, that if the proper length λ of a metric fluctuation (for instance, a gravitational wave) has to be of sub-horizon size at the beginning of inflation (in order to represent a causal perturbation present today inside our Hubble radius), then $\lambda(t_i) < H_i^{-1}$ so that, for $H_i^{-1} \sim \lambda_P$, its initial proper energy was larger than Planckian, $\omega(t_i) = \lambda^{-1}(t_i) > H_i \sim M_P$. On the other hand, the extrapolation to this regime of known, sub-Planckian physics is problematic, and the cosmological predictions one obtains in this way are questionable. Toy model calculations have shown that the inflationary results on the spectrum of perturbations may indeed be strongly sensitive to the details of the trans-Planckian initial conditions [17]. This problem is absent if $H_i^{-1} \gg \lambda_P$ since, in that case, the ratio $\omega/M_P = \lambda_P \omega \sim \tilde{a}^{-1}$ decreases as we go back in time during inflation, so that the initial fluctuations are always normalized well inside the perturbative regime (see also Chapters 7 and 8).

Let us finally comment on the "naturalness" of the choice of initial conditions for a phase of inflationary evolution. The Planck (or string) length certainly provides a natural standard [18] for the size of the initial homogeneous patches when initial conditions are imposed on a cosmological configuration approaching the high-curvature, quantum gravity regime (as in models of post-big bang inflation). In models of pre-big bang inflation, however, initial conditions are to be imposed when the Universe is deeply inside the low-curvature, weak-coupling, highly classical regime. In that case, the Planck or string length is certainly not a typical scale for the background geometry, while the classical horizon scale H^{-1} seems to provide the relevant standard for a natural homogeneity scale [14].

It should be noted, also, that if we assume the saturation of the holographic bound applied to a cosmological metric [19–22], then a large (homogeneous) Hubble horizon should imply a large initial entropy, at least if $S_i \sim$ (horizon area in Planck units). This should correspond, in a quantum context, to a small probability P that such a configuration be obtained through a process of quantum tunneling, since $P \sim \exp[-S]$. In models of pre-big bang inflation, however, quantum effects such as tunneling or reflection of the Wheeler–De Witt wave function [23, 24, 25] are expected to become important possibly *towards the end* of inflation, not the beginning (see Appendix 6A). They may be effective *to exit* from the inflationary regime, *not to enter* it, and to explain the origin of the initial state. A large entropy of the initial state, in the weakly coupled, highly classical regime, can only correspond to a large probability of such a state which, for classical and macroscopic configurations, is expected to be proportional to $\exp[+S]$.

In any case, an initial state characterized by a set of large (or small) dimensionless parameters, associated with a large value of the initial horizon scale H_i^{-1}, is an unavoidable aspect of all models in which inflation starts at scales much smaller than Planckian. In the post-big bang inflationary scenario, where the observational tracks of any pre-Planckian epoch are washed out, one might regard as unnatural [18] having an initial homogeneity scale of order H^{-1}, whenever H is small in Planck (or string) units. In the context of models in which inflation

Table 5.1 Kinematical and dynamical properties of the various inflationary classes

	Class I		Class II	
	power-inflation	de Sitter	super-inflation	accelerated contraction
cosmic time, $a = |t|^{\beta}$	$\beta > 1, t > 0$	$\beta = \infty, t > 0$	$\beta < 0, t < 0$	$0 < \beta < 1, t < 0$
conformal time, $a = |\eta|^{\alpha}$	$\alpha < -1, \eta < 0$	$\alpha = -1, \eta < 0$	$-1 < \alpha < 0, \eta < 0$	$\alpha > 0, \eta < 0$
kinematics	accelerated expansion	accelerated expansion	accelerated expansion	accelerated contraction
curvature H^2	decreasing	constant	growing	growing
horizon $d_e(t)$	growing	constant	shrinking	shrinking

precedes the Planck era, however, the phenomenological imprints of the Planck epoch are not necessarily washed out by a long and subsequent inflationary phase. The pre-Planckian history of the Universe may become visible [26], and sub-Planckian initial conditions accessible (in principle) to observational tests, so that their naturalness could also be analyzed with a Bayesian approach, in terms of a-posteriori probabilities, as discussed in [15] (see also [27] for a more detailed comparison of pre-big bang versus post-big bang inflation).

We conclude this chapter by reporting, in Table 5.1 a schematic summary of the four possible types of accelerated evolution, and of their main properties. The powers α and β used in the table are related by Eq. (5.18), and the event horizon is defined in Eqs. (5.35)–(5.37).

References

[1] F. Lucchin and S. Matarrese, *Phys. Lett.* **B164** (1985) 282.
[2] A. Guth, *Phys. Rev.* **D23** (1981) 347.
[3] D. Shadev, *Phys. Lett.* **B317** (1984) 155.
[4] R. B. Abbott, B. Bednarz and S. D. Ellis, *Phys. Rev.* **D33** (1986) 2147.
[5] E. W. Kolb, D. Lindley and D. Seckel, *Phys. Rev.* **D30** (1984) 1205.
[6] M. Gasperini and G. Veneziano, *Mod. Phys. Lett.* **A8** (1993) 3701.
[7] A. D. Linde, *Phys. Lett.* **B129** (1983) 177.
[8] J. Khouri, B. A. Ovrut, P. J. Steinhardt and N. Turok, *Phys. Rev.* **D64** (2001) 123522.
[9] J. Khouri, B. A. Ovrut, N. Seiberg, P. J. Steinhardt and N. Turok, *Phys. Rev.* **D65** (2002) 086007.
[10] P. J. Steinhardt and N. Turok, *Phys. Rev.* **D65** (2002) 126003.
[11] L. A. Boyle, P. J. Steinhardt and N. Turok, *Phys. Rev.* **D69** (2004) 127302.
[12] R. R. Caldwell, M. Kamionkowski and N. N. Weinberg, *Phys. Rev. Lett.* **91** (2003) 07130.
[13] M. Gasperini and G. Veneziano, *Phys. Rev.* **D50** (1994) 2519.
[14] M. Gasperini, *Phys. Rev.* **D61** (2000) 87301.
[15] A. Buonanno, T. Damour and G. Veneziano, *Nucl. Phys.* **B543** (1999) 275.
[16] R. H. Brandenberger and J. Martin, *Mod. Phys. Lett.* **A16** (2001) 999.
[17] J. A. Martin and R. Brandenberger, *Phys. Rev.* **D68** (2003) 063513.
[18] N. Kaloper, A. Linde and R. Bousso, *Phys. Rev.* **D59** (1999) 043508.
[19] R. Easther and D. A. Lowe, *Phys. Rev. Lett.* **82** (1999) 4967.
[20] G. Veneziano, *Phys. Lett.* **B454** (1999) 22.
[21] N. Kaloper and A. Linde, *Phys. Rev.* **D60** (1999) 103509.
[22] R. Brustein and G. Veneziano, *Phys. Rev. Lett.* **84** (2000) 5695.
[23] M. Gasperini and G. Veneziano, *Gen. Rel. Grav.* **28** (1996) 1301.
[24] M. Gasperini, J. Maharana and G. Veneziano, *Nucl. Phys.* **B472** (1996) 349.
[25] M. Gasperini, *Int. J. Mod. Phys.* **D10** (2001) 15.
[26] M. Gasperini, *Mod. Phys. Lett.* **A14** (1999) 1059.
[27] M. Gasperini, *Class. Quantum Grav.* **17** (2000) R1.

6

The string phase

From the solutions of the low-energy string cosmology equations, presented in the previous chapters, we have learned that the traditional picture of a Universe which emerges from the inflation of a very small and curved space-time patch is a *possibility*, not a *necessity*: quite different initial conditions are possible, and not necessarily unlikely. In particular, there are scenarios suggesting an overturning of the traditional scheme of slow-roll inflation, describing a Universe which starts inflating from an initial state characterized by a perturbatively small space-time curvature. Such a state is unstable and, pushed by the dilaton, tends to decay with an accelerated evolution towards higher and higher curvature configurations.

The growth of the curvature during this accelerated regime is possibly accompanied by the growth of the dilaton, and thus of the string coupling $g_s = \exp(\phi/2)$. As time goes on, in such a context, the Universe is thus approaching the two limits marking the validity of the lowest-order string effective action: (*i*) the string curvature scale, reached when $\lambda_s H \sim 1$ and/or $\lambda_s \dot{\phi} \sim 1$, and (*ii*) the strong coupling regime, reached when $g_s \sim 1$. Reaching the first boundary requires the inclusion of α' corrections (higher powers of the curvature and higher derivatives of all background fields), while the second boundary requires the inclusion of quantum loop corrections (higher genus corrections and higher powers of g_s^2). Crossing both limits would necessarily imply a Universe entering the non-perturbative regime of the (still) largely unknown M-theory (see for instance [1, 2]).

In the context of the inflationary solutions of the higher-dimensional, tree-level, gravi-dilaton effective action there is no way to stop the accelerated growth of the curvature and of the string coupling. We may thus expect that the damping of inflation, the stabilization of the extra $d - 3$ spatial dimensions in some compactified configuration compatible with the standard-model symmetries, as well as the eventual transition to a phase of decelerated expansion of three spatial dimensions, be eventually induced by the effects of the α' and loop corrections, with the possible contribution of other fields. In particular, this could include the contribution of the R–R and/or NS–NS p-form fields, typical of the superstring effective actions: we should recall that the regular bouncing model of Section 4.3, obtained by "boosting" the Milne solution, was based on the presence of the NS–NS two-form B. Higher-order corrections seem to be generally required, however, to contrast the possible tendency of the inflationary Kasner-like solutions to enter the chaotic regime [3].

Concerning the relative importance of the various corrections we may, in principle, distinguish two cases, depending on the parameters characterizing the initial configuration. A first possibility is that the string loop corrections become important at an epoch in which the curvature and the gradients of all background fields are still small (in string units), so that the α' corrections can be safely neglected. To estimate the importance of the loop corrections on the cosmological evolution we can compute the backreaction of the particles produced by the mechanism of parametric amplification of the vacuum fluctuations (see Chapter 7). The total energy density of such particles, produced at a curvature scale H, is of the order of $\rho_q \sim NH^4$, where N is the total number of effective degrees of freedom participating in the amplification process. The backreaction starts to become important when ρ_q approximates the critical energy density, $\rho_c \sim M_P^2 H^2 \sim (g_s \lambda_s)^{-2} H^2$, namely when

$$N g_s^2 \lambda_s^2 H^2 \sim 1. \tag{6.1}$$

When the above condition is satisfied inflation may be stopped by the quantum corrections, and the background may experience an almost immediate "bounce" towards the decreasing-curvature regime (see trajectory $\boxed{1} \rightarrow \boxed{1a}$ of Fig. 6.1), without any contribution from the α' corrections. Examples of this type have been reported in Chapter 4, with the quantum backreaction simulated by effective sources with negative energy density, or by effective perturbative potentials (see [4] for other possible examples). The "big bang", in this case, may be represented as the process of radiation production due to the amplification of the vacuum fluctuations associated with the curvature bounce. It should be stressed that the condition (6.1) also signals the saturation of the so-called "Hubble entropy bound" [5] for the radiation produced.

The condition (6.1), however, marks the beginning of a phase characterized by a copious, non-perturbative production of strings and higher-dimensional branes (see for instance [6]). If the loop corrections do not induce an immediate bounce, the Universe necessarily enters a truly "stringy" phase, which is higher-dimensional, strongly coupled, populated by strings, branes and antibranes, possibly winding and wrapping around the compact dimensions [7, 8, 9], eventually colliding and annihilating among each other. The "big bang", in this case, could be represented by a collision of branes [10]. Also, the interactions of the branes with the background fields could be responsible for the stabilization of the

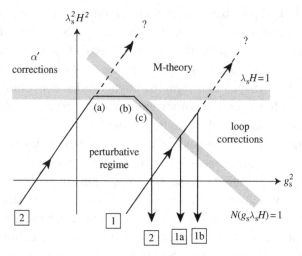

Figure 6.1 Qualitative sketch of the possible evolution of regular (bouncing) cosmologies in the plane $\{g_s^2, \lambda_s^2 H^2\}$ (in logarithmic scale). For trajectory $\boxed{1}$ the string loop corrections become important before the α' corrections (in case $\boxed{1a}$ there is an immediate bounce due to the loop backreaction, in case $\boxed{1b}$ the model goes through a phase of string/brane-dominated evolution). Trajectory $\boxed{2}$, on the contrary, leads the model to a regime of high, constant curvature, and only later may quantum loop effects come into play, eventually inducing the bounce of the curvature.

background in a standard-model configuration [11], possibly inducing an additional phase of slow-roll, post-big bang inflation [12, 13, 14] on a brane identified with our Universe, which subsequently evolves through the steps of the standard cosmological picture (trajectory $\boxed{1} \rightarrow \boxed{1b}$ of Fig. 6.1). Such a phase of brane inflation (see Chapter 10), however, is not an unavoidable necessity if this stringy phase is preceded by a sufficiently long period of pre-big bang inflation.

The second possibility is that the high curvature limit is reached when the background fields are still in the weak coupling regime, $g_s \ll 1$. In this case, when $\lambda_s H \sim 1$, the Universe enters a phase which is again typically "stringy", but which has different properties with respect to the previous one. The string effects, represented by the α' corrections, now tend to stabilize the curvature at a constant value near the string scale [15], while the dilaton keeps growing (see trajectory $\boxed{2}$ of Fig. 6.1, from (a) to (b)). The metric evolution remains accelerated, so that the Universe is still inside the regime of pre-big bang inflation: if this phase is long enough it may affect in a significant way the final spectra of cosmological pertubations, as we shall see in the following chapters.

The discussion of this chapter will concentrate, first of all, on the dynamical properties of the high-curvature, weak-coupling regime. It will be shown in Section 6.1 that, when the quantum loop corrections are negligible, a background configuration with constant curvature and linearly evolving dilaton may represent a solution of the string effective action to all orders in α' (and thus correspond to some exact conformal field theory model). Such a solution represents a "fixed point" of the string cosmology equations which, in some models, is also smoothly connected to the perturbative vacuum. It is thus possible that the Universe, starting from an initial, perturbative configuration, is attracted at late times towards such a (typically stringy) high-curvature configuration.

The extension in time of such a phase, even if long, is finite because – unless one introduces some extra mechanism of dilaton stabilization – the string coupling keeps growing until the loop corrections become important (at the scale $N g_s^2 \sim 1$), and move the gravi-dilaton system away from the fixed-point configuration. Then, if the loop perturbations are pushing towards unbounded growth of the curvature and of the coupling, the system necessarily enters the full M-theory regime (dashed trajectories of Fig. 6.1). If, on the contrary, only the coupling is growing (but the curvature is frozen), then the system is led to the same phase populated by strings and branes as in the previous case, with the complication that α' corrections are also necessary (see [16] for recent studies in this regime). Finally, if the loop corrections are appropriate, the Universe may bounce instead towards the standard cosmological regime as illustrated by trajectory $\boxed{2}$ of Fig. 6.1, and as will be discussed in Section 6.2 (see in particular the discussion after Eq. (6.32) for a precise definition of the physical effects coming into play at the points marked by (a), (b) and (c) in Fig. 6.1).

In Section 6.3 we will finally consider the possibility of a Universe which, evolving from the weak coupling along trajectory $\boxed{1}$, does not immediately bounce when hitting the borderline of loop backreaction, but enters a phase where the strong coupling effects are associated with the production of a gas of strings and branes. We will discuss, in particular, the case in which the Universe is filled by a string gas as the dominant source of the background geometry. It will be shown that the presence of "winding modes", in backgrounds with compact topology, is not an efficient source of inflation but can induce the transition to a final configuration with only three dimensions expanding in a decelerated way, and the other spatial dimensions remaining compact at the string scale [7, 8] (see [17] for a recent review of this string-gas cosmological scenario).

6.1 High-curvature fixed points of the string cosmology equations

In the context of higher-derivative models of gravity, it is known that the presence of non-minimally coupled scalar fields tends to remove any exact solution at constant curvature [18, 19] (which generically exists, instead, in models of pure gravity containing higher powers of the curvature invariants in the action [20]). In the context of the S-frame string effective action, however, such solutions may exist, even to all orders in α', provided they are associated with a non-trivial (time-dependent) dilaton background.

Let us consider, in fact, a spatially flat, homogeneous but anisotropic gravi-dilaton background, parametrized by

$$g_{00} = N^2(t), \qquad g_{ij} = -a_i^2(t)\delta_{ij}, \qquad a_i = e^{\beta_i}, \qquad \phi = \phi(t), \qquad (6.2)$$

where $i, j = 1, \ldots, d$. Defining

$$\overline{\phi} = \phi - \sum_i \beta_i \qquad (6.3)$$

and using the fact that, to the lowest order of the g_s^2 expansion, the dependence of the S-frame action on a constant dilaton field is fixed [21], we can write the exact S-frame gravi-dilaton action, to all orders in α', as [15]

$$S = \int dt \, N e^{-\overline{\phi}} L\left(\beta_i^{(n)}, \overline{\phi}^{(n)}\right) \qquad (6.4)$$

(modulo a spatial volume factor). Here the effective Lagrangian L is a general function of the (covariantized) time derivative of the fields $\overline{\phi}$, β_i, at all orders $n \geq 1$:

$$\beta_i^{(n)} = \prod_{k=1}^{n} \left(\frac{1}{N}\frac{d}{dt}\right)^k \beta(t), \qquad \overline{\phi}^{(n)} = \prod_{k=1}^{n} \left(\frac{1}{N}\frac{d}{dt}\right)^k \overline{\phi}(t). \qquad (6.5)$$

Note that we are excluding from this discussion the possible presence of a constant cosmological term, which means that we are considering a critical number of dimensions or, for $d \neq d_c$, that the constant contribution $(d - d_c)/\alpha'$ is canceled by the appropriate contribution of some "passive" sector of the model we are considering.

The variation of the action with respect to N, $\overline{\phi}$ and β_i, and the subsequent choice of the cosmic-time gauge $N = 1$, gives, respectively, the equations

$$L - \dot{\beta}_i \frac{\partial L}{\partial \dot{\beta}_i} - \dot{\overline{\phi}}\frac{\partial L}{\partial \dot{\overline{\phi}}} - 2\ddot{\beta}_i\frac{\partial L}{\partial \ddot{\beta}_i} - 2\ddot{\overline{\phi}}\frac{\partial L}{\partial \ddot{\overline{\phi}}} + \cdots$$

$$+ e^{\overline{\phi}}\frac{d}{dt}\left(e^{-\overline{\phi}}\dot{\beta}_i\frac{\partial L}{\partial \ddot{\beta}_i} + e^{-\overline{\phi}}\dot{\overline{\phi}}\frac{\partial L}{\partial \ddot{\overline{\phi}}}\right) + \cdots = 0, \qquad (6.6)$$

$$-L - e^{\overline{\phi}}\frac{d}{dt}\left(e^{-\overline{\phi}}\frac{\partial L}{\partial \dot{\overline{\phi}}}\right) + e^{\overline{\phi}}\frac{d^2}{dt^2}\left(e^{-\overline{\phi}}\frac{\partial L}{\partial \ddot{\overline{\phi}}}\right) + \cdots = 0, \qquad (6.7)$$

$$\frac{\partial L}{\partial \dot{\beta}_i} - e^{\overline{\phi}}\frac{d}{dt}\left(e^{-\overline{\phi}}\frac{\partial L}{\partial \ddot{\beta}_i}\right) + \cdots = e^{\overline{\phi}}Q_i, \qquad (6.8)$$

where Q_i are d integration constants. These equations have been multiplied by $\exp\overline{\phi}$, and the last equation for β_i has been integrated a first time, using the fact that β_i are cyclic variables for the action (6.4), and the associated conjugate momenta are conserved. If $\dot{\overline{\phi}} \neq 0$ then the above $d+2$ equations are not all independent, because of the condition [15]

$$\dot{\beta}_i \frac{\delta S}{\delta \beta_i} + \dot{\overline{\phi}} \frac{\delta S}{\delta \overline{\phi}} = N \frac{\mathrm{d}}{\mathrm{d}t} \left(\frac{\delta S}{\delta N} \right), \qquad (6.9)$$

following from the general covariance of the action (the validity of this condition can also be checked directly using Eqs. (6.6)–(6.8)). We are thus led to a system of $d+1$ independent equations for the $d+1$ variables $\beta_i, \overline{\phi}$.

Let us look for solutions of the form

$$H_i = \dot{\beta}_i = x_i = \text{const}, \qquad \dot{\overline{\phi}} = x_0 = \text{const}, \qquad (6.10)$$

describing, in cosmic time, a phase of exponential evolution of the scale factors, $a_i = \exp(x_i t)$, and linear evolution of the dilaton, $\phi = x_0 t + \text{const}$. The background scalar curvature is constant, even if the space-time manifold is not maximally symmetric. Such a configuration represents a fixed point for the system of Eqs. (6.6)–(6.8), in the sense that if the gravi-dilaton system at some time (for instance the initial time) is there, then it is trapped for ever in such a configuration (modulo the introduction of external perturbations). This behavior is similar to that described by the renormalization group equations [15], in a different context.

Imposing the conditions (6.10), the background equations become a system of $d+1$ algebraic equations for the $d+1$ unknown variables $\{x_0, x_i\}$. There is always the trivial solution, describing a constant dilaton in Minkowski space, and representing a well-known (exact) string theory solution, to all orders in α', in critical dimensions (if $d \neq d_c$ the corresponding solution is flat space-time with linear dilaton [22]). But also non-trivial geometries (at constant scalar curvature) are allowed, in principle to all orders, provided the algebraic system admits a real (non-trivial) set of values for the constant x_0, x_i. Assuming that such a solution exists, an interesting question is whether this phase at constant curvature may represent (or not) an unavoidable stage of the cosmic evolution for models starting from perturbative initial conditions.

Let us note, to this purpose, that in the limit in which Eq. (6.10) is satisfied, the left-hand side of Eq. (6.8) becomes a constant. To obtain a constant also on the right-hand side, a first possibility (for $Q_i \neq 0$) is that, in the same limit, also $\dot{\overline{\phi}} = 0$. In this case, however, the dilaton equation (6.7) is no longer a consequence of the other equations, and must be imposed as an additional condition. One obtains, in this way, additional constraints on the values of the conserved moment Q_i. These integration constants stay fixed at the value they had in the initial configuration: the system may thus evolve towards the configuration (6.10) only if the parameters of the initial state are appropriately fine-tuned to satisfy the required constraints.

A more interesting possibility is the case in which the solution (6.10) satisfies the condition

$$\dot{\bar{\phi}} = \dot{\phi} - \sum_i \dot{\beta}_i < 0. \tag{6.11}$$

In such a case the right-hand side of Eq. (6.8) tends to go to zero at large enough times, and the equation can be automatically satisfied by the constant-curvature configuration quite irrespective of the initial values (even non-zero) of the momenta Q_i, independently assigned by the low-energy initial conditions. A background configuration satisfying Eqs. (6.10) and (6.11) may thus play the role of "late-time attractor" for the string cosmology equations.

However, the condition (6.11) is necessary, but not sufficient for this purpose. Indeed, it only implies that it will attract towards the configuration (6.10) all phase-space trajectories passing *sufficiently near* to it. Low-energy solutions, starting near the trivial fixed point with $\dot{\beta}_i = 0 = \dot{\phi}$, will smoothly evolve towards the high-curvature fixed point (6.10) only in the absence of a singularity disconnecting the perturbative vacuum from the high-curvature string phase (and/or in the absence of other fixed points).

An example of such a smooth evolution can be obtained, to first order in α', by considering the effective S-frame action introduced in Chapter 2, that we repeat here for the reader's convenience:

$$S = -\frac{1}{2\lambda_{\rm s}^{d-1}} \int d^{d+1}x \sqrt{|g|}\, e^{-\phi} \left[R + (\nabla\phi)^2 - \frac{\alpha'}{4} R_{\rm GB}^2 + \frac{\alpha'}{4}(\nabla\phi)^4 \right]. \tag{6.12}$$

Recall that the higher-curvature corrections have been represented by the Euler–Gauss–Bonnet invariant $R_{\rm GB}^2 = R^2 - 4R_{\mu\nu}^2 + R_{\mu\nu\alpha\beta}^2$, in order to eliminate terms with higher than second derivatives from the field equations. The associated field equations have been presented, in fully covariant form, in Section 2.3. For our applications, however, it is useful to derive here the equations governing the Bianchi-I-type background (6.2) starting directly from the action, written in terms of N, ϕ, β_i and of their derivatives. For such a background metric, in particular, the connection and the scalar curvature have already been computed in Eq. (5.23). For the computation of $G_{\rm GB}^2$ we also need the Ricci-squared invariant,

$$N^4 R_{\mu\nu} R^{\mu\nu} = F^2 \left[\sum_i H_i^2 + \left(\sum_i H_i \right)^2 \right]$$

$$- 2F \left[\sum_i H_i \sum_i \dot{H}_i + \sum_i H_i \dot{H}_i + 2 \sum_i H_i \sum_i H_i^2 \right]$$

$$+\sum_i \dot{H}_i^2 + \left(\sum_i \dot{H}_i\right)^2 + \left(\sum_i H_i^2\right)^2 + \left(\sum_i H_i^2\right)\sum_i H_i^2$$

$$+2\sum_i \dot{H}_i \sum_i H_i^2 + 2\sum_i H_i \sum_i H_i \dot{H}_i, \tag{6.13}$$

and the Riemann-squared invariant,

$$N^4 R_{\mu\nu\alpha\beta}R^{\mu\nu\alpha\beta} = 4F^2 \sum_i H_i^2 - 8F\sum_i H_i^3 - 8F\sum_i H_i \dot{H}_i + 2\sum_i H_i^4 + 4\sum_i \dot{H}_i^2$$

$$+2(\sum_i H_i^2)^2 + 8\sum_i \dot{H}_i H_i^2. \tag{6.14}$$

Their combination with $N^4 R^2$ leads to the following explicit expression for the Gauss–Bonnet invariant:

$$N^4 R_{\mathrm{GB}}^2 = 12F\sum_i H_i \sum_i H_i^2 - 8F\sum_i H_i^3 - 4F\sum_i H_i \left(\sum_i H_i\right)^2 + 2\sum_i H_i^4 + \left(\sum_i H_i\right)^4$$

$$-\left(\sum_i H_i^2\right)^2 + 8\sum_i \dot{H}_i H_i^2 + 4\sum_i \dot{H}_i\left(\sum_i H_i\right)^2 - 4\sum_i \dot{H}_i \sum_i H_i^2$$

$$-2\sum_i H_i^2\left(\sum_i H_i\right)^2 - 8\sum_i H_i \sum_i H_i \dot{H}_i. \tag{6.15}$$

For the discussion of this chapter it will be enough to assume that the spatial sections of the Bianchi-I manifold can be factorized as the product of two conformally flat spaces, with d and n dimensions, and scale factors $\exp\beta(t)$ and $\exp\gamma(t)$, respectively. Integrating the action by parts, so as to eliminate all terms with higher than first derivatives, we also automatically eliminate all terms containing $F = \dot{N}/N$ (as expected, as N is only an auxiliary field whose variation provides the Hamiltonian constraint). The action (6.12) (modulo the spatial volume factor, in string units) then takes the form

$$S = \frac{\lambda_{\mathrm{s}}}{2}\int dt\, e^{d\beta + n\gamma - \phi}\left[\frac{1}{N}\left(-\dot{\phi}^2 - d(d-1)\dot{\beta}^2 - n(n-1)\dot{\gamma}^2 - 2dn\dot{\beta}\dot{\gamma}\right.\right.$$

$$\left. + 2d\dot{\beta}\dot{\phi} + 2n\dot{\gamma}\dot{\phi}\right) + \frac{k\alpha'}{4N^3}\left(c_1\dot{\beta}^4 + c_2\dot{\gamma}^4 + c_3\dot{\phi}\dot{\beta}^3 + c_4\dot{\phi}\dot{\gamma}^3 + c_5\dot{\phi}\dot{\beta}\dot{\gamma}^2\right.$$

$$\left.\left. + c_6\dot{\phi}\dot{\beta}^2\dot{\gamma} + c_7\dot{\beta}^2\dot{\gamma}^2 + c_8\dot{\beta}\dot{\gamma}^3 + c_9\dot{\beta}^3\dot{\gamma} - \dot{\phi}^4\right)\right], \tag{6.16}$$

where

$$c_1 = -\frac{d}{3}(d-1)(d-2)(d-3),$$

$$c_2 = -\frac{n}{3}(n-1)(n-2)(n-3),$$

$$c_3 = \frac{4}{3}d(d-1)(d-2),$$

$$c_4 = \frac{4}{3}n(n-1)(n-2),$$

$$c_5 = 4dn(n-1),$$

$$c_6 = 4dn(d-1),$$ \hfill (6.17)

$$c_7 = -2dn(d-1)(n-1),$$

$$c_8 = -\frac{4}{3}dn(n-1)(n-2),$$

$$c_9 = -\frac{4}{3}dn(d-1)(d-2).$$

Note that we have re-inserted the coefficient k as a pre-factor multiplying the α' corrections, to take into account possible different contributions from different string models: for instance, $k = 1$ for the bosonic string, $k = 1/2$ for the heterotic superstring [21].

We can now easily obtain the field equations, by varying the action with respect to ϕ, N, β and γ. For simplicity, we report here the equations for the d-dimensional isotropic case only, setting everywhere $n = 0$ (see [15] for a more general discussion). The variation with respect to N then gives

$$\dot{\phi}^2 + d(d-1)\dot{\beta}^2 - 2d\dot{\beta}\dot{\phi} - \frac{3}{4}k\alpha'\left(c_1\dot{\beta}^4 + c_3\dot{\phi}\dot{\beta}^3 - \dot{\phi}^4\right) = 0. \quad (6.18)$$

The variation with respect to ϕ gives

$$-2\ddot{\phi} + 2d\ddot{\beta} + \dot{\phi}^2 + d(d+1)\dot{\beta}^2 - 2d\dot{\phi}\dot{\beta}$$

$$+k\frac{\alpha'}{4}\left[3c_3\dot{\beta}^2\ddot{\beta} - 12\dot{\phi}^2\ddot{\phi} + 3\dot{\phi}^4 + (dc_3 + c_1)\dot{\beta}^4 - 4d\dot{\beta}\dot{\phi}^3\right] = 0. \quad (6.19)$$

The variation with respect to β gives

$$-2d(d-1)\ddot{\beta} + 2d\ddot{\phi} - d^2(d-1)\dot{\beta}^2 - d\dot{\phi}^2 + 2d(d-1)\dot{\phi}\dot{\beta}$$

$$+k\frac{\alpha'}{4}\left[12c_1\dot{\beta}^2\ddot{\beta} + 3c_3\ddot{\phi}\dot{\beta}^2 + 6c_3\dot{\phi}\dot{\beta}\ddot{\beta} + 3dc_1\dot{\beta}^4\right.$$

$$\left. +(2dc_3 - 4c_1)\dot{\phi}\dot{\beta}^3 - 3c_3\dot{\phi}^2\dot{\beta}^2 + d\dot{\phi}^4\right] = 0. \quad (6.20)$$

For the constant-curvature configuration specified by Eq. (6.10), all the above equations become algebraic equations. By setting $\dot{\beta} = x_1 = \text{const}$, $\dot{\phi} = x_0 = \text{const}$, we obtain, from Eqs. (6.18) and (6.19) (in units $k\alpha' = 1$),

$$x_0^2 + d(d-1)x_1^2 - 2dx_0x_1 - \frac{3}{4}\left(c_1x_1^4 + c_3x_0x_1^3 - x_0^4\right) = 0,$$

$$x_0^2 + d(d+1)x_1^2 - 2dx_0x_1 + \frac{1}{4}\left[3x_0^4 + (c_1 + dc_3)x_1^4 - 4dx_1x_0^3\right] = 0. \tag{6.21}$$

One can easily check that, for any value of d from 1 to 9, there is a non-trivial pair of real solutions defining a fixed point for the (isotropic) gravi-dilaton system in the plane $\{\dot{\phi}, \dot{\beta}\}$. Anisotropic fixed points seem to be absent for the simple example we have considered, but they can be obtained, for instance, by adding to the action (6.12) the contributions of the NS–NS two-form [23], with the appropriate α' corrections. For the isotropic case we report here the solutions of the above equations for $d = 3, 6, 9$:

$$d = 3, \qquad x_0 = \pm 1.40..., \qquad x_1 = \pm 0.616...,$$

$$d = 6, \qquad x_0 = \pm 1.37..., \qquad x_1 = \pm 0.253..., \tag{6.22}$$

$$d = 9, \qquad x_0 = \pm 1.38..., \qquad x_1 = \pm 0.163...,$$

where the same sign has to be taken for x_0 and x_1 in any pair (the fact that there are solutions of opposite sign is due to the time-reversal symmetry of the string cosmology equations).

It is important to stress that, for the solutions (6.22), the third equation (6.20) also turns out to be automatically satisfied, as expected, and that the same solutions can also be obtained from the system of equations (6.18) and (6.20), or from Eqs. (6.19) and (6.20). In this last case one obtains an additional solution which satisfies the condition $\ddot{\phi} = x_0 - dx_1 = 0$. In that case, however, the three equations are all independent, and one has to impose that the third equation is also satisfied: one then finds that the third condition completely eliminates the new solution.

The pairs of positive values in the above solution (6.22) describe expansion and, for $d \geq 3$, satisfy $\ddot{\phi} = \dot{\phi} - d\dot{\beta} < 0$: thus, according to the previous discussion, they can in principle represent late-time attractors of the low-energy inflationary evolution (in the absence of obstructions). This is indeed what happens for the action (6.12), as clearly shown by the numerical integration of Eqs. (6.19) and (6.20) reported in Fig. 6.2. The initial curvature scale has been fixed by setting $\dot{\beta} = 0.05$ at $t = -1$ (in string units $k\alpha' = 1$), and the associated value of $\dot{\phi}$ has been obtained, for any d, using Eq. (6.18) as a constraint on the set of initial data. The numerical integration from $t = -100$ to $t = 100$ then leads to the results illustrated in Fig. 6.2.

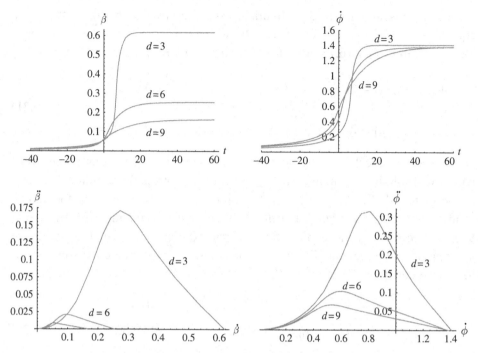

Figure 6.2 Numerical solutions of the isotropic equations (6.18)–(6.20) (in units $k\alpha' = 1$) for $d = 3, 6, 9$, and with initial conditions on the expanding pre-big bang branch of the tree-level solutions (approaching from positive values the trivial fixed point $\dot{\beta} = 0 = \dot{\phi}$). The upper panels illustrate the time evolution towards a final string phase at constant curvature and linear dilaton. The lower panels illustrate the smooth flow of the curvature parameters between the two fixed points.

Starting from initial conditions satisfying $\dot{\beta} > 0$ and $\dot{\bar{\phi}} > 0$, and thus compatible with the expanding, pre-big bang branch of the low-energy solutions (see Section 4.1), the gravi-dilaton system smoothly evolves in time to reach the expanding, fixed-point configurations of Eq. (6.22) (Fig. 6.2, upper panels). This is possible because, for the action (6.12), the high-curvature fixed points (i.e. the string phase) and the trivial fixed point (i.e. the perturbative vacuum), both characterized by $\ddot{\beta} = 0$ and $\ddot{\phi} = 0$, are connected by a smooth "renormalization group" flow of the curvature parameters $\ddot{\beta}$, $\ddot{\phi}$ (Fig. 6.2, lower panels). It is worth noticing that these isotropic, high-curvature fixed points can be reached even starting from slightly anisotropic initial configurations, provided all the variables β_i have the same sign [15].

Real solutions of the type (6.22) – with constant curvature, linear dilaton, and properties of asymptotic attractors – may exist for scalar-tensor models of gravity with higher-derivative corrections, whether or not formulated in a string theory context. They are not, however, a property of *any* model with quadratic curvature corrections: if we consider, for instance, the action (2.77) (related to the action (6.12) by a truncated field redefinition), and we repeat the previous discussion, we find that there are no (non-trivial) real solutions to the system of algebraic equations for any reasonable number of dimensions.

In addition, even assuming that the equations admit non-trivial fixed points, they are not necessarily connected to the perturbative vacuum as in the previous example: there are examples in which the vacuum and the string phase are disconnected by a curvature singularity, or by a non-physical region of phase space where the curvature becomes imaginary.

An example of this "disconnected" situation can be obtained by starting from the action (2.77), and performing the field redefinition

$$g_{\mu\nu} \to g_{\mu\nu} + 4\alpha' R_{\mu\nu}, \qquad \phi \to \phi + \alpha' \left[R - (\nabla\phi)^2 \right]. \qquad (6.23)$$

Truncating the action to order α', and applying the results of Chapter 2 (in particular Eq. (2.75)) we are led to the action

$$
S = -\frac{1}{2\lambda_s^{d-1}} \int d^{d+1}x \sqrt{|g|} e^{-\phi} \Bigg\{ R + (\nabla\phi)^2
$$
$$
- \frac{k\alpha'}{4} \left[R_{GB}^2 - 4 G^{\mu\nu} \nabla_\mu \phi \nabla_\nu \phi + 2\nabla^2 \phi (\nabla\phi)^2 - (\nabla\phi)^4 \right] \Bigg\}, \qquad (6.24)
$$

where $G_{\mu\nu}$ is the Einstein tensor. It is worth noticing that this action is compatible with a symmetry [24, 25] which may be regarded as an extension to order α' of the scale-factor duality symmetry typical of the tree-level action. This new invariance property can be easily displayed by considering the background geometry (6.2) in the cosmic-time gauge $N = 1$, by introducing the matrix G associated with the spatial metric tensor g_{ij}, the shifted dilaton $\overline{\phi}$ according to Eq. (4.119), and finally defining the $2d \times 2d$ matrix

$$
M_{\alpha'} = \begin{pmatrix} G^{-1} - \frac{k\alpha'}{4} G^{-1} \dot{G} G^{-1} \dot{G} G^{-1} & 0 \\ 0 & G + \frac{k\alpha'}{4} \dot{G} G^{-1} \dot{G} \end{pmatrix}, \qquad (6.25)
$$

which generalizes to first order in α' (and for $B_{\mu\nu} = 0$) the $O(d, d)$ matrix M of Eq. (4.124). Using the results of Section 4.3, the action (6.24) can then be rewritten in terms of $M_{\alpha'}$, $\overline{\phi}$ and of the $O(d, d)$ metric η, and turns out to be invariant (to first order in α') under the transformation [24, 25]

$$
M_{\alpha'} \to \eta M_{\alpha'} \eta, \qquad \overline{\phi} \to \overline{\phi}, \qquad (6.26)
$$

generalizing the tree-level transformation associated with the inversion of the scale factor (see Eq. (4.137)).

The field equations for the action (6.24) are still second-order equations. If we repeat the previous analysis we find that they admit high-curvature fixed points of the type (6.10), for any d (at least up to $d = 9$). However, such points cannot be smoothly approached by low-energy trajectories emerging from the trivial fixed point: all low-energy solutions describing expanding pre-big bang inflation necessarily evolve towards a curvature singularity, which prevents their continuous convergence towards a stable string phase.

As shown by the quoted examples, the nice property of the action (6.12) to admit solutions able to damp the initial growth of the curvature *is not* invariant, in general, under field redefinitions of the type (2.74), truncated to first order in α'. This ambiguity affects all models truncated to any given *finite* order of the α' expansion and can be resolved, in principle, only by considering an exact conformal model which automatically includes the corrections to all orders (see [26, 27] for some specific string cosmology examples of exact conformal models).

Even with the choice of an appropriate action, compatible with a smooth connection of the perturbative vacuum with the string phase, a model of pre-big bang inflation based only on the α' corrections is still incomplete. Indeed, in a possibly realistic picture, the Universe cannot stay frozen for ever in the constant-curvature regime, but has to evolve towards the standard, decelerated, decreasing-curvature regime. In other words, and with reference to the four asymptotic branches of the low-energy solutions, a complete transition (in the absence of matter sources and/or of other background fields) should connect branch $\boxed{4}$ to branch $\boxed{1}$ of Fig. 4.3.

This requires, as already stressed, that the initial configuration with $\dot{\phi} > 0$ may evolve towards a final one with $\dot{\phi} < 0$. This requirement is indeed satisfied by the solutions of the action (6.12), as illustrated in Fig. 6.3 (left panel), where we have plotted the same numerical solution as in Fig. 6.2, for $d = 3$, in the plane $(\dot{\phi}, \sqrt{3}\dot{\beta})$. The final fixed points, however, are not placed on the post-big bang, asymptotic branches of the low-energy solutions, represented by the dashed lines. Thus, for the action (6.12), it it impossible to perform a "phase-space loop" leading the system back to the origin (as was the case for the examples presented in Appendix 4B).

Such an impossibility, which appears as a clear asymmetry in the plane $(\dot{\phi}, \sqrt{3}\dot{\beta})$, is also due in part to the fact that we are considering an action which is invariant for time reflections but is not invariant for scale-factor duality transformations: thus, there are no "self-dual" solutions. Indeed, a smooth evolution from the perturbative vacuum to the string phase (or vice versa) is allowed for the expanding pre-big bang branch of the lowest-order solution (curve $\boxed{4}$), and for its time-symmetric partner, the contracting post-big bang branch (curve $\boxed{3}$),

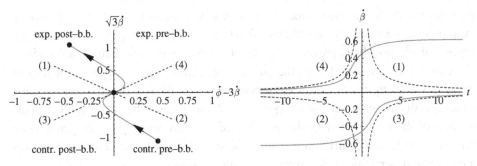

Figure 6.3 The same numerical solution as in Fig. 6.2, for the case $d = 3$. The dashed curves represent the (singular) branches of the tree-level, vacuum solutions (see also Figs. 4.1 and 4.3). The left panel shows the smooth connection of the initial asymptotic state with $\dot{\beta} = 0 = \dot{\phi}$ to the fixed points of Eq. (6.21). The right panel shows the time-symmetric evolution of the regularized Hubble parameter.

as illustrated in Fig. 6.3, right panel. The expanding post-big bang branch (curve $\boxed{1}$), on the contrary, stays singular, and cannot be smoothly connected to the regularized pre-big bang branch.

It is possible, as we have seen, to restore the duality invariance at the level of the truncated action. In that case the non-trivial fixed points may acquire a self-dual distribution in the phase-space plane: however, this may not be enough for implementing a regular and complete transition, since the points could be placed in phase-space regions physically disconnected from the origin (as in the example (6.24)). The existence and the location of the fixed points, with reference to a possible duality symmetry of the higher-order action, has been discussed in general in [28, 29].

What seems to be crucial in order to shift the cosmological evolution away from the fixed point, towards the standard post-big bang regime (even without respecting self-duality), is the effect of the quantum loop corrections. Indeed, as long as the Universe is anchored to the high-curvature string phase, the dilaton keeps growing linearly in cosmic time, and the string coupling g_s^2 keeps growing exponentially, so that the loop corrections are doomed to become eventually important, quite independently from the initial value of the dilaton. The possible effect of such corrections will be discussed in the following section.

6.2 Strong coupling corrections and the curvature "bounce"

The backreaction of the quantum loop corrections on the background geometry may be described, in an appropriate limit, as the contribution of an effective source with negative energy density [30, 31], whose presence tends to contrast the growth of both the curvature and the dilaton kinetic energy [32]. Some particular examples, represented by an effective fluid or by a non-local potential containing powers of g_s^2, have been presented in Chapter 4 within the small curvature regime. Another effect associated with the string loop corrections is the backreaction of the massive string modes: they are produced with a density distribution which is exponentially rising with the energy [33, 34], and could be responsible for the bounce of H and $\dot{\phi}$ even in the absence of α' corrections [35]. In general, all types of effects and contributions should be simultaneously taken into account for a complete description of the strong coupling regime.

In this section we discuss, in particular, the effect of the quantum loop corrections on a background which is already inside the high-curvature regime, and which is temporarily stabilized in the string-phase configuration, at constant curvature. It is clear that the inclusion in the effective action of terms growing with ϕ can easily induce an acceleration of the gravi-dilaton system, removing it from the high-curvature fixed point at late enough times. It should be immediately stressed, however, that a successful transition to the standard, post-big bang regime – the so-called "graceful exit" [36] – must fulfil various non-trivial requirements.

First of all, a true smoothing out of the low-energy singularities requires that the curvature be regularized in all frames, not just in the S-frame as in the case of the example presented in the previous section. The physical motivation underlying this condition is that different test particles, associated with fundamental fields differently coupled to the dilaton, follow the geodesics of different metric frames. The gravitons, for instance, evolve according to the E-frame geodesics [37]; other fields, for instance those belonging to the R–R sector of type IIA and type IIB superstrings, are canonically coupled to the S-frame metric background, and follow instead the S-frame geodesics. The regularity and completeness of a classical space-time manifold, on the other hand, are fully determined by the properties of its geodesic network [38]. If the regularization of the curvature is not frame independent, the singularity might disappear for some types of test particles, but would persist for others [39].

A simple example of this effect can be obtained by considering the isotropic S-frame solution with constant H and $\dot{\phi}$, derived in the previous section. The dilaton is monotonically growing, so that the curvature certainly diverges in the E-frame representation, where the Hubble parameter is given by

$$H_{\mathrm{E}} = \left(H - \frac{\dot{\phi}}{d-1} \right) e^{\phi/(d-1)} \qquad (6.27)$$

(see also Eq. (4B.11)), where the dot denotes differentiation with respect to the S-frame cosmic time. Indeed, for constant H and $\dot{\phi}$, the E-frame curvature avoids divergences at $t \to +\infty$ only if ϕ is constant or decreasing. If we have, on the contrary, a regular S-frame solution in which H and $\dot{\phi}$ are decreasing (fast enough) as $t \to +\infty$, then the dilaton could be growing and would still avoid the curvature singularity in the E-frame, as well as in other frames (this is what happens, for instance, in the solution of Eq. (4B.8)). For a growing dilaton ($\dot{\phi} > 0$), however, Eq. (6.27) also implies the constraint

$$\dot{\phi} < (d-1)H, \qquad (6.28)$$

(stronger than the condition (6.11)), which is required to implement a realistic scenario in which the metric is expanding also in the E-frame ($H_E > 0$), after the transition. Both the constraints (6.11) and (6.28) can be translated into energy conditions, to be satisfied by the stress tensor of the effective matter sources controlling the background evolution [40, 41].

However, a dilaton which keeps growing asymptotically is doomed to conflict with the perturbative approach and with the truncated expansion of the string effective action: sooner or later the background will be led to the full non-perturbative M-theory regime – or at least to the regime where the quantum loop corrections are to be included, to all orders. In that case, a possibly realistic scenario could be implemented by invoking a "saturation" mechanism (see Section 9.3) by which, in spite of the growth of ϕ and of the "bare" coupling $\exp(\phi)$, the effective string coupling g_s^2 tends to become frozen on moderate values, compatible with the condition $g_s^2 = M_s^2/M_P^2 \lesssim 1$.

In the absence of a saturation mechanism, the stabilization of g_s seems to require the stabilization of the dilaton itself, $\phi \simeq$ const. In that case, if the growth of the dilaton leads to the phase of high (but constant) curvature and strong coupling, where the Universe becomes populated by higher-dimensional branes, an interesting possibility is that the dilaton (together with the volume of the extra dimensions) becomes stabilized by appropriate fluxes of R–R and NS–NS p-form fields [42, 11], or by the mechanism recently proposed in [43]. If stabilization is not achieved in this way, because the bounce to the decreasing-curvature regime is induced very early by the loop corrections, there is still the possibility of a "late-time" dilaton stabilization if, after the transition to the radiation-dominated Universe, some attracting minimum is developed in the non-perturbative dilaton potential.

Let us consider, for instance, a late-time picture of cosmological evolution where the Universe has already entered the decelerated, decreasing curvature regime (thanks to the effects of the loop corrections), and the value of ϕ has grown large enough to justify the introduction of a non-perturbative potential $V(\phi)$. Let us suppose, also, that the transition from the string phase to the decelerated regime has largely amplified the quantum fluctuations of all background fields, which after re-entering inside the horizon behave as an effective radiation fluid, and tend to become the dominant source of the background geometry. The α' corrections are going to become more and more negligible, so that the background evolution is determined by the effective low-energy equations, with the possible inclusion of dilaton loop contributions due to the strong coupling regime.

Including such contributions into the (S-frame) dilaton equation (2.38), for a generic matter source with equation of state $p = p(\rho)$ and zero dilaton charge, one then obtains the evolution equation [44]:

$$A(\phi)(\ddot{\phi} + dH\dot{\phi}) - B(\phi)\dot{\phi}^2 + V + \frac{1}{2}(d-1)V' + \frac{1}{2}C(\phi)\,e^{\phi}(\rho - dp) = 0 \quad (6.29)$$

(we have used units $2\lambda_s^{d-1} = 1$, and $V' = \partial V/\partial\phi$). Here A, B, C are the "form factors" containing the loop corrections (and reducing to 1 in the weak coupling limit $g_s \ll 1$); other quantum corrections are possibly included in the potential $V(\phi)$. In such a context, a stable solution with $\phi = \phi_0 = $ const is thus possible only if

$$dp - \rho = 2C^{-1}(\phi_0)\,e^{-\phi_0}\left(V + \frac{d-1}{2}V'\right) = \text{const.} \quad (6.30)$$

Combining this condition with the conservation of the matter stress tensor, we are left with three possible configurations compatible with a constant dilaton [45]:

(1) vacuum, $\rho = p = 0$, and $V + V'(d-1)/2 = 0$;
(2) cosmological constant, $\rho = -p = \rho_0 = $ const, and $V + V'(d-1)/2$
    $= -C(\phi_0)\rho_0\,e^{\phi_0}(d+1)/2 = \text{const} \neq 0$;
(3) radiation, $\rho = dp$, and $V + V'(d-1)/2 = 0$.

The first two cases are still peculiar to the inflationary regime. When the background becomes decelerated and radiation dominated, however, it becomes compatible with a dilaton stabilized at a value ϕ_0 such that

$$V(\phi_0) + \frac{d-1}{2}V'(\phi_0) = 0. \quad (6.31)$$

Note that this value exactly coincides with an extremum of the E-frame potential $\widetilde{V}$, related to the S-frame potential V by Eq. (2.47) (the above condition is equivalent, indeed, to $\widetilde{V}' = 0$). Note also that the radiation-dominated phase is compatible with a constant dilaton even in the limiting case of vanishing potential, as illustrated by the examples of Chapter 4 in which the dilaton is asymptotically driven to a constant: in that case, the final value of the dilaton depends on the parameters of the initial post-big bang configuration.

In this section we present an example of transition to the post-big bang regime with a dilaton which tends to be stable, asymptotically; the case of a dilaton which keeps (slowly) varying even after inflation will be discussed in Section 9.3.

According to the previous discussion, an effective action possibly accounting for a graceful exit from the high-curvature string phase, and preventing a significant trespass into the strong coupling regime, should contain the following terms:

$$S = S_0 + S_{\alpha'} + S_{\text{loop}} + S_m. \tag{6.32}$$

Here S_0 is the lowest-order gravi-dilaton action, describing the initial inflationary evolution away from the string perturbative vacuum; $S_{\alpha'}$ contains the high-curvature corrections damping the acceleration, and leading the system to a (temporary) stabilization around the fixed-point configuration; S_{loop} contains the quantum loop corrections inducing the curvature bounce; S_m finally represents the matter sources, possibly interacting with the dilaton and contributing to its final stabilization. With reference to the trajectory labelled $\boxed{2}$ in Fig. 6.1, we can say that $S_{\alpha'}$ comes into play at point (a), and is responsible for the first correction to the initial trajectory; S_{loop} comes into play at point (b), where the background is shifted away from the fixed point, towards the low-curvature regime; finally, the last and decisive correction is due to S_m, stopping the growth of the dilaton at point (c) of Fig. 6.1.

Models evolving in this way have been explicitly constructed in [29, 40, 41], using in particular loop corrections which become sub-leading at large times, after their contribution to the graceful exit: in that case, the final asymptotic regime is still described by the solutions of the lowest-order effective action (this scenario is consistent provided the final growth of the dilaton is stopped, of course). Here we follow the particular example discussed in [29, 46], based on the most general (first-order) α' corrections leading to equations containing at most second derivatives. Limiting ourselves to the four-dimensional case we consider the action

$$S_{\alpha'} = -\frac{1}{2\lambda_s^2} \frac{k\alpha'}{4} \int d^4x \sqrt{-g}\, e^{-\phi} [\, a_0 R_{\text{GB}}^2 + b_0 G^{\mu\nu} \nabla_\mu \phi \nabla_\nu \phi$$

$$+ c_0 \nabla^2 \phi (\nabla\phi)^2 + d_0 (\nabla\phi)^4 \,], \tag{6.33}$$

where the coefficients a_0, b_0, c_0, d_0 are fixed in such a way as to be compatible with the conformal invariance of the sigma model action, to order α' (see Chapter 3). In particular, they must satisfy the conditions

$$a_0 = -1, \qquad b_0 + 2(c_0 + d_0) = 2, \tag{6.34}$$

ensuring that the above action can be obtained by performing a suitable field redefinition (of the type (2.74), (2.75)) from the Riemann-squared term of the action (2.77), which correctly reproduces the string scattering amplitudes [21]. For $b_0 = c_0 = 0$ one recovers, in particular, the action (6.12) used in the previous section, while for $b_0 = 4$ and $c_0 = -2$ one gets $d_0 = 1$, and recovers the duality-invariant action (6.24).

Concerning S_{loop}, it should be mentioned that the existing computations for the gravi-dilaton and the moduli sectors are strongly dependent on the considered model of compactification (see for instance [47, 48, 49]). In the absence of definite results, generally valid for any model and for the full string-loop expansion, we follow here the semi-phenomenological approach of [28, 29] where the loop expansion is simulated by the inclusion of additional α' corrections, weighted by higher powers of $g_s^2 = \exp(\phi)$. We limit the expansion to the two-loop order, and set

$$S_{\text{loop}} = -\frac{1}{2\lambda_s^2}\frac{k\alpha'}{4}\int d^4x\sqrt{-g}\,e^{-\phi}$$

$$\times \left\{ A_1 e^{\phi}\left[a_1 R_{\text{GB}}^2 + b_1 G^{\mu\nu}\nabla_\mu\phi\nabla_\nu\phi + c_1\nabla^2\phi(\nabla\phi)^2 + d_1(\nabla\phi)^4\right] \right.$$

$$\left. + A_2 e^{2\phi}\left[a_2 R_{\text{GB}}^2 + b_2 G^{\mu\nu}\nabla_\mu\phi\nabla_\nu\phi + c_2\nabla^2\phi(\nabla\phi)^2 + d_2(\nabla\phi)^4\right] \right\}.$$

(6.35)

Here the one-loop and two-loop terms have been parametrized so as to contain the higher-curvature corrections in the same form as in $S_{\alpha'}$ (to avoid higher-than-second derivatives of the background in the field equations); however, the coefficients $a_1, b_1, \ldots$ and $a_2, b_2, \ldots$ are in principle different from the tree-level coefficients $a_0, b_0, \ldots$, and are not subject to the constraint (6.34). The constant parameters A_1 and A_2 control the onset of the loop corrections, and their late-time suppression for a successful exit. Their precise values, as well as the values of the other coefficients, should be the outcome of a specific computation performed within a given compactification model.

For the choice of S_m we note, finally, that the stabilization of ϕ at a minimum of its effective potential energy requires some level of fine-tuning on the parameters of both the potential and the cosmological state at the beginning of the decelerated regime [50]. We thus assume that the dilaton damping is simply due to its coupling to the radiation, and that the radiation fluid, described by S_m, is quite negligible in the initial phase of pre-big bang evolution, but is copiously produced by the curvature bounce and becomes important in the post-big bang regime.

Summing up all terms, and considering a homogeneous, spatially flat metric background, with scale factor $a(t) = \exp\beta(t)$, we can write the action (6.32) (modulo a total derivative, in units $2\lambda_s^2 = 1$) as follows:

$$S = \int \frac{dt}{N}e^{3\beta-\phi}(6\dot\beta\dot\phi - \dot\phi^2 - 6\dot\beta^2) - \frac{k\alpha'}{4}\sum_{n=0}^{2}\int \frac{dt}{N^3}e^{3\beta+(n-1)\phi}L_n + S_m, \quad (6.36)$$

where (setting $A_0 = 1$, and using Eqs. (6.33) and (6.35))

$$L_n = A_n \left[a_n \dot{\phi} \dot{\beta}^3 + 3b_n \dot{\phi}^2 \dot{\beta}^2 + 2c_n \dot{\beta} \dot{\phi}^3 + \left(\frac{c_n}{3} + d_n \right) \dot{\phi}^4 \right]. \quad (6.37)$$

The variation with respect to N, β and ϕ leads, respectively, to the equations

$$\dot{\phi}^2 + 6\dot{\beta}^2 - 6\dot{\beta}\dot{\phi} + \frac{3}{4}k\alpha' \sum_{n=0}^{2} L_n e^{n\phi} = e^{\phi}\rho, \quad (6.38)$$

$$2\ddot{\phi} - 4\ddot{\beta} + 4\dot{\beta}\dot{\phi} - \dot{\phi}^2 - 6\dot{\beta}^2$$

$$+ \frac{k\alpha'}{4} \sum_{n=0}^{2} e^{n\phi} \left[L_n + \frac{d}{dt} \left(\frac{\partial L_n}{\partial \dot{\beta}} \right) + (3\dot{\beta} + (n-1)\dot{\phi}) \frac{\partial L_n}{\partial \dot{\beta}} \right] = \frac{1}{3} e^{\phi}\rho, \quad (6.39)$$

$$6\ddot{\beta} - 2\ddot{\phi} - 6\dot{\beta}\dot{\phi} + 12\dot{\beta}^2 + \dot{\phi}^2$$

$$+ \frac{k\alpha'}{4} \sum_{n=0}^{2} e^{n\phi} \left[(n-1)L_n + \frac{d}{dt} \left(\frac{\partial L_n}{\partial \dot{\phi}} \right) + (3\dot{\beta} + (n-1)\dot{\phi}) \frac{\partial L_n}{\partial \dot{\phi}} \right] = 0 \quad (6.40)$$

(in the equation for β we have divided by 3 and we have used the equation of state of the radiation fluid contained in S_m).

A detailed analysis of the fixed points of this system of equations, and of the subsequent transition to the decelerated regime, shows that there is a wide region of parameter space allowing fixed points with $\dot{\phi} < 0$, smoothly connected to the vacuum and compatible with the graceful exit [29], even without the inclusion of the radiation contribution. To give an explicit example we present here a numerical integration with the following values of the non-zero loop coefficients,

$$a_0 = a_1 = a_2 = -d_0 = -d_1 = -d_2 = -1, \quad (6.41)$$

chosen to preserve the simple form of the action (6.12) (but the bounce is implemented even if b_0, c_0 are non-zero, and different from b_1, b_2 and c_1, c_2). We also choose, for our example, $A_1 = 1$ and $A_2 < 0$, with $|A_2| \ll 1$ (the opposite sign of the two terms seems to be required for a successful transition [28, 29], at least when the loop corrections are truncated to second order). With this choice of parameters the numerical integration of Eqs. (6.38)–(6.40) shows that the background can smoothly evolve from the string perturbative vacuum to the high-curvature string phase, and then decay (thanks to the loop corrections) towards the decelerated, decreasing curvature regime, even if $\rho = 0$. The final dilaton damping, however, requires the coupling to the radiation fluid, which is here made more effective by allowing a possible decay into radiation, with decay width Γ: the matter conservation equation is then modified as follows:

$$\dot{\rho} + 4H\rho - \frac{1}{2}\Gamma\dot{\phi}^2 = 0, \quad (6.42)$$

and the dilaton equation (6.40) has to be completed by adding, on the right-hand side, the friction term $-\Gamma(\dot\phi/2)\exp(\phi)$ required for the covariant conservation of the total stress tensor.

The resulting system of equations (6.39), (6.40) and (6.42) has been numerically integrated from $t=-40$ to $t=100$, for the particular values $k\alpha'=1$, $A_2=-2\times 10^{-3}$, $\Gamma=5.63\times 10^{-4}$, using Eq. (6.38) as a constraint on the initial data, specified (in string units) as follows: $\phi=-10$, $\rho=0.005$, $\dot\beta=0.05$, $\dot\phi=0.22566$ at $t=-1$. The results are illustrated in Fig. 6.4, where the curvature bounce and the dilaton damping are shown in the left panel, while the closure of the phase-space cycle (i.e. the exit completion) is shown in the right panel.

In spite of the possible existence of phenomenological examples of smooth bouncing transitions and a graceful exit from the phase of high-curvature inflation, it should be stressed that it is at present unclear whether such a scenario could be the outcome of a reliable string theory computation. A perturbative approach to the two-loop order, in particular, may be self-consistent only for a sufficiently small value of the final coupling, $g_s \ll 1$. In the numerical example reported in Fig. 6.4, on the contrary, the coupling seems to approach a final value which is not sufficiently small. It is possible, of course, to choose a different set of values for the parameters of the numerical solution, and to end up with smaller asymptotic values for the final dilaton. In that case, however, a smooth transition seems to require values of A_1 and A_2 larger than one (in modulus), and of the same order [29, 46], in contrast with the expectations for the coefficients of a perturbative expansion (see Eq. (6.35)). This result might signal a natural tendency of the background to enter the strong coupling regime and the phase populated by strings and higher-dimensional branes, thus probably implying that only in that

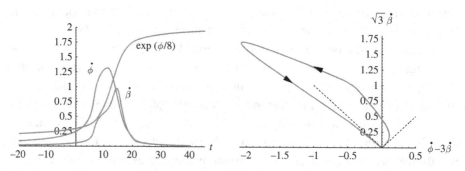

Figure 6.4 Numerical integration of Eqs. (6.38)–(6.42) (the values of the parameters and of the initial conditions used in these plots are given in the text). The left panel shows the curvature bounce, the right panel shows the smooth evolution from the vacuum to the fixed point of Fig. 6.3, and the subsequent "exit" induced by the quantum corrections (the dashed lines are the low-energy solutions describing pre- and post-big bang evolution).

phase is an efficient and realistic dilaton stabilization achieved, together with the stabilization of the internal moduli [11, 42].

6.3 String gas cosmology

Assuming that the Universe does not immediately bounce back when reaching the regime of (moderately) strong coupling, then the cosmological evolution enters a new phase in which the Dirichlet branes, or D_p-branes (see Appendix 3A) – the fundamental, p-dimensional string theory objects [51], with effective mass $\sim g_s^{-1}(\alpha')^{-(p+1)/2}$ – start to become light, and can be gravitationally produced in pairs from the vacuum [6]. In this phase, the Universe is potentially large and filled with a "hot soup" of strings, branes and antibranes (i.e. branes with opposite spatial orientation), of all possible sizes and of all dimensions p allowed in a string theory context, possibly wrapping around the compact dimensions. Such a configuration has been invoked as the initial state of the "string gas" [7, 8] or "brane gas" [9] cosmological scenario, as well as of other scenarios [52, 53, 54] trying to explain why our effective macroscopic world is three-dimensional (see Chapter 10 for a specific introduction to brane cosmology).

A possible way to take into account the loop corrections arising from higher-genus topologies, in such a context, is to add to the tree-level action of the background fields, Eq. (2.21), the sigma model action of fundamental strings, which has no dilaton pre-factor in it, and is thus of order g_s^2 with respect to the usual tree-level action [55]. Summing over all components of the string gas (see e.g. Eq. (4.151)), and using the corresponding energy-momentum tensor as the source of the "bulk" geometry, one then arrives at the so-called string gas approach to early cosmological evolution, which will be briefly illustrated in this section (a similar approach can be adopted for a gas of higher-dimensional branes).

There are two main assumptions in this approach: (*i*) the toroidal topology of the spatial dimensions, and (*ii*) the "adiabatic" evolution of the string gas components (see [17] for a detailed discussion of this scenario). Thanks to this second assumption, in particular, one can neglect the gradients of the metric and of the other background fields in the string evolution equations: thus, locally, each single string will not be influenced by the time evolution of the cosmological background, and one can still apply the results for the energy spectrum obtained in a flat, static background.

Let us then recall that, for a string wound around one compact dimension of radius R, the energy spectrum is fixed by the mass-shell condition (4.5), which can be rewritten as

$$E^2 = \left|\vec{P}_{\mathrm{nc}}\right|^2 + \frac{n^2}{R^2} + m^2\frac{R^2}{\alpha'^2} + \frac{2}{\alpha'}(N + \tilde{N} - \delta - \tilde{\delta}). \tag{6.43}$$

Here $\vec{P}_{nc}$ is the momentum associated with translations along the flat non-compact directions; n and m are, respectively, the excitation levels of the momentum and winding modes in the compact dimension, and N, $\tilde{N}$ are the eigenvalues of the number operators for right- and left-moving modes, respectively. Note that we have not specified the constants δ, $\tilde{\delta}$ whose explicit values are fixed by the normal ordering prescription, but depend on the model of string that we adopt (for the bosonic string $\delta = \tilde{\delta} = 1$, see Appendix 3A).

If all spatial dimensions are compact, as assumed by the string gas cosmology, then $\vec{P}_{nc} = 0$. Let us call a_i, $i = 1, \ldots, d$, the scale factors along the various spatial dimensions, and replace the constant radii R_i with the "rolling radii" $R_i \rightarrow \sqrt{\alpha'} a_i$, by exploiting the adiabatic assumption. The previous equation for the energy spectrum can then be generalized as follows:

$$E^2 = \frac{1}{\alpha'} \sum_i \left(\frac{n_i^2}{a_i^2} + m_i^2 a_i^2 \right) + \frac{2}{\alpha'} (N + \tilde{N} - \delta - \tilde{\delta}). \tag{6.44}$$

Summing over all strings, the energy density of the string gas is given by

$$\rho = \sum_s n_s E_s, \tag{6.45}$$

where $n_s = N_s V^{-1}$ is the number density of string states in a (proper) spatial volume $V \sim \prod_i a_i$, with energy E_s specified by a given set of quantum numbers, i.e. $E_s = E_s(n, m, N, \tilde{N})$.

To illustrate the dynamical effects of the winding modes on the background geometry we concentrate our discussion on the massless levels of the string spectrum (for closed bosonic strings, for instance, on the case $N = \tilde{N} = 1 = \delta = \tilde{\delta}$), and we consider an isotropic initial situation in which the gas of winding and momentum modes is uniformly distributed in all spatial dimensions: thus $a_i = a$, $\forall i$, and $V \sim a^d$. The equation of state of the gas can then be obtained using again the adiabatic assumption, $T\,dS = d(\rho V) + p\,dV = 0$, which gives for the pressure

$$p = -\left(\frac{\partial \rho V}{\partial V} \right)_{S=\text{const}}. \tag{6.46}$$

For the winding modes, taking the square root of the spectrum (6.44), we obtain $E_w \sim \rho_w a^d \sim a$, from which $\rho_w \sim a^{1-d}$ and

$$p_w = -\frac{\partial E_w}{\partial a} \frac{da}{dV} = -\frac{E_w}{dV} = -\frac{1}{d} \rho_w. \tag{6.47}$$

For the momentum modes we have, instead, $E_m \sim \rho_m a^d \sim a^{-1}$, from which $\rho_m \sim a^{-1-d}$ and

$$p_m = -\frac{\partial E_m}{\partial a} \frac{da}{dV} = \frac{E_m}{dV} = \frac{1}{d} \rho_m. \tag{6.48}$$

The gas of momentum modes is thus characterized by a radiation-like equation of state, while for the winding modes we find the "dual" equation of state, according to the terminology of Section 4.2 (the same equation of state, $p/\rho = \gamma = -1/d$, is also obtained for a gas of long, stretched strings, frozen outside the horizon [57, 58]). These results are automatically consistent with the (separate) covariant conservation of both winding and momentum gases, which implies $\rho \sim a^{-d(1+\gamma)}$ (see Eq. (1.29)), and which separately reproduces the same scale-factor dependence of ρ_w and ρ_m as that directly obtained from the spectrum (6.44).

Let us now consider the effects of these gases on the evolution of the gravidilaton background, with particular attention to the possible production of a final configuration with three spatial dimensions much larger than the others. We start from the background equations (4.39)–(4.41), which we repeat here, for the reader's convenience, for an isotropic d-dimensional space and for sources with zero dilaton charges ($\sigma = 0$):

$$\dot{\overline{\phi}}^2 - dH^2 = e^{\overline{\phi}}\overline{\rho}, \qquad \dot{H} - H\dot{\overline{\phi}} = \frac{1}{2}e^{\overline{\phi}}\overline{p},$$

$$\ddot{\overline{\phi}} - dH^2 = \frac{1}{2}e^{\overline{\phi}}\overline{\rho} \tag{6.49}$$

(in the last equation we have eliminated $\dot{\overline{\phi}}^2$ through Eq. (4.39)). The shifted variables, $\overline{\rho}$, $\overline{p}$ are defined as in Eq. (4.38). We study the solutions of these equations in the phase-space plane spanned by the coordinates $\{\dot{\overline{\phi}}, \sqrt{d}H\}$, using as sources a mixture of winding and momentum modes,

$$\overline{\rho} = \overline{\rho}_w + \overline{\rho}_m, \qquad \overline{p} = -\frac{\overline{\rho}_w}{d} + \frac{\overline{\rho}_m}{d}, \tag{6.50}$$

and exploiting the solutions of the conservation equation, $\overline{\rho}_w \sim a$, $\overline{\rho}_m \sim a^{-1}$. A possible dynamical explanation of the observed large-scale dimensionality can then be obtained [8] starting from an initial configuration in which the background has already reached the post-big bang regime, characterized by decelerated expansion with $H > 0$ and $\dot{\overline{\phi}} < 0$. The condition of positive energy density implies, in this case, $|\dot{\overline{\phi}}| > \sqrt{d}H$ (from the first equation of the system (6.49)), and locates the initial state in the upper-left quadrant of phase space (see Fig. 6.3, left panel), *below* the dashed line labeled by (1) (representing the vacuum solution $\dot{\overline{\phi}} = -\sqrt{d}H$) but *above* the horizontal axis $H = 0$. The background may have reached such a state, for instance, driven by the effects of the loop backreaction, as illustrated by the numerical solution plotted in the right panel of Fig. 6.4.

Starting from an initial configuration in which the background is isotropically expanding, it is clear that the density of the winding gas, growing like $\bar{\rho}_w \sim a$, will rapidly become dominant with respect to the radiation gas, which dilutes as $\rho_m \sim a^{-1}$. On the other hand, the low-energy gravi-dilaton equations sourced by a perfect barotropic fluid have been exactly integrated in Chaper 4, for any equation of state $p/\rho = \gamma = $ const (see Eqs. (4.75) and (4.76)). From the general solution we can see that the background, starting from the given initial conditions, tends to reach, at late times, the asymptotic configuration satisfying the condition

$$H = -\gamma \dot{\phi} \qquad (6.51)$$

(obtained by combining Eqs. (4.75) and (4.76), in the limit $x \to +\infty$). The winding-dominated background thus asymptotically approaches the state $\dot{\phi} = dH$, represented in the phase-space plane by a line crossing the origin, with positive angular coefficient $1/\sqrt{d}$. This effect is illustrated by the plots reported in Fig. 6.5, and corresponding to a numerical integration of the equations (6.49) for $d = 3$ and for a mixture of momentum and winding modes in the initial ratio 1:100. The three plotted trajectories have been obtained by varying the initial conditions of $\dot{\phi}$ (-1.25, -1 and -0.75) at fixed initial values of all the other parameters.

As clearly illustrated in the figure, all the trajectories we are considering evolve towards the origin in the region of negative $\dot{\phi}$, and may thus join the asymptotic regime $\dot{\phi} = dH$ only when H is also negative [8, 9]. This implies that, starting from an initial expanding configuration, the background is unavoidably driven by the winding modes towards a final contracting state. In particular, for H and $\dot{\phi}$ approaching zero from negative values, the Universe enters the so-called "loitering" regime [9, 59] (see Fig. 6.5). The same regime, characterized by decreasing dilaton and by the decelerated contraction of the internal dimensions, filled with a negative pressure fluid, was also obtained in the context of the

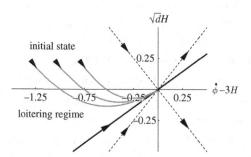

Figure 6.5 Phase space trajectories describing the winding-driven evolution, and the asymptotic approach to the loitering regime (for $d = 3$). The dashed lines represent the same vacuum solutions as in Fig. 6.3; the solid line crossing the origin represents the asymptotic solution (6.51).

phenomenological examples presented in Appendix 4B (see Eqs. (4B.54) and (4B.55), and also Fig. 4.8).

Thus, if the initial Universe has reached a phase of decreasing dilaton, and has compact spatial dimensions isotropically filled by a gas of string winding modes, it cannot expand, and its size remains confined at the initial scale, naturally identified with the string scale – with possible oscillations due to the alternate dominance of winding modes (preventing expansion) and momentum modes (preventing contraction). Can winding modes disappear, allowing the Universe to expand?

In a background with stable cycles, such as a torus, the expansion becomes possible only if winding modes annihilate with antiwinding modes (i.e. with modes wound around the cycles with the opposite orientation). Assuming a symmetric initial configuration with equal numbers of winding and antiwinding modes (for instance, a bath in thermal equilibrium), using the fact that the world-volumes of extended objects with p spatial dimensions have measure zero probability of intersecting in more than $2p + 1$ spatial dimensions, and following the intuition that *interactions* are due to *intersections*, it was argued in [7] that string ($p = 1$) winding modes can annihilate only in $d \leq 2 + 1 = 3$ spatial dimensions. Thus, at most three spatial dimensions may become permanently dominated by momentum modes (i.e. by radiation) and can expand, hence explaining the observed space-time dimensionality. The remaining $d - 3$ spatial dimensions, on the contrary, are kept confined at the string scale by the presence of the winding gas and by their dynamical backreaction.

This qualitative counting argument has been verified numerically [60] using the fact that, in $d = 3$, the interaction of winding states and their energy transfer to non-winding states can be described in close analogy with the cosmic string case. The intersection of two winding strings with the opposite orientation, in particular, produces closed loops with vanishing winding number, characterized by the same equation of state of non-relativistic (dust) matter. A quantitative, numerical analysis of the background evolution in the presence of "unwinding", due to loop production, has also shown [59] that the annihilation of the winding modes leads the background to exit from the loitering phase, and to approach the origin of the phase-space plane from positive values of H, in a state of three-dimensional, loop-driven isotropic expansion. These conclusions, however, have been challenged by a series of numerical works [61, 62, 63] showing that, because of the dilaton dependence of the string coupling, the annihilation mechanism and the resulting liberation of three growing dimensions only works if the initial value of the dilaton is sufficiently large. Otherwise, the compactification of all or none of the spatial dimensions is the most probable final configuration.

The above scenario, in which the background evolution is controlled by the presence of string winding modes, can be easily generalized to the case in which all spatial dimensions are isotropically filled with a gas of D_p-branes, wrapping around the cycles of the toroidal background, and possibly oscillating in the directions transverse to the brane. The dynamics of these higher-dimensional objects, extended along p spatial dimensions, is described by the Dirac–Born–Infeld (DBI) action (see e.g. [51]),

$$S_p = -T_p \int d^{p+1}\xi\, e^{-\phi/2} \left| \det(h_{ab} + B_{ab} + 2\pi\alpha' F_{ab}) \right|^{1/2}, \qquad (6.52)$$

which generalizes the Nambu–Goto action of a string, see Eq. (3A.1). Here $T_p \sim (\alpha')^{-(p+1)/2}$ is the tension (or mass per unit volume) of the brane, h_{ab} the induced metric on the brane, B_{ab} the induced antisymmetric tensor, and F_{ab} the possible gauge field confined on the brane. Finally, the coordinates ξ^a, $a = 1, \ldots, p+1$, span the brane world-volume.

As in the case of the string gas we limit ourselves to a torsionless background, $B = 0$, and we concentrate the discussion on the degrees of freedom leading to a gas with an effective negative pressure, thus neglecting gauge fields and other matter fields possibly living on the brane, as well as the oscillations transverse to the brane (they are all degrees of freedom contributing an "ordinary", positive pressure term to the brane gas, as can be seen by expanding to second order the above action [9]). We are then left with the action for the brane winding modes, which in the adiabatic approximation has the same form as the action for a cosmological term $\Lambda_p \sim g_s^{-1} T_p$ confined on the $(p+1)$-dimensional world-volume of the brane. In an isotropic, $(d+1)$-dimensional bulk manifold, with scale factor a, we have thus a "wrapping" energy $E_p \sim g_s^{-1} T_p a^p$ localized on the brane, with effective pressure (in a bulk volume $V \sim a^d$) given, in the adiabatic approximation, by

$$\widetilde{p} = -\frac{\partial E_p}{\partial V} = -\frac{p E_p}{dV} = -\frac{p}{d}\rho_p. \qquad (6.53)$$

The equation of state is thus controlled by the barotropic parameter $\gamma = -p/d$ (for $p = 1$ we recover the same result as in the case of the string winding modes). Using the conservation equation we eventually obtain

$$\overline{\rho}_p = \rho_p a^d \sim a^{-d\gamma} = a^p, \qquad (6.54)$$

generalizing to any p the scaling behavior of string winding modes, $\overline{\rho} \sim a$.

We can now repeat the analysis of the background equations (6.49), to find qualitatively the same results as before: the contribution of the winding modes of the brane becomes dominant with respect to that of the positive-pressure sources, and prevents the expansion of the spatial dimensions, leading the background towards an asymptotic regime of loitering contraction along the phase-space trajectory $\overset{\bullet}{\overline{\phi}} = -H/\gamma = dH/p$. The annihilation of these modes in $d \leq 2p+1$

dimensions leads eventually to the same conclusions as the string gas cosmology, but with the additional possible introduction of an interesting hierarchy in the size of the compact internal dimensions [9].

Suppose, in fact, we start with a hot, thermal mixture of all possible types of p branes, with comparable number densities for all species of branes, and with equal numbers of winding and antiwinding modes. Suppose also that, initially, the Universe expands isotropically: as the gravitational contribution of the branes grows as $\bar{\rho} \sim a^p$, we will have a series of phases, starting with the phase in which the component of the mixture with the largest dimensionality (say p_1) becomes dominant first. The annihilation of these modes will leave $2p_1 + 1$ dimensions free to expand, but this expansion will stop when other higher-dimensional objects (say p_2-branes, with $p_2 < p_1$) will become dominant. Their annihilation will leave a new subset of $2p_2 + 1$ expanding dimensions, and so on, down to $p = 1$. The domain-wall problem, possibly arising at each stage of brane-dominated evolution, should be solved by the loitering phase preceding the annihilation [59].

In the realistic case of a primordial Universe living in critical ($d = 9$) superstring dimensions we may note, first of all, that there is no difficulty for the intersection, the interaction and the self-annihilation of all branes with $p \geq 4$ (i.e. for all p such that $9 \leq 2p + 1$). The isotropic expansion of the nine spatial dimensions is thus possibly prevented by $p = 3, 2, 1$ branes only. The $p = 3$ branes (if they exist) are the first to become dominant, and when they annihilate will first allow a seven-dimensional spatial section (say, a seven-dimensional torus T^7) freedom to expand. The size of this (hypothetical) T^7, however, cannot grow too large, because D_2-branes, sooner or later, will become dominant, halting the expansion. The annihilation will allow a T^5 subspace of this T^7 to expand, until the winding modes of D-strings ($p = 1$ branes) or fundamental strings dominate. Within this T^5, the disappearance of the string winding modes will finally allow a T^3 subspace to expand, and become large without further obstructions.

Thus, we eventually recover the same result as before, with a possible hierarchy of size of the internal $d - 3$ dimensions: after the above phase transitions, the nine-dimensional spatial manifold M^9 will acquire the structure:

$$M^9 \to T^2 \times T^7 \to T^2 \times T^2 \times T^5 \to T^2 \times T^2 \times T^2 \times T^3. \qquad (6.55)$$

Interestingly enough one obtains, in this context, a configuration in which two dimensions may be larger, in principle, than the remaining internal ones, suggesting a possible connection with proposed scenarios with large internal dimensions [64, 65].

A final, important comment on string gas cosmology must concern the stabilization of the volume of the internal dimensions and, eventually, of the dilaton, both required for a phenomenologically consistent late-time cosmology.

The interactions of winding modes, and their backreaction on the geometry, suggest, as we have seen, a possible explanation of the large-scale dimensionality of the space-time in which we live (provided we accept some level of fine-tuning on the initial value of the dilaton). In the same way, the interplay of winding and momentum modes can stabilize the radius of the internal dimensions in an oscillating (S-frame) configuration [66]. However, this is not enough to stabilize the extra dimensions in all frames since, in this context, the dilaton keeps running. The inclusion of additional string states that are massless at the self-dual radius $R \sim \sqrt{\alpha'}$ seems to be required, in particular, to stabilize the extra dimensions in the E-frame [67] (but, again, the whole process is sensitive to the initial conditions [17]).

An alternative approach, possibly free from fine-tuning problems, is based on the mechanism of "quantum moduli trapping" [43]: when the background reaches an "enhanced symmetry point" like the self-dual radius, there is an additional production of massive states which become light near that point, and which should generate a confining potential preventing the modulus departing from it (see [17] for an application of this mechanism to a bosonic string model embedded in a five-dimensional space-time compactified on a circle).

Appendix 6A
Birth of the Universe in quantum string cosmology

In the standard cosmological scenario the Universe is expected to emerge from a space-time singularity, and to evolve initially through a phase of very high curvature and density, well inside the quantum gravity regime. In string cosmology, instead, there are scenarios which avoid the singularity, and in which the Universe emerges from a state of perturbative vacuum: in that case the initial phase is classical, with a curvature and a density very small in string (or Planck) units.

Even for those scenarios, however, the transition to the decelerated, radiation-dominated regime seems to occur only after the establishment of the high-curvature and strong coupling regime, as discussed in this chapter. The "birth of our Universe", regarded as the beginning of the present (Friedman-like) cosmological state, corresponds in that case to the transition (or "bounce") from the phase of growing to decreasing curvature, and also in that case can be described using quantum cosmology methods, as for a Universe born from an initial singularity.

There is, however, a crucial difference between a quantum description of the "big bang" and of the "big bounce": indeed, the bounce is preceded by a long period of classical evolution, while the standard big bang picture assumes that there is an abrupt truncation of the space-time dynamics at the singularity, with no classical description allowed at previous epochs (actually, there are no "previous" epochs, as the time coordinate itself ends at the singularity with no further allowed extension). It follows that in the standard scenario the initial state of the Universe is unknown, in principle, and has to be fixed through some ad hoc prescription: there are indeed various possible choices for the initial boundary conditions [68–73], leading in general to different quantum pictures for the early cosmological evolution.

In string cosmology models of bouncing, on the contrary, the initial state is fixed by the given pre-big bang (or pre-bounce) evolution, which, for instance, approaches asymptotically the string perturbative vacuum: this unambiguously determines the initial "wave function" of the Universe [74, 75], and the subsequent transition probabilities.

In a quantum cosmology context the Universe is described by a wave function evolving in *superspace*, according to the so-called Wheeler–De Witt (WDW) equation [76, 77], in much the same way as in ordinary quantum mechanics a particle is described by a wave function evolving in Hilbert space, according to the Schrödinger equation. Each point of Hilbert space corresponds to a state, and the quantum dynamics may allow transitions even in the case in which such transitions are forbidden in the context of the classical dynamics. In the same way, each point of superspace corresponds to a possible geometric configuration of the space-like sections of our space-time manifold, and the quantum cosmology dynamics may allow transitions between different geometrical states (for instance, from contraction to expansion, or from growing to decreasing curvature), even in the case in which such configurations are classically disconnected by a singularity.

It is probably appropriate to recall here that the quantum cosmology approach is affected by various problems, already present also in the context of the standard cosmological scenario: we can mention, in particular, the meaning of the probabilistic interpretation [78], the existence (and the possible meaning) of a semiclassical limit [79], the unambiguous identification of a time-like coordinate in superspace [80]. The problems of the boundary conditions and of the operator ordering in the WDW equation disappear, as we shall see, in a string cosmology context: the other problems, however, remain. Nevertheless, if we accept giving up a deterministic description of the bouncing transition, the quantum cosmology approach may allow a precise formulation of the question concerning the birth of our present cosmological state, and may provide a quantitative answer to this question.

For an elementary illustration of this possibility we discuss in this appendix the simplest example of "low-energy" quantum string cosmology, based on the following gravi-dilaton effective action:

$$S = -\frac{1}{2\,\lambda_s^{d-1}} \int \mathrm{d}^{d+1}x \sqrt{|g|}\, \mathrm{e}^{-\phi} \left[R + (\nabla\phi)^2 + V(\phi, g) \right]. \tag{6A.1}$$

We neglect higher-order (α' and loop) corrections, except those encoded in the (possibly non-local and non-perturbative) dilaton potential: the approach is similar to that of low-energy quantum mechanics, where one neglects relativistic and higher-order corrections. This first approximation is already sufficient, however, to take into account the quantum geometric effects we are interested in (see [81, 82] for higher-curvature contributions to the WDW equation).

We use, for our examples, the background (6.2) in the isotropic limit. Thus, our cosmological system is characterized by two degrees of freedom, the scale factor a and the dilaton ϕ (the "lapse" function $N = \sqrt{g_{00}}$ can be fixed to arbitrary values by choosing an appropriate gauge). The quantum evolution of this system can then be described by the WDW equation in a two-dimensional "minisuperspace", which we parametrize using the following convenient coordinates:

$$\beta = \sqrt{d}\,\ln a, \qquad \overline{\phi} = \phi - \sqrt{d}\,\beta - \ln \int \mathrm{d}^d x\, \lambda_s^{-d} \tag{6A.2}$$

(we are assuming spatial section of finite volume, $\int \mathrm{d}^d x < \infty$). Each given "point" $\{\beta(t), \overline{\phi}(t)\}$ of such minisuperspace represents a classical solution of the action (6A.1).

To obtain the WDW equation we now rewrite the action in terms of the variables β and $\overline{\phi}$, without fixing the temporal gauge, and using for the background the previous results (5.23)–(5.25). Defining

$$S = \int \mathrm{d}t\, L(N, \beta, \overline{\phi}) \tag{6A.3}$$

we obtain (after integration by parts)

$$L(N, \beta, \overline{\phi}) = \lambda_s \frac{\mathrm{e}^{-\overline{\phi}}}{2N} \left[\dot{\beta}^2 - \dot{\overline{\phi}}^2 - N^2\, V(\beta, \overline{\phi}) \right]. \tag{6A.4}$$

The variation with respect to N defines the total energy density of the system, and leads to the so-called Hamiltonian constraint. In the cosmic-time gauge $N = 1$,

$$\left(\frac{\delta S}{\delta N} \right)_{N=1} = 0 \quad \Rightarrow \quad \dot{\overline{\phi}}^2 - \dot{\beta}^2 - V = 0. \tag{6A.5}$$

The two momenta, canonically conjugated to the minisuperspace coordinates β and $\overline{\phi}$, are given, respectively, by

$$\Pi_\beta = \left(\frac{\delta L}{\delta \dot{\beta}}\right)_{N=1} = \lambda_s \dot{\beta} e^{-\overline{\phi}},$$

$$\Pi_{\overline{\phi}} = \left(\frac{\delta L}{\delta \dot{\overline{\phi}}}\right)_{N=1} = -\lambda_s \dot{\overline{\phi}} e^{-\overline{\phi}}. \tag{6A.6}$$

The Hamiltonian constraint can thus be rewritten in terms of the momenta as

$$\Pi_\beta^2 - \Pi_{\overline{\phi}}^2 + \lambda_s^2 V(\beta, \overline{\phi}) e^{-2\overline{\phi}} = 0. \tag{6A.7}$$

The WDW equation is finally obtained as the differential implementation of the Hamiltonian constraint through the gradient representation of the momenta, $\Pi \to i\nabla$:

$$\left[\partial_{\overline{\phi}}^2 - \partial_\beta^2 + \lambda_s^2 V(\beta, \overline{\phi}) e^{-2\overline{\phi}} \right] \Psi(\beta, \overline{\phi}) = 0. \tag{6A.8}$$

In the absence of dilaton potential this equation reduces to the free d'Alembert equation in a flat, two-dimensional manifold, and the solution can be factorized in plane waves as follows:

$$\Psi(\beta, \overline{\phi}) = \psi_\beta^\pm \psi_{\overline{\phi}}^\pm \sim e^{\pm ik\beta} e^{\pm i\overline{\phi}}. \tag{6A.9}$$

Here k is a positive constant, and $\psi_\beta^\pm$, $\psi_{\overline{\phi}}^\pm$ are free momentum eigenstates, satisfying the eigenvalue equations

$$\Pi_\beta \psi_\beta^\pm = \pm k \, \psi_\beta^\pm,$$

$$\Pi_{\overline{\phi}} \psi_{\overline{\phi}}^\pm = \pm k \, \psi_{\overline{\phi}}^\pm. \tag{6A.10}$$

Recalling that $\Pi_\beta \sim \dot{\beta}$ and $\Pi_{\overline{\phi}} \sim -\dot{\overline{\phi}}$ (according to their definitions), one can immediately check that the four particular solutions of type (6A.9) – corresponding to the four possible combinations of positive and negative eigenvalues – provide a plane-wave representation of the four asymptotic branches of the classical low-energy solutions, satisfying the condition

$$\dot{\overline{\phi}} = \pm\sqrt{d}H = \pm\dot{\beta} \quad \Rightarrow \quad \Pi_\beta = \pm\Pi_{\overline{\phi}}, \tag{6A.11}$$

and corresponding to the bisecting lines of the "phase space" plane of Fig. 4.3. With reference to Fig. 4.3 we have the following correspondence between the classical solutions and the plane-wave representation in minisuperspace:

- expansion $\longrightarrow \dot{\beta} > 0 \longrightarrow \psi_\beta^+$,
- contraction $\longrightarrow \dot{\beta} < 0 \longrightarrow \psi_\beta^-$,
- pre-big bang $\longrightarrow \dot{\overline{\phi}} > 0 \longrightarrow \psi_{\overline{\phi}}^-$,
- post-big bang $\longrightarrow \dot{\overline{\phi}} < 0 \longrightarrow \psi_{\overline{\phi}}^+$.

In this representation, the transition from an initial *expanding pre-big bang* phase to a final *expanding post-big bang* phase (represented by the upper dashed curve of Fig. 4.3) becomes a transition between the initial state

$$\Psi_{\text{in}} = \psi_\beta^+ \psi_{\overline{\phi}}^- \sim e^{ik\overline{\phi}-ik\beta}, \qquad \Pi_\beta > 0, \qquad \Pi_{\overline{\phi}} < 0, \qquad (6A.12)$$

and the final state

$$\Psi_{\text{out}} = \psi_\beta^+ \psi_{\overline{\phi}}^+ \sim e^{-ik\overline{\phi}-ik\beta}, \qquad \Pi_\beta > 0, \qquad \Pi_{\overline{\phi}} > 0. \qquad (6A.13)$$

The corresponding minisuperspace trajectory describes a monotonic evolution along β and a reflection along $\overline{\phi}$: thus, it can be represented as a scattering process of the WDW wave function in the plane spanned by β and $\overline{\phi}$, where β plays the role of a time-like coordinate, and $\overline{\phi}$ of space-like coordinate [74, 75] (see Fig. 6.6).

The boundary conditions for this process are uniquely fixed by the choice of the string perturbative vacuum as the initial state, corresponding to an incoming wave emerging from the asymptotic region $\beta \to -\infty$, $\overline{\phi} \to -\infty$, associated with a positive eigenvalue of Π_β and apposite eigenvalue of $\Pi_{\overline{\phi}}$. As illustrated in the figure, such an incident wave is partially transmitted towards the singularity ($\beta \to +\infty$, $\overline{\phi} \to +\infty$), and partially reflected back towards a post-big bang configuration, still expanding, but with decreasing curvature. The transition probability is fixed by the reflection coefficient

$$R_k = \frac{|\Psi_k^+|^2}{|\Psi_k^-|^2}, \qquad (6A.14)$$

given by the ratio between the left-moving and the right-moving parts of the asymptotic solution along $\overline{\phi}$.

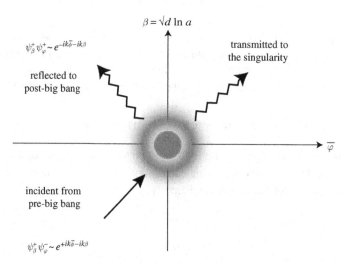

Figure 6.6 Birth of the Universe from the string perturbative vacuum, as a process of scattering and reflection of the WDW wave function in minisuperspace.

6A.1 Tunneling from the string perturbative vacuum

Without potential in the WDW equation there is no scattering, and then no transition, of course. With appropriate potentials, allowing a smooth evolution from the pre- to the post-big bang regime at the level of the classical solution, the transition probability R_k tends to unity; such a probability may be non-zero, however, even if the two branches of the classical solution are causally disconnected by a singularity [74]. We present here a very simple illustration of this effect, using as dilaton potential a cosmological constant, $V(\beta, \overline{\phi}) = \Lambda = \text{const}$ [75].

In this case the classical solution satisfies the condition

$$\dot{\overline{\phi}}^2 - dH^2 \equiv \dot{\overline{\phi}}^2 - \dot{\beta}^2 = \Lambda, \tag{6A.15}$$

and is represented by a hyperbola in the plane $\{\dot{\overline{\phi}}, \dot{\beta}\}$ of Fig. 4.3 (in the limit $\dot{\overline{\phi}} \to \pm\infty$, $\dot{\beta} \to \pm\infty$ one recovers the four distinct branches of the free solution (6A.11)). The explicit form of this hyperbolic solution, obtained by integrating the low-energy equations (4A.29) (with $V = \Lambda$ and no sources), is well known [83], and can be written as

$$a(t) = a_0 \left(\tanh \left| \frac{\sqrt{\Lambda}}{2} t \right| \right)^{\pm 1/\sqrt{d}},$$

$$\overline{\phi} = \phi_0 - \ln \sinh \left| \sqrt{\Lambda} t \right|, \tag{6A.16}$$

where a_0 and ϕ_0 are integration constants. This solution contains two branches, of the pre- and post-big bang type, defined respectively for $t < 0$ and $t > 0$, and separated by a singularity of the curvature invariants and of the effective string coupling at $t = 0$. Both branches are characterized by a conserved momentum along the β axis, defined by Eq. (6A.6) as

$$\lambda_s \dot{\beta} e^{-\overline{\phi}} = \pm \lambda_s \sqrt{\Lambda} e^{-\phi_0} \equiv \pm k = \text{const.} \tag{6A.17}$$

For this potential the WDW equation reduces to

$$\left(\partial_{\overline{\phi}}^2 - \partial_{\beta}^2 + \lambda_s^2 \Lambda e^{-2\overline{\phi}} \right) \Psi(\beta, \overline{\phi}) = 0, \tag{6A.18}$$

and can be solved by exploiting the conservation of Π_β, separating the variables as

$$\Psi(\beta, \overline{\phi}) = \psi_k(\overline{\phi}) e^{-ik\beta}, \tag{6A.19}$$

where

$$\left(\partial_{\overline{\phi}}^2 + k^2 + \lambda_s^2 \Lambda e^{-2\overline{\phi}} \right) \psi_k(\overline{\phi}) = 0. \tag{6A.20}$$

The general solution of this equation is a linear combination of Bessel functions [84] $J_\nu(z)$ and $J_{-\nu}(z)$, of index $\nu = ik$ and argument $z = \lambda_s \sqrt{\Lambda} \exp(-\overline{\phi})$. Assuming a flat and perturbative geometric configuration as the initial condition, i.e. choosing the initial wave function incoming from the asymptotic regime $\overline{\phi} \to -\infty$, $\beta \to -\infty$ (see Fig. 6.6), we can impose that there are only right-moving waves (along $\overline{\phi}$) approaching the singularity $\overline{\phi} \to +\infty$ (namely waves representing a state with $\dot{\overline{\phi}} > 0$, or $\Pi_{\overline{\phi}} < 0$). Using the small argument limit of the Bessel functions,

$$\lim_{\overline{\phi} \to +\infty} J_{\pm ik} \left(\lambda_s \sqrt{\Lambda} e^{-\overline{\phi}} \right) \sim e^{\mp ik\overline{\phi}}, \tag{6A.21}$$

we can then uniquely fix the WDW solution (modulo an arbitrary normalization factor N_k) as follows:

$$\Psi_k(\beta, \overline{\phi}) = N_k J_{-ik}\left(\lambda_s \sqrt{\Lambda}\, e^{-\overline{\phi}}\right) e^{-ik\beta}. \tag{6A.22}$$

After imposing the boundary conditions we expand the normalized solution in the opposite, perturbative limit $\overline{\phi} \to -\infty$, where we find

$$\lim_{\overline{\phi} \to -\infty} \Psi_k(\beta, \overline{\phi}) = \frac{N_k e^{-ik\beta}}{(2\pi z)^{1/2}} \left[e^{-i(z-\pi/4)} e^{k\pi/2} + e^{i(z-\pi/4)} e^{-k\pi/2} \right]$$

$$\equiv \Psi_k^- + \psi_k^+. \tag{6A.23}$$

This expression contains a superposition of the initial incoming state (Ψ_k^-, characterized by $\Pi_{\overline{\phi}} < 0$, i.e. by growing dilaton), and of the reflected component (Ψ_k^+, characterized by $\Pi_{\overline{\phi}} > 0$, i.e. by decreasing dilaton). The corresponding transition probability is then determined by the reflection coefficient (6A.14) as [75]

$$R_k = e^{-2\pi k}. \tag{6A.24}$$

Therefore, the quantum probability for this process is non-zero even if this transition is classically forbidden. We find, in particular, the same exponential suppression appearing also in the "tunneling" processes typical of the standard quantum cosmology scenario, where the quantum effects are expected to generate a Universe in the appropriate inflationary state. There is, however, an important difference due to the fact that, in the string cosmology scenario we are considering, the quantum gravity (Planckian) regime is reached *at the end* of a long phase of accelerated evolution, when the Universe is expected to *exit* (not to enter) the inflationary regime. Thus, quantum effects have not to be responsible for inflationary initial conditions (indeed, the Universe emerges from the scattering process into a state of standard, decelerated expansion).

In spite of these differences, the result (6A.24) is formally very similar to the probability that our Universe may emerge from the Planckian regime according to the "*tunneling from nothing*" (and other similar) inflationary scenarios [70–73]. The explanation of this formal analogy is simple, if we recall that the choice of the string perturbative vacuum as the initial state implies that there are *only outgoing (right-moving) waves approaching the singularity* at $\overline{\phi} \to \infty$. This is exactly equivalent to imposing tunneling boundary conditions, which select "*only outgoing modes at the singular space-time boundary*" [78, 85]. In this sense, the quantum reflection that we have illustrated can also be interpreted as a tunneling process, not "from nothing" but "from the string perturbative vacuum".

6A.2 Operator ordering

It seems appropriate, at this point, to stress that the WDW equation of string cosmology is not affected by operator ordering ambiguities, thanks to the symmetry properties of the string effective action [74]. The presence of the duality symmetry, in particular, implies that the minisuperspace geometry is globally flat: thus, it is always possible to choose a convenient parametrization associated with a flat metric in momentum space, and to a Hamiltonian manifestly free from operator ordering problems. Conversely, if we introduce curvilinear coordinates in minisuperspace, the ordering imposed by duality is equivalent to the requirement of general reparametrization invariance.

For a general illustration of this property we may start from the $O(d, d)$-covariant form of the low-energy effective action introduced in Chapter 4, considering in particular the Lagrangian associated with the action (4.132):

$$L(M, \overline{\phi}) = -\frac{\lambda_s}{2} e^{-\overline{\phi}} \left[\dot{\overline{\phi}}^2 + \frac{1}{8} \mathrm{Tr}(\dot{M} \dot{M}^{-1}) + V \right].$$ (6A.25)

The canonical momentum Π_M for the torsion-graviton background,

$$\Pi_M = \frac{\delta L}{\delta \dot{M}} = \frac{\lambda_s}{8} e^{-\overline{\phi}} M^{-1} \dot{M} M^{-1},$$ (6A.26)

leads to the classical Hamiltonian density

$$H = \frac{4}{\lambda_s} e^{\overline{\phi}} \mathrm{Tr}(M \Pi_M M \Pi_M),$$ (6A.27)

which would seem to have ordering problems, since $[M, \Pi_M] \neq 0$. However, thanks to the $O(d, d)$ properties of the matrix M, we can always rewrite the torsion-graviton kinetic terms in the form (4.131)

$$\frac{1}{8} \mathrm{Tr}(\dot{M} \dot{M}^{-1}) = \frac{1}{8} \mathrm{Tr}(\dot{M} \eta)^2,$$ (6A.28)

with a corresponding momentum

$$\Pi_M = -\frac{\lambda_s}{8} e^{-\overline{\phi}} \eta \dot{M} \eta.$$ (6A.29)

The associated Hamiltonian density,

$$H = -\frac{4}{\lambda_s} e^{\overline{\phi}} \mathrm{Tr}(\eta \Pi_M \eta \Pi_M),$$ (6A.30)

has a flat metric in momentum space, and no ordering problems for the WDW equation.

A flat minisuperspace metric also manifestly appears in the gravi-dilaton Hamiltonian (6A.7). Let us discuss, however, the effects of changing the chosen parametrization, using, for instance, the new pair of coordinates $\{a, \overline{\phi}\}$ different from those of Eq. (6A.4). The action becomes

$$S = \frac{\lambda_s}{2} \int dt \, \frac{e^{-\overline{\phi}}}{N} \left(d \frac{\dot{a}^2}{a^2} - \dot{\overline{\phi}}^2 - N^2 V \right),$$ (6A.31)

and the new conjugate momenta,

$$\Pi_a = d \lambda_s \frac{\dot{a}}{a^2} e^{-\overline{\phi}}, \qquad \Pi_{\overline{\phi}} = -\lambda_s \dot{\overline{\phi}} e^{-\overline{\phi}},$$ (6A.32)

lead to the following (classical) Hamiltonian constraint:

$$\frac{a^2}{d} \Pi_a^2 - \Pi_{\overline{\phi}}^2 + \lambda_s^2 V e^{-2\overline{\phi}} = 0.$$ (6A.33)

The kinetic part of this Hamiltonian has a non-trivial 2×2 metric γ_{AB} in minisuperspace:

$$H_\gamma = \frac{a^2}{d}\Pi_a^2 - \Pi_{\bar\phi}^2 = \gamma^{AB}\Pi_A\Pi_B,$$

$$\gamma_{AB} = \mathrm{diag}\left(\frac{d}{a^2}, -1\right), \tag{6A.34}$$

and we now encounter an ordering problem for the quantum system, since $[a, \Pi_a] \neq 0$.

Using the differential representation of the momentum operators we can write, in general, the Hamiltonian operator in the form

$$H_\gamma = \frac{\partial}{\partial\bar\phi^2} - \frac{a^2}{d}\frac{\partial^2}{\partial a^2} - \epsilon\frac{a}{d}\frac{\partial}{\partial a}, \tag{6A.35}$$

where ϵ is a (real) c-number parameter, depending on the ordering (there are no additional contributions to the ordered Hamiltonian from the scalar curvature $\mathcal{R}$ of superspace [86], since $\mathcal{R}$ is vanishing for the metric (6A.34)). On the other hand, if we want to reproduce the general result (6A.30) – which is valid also in the isotropic case with $B = 0$ and $G = -a^2 I$ – one easily finds that the operators $a^2\partial_a^2$ and $a\partial_a$ must appear in the Hamiltonian with the same numerical coefficient, which forces us to the choice $\epsilon = 1$. Otherwise stated, the ordering specified by $\epsilon = 1$ is the only one compatible with the invariance of the action under the scale-factor duality transformation $a \to \tilde{a} = a^{-1}$, $\bar\phi \to \bar\phi$. By applying such a transformation to the ordered Hamiltonian (6A.35) one obtains, in fact,

$$H_\gamma(a) = H_\gamma(\tilde{a}) + \frac{2}{d}(\epsilon - 1)\tilde{a}\frac{\partial}{\partial\tilde{a}}, \tag{6A.36}$$

so that only for $\epsilon = 1$ is the form of the Hamiltonian preserved.

It may be observed, finally, that this ordering prescription is exactly equivalent to the requirement of reparametrization invariance in minisuperspace, which imposes on the kinetic part of the Hamiltonian the covariant d'Alembert form [86]:

$$\mathcal{H} = -\nabla_A\nabla^A = -\frac{1}{\sqrt{-\gamma}}\partial_A(\sqrt{-\gamma}\gamma^{AB}\partial_B). \tag{6A.37}$$

For the metric (6A.34), on the other hand,

$$-\nabla_A\nabla^A = \frac{\partial}{\partial\bar\phi^2} - \frac{a}{d}\frac{\partial}{\partial a} - \frac{a^2}{d}\frac{\partial^2}{\partial a^2}, \tag{6A.38}$$

which again implies $\epsilon = 1$, when compared to Eq. (6A.35).

6A.3 Scattering of the wave function in minisuperspace

Let us conclude this appendix by noting that the scattering process illustrated in Fig. 6.6 is not the only process of quantum string cosmology which may occur – and possibly describe the birth of our present cosmological state – in the two-dimensional minisuperspace of Eq. (6A.8). Even after fixing the boundary conditions with the choice of the perturbative vacuum (represented, asymptotically, by the wave function $\psi_\beta^+\psi_{\bar\phi}^-$), there

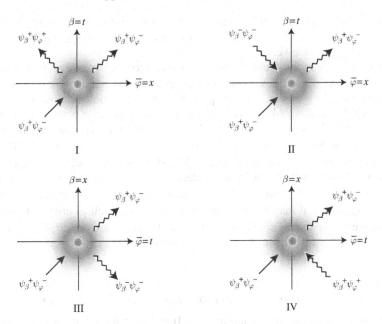

Figure 6.7 The four possible channels of vacuum decay in the two-dimensional WDW minisuperspace spanned by β and $\overline{\phi}$.

are indeed four different types of processes, associated with four different configurations $\psi_\beta^\pm \psi_{\overline{\phi}}^\pm$ of the wave function in the final state, and corresponding to four possible "decay channels" of the string perturbative vacuum [87].

Such a range of possibilities is due to the arbitrary choice of the time-like coordinate in minisuperspace [88], and also to the fact that, with appropriate assumptions about the scattering potential (supposed to be localized in a finite region of superspace), one can obtain asymptotical $|\text{out}\rangle$ states which are superpositions of waves with the same Π_β and opposite $\Pi_{\overline{\phi}}$, *or* with the same $\Pi_{\overline{\phi}}$ and opposite Π_β. Taking into account that one component of the $|\text{out}\rangle$ state must correspond to the transmitted part of the incident wave function, $\Psi_{\text{in}} = \psi_\beta^+ \psi_{\overline{\phi}}^-$, the possible components of the outgoing wave function will be of the type $\Psi_{\text{out}} \sim \psi_\beta^+ \psi_{\overline{\phi}}^\pm$ and $\Psi_{\text{out}} \sim \psi_{\overline{\phi}}^- \psi_\beta^\pm$, depending on momentum conservation along β or along $\overline{\phi}$, respectively. A graphic illustration of the possible processes is shown in Fig. 6.7.

Two of these processes (I and IV) describe a monotonic evolution along β, the other two (II and III) along $\overline{\phi}$. The time-like coordinate, however, coincides with β for cases I and II, and with $\overline{\phi}$ for cases III and IV. As a consequence, only cases I and III represent a true "reflection" of the momenta along a spatial axis: the other two processes, II and IV, are qualitatively different, as the final state is a superposition of modes of positive and negative frequency with respect to the time-like axis (namely, they are positive and negative "energy" eigenstates). In the context of the so-called "third quantization" formalism [89–92] (i.e. second quantization of the WDW wave function), they represent a Bogoliubov mixing describing the production of a "pair of universes" from the vacuum. The wave moving backward with respect to the time axis has to be "reinterpreted" (as in quantum field theory for the antiparticles) as an "antiuniverse" of positive energy and opposite spatial momentum (in superspace). In our case such a superspace inversion corresponds to a reflection of $\dot{\beta}$, so that this reinterpretation principle changes expansion into contraction, and vice versa.

Let us discuss in more detail the four possible processes, starting from the spatial reflections I and III. Case I describes a reflection along the spatial axis $\overline{\phi}$, and has been discussed previously (see Fig. 6.6). In particular, the evolution is monotonic along β, so that the Universe always keeps expanding. The incident wave is partially transmitted towards the big bang singularity ($\overline{\phi} \to +\infty$), and partially reflected back towards the low-energy, expanding, post-big bang regime ($\beta \to \infty$, $\overline{\phi} \to -\infty$). In case III, on the contrary, the evolution is monotonic along $\overline{\phi}$, and the incident wave is reflected along β, interpreted as the spatial axis. The incident wave is totally transmitted towards the singular region ($\overline{\phi} \to +\infty$), but in part as an expanding ($\dot{\beta} > 0$) and in part as a contracting ($\dot{\beta} < 0$) configuration. It may be noted that this process (as well as process II) requires a duality breaking potential $V = V(\beta, \overline{\phi})$, otherwise Π_β is conserved, $[H, \Pi_\beta] = 0$, and the reflection along β is forbidden, even at the quantum level.

6A.4 Antitunneling from the string perturbative vacuum

Consider now the third-quantization processes, II and IV. Case IV, in particular, describes the production of universe–antiuniverse pairs (one expanding, the other contracting) from the string perturbative vacuum [93]. However, the pairs evolve towards the strong coupling regime ($\overline{\phi} \to +\infty$), so that all members of the pairs fall inside the big bang singularity. Case II is probably more interesting, from a phenomenological point of view, since there the universe and the antiuniverse of the pair are both expanding: one falls inside the singularity, but the other expands towards the low-energy, post-big bang regime ($\overline{\phi} \to -\infty$, $\beta \to +\infty$), and may represent the birth of a universe like ours in a standard Friedman-like configuration [94].

In this second case, the production of pairs from the vacuum – just as in the particle case – is associated with the parametric amplification of the wave function in the minisuperspace. The birth of the Universe, in this case, may be seen as a process of *antitunneling from the string perturbative vacuum*, and the probability of this process is no longer exponentially suppressed, as in the case of the tunneling transitions of type I and III.

This type of process requires an effective potential containing both the metric and the dilaton, but *not* in the combination $\overline{\phi}$: otherwise, $[V, \Pi_\beta] = 0$, the Hamiltonian is translationally invariant along the β axis, and the $|\text{out}\rangle$ state cannot be a mixture of states with positive and negative eigenvalues of Π_β. A simple example of a potential of this type is the two-loop potential

$$V(\beta, \overline{\phi}) = \Lambda\theta(-\beta)e^{2\phi} \equiv \Lambda\theta(-\beta)e^{2(\overline{\phi}+\sqrt{d}\beta)}, \tag{6A.39}$$

which has been studied in [94]. Here Λ is a positive constant, and θ is the Heaviside step function, introduced to mimic an efficient damping of the potential outside the interaction regime (in particular, in the large-radius limit $\beta \to \infty$ of the expanding, post-big bang configuration).

With this potential, the WDW equation (6A.8) can be conveniently separated in terms of the eigenstates of $\Pi_{\overline{\phi}}$, looking for solutions of the form

$$\Psi(\beta, \overline{\phi}) = \Psi_k(\beta)e^{ik\overline{\phi}}, \tag{6A.40}$$

where

$$\left[\partial_\beta^2 + k^2 - \lambda_s^2 \Lambda \theta(-\beta) e^{2\sqrt{d}\beta}\right] \Psi_k(\beta) = 0. \qquad (6A.41)$$

In the region $\beta > 0$ the potential is vanishing, and the solution is a superposition of eigenstates of Π_β represented by the free (positive and negative) frequency modes $\psi_\beta^\pm$. In the region $\beta < 0$ the solution is a combination of Bessel functions $J_\nu(z)$, of imaginary index $\nu = \pm ik/\sqrt{d}$ and argument $z = i\lambda_s \sqrt{\Lambda/d} \exp(\sqrt{d}\beta)$. The boundary conditions are fixed at $\beta \to -\infty$, imposing that the universe starts expanding from the string perturbative vacuum,

$$\lim_{\beta \to -\infty} \Psi(\beta, \overline{\phi}) \sim e^{ik(\overline{\phi} - \beta)}. \qquad (6A.42)$$

Imposing the matching of Ψ and Ψ' at $\beta = 0$ one then computes the Bogoliubov coefficients determining, in the third-quantization formalism, the probability distribution of the pairs of universes produced from the vacuum, for each mode k (where k represents a given configuration in the space of the initial parameters).

One finds [93, 94], in particular, that the initial wave function is parametrically amplified provided $k < \lambda_s \sqrt{\Lambda}$. In that case the birth of our present, post-big bang cosmological phase may proceed efficiently as a forced production of pairs of universes from the quantum fluctuations of the string perturbative vacuum. Since $k \sim g_s^{-2}$, the process is strongly favored for configurations of large enough coupling g_s and/or large enough cosmological constant Λ (in string units). For $k \gg \lambda_s \sqrt{\Lambda}$, on the contrary, there is no parametric amplification, and the initial state runs almost undisturbed towards the singularity. Only a small, exponentially suppressed fraction of it is able to emerge in the post-big bang regime (just as in the case of tunneling, or reflections): the number of produced pairs is exponentially damped, $n_k \sim \exp(-k/\lambda_s \sqrt{\Lambda})$, and the resulting distribution describes a "thermal bath" (or "foam") of "baby universes" at an effective temperature $T \sim \sqrt{\Lambda}$ in superspace (see [94, 95] for a more detailed discussion).

References

[1] E. Witten, *Nucl. Phys.* **B443** (1995) 85.
[2] P. Horava and E. Witten, *Nucl. Phys.* **B460** (1996) 506.
[3] T. Damour and M. Henneaux, *Phys. Lett.* **B488** (2000) 108.
[4] M. Gasperini and G. Veneziano, *Phys. Rep.* **373** (2003) 1.
[5] G. Veneziano, *Phys. Lett.* **B454** (1999) 22.
[6] M. Maggiore and A. Riotto, *Nucl. Phys.* **B548** (1999) 427.
[7] R. Brandenberger and C. Vafa, *Nucl. Phys.* **B316** (1989) 391.
[8] A. A. Tseytlin and C. Vafa, *Nucl. Phys.* **B372** (1992) 443.
[9] S. Alexander, R. Brandenberger and D. Easson, *Phys. Rev.* **D62** (2000) 103509.
[10] J. Khouri, B. A. Ovrut, P. J. Steinhardt and N. Turok, *Phys. Rev.* **D64** (2001) 123522.
[11] S. Kachru, R. Kallosh, A. Linde and S. P. Trivedi, *Phys. Rev.* **D68** (2003) 046005.
[12] G. Dvali and S. H. H. Tye, *Phys. Lett.* **B450** (1999) 72.
[13] C. Burgess, M. Majumdar, D. Nolte *et al.*, *JHEP* **0107** (2001) 047.
[14] S. Kachru, R. Kallosh, A. Linde *et al.*, *JCAP* **0310** (2003) 013.
[15] M. Gasperini, M. Maggiore and G. Veneziano, *Nucl. Phys.* **B494** (1997) 315.
[16] M. Berg, M. Haack and B. Kors, *Phys. Rev. Lett.* **96** (2006) 021601.
[17] T. Battenfeld and S. Watson, *Rev. Mod. Phys.* **78** (2006) 435.
[18] G. Boulware and S. Deser, *Phys. Rev. Lett.* **55** (1985) 2656.
[19] S. Kalara and K. A. Olive, *Phys. Lett.* **B218** (1989) 148.
[20] M. S. Madsen and J. D. Barrow, *Nucl. Phys.* **B323** (1989) 242.
[21] R. R. Metsaev and A. A. Tseytlin, *Nucl. Phys.* **B293** (1987) 385.
[22] C. Myers, *Phys. Lett.* **B199** (1987) 371.
[23] S. Foffa and M. Maggiore, *Phys. Rev.* **D58** (1998) 023505.
[24] K. A. Meissner, *Phys. Lett.* **B392** (1997) 298.
[25] N. Kaloper and K. A. Meissner, *Phys. Rev.* **D56** (1997) 7940.
[26] E. Kiritsis and C. Kounnas, *Phys. Lett.* **B331** (1994) 51.
[27] H. J. de Vega, A. L. Larsen and N. Sanchez, *Phys. Rev.* **D61** (2000) 066003.
[28] R. Brustein and R. Madden, *JHEP* **9907** (1999) 006.
[29] C. Cartier, E. J. Copeland and R. Madden, *JHEP* **0001** (2000) 035.
[30] Y. Nambu, *Phys. Rev.* **D63** (2001) 044013.
[31] Y. Nambu, *Phys. Rev.* **D65** (2002) 104013.
[32] A. Ghosh, R. Madden and G. Veneziano, *Nucl. Phys.* **B570** (2000) 207.
[33] A. Lawrence and E. Martinec, *Class. Quantum Grav.* **13** (1996) 63.
[34] S. S. Gubser, *Phys. Rev.* **D69** (2004) 123507.
[35] M. Maggiore, *Nucl. Phys.* **B525** (1998) 413.

[36] R. Brustein and G. Veneziano, *Phys. Lett.* **B329** (1994) 429.
[37] N. Kaloper and K. A. Olive, *Phys. Rev.* **D57** (1998) 811.
[38] S. W. Hawking and G. F. R. Ellis, *The Large Scale Structure of Spacetime* (Cambridge: Cambridge University Press, 1973).
[39] M. Gasperini, *Int. J. Mod. Phys.* **D13** (2004) 2267.
[40] R. Brustein and R. Madden, *Phys. Lett.* **B410** (1997) 110.
[41] R. Brustein and R. Madden, *Phys. Rev.* **D57** (1998) 712.
[42] S. B. Giddings, S. Kachru and J. Polchinski, *Phys. Rev.* **D66** (2002) 106006.
[43] L. Kofman, A. Linde, X. Liu *et al.*, *JHEP* **0405** (2004) 030.
[44] M. Gasperini, F. Piazza and G. Veneziano, *Phys. Rev.* **D65** (2002) 023508.
[45] M. Gasperini, *Phys. Rev.* **D64** (2001) 043510.
[46] C. Cartier, E. J. Copeland and M. Gasperini, *Nucl. Phys.* **B607** (2001) 406.
[47] I. Antoniadis, E. Gava and K. S. Narain, *Phys. Lett.* **B283** (1992) 209.
[48] I. Antoniadis, J. Rizos and K. Tamvakis, *Nucl. Phys.* **B415** (1994) 497.
[49] S. Foffa, M. Maggiore and R. Sturani, *Nucl. Phys.* **B552** (1999) 395.
[50] M. Gasperini and G. Veneziano, *Phys. Lett.* **B387** (1996) 715.
[51] J. Polchinski, "Tasi Lectures on D-Branes", in *Boulder 1996, Fields, Strings and Duality* (Singapore: World Scientific, 1997), p. 293.
[52] A. C. Davis and M. Majumdar, *JHEP* **0203** (2002) 056.
[53] R. Durrer, M. Kunz and M. Sakellariadou, *Phys. Lett.* **B614** (2005) 125.
[54] A. Karch and L. Randall, *Phys. Rev. Lett.* **95** (2005) 161601.
[55] A. A. Tseytlin, *Class. Quantum Grav.* **9** (1992) 979.
[56] S. P. de Alvis and K. Sato, *Phys. Rev.* **D53** (1996) 7187.
[57] M. Gasperini, N. Sanchez and G. Veneziano, *Nucl. Phys.* **B364** (1991) 365.
[58] M. Gasperini, M. Giovannini, K. A. Meissner and G. Veneziano, in *String Theory in Curved Space Times*, ed. N. Sanchez (Singapore: World Scientific, 1998), p. 49.
[59] R. Brandenberger, D. A. Easson and D. Kimberly, *Nucl. Phys.* **B623** (2002) 421.
[60] M. Sakellariadou, *Nucl. Phys.* **B468** (1996) 319.
[61] R. Easther, M. R. Greene, M. G. Jackson and D. Kabat, *Phys. Rev.* **D67** (2003) 123501.
[62] R. Easther, M. R. Greene, M. G. Jackson and D. Kabat, *JCAP* **0401** (2004) 006.
[63] R. Easther, M. R. Greene, M. G. Jackson and D. Kabat, *JCAP* **0502** (2005) 009.
[64] N. Arkani-Hamed, S. Dimopoulos and G. Dvali, *Phys. Lett.* **B429** (1998) 263.
[65] I. Antoniadis, N. Arkani-Hamed, S. Dimopoulos and G. Dvali, *Phys. Lett.* **B436** (1998) 257.
[66] S. Watson and R. Brandenberger, *JCAP* **0311** (2003) 008.
[67] S. P. Patil and R. Brandenberger, *JCAP* **0601** (2006) 005.
[68] J. B. Hartle and S. W. Hawking, *Phys. Rev.* **D28** (1983) 2960.
[69] S. W. Hawking, *Nucl. Phys.* **B239** (1984) 257.
[70] A. Vilenkin, *Phys. Rev.* **D30** (1984) 509.
[71] A. D. Linde, *Sov. Phys. JETP* **60** (1984) 211.
[72] Y. Zeldovich and A. A. Starobinski, *Sov. Astron. Lett.* **10** (1984) 135.
[73] V. A. Rubakov, *Phys. Lett.* **B148** (1984) 280.
[74] M. Gasperini, J. Maharana and G. Veneziano, *Nucl. Phys.* **B472** (1996) 349.
[75] M. Gasperini and G. Veneziano, *Gen. Rel. Grav.* **28** (1996) 1301.
[76] B. S. De Witt, *Phys. Rev.* **160** (1967) 1113.
[77] J. A. Wheeler, in *Battelle Rencontres*, eds C. De Witt and J. A. Wheeler (New York: Benjamin, 1968).
[78] A. Vilenkin, *Phys. Rev.* **D33** (1986) 3650.
[79] M. G. Bento and O. Bertolami, *Class. Quantum Grav.* **12** (1995) 1919.
[80] M. Cavaglià and V. De Alfaro, *Gen. Rel. Grav.* **29** (1997) 773.

[81] M. D. Pollock, *Nucl. Phys.* **B324** (1989) 187.
[82] M. D. Pollock, *Int. J. Mod. Phys.* **A7** (1992) 4149.
[83] M. Muller, *Nucl. Phys.* **B337** (1990) 37.
[84] M. Abramowitz and I. A. Stegun, *Handbook of Mathematical Functions* (New York: Dover, 1972).
[85] A. Vilenkin, *Phys. Rev.* **D37** (1988) 888.
[86] A. Ashtekar and R. Geroch, *Rep. Prog. Phys.* **37** (1974) 1211.
[87] M. Gasperini, *Int. J. Mod. Phys.* **A13** (1998) 4779.
[88] M. Cavaglià and C. Ungarelli, *Class. Quantum Grav.* **16** (1999) 1401.
[89] V. A. Rubakov, *Phys. Lett.* **B214** (1988) 503.
[90] M. McGuigan, *Phys. Rev.* **D38** (1988) 3031.
[91] M. McGuigan, *Phys. Rev.* **D39** (1989) 2229.
[92] M. McGuigan, *Phys. Rev.* **D41** (1990) 418.
[93] A. Buonanno, M. Gasperini, M. Maggiore and C. Ungarelli, *Class. Quantum Grav.* **14** (1997) L97.
[94] M. Gasperini, *Int. J. Mod. Phys.* **D10** (2001) 15.
[95] M. Gasperini and G. Veneziano, *Phys. Rep.* **373** (2003) 1.

7

The cosmic background of relic gravitational waves

In Chapter 5 we have shown that there are various classes of inflationary kinematics, and we have stressed that the kinematic properties of the string cosmology backgrounds may be very different from the "standard" kinematic properties typical of field theory models of inflation. A question which naturally arises is whether such kinematic differences may correspond (at least in principle) to observable phenomenological differences, suitable to provide a clear signature of the various primordial scenarios.

The answer to this question is positive, as the transition from accelerated to decelerated (i.e. from inflationary to standard) evolution amplifies the quantum fluctuations of the various background fields, and may produce a large amount of various species of radiation. The spectral properties of this radiation, on the other hand, are strongly correlated to the kinematics of the inflationary phase, as we shall see in this chapter. A direct (or indirect) observation of such a primordial component of the cosmic radiation may thus give us important information on the inflationary dynamics, testing the predictions of the various cosmological scenarios.

This chapter is devoted to the study of the direct inflationary production of gravitational radiation, starting from the amplification of the tensor part of the metric fluctuations and the subsequent formation of a cosmic background of relic gravitons. Such a background, if produced at curvature scales $H \lesssim M_\mathrm{P}$, is expected to survive almost unchanged till the present epoch, thus transmitting to us (encoded inside its spectral structure) a faithful imprint of the primordial kinematics [1].

We should recall that the Universe becomes "transparent" to the gravitational radiation just below the Planck scale. For the electromagnetic radiation such a transparence is reached only at exceedingly lower scales, i.e. at the "decoupling" scale $H_{\mathrm{dec}} \sim 10^{-57} M_\mathrm{P}$ when the temperature drops below 3000 K, roughly corresponding to the temperature of recombination of hydrogen atoms (see for instance [2]). This big difference of scales underlines the exceptional relevance of a direct observation of the primordial graviton background for a possible reconstruction of the early history of our Universe.

7.1 Propagation of tensor perturbations

In order to discuss the production of gravitational waves in a generic inflationary scenario we need, first of all, the equation describing the propagation of tensor perturbations on a cosmological background (generalizing the usual d'Alembert wave equation, valid in Minkowski space). Such an equation can be deduced in two different (but equivalent) ways.

The first approach is based on the direct perturbation of the field equations, and on the expansion of the metric to first order around a given zeroth-order solution $g_{\mu\nu}$,

$$g_{\mu\nu} \to g_{\mu\nu} + \delta g_{\mu\nu}, \qquad \delta g_{\mu\nu} \equiv h_{\mu\nu}. \tag{7.1}$$

Using the unperturbed equations for $g_{\mu\nu}$ one then obtains a linearized equation for the first-order perturbation $h_{\mu\nu}$. This is a convenient approach if we are only interested in the classical propagation of tensor perturbations, but is *not* appropriate if we want to study the evolution of the quantum fluctuations of the metric background.

The second approach is based on the same metric expansion as before, but the expansion is now applied to the action, which is perturbed up to terms quadratic in $h_{\mu\nu}$. Using the unperturbed equations for $g_{\mu\nu}$ one obtains a quadratic action, $\delta^{(2)}S \equiv S(h^2)$, whose variation with respect to $h_{\mu\nu}$ provides a linear propagation equation – exactly the same equation as that obtained with the previous method. Given $\delta^{(2)}S$, however, it is possible to diagonalize the kinetic term of the fluctuations, and define the so-called "normal modes" of oscillations: such modes can be quantized by imposing canonical commutation relations, and can be used to normalize the perturbed solutions to a spectrum of quantum fluctuations of the vacuum [3], as we shall see in this section.

We apply this second procedure to the lowest-order, gravi-dilaton effective action (already introduced in Chapters 2 and 3),

$$S = -\frac{1}{2\lambda_s^{d-1}} \int d^{d+1}x \sqrt{|g|}\, e^{-\phi} \left[R + (\nabla\phi)^2 + V(\phi) \right], \tag{7.2}$$

where we have also included a scalar potential for possible applications to generic inflationary scenarios. In order to study the pure tensor part of the metric fluctuations (see Chapter 8 for scalar metric perturbations) we do not perturb the scalar dilaton field, setting $\delta\phi = 0$ (indeed, scalar and tensor perturbations are decoupled in the linear approximation). We can then parametrize the metric perturbation with a transverse and traceless tensor field, as follows:

$$\delta g_{\mu\nu} = h_{\mu\nu}, \qquad \nabla_\nu h_\mu{}^\nu = 0, \qquad g^{\mu\nu} h_{\mu\nu} = 0, \tag{7.3}$$

where ∇_ν represents the unperturbed covariant derivative, and the indices of $h_{\mu\nu}$ are raised and lowered with the unperturbed metric $g_{\mu\nu}$.

In order to expand the action up to terms quadratic in the fluctuations variable $h_{\mu\nu}$, let us start considering the contravariant components of the metric and the determinant of the metric tensor. We obtain

$$\delta^{(1)} g^{\mu\nu} = -h^{\mu\nu}, \qquad \delta^{(2)} g^{\mu\nu} = h^{\mu\alpha} h_\alpha{}^\nu, \tag{7.4}$$

$$\delta^{(1)} \sqrt{-g} = 0, \qquad \delta^{(2)} \sqrt{-g} = -\frac{1}{4} \sqrt{-g}\, h_{\mu\nu} h^{\mu\nu}, \tag{7.5}$$

where we are using the notation in which $\delta^{(k)} A$ denotes the kth term in the expansion of the variable A in powers of h. As usual, the signs of the contravariant components are fixed by the condition

$$(g_{\mu\nu} + h_{\mu\nu})(g^{\nu\alpha} + \delta^{(1)} g^{\nu\alpha} + \delta^{(2)} g^{\nu\alpha}) = \delta_\mu^\alpha. \tag{7.6}$$

For the Christoffel connection we obtain

$$\delta^{(1)} \Gamma_{\mu\nu}{}^\alpha = \frac{1}{2} g^{\alpha\beta} \left(\partial_\mu h_{\nu\beta} + \partial_\nu h_{\mu\beta} - \partial_\beta h_{\mu\nu} \right)$$
$$- \frac{1}{2} h^{\alpha\beta} \left(\partial_\mu g_{\nu\beta} + \partial_\nu g_{\mu\beta} - \partial_\beta g_{\mu\nu} \right), \tag{7.7}$$

$$\delta^{(2)} \Gamma_{\mu\nu}{}^\alpha = -\frac{1}{2} h^{\alpha\beta} \left(\partial_\mu h_{\nu\beta} + \partial_\nu h_{\mu\beta} - \partial_\beta h_{\mu\nu} \right)$$
$$+ \frac{1}{2} h^{\alpha\rho} h_\rho^\beta \left(\partial_\mu g_{\nu\beta} + \partial_\nu g_{\mu\beta} - \partial_\beta g_{\mu\nu} \right), \tag{7.8}$$

and so on for the components of the Ricci tensor, $\delta^{(1)} R_{\mu\nu}$, $\delta^{(2)} R_{\mu\nu}$. We may note that the first-order perturbation $\delta^{(1)} \Gamma$, using the explicit form of the unperturbed connection, can also be rewritten in a useful covariant form as

$$\delta^{(1)} \Gamma_{\mu\nu}{}^\alpha = \frac{1}{2} g^{\alpha\beta} \left(\nabla_\mu h_{\nu\beta} + \nabla_\nu h_{\mu\beta} - \nabla_\beta h_{\mu\nu} \right). \tag{7.9}$$

Applying this procedure to Eq. (7.2), we are led to consider the following perturbed action:

$$\delta^{(2)} S = -\frac{1}{2\lambda_s^{d-1}} \int d^{d+1} x\, e^{-\phi} \left[\sqrt{|g|} \left(\delta^{(1)} g^{\mu\nu} \delta^{(1)} R_{\mu\nu} + R_{\mu\nu} \delta^{(2)} g^{\mu\nu} + g^{\mu\nu} \delta^{(2)} R_{\mu\nu} \right. \right.$$
$$\left. \left. + \delta^{(2)} g^{\mu\nu} \partial_\mu \phi \partial_\nu \phi \right) + \left(g^{\mu\nu} R_{\mu\nu} + g^{\mu\nu} \partial_\mu \phi \partial_\nu \phi + V \right) \delta^{(2)} \sqrt{|g|} \right]. \tag{7.10}$$

Up to now the formalism is completely covariant, and all equations are valid quite independently of the specific form of the unperturbed metric. Since we are interested in the perturbation of a cosmological background, however, it is possible (and more convenient) to compute the perturbed action using directly the synchronous gauge, where $g_{00} = 1$. In particular, in order to illustrate the possible dynamical contribution of the internal moduli [4–7], we work with a simple example of higher-dimensional manifold whose spatial sections can be factorized as the product of two conformally flat spaces, with d and n dimensions, and scale factors $a(t)$ and $b(t)$, respectively. We thus set, in the synchronous gauge,

$$g_{00} = 1, \qquad g_{0i} = 0, \qquad g_{ij} = -a^2(t)\delta_{ij},$$
$$g_{ia} = 0, \qquad g_{ab} = -b^2(t)\delta_{ab}, \tag{7.11}$$

where $i, j = 1, \ldots, d$ are "external space" indices, $a, b = d+1, \ldots, d+n$ are "internal space" indices, and $\mu, \nu = 0, \ldots, d+n$. For this background, and in this gauge,

$$\Gamma_{0i}{}^{j} = H\delta_i^j, \qquad \Gamma_{ij}{}^{0} = -Hg_{ij},$$
$$\Gamma_{0a}{}^{b} = F\delta_a^b, \qquad \Gamma_{ab}{}^{0} = -Fg_{ab},$$
$$R_{ij} = -g_{ij}\left(\dot{H} + dH^2 + nHF\right), \tag{7.12}$$
$$R_{ab} = -g_{ab}\left(\dot{F} + nF^2 + dHF\right),$$
$$R_{00} = -d\dot{H} - n\dot{F} - dH^2 - nF^2,$$

where $H = \dot{a}/a$ and $F = \dot{b}/b$.

We concentrate our study on the (transverse and traceless) tensor fluctuations of the external d-dimensional space, which in the linear approximation are decoupled from the possible fluctuations of the internal metric background. Also, we assume that the translations along all internal dimensions are exact isometries not only of the unperturbed metric (7.11), but also of its perturbations. As a consequence, $h_{\mu\nu}$ only depends on the external space-time coordinates, and Eq. (7.3) (for our background, in the synchronous gauge) reduces to

$$h_{0\mu} = 0 = h_{a\mu}, \qquad h_{ij} = h_{ij}(t, \vec{x}),$$
$$g^{ij}h_{ij} = 0, \qquad \partial_j h_i^j = 0 \tag{7.13}$$

(see [8] for a more general discussion in which also the internal gradients of h_{ij} are non-vanishing). The quadratic, perturbed action can then be written explicitly as

$$\delta^{(2)}S = -\frac{1}{2\lambda_s^{d-1}} \int d^{d+1}x \, a^d b^n e^{-\phi} \left\{ -h^{ij}\delta^{(1)}R_{ij} - h_i^j h_j^i \left(\dot{H} + dH^2 + nHF\right) \right.$$
$$\left. + g^{\mu\nu}\delta^{(2)}R_{\mu\nu} - \frac{1}{4}h_i^j h_j^i \left[\dot{\phi}^2 + V - 2d\dot{H} - 2n\dot{F} - d(d+1)H^2 - n(n+1)F^2\right.\right.$$
$$\left.\left. - 2dnHF\right]\right\}, \tag{7.14}$$

where we have integrated over the trivial coordinates of the compact internal space, and we have omitted the dimensionless volume factor $\lambda_s^{-n} \int d^n x$.

We now need the perturbed expression of the Ricci tensor, at first and second order. From the definitions (7.7) we find, first of all, the first-order components of the perturbed connection:

$$\delta^{(1)}\Gamma_{ij}{}^0 = -\frac{1}{2}\dot{h}_{ij},$$

$$\delta^{(1)}\Gamma_{0i}{}^j = \frac{1}{2}g^{jk}\dot{h}_{ik} - Hh_i^j, \tag{7.15}$$

$$\delta^{(1)}\Gamma_{ij}{}^k = \frac{1}{2}\left(\partial_i h_j^k + \partial_j h_i^k - \partial^k h_{ij}\right)$$

(the dot denotes differentiation with respect to cosmic time, and $\partial^k \equiv g^{jk}\partial_j$). Using the identity

$$\nabla_0 h_i^j = \dot{h}_i^j = \nabla_0 \left(g^{jk}h_{ik}\right) = g^{jk}\left(\nabla_0 h_{ik}\right) = g^{jk}\dot{h}_{ik} - 2Hh_i^j, \tag{7.16}$$

the second term of Eq. (7.15) can be rewritten in the form

$$\delta^{(1)}\Gamma_{0i}{}^j = \frac{1}{2}\dot{h}_i^j, \tag{7.17}$$

more convenient for further applications. The first-order perturbation of the Ricci tensor,

$$\delta^{(1)}R_{ij} = \partial_\mu\left(\delta^{(1)}\Gamma_{ij}{}^\mu\right) - \partial_i\left(\delta^{(1)}\Gamma_{j\mu}{}^\mu\right) + \delta^{(1)}\Gamma_{\mu\rho}{}^\mu\Gamma_{ij}{}^\rho$$

$$+ \Gamma_{\mu\rho}{}^\mu\delta^{(1)}\Gamma_{ij}{}^\rho - \delta^{(1)}\Gamma_{i\rho}{}^\mu\Gamma_{\mu j}{}^\rho - \Gamma_{i\rho}{}^\mu\delta^{(1)}\Gamma_{\mu j}{}^\rho, \tag{7.18}$$

then becomes

$$\delta^{(1)}R_{ij} = -\frac{1}{2}\ddot{h}_{ij} + \frac{1}{2}\frac{\nabla^2}{a^2}h_{ij} - \frac{1}{2}\dot{h}_{ij}[(d-2)H + nF] + \frac{H}{2}\left(g_{kj}\dot{h}_i^k + g_{ik}\dot{h}_j^k\right), \tag{7.19}$$

where ∇^2 is the flat-space Laplace operator, $\nabla^2 \equiv \delta^{ij}\partial_i\partial_j$.

The final perturbation equations will be simpler if we use, as gravitational variables, the mixed components $h_i^j \equiv g^{jk}h_{ik}$ of the fluctuation tensor. Let us compute for this purpose the covariant time-derivative of Eq. (7.16):

$$\nabla_0\nabla_0 h_i^j = \nabla_0 \dot{h}_i^j = \ddot{h}_i^j = g^{jk}\ddot{h}_{ik} - (2\dot{H} + 4H^2)h_i^j - 4H\dot{h}_i^j. \tag{7.20}$$

The inversion of Eqs. (7.16) and (7.20) leads to the identities

$$\dot{h}_{ij} = g_{jk}\left(\dot{h}_i^k + 2Hh_i^k\right),$$

$$\ddot{h}_{ij} = g_{jk}\left(\ddot{h}_i^k + 4H\dot{h}_i^k + 2\dot{H}h_i^k + 4H^2 h_i^k\right), \tag{7.21}$$

which can be used to eliminate $\ddot{h}_{ij}$ and $\dot{h}_{ij}$ in Eq. (7.19), and to obtain

$$\delta^{(1)}R_{ij} = -\frac{1}{2}g_{jk}\left[\ddot{h}_i^k - \frac{\nabla^2}{a^2}h_i^k + (dH+nF)\dot{h}_i^k + (2\dot{H}+2dH^2+2nHF)h_i^k\right].$$

$$(7.22)$$

This result can be used to compute the mixed form of the perturbed Ricci tensor, $\delta^{(1)}R_i{}^j$, and to check that it reduces, as usual, to the covariant d'Alembert form for the higher-dimensional background (7.11):

$$\delta^{(1)}R_i{}^j = g^{jk}\delta^{(1)}R_{ik} - h^{jk}R_{ik} = g^{jk}\delta^{(1)}R_{ik} + \left(\dot{H}+dH^2+nHF\right)h_i^j$$

$$= -\frac{1}{2}\left[\ddot{h}_i^j + (dH+nF)\dot{h}_i^j - \frac{\nabla^2}{a^2}h_i^j\right]$$

$$\equiv -\frac{1}{2}\left(\nabla_\mu\nabla^\mu\right)h_i^j. \qquad (7.23)$$

Let us now consider second-order perturbations. From Eqs. (7.8), (7.13) and (7.16), we first compute the perturbed connection:

$$\delta^{(2)}\Gamma_{0i}{}^j = -\frac{1}{2}\dot{h}_i^k h_k^j,$$

$$(7.24)$$

$$\delta^{(2)}\Gamma_{ij}{}^k = -\frac{1}{2}h^{kl}\left(\partial_i h_{jl} + \partial_j h_{il} - \partial_l h_{ij}\right).$$

The perturbed Ricci tensor,

$$\delta^{(2)}R_{\nu\alpha} = \partial_\mu\left(\delta^{(2)}\Gamma_{\nu\alpha}{}^\mu\right) - \partial_\nu\left(\delta^{(2)}\Gamma_{\mu\alpha}{}^\mu\right) + \delta^{(1)}\Gamma_{\mu\rho}{}^\mu\delta^{(1)}\Gamma_{\nu\alpha}{}^\rho - \delta^{(1)}\Gamma_{\nu\rho}{}^\mu\delta^{(1)}\Gamma_{\mu\alpha}{}^\rho$$

$$+ \delta^{(2)}\Gamma_{\mu\rho}{}^\mu\Gamma_{\nu\alpha}{}^\rho + \Gamma_{\mu\rho}{}^\mu\delta^{(2)}\Gamma_{\nu\alpha}{}^\rho - \delta^{(2)}\Gamma_{\nu\rho}{}^\mu\Gamma_{\mu\alpha}{}^\rho - \Gamma_{\nu\rho}{}^\mu\delta^{(2)}\Gamma_{\mu\alpha}{}^\rho,$$

$$(7.25)$$

using the previous expressions for Γ, $\delta^{(1)}\Gamma$ and $\delta^{(2)}\Gamma$, can be written explicitly as

$$\delta^{(2)}R_{00} = \frac{1}{2}\ddot{h}_i^j h_j^i + \frac{1}{4}\dot{h}_i^j \dot{h}_j^i + H\,\dot{h}_i^j h_j^i,$$

$$\delta^{(2)}R_{ij} = \frac{1}{2}\partial_l h_j^k \partial^l h_{ik} + \frac{1}{4}h_i^k \partial_i \partial_j h_k^l + \frac{H}{2}g_{ij}\dot{h}_i^k h_l^l + \frac{1}{4}g_{kl}\dot{h}_j^k h_i^l + \frac{1}{4}g_{jl}\dot{h}_i^k h_k^l, \quad (7.26)$$

$$\delta^{(2)}R_{ab} = \frac{F}{2}g_{ab}\,\dot{h}_l^k h_k^l$$

(we have used Eq. (7.21) for $\dot{h}_{ik}$, and we have neglected all terms that, after integration by parts, do not contribute to the perturbed action because of the gauge condition $\partial_j h_i^j$).

We now insert the Ricci perturbations in the action (7.14), and sum up all contributions. Adopting a matrix notation for h_i^j we write

$$h_i^j h_j^i = \text{Tr}(h^2), \qquad \dot{h}_i^j h_j^i = \text{Tr}(\dot{h}h) \tag{7.27}$$

(and so on), and we arrive at the following quadratic action:

$$\delta^{(2)}S = -\frac{1}{2\lambda_s^{d-1}} \int d^{d+1}x \, a^d b^n e^{-\phi} \text{Tr}\left\{ \ddot{h}h - \frac{1}{4}h\frac{\nabla^2}{a^2}h + \frac{3}{4}\dot{h}^2 + \dot{h}h\left[(d+1)H + nF\right] \right.$$
$$\left. + \frac{h^2}{2}\left[d\dot{H} + n\dot{F} + \frac{d}{2}(d+1)H^2 + \frac{n}{2}(n+1)F^2 + dnHF - \frac{\dot{\phi}^2}{2} - \frac{V}{2} \right] \right\}, \tag{7.28}$$

where the terms containing $\ddot{h}h$ and $\dot{h}h$ can be integrated by parts, using the identity

$$a^d b^n e^{-\phi} \text{Tr}\left[\ddot{h}h + \dot{h}h(dH + H + nF) \right] = \frac{d}{dt}\left[a^d b^n e^{-\phi} \text{Tr}\left(\dot{h}h + \frac{H}{2}h^2 + \frac{\dot{\phi}}{2}h^2 \right) \right]$$
$$- a^d b^n e^{-\phi} \text{Tr}\left\{ \dot{h}^2 + \frac{1}{2}h^2[\dot{H} + dH^2 + nHF + \ddot{\phi} - \dot{\phi}^2 + (d-1)H\dot{\phi} + nF\dot{\phi}] \right\}. \tag{7.29}$$

We obtain (modulo a total derivative)

$$\delta^{(2)}S = \frac{1}{2\lambda_s^{d-1}} \int d^{d+1}x \, a^d b^n e^{-\phi} \text{Tr}\left\{ \frac{1}{4}\dot{h}^2 + \frac{1}{4}h\frac{\nabla^2}{a^2}h + \frac{h^2}{2}\left[\ddot{\phi} - \frac{\dot{\phi}^2}{2} + \frac{V}{2} - (d-1)\dot{H} \right.\right.$$
$$\left.\left. - n\dot{F} + (d-1)H\dot{\phi} + nF\dot{\phi} - \frac{d}{2}(d-1)H^2 - \frac{n}{2}(n+1)F^2 - (d-1)nHF \right] \right\}. \tag{7.30}$$

A last simplification is due to the vanishing of the coefficient of the quadratic "mass term" $h^2/2$, thanks to the unperturbed gravitational equations for the homogeneous background (7.11) (see in particular the spatial equation, obtained by varying with respect to β the action (6.16) without α' corrections and with the addition of a potential $V(\phi)$). This leads us to the final result

$$\delta^{(2)}S = \frac{1}{2\lambda_s^{d-1}}\frac{1}{4} \int d^d x \, dt \, a^d b^n e^{-\phi} \left(\dot{h}_i^j \dot{h}_j^i + h_i^j \frac{\nabla^2}{a^2}h_j^i \right), \tag{7.31}$$

which, using conformal time $(d\eta = dt/a)$, can also be rewritten

$$\delta^{(2)}S = \frac{1}{2\lambda_s^{d-1}}\frac{1}{4} \int d^d x \, d\eta \, a^{d-1} b^n e^{-\phi} \left(h_i'^j h_j'^i + h_i^j \nabla^2 h_j^i \right), \tag{7.32}$$

where $h' = a\dot{h}$, and the prime denotes differentiation with respect to η.

The variation of this action with respect to h provides the linearized equation for the tensor perturbations in the higher-dimensional background (7.11), and in the string frame. Such an equation generalizes the results first obtained and studied in [9, 10, 11], as it includes contributions from the geometry of the internal dimensions (the "modular" dynamics described by $b(t)$), and from the dilaton. In the conformal-time gauge, in particular, the variation of the action (7.32) leads to

$$h_i''^j + \left[(d-1)\frac{a'}{a} + n\frac{b'}{b} - \phi' \right] h_i'^j - \nabla^2 h_i^j = 0. \tag{7.33}$$

In cosmic time, from the action (7.31),

$$\ddot{h}_i^j + (dH + nF - \dot{\phi})\dot{h}_i^j - \frac{\nabla^2}{a^2} h_i^j = 0. \tag{7.34}$$

In a more general, fully covariant form, the perturbation equation can be expressed as

$$\frac{1}{\sqrt{|g|}} \partial_\mu \left(\sqrt{|g|} \, e^{-\phi} g^{\mu\nu} \partial_\nu \right) h_i^j = 0. \tag{7.35}$$

The components of h_i^j are decoupled in the linear approximation, and satisfy an evolution equation which in the Einstein frame (where the factor $\exp(-\phi)$ disappears) is exactly the same equation describing the evolution of a free, minimally coupled scalar field.

It should be stressed that the above equations are obtained from the lowest-order string effective action, and are valid in the low-energy limit. When the curvature of the background is high in string units ($\lambda_s H \sim 1$) these equations should be improved, possibly including all relevant higher-derivative contributions. Such contributions can be in principle determined by perturbing the full action containing the α' corrections (see Appendix A7 for an explicit computation to first order in α' [12]).

It is also worth noticing that the equations we have obtained do not contain explicit contributions either from the scalar potential V, or from other matter fields (possibly present in the action, and minimally coupled to the metric). Both the potential and the matter fields, however, do indirectly affect the evolution of h_i^j through their contributions to the background solutions for $a(t)$, $b(t)$ and $\phi(t)$.

7.1.1 Frame independence

Up to now all computations have been performed in the string frame, where the gravi-dilaton action has the form (7.2). In other frames the functional dependences of the action on the dilaton may be different, with a resulting different equation for the evolution of tensor perturbations. The dynamical properties of tensor perturbations (and, in particular, their spectral energy distribution) are however "*frame independent*", at least in the case in which the metrics of the various frames differ by a conformal rescaling which is only dilaton dependent. This important property can be illustrated by various simple arguments.

One may consider, for instance, a generic dilaton-dependent metric rescaling,

$$g_{\mu\nu} = \tilde{g}_{\mu\nu} e^{\psi(\phi)}, \qquad g^{\mu\nu} = \tilde{g}^{\mu\nu} e^{-\psi(\phi)}, \tag{7.36}$$

and compute the transformed metric perturbations, with unperturbed dilaton. One obtains

$$\delta\phi = 0, \qquad \delta g_{\mu\nu} \equiv h_{\mu\nu} = \delta\tilde{g}_{\mu\nu} e^{\psi} \equiv \tilde{h}_{\mu\nu} e^{\psi}, \tag{7.37}$$

so that

$$h_\mu^\nu \equiv g^{\nu\alpha} h_{\mu\alpha} = \tilde{g}^{\nu\alpha} e^{-\psi} \tilde{h}_{\mu\alpha} e^{\psi} \equiv \tilde{h}_\mu^\nu, \tag{7.38}$$

which shows that the tensor perturbation variable, in mixed form, is the same in both frames.

One can also check that the evolution equations for h and $\tilde{h}$ are formally different in the two frames, and the background solutions around which the expansion is performed are different, but these two differences exactly compensate each other to give the same effective equation for h and $\tilde{h}$.

A simple example of this effect is obtained by considering the E-frame metric $\tilde{g}_{\mu\nu}$, related to the S-frame metric $g_{\mu\nu}$ by the rescaling

$$g_{\mu\nu} = \left(\frac{\lambda_s}{\lambda_P}\right)^2 \tilde{g}_{\mu\nu} e^{2\phi/(d+n-1)}, \qquad \phi = \tilde{\phi}\left[\lambda_P^{d+n-1}(d+n-1)\right]^{1/2}, \tag{7.39}$$

appropriate to a $(d+n+1)$-dimensional manifold. The action (7.2) becomes, in Planck units,

$$S = -\int d^{d+n+1}x \sqrt{|\tilde{g}|} \left[\frac{\tilde{R}}{2\lambda_P^{d+n-1}} - \frac{1}{2}\left(\tilde{\nabla}\tilde{\phi}\right)^2 + \tilde{V}(\tilde{\phi})\right] \tag{7.40}$$

(see Section 2.2). Expanding the metric $\tilde{g}_{\mu\nu}$ around a background factorized as in Eq. (7.11), with scale factors $\tilde{a}$ and $\tilde{b}$, and following the same procedure as before, we obtain the perturbed action

$$\delta^{(2)} S = \frac{1}{2\lambda_P^{d-1}} \frac{1}{4} \int d^d x \, d\eta \, \tilde{a}^{d-1} \tilde{b}^n \left(\tilde{h}_i^{\prime j} \tilde{h}_j^{\prime i} + \tilde{h}_i^j \nabla^2 \tilde{h}_j^i\right). \tag{7.41}$$

We have used the conformal time coordinate, which is the same in both frames (see Eq. (4.106)). The corresponding E-frame perturbation equation

$$\widetilde{h}_i^{\prime\prime j} + \left[(d-1)\frac{\widetilde{a}'}{\widetilde{a}} + n\frac{\widetilde{b}'}{\widetilde{b}} \right] \widetilde{h}_i^{\prime j} - \nabla^2 \widetilde{h}_i^j = 0, \tag{7.42}$$

is thus a "pure" (covariant) d'Alembert equation for the metric $\widetilde{g}_{\mu\nu}$, different from Eq. (7.33). According to the transformation (7.39), however,

$$\widetilde{a} = a\left(\frac{\lambda_P}{\lambda_s}\right) e^{-\phi/(d+n-1)}, \qquad \widetilde{b} = b\left(\frac{\lambda_P}{\lambda_s}\right) e^{-\phi/(d+n-1)}, \tag{7.43}$$

from which

$$(d-1)\frac{\widetilde{a}'}{\widetilde{a}} + n\frac{\widetilde{b}'}{\widetilde{b}} = (d-1)\frac{a'}{a} + n\frac{b'}{b} - \phi', \tag{7.44}$$

so that the S-frame and E-frame equations for h and $\widetilde{h}$ actually coincide, in agreement with the identity (7.38).

7.1.2 Canonical normalization

In the last part of this section we show how to define, starting from the perturbed action, the canonical variable associated with the dynamics of tensor perturbations. Such a variable is a necessary ingredient for the normalization of the initial tensor spectrum, and for studying the evolution of the quantum fluctuations of the vacuum.

Let us come back to the action (7.32), and set $h_i^j = h_A(\epsilon^A)_i^j$, where ϵ^A is the polarization tensor representing a given polarization state h_A. The sum over A runs over all possible independent polarizations, which are in general $(d+1)$ $(d-2)/2$ for transverse and traceless tensor fluctuations, defined on a $D = (d+1)$-dimensional, spatially flat manifold. Tensor fluctuations are represented by a D-dimensional symmetric matrix, which has in general $D(D+1)/2$ independent components. On these components we can always impose D gauge conditions (choosing, for instance, the harmonic gauge [2]) and, in addition, we can eliminate other D components through an appropriate coordinate transformation, using the reparametrization invariance of the gravitational theory. We are thus left with

$$\frac{1}{2}D(D+1) - 2D = \frac{1}{2}D(D-3) \equiv \frac{1}{2}(d+1)(d-2) \tag{7.45}$$

components, corresponding to the number of independent polarization states (in $d = 3$, for instance, we have the two well-known "*plus*" and "*cross*" polarizations

[13], h_+ and $h_\times$). By exploiting the properties of the spin-two polarization tensor, $\text{Tr}(\epsilon^A \epsilon^B) = 2\delta^{AB}$, one obtains $h_i^j h_j^i = 2\sum_A h_A^2$. For each polarization mode $h_A(t, x_i)$ the action (7.32) can thus be rewritten as

$$\delta^{(2)} S = \frac{1}{2} \int d^d x \, d\eta \, z^2(\eta) \left(h'^2 + h \nabla^2 h \right) \tag{7.46}$$

(we have omitted, for simplicity, the polarization index), where

$$z(\eta) = \frac{M_s^{(d-1)/2}}{\sqrt{2}} a^{(d-1)/2} b^{n/2} e^{-\phi/2}. \tag{7.47}$$

The above action describes the dynamics of a scalar variable h, non-minimally coupled to a time-dependent background field $z(\eta)$ (also called the "pump field"), represented in this case by the dilaton and by the external and internal scale factors (the so-called "moduli" fields). The canonical variable u, which diagonalizes the kinetic part of the action and describes, asymptotically, a freely oscillating field, can now be defined as

$$u = zh. \tag{7.48}$$

This variable has the correct dimensions $[u] = [M^{(d-1)/2}]$ appropriate to a scalar field in a D-dimensional manifold; the perturbed action $\delta^{(2)} S$, written in terms of u, and after integration by parts, takes the canonical (diagonalized) form

$$\delta^{(2)} S = \frac{1}{2} \int d^d x \, d\eta \left(u'^2 + u \nabla^2 u + \frac{z''}{z} u^2 \right); \tag{7.49}$$

the variation with respect to u eventually leads to the canonical (Schrödinger-like) evolution equation

$$u'' - \left[\nabla^2 + U(\eta) \right] u = 0, \qquad U = \frac{z''}{z}. \tag{7.50}$$

Note that, if we write z in terms of the E-frame variables $\tilde{a}, \tilde{b}$ of Eq. (7.43), we obtain

$$z = \frac{M_P^{(d-1)/2}}{\sqrt{2}} \tilde{a}^{(d-1)/2} \tilde{b}^{n/2} \equiv \tilde{z}, \tag{7.51}$$

where we have used (7.43) *without* rescaling the internal modulus $\tilde{b}$ with the factor λ_P/λ_s, because such a factor has already been absorbed into the internal volume factor $\lambda_P^{-n} \int d^n x$ (not explicitly written in the E-frame action (7.41)). Since $\tilde{z}(\eta)$ exactly represents the E-frame pump field (see Eq. (7.41)), it follows that the canonical transformation (7.48) also diagonalizes the E-frame action. In other words, the canonical action and the evolution equation are frame independent.

We can notice, at this point, that the effective action (7.49) is formally the same as the action for a free scalar field in Minkowski space, with time-dependent "mass term" $-z''/z$, and effective Lagrangian

$$\mathcal{L} = \frac{1}{2}\left[u'^2 - (\nabla_i u)^2 + \left(\frac{z''}{z}\right)u^2\right]. \tag{7.52}$$

This field can be quantized starting from the definition of the usual momentum density $\mathcal{P}$, canonically conjugate to u,

$$\mathcal{P} = \frac{\partial \mathcal{L}}{\partial u'} = u', \tag{7.53}$$

and imposing (equal-time) canonical commutation relations

$$\left[u(x_i, \eta), \mathcal{P}(x_i', \eta)\right] = i\delta^d(x - x') \tag{7.54}$$

on the $\eta = $ const hypersurface. The classical variable u is then promoted to a field operator, and can be expanded over a complete set of solutions of the classical equation (7.50). Such an equation can be separated using the Fourier modes ψ_k satisfying the eigenvalue equation $\nabla^2 \psi_k = -k^2 \psi_k$, so that the field u can be expanded in plane waves as follows:

$$u(x_i, \eta) = \int \frac{\mathrm{d}^d k}{(2\pi)^{d/2}}\left[a_k u_k(\eta)\mathrm{e}^{ik_i x^i} + a_k^\dagger u_k^*(\eta)\mathrm{e}^{-ik_i x^i}\right]. \tag{7.55}$$

The modes $u_k(\eta)$ satisfy the eigenvalue equation

$$u_k'' + \left(k^2 - \frac{z''}{z}\right)u_k = 0, \tag{7.56}$$

and are chosen to be positive frequency modes with respect to η, i.e. $iu_k' = \alpha_k u_k$, with $\alpha_k > 0$, on a given (initial) hypersurface $\eta = $ const. Notice that we have already imposed on the Fourier expansion (7.55) the condition following from the reality of the classical solution, which now becomes a hermiticity condition $(u = u^\dagger)$ for the corresponding quantum operator.

Using the mode expansion, the commutation relation (7.54) can be written explicitly as

$$\frac{1}{(2\pi)^d}\int \mathrm{d}^d k\, \mathrm{d}^d k'\left\{u_k u_{k'}'[a_k, a_{k'}]\mathrm{e}^{i(kx+k'x')} + u_k u_{k'}'^*[a_k, a_{k'}^\dagger]\mathrm{e}^{i(kx-k'x')}\right.$$

$$\left. + u_k^* u_{k'}'[a_k^\dagger, a_{k'}]\mathrm{e}^{-i(kx-k'x')} + u_k^* u_{k'}'^*[a_k^\dagger, a_{k'}^\dagger]\mathrm{e}^{-i(kx+k'x')}\right\} = i\delta^d(x - x').$$

$$\tag{7.57}$$

It is convenient, on the other hand, to normalize the modes u_k in such a way that the commutation relations for a and $a^\dagger$ keep the usual canonical form,

$$[a_k, a_{k'}] = 0 = [a_k^\dagger, a_{k'}^\dagger], \qquad [a_k, a_{k'}^\dagger] = \delta^d(k - k'), \qquad (7.58)$$

so that we can apply the usual interpretation of a_k and $a_k^\dagger$ as annihilation and creation operators. By integrating over k, k', and using the integral representation of the delta distribution, one then obtains that Eqs. (7.57) and (7.58) are compatible provided the modes u_k satisfy the normalization condition

$$u_k u_k'^* - u_k' u_k^* = \mathrm{i}. \qquad (7.59)$$

It can be easily checked, in particular, that this condition is exactly equivalent to the orthonormality relation for the set of modes $\bar{u}_k(x, \eta) = u_k(\eta) e^{\mathrm{i}kx}/(2\pi)^{d/2}$ (used for the expansion of the u operator), where the orthonormality is referred to the so-called Klein–Gordon scalar product $\langle \bar{u}_k | \bar{u}_{k'} \rangle$, defined as

$$\langle \bar{u}_k | \bar{u}_{k'} \rangle \equiv -\mathrm{i} \int \mathrm{d}^d x \left(\bar{u}_k \bar{u}_{k'}'^* - \bar{u}_k' \bar{u}_{k'}^* \right) = \delta^d(k - k'). \qquad (7.60)$$

The normalization condition (7.59) can now be applied to fix the initial amplitude of the quantum fluctuations of the metric in a typical inflationary background where, as we shall see in the next section, the effective potential $z''/z \to 0$ as $\eta \to -\infty$. In such an initial asymptotic regime the canonical equation (7.56) reduces to the free-field equation,

$$u_k'' + k^2 u_k = 0, \qquad \eta \to -\infty, \qquad (7.61)$$

with oscillating solutions. Imposing on the solution to represent a positive frequency mode on the initial hypersurface at $\eta \to -\infty$, namely $u_k(\eta) = N_k e^{-\mathrm{i}k\eta}$, and using the canonical normalization (7.59), one obtains $2k|N|^2 = 1$, i.e.

$$u_k = \frac{e^{-\mathrm{i}k\eta}}{\sqrt{2k}}, \qquad \eta \to -\infty. \qquad (7.62)$$

The initial amplitude may contain, in general, an arbitrary phase factor, $u_k \to u_k e^{\mathrm{i}\alpha_k}$ (which we have omitted here for simplicity), possibly associated with random (or chaotic) initial conditions. Quite irrespective of such an initial phase, the important property of the normalized, positive frequency solution (7.62) is that it identifies the initial vacuum state of the tensor fluctuations as the incoming state of lowest energy associated with the action (7.49). In a time-dependent background, however, the notion of positive frequency *is not* time independent, because of the evolution of the effective mass term $-z''/z$. Hence, the definition of the vacuum state is neither unique nor globally defined on the whole cosmological manifold [14]. This implies, in particular, that a state which is initially "empty" of particles (i.e. of *quanta* of the given fluctuations) may become non-empty later on, at subsequent epochs. This effect is the basis of the mechanism of cosmological particle production, which we shall discuss in the next section.

It may be useful to recall, finally, that the normalization (7.62) of the canonical variable u_k automatically fixes the normalization of the Fourier component of the metric fluctuation variable, $h_k = u_k/z$. The equation for h_k can be directly obtained from the canonical equation (7.56) and can be written, in terms of z, as

$$h_k'' + 2\frac{z'}{z}h_k' + k^2 h_k = 0. \tag{7.63}$$

For the phenomenological applications in this book we use the following convenient definition of Fourier transform,

$$h(x_i, \eta) = \frac{\sqrt{V}}{(2\pi)^d} \int d^d k \, h_k(\eta) e^{ik\cdot x}, \tag{7.64}$$

corresponding to the continuum limit of the discrete Fourier expansion over a comoving (d-dimensional) normalization volume V:

$$h(x_i, \eta) = \frac{1}{\sqrt{V}} \sum_k h_k(\eta) e^{ik\cdot x}. \tag{7.65}$$

With such a definition, the variable h_k has dimensions $[h_k] = [M^{-d/2}]$, so that both $h(x)$ and $k^{d/2}h_k$ are dimensionless quantities. The momentum variable, canonically conjugate to h according to the action (7.46), is given by

$$\Pi_k = z^2 h_k', \tag{7.66}$$

and satisfies the equation

$$\Pi_k'' - 2\frac{z'}{z}\Pi_k' + k^2 \Pi_k = 0. \tag{7.67}$$

The coupled analysis of the equations of motion of h and Π turns out to be very useful for discussing the amplification of tensor perturbations and the computation of their final spectrum [15], as we shall see in the next section.

7.2 Parametric amplification and spectral distribution

In the previous section we have shown how to derive the equation governing the evolution of tensor perturbations on a cosmological background, and how to impose the canonical normalization on their initial spectrum. Here we discuss the possible amplification of such perturbations, which is typically (but not necessarily) triggered by the accelerated evolution in time of the "pump field" $z = \tilde{z}$. In a string cosmology context z may contain, in general, contributions from the external space-time geometry, from the internal moduli fields, and from the dilaton: as a consequence, tensor perturbations may be amplified by the inflationary expansion of the external space [11, 16–20], and/or by the accelerated contraction of the internal dimensions [4, 5, 6], and/or by the accelerated variation in time of the coupling constants [7].

In order to illustrate the amplification mechanism it is convenient to parametrize the evolution of z in the negative range of the conformal time coordinate, with a power α, as follows:

$$z(\eta) = \frac{M_P^{(d-1)/2}}{\sqrt{2}} \left(-\frac{\eta}{\eta_1}\right)^{\alpha}, \qquad -\infty < \eta < 0. \tag{7.68}$$

Here η_1 is an (appropriate) reference time scale, and the mass factor is required for the correct normalization of the canonical variable (see Eq. (7.47)). In the standard cosmological picture in which there are no contributions from internal dimensions, and the external space is three-dimensional ($n = 0$, $d = 3$), the pump field is proportional to the E-frame scale factor, namely $z \sim \tilde{a} \sim (-\eta)^{\alpha}$ (see Eq. (7.51)). As α is varying from $-\infty$ to $+\infty$, one then reproduces all possible types of inflationary backgrounds (see Chapter 5, and in particular Fig. 5.1): power-inflation ($\alpha < -1$), de Sitter ($\alpha = -1$), super-inflation ($-1 < \alpha < 0$) and accelerated contraction ($\alpha > 0$). In all these cases the evolution is accelerated, in the sense that

$$\text{sign}\left\{\frac{d\tilde{a}}{d\tilde{t}}\right\} = \text{sign}\left\{\frac{d^2\tilde{a}}{d\tilde{t}^2}\right\}. \tag{7.69}$$

Given the power-law behavior (7.68), the effective potential takes the form $U = z''/z = \alpha(\alpha-1)/\eta^2$, and the canonical perturbation equation (7.56) becomes an exact Bessel equation,

$$u_k'' + \left[k^2 - \frac{\alpha(\alpha-1)}{\eta^2}\right] u_k = 0, \tag{7.70}$$

valid in the negative-time range $\eta < 0$. At early enough times ($\eta \to -\infty$) the potential $U(\eta)$ goes to zero (as anticipated in the previous section), and the kinetic energy tends to become dominant with respect to the potential energy. A given massless mode of comoving frequency k enters the "kinetic-dominated" regime for values of η such that $|k\eta| \gg 1$. Inside such a regime the proper frequency $\omega = k/a$ is much greater than the Hubble scale,

$$\omega = \frac{k}{a} \gg |a\eta|^{-1} \sim |H|, \tag{7.71}$$

or – equivalently – the proper wavelength $\lambda = a/k$ is much smaller than the Hubble horizon H^{-1}, i.e. $|\lambda H| \ll 1$. One says that, in the limit $\eta \to -\infty$, all modes tend to be "*inside the horizon*". In this limit the canonical variable u_k is freely oscillating with constant amplitude, according to Eqs. (7.61) and (7.62), and the metric fluctuation $h_k(\eta)$ is also oscillating, but with an "adiabatically" evolving amplitude,

$$h_k(\eta) = \frac{e^{-ik\eta}}{z(\eta)\sqrt{2k}}. \qquad (7.72)$$

The associated canonical momentum (7.66), in this limit, is given by

$$\Pi_k(\eta) = -i\sqrt{\frac{k}{2}}z(\eta)e^{-ik\eta}. \qquad (7.73)$$

Consider now the opposite limit $\eta \to 0_-$, corresponding to the asymptotic future of the cosmological phase associated with the pump field (7.68). At late enough times all modes tend to be "*outside the horizon*", namely tend to be in a configuration in which the kinetic energy is negligible with respect to the potential energy, $|k\eta| \ll 1$, the proper frequency is much smaller than the Hubble scale, and the proper wavelength is much larger than the Hubble horizon

$$\omega = \frac{k}{a} \ll |a\eta|^{-1} \sim |H|, \qquad |\lambda H| \gg 1. \qquad (7.74)$$

In this regime the tensor perturbation equation (7.63) can be solved by the following asymptotic expansion [3, 15, 21]:

$$h_k(\eta) = A_k\left[1 - k^2\int_{\eta_{ex}}^{\eta} d\eta' z^{-2}(\eta')\int_{\eta_{ex}}^{\eta'} d\eta'' z^2(\eta'') + \cdots\right]$$
$$+ B_k\left[k\int_{\eta_{ex}}^{\eta} d\eta' z^{-2}(\eta') + \cdots\right], \qquad (7.75)$$

valid for the mode k and for times $\eta_{ex} < \eta < 0$, where $|\eta_{ex}| = k^{-1}$ is the time scale of horizon crossing of the given mode. For the conjugate momentum Π_k, satisfying Eq. (7.67), we can write a similar expansion, using however an "inverted" (or "dual") pump field, $z \to z^{-1}$:

$$\Pi_k(\eta) = z^2 h'_k = kB_k\left[1 - k^2\int_{\eta_{ex}}^{\eta} d\eta' z^2(\eta')\int_{\eta_{ex}}^{\eta'} d\eta'' z^{-2}(\eta'') + \cdots\right]$$
$$- kA_k\left[k\int_{\eta_{ex}}^{\eta} d\eta' z^2(\eta') + \cdots\right]. \qquad (7.76)$$

The coefficients A_k and B_k are integration constants, to be determined through the initial normalization at $\eta = \eta_{ex}$:

$$A_k = h_k(\eta_{ex}), \qquad B_k = k^{-1}\Pi_k(\eta_{ex}). \qquad (7.77)$$

With the power-law pump field of Eq. (7.68) the integration can be performed exactly and we obtain, to leading order as $\eta \to 0_-$,

$$
h_k(\eta) = \begin{cases} A_k + \frac{2B_k}{M_{\mathrm{P}}^{d-1}} \frac{k\eta_1}{(1-2\alpha)} \left(\left| \frac{\eta_{\mathrm{ex}}}{\eta_1} \right|^{1-2\alpha} - \left| \frac{\eta}{\eta_1} \right|^{1-2\alpha} \right) + \cdots, & \alpha \neq 1/2, \\[2ex] A_k - \frac{2B_k}{M_{\mathrm{P}}^{d-1}} k\eta_1 \ln \left| \frac{\eta}{\eta_{\mathrm{ex}}} \right| + \cdots, & \alpha = 1/2, \end{cases}
$$

$$
\Pi_k(\eta) = \begin{cases} k \left[B_k - \frac{A_k M_{\mathrm{P}}^{d-1}}{2} \frac{k\eta_1}{(1+2\alpha)} \left(\left| \frac{\eta_{\mathrm{ex}}}{\eta_1} \right|^{1+2\alpha} - \left| \frac{\eta}{\eta_1} \right|^{1+2\alpha} \right) + \cdots \right], & \alpha \neq -1/2, \\[2ex] k \left(B_k + \frac{A_k M_{\mathrm{P}}^{d-1}}{2} k\eta_1 \ln \left| \frac{\eta}{\eta_{\mathrm{ex}}} \right| + \cdots \right), & \alpha = -1/2. \end{cases}
$$

$$(7.78)$$

It is important to notice that these asymptotic solutions are always characterized by a constant term and by a time-dependent one. This second term may be growing or decreasing as $\eta \to 0_-$, depending on the value of the power α, i.e. on the particular kinematic behavior of the background fields a, b, ϕ. Thus, at late enough times, tensor perturbations are either frozen – if the time-dependent term decays away – or growing – if the time-dependent term is growing.

In any case, the behavior in time of the amplitude is different from the adiabatic evolution (7.72), (7.73), typical of modes inside the horizon. Comparing with the adiabatic solutions, we find that for a growing pump field ($\alpha < 0$) the evolution outside the horizon always tends to enhance the amplitude, i.e.

$$
\frac{h_k(|k\eta| > 1)}{h_k(|k\eta| < 1)} \to 0, \qquad \eta \to 0_-. \tag{7.79}
$$

For a decreasing pump field ($\alpha > 0$), on the contrary, what is enhanced is the amplitude of the conjugate momentum, i.e.

$$
\frac{\Pi_k(|k\eta| > 1)}{\Pi_k(|k\eta| < 1)} \to 0, \qquad \eta \to 0_-. \tag{7.80}
$$

For $\alpha > 1$ both amplitudes are enhanced, and the background may become gravitationally unstable [22, 23], since the initial homogeneity may be completely destroyed by the growth of the quantum fluctuations of the metric.

Therefore, it is a long enough phase of "super-horizon" evolution (also called a phase of "stretching" outside the horizon) that is responsible for the cosmological amplification of the metric fluctuations. Such an amplification occurs, according to the canonical equation (7.70), whenever the fluctuations evolve from an initial adiabatic and oscillating regime to a final regime dominated by the "geometric" potential energy, $U \gg k^2$. In all cases, it is important to stress that the kinematical amplification of the amplitude always corresponds to a physical amplification of the energy density stored in the fluctuations.

For a concrete illustration of this point let us consider the Hamiltonian density H associated with the perturbed action (7.46). For each mode k, using the canonical momentum (7.66), we have

$$H_k = \frac{1}{2} \left(z^{-2} |\Pi_k|^2 + k^2 z^2 |h_k|^2 \right). \tag{7.81}$$

The total Hamiltonian is obtained by integrating over all modes. In the initial, oscillating regime, described by the solutions (7.72) and (7.73), we find that $z|h|$ and $z^{-1}|\Pi|$ are time independent, so that the energy density stored in a given oscillating mode stays frozen, at a constant value fixed by its canonical normalization. In the final super-horizon regime, on the contrary, we must apply the asymptotic solutions (7.78), and we find that both terms in the Hamiltonian are in general time dependent:

$$z h_k = \left(\frac{M_P^{d-1}}{2} \right)^{1/2} \widehat{A}_k \left| \frac{\eta}{\eta_1} \right|^{\alpha} - \left(\frac{2}{M_P^{d-1}} \right)^{1/2} B_k \frac{k\eta_1}{(1-2\alpha)} \left| \frac{\eta}{\eta_1} \right|^{1-\alpha},$$

$$z^{-1} \Pi_k = k \left[\left(\frac{2}{M_P^{d-1}} \right)^{1/2} \widehat{B}_k \left| \frac{\eta}{\eta_1} \right|^{-\alpha} \right. \tag{7.82}$$

$$\left. + \left(\frac{M_P^{d-1}}{2} \right)^{1/2} A_k \frac{k\eta_1}{(1+2\alpha)} \left| \frac{\eta}{\eta_1} \right|^{1+\alpha} \right],$$

where we have represented by $\widehat{A}_k$ and $\widehat{B}_k$ the appropriate combinations of constants containing the frozen part of the solution (7.78) (in the case $\alpha = \pm 1/2$ we have to include the logarithmic corrections that we are neglecting here is assuming $\alpha \neq \pm 1/2$).

With a simple analysis of the time evolution of h_k and Π_k as $\eta \to 0_-$ it can be easily checked that for $\alpha > 1/2$ the metric fluctuation h is growing, but the momentum variable Π is frozen, and its contribution $z^{-1}\Pi$ asymptotically dominates the Hamiltonian. For $\alpha < -1/2$, on the contrary, the momentum Π is growing, while the variable h is frozen, and its contribution zh dominates the asymptotic Hamiltonian. Finally, for $-1/2 < \alpha < 1/2$, both variables can dominate the Hamiltonian, but they are always represented by the frozen part of the asymptotic solution (7.78). Thus, quite irrespective of the value (and sign) of α, the energy density of the super-horizon fluctuations can be estimated by inserting into the Hamiltonian (7.81) the frozen part of the solutions for h_k and Π_k, namely [15]

$$H_k = \frac{k^2}{2} \left(z^2 \left| \widehat{A}_k \right|^2 + z^{-2} \left| \widehat{B}_k \right|^2 \right), \qquad \eta \to 0_-. \tag{7.83}$$

This quadratic form is always growing as $\eta \to 0_-$, and its growth represents, as already anticipated, the amplification of the energy density stored in each fluctuation mode. Such an energy growth may be associated either with the amplification of the metric amplitude h, if z is growing, or with the amplification of the momentum amplitude Π, if z is decreasing, as illustrated by the previous discussion.

7.2.1 Spectral amplitude

For later phenomenological applications it is now convenient to establish a precise connection between the spectral distribution of the amplified fluctuations and the kinematic parameters of the cosmological background. Let us introduce the so-called spectral amplitude, which can be defined starting from the two-point correlation function $\xi_h(r)$ which characterizes the fluctuations on a comoving scale of distances r:

$$\xi_h(\vec{r}) = \langle h(\vec{x}) h(\vec{x} + \vec{r}) \rangle. \tag{7.84}$$

The brackets here denote quantum expectation values if the perturbations are quantized, and are expanded into annihilation and creation operators (see the previous section); if we are working in the classical limit, instead, the brackets may denote an *ensemble* average represented by a spatial integral over some given volume V:

$$\xi_h(\vec{r}) = \frac{1}{V} \int d^d x \, h(\vec{x}) h(\vec{x} + \vec{r}). \tag{7.85}$$

In this last case, using the Fourier transform (7.64), and the reality condition $h_{\vec{k}} = h^*_{-\vec{k}}$ for the metric perturbation field, we obtain

$$\xi_h(\vec{r}) = \frac{1}{(2\pi)^d} \int d^d k \, |h_k|^2 e^{-i\vec{k}\cdot\vec{r}}. \tag{7.86}$$

The same result can be obtained, of course, by evaluating the expectation value for the quantum field h.

We now assume that the amplified fluctuations satisfy the so-called "isotropy condition", i.e. that $|h_k|$ is only a function of $|\vec{k}|$ (see for instance [24]). Using polar coordinates, and integrating over the angles of the $(d-1)$-dimensional unit sphere, we are led to

$$\xi_h(r) = \frac{4\pi^{d/2}}{(2\pi)^d \Gamma(d/2)} \int_0^\infty \frac{dk}{k} \frac{\sin kr}{kr} |\delta_h(k)|^2, \qquad |\delta_h(k)|^2 = k^d |h_k|^2, \tag{7.87}$$

where Γ is the Euler Gamma function (we are assuming that $|h_k|^2$ satisfies the appropriate cut-off at high and low frequency, so as to make the above integral convergent). The quantity $|\delta_h(k)|^2$ is called "spectral amplitude" (or "power spectrum"), and represents the typical (dimensionless) amplitude of a tensor fluctuation on the comoving length scale $r = k^{-1}$. Indeed, for a power-law spectrum, $|\delta_h|^2$ is proportional to the two-point correlation function evaluated at $r = k^{-1}$:

$$|\delta_h(k)|^2 = k^d |h_k|^2 \sim [\xi(r)]_{r=k^{-1}} . \tag{7.88}$$

Let us now compute the spectral amplitude outside the horizon, in the limit $\eta \to 0$, using the asymptotic solution (7.78). We have to fix, first of all, the integration constants A_k, B_k, by exploiting the continuity with the oscillating solutions at $\eta = \eta_{ex} = -k^{-1}$. If we are interested in the amplification of the quantum fluctuations of the vacuum we can adopt, in particular, the canonical normalization (7.72), (7.73), which leads to

$$h_k(\eta_{ex}) = \frac{e^{i\varphi_k}}{z_{ex}\sqrt{2k}} \equiv \frac{(k\eta_1)^\alpha}{\sqrt{kM^{d-1}}} e^{i\varphi_k}$$

$$\Pi_k(\eta_{ex}) = z_{ex}\sqrt{\frac{k}{2}} e^{i\theta_k} \equiv \frac{1}{2}\sqrt{kM^{d-1}}(k\eta_1)^{-\alpha} e^{i\theta_k}, \tag{7.89}$$

where φ_k and θ_k are arbitrary (constant) phase factors. Determining A_k, B_k according to Eqs. (7.77), and inserting their values into (7.78), we obtain the asymptotic normalized solution for $\eta_{ex} < \eta < 0$:

$$h_k(\eta) = \begin{cases} \dfrac{1}{\sqrt{kM^{d-1}}}\left[(k\eta_1)^\alpha \left(e^{i\varphi_k} + \dfrac{e^{i\theta_k}}{1-2\alpha}\right) - \dfrac{(k\eta_1)^{1-\alpha}}{1-2\alpha}\left|\dfrac{\eta}{\eta_1}\right|^{1-2\alpha} e^{i\theta_k}\right], & \alpha \neq 1/2, \\[4mm] \dfrac{(k\eta_1)^{1/2}}{\sqrt{kM^{d-1}}}\left[e^{i\varphi_k} - e^{i\theta_k}\ln|k\eta|\right], & \alpha = 1/2. \end{cases}$$

$$\tag{7.90}$$

When computing the final spectral amplitude, $|\delta_h|^2 = k^d |h_k|^2$, we may thus conveniently distinguish three cases.

(1) In the case $\alpha < 1/2$ the asymptotic amplitude of h_k is dominated by the constant term of the solution (7.90), and we obtain the spectral amplitude

$$|\delta_h(k)|^2 \sim \left(\frac{k_1}{M_P}\right)^{d-1}\left(\frac{k}{k_1}\right)^{d-1+2\alpha}, \tag{7.91}$$

where we have defined $k_1 \equiv \eta_1^{-1}$, and we have omitted an unimportant (α-dependent) constant coefficient of order one. Using the definitions (7.51) and (7.68), and evaluating the pump field z at $\eta = -\eta_1$, we find $\tilde{a}_1^{d-1}\tilde{b}_1^n = 1$. We may thus separately normalize at η_1 the two scale factors by setting

$$\tilde{a} = (-\eta/\eta_1)^\beta, \qquad \tilde{b} = (-\eta/\eta_1)^\gamma, \tag{7.92}$$

where, according to the parametrization (7.68),

$$(d-1)\beta + n\gamma = 2\alpha. \tag{7.93}$$

In this case $\tilde{a}_1 \equiv \tilde{a}(-\eta_1) = 1$, and the spectral amplitude of the mode k_1 can be directly referred to the E-frame curvature scale of the d-dimensional external space, evaluated at the time of horizon crossing, and defined by $H_1 \equiv k_1/\tilde{a}_1 = k_1$. The previous spectrum (7.91) may thus be rewritten as

$$|\delta_h(k)|^2 \sim \left(\frac{H_1}{M_P}\right)^{d-1} \left(\frac{k}{k_1}\right)^{d-1+2\alpha}, \tag{7.94}$$

or, in terms of the E-frame parameters β and γ,

$$|\delta_h(k)|^2 \sim \left(\frac{H_1}{M_P}\right)^{d-1} \left(\frac{k}{k_1}\right)^{(d-1)(1+\beta)+n\gamma}. \tag{7.95}$$

(2) In the case $\alpha > 1/2$ the asymptotic amplitude of h_k is dominated by the time-dependent term of the solution (7.90). Using the same arguments as before we obtain the spectrum

$$|\delta_h(k, \eta)|^2 \sim \left(\frac{H_1}{M_P}\right)^{d-1} \left(\frac{k}{k_1}\right)^{d+1-2\alpha} (k_1\eta)^{2(1-2\alpha)}, \tag{7.96}$$

or, in terms of the E-frame parameters β and γ:

$$|\delta_h(k, \eta)|^2 \sim \left(\frac{H_1}{M_P}\right)^{d-1} \left(\frac{k}{k_1}\right)^{d+1-(d-1)\beta-n\gamma} (k_1\eta)^{2(1-2\alpha)}. \tag{7.97}$$

It is important to note that in this case the spectral amplitude is time dependent, and tends to diverge as $\eta \to 0_-$, since the power $1-2\alpha$ is negative. The perturbative analysis remains valid provided the amplification phase is not extended much beyond the reference time $\eta_1 = 1/k_1$: indeed, the time-dependent factor present in δ_h remains smaller than one, as long as $|\eta/\eta_1| > 1$. Even so, the time dependence of the spectral amplitude is important, as it may change the final distributions of the modes when they re-enter the horizon, after the end of the accelerated epoch.

(3) Finally, in the limiting case $\alpha = 1/2$, there is only a logarithmic growth of the perturbations, and we obtain the spectral amplitude [21, 25]

$$|\delta_h(k, \eta)|^2 \sim \left(\frac{H_1}{M_P}\right)^{d-1} \left(\frac{k}{k_1}\right)^{d} (\ln|k\eta|)^2. \tag{7.98}$$

This particular case corresponds to the vacuum solutions of the tree-level string cosmology equations (see Eqs. (4.31)–(4.34) for the more general, fully aniso-tropic version). For a $(1+d+n)$-dimensional background, such solutions, written in conformal time, lead in fact to the pump field

$$z^2 \sim (-\eta)^{2\alpha} \sim a^{d-1} b^n e^{-\phi} \sim (-\eta), \tag{7.99}$$

corresponding to $\alpha = 1/2$ for any value of d and n, and for all background kinematics satisfying the conditions (4.33) and (4.34).

A few remarks on the spectral distributions (7.94), (7.96) and (7.98) are now in order. We should note, first of all, that for any realistic phenomenological application the model of the background should be completed by extending the cosmological solutions to the region $\eta > 0$, so as to possibly include all (accelerated and decelerated) phases up to the present epoch. The final distribution of the amplified perturbations, computed after the fluctuation modes have re-entered the horizon, is affected in general by all background transitions, and may be different from the spectrum we have just computed outside the horizon, after a single amplification. Nevertheless, there is a close connection between the frequency distribution of the spectrum and the kinematic behavior of the background fields, which clearly emerges even at the level of this simple example, and which we now illustrate in some detail.

We should consider the so-called "spectral index" n_T (not to be confused with the number of internal dimensions!), which characterizes the spectral distribution as a power of k, and which is defined by

$$n_T = \frac{d \ln |\delta_h(k)|^2}{d \ln k}. \tag{7.100}$$

From Eqs. (7.94), (7.96) and (7.98) we obtain, modulo logarithmic corrections,

$$n_T = \begin{cases} d - 1 + 2\alpha, & 2\alpha \leq 1, \\ d + 1 - 2\alpha, & 2\alpha \geq 1. \end{cases} \tag{7.101}$$

The spectrum is said to be "flat" if it is k-independent (i.e. if $n_T = 0$), growing (or "blue") if $n_T > 0$, and decreasing (or "red") if $n_T < 0$. As clearly shown by the previous equation, the spectrum that we have obtained is characterized by a "maximal" spectral index $n_T = d$, corresponding to a pump field with power $\alpha = 1/2$. In particular, the tensor spectrum is growing if $1 - d < 2\alpha < 1 + d$, decreasing if $2\alpha < 1 - d$ or $2\alpha > 1 + d$, and flat in the limiting case in which $2\alpha = 1 \pm d$ (see Fig. 7.1).

Let us concentrate on the case $2\alpha \leq 1$, which contains the range $\alpha < 0$ and thus includes the conventional scenarios where one neglects the contributions of the moduli fields: in this case, the fluctuations are only amplified by the inflationary expansion of the external scale factor $a \sim (-\eta)^\beta$, with $\beta = 2\alpha/(d-1)$. The corresponding spectrum is given by Eq. (7.95). With $n = 0$, in particular, we obtain

$$|\delta_h(k, \eta)|^2 \sim k^{(d-1)(1+\beta)}, \qquad \beta(d-1) \leq 1, \tag{7.102}$$

and we can immediately establish a close connection between the frequency behavior of this spectrum and the time behavior of the background curvature

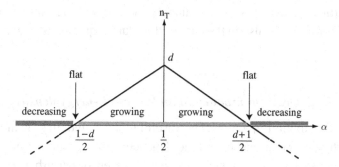

Figure 7.1 The spectral index n_T as a function of the pump field kinematics, specified by the power α of Eq. (7.68). The spectrum is flat for $\alpha = (1 \pm d)/2$, and the maximal power k^d (modulo log corrections) is obtained for $\alpha = 1/2$.

scale [26]. The spectrum is decreasing, flat or increasing depending on whether the E-frame scale factor describes power-inflation, de Sitter inflation or super-inflation, respectively:

$$
\begin{array}{llcl}
\text{power-inflation,} & \beta < -1 & \Rightarrow & \text{red spectrum,} \\
\text{de Sitter inflation,} & \beta = -1 & \Rightarrow & \text{flat spectrum,} \qquad (7.103) \\
\text{super-inflation,} & \beta > -1 & \Rightarrow & \text{blue spectrum.}
\end{array}
$$

A red spectrum is thus associated with a background with decreasing curvature, a flat spectrum with a background with constant curvature, and a blue spectrum with a background with growing curvature, as can be easily deduced from Table 5.1, which summarizes the main properties of the various kinematic classes of inflation.

It must be noted that such a direct connection between the spectrum and the curvature cannot be applied, in general, in the presence of dynamical internal dimensions, contributing with the factor $n\gamma \neq 0$ to the full spectral power. In that case the spectral behavior is strongly dependent on the number of internal and external dimensions [6], and it is possible, for instance, to obtain non-blue spectra even from a background whose overall curvature is growing.

Also, we cannot generically apply (7.103) if the pump field has a power $\alpha > 1/2$. Such a power can be easily obtained in a string cosmology context: we may recall here, as a typical example, the isotropic, d-dimensional background driven by perfect fluid sources, presented in Section 4.2, and described by the E-frame scale factor of Eq. (4.108). Setting to zero the dilatonic charge ($\gamma_0 = 0$) we obtain, in this case,

$$
2\alpha = \frac{2(1 - \gamma)}{1 - 2\gamma + 2d\gamma}. \qquad (7.104)
$$

Thus, in particular, $\alpha = 1$ in the radiation case $(\gamma = 1/d)$, and $\alpha = (d+1)/(d+3) > 1/2$ (for $d > 2$) also in the case of the "dual" equation of state, $\gamma = -1/d$.

7.2.2 Graviton production and spectral energy density

The spectral distribution δ_h represents a useful tool for studying the amplification of the quantum fluctuations in a time-dependent background. For many phenomenological applications, related in particular to tensor perturbations and to the study of the cosmic background of relic gravitational waves, another convenient variable is represented by the so-called "spectral energy density" $\Omega_g(k)$, which measures the energy density (in critical units) stored inside the fluctuations, per logarithmic interval of frequency:

$$\Omega_g(k) = \frac{\mathrm{d}(\rho/\rho_c)}{\mathrm{d}\ln k} = \frac{k}{\rho_c}\frac{\mathrm{d}\rho}{\mathrm{d}k}. \tag{7.105}$$

The present value of this spectral parameter, $\Omega_g(k, \eta_0)$, refers to all fluctuation modes which are inside our present horizon $(k\eta_0 \gg 1)$, and thus estimates the available energy density for producing signals in a gravitational detector, as a function of the parameters of the given inflationary scenario.

For a basic introduction to the computation of Ω_g we start from the canonical evolution equation (7.56). The effective potential $U = z''/z$, in a realistic cosmological scenario, should approach zero at the beginning of the primordial accelerated phase, $\eta \to -\infty$, and should be decreasing towards zero also in the opposite limit $\eta \to +\infty$, during the phase of standard decelerated evolution. This is certainly the case if we consider a standard, isotropic, $d = 3$ background, describing a Universe which in the E-frame evolves from inflation $(z \sim \tilde{a} \sim (-\eta)^\alpha, \eta < 0)$ to radiation-dominated $(z \sim \tilde{a} \sim \eta, \eta > 0)$ and matter-dominated $(z \sim \tilde{a} \sim \eta^2, \eta > 0)$ decelerated expansion. This implies that the global evolution of the tensor fluctuations across this type of background can be described as a process of potential scattering of the canonical variable u_k, which is freely oscillating in the initial configuration, interacts with an effective potential localized in a finite, intermediate region, and tends to be oscillating again in the asymptotic region $\eta \to +\infty$ (see Fig. 7.2).

There is, however, an important difference between such a cosmological evolution of the perturbations and an ordinary problem of quantum mechanical scattering of the Schrödinger wave function: in our case, the differential variable of the canonical equation (7.56) is a *time-like* parameter, not a space-like coordinate. As a consequence, the oscillation frequency corresponds to *energy*, not to

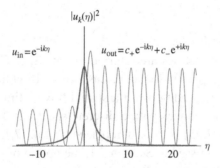

Figure 7.2 Parametric amplification of the canonical variable u_k in a cosmo-logical background characterized by the bell-like potential $U(\eta) = 2/(2+\eta^2)$ (bold curve), describing a smooth transition from de Sitter inflation, $a \sim (-\eta)^{-1}$, to matter-dominated evolution, $a \sim \eta^2$. The oscillating curve shows a numerical solution of the canonical equation (7.56) with $k = 1.1$, and initial conditions $u(\eta_i) = 0$, $u'(\eta_i) = -0.74$ at $\eta_i = -10$. Such an amplification can be described, in a semiclassical way, as an "anti-tunneling" effect [27].

momentum. So, even choosing the initial solution canonically normalized as a positive energy mode,

$$\eta \to -\infty, \qquad u_k^{\text{in}} = \frac{1}{\sqrt{2k}} e^{-ik\eta + i\phi_{\text{in}}} e^{ik \cdot x} \tag{7.106}$$

(ϕ_{in} is an arbitrary initial phase), we obtain a final asymptotic solution which is in general a superposition of *positive* and *negative* energy modes,

$$\eta \to +\infty, \qquad u_k^{\text{out}} = \frac{1}{\sqrt{2k}} \left[c_+(k) e^{-ik\eta} + c_-(k) e^{+ik\eta} \right] e^{ik \cdot x}. \tag{7.107}$$

It is well known, in a quantum field theory context, that such a frequency mixing describes a process of pair production from the vacuum (see for instance [14]), where the complex coefficients $c_\pm(k)$ parametrize the so-called "Bogoliubov transformation" connecting the annihilation and creation operators of the $|\text{in}\rangle$ states to the corresponding set of operators of the $|\text{out}\rangle$ ones [1, 19, 20]. Such a mechanism of particle production is a crucial ingredient for the discussion of the cosmic background of relic gravitational waves produced by inflation, and will be introduced in some detail in the following.

Consider an orthonormal and complete set of solutions $v_k(\eta, x)$ of Eq. (7.56), which are positive-frequency modes on a given initial hypersurface $\eta = \eta_i$. The field u can then be expanded in this basis as

$$u = \sum_k \left(b_k v_k + b_k^\dagger v_k^* \right), \tag{7.108}$$

where

$$\langle v_k | v_{k'} \rangle = \delta_{kk'} = -\langle v_k^* | v_{k'}^* \rangle, \qquad \langle v_k | v_{k'}^* \rangle = 0, \qquad (7.109)$$

and the orthonormality relations are referred to the Klein–Gordon scalar product (7.60). Note that, for simplicity, we are discussing the Bogoliubov transformations using periodic boundary conditions on a spatial box of finite comoving volume, corresponding to Fourier modes with a discrete spectrum. In such a context, the continuous expansion (7.55) is exactly replaced by Eq. (7.108).

As we have already stressed in Section 7.1, the notion of positive frequency, in general, is not time invariant, even if the cosmological background is characterized by a distinguished time direction (i.e. by a globally defined time-like Killing vector). So, given a final hypersurface $\eta = \eta_f$ (different from the initial one), the field u can in principle be expanded over a different set of mode solutions u_k, which are positive frequency modes at $\eta = \eta_f$:

$$u = \sum_k \left(a_k u_k + a_k^\dagger u_k^* \right), \qquad (7.110)$$

where

$$\langle u_k | u_{k'} \rangle = \delta_{kk'} = -\langle u_k^* | u_{k'}^* \rangle, \qquad \langle u_k | u_{k'}^* \rangle = 0. \qquad (7.111)$$

When the field u is quantized the Fourier coefficients $\{b, b^\dagger\}$ and $\{a, a^\dagger\}$ are promoted to annihilation and creation operators satisfying canonical commutation relations, and determining two Fock representations (in principle different) of the Hilbert space, for the system of quantum fluctuations we are considering.

The relation between the two sets of operators a and b can be easily obtained by equating the scalar products $\langle u_k | u \rangle$ obtained from Eqs. (7.108) and (7.110), and using the properties $\langle u^* | v^* \rangle = -\langle u | v \rangle^* = -\langle v | u \rangle$ following from the definition (7.60). We obtain

$$a_k = \sum_{k'} \left(\alpha_{kk'} b_{k'} + \beta_{kk'}^* b_{k'}^\dagger \right), \qquad (7.112)$$

where

$$\alpha_{kk'} = \langle u_k | v_{k'} \rangle, \qquad \beta_{kk'}^* = \langle u_k | v_{k'}^* \rangle, \qquad (7.113)$$

are the so-called Bogoliubov coefficients. In the same way, the product $\langle v_k | u \rangle$ leads to the inverse relation

$$b_{k'} = \sum_k \left(a_k \alpha_{kk'}^* - a_k^\dagger \beta_{kk'}^* \right). \qquad (7.114)$$

Equations (7.112) and (7.114) are compatible with the canonical commutation relations of a and b provided the Bogoliubov coefficients satisfy the conditions

$$\sum_{k'} \left(\alpha_{kk'} \alpha_{k'i}^* - \beta_{kk'}^* \beta_{k'i} \right) = \delta_{ik}, \qquad (7.115)$$

$$\sum_{k'}(-\alpha_{kk'}\beta_{ik'}^* + \beta_{kk'}^*\alpha_{ik'}) = 0. \tag{7.116}$$

Suppose now that the spatial dependence of both modes v_k and u_k may be expanded in plane waves, namely that

$$v_k = v_k(\eta)e^{ik\cdot x}, \qquad u_k = u_k(\eta)e^{ik\cdot x}, \tag{7.117}$$

just as in the case of the asymptotic solutions (7.106) and (7.107). In such a case the Bogoliubov transformations become diagonal and isotropic in momentum space, and the Bogoliubov coefficients may be written in the general form

$$\alpha_{kk'} = c_+(k)\delta_{kk'}, \qquad \beta_{kk'}^* = c_-^*(k)\delta_{-kk'} \tag{7.118}$$

(we have used (7.113), and the definition of the Klein–Gordon scalar product in the discrete representation). The condition (7.115) implies

$$|c_+(k)|^2 - |c_-(k)|^2 = 1 \tag{7.119}$$

for all k (which also guarantees that the asymptotic modes (7.107) still satisfy the canonical normalization (7.59)). Inserting (7.118) into (7.112), and using the adjoint relation for $a^\dagger$, we can finally express the Bogoliubov transformation in matrix form as

$$\begin{pmatrix} a_k \\ a_{-k}^\dagger \end{pmatrix} = \begin{pmatrix} c_+(k) & c_-^*(k) \\ c_-(k) & c_+^*(k) \end{pmatrix} \begin{pmatrix} b_k \\ b_{-k}^\dagger \end{pmatrix}. \tag{7.120}$$

It may be useful, for specific cosmological applications, to note that the above Bogoliubov transformation can also be represented as a unitary transformation generated by the "two-mode squeezing operator" S_k [28–31],

$$S_k = \exp\left(-s_k^* b_k b_{-k} + s_k b_k^\dagger b_{-k}^\dagger\right), \tag{7.121}$$

where $S_k = S_{-k}$, $S_k^\dagger(s) = S_k^{-1}(s) = S_k(-s)$, and where $s_k = r_k \exp(2i\theta_k)$ is the so-called "squeezing parameter", related to $c_\pm(k)$ by

$$c_+(k) = \cosh r_k, \qquad c_-^*(k) = e^{2i\theta_k}\sinh r_k. \tag{7.122}$$

Consider in fact the transformation $S_k^\dagger b_k S_k$, where we define

$$S_k = e^{A_k}, \qquad A_k = r_k\left(e^{2i\theta_k}b_k^\dagger b_{-k}^\dagger - e^{-2i\theta_k}b_k b_{-k}\right), \tag{7.123}$$

and apply the well-known BHC operator expansion [32],

$$e^{-A}be^A = b - [A, b] + \frac{1}{2}[A, [A, b]] + \cdots + \frac{(-1)^n}{n!}[A, [\cdots [A, b]\cdots]] + \cdots. \tag{7.124}$$

Using the canonical commutation relations for b and $b^\dagger$,

$$[A_k, b_k] = -re^{2i\theta_k}b^\dagger_{-k},$$

$$[A_k, [A_k, b_k]] = r^2 b_k, \tag{7.125}$$

$$[A_k, [A_k, [A_k, b_k]]] = -r^3 e^{2i\theta_k}b^\dagger_{-k}, \cdots,$$

one is led to

$$S^\dagger_k b_k S_k = b_k \left(1 + \frac{r^2}{2!} + \cdots\right) + b^\dagger_{-k}e^{2i\theta_k}\left(r + \frac{r^3}{3!} + \cdots\right)$$

$$\equiv b_k \cosh r_k + b^\dagger_{-k}e^{2i\theta_k}\sinh r_k. \tag{7.126}$$

With the same procedure one obtains the transformation of $b^\dagger_{-k}$,

$$S^\dagger_k b^\dagger_{-k} S_k = b_k e^{-2i\theta_k}\sinh r_k + b^\dagger_{-k}\cosh r_k. \tag{7.127}$$

Comparing with Eqs. (7.120) and (7.122) one can finally rewrite the Bogoliubov transformations as

$$a_k = S^\dagger_k b_k S_k, \qquad a^\dagger_{-k} = S^\dagger_k b^\dagger_{-k} S_k. \tag{7.128}$$

In the Heisenberg representation the unitary operator S_k thus transforms the initial operators $\{b, b^\dagger\}$ into the final ones $\{a, a^\dagger\}$. In the Schrödinger representation, in which the operators are fixed, the operator S_k transforms instead the initial state of the tensor fluctuations into a final "squeezed state".

We may suppose that our initial state coincides with the vacuum state $|0\rangle$, annihilated by all b_k operators, and characterized by zero expectation value of the number operator $n = b^\dagger b$, for all modes k,

$$b_k|0\rangle = 0 = \langle 0|b^\dagger_k, \qquad \langle n_k\rangle_{\rm in} = \langle 0|b^\dagger_k b_k|0\rangle = 0. \tag{7.129}$$

In the final "squeezed vacuum" state, $|s_k\rangle = S_k|0\rangle$, the expectation value of the number operator is $\langle n_k\rangle_{\rm out} = \langle s_k|b^\dagger_k b_k|s_k\rangle$, which obviously coincides with the corresponding one for the Heisenberg representation, $\langle n_k\rangle_{\rm out} = \langle 0|a^\dagger_k a_k|0\rangle$ (see Eq. (7.128)). The important point to be stressed is that, independently of the chosen representation, the final asymptotic configuration is in general characterized by a non-zero expectation number of (pairs of) particles in each mode $|\vec{k}|$ (particles are created in pairs because of momentum conservation). Such a number is determined by the squeezing parameter $r = |s_k|$ or, equivalently, by the Bogoliubov coefficient $c_-(k)$ controlling the presence of negative frequency modes in the $|{\rm out}\rangle$ solution (7.107). Indeed, for each mode k,

$$\langle n_k + n_{-k} \rangle_{\text{out}} = \langle s | b_k^\dagger b_k + b_{-k}^\dagger b_{-k} | s \rangle = \langle 0 | S_k^\dagger \left(b_k^\dagger b_k + b_{-k}^\dagger b_{-k} \right) S_k | 0 \rangle$$

$$= \langle 0 | a_k^\dagger a_k + a_{-k}^\dagger a_{-k} | 0 \rangle = \langle 0 | c_- b_{-k} c_-^* b_{-k}^\dagger + c_- b_k c_-^* b_k^\dagger | 0 \rangle$$

$$= 2 |c_-(k)|^2 = 2 \sinh^2 r_k. \tag{7.130}$$

One can also start, more generally, from an initial state different from the vacuum: a non-trivial number state, for instance, or a statistical mixture of number states. The final configuration will then describe a "squeezed number" state, or a squeezed statistical mixture [33, 34], with total number of particles which is always increased with respect to the initial configuration. The squeezed-state formalism turns out to be useful, in particular, for the analysis of the statistical properties of the radiation produced, and for the study of the entropy growth associated with particle production [35, 36].

By applying the formalism of the Bogoliubov transformations one can then describe the amplification of tensor perturbations, in a second-quantization language, as the production of pairs of gravitons from the given initial state. For each mode $\vec{k}$, the differential energy density of the amplified perturbations (in d spatial dimensions) is given by

$$d\rho_k = \frac{k}{2}(d+1)(d-2)\langle n_k \rangle \frac{d^d k}{(2\pi)^d}, \tag{7.131}$$

where k is the energy of the given (massless) mode, $(d+1)(d-2)/2$ is the number of polarization states, and $\langle n_k \rangle$ the number density of produced gravitons. Assuming that the final graviton distribution satisfies the isotropy condition (i.e. that $\langle n_k \rangle$ depends only on $k = |\vec{k}|$), and integrating over the whole $(d-1)$-dimensional angular sphere, one obtains

$$k \frac{d\rho_k}{dk} = \frac{(d+1)(d-2)\pi^{d/2}}{(2\pi)^d \Gamma(d/2)} k^{d+1} \langle n_k \rangle. \tag{7.132}$$

The computation of the spectral energy density requires knowledge of $\langle n_k \rangle = |c_-(k)|^2$, and then requires the solution of the perturbation equation in the asymptotic limit of large positive times (see Eq. (7.107)).

7.2.3 Matching conditions

Before concluding this section we present an explicit example of computation of the spectrum for a very simple model of background, which includes a first phase of accelerated evolution from $-\infty$ to $-\eta_1$, and a second decelerated phase from $-\eta_1$ to $+\infty$. After solving the perturbation equation, separately in the two phases, we fix the integration constants by imposing (*i*) the canonical normalization, and (*ii*) the continuity of the metric fluctuation h and of its first derivative h' (or, equivalently, of its conjugate momentum) at the transition epoch $\eta = -\eta_1$. For a more general background characterized by $n+1$ distinct phases, such a procedure can be iterated n times, and it is also possible to provide diagrammatic prescriptions allowing an automatic (and fast) estimate of the final spectrum [37]).

We work in the so-called "sudden approximation" [38], considering a background characterized by the following graviton pump field $z(\eta)$:

$$
z(\eta) = \begin{cases} \left(\dfrac{M_P^{d-1}}{2}\right)^{1/2} \left(-\dfrac{\eta}{\eta_1}\right)^{\alpha_1}, & -\infty \le \eta \le -\eta_1, \\[3mm] \left(\dfrac{M_P^{d-1}}{2}\right)^{1/2} \left(\dfrac{\eta+2\eta_1}{\eta_1}\right)^{\alpha_2}, & -\eta_1 \le \eta \le +\infty \end{cases}
$$
(7.133)

(we have assumed the continuity of z in η_1). The canonical equation becomes, in the two phases,

$$
u_k'' + \left[k^2 - \frac{\alpha_1(\alpha_1-1)}{\eta^2}\right] u_k = 0, \qquad \eta \le -\eta_1,
$$

$$
u_k'' + \left[k^2 - \frac{\alpha_2(\alpha_2-1)}{(\eta+2\eta_1)^2}\right] u_k = 0, \qquad \eta \ge -\eta_1.
$$
(7.134)

The general solution can be written in terms of the first- and second-kind Hankel functions [39],

$$
u_k^1 = \eta^{1/2} \left[A_+^1(k) H_{\nu_1}^{(2)}(k\eta) + A_-^1(k) H_{\nu_1}^{(1)}(k\eta) \right], \qquad \eta \le -\eta_1,
$$

$$
u_k^2 = (\eta+2\eta_1)^{1/2} \left[A_+^2(k) H_{\nu_2}^{(2)}(k\eta+2k\eta_1) \right.
$$

$$
\left. + A_-^2(k) H_{\nu_2}^{(1)}(k\eta+2k\eta_1) \right], \qquad \eta \ge -\eta_1, \qquad (7.135)
$$

where $A_\pm^{1,2}$ are integration constants, and the Bessel indices ν_1, ν_2 are determined by the pump-field kinematics as

$$
\nu_1 = 1/2 - \alpha_1, \qquad \nu_2 = 1/2 - \alpha_2. \tag{7.136}
$$

We now use the canonical normalization at $\eta \to -\infty$ to fix the integration constants $A_\pm^1$. Using the large-argument limit of the Hankel functions we have [39]

$$
H_\nu^{(2)}(k\eta) = \sqrt{\frac{2}{\pi k\eta}} e^{-ik\eta - i\epsilon_\nu}, \qquad H_\nu^{(1)}(k\eta) = \sqrt{\frac{2}{\pi k\eta}} e^{ik\eta + i\epsilon_\nu}, \qquad \eta \to -\infty
$$
(7.137)

($\epsilon_\nu = -\nu\pi/2 - \pi/4$ is a ν-dependent phase factor), and we can thus impose, according to Eq. (7.106),

$$A^1_+ = \sqrt{\pi/4}, \qquad A^1_- = 0. \tag{7.138}$$

By applying the large argument behavior to the second solution, u^2_k, we can then identify the Bogoliubov coefficients in the asymptotic limit $\eta \to +\infty$. Comparing with Eq. (7.107) we obtain

$$c_\pm(k) = \sqrt{\frac{4}{\pi}} A^2_\pm(k) \tag{7.139}$$

(modulo a constant phase factor absorbed into $c_\pm$). With such a normalization, the general solution for the metric fluctuation field, $h_k = u_k/z$, can be written eventually in the form

$$h^1_k = e^{i\pi\alpha_1} \left(\frac{\pi\eta_1}{2M_P^{d-1}} \right)^{1/2} \left(\frac{\eta}{\eta_1} \right)^{\nu_1} H^{(2)}_{\nu_1}(k\eta), \qquad \eta \le -\eta_1,$$

$$h^2_k = \left(\frac{\pi\eta_1}{2M_P^{d-1}} \right)^{1/2} \left(\frac{\eta + 2\eta_1}{\eta_1} \right)^{\nu_2} \left[c_+ H^{(2)}_{\nu_2}(k\eta + 2k\eta_1) \right. \tag{7.140}$$

$$\left. + c_- H^{(1)}_{\nu_2}(k\eta + 2k\eta_1) \right], \qquad \eta \ge -\eta_1,$$

It may be instructive, at this point, to compare this exact result with our previous approximate solutions. To this purpose we exploit the small argument limit of the Hankel functions, which gives, for $\nu \ne 0$ [39],

$$H^{(1)}_\nu(x) = p_\nu x^\nu + iq_\nu x^{-\nu} + \cdots ,$$

$$H^{(2)}_\nu(x) = p^*_\nu x^\nu - iq_\nu x^{-\nu} + \cdots , \tag{7.141}$$

where q and p are ν-dependent coefficients, with modulus of order one (when $\nu = 0$ there are additional logarithmic corrections). By applying such an expansion to the h^1_k solution (i.e. considering metric fluctuations well outside the horizon, with $|k\eta| \ll 1$), and setting $\nu = \nu_1 = 1/2 - \alpha_1$, one can easily reproduce the asymptotic solution (7.90), characterized by a constant part and a part which depends on time as $|\eta|^{1-2\alpha_1}$. One also exactly recovers the same k-dependence, with two terms proportional to $k^{1/2-\alpha_1}$ and $k^{\alpha_1-1/2}$.

We are now in a position to determine the coefficients $c_\pm$ by imposing the continuity conditions at the transition time $\eta = -\eta_1$. It should be noted that the requirement of continuity of the metric fluctuation h is not equivalent, in general, to the continuity of the canonical variable u, if the matching conditions are imposed on a discontinuous background [40]. The continuity of h, on the other hand, is needed to guarantee the continuity of the total energy density across the matching hypersurface [41]. In order to adhere to the standard matching prescriptions at constant energy density, we thus impose

$$h^1(-\eta_1) = h^2(-\eta_1), \qquad h'^1(-\eta_1) = h'^2(-\eta_1). \qquad (7.142)$$

We also use the following properties of the Hankel functions,

$$H_\nu^{(2)}(-x) = -e^{i\pi\nu} H_\nu^{(1)}(x),$$

$$\frac{d}{dx} H_\nu^{(2)}(-x) = e^{i\pi\nu} \frac{d}{dx} H_\nu^{(1)}(x), \qquad (7.143)$$

which are useful to express the final result in terms of the Hankel functions evaluated in $x_1 \equiv k\eta_1$. The continuity of the solution (7.140) then gives

$$c_+ H_{\nu_2}^{(2)}(x_1) + c_- H_{\nu_2}^{(1)}(x_1) = e^{i\phi_1} H_{\nu_1}^{(1)}(x_1), \qquad (7.144)$$

where ϕ is a constant phase factor, $e^{i\phi_1} = (-1)^{\nu_1+1} e^{i\pi(\nu_1+\alpha_1)}$. The continuity of h' provides a second condition which, by using Eq. (7.144), can be written as

$$c_+ H_{\nu_2}'^{(2)}(x_1) + c_- H_{\nu_2}'^{(1)}(x_1) = -e^{i\phi_1} \left[H_{\nu_1}'^{(1)}(x_1) + \frac{\nu_1+\nu_2}{x_1} H_{\nu_1}^{(1)}(x) \right], \qquad (7.145)$$

where the prime denotes the derivative of $H_\nu(x)$ with respect to its argument.

The solution of the system of algebraic equations (7.144) and (7.145), for the two unknown variables $c_\pm$, can be simplified by using the Wronskian properties of the Hankel functions [39],

$$H_\nu^{(2)}(x) H_\nu'^{(1)}(x) - H_\nu'^{(2)}(x) H_\nu^{(1)}(x) = \frac{4i}{\pi x}, \qquad (7.146)$$

which immediately provides the determinant of the coefficients of the above algebraic system. The exact result for $c_\pm$ is, finally,

$$c_+ = -\frac{i\pi}{4} x_1 e^{i\phi_1} \left[H_{\nu_1}^{(1)} H_{\nu_2}'^{(1)} + H_{\nu_1}'^{(1)} H_{\nu_2}^{(1)} + \frac{\nu_1+\nu_2}{x_1} H_{\nu_1}^{(1)} H_{\nu_2}^{(1)} \right]_{x_1},$$

$$c_- = \frac{i\pi}{4} x_1 e^{i\phi_1} \left[H_{\nu_2}^{(2)} H_{\nu_1}'^{(1)} + H_{\nu_2}'^{(2)} H_{\nu_1}^{(1)} + \frac{\nu_1+\nu_2}{x_1} H_{\nu_1}^{(1)} H_{\nu_2}^{(2)} \right]_{x_1}, \qquad (7.147)$$

where all the Hankel functions are evaluated in $x_1 = k\eta_1$. One can immediately check, by repeated application of the Wronskian condition (7.146), and by using the property $[H_\nu^{(1)}(x)]^* = H_\nu^{(2)}(x^*)$ holding for all real ν, that the Bogoliubov coefficients satisfy the normalization condition $|c_+|^2 - |c_-|^2 = 1$.

The value of $|c_-(k)|^2$ represents the expectation number of gravitons produced in the transition between the two cosmological phases. When inserted into Eq. (7.132), it determines the associated energy density distribution. The above result, however, has been obtained using the sudden approximation for the background transition at η_1, and is thus valid in the limit of low enough frequency modes, $k\eta_1 \ll 1$. In the opposite regime $k\eta_1 \gg 1$, using the large argument limit of the Hankel functions, one would find from Eq. (7.147) a k-independent result, with $|c_-| \sim 1$, which would correspond to an ultraviolet divergence of the spectrum, $\Omega(k) \sim k^{d+1}$ for $k \to \infty$. Such a divergence is due to the fact that the sudden approximation is inadequate for modes with an energy k much larger than (or of the order of) the peak of the effective potential, $|U(\eta_1)|^{1/2} \sim \eta_1^{-1}$ – namely, for modes which do not "hit" the potential barrier. For such modes, the mixing coefficients should be computed by replacing the potential step with a smooth transition of $U(\eta)$, and one finds that the coefficient c_- is exponentially suppressed as $\exp(-k\eta_1)$, thus avoiding the ultraviolet divergence [14, 19, 20, 38]. Thus, if we are not interested in a precise calculation of the high-frequency tail of the spectrum, we may avoid constructing a detailed (and smooth) model of background in the transition region, limiting ourselves to the frequency band with $k\eta_1 \ll 1$.

In this chapter we choose this option, replacing the exponential decay of the spectrum at high frequency with the sharp cut-off

$$|c_+(k)| \simeq 1, \qquad |c_-(k)| \simeq 0, \qquad k\eta_1 \gg 1. \qquad (7.148)$$

In the opposite regime, $k\eta_1 \ll 1$, the result (7.147) is valid, and the form of c_- can be simplified using the derivative property of the Hankel functions,

$$H'_\nu = -H_{\nu+1}(x) + \frac{\nu}{x}H_\nu(x), \qquad (7.149)$$

and the small argument limit (7.141) where, for $\nu \neq 0$ [39],

$$p_\nu = \frac{2^{-\nu}}{\Gamma(1+\nu)}(1 \mp \mathrm{i}\cot\nu\pi), \qquad q_\nu = -\frac{2^\nu}{\Gamma(1-\nu)}\csc\nu\pi \qquad (7.150)$$

(the case $\nu = 0$ includes a logarithmic term, and will be considered at the end of this section). Keeping the leading terms for $x_1 = k\eta_1 \ll 1$ we arrive at

$$c_-(x_1) = \beta_{+-}x_1^{\alpha_1-\alpha_2} + \beta_{-+}x_1^{-\alpha_1+\alpha_2} + \beta_{++}x_1^{\alpha_1+\alpha_2-1} + \beta_{--}x_1^{1-\alpha_1-\alpha_2}, \qquad (7.151)$$

where

$$\beta_{+-} = -\frac{\pi}{4}\mathrm{e}^{\mathrm{i}\phi_1}p^*_{\nu_2}[2(\nu_1+\nu_2)q_{\nu_1} - q_{\nu_1+1}],$$

$$\beta_{-+} = -\frac{\pi}{4}\mathrm{e}^{\mathrm{i}\phi_1}p_{\nu_1}[q_{\nu_2+1} - 2q_{\nu_2}(\nu_1+\nu_2)],$$

$$\beta_{--} = \mathrm{i}\frac{\pi}{2}\mathrm{e}^{\mathrm{i}\phi_1}(\nu_1+\nu_2)p_{\nu_1}p^*_{\nu_2}, \qquad (7.152)$$

$$\beta_{++} = \mathrm{i}\frac{\pi}{4}\mathrm{e}^{\mathrm{i}\phi_1}[2(\nu_1+\nu_2)q_{\nu_1}q_{\nu_2} - q_{\nu_1}q_{\nu_2+1} - q_{\nu_2}q_{\nu_1+1}].$$

Table 7.1 *Spectral distribution of the Bogoliubov coefficients and of the graviton energy density as a function of the kinematic parameters of the two pump fields and of the number* d *of spatial dimensions*

Pump fields $z_i \sim \eta^{\alpha_i}$	Bogoliubov coefficient $\lvert c_- \rvert$	Spectral density $\Omega \sim k^{d+1}\lvert c_- \rvert^2$
$\alpha_1 < 1/2,\ \alpha_2 > 1/2$	$k^{\alpha_1 - \alpha_2}$	$k^{d+1+2\alpha_1 - 2\alpha_2}$
$\alpha_1 > 1/2,\ \alpha_2 < 1/2$	$k^{-\alpha_1 + \alpha_2}$	$k^{d+1-2\alpha_1 + 2\alpha_2}$
$\alpha_1 < 1/2,\ \alpha_2 < 1/2$	$k^{\alpha_1 + \alpha_2 - 1}$	$k^{d-1+2\alpha_1 + 2\alpha_2}$
$\alpha_1 > 1/2,\ \alpha_2 > 1/2$	$k^{1-\alpha_1 - \alpha_2}$	$k^{d+3-2\alpha_1 - 2\alpha_2}$

It is now evident how the pump-field kinematics (i.e. the explicit values of α_1, α_2) plays a crucial role in determining the leading k-behavior of c_- for $x_1 \ll 1$, and then in fixing the final spectral energy distribution $\Omega(k) \sim k^{d+1}\lvert c_- \rvert^2$ (see Table 7.1). In the next section we will present some typical examples of graviton spectra associated with different inflationary models. It may be instructive, however, to discuss immediately the particular case in which the pump field of the second (decelerated) phase evolves linearly in conformal time, so that $\alpha_2 = 1$ (for an isotropic, $d = 3$ manifold, this means that we are then in the presence of a radiation-dominated phase). In such a case the energy density of the produced gravitational radiation, in its final state inside the horizon, exhibits exactly the same spectral distribution as the primordial metric fluctuations in the inflationary, super-horizon regime.

In fact, by setting $\alpha_1 = \alpha$, $\alpha_2 = 1$ in Eq. (7.151), we find, to leading order as $x_1 \to 0$,

$$c_-(x_1) = \beta_{+-}x_1^{\alpha-1} + \beta_{--}x_1^{-\alpha}, \qquad (7.153)$$

and we can distinguish two possibilities: the case $\alpha < 1/2$, in which the first term is dominant, and the case $\alpha > 1/2$, in which the second term is dominant. Inserting this result into the definition (7.105) we find

$$\Omega(k) \sim k^{d+1}\lvert c_-(k) \rvert^2 \sim \begin{cases} k^{d-1+2\alpha}, & 2\alpha < 1, \\ k^{d+1-2\alpha}, & 2\alpha > 1, \end{cases} \qquad (7.154)$$

and we reproduce exactly the same spectral distribution as that obtained for $\lvert \delta_h(k) \rvert^2$, in the case of super-horizon tensor perturbations (see Eq. (7.101) and Fig. 7.1). For $2\alpha < 1$, and in the absence of contributions from internal dimensions, one recovers, in particular, the close connection between spectral behavior (in frequency) and curvature behavior (in time), already illustrated in Eq. (7.103). The spectral variable $\Omega(k)$, on the other hand, is directly measurable (at least in principle) by a gravity-wave detector, and this explains the claim – made at the beginning of this chapter – about the possibility of obtaining direct information on the primordial cosmological dynamics through the study of its gravitational relics.

Let us finally compute the spectrum in the limiting case in which one of the two pump fields has a power $\alpha = 1/2$, corresponding to a vanishing Bessel index $\nu = 1/2 - \alpha = 0$ in Eq. (7.147). We may consider, for instance, the case $\nu_1 = 0$, namely $\alpha_1 = 1/2$, which corresponds to a particular example of pre-big bang kinematics (see Eq. (7.99)). The computation of c_- then requires the small argument expansion of the Hankel function $H_0^{(1)}$, which is given by

$$H_0^{(1)}(x_1) \simeq p_0 + iq_0 \ln x_1 + ir_0 x_1^2 \ln x_1 + s_0 x_1^2 + \cdots, \tag{7.155}$$

where, in particular,

$$p_0 = 1 + (2i/\pi)(\gamma - \ln 2), \qquad q_0 = 2/\pi, \tag{7.156}$$

and $\gamma = 0.5772\ldots$ is the Euler–Mascheroni constant. Equation (7.147) then gives, to leading order as $x_1 \to 0$,

$$c_-(x_1) = \beta_+ x_1^{1/2-\alpha_2} \ln x_1 + \beta_- x_1^{\alpha_2-1/2} \ln x_1, \tag{7.157}$$

where

$$\beta_+ = -\frac{\pi}{2} e^{i\phi_1} \nu_2 q_0 p_{\nu_2}^*, \qquad \beta_- = i\frac{\pi}{4} e^{i\phi_1} q_0 \left(2\nu_2 q_{\nu_2} - q_{\nu_2+1}\right). \tag{7.158}$$

This result agrees with Eq. (7.151) for $\alpha_1 = 1/2$, modulo the expected logarithmic corrections (see also Eq. (7.98)). In the particular case $\alpha_2 = 1$, one recovers the typical spectrum of the so-called "minimal" pre-big bang models [25, 42],

$$|c_-|^2 \sim x_1^{-1} \ln^2 x_1, \qquad \Omega(k) \sim k^d \ln^2(k\eta_1), \tag{7.159}$$

which will be discussed in more detail in the next section.

7.3 Expected relic gravitons from inflation

The aim of this section is to present an explicit computation of the graviton spectra predicted by the most typical classes of inflationary models, in view of subsequent estimates of the experimental sensitivity required for their possible detection. The maximum expected signal seems to be outside the present operating range of existing gravitational antennas, but it could be accessible to the advanced versions of these instruments (hopefully, of near-future implementation), especially in the case of a *growing* spectrum, which is indeed typical of string cosmology (see Section 7.4).

For the purpose of this section it is sufficient to consider a homogeneous and isotropic cosmological background, without dynamical contributions from the internal dimensions (i.e. with $d = 3$, $n = 0$). Also, to take into account the temporal evolution of the graviton energy density, it is convenient to adopt as spectral variable the proper frequency, $\omega(t) = k/a(t)$. By using the explicit form of the critical energy density, $\rho_c = 3M_P^2 H^2(t)$, we then obtain from Eqs. (7.105) and (7.132):

$$\Omega_g(\omega, t) = \frac{\omega^4 |c_-(\omega)|^2}{3\pi^2 M_P^2 H^2}. \tag{7.160}$$

Finally, we use a simple model of background which undergoes a first transition (at $\eta = -\eta_1$) from inflation to radiation-dominated expansion, and a second transition (at $\eta = \eta_{eq}$) to the final matter-dominated regime (see Fig. 7.3). Where string cosmology models are concerned we assume that the dilaton field is possibly dynamical during inflation, and that it becomes frozen (trapped, for instance, at the minimum of an appropriate potential) during the subsequent decelerated epochs, consistent with the standard cosmological evolution. For $d = 3$ and $n = 0$ the pump field of tensor perturbations will always coincide with the E-frame scale factor, and its behavior in time during the various phases is summarized in Fig. 7.3.

We start by considering the standard inflationary scenario, with a pump field describing a phase of (E-frame) accelerated expansion, $\alpha < 0$. For the first transition

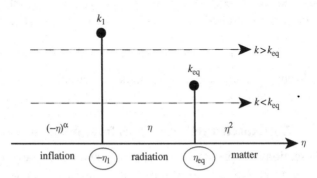

Figure 7.3 Schematic view of the simplest, "quasi-realistic" example of inflationary scenario. We have specified, in each phase, the (conformal) time dependence of the pump field $z(\eta) \sim \tilde{a}(\eta)$. High-frequency modes with $k > k_{eq} = \eta_{eq}^{-1}$ are only affected by the first transition, while lower-frequency modes, $k < k_{eq}$, are affected by both transitions.

at $\eta = -\eta_1$ we can apply the results of the previous section, and from Eqs. (7.151) and (7.153) we obtain, for $\alpha_2 = 1$, $\alpha_1 = \alpha < 0$,

$$c_-(\omega) = \beta_{+-} \left(\frac{\omega}{\omega_1}\right)^{\alpha-1}, \qquad (7.161)$$

where $\omega_1 = k_1/a$, and $k_1 = \eta_1^{-1}$. Thus,

$$\Omega_g(\omega, t) = \frac{|\beta_{+-}|^2 H_1^4}{3\pi^2 M_P^2 H^2} \left(\frac{a_1}{a}\right)^4 \left(\frac{\omega}{\omega_1}\right)^{2+2\alpha}, \qquad \omega < \omega_1, \qquad (7.162)$$

where we have explicitly introduced the useful parameter $H_1 = k_1/a_1 \equiv \omega_1(t_1)$, controlling the curvature scale of the background at the epoch of horizon-crossing of the cut-off frequency ω_1. Note that, for a power-law scale factor, the parameter H_1 roughly corresponds to the E-frame curvature scale at the end of the inflationary period. We can also use the fact that the Universe becomes radiation dominated at $\eta = -\eta_1$, so that

$$H_1^2 \left(\frac{a_1}{a}\right)^4 = \frac{\rho_1}{3M_P^2} \left(\frac{a_1}{a}\right)^4 = \frac{\rho_r(t)}{3M_P^2} = \Omega_r(t)H^2, \qquad (7.163)$$

where $\Omega_r(t) = \rho_r/\rho_c$ is the instantaneous fraction of critical energy density associated with the radiation produced at η_1. Equation (7.162) can then be rewritten as

$$\Omega_g(\omega, t) = \frac{|\beta_{+-}|^2}{3\pi^2} \left(\frac{H_1}{M_P}\right)^2 \Omega_r(t) \left(\frac{\omega}{\omega_1}\right)^{2+2\alpha}, \qquad \omega < \omega_1, \qquad t_1 < t.$$

$$\tag{7.164}$$

Taking into account that $|\beta_{+-}|^2$ is only an (α-dependent) numerical factor of order one we can already deduce, from this expression, the three main primordial properties of the inflation-generated background of relic gravitons:

(1) the spectral distribution tends to follow the behavior in time of the curvature scale during the inflationary regime, i.e. the spectrum is flat for de Sitter, $\alpha = -1$, decreasing for power-inflation, $\alpha < -1$, and growing for super-inflation, $-1 < \alpha < 0$ (as repeatedly stressed also in the previous section);
(2) the energy density scales in time like the radiation energy density, $\rho_g \sim a^{-4}$;
(3) the overall intensity is controlled by the value of the curvature scale (in Planck units) characterizing the background at the end of the inflationary evolution.

We should remember, however, that the spectrum (7.164) is valid for all modes only during the radiation phase, i.e. *before* the second background transition at $\eta = \eta_{eq}$. If we are interested in the full graviton spectrum at the present epoch, $\eta_0 > \eta_{eq}$, we must also take into account the second transition that affects the lower-frequency modes with $k\eta_{eq} < 1$.

To this purpose – and also in view of further applications – it is convenient to compute the Bogoliubov coefficients for a general case in which the metric fluctuations asymptotically contain a mixture of positive and negative frequency modes both *before* and *after* the transition. We thus extend the computation of Section 7.2 by assuming that, at a given time $\eta = \eta_2 > 0$, there is a background transition to a new phase, whose pump field is controlled by the generic power α_3, and is defined by

$$z(\eta) = \frac{M_{\mathrm{P}}}{\sqrt{2}} \left(\frac{\eta + 2\eta_1}{\eta_1} \right)^{\alpha_3} \left(\frac{\eta_2 + 2\eta_1}{\eta_1} \right)^{\alpha_2 - \alpha_3}, \qquad 0 < \eta_2 \leq \eta, \qquad (7.165)$$

in such a way as to match continuously at η_2 with the previous definition (see Eq. (7.133)).

The canonical solutions describing the evolution of tensor perturbations in the first, inflationary phase, $\eta \leq -\eta_1$, and in the second, post-inflationary phase, $-\eta_1 \leq \eta \leq \eta_2$, are still valid, and are given by Eq. (7.140). Putting $d = 3$, and defining the convenient variable

$$y = k\eta + 2k\eta_1 \equiv x + 2x_1, \qquad (7.166)$$

we can write such solutions as

$$h_k^1(\eta) = e^{i\pi\alpha_1} \left(\frac{\pi\eta_1}{2M_{\mathrm{P}}^2} \right)^{1/2} \left(\frac{x}{x_1} \right)^{\nu_1} H_{\nu_1}^{(2)}(x), \qquad \eta \leq -\eta_1,$$

$$h_k^2(\eta) = \left(\frac{\pi\eta_1}{2M_{\mathrm{P}}^2} \right)^{1/2} \left(\frac{y}{x_1} \right)^{\nu_2} \left[c_+^2 H_{\nu_2}^{(2)}(y) + c_-^2 H_{\nu_2}^{(1)}(y) \right], \qquad -\eta_1 \leq \eta \leq \eta_2.$$

$$(7.167)$$

In the third cosmological phase, the solution corresponding to the pump field (7.165) is

$$h_k^3(\eta) = \left(\frac{\pi\eta_1}{2M_{\mathrm{P}}^2} \right)^{1/2} \left(\frac{y}{x_1} \right)^{\nu_3} \left(\frac{y_2}{x_1} \right)^{\nu_2 - \nu_3} \left[c_+^3 H_{\nu_3}^{(2)}(y) + c_-^3 H_{\nu_3}^{(1)}(y) \right], \qquad \eta_2 \leq \eta,$$

$$(7.168)$$

where $\nu_3 = 1/2 - \alpha_3$, and $y_2 = x_2 + 2x_1$, with $x_2 = k\eta_2$.

The coefficients $c_\pm^2$, describing the amplification associated with the first background transition, have been already reported in Eq. (7.147). By eliminating the derivatives of the Hankel functions (through Eq. (7.149)), such coefficients can be conveniently rewritten as

$$
c_+^2 = -\frac{i\pi}{4} x_1 e^{i\phi_1} \left[-H_{\nu_1}^{(1)} H_{\nu_2+1}^{(1)} - H_{\nu_2}^{(1)} H_{\nu_1+1}^{(1)} + \frac{2}{x_1} (\nu_1 + \nu_2) H_{\nu_1}^{(1)} H_{\nu_2}^{(1)} \right]_{x_1} ,
$$
$$
c_-^2 = \frac{i\pi}{4} x_1 e^{i\phi_1} \left[-H_{\nu_2}^{(2)} H_{\nu_1+1}^{(1)} - H_{\nu_1}^{(1)} H_{\nu_2+1}^{(2)} + \frac{2}{x_1} (\nu_1 + \nu_2) H_{\nu_1}^{(1)} H_{\nu_2}^{(2)} \right]_{x_1}
$$
(7.169)

(all functions are evaluated in $x_1 = k\eta_1$). For the computation of the coefficients $c_\pm^3$, associated with the final phase, we should recall that the process of graviton production is exponentially suppressed for all modes with high enough frequency, i.e. $x_2 = k\eta_2 \gg 1$. If we are only concerned with an estimate of the leading contributions to the graviton spectrum we can safely assume that all modes which do not "hit" the effective potential barrier at η_2 are not affected by the transition, so that $|c_\pm^2(k > k_2)| \simeq |c_\pm^3(k > k_2)|$, and their spectral distribution is transmitted almost unchanged from the second to the third cosmological phase. Low-frequency modes, with $x_2 \leq 1$, are instead strongly influenced by the second transition, and the corresponding coefficients $c_\pm^3$ can be computed in the "sudden" approximation, imposing the continuity of h and h' at $\eta = \eta_2$.

Performing the matching, and exploiting the properties of the Hankel functions, one arrives at the following expression for the coefficient c_-^3, which controls the final number density of produced gravitons:

$$
c_-^3 = -\frac{i\pi}{4} y_2 \left[-H_{\nu_3}^{(2)}(y_2) \left(c_+^2 H_{\nu_2+1}^{(2)} + c_-^2 H_{\nu_2+1}^{(1)} \right)_{y_2} \right.
$$
$$
\left. + \left(H_{\nu_3+1}^{(2)} + \frac{2}{y_2}(\nu_2 - \nu_3) H_{\nu_3}^{(2)} \right)_{y_2} \left(c_+^2 H_{\nu_2}^{(2)} + c_-^2 H_{\nu_2}^{(1)} \right)_{y_2} \right]
$$
(7.170)

(all the Hankel functions, except those contained inside $c_\pm^2$, are evaluated at $y_2 \ll 1$). This result, in combination with Eq. (7.169), determines the graviton spectrum for a double transition with generic (but $\neq 1/2$) values of $\alpha_1, \alpha_2, \alpha_3$. In the limit $c_-^2 \to 0$ and $c_+^2 \to 1$ the solution for h^2 acquires the canonical normalization associated with a vacuum fluctuation spectrum, and one recovers for c_-^3 the result (7.147) – with the only differences due to a matching performed at another (positive) value of η.

Let us now explicitly compute c_-^3 for the simple model in which the second cosmological phase corresponds to the radiation-dominated regime, so that $\alpha_2 = 1$, $\nu_2 = -1/2$, with associated Hankel functions

$$H^{(2)}_{-\frac{1}{2}}(x) = \sqrt{\frac{2}{\pi x}}\, e^{-ix}, \qquad H^{(2)}_{\frac{1}{2}}(x) = i\sqrt{\frac{2}{\pi x}}\, e^{-ix},$$

$$H^{(1)}_{-\frac{1}{2}}(x) = \sqrt{\frac{2}{\pi x}}\, e^{ix}, \qquad H^{(1)}_{\frac{1}{2}}(x) = -i\sqrt{\frac{2}{\pi x}}\, e^{ix}. \tag{7.171}$$

The first term multiplying $H^{(2)}_{\nu_3}$ in Eq. (7.170) becomes

$$c_+^2 H^{(2)}_{1/2}(y_2) + c_-^2 H^{(1)}_{1/2}(y_2)$$

$$= e^{i\phi_1}\left(\frac{x_1}{y_2}\right)^{1/2}\left[\left(-H^{(1)}_{\nu_1+1} + \frac{2}{x_1}(\nu_1 - \frac{1}{2})H^{(1)}_{\nu_1}\right)_{x_1} \cos(y_2 - x_1)\right.$$

$$\left. + H^{(1)}_{\nu_1}(x_1)\sin(y_2 - x_1)\right]. \tag{7.172}$$

For a third phase dominated by dust matter we also have $\alpha_3 = 2$ and $\nu_3 = -3/2$. One can then easily check that the contribution of the term multiplying $H^{(2)}_{\nu_3+1}$, in the second part of Eq. (7.170), is sub-leading with respect to the term multiplying $H^{(2)}_{\nu_3}$, whose explicit form is

$$\frac{2}{y_2}(\nu_2 - \nu_3)\left(c_+^2 H^{(2)}_{-1/2} + c_-^2 H^{(1)}_{-1/2}\right)_{y_2}$$

$$= \frac{e^{i\phi_1}}{y_2}\left(\frac{x_1}{y_2}\right)^{1/2}\left[2H^{(1)}_{\nu_1}(x_1)\cos(y_2 - x_1)\right.$$

$$\left. + \left(2H^{(1)}_{\nu_1+1} - \frac{4}{x_1}(\nu_1 - \frac{1}{2})H^{(1)}_{\nu_1}\right)_{x_1}\sin(y_2 - x_1)\right]. \tag{7.173}$$

On the other hand, in the limit $x_1 \ll x_2 \ll 1$, one can use the approximations $y_2 \simeq x_2$, $\cos(x_2 - x_1) \simeq 1$ and $\sin(x_2 - x_1) \simeq x_2$; comparing, in this limit, the two previous equations, and summing up all leading terms, one is left with

$$c_-^3(k) = -\frac{i\pi}{4}e^{i\phi_1}x_2\left(\frac{x_1}{x_2}\right)^{1/2}\left[3H^{(1)}_{\nu_1+1} - \frac{6}{x_1}(\nu_1 - \frac{1}{2})H^{(1)}_{\nu_1}\right]_{x_1} H^{(2)}_{-3/2}(x_2).$$

$$\tag{7.174}$$

One can finally exploit the small argument expansion of the Hankel functions, identifying ν_1 with $1/2 - \alpha$, and η_2 with η_{eq}. The Bogoliubov coefficient for a mode k undergoing the double transition can then be written as

$$c_-^3(k) = \beta_\nu \, x_{eq}^{-1} \, x_1^{-|1/2-\alpha|-1/2}, \qquad k < k_{eq}. \qquad (7.175)$$

Here $x_{eq} \equiv k\eta_{eq}$, and β_ν is a numerical factor (with modulus of order one) determined by the expansion (7.141):

$$\beta_\nu = \begin{cases} \dfrac{3\pi}{4} e^{i\phi_1} p_{-3/2}^* \left[q_{\nu+1} + (2\nu - 1)q_\nu \right], & \nu > 0, \\[4mm] -i\dfrac{3\pi}{4} e^{i\phi_1} p_{-3/2}^* (2\nu - 1)p_\nu, & \nu < 0. \end{cases} \qquad (7.176)$$

We are now in a position to provide the complete energy spectrum for the graviton production associated with our model of background. Such a spectrum is characterized by two distinct branches, since the number density of high-frequency modes $(k > k_{eq})$ is controlled by the Bogoliubov coefficient (7.161), while the number density of low-frequency modes $(k < k_{eq})$ is controlled by the coefficient (7.175). Thus,

$$|c_-(\omega)| \sim \begin{cases} \left(\dfrac{\omega}{\omega_1} \right)^{\alpha-1}, & \omega_{eq} < \omega < \omega_1, \\[4mm] \left(\dfrac{\omega}{\omega_1} \right)^{\alpha-1} \left(\dfrac{\omega}{\omega_{eq}} \right)^{-1}, & \omega_0 < \omega < \omega_{eq} \end{cases} \qquad (7.177)$$

(we have assumed $\alpha < 1/2$, which is the typical case for the conventional inflationary scenario).

The above spectrum is valid in the frequency band ranging from the upper limit ω_1 (beyond which graviton production is exponentially suppressed), to the lower limit $\omega_0 = H_0$, fixed by the scale of the present Hubble horizon (lower-frequency modes are still outside the horizon, and are characterized by a different spectrum). Inserting the Bogoliubov coefficients in Eq. (7.160), using the definition (7.163) of Ω_r, and absorbing into H_1^2 all numerical factors of order one appearing in the computation of Ω and c_-, we obtain the following result for the graviton energy density, at the present time t_0:

$$\Omega_g(\omega, t_0) = \begin{cases} \left(\dfrac{H_1}{M_P} \right)^2 \Omega_r(t_0) \left(\dfrac{\omega}{\omega_1} \right)^{2+2\alpha}, & \omega_{eq} < \omega < \omega_1, \\[4mm] \left(\dfrac{H_1}{M_P} \right)^2 \Omega_r(t_0) \left(\dfrac{\omega}{\omega_1} \right)^{2+2\alpha} \left(\dfrac{\omega}{\omega_{eq}} \right)^{-2}, & \omega_0 < \omega < \omega_{eq}. \end{cases} \qquad (7.178)$$

This spectral distribution applies to the case $\alpha < 1/2$, but it can be easily extended to include the case $\alpha > 1/2$, according to the general result (7.153), simply by replacing $2+2\alpha$ with $3-2|\nu|$, where $\nu = 1/2 - \alpha$.

There are three important parameters, closely related to inflation, in the above spectrum: (*i*) the curvature scale H_1, which controls the amplitude of the graviton background; (*ii*) the kinematic power α, which controls the slope of the spectrum; and (*iii*) the cut-off scale $\omega_1(t)$, which controls the position of the so-called "end-point" of the spectrum, beyond which the produced radiation becomes exponentially small. For the model represented in Fig. 7.3 the present value of the end-point frequency, ω_1, is obtained by rescaling $\omega_1(t_1) \equiv H_1$ through the radiation and matter epoch, down to the present time t_0. The result (in Planck units) is

$$\omega_1(t_0) = \omega_1(t_1)\left(\frac{a_1}{a_0}\right) = H_1\left(\frac{a_1}{a_{eq}}\right)\left(\frac{a_{eq}}{a_0}\right) = \left(\frac{H_1}{M_P}\right)^{1/2}\left(\frac{H_{eq}}{M_P}\right)^{1/2}\left(\frac{H_0}{H_{eq}}\right)^{2/3}M_P.$$

(7.179)

By using the values of H_0 and H_{eq} reported in Eqs. (1.78) and (1.79) one obtains

$$\omega_1(t_0) \simeq 4\left(\frac{H_1}{M_P}\right)^{1/2}\left(\frac{\Omega_m}{0.3}\right)^{-1/3}10^{11}\,\text{Hz}.$$

(7.180)

Thus, the transition scale H_1 in principle controls not only the *intensity* of the spectrum, but also its *extension* in frequency.

If the kinematical details of the post-inflationary evolution were known, from t_1 down to the present epoch, then the spectrum would contain only two unknown parameters (H_1 and α) carrying the direct imprint of the inflationary epoch. All other parameters are indeed determined by the subsequent evolution. In the model of Fig. 7.3, for instance, the two scales ω_0 and ω_{eq}, corresponding to a mode crossing the Hubble radius today and at the equality epoch, respectively, are fixed by

$$\omega_0 = H_0 \simeq 3.2 \times 10^{-18}\,h\,\text{Hz},$$

$$\omega_{eq}(t_0) = \omega_{eq}(t_{eq})\left(\frac{a_{eq}}{a_0}\right) = \frac{H_{eq}}{H_0}\left(\frac{H_0}{H_{eq}}\right)^{2/3}\omega_0 \simeq 75\,h\left(\frac{\Omega_m}{0.3}\right)^{2/3}\omega_0.$$

(7.181)

In that model, also, the radiation which starts dominating the Universe at the epoch t_1 is the same radiation today filling our present Universe. Using the value reported in Eq. (1.34), we have

$$\Omega_r(t_0) \simeq 4 \times 10^{-5}\,h^{-2},$$

(7.182)

and the graviton spectrum (7.178) is then completely determined for any given value of α and H_1.

7.3.1 Phenomenological bounds on the graviton background

For a comparison of the inflationary predictions with the existing phenomenological constraints we should note, first of all, that an inflationary model of conventional type is typically characterized by a phase of accelerated expansion with slightly decreasing (or constant) curvature: in particular, by $\alpha = -1$ for de Sitter, and by $\alpha < -1$ for power-law inflation (see Chapter 5). The spectrum (7.178) associated with a phase of de Sitter inflation is thus flat (i.e. "scale-invariant") from ω_{eq} to ω_1, and decreasing from ω_0 to ω_{eq}. In the case of power-law inflation both branches of the spectrum are decreasing. In any case, the maximum intensity is reached at the lower end of the spectrum, and it is thus clear that the most constraining bounds are expected to emerge from large-scale observations, possibly in the range of the lowest frequency ω_0.

We must recall, in this respect, the important bound obtained from the present measurements of the CMB anisotropy. In fact, a cosmic background of relic gravitational waves perturbs the large-scale homogeneity of the cosmological metric, and induces a corresponding distortion in the temperature of the cosmic background of electromagnetic radiation (see the discussion of the next chapter). One finds, in general, that the temperature fluctuations $|\Delta T/T|$, induced over a proper-length scale (a/k), are of the same order of magnitude as the amplitude of the tensor fluctuations, $|\delta_h|$, evaluated on the same length scale [16, 17, 18]. At the level of the present Hubble horizon

$$|\Delta T/T|_{H_0} \sim |\delta_h|_{k_0 = a_0 H_0} \tag{7.183}$$

(see Section 8.2 for a precise relation between the multipole expansion of $\Delta T/T$ and the tensor perturbation spectrum). Large-scale measurements [44, 45], on the other hand, imply

$$|\Delta T/T|_{H_0} \sim 10^{-5}. \tag{7.184}$$

These two conditions can be easily translated into a bound on the graviton energy density Ω_g present on large scales, by computing the stress tensor $\tau_{\mu\nu}$ associated with the tensor fluctuations. We can conveniently start from the E-frame action (7.41), written in the case $d = 3$, $n = 0$. In the cosmic time gauge we obtain

$$\tau_0^0 = \frac{M_P^2}{4} \sum_A \left[\dot{h}_A^2 + \left(\frac{\nabla h_A}{a} \right)^2 \right], \tag{7.185}$$

where the sum ranges over the two polarizations. By averaging over the spatial coordinates, using the Fourier transform (7.64) and the reality condition $h_{-k} = h_k^*$, we obtain

$$\rho_g = \langle \tau_0^0 \rangle = \frac{M_P^2}{2} \sum_A \int \frac{d^3 k}{(2\pi)^3} |h_A(k)|^2 \frac{k^2}{a^2} \tag{7.186}$$

(see also Section 7.4). Assuming that we are dealing with a stochastic, isotropic and unpolarized graviton background, we can repeat the same steps as in the computation of the spectral amplitude (7.87). Dividing by $\rho_c = 3M_P^2 H^2$ we finally obtain

$$\frac{d\rho_g}{\rho_c} = \frac{|\delta_h|^2}{6\pi^2}\left(\frac{k}{aH}\right)^2\frac{dk}{k},$$
(7.187)

from which, at the present time t_0,

$$\Omega_g(\omega, t_0) = \frac{1}{6\pi^2}|\delta_h(t_0)|^2\left(\frac{\omega}{\omega_0}\right)^2.$$
(7.188)

Taking into account that scalar metric perturbations also contribute to $\Delta T/T$ (and are possibly dominant, as we shall discuss in Section 8.2), from the combination of Eqs. (7.183), (7.184) and (7.188) we can obtain the following upper bound on the present energy density of the relic graviton background at the Hubble horizon scale:

$$h^2\Omega_g(\omega_0, t_0) \lesssim 10^{-10}, \qquad \omega_0 \simeq 10^{-18}\,\text{Hz}.$$
(7.189)

Using this bound, and the numerical values (7.180) and (7.181), it follows that the highest allowed graviton spectrum predicted by conventional inflationary models can be represented as in Fig. 7.4, where we have inserted the de Sitter case, $\alpha = -1$, and two examples of power-law inflation, with $\alpha = -1.1$ and $\alpha = -1.2$. Above the cut-off frequency we have assumed that the spectrum is exponentially suppressed as $\Omega_g(\omega) = \Omega_g(\omega_1)\exp[-(\omega - \omega_1)/\omega_1]$.

It should be noted that the condition (7.189) provides a very stringent constraint on the final inflation scale H_1. Indeed, by using Eq. (7.181) for ω_{eq}/ω_0, and deducing from Eq. (7.179) the relation

$$\frac{\omega_1}{\omega_0} = \left(\frac{H_1}{M_P}\right)^{1/2}\left(\frac{M_P}{H_0}\right)^{1/2}\left(\frac{H_0}{H_{eq}}\right)^{1/6},$$
(7.190)

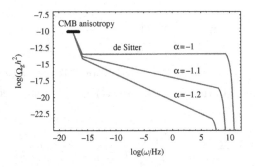

Figure 7.4 Typical examples of graviton spectra in standard models of de Sitter and power-law inflation. The cut-off frequency ω_1, beyond which the spectrum is exponentially suppressed, saturates the upper limit (7.193).

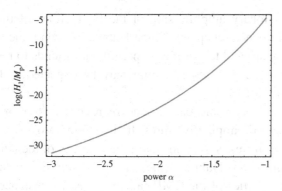

Figure 7.5 Maximal allowed value of the transition scale H_1 as a function of the power α, determining (in conformal time) the rate of inflationary expansion. The curve is obtained from Eq. (7.191) with $h = 0.7$ and $\Omega_m = 0.3$.

we can rewrite the isotropy bound for the spectrum (7.178) in the form

$$\left(\frac{H_1}{M_P}\right)^{1-\alpha} \lesssim 10^{-10}\,\Omega_r(t_0)^{-1}\left(\frac{M_P}{H_0}\right)^{1+\alpha}\left(\frac{H_0}{H_{eq}}\right)^{1+\alpha/3} h^{-2}. \qquad (7.191)$$

Inserting the numerical values H_0, H_{eq} and Ω_r, and choosing the typical values $h = 0.7$, $\Omega_m = 0.3$, we obtain the upper limit on the curvature scale H_1 plotted in Fig. 7.5. The highest allowed scale corresponds to the case of de Sitter inflation ($\alpha = -1$), which implies

$$H_1 \lesssim 10^{-5} M_P. \qquad (7.192)$$

The bound becomes more and more stringent as α becomes smaller than -1 in models of power-law inflation.

One may also note that the constraint on the scale H_1 induces a corresponding constraint on the allowed extension in frequency of the spectrum, since ω_1 is determined by H_1, according to Eq. (7.180). Using Eq. (7.191) one obtains

$$\log\left(\frac{\omega_1}{Hz}\right) \lesssim 11.6 + \frac{24.4 + 29.1\alpha}{1-\alpha} - \frac{2+\alpha}{1-\alpha}\log h + \frac{4}{3(1-\alpha)}\log\left(\frac{0.3}{\Omega_m}\right). \qquad (7.193)$$

This is the reason why *steeper* spectra are also *shorter* spectra, as apparent in Fig. 7.4, where all the spectra are plotted up to the maximal value of ω_1 allowed by Eq. (7.193). Assuming $\Omega_m = 0.3$ and $h = 0.7$, the previous bound is saturated by $\log(\omega_1/Hz) \simeq 9.3$ for $\alpha = -1$, by $\log(\omega_1/Hz) \simeq 8$ for $\alpha = -1.1$ and by $\log(\omega_1/Hz) \simeq 6.9$ for $\alpha = -1.2$, as illustrated in Fig. 7.4.

We should finally stress that the hypothesis of saturation of the CMB limit used in the plots of Fig. 7.4 is not completely ad hoc, being motivated by the standard inflationary mechanism for producing the observed CMB anisotropy, based on the inflationary amplification of the fluctuations of the metric tensor (see the details in the next chapter). At the Hubble horizon scale one obtains that the amplitude of scalar and tensor metric perturbations tends to be of comparable magnitude, so that the graviton spectrum may be expected to be close to the saturation of the limit (7.189).

This means that the cosmic background of relic gravitons might even have already been detected, implicitly, through the observations of the large-scale anisotropy [44, 45]. If this is true, however, the intensity of the tensor background would be so high, at the Hubble scale, that it should significantly also affect the polarization properties of the CMB radiation (see for instance [46, 47]). The planned measurements of the CMB anisotropy should clarify this point in the near future.

The situation is very different if the relic graviton background is characterized by a growing spectral distribution, which is typically the case for string cosmology models [26] like those associated with the pre-big bang [48] and with the ekpyrotic [49] scenarios (but also in the case of "quintessential inflation", see e.g. [50]). A growing spectrum reaches the peak in the high-frequency regime, and seems to be more accessible (at least in principle) to a possible direct detection by the existing gravitational antennas.

If the growth is very fast, however, the intensity of the spectrum at low frequencies is likely to be very small, in particular much smaller than the saturation limit (7.189). This implies that the *direct* inflationary amplification of tensor metric fluctuations is not at the level of contributing to the observed large-scale anisotropy. This is a positive result, on one hand, since it allows evasion of the stringent limit on the inflationary curvature scale imposed by the CMB anisotropy. On the other hand, a similar result should also be expected for the spectrum of the scalar metric perturbations, and this complicates the inflationary explanation of the observed anisotropy, requiring the introduction of an additional auxiliary field (the curvaton), as will be discussed in the next chapter.

Before presenting explicit examples of graviton spectra typical of string cosmology we should recall that, for growing spectra, there are additional phenomenological bounds that can constrain the intensity of the relic background at frequency scales much higher than those relevant to the CMB anisotropy.

A first bound follows from almost a decade of monitoring the radio pulse arriving from a number of millisecond pulsars: at present, no detectable distortion of pulsar timing, due to the presence of a background of relic gravitational waves, has been found. Consistency of these observations requires that the energy density of the background be small enough, at frequency scales of the order of the inverse of the observation time: more precisely [51],

$$h^2 \Omega_g(\omega_p, t_0) \lesssim 10^{-8}, \qquad \omega_p \simeq 10^{-8} \text{Hz}. \tag{7.194}$$

A second bound comes from the standard nucleosynthesis analysis [52]. In order to avoid too fast an expansion-rate of the Universe at the time of nucleosynthesis (which would spoil the remarkably accurate predictions of the observed abundance of light elements), the total energy density of the radiation that dominates the Universe at that time has to be bounded. This in turn constrains the energy density of the graviton background: its total energy, integrated over all modes, cannot exceed, roughly, the energy density of one massless degree of freedom in thermal equilibrium at the nucleosynthesis epoch (namely, about one-tenth of the total energy density). Using Eq. (7.182), this gives the following bound on the total integrated spectrum:

$$h^2 \int d\ln \omega \, \Omega_g(\omega, t_0) \lesssim 0.5 \times 10^{-5} \qquad (7.195)$$

(see [53] for an accurate computation). This number also provides a crude upper limit on the peak value of the spectrum, irrespective of its position in frequency.

7.3.2 Primordial gravitons from pre-big bang inflation

In order to present a typical example of a growing spectrum we can start again from the sketch of background evolution illustrated in Fig. 7.3, considering, however, an inflationary phase characterized by growing curvature (i.e. considering a phase of type II inflation, according to the classification of Chapter 5). In this case the inflationary pump field has a power $\alpha > -1$, and such a power may even be positive, in principle. Indeed, the pump field of an isotropic background corresponds to the E-frame scale factor, and we know that pre-big bang models are represented in the E-frame by a contracting phase (see Chapter 5), while ekpyrotic models are contracting even in the string frame [54] (see Chapter 10).

According to our previous results we may expect that a phase of inflation at growing curvature be associated with a growing graviton spectrum. In the simplest case of the low-energy, dilaton-dominated pre-big bang evolution, for instance, the power of the pump field is $\alpha = 1/2$ (see Eq. (7.99)), and one obtains a spectral distribution with the "maximal" slope $\Omega_g \sim \omega^3$ (modulo logarithmic corrections). The same is true for the so-called kinetic phase (dominated by a scalar modulus) of the ekpyrotic/cyclic scenario [55] that we will illustrate in Section 10.4. For all these models, the slope is obviously different in the low-energy "tail" of the spectrum ($\omega < \omega_{eq}$), which is also affected by the radiation-to-matter transition, according to Eq. (7.178); for such a rapidly growing spectrum, however, the amplitude of the low-frequency part is so damped that its particular slope becomes irrelevant in any realistic phenomenological application.

A model of string cosmology inflation characterized by only one kinematical power, that remains unchanged as the Universe evolves from an initial state (possibly approaching the string perturbative vacuum), to a final, radiation-dominated state, is probably much too simple to be realistic, however. Indeed, as discussed in Chapter 6, the transition to the standard cosmological phase is likely to occur in the high-curvature and strong-coupling regime, where the background is still characterized by an accelerated evolution, but the kinematics is in general different from that predicted by the tree-level action.

In order to take this effect into account we consider here a two-phase model of pre-big bang inflation. In the initial phase, extending in time from $\eta = -\infty$ to $\eta = -\eta_s$, the accelerated growth of the curvature and of the dilaton is described by the solutions of the lowest-order string effective action, with corresponding pump field $z \sim (-\eta)^{1/2}$. This low-energy evolution eventually leads to a second, high-curvature (string) phase, during which the S-frame curvature stays approximately frozen at a value controlled by the string scale M_s, while the dilaton keeps growing until the strong-coupling regime $g_s^2 = \exp\phi \sim 1$ is reached, and the curvature approaches the string scale also in the Einstein frame, $H_1 \sim M_s$. This second phases ranges in time from $\eta = -\eta_s$ to $\eta = -\eta_1$, may include higher-derivative effects, and the corresponding pump field will be parametrized by a generic power-law behavior, $z \sim (-\eta)^{\alpha}$. At the end of this "stringy" phase the Universe is expected to enter the standard radiation-dominated regime, and to follow all subsequent steps of a standard cosmological evolution (see Fig. 7.6). Note that the power α characterizing the string phase is not completely arbitrary for this model, because the graviton pump field, expressed through the S-frame scale factor a_s, can be written as $z \sim a_s \exp(-\phi/2)$. On the other hand, if the S-frame curvature is nearly constant, and the dilaton is growing for $\eta \to 0_-$, then, in the string phase,

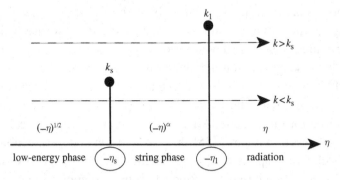

Figure 7.6 "Minimal" example of pre-big bang scenario. The S-frame curvature scale grows at low energies, and stays constant in the string phase. The final E-frame curvature scale is of the order of the string mass M_s. The string coupling is small at the beginning of the string phase, $g_s(\eta_s) \ll 1$, and becomes strong at the inflation-radiation transition, $g_s(\eta_1) \sim 1$. Also illustrated in the figure is the fact that high-frequency modes $(k > k_s)$ are only affected by the second background transition, while low-frequency modes $(k < k_s)$ are affected by both transitions.

$$a_{\rm s} \sim (-\eta)^{-1}, \qquad \exp(-\phi/2) \sim (-\eta)^{\beta}, \qquad \beta > 0,$$

$$z \sim (-\eta)^{\alpha} \quad \Rightarrow \quad \alpha = \beta - 1, \tag{7.196}$$

from which $\alpha \geq -1$.

The model of background evolution we have just described is called the "minimal" pre-big bang model [27], since it is the simplest example of consistent, inflationary evolution from the string perturbative vacuum, compatible with the dynamics of the string effective action. The complete model of the background includes, of course, the final transition to the matter-dominated epoch. However, such a phase is not explicitly shown in Fig. 7.6 because we confine ourselves to the branch $\omega > \omega_{\rm eq}$ of the spectrum, i.e. to modes which are not affected by this last transition. Indeed, as the spectrum is rapidly growing, its lowest-frequency part is very strongly suppressed and seems to be of little phenomenological relevance.

For the computation of the spectrum we follow the usual procedure, defining the pump field in the "sudden" approximation,

$$z(\eta) = \begin{cases} (M_{\rm P}/\sqrt{2})\,(\eta_{\rm s}/\eta_1)^{\alpha}\,(-\eta/\eta_{\rm s})^{1/2}, & \eta \leq -\eta_{\rm s}, \\ (M_{\rm P}/\sqrt{2})\,(-\eta/\eta_1)^{\alpha}, & -\eta_{\rm s} \leq \eta \leq -\eta_1, \\ (M_{\rm P}/\sqrt{2})\,((\eta+2\eta_1)/\eta_1), & -\eta_1 \leq \eta, \end{cases} \tag{7.197}$$

and matching the perturbations h, h' at $-\eta_{\rm s}$ and $-\eta_1$. We obtain a spectrum with two branches, since modes of high enough frequency ($k > k_{\rm s} = \eta_{\rm s}^{-1}$) will be affected only by the second transition, while modes with $k < k_{\rm s}$ will be affected by both transitions (see Fig. 7.6). We note that the evolution of tensor fluctuations in the high-curvature regime, $\eta > -\eta_{\rm s}$, should be described by perturbing the full background equations, possibly including all (higher-derivative) α' corrections (see Appendix 7A). The inclusion of such corrections, however, does not change in a significant, qualitative way the results obtained from the tree-level perturbation equation [12, 56] (at least until the string-phase curvature stays constant in the S-frame); this justifies a first estimate of the spectrum in terms of the low-energy equation (7.56).

For the high-frequency modes with $k > k_{\rm s}$ we can directly apply the result of Eq. (7.153). Taking into account both possibilities, $\alpha > 1/2$ and $\alpha < 1/2$, the frequency dependence of the energy density, for this branch of the spectrum, can then be parametrized as

$$\left(\frac{\omega}{\omega_1}\right)^4 |c_-(\omega)|^2 \sim \left(\frac{\omega}{\omega_1}\right)^{3-2|\nu|}, \qquad \nu = \frac{1}{2} - \alpha, \qquad \omega_{\rm s} < \omega < \omega_1 \tag{7.198}$$

(in agreement with Eq. (7.154) for $d = 3$).

In the case $k < k_s$, on the contrary, we must include the effects of both transitions. The solution of the canonical equation, in the three different phases, is

$$h_k^1(\eta) = e^{i\pi/2} \left(\frac{\pi\eta_s}{2M_P^2}\right)^{1/2} \left(\frac{x_1}{x_s}\right)^{\alpha} H_0^{(2)}(x), \qquad\qquad \eta \le -\eta_s,$$

$$h_k^2(\eta) = e^{i\pi\alpha} \left(\frac{\pi\eta_1}{2M_P^2}\right)^{1/2} \left(\frac{x}{x_1}\right)^{\nu} \left[c_+^2 H_\nu^{(2)}(x) + c_-^2 H_\nu^{(1)}(x)\right], \qquad -\eta_s \le \eta \le -\eta_1,$$

$$\tag{7.199}$$

$$h_k^3(\eta) = \left(\frac{\pi\eta_1}{2M_P^2}\right)^{1/2} \left(\frac{y}{x_1}\right)^{-1/2} \left[c_+^3 H_{-1/2}^{(2)}(y) + c_-^3 H_{-1/2}^{(1)}(y)\right], \qquad -\eta_1 \le \eta,$$

where $x = k\eta$ and $y = k\eta + 2k\eta_1$. By matching the solutions at $-\eta_s$ and $-\eta_1$, and by exploiting the Wronskian properties and derivative properties of the Hankel functions, we obtain

$$c_-^3 = \frac{\pi}{4} x_1 \left[-H_{-1/2}^{(2)}(x_1) \left(c_+^2 e^{i\pi\nu} H_{\nu+1}^{(1)} + c_-^2 e^{-i\pi\nu} H_{\nu+1}^{(2)}\right)_{x_1} \right.$$

$$\left. + \left(\frac{2\nu-1}{x_1} H_{-1/2}^{(2)} - H_{1/2}^{(2)}\right)_{x_1} \left(c_+^2 e^{i\pi\nu} H_\nu^{(1)} + c_-^2 e^{-i\pi\nu} H_\nu^{(2)}\right)_{x_1} \right]$$

$$\tag{7.200}$$

where $x_1 = k\eta_1$, and

$$c_+^2 e^{i\pi\nu} = \frac{i\pi}{4} x_s \left[-H_0^{(1)} H_{\nu+1}^{(2)} + H_\nu^{(2)} H_1^{(1)} + \frac{2\nu}{x_s} H_0^{(1)} H_\nu^{(2)} \right]_{x_s},$$

$$c_-^2 e^{-i\pi\nu} = \frac{i\pi}{4} x_s \left[-H_\nu^{(1)} H_1^{(1)} + H_0^{(1)} H_{\nu+1}^{(1)} - \frac{2\nu}{x_s} H_\nu^{(1)} H_0^{(1)} \right]_{x_s},$$

$$\tag{7.201}$$

where $x_s = k\eta_s$. Notice that in the limit $c_+^2 = 1$ and $c_-^2 = 0$ (in which one neglects the production of particles induced by the first transition) one precisely recovers the Bogoliubov coefficient of Eq. (7.169) (modulo a phase factor, due to a different choice of the initial phase).

Using the exact analytical expression (7.200) one can plot the spectral distribution of gravitons in the various frequency sectors (see for instance [57]). An approximate, analytical estimate for the frequency range $k \ll k_s < k_1$ can be easily obtained, however, by recalling the explicit form (7.171) of the Hankel functions for the special case $\nu = \pm 1/2$, and by exploiting the small argument expansions (7.141) and (7.155). The final result, to leading order in $x_1 \ll 1$, $x_s \ll 1$, is

$$c_-^3(k) = x_1^{-1/2} \left[\beta_1 \left(\frac{x_1}{x_s}\right)^{-\alpha+1/2} + (\beta_2 + \beta_3 \ln x_s) \left(\frac{x_1}{x_s}\right)^{\alpha-1/2} \right], \tag{7.202}$$

where β_1, β_2, β_3 are numerical coefficients determined by the asymptotic expansion of the Hankel functions:

$$\beta_1 = 2Aq_1, \qquad \beta_2 = A(2q_1 - 4i\nu p_0),$$

$$\beta_3 = 4\nu Aq_0, \qquad A = i\frac{\pi}{4}e^{-ix_1}\frac{(\pi/2)^{1/2}}{2^\nu\Gamma(1+\nu)}(2\nu - 1)q_\nu. \tag{7.203}$$

The frequency dependence of the energy density for this low-energy branch, neglecting the small logarithmic corrections, and taking into account that $\omega_s < \omega_1$, can then be written as

$$\left(\frac{\omega}{\omega_1}\right)^4 |c_-^3(\omega)|^2 \sim \left(\frac{\omega}{\omega_1}\right)^3 \left(\frac{\omega_s}{\omega_1}\right)^{-2|\alpha - 1/2|}, \qquad \omega < \omega_s, \tag{7.204}$$

which consistently matches the behavior of the high-frequency branch (7.198) at the transition scale $\omega = \omega_s$.

Putting the two branches together into the spectral energy density (7.160), and using the definition (7.163) of Ω_r, we can finally write the full graviton spectrum for this class of minimal pre-big bang models as

$$\Omega_g(\omega, t_0) = \left(\frac{H_1}{M_P}\right)^2 \Omega_r(t_0)\left(\frac{\omega}{\omega_1}\right)^{3-2|1/2-\alpha|}, \qquad \omega_s < \omega < \omega_1,$$

$$= \left(\frac{H_1}{M_P}\right)^2 \Omega_r(t_0)\left(\frac{\omega_s}{\omega_1}\right)^{-2|1/2-\alpha|}\left(\frac{\omega}{\omega_1}\right)^3, \qquad \omega_{eq} < \omega < \omega_s. \tag{7.205}$$

As before, we have absorbed inside the parameter H_1 the numerical factors of order one associated with the computation of the Bogoliubov coefficients. Also, as in the previous cases, the present value of the cut-off frequency $\omega_1(t_0)$ is determined by H_1 and by the post-inflationary evolution. For the simple post-inflationary scenario of Fig. 7.3, in particular, the result (7.180) is still valid. In this class of string cosmology models, however, the value of H_1 is not an arbitrary parameter as in the context of the standard inflationary models, but is closely related to the fundamental string mass scale, $H_1 \simeq M_s$. As a consequence, the end-point coordinates of the spectrum are basically controlled by the fundamental ratio between string and Planck mass: using Eqs. (7.180) and (7.182), with $\Omega_m = 0.3$ and $h = 0.7$, one finds

$$\omega_1(t_0) \simeq 4 \times 10^{11}\text{Hz}\left(\frac{M_s}{M_P}\right)^{1/2},$$

$$\Omega_g(\omega_1, t_0) \simeq 8 \times 10^{-5}\left(\frac{M_s}{M_P}\right)^2. \tag{7.206}$$

The most striking difference from the conventional inflationary predictions is the fast growth in frequency of the spectrum. The low-energy band $\omega < \omega_s$, in particular, is characterized by a nearly thermal (i.e. Rayleigh–Jeans) behavior, $\Omega_g \sim \omega^3$, that simulates the low-energy part of the black-body spectrum. At higher frequencies the slope is still growing, but it is in general flatter, as $\Omega_g \sim \omega^{3-2\nu}$, with $\nu > 0$ (see Fig. 7.7). These important spectral properties remain valid even when the graviton distribution is correctly computed including in the perturbation equations the required higher-curvature corrections (at least when they are truncated to first-order in α' [12, 56], see Appendix 7A).

It should be noted, also, that for the minimal models we are considering the spectrum associated with the string phase may be flat (in the limiting case in which $\alpha = -1$, corresponding to a frozen dilaton), but it cannot be decreasing. A decreasing spectrum would require $2\nu = |1 - 2\alpha| > 3$, namely $\alpha < -1$ or $\alpha > 2$. The first possibility is excluded by the assumption of non-decreasing dilaton, Eq. (7.196)), while the second is excluded by the fact that the cosmological background would become unstable, as already stressed in Section 7.2 (see, in particular, the comments following Eq. (7.80)). On the other hand, a decreasing high-frequency branch of the spectrum would become rapidly inconsistent with the existing phenomenological upper bounds, given the fixed value of the end-point of the spectrum (controlled by the ratio M_s/M_P according to Eq. (7.206)). For realistic values of this ratio, $M_s \sim 0.1 - 0.01 M_P$, consistent with string models of unified gravitational and gauge interactions [58], the end-point value $\Omega_g(\omega_1)$ is in fact automatically compatible with the nucleosynthesis bound (7.195), but is already quite close to the maximal allowed value, as illustrated in Fig. 7.7.

The spectra plotted in Fig. 7.7 have been obtained from Eqs. (7.205) and (7.206) using the maximum theoretically expected value, $M_s \simeq 0.1 M_P$, and choosing different values for the two (independent and arbitrary) parameters α and ω_s. In particular: $\alpha = -1, -0.9, -0.75$, and $\log(\omega_s/\omega_1) = -8, -11, -5$, respectively. Above the cut-off frequency we have assumed the usual exponential suppression, controlled by the factor $\exp[-(\omega - \omega_1)/\omega_1]$. It is clear from the figure that, once the end-point has been fixed at the string scale, the resulting graviton distribution (7.205) is fully determined by the two parameters α and ω_s, which control the rate of kinematical evolution and the extension in time of the string phase, or – equivalently – the coordinates of the break-point of the spectrum in the plane $\{\omega, \Omega_g\}$.

From a physical point of view such parameters can be related to the value of the string coupling g_s at the time when the S-frame curvature reaches the string scale, and to the rate of growth of g_s during the string phase (see the discussion of the next chapter). They depend, in principle, on the initial conditions and on the dynamical details of the high-curvature regime, and thus remain completely

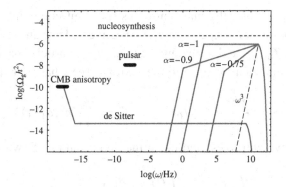

Figure 7.7 Three examples of spectra for the minimal pre-big bang scenario, corresponding to different values of α $(-1, -0.9, -0.75)$, and ω_s/ω_1 $(10^{-8}, 10^{-11}, 10^{-5})$. For the peak position we have used Eq. (7.206) with $M_s = 0.1 M_P$. The thin, dashed line labeled ω^3 corresponds to a low-energy model with negligible extension in time of the string phase (i.e. with $\omega_s \simeq \omega_1$). Also shown, for comparison, is the scale-invariant spectrum associated with standard de Sitter inflation, and the three phenomenological bounds imposed by the CMB anisotropy (7.189), pulsar-timing data (7.194) and nucleosynthesis (7.195).

arbitrary in the context of a minimal model which only fixes the peak value of the spectrum and its position in frequency. It follows, in particular, that the string branch of the spectrum could also be extended to arbitrarily small frequency scales – provided the slope is not too flat, to avoid conflicting with the pulsar and CMB bounds on the relic intensity at large scales. It should be recalled, also, that the nucleosynthesis bound (7.195) applies to the total integrated energy density, and thus becomes more and more stringent as the spectrum flattens and the string phase lengthens.

The precision attained by the minimal model in fixing the present coordinates of the end-point of the spectrum might be reduced, however, if we take explicitly into account that the present values of the spectral parameters are also dependent on the details of the post-inflationary evolution. The actual peak value, in particular, could be depressed with respect to the predictions of Eq. (7.206), if the critical fraction of radiation energy density that we are presently observing has not been entirely produced at the transition epoch t_1, but contains contributions from later epochs.

This effect can be appropriately illustrated by introducing the phenomenological parameter δS, representing the fraction of present (black-body) entropy due to all reheating processes possibly occurring after the end of the string phase. Such a parameter is defined by [53]

$$\delta S = \frac{S_0 - S_1}{S_0}, \tag{7.207}$$

where S_0 and S_1 are the (thermal) entropies (per unit comoving volume) of the radiation background at $t = t_0$ and $t = t_1$, respectively. They are defined in terms of the temperatures T_0, T_1, and of the numbers n_0, n_1 of particle species contributing (with their own statistical weight) to the entropy (see Chapter 1, Eq. (1.67)):

$$S_0 = \frac{2\pi^2}{45} n_0 (a_0 T_0)^3, \qquad S_1 = \frac{2\pi^2}{45} n_1 (a_1 T_1)^3. \tag{7.208}$$

Combining the two entropies, and using the definition of δS, we are led to

$$\omega_1(t_0) \equiv H_1 \left(\frac{a_1}{a_0} \right) = T_0 \left(\frac{H_1}{T_1} \right) \left(\frac{n_0}{n_1} \right)^{1/3} (1 - \delta S)^{1/3}. \tag{7.209}$$

Let us now assume that the radiation which starts dominating the Universe at the epoch t_1 is produced in a state of thermal equilibrium, so that

$$H_1^2 = \frac{\pi^2 N_1}{90 M_P^2} T_1^4, \tag{7.210}$$

according to Eq. (1.60). Dividing by T_1^2 and M_P^2, respectively, and combining the two corresponding equations, we obtain the useful condition

$$\frac{H_1}{T_1} = \left(\frac{\pi^2 N_1}{90} \right)^{1/4} \left(\frac{H_1}{M_P} \right)^{1/2}, \tag{7.211}$$

which we can insert into Eq. (7.209). Assuming $H_1 \simeq M_s$, $N_1 \simeq n_1 \sim 10^3$, and using the standard numerical values associated with the present black-body radiation (composed of photons and three neutrino types, see Chapter 1),

$$n_0 = \sum_b N_b + \frac{7}{8} \sum_f N_f \left(\frac{T_\nu}{T_\gamma} \right)^3 = 2 + \frac{7}{8} \times 6 \times \frac{4}{11} \simeq 3.9,$$

$$T_0 = T_\gamma \simeq 2.3 \times 10^{-4} \, \text{eV} \simeq 3.5 \times 10^{11} \, \text{Hz}, \tag{7.212}$$

we finally obtain

$$\omega_1(t_0) \simeq 1.5 \left(\frac{M_s}{M_P} \right)^{1/2} \left(\frac{10^3}{n_1} \right)^{1/12} (1 - \delta S)^{1/3} 10^{11} \, \text{Hz}. \tag{7.213}$$

In the absence of models providing a detailed (and reliable) description of the transition from the string to the radiation phase, and allowing a precise geometric definition of the maximum amplified frequency ω_1, we can reasonably identify the end-point of the spectrum with the limiting frequency corresponding to the production of one graviton per polarization state and per unit phase-space volume [53]. In that case $|c_-(\omega_1)|^2 = 1$, and the associated energy density, from Eqs. (7.160) and (7.213), is

$$\Omega_g(\omega_1, t_0) = \frac{\omega_1^4(t_0)}{\pi^2 \rho_c} \simeq 1.2 \times 10^{-6} \, h^{-2} \left(\frac{M_s}{M_P}\right)^2 \left(\frac{10^3}{n_1}\right)^{1/3} (1 - \delta S)^{4/3}.$$

(7.214)

In spite of the different definition of the end-point parameters, we find that the values of ω_1 and $\Omega_g(\omega_1)$ that we have obtained are in good agreement with the previous estimates (7.206), in the limit $\delta S = 0$.

If $\delta S \neq 0$, i.e. if we are in the presence of additional, post-inflationary processes of radiation production (due to some reheating phase occurring at epochs much later than t_1), then the present peak values of ω_1 and $\Omega_g(\omega_1)$ turn out to be lower than the predictions of the minimal model, because of an effective dilution (in critical units) of the graviton energy density. Suppose, for instance, that only a very small fraction (say, 0.1%) of the present black-body entropy is a true relic of the transition between inflation and the radiation phase. In that case $1 - \delta S = 10^{-3}$, so that – according to Eq. (7.214) – the height of the peak has to be lowered by four orders of magnitude with respect to the value reported in Fig. 7.7. Even then, however, the intensity of the graviton background would stay well above the flat, de Sitter spectrum, which represents the most optimistic prediction of the standard inflationary scenario. In addition, one should not forget that the above dilution of the energy density applies to any given primordial graviton background, and then also to the graviton background produced in the context of standard inflation.

Another possible source of uncertainty about the end-point values of the spectrum, in the context of the pre-big bang scenario, arises from the fact that the fundamental ratio M_s/M_P is expected to control the *height* of the peak, but not necessarily its *position* in frequency. There are indeed "non-minimal" models where the strong correlation between the end-point frequency and the peak of the spectrum is lost, and the peak may be located at frequency scales much lower than ω_1.

In fact, in the minimal scenario illustrated in Fig. 7.6, and corresponding to the spectrum (7.205), the end of the high-curvature phase coincides with the freezing of the string coupling parameter g_s^2 and with the beginning of the standard radiation era. However, we may also consider a different, non-minimal scenario in which the coupling is still small at the end of the string phase, $g_s^2(t_1) \ll 1$, so that the dilaton keeps growing (in a decelerated way) also for $t > t_1$ while the curvature is already decreasing, and the radiation produced in the transition at $t = t_1$ becomes dominant only at much later times [27].

The main difference between these two possibilities is that in the non-minimal case the effective potential appearing in the canonical perturbation equation is non-monotonic even before the beginning of the standard radiation epoch: as a consequence, the highest-frequency modes re-enter inside the horizon during the intermediate, dilaton-dominated phase which follows the string phase and precedes the beginning of the radiation era. This modifies the slope of the spectrum in the high-frequency sector, with the possible appearance of a negative power of ω: the full graviton spectrum may become non-monotonic, and the peak may no longer be coincident with the end-point [27, 40, 59, 60] (see also [61]).

This is good news, from an experimental point of view, because it suggests the possibility of a maximum signal even at frequencies much lower than the gigahertz band, predicted by Eq. (7.206). However, it also provides a warning against a too naive interpretation (and extrapolation) of possible experimental data, in view of the complexity of the parameter space of string cosmology models. In spite of these theoretical uncertainties it is still possible to define, with reasonable accuracy, the maximal allowed region for the expected graviton background in the phenomenological plane $\{\omega, \Omega_g\}$.

The allowed region is shown in Fig. 7.8 for the frequency range $\omega > 10^{-4}$ Hz, that seems to be phenomenologically relevant for the present gravitational detectors (see next section). The maximal allowed intensity coincides with the peak (7.206) of the minimal pre-big bang models, evaluated for the theoretical upper limit of the string-to-Planck mass ratio $M_s = 0.1 M_P$. This leads to the border value

$$\Omega_g \lesssim 10^{-6} h^{-2}, \tag{7.215}$$

which also includes the small uncertainties associated with the identification $H_1 \sim M_s$. This value is automatically compatible with the nucleosynthesis bound (7.195), even if the spectrum is nearly flat from the gigahertz down to the millihertz scale. We have also reported in the figure three lines of constant strain density, $\sqrt{S_h} = 10^{-19}, 10^{-23}, 10^{-25}$ Hz$^{-1/2}$. The strain density, which we shall introduce in the next section, is an alternative variable for characterizing the intensity of gravity-wave backgrounds, more convenient for a direct comparison with the experimental sensitivities of the gravitational antennas.

Also shown in the figure is the scale-invariant de Sitter spectrum, in order to emphasize the relatively large enhancement (up to eight orders of magnitude) of the relic background expected in models of pre-big bang inflation, with respect to that expected in the standard inflationary scenario. Such an enhancement is basically due to the fact that the standard spectrum is decreasing, the normalization is imposed at the low-frequency end of the spectrum, and the peak value is controlled by the anisotropy of the CMB radiation, which imposes (for $\omega \gtrsim \omega_{eq}$) $\Omega_g \lesssim \Omega_r(t_0)(\Delta T/T)^2 \lesssim 10^{-14}$ (see Eq. (7.178)). The pre-big bang spectrum, on the contrary, is growing, the normalization is imposed at the high-frequency end of the spectrum, and the peak value is controlled by the string-to-Planck

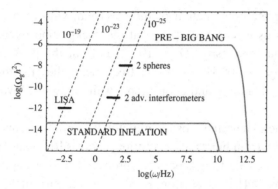

Figure 7.8 The allowed region for the relic graviton spectrum is below the upper border (marked *pre-big bang*) for models of string cosmology inflation, and below the lower border for the standard inflationary scenario. The three dashed lines correspond to three different, constant values of the strain density (see Eq. (7.224)). The figure also shows the level of sensitivity expected to be reached through the cross-correlation of two resonant spherical detectors, two advanced (ground-based) interferometers, and two space-interferometers such as LISA (see Section 7.4).

mass ratio, which imposes $\Omega_g \lesssim \Omega_r(t_0)(M_s/M_P)^2 \lesssim 10^{-6}$ (see Eq. (7.205)). This explains the eight orders of magnitude between the two allowed regions of Fig. 7.8.

It should be stressed that a blue graviton spectrum may be regarded as a rather typical prediction of string cosmology. However, such a high level of intensity, like that obtained in the context of the pre-big bang scenario, is not necessarily a property of all string cosmology models. We can mention, as an important counter-example, the case of the ekpyrotic/cyclic scenario where, as in the pre-big bang case, there are three main cosmological phases: (1) an initial "ekpyrotic" phase, in which two colliding branes are slowly approaching one another along an external spatial dimension; (2) a "kinetic" phase, dominated by a modulus scalar field which determines the interbrane distance, and in which the scale factor is rapidly contracting, leading the Universe to bounce, and then re-expand; (3) the standard radiation-dominated phase (see Section 10.4 for a detailed discussion). The corresponding graviton spectrum is similar, in many respects, to the pre-big bang spectrum: it is flat at very low frequencies, and growing at high frequencies, with a monotonic slope. In particular [55]:

$$\Omega_g \sim \begin{cases} \omega^3, & \omega_r < \omega < \omega_{\text{end}}, \\ \omega^2, & \omega_{\text{eq}} < \omega < \omega_r, \\ \text{const}, & \omega < \omega_{\text{eq}}, \end{cases} \qquad (7.216)$$

where the frequency scales ω_r and ω_{end} correspond to modes crossing the horizon at the onset of the radiation era and at the end of the ekpyrotic phase, respectively.

However, a complete computation of the spectrum, with realistic values of the parameters, leads us to conclude [55] that the background intensity reaches the standard inflationary intensity ($\Omega_g \sim 10^{-14}$) only at the peak of the spectrum, located around $\omega_{end} \sim 10^8$ Hz. At lower frequencies the graviton distribution is suppressed by the quadratic and cubic slope, well below the level of standard de Sitter inflation.

Stronger (but non-Gaussian) cosmic graviton backgrounds can arise, in a string cosmology context, from bursts of gravitational waves [62] generated by a cosmic network of fundamental strings, possibly produced [63] in models of D-brane–antibrane inflation [64] (see Section 10.5). The intensity of such backgrounds might be comparable to the maximal intensity predicted by pre-big bang models, even for relatively small values of the effective tension of the produced strings. They are thus in principle accessible to the expected sensitivities of near-future interferometric detectors, as will be discussed in the next section.

Primordial gravitational backgrounds of inflationary origin, generated by mechanisms other than the direct amplification of tensor perturbations, can also appear in the context of the conventional inflationary models. These backgrounds are possibly due to graviton radiation from cosmic strings and topological defects [65], from bubble collisions at the end of a first-order phase transition [66], from parametric resonance of the inflaton oscillations [67], etc. Their intensity can easily exceed the de Sitter bound of Fig. 7.8, but tends to stay lower than (or at most equal to) the maximal allowed string cosmology level (see for instance [68]).

We should mention, finally, the possibility of graviton "foregrounds" of astrophysical (non-primordial) origin, which are the superposition of gravitational waves produced by a large number of individual sources. These backgrounds can be stochastic, and may in principle overcome even the nucleosynthesis limit (as they are produced later), but in that case they are peaked around a rather narrow frequency band. In any case, their presence has the important effect of determining a "fundamental" sensitivity limit for any gravitational antennas looking for primordial background in the same frequency range [69].

7.4 Sensitivities and cross-correlation of gravitational detectors

The intensity of a cosmic background of gravitational radiation can be parametrized not only by its spectral energy density, $\Omega_g(\omega)$, but also by the so-called "strain density", $S_h(\omega)$. This second variable is more appropriate for a direct comparison of the induced signal with the noise power-spectrum of a gravitational detector, and can be conveniently used in order to define the experimental sensitivity required to detect a given background.

The precise relation between Ω_g and S_h can be deduced starting from the quadratic (E-frame) action (7.41) for tensor perturbations, which can be written in covariant form (and in $d = 3$) as

$$\delta^{(2)}S = \frac{M_P^2}{2}\frac{1}{4}\int d^4x\sqrt{-g}\, g^{\mu\nu}\partial_\mu h_i^j \partial_\nu h_j^i. \tag{7.217}$$

Decomposing h_i^j into the two physical polarization modes, $h_i^j = h_A(\epsilon^A)_i^j$, $A = 1, 2$, with $\mathrm{Tr}(\epsilon^A\epsilon^B) = 2\delta^{AB}$, the variation of the above action with respect to the unperturbed metric, according to the standard definition (1.3), leads to the energy-momentum tensor

$$T_{\mu\nu} = \frac{M_P^2}{2}\sum_A\left[\partial_\mu h_A \partial_\nu h_A - \frac{1}{2}g_{\mu\nu}(\partial_\alpha h_A)^2\right]. \tag{7.218}$$

The energy density of a cosmic graviton background, consisting of a stochastic collection of standing waves of frequency ω and wave-number $\vec{k}/a = \omega\hat{n}$, with $|\hat{n}| = 1$, is then obtained by taking the spatial (or *ensemble*) average of the τ_0^0 component,

$$\rho_g = \langle\tau_0^0\rangle = \frac{M_P^2}{2}\sum_A\langle\dot{h}_A^2\rangle = \frac{M_P^2}{4}\langle\dot{h}_{ij}\dot{h}^{ij}\rangle \tag{7.219}$$

(see e.g. [13]), in agreement with the results already presented in Eq. (7.186). In a quantum field theory context, in which the metric fluctuations are quantized, the brackets denote the quantum expectation value in a generic n-particle state.

For a better comparison with the experimental variables we now expand the fluctuations in Fourier modes, introducing explicitly the variable ν such that $\omega = 2\pi\nu$, and working in units $h = 1$ (i.e., $\hbar = 1/2\pi$):

$$h_A(\vec{x}, t) = \int_{-\infty}^{+\infty}d\nu\, h_A(\vec{x}, \nu)e^{-2\pi i\nu t}, \tag{7.220}$$

where h_A satisfies the reality condition $h_A(-\nu) = h_A^*(\nu)$. The energy density becomes

$$\rho_g = \frac{M_P^2}{2}4\pi^2\sum_A\int_{-\infty}^{+\infty}d\nu\, d\nu'\,\nu\nu'\,\langle h_A(\nu)h_A^*(\nu')\rangle e^{-2\pi i(\nu-\nu')t}. \tag{7.221}$$

Let us also define, for a stochastic background of gravitational waves (see for instance [24]),

$$\langle h_A(\nu)h_{A'}^*(\nu')\rangle = \delta_{AA'}\delta(\nu-\nu')\frac{1}{2}S_h^A(|\nu|), \tag{7.222}$$

where $S_h^A(|\nu|)$ is the so-called one-sided strain density, defined in the positive frequency range, with dimensions Hz^{-1}. Inserting this definition into Eq. (7.221), and summing over polarizations (which gives a factor of 2 for unpolarized backgrounds with $S_h^1 = S_h^2 = S_h$), we obtain

$$\rho_g = 4\pi^2 M_P^2 \int_0^\infty d\nu \, \nu^2 S_h(|\nu|) \equiv \int d\rho_g, \qquad (7.223)$$

from which

$$\Omega_g(\nu, t_0) = \frac{1}{\rho_c} \frac{d\rho_g}{d\ln\nu} = \frac{4\pi^2 |\nu|^3}{3H_0^2} S_h(|\nu|). \qquad (7.224)$$

Numerically,

$$h^2 \Omega_g(\nu, t_0) \simeq 1.28 \times 10^{45} S_h(|\nu|) \left(\frac{\nu}{kHz}\right)^3 Hz. \qquad (7.225)$$

The curves $S_h = $ const, therefore, are lines of angular coefficient 3 in the plane $\{\log\omega, \log\Omega_g\}$ (see Fig. 7.8).

The strain-density variable S_h is a useful indicator of the minimal level of instrumental noise of a gravitational antenna required for the successful detection of a given gravitational background. The instrumental noise, in fact, is described by the so-called one-sided noise power spectrum, $P|(\nu|)$, defined by

$$\langle n(\nu)n^*(\nu')\rangle = \delta(\nu - \nu')\frac{1}{2}P(|\nu|) \qquad (7.226)$$

(see for instance [24, 70]), where $P(|\nu|)$ is a real function defined over all positive frequencies, and $n(\nu)$ is the Fourier component of the noise $n(t)$. In particular,

$$\langle n^2(t)\rangle = \int_0^\infty d\nu \, P(|\nu|). \qquad (7.227)$$

The comparison of Eqs. (7.226) and (7.222) shows that the signal produced by a stochastic background of gravitational waves – which manifests itself as an excess of noise in the gravitational antenna – will in principle be detectable (by a single instrument) only if $S_h > P$. The noise power spectrum, $P(|\nu|)$, of a given instrument thus determines the minimal intensity of a graviton background detectable by that instrument. Vice versa, the spectral intensity $S_h(|\nu|)$ of a graviton background specifies the *maximal noise* $P(|\nu|)$ (i.e. the *minimal sensitivity*) required for its possible detection.

We have seen, for instance, that the relic background of cosmic gravitons produced in the context of the pre-big bang scenario is characterized by the *maximal* expected intensity

$$\Omega_g^{max} \simeq 10^{-6}h^{-2}, \qquad (7.228)$$

corresponding to the upper border of the allowed region of Fig. 7.8. It follows, from Eq. (7.224), that the *minimal* sensitivity required for its possible detection is [27]

$$\left(\sqrt{S_h}\right)^{\text{min}} = \nu^{-3/2} \left(\frac{3H_0^2}{4\pi^2}\right)^{1/2} \left(\sqrt{\Omega_g}\right)^{\text{max}} \simeq 2.8 \times 10^{-26} \left(\frac{\text{kHz}}{\nu}\right)^{3/2} \text{Hz}^{-1/2}.$$

(7.229)

It is important to notice that the minimal sensitivity level required to cross the border of the allowed region grows with the frequency band to which the detector is tuned – even if such a border is the same at all frequency scales – as clearly illustrated in Fig. 7.8. As a consequence, detectors working (or planned to work) at lower frequencies are strongly favored, from an experimental point of view, with respect to high-frequency detectors. Unfortunately, from a theoretical point of view, the probability of a large background intensity seems to be greater at higher frequencies, at least in the context of the minimal pre-big bang scenario (see Fig. 7.7).

In any case, the direct detection of a primordial graviton background by means of a unique gravitational antenna is very unlikely, at present, because of two important problems.

The first problem is the fact that the sensitivity of presently operating detectors is rather far from the limiting value (7.229). For the cryogenic, resonant-mass gravitational antennas (ALLEGRO, AURIGA, EXPLORER, NAUTILUS and NIOBE) the best available sensitivity (see for instance [71] for NAUTILUS) is at present around $\sqrt{S_h} \simeq 3 \times 10^{-22}$ Hz$^{-1/2}$, with resonant frequencies in the kHz range (the typical noise spectrum of a resonant-bar detector is reported in Fig. 7.9). The corresponding detectable level of Ω_g, from Eq. (7.225), is only $h^2 \Omega_g \sim 10^2$. For the present interferometric antennas (GEO 600, LIGO, TAMA 300, VIRGO) the best sensitivity of the first-generation detectors [72, 73] is also around $\sqrt{S_h} \sim 10^{-22}$ Hz$^{-1/2}$. Such a sensitivity, however, is available in a lower frequency range, $\nu \sim 10^2$ Hz, and thus corresponds to a lower value of the detectable energy density, $h^2 \Omega_g \sim 10^{-1}$.

In all cases we are dealing with sensitivities well above the border of the allowed region, specified by Eq. (7.228). In order to approach the border we must wait, for instance, for the second (and higher) generations of interferometric detectors. In particular, the Advanced LIGO and Enhanced LIGO projects (also called LIGO II and LIGO III, see [74]), are expected to improve the present LIGO I sensitivity by two orders of magnitude, thus reaching the borderline (7.228). For a better illustration of the expected sensitivities we have plotted in Fig. 7.10 the analytical fits [75] of the noise power spectrum of LIGO I,

$$P(|\nu|) = \frac{3}{2}10^{-46} \left[\left(\frac{\nu}{200\,\text{Hz}}\right)^{-4} + 2 + 2\left(\frac{\nu}{200\,\text{Hz}}\right)^2\right] \text{Hz}^{-1}, \qquad \nu > 40\,\text{Hz},$$

(7.230)

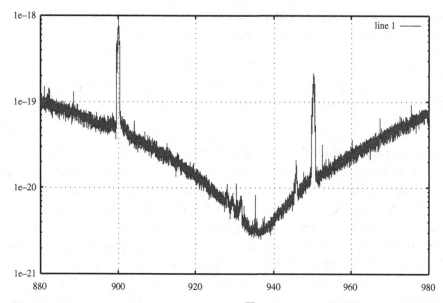

Figure 7.9 Typical noise spectral density $\sqrt{P}$ (in units $Hz^{-1/2}$) versus frequency (in Hz) for the resonant-bar NAUTILUS. The plot refers to the run of the year 2003, and to a bar temperature of 2 K (courtesy of the NAUTILUS group; see also www.roma1.infn.it/rog).

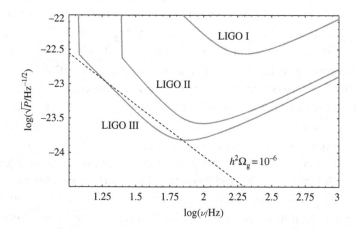

Figure 7.10 The square root of the noise power spectrum of Eqs. (7.230), (7.231) and (7.232). The noise is assumed to go to infinity below the seismic cut-off frequencies at 40 Hz, 25 Hz and 12 Hz, respectively. The dashed line is obtained from Eq. (7.229) by identifying $S_h(|\nu|)$ with $P(|\nu|)$. It represents the maximal level of noise possibly compatible with a single-instrument detection of pre-big bang gravitons.

of LIGO II,

$$P(|\nu|) = \frac{7.9}{11} 10^{-48} \left[\left(\frac{\nu}{110\,\text{Hz}} \right)^{-9/2} + \frac{9}{2} + \frac{9}{2} \left(\frac{\nu}{110\,\text{Hz}} \right)^2 \right] \text{Hz}^{-1}, \qquad \nu > 25\,\text{Hz},$$

(7.231)

and of LIGO III,

$$P(|\nu|) = \frac{2.3}{5} 10^{-48} \left[\left(\frac{\nu}{75\,\text{Hz}} \right)^{-4} + 2 + 2 \left(\frac{\nu}{75\,\text{Hz}} \right)^2 \right] \text{Hz}^{-1}, \qquad \nu > 12\,\text{Hz}.$$

(7.232)

The infrared cut-off of the above spectra is a seismic cut-off, below which the effective noise can be treated as infinite. Also shown in Fig. 7.10 is the limiting noise that should be reached in order to detect (with a single instrument) a graviton background with $\Omega_g \leq 10^{-6} h^{-2}$.

It may be interesting to note that such a limit may become easily accessible to gravitational antennas like the space interferometers of the LISA project (see for instance [76]), because in space the detectors are not affected by seismic noise, which is crucial in limiting the sensitivity band of the Earth-based antennas. The goal of LISA is to reach the sensitivity $\sqrt{S_h} \simeq 4 \times 10^{-21}\,\text{Hz}^{-1/2}$, in the frequency band from 3 to 10 mHz. The corresponding value of detectable energy density, in this frequency band, is the remarkably small value $h^2 \Omega_g \sim 10^{-11}$, well below the upper bound (7.228). Another interferometric antenna in space, which should reach the maximal sensitivity in a frequency band intermediate between LISA and LIGO, is the so-called DECIGO project [77]. The planned sensitivity is $\sqrt{S_h} \simeq 10^{-23}\,\text{Hz}^{-1/2}$, in the frequency band from 0.1 to 1 Hz, which would correspond to $h^2 \Omega_g \sim 10^{-13}$. We should also quote another project, consisting of a "constellation" of four spatial interferometers (operating in the same range as DECIGO): the so-called "big bang observer" (BBO), which is presently begin investigated by NASA [78]. All these space interferometer projects seem to offer promising possibilities for exploring, with a single detector, a region of parameter space physically relevant for models of string cosmology inflation.

At higher frequencies there are other interesting projects based on resonant-mass detectors with spherical (or truncated icosahedron) geometry, such as TIGA [79] or SFERA [80]. Hollow spheres, pushed to their extreme quantum limit, are also particularly promising (see for instance [81]), as they could reach a strain sensitivity $\sqrt{S_h} \simeq 3 \times 10^{-24}\,\text{Hz}^{-1/2}$ in the kHz frequency range (the resonant frequency can be tuned, in principle, by varying the radius of the sphere). Finally, there is work in progress on the possible use of two coupled electromagnetic microwave cavities as a high-frequency gravitational detector. It seems possible, at present, to reach a strain sensitivity $\sqrt{S_h} \simeq 10^{-20}\,\text{Hz}^{-1/2}$ for frequencies in the kHz range [82] (which is, however, well above the required sensitivity (7.229)).

As explicitly illustrated by the above examples, the limiting sensitivity (7.229) would seem to be out of reach for both present and near-future detectors (with the possible important exception of space interferometers). There is, in addition, a second difficulty in the case of single-instrument detection, from which even space interferometers are not free. Even assuming that the required sensitivity is reached, an unambiguous detection of a stochastic background of primordial gravitational radiation with a single experimental apparatus would require a complete and exact knowledge of all the intrinsic experimental noises, as well as of all (non-primordial) gravitational backgrounds of different origin which could interact with the given detector.

Fortunately, an efficient answer to both difficulties is known. The answer consists in the search for the background signal not through a *single* antenna, but through the cross-correlation of the outputs of *two or more* antennas [24, 83, 84].

7.4.1 Correlated response of two detectors

This important technique is based on the fact that the generic outputs $s_1(t)$ and $s_2(t)$ of two different detectors (where $s(t)$ represents, in practice, the microscopic oscillations of the detector as a function of time), can be decomposed as

$$s_{1,2}(t) = h_{1,2}(t) + n_{1,2}(t), \tag{7.233}$$

where $h(t)$ is the so-called "physical strain", i.e. the signal induced by the incident gravitational radiation, while $n(t)$ represents the noise. The noise component depends on both the instrumental properties of the detector and the possible local disturbance. The signal component depends on both the properties of the incident radiation (intensity, polarization, propagation direction) specified by the metric fluctuation tensor h_{ij}, and the geometrical properties of the antenna itself (shape, orientation) specified by the so-called "response tensor" D_{ij} (see below).

For an explicit definition of the signal it is convenient to expand the incident radiation in plane waves of proper frequency (or energy) ν and proper wave-number $\vec{p} = \vec{k}/a = 2\pi\nu\widehat{n}$, where $\widehat{n}$ is a unit vector specifying the propagation direction on the two-sphere Ω_2. We then define

$$h_{ij}(t, \vec{x}) = \int_{-\infty}^{+\infty} d\nu \int_{\Omega_2} d^2\widehat{n}\, h_A(\nu, \widehat{n})\, \epsilon_{ij}^A(\widehat{n}) e^{2\pi i\nu(\widehat{n}\cdot\vec{x}-t)}, \tag{7.234}$$

where each polarization component h_A satisfies the reality condition $h_A(\nu) = h_A^*(-\nu)$. The physical strains $h_{1,2}(t)$, induced in each of the two antennas, are then obtained by projecting the incident radiation onto the response tensor of the detectors, $D^{ij}_{1,2}$, namely,

$$h_{1,2}(t) = \int_{-\infty}^{+\infty} d\nu \int_{\Omega_2} d^2\widehat{n} \, h_A(\nu, \widehat{n}) \, F^A_{1,2}(\widehat{n}) \, e^{2\pi i\nu(\widehat{n}\cdot\vec{x}_{1,2}-t)}. \tag{7.235}$$

Here $\vec{x}_{1,2}$ are the (known and constant) positions of the centers of mass of the detectors, and

$$F^A_{1,2}(\widehat{n}) = \frac{1}{2}\epsilon^A_{ij}(\widehat{n})D^{ij}_{1,2} \tag{7.236}$$

are the so-called "pattern functions". They parametrize, for each antenna, the response to a given polarization mode and to a given propagation direction of the incident radiation, as a function of the spatial orientation of the axes of the detector [24, 70].

It should be noted that a gravitational antenna may be characterized by more than one response mode, associated with different response tensors. The response of the interferometric antennas, for instance, may be described by the so-called *differential mode*, represented by the symmetric, trace-free tensor

$$D^{ij} = u^i u^j - v^i v^j, \tag{7.237}$$

where $\widehat{u}$ and $\widehat{v}$ are two unit vectors specifying the directions (not necessarily orthogonal) of the two arms of the interferometer. But there is also the *common mode*, represented by the tensor

$$D^{ij} = u^i u^j + v^i v^j. \tag{7.238}$$

The main response of a cylindrical, resonant-bar detector, with axis along the $\widehat{u}$ direction, is represented by the tensor $D^{ij} = u^i u^j$. A spherical, resonant-mass detector has many response modes: the simplest one, the so-called monopole mode, is represented by the tensor $D^{ij} = \delta^{ij}$. Such a response mode, in particular, is irrelevant for gravitational radiation described by the traceless field h_{ij}, since the corresponding pattern function is vanishing. It may be relevant, however, for representing the response to a possible scalar (or dilatonic) component of the gravitational radiation, as we shall see in Chapter 9.

The cross-correlation of the outputs of the two detectors is based on a series of assumptions, on both the signals $h_{1,2}$ and the noises $n_{1,2}$, that we now list in detail. The two noises are assumed to be statistically uncorrelated [24], namely

$$\langle n_1(t)n_2(t')\rangle = 0, \tag{7.239}$$

where the brackets denote ensemble average. This assumption is justified by the fact that the two detectors are always different, in principle (even if the type is the same); in addition, they are placed at different locations, usually with large spatial separations.

The noises are also assumed to be statistically independent of the physical strains, namely,

$$\langle n_{1,2}(t) h_{1,2}(t') \rangle = 0. \tag{7.240}$$

Finally, we assume we are dealing with the most unfavorable case in which, for each detector, the noise is much larger in magnitude than the physical strain,

$$|n_{1,2}(t)| \gg |h_{1,2}(t)| \tag{7.241}$$

(in the opposite case, the signal would be easily detected even by a single gravitational antenna).

Concerning the signal, one assumes, as already mentioned in this chapter, that the inflationary amplification of the metric fluctuations produces an unpolarized, isotropic and stationary background of primordial gravitons, represented by a stochastic collection of standing waves, whose average energy-density distribution can be adequately described in terms of the spectral variable S_h, defined by Eqs. (7.222)–(7.224). In other words, the Fourier amplitudes $h_A(\nu, \widehat{n})$ are assumed to be represented by Gaussian variables satisfying the so-called stochastic conditions

$$\langle h_A(\nu, \widehat{n}) \rangle = 0,$$

$$\langle h_A(\nu, \widehat{n})\, h_{A'}^*(\nu', \widehat{n}') \rangle = \frac{1}{4\pi} \delta_{AA'} \delta(\nu - \nu') \delta^2(\widehat{n}, \widehat{n}') \frac{1}{2} S_h(|\nu|), \tag{7.242}$$

where $\delta^2(\widehat{n}, \widehat{n}') = \delta(\phi - \phi') \delta(\cos\theta - \cos\theta')$ (note that the integration of the second equation over the angular variables exactly reproduces the definition (7.222) of the strain density S_h). Using Eq. (7.224) the last equation can be rewritten as

$$\langle h_A(\nu, \widehat{n})\, h_{A'}^*(\nu', \widehat{n}') \rangle = \delta_{AA'} \delta(\nu - \nu') \delta^2(\widehat{n}, \widehat{n}') \frac{3 H_0^2 \Omega_g(\nu)}{32\pi^3 |\nu|^3} \tag{7.243}$$

(in agreement with the notation of [24]).

We are now in a position to present a precise computation of the signal-to-noise ratio resulting from the cross-correlation of the two detectors. Given the two outputs $s_{1,2}$, defined over a total observation time T, we can define an integrated "signal" S as

$$S = \int_{-T/2}^{T/2} dt\, dt'\, s_1(t) s_2(t') Q(t - t'). \tag{7.244}$$

Here $Q(t)$ is a real "filter" function, which will be appropriately chosen so as to maximize the signal-to-noise ratio (SNR), defined by the statistical average as

$$\text{SNR} = \frac{\langle S \rangle}{\Delta S}, \qquad \Delta S = \left(\langle S^2 \rangle - \langle S \rangle^2 \right)^{1/2}. \tag{7.245}$$

In order to compute $\langle S \rangle$ we start by applying the assumptions (7.239) and (7.240), which lead to

$$\langle S \rangle = \int_{-T/2}^{T/2} dt\, dt'\, \langle h_1(t)h_2(t')\rangle Q(t-t'). \tag{7.246}$$

We expand the strain as in Eq. (7.235), and perform the average using the stochastic condition (7.242). We obtain

$$\langle S \rangle = \frac{N}{8\pi} \int_{-T/2}^{T/2} dt\, dt' \int_{-\infty}^{+\infty} d\nu\, S_h(|\nu|)\gamma(|\nu|)Q(t-t')\, e^{-2\pi i\nu(t-t')}, \tag{7.247}$$

where

$$\gamma(\nu) = \frac{1}{N} \sum_A \int_{\Omega_2} d^2\hat{n}\, F_1^A(\hat{n})F_2^A(\hat{n})e^{2\pi i\nu\hat{n}\cdot(\vec{x}_1-\vec{x}_2)} \tag{7.248}$$

is the so-called "overlap reduction function" [24, 83, 84], which determines the signal resulting from the cross-correlation of the two outputs, taking into account the relative distance and orientation of the two detectors. The constant factor N is an overall normalization coefficient, which can be conveniently chosen in such a way that γ equals unity for coincident $(\vec{x}_1 = \vec{x}_2)$ and coaligned $(F_1^A = F_2^A)$ detectors. Assuming that the observation time T is much larger than the time intervals $t - t'$ over which $Q \neq 0$, we can approximate one of the time integrals of Eq. (7.247) with the Fourier transform $Q(\nu)$ of $Q(t)$:

$$Q(\nu) = \int_{-\infty}^{+\infty} d\nu\, Q(t-t')e^{-2\pi i\nu(t-t')}. \tag{7.249}$$

Replacing S_h with the spectral energy density Ω_g, according to Eq. (7.224), we finally obtain

$$\langle S \rangle = \frac{3NTH_0^2}{32\pi^3} \int_{-\infty}^{+\infty} \frac{d\nu}{|\nu|^3}\, Q(|\nu|)\gamma(|\nu|)\Omega_g(|\nu|). \tag{7.250}$$

We now need to compute the variance ΔS which, for uncorrelated noises much larger than the physical strain (see Eqs. (7.239)–(7.241)), can be written as

$$\Delta S^2 = \langle S^2 \rangle = \int_{-T/2}^{T/2} dt\, dt'\, d\tau\, d\tau'\, \langle n_1(t)n_2(t')n_1(\tau)n_2(\tau')\rangle\, Q(t-t')Q(\tau-\tau'). \tag{7.251}$$

Since $\langle n_1 n_2 \rangle = 0$, the only contribution to the above integral comes from the term $\langle n_1(t)n_1(\tau)\rangle\langle n_2(t')n_2(\tau')\rangle$. The definition (7.226) of the noise power spectrum, on the other hand, can be inverted to give

$$\langle n(t)n(\tau)\rangle = \frac{1}{2} \int_{-\infty}^{+\infty} d\nu P(|\nu|)e^{-2\pi i\nu(t-\tau)}. \tag{7.252}$$

Inserting the spectral noises P_1, P_2 into Eq. (7.251), and using the definition of $Q(\nu)$, we obtain

$$
\Delta S^2 = \frac{1}{4} \int_{-T/2}^{T/2} dt\, dt'\, d\tau\, d\tau' \int_{-\infty}^{+\infty} d\nu_1\, d\nu_2\, d\nu\, d\nu'\, Q(\nu_1)Q(\nu_2)P_1(\nu)P_2(\nu')
$$

$$
\times \exp\{2\pi i\left[t(\nu_1 - \nu) - t'(\nu_1 + \nu') + \tau(\nu_2 + \nu) - \tau'(\nu_2 - \nu')\right]\}
$$

$$
= \frac{1}{4} \int_{-\infty}^{+\infty} d\nu_1\, d\nu_2\, d\nu\, d\nu'\, Q(\nu_1)Q(\nu_2)P_1(\nu)P_2(\nu')
$$

$$
\times \delta_T(\nu_1 - \nu)\delta(\nu_1 + \nu')\delta(\nu_2 + \nu)\delta(\nu_2 - \nu'), \tag{7.253}
$$

where we have extended the integrals in dt', $d\tau$, $d\tau'$ from $-\infty$ to $+\infty$, assuming as before that T is much larger than the time intervals over which $Q \neq 0$. Also, we have defined

$$
\delta_T(f) = \int_{-T/2}^{T/2} dt\, e^{-2\pi i ft} = \frac{\sin(\pi ft)}{\pi f}. \tag{7.254}
$$

The integration over ν_1, ν_2, ν', using the reality condition $Q(-\nu) = Q^*(\nu)$, leads to the final result

$$
(\Delta S)^2 = \langle S^2 \rangle = \frac{T}{4} \int_{-\infty}^{+\infty} d\nu\, P_1(|\nu|)P_2(|\nu|)\, |Q(\nu)|^2. \tag{7.255}
$$

We can now choose the function Q in such a way as to maximize the ratio

$$
\text{SNR} = \frac{\langle S \rangle}{\Delta S} = NT^{1/2} \frac{6H_0^2}{32\pi^3} \frac{\int_{-\infty}^{+\infty} d\nu\, |\nu|^{-3}Q(\nu)\gamma(|\nu|)\Omega_g(|\nu|)}{\left[\int_{-\infty}^{+\infty} d\nu\, P_1(|\nu|)P_2(|\nu|)\, |Q(\nu)|^2\right]^{1/2}}. \tag{7.256}
$$

We note that the following combination of any two frequency-dependent variables, A and B,

$$
(A, B) = \int_{-\infty}^{+\infty} d\nu\, A^*(\nu)B(\nu)P_1(|\nu|)P_2(|\nu|), \tag{7.257}
$$

satisfies all the properties of a positive-definite inner product in ordinary Euclidean space [24]. In terms of such a product, the SNR expression can be rewritten as

$$
(\text{SNR})^2 = TN^2 \left(\frac{3H_0^2}{16\pi^3}\right)^2 \frac{(A, Q)^2}{(Q, Q)}, \tag{7.258}
$$

where

$$
A = \frac{\gamma(|\nu|)\Omega_g(|\nu|)}{|\nu|^3 P_1(|\nu|)P_2(|\nu|)}. \tag{7.259}
$$

Written in terms of a scalar product, it is easy to check that the above ratio can be maximized by choosing Q proportional to A, i.e. $Q = \lambda A$, with λ a real (arbitrary) normalization constant, so that $(A, Q)^2/(Q, Q) = (A, A)$. This "optimal filtering" prescription then leads to the final result of this cross-correlation analysis:

$$\text{SNR} = N \frac{3H_0^2}{16\pi^3} \left[T \int_{-\infty}^{+\infty} \frac{d\nu}{\nu^6} \frac{\gamma^2(|\nu|)\Omega_g^2(|\nu|)}{P_1(|\nu|)P_2(|\nu|)} \right]^{1/2}. \qquad (7.260)$$

As clearly shown by this expression, the signal induced by a given cosmic background Ω_g grows with the square root of the correlation time T, and is larger for larger overlap γ and smaller noises P_1, P_2, of the two detectors.

The important consequence of the above analysis is that the experimental sensitivity of two cross-correlated detectors to a graviton background is greatly improved with respect to the sensitivity of a single detector. For a quantitative estimate of such improvement one should first provide a precise definition of the level of signal required for an unambiguous detection. Following [24] we can say, in particular, that a stochastic background can be detected, with a detection rate β and a false-alarm rate α, if

$$\text{SNR} \geq \sqrt{2}\left(\text{erfc}^{-1}2\alpha - \text{erfc}^{-1}2\beta\right). \qquad (7.261)$$

We refer to the existing literature for a complete list of results on the cross-correlation of all possible pairs of detectors. The level of sensitivity reached by the correlation of two interferometers, in particular, has been studied in [24, 70, 85]. We recall here that for an observation time $T = 4$ months, a detection rate of 0.95%, a false-alarm rate of 0.05%, and with the noise level of the first generation of interferometers, the minimum detectable graviton background (with a flat spectrum, $\Omega_g = \text{const}$) is $h^2\Omega_g \simeq 5 \times 10^{-6}$, just about at the borderline of the allowed region. With the planned sensitivities of Advanced LIGO, however, one expects a much better limit [24],

$$h^2\Omega_g \simeq 5 \times 10^{-11}, \qquad (7.262)$$

well inside the allowed region.

The cross-correlation between two resonant bars, and between a bar and an interferometer, has been studied in [86, 87]. The minimum detectable value of Ω_g, in those cases, is at present [70] $h^2\Omega_g \sim 10^{-4}$ for a flat spectrum, one year of observation time, at the 90% confidence level. The correlation of the spherical resonant detectors seems to be more promising, however. Two resonant spheres of 3 m diameter, located at the same site, could reach a sensitivity [88] corresponding to $h^2\Omega_g \simeq 4 \times 10^{-7}$. Also, studies reported in [89] suggest that the correlation of two hollow spheres could reach $h^2\Omega_g \sim 10^{-9}$. Both predictions are well inside the allowed region (see Fig. 7.8). Finally, the cross-correlation for the spatial interferometer LISA has been discussed in [69] by taking into account that, at low frequencies, the sensitivity to a primordial graviton background is fundamentally limited by the possible presence of other stochastic backgrounds, of astrophysical origin, generated at much later times.

A detailed analysis, performed in the case of two identical LISAs, leads to a minimum detectable background $h^2\Omega_g \sim 10^{-12}$, in the mHz range, for a flat spectrum.

It should be stressed, at this point, that all the numerical results we have reported have been obtained by assuming that the graviton spectrum is flat ($\Omega_g = $ const) in the whole frequency band where $\gamma \neq 0$ and the noises $P_{1,2}$ are near the minimum. In principle, however, the value of SNR depends not only on the intensity but also on the slope of the graviton spectrum. The expected level of sensitivity to a growing spectral distribution has been computed in [90, 91] and [60, 92] for a pair of LIGO and a (virtual) pair of VIRGO interferometers, respectively. The minimum detectable intensity, in that case, does not seem to present any substantial improvement with respect to a flat graviton spectrum, apart from a weak frequency dependence, which tends to enhance the sensitivity at the lower end of the resonant band of the two detectors.

We conclude this chapter by quoting the best direct experimental upper limit existing at present on the energy density of a stochastic background of cosmic gravitons, obtained from the cross-correlated analysis of the data of the two LIGO interferometers [93]:

$$h^2\Omega_g \lesssim 8 \times 10^{-4} \left(\frac{\nu}{100\mathrm{Hz}}\right)^\alpha, \qquad h = 0.72, \qquad 69\ \mathrm{Hz} \leq \nu \leq 156\ \mathrm{Hz}, \tag{7.263}$$

valid for a flat ($\alpha = 0$), quadratic ($\alpha = 2$) and cubic ($\alpha = 3$) spectrum, and relative to an integration time $T = 200$ hours. The integration time can be easily improved, of course, and with one year of data at design sensitivity the LIGO detector will reach a limit several times below the nucleosynthesis bound [93]. We may thus expect that, in the near future, the cross-correlation of advanced (ground-based) interferometers, space interferometers, resonant spheres and hollow spheres will be able to explore (and to set precise constraints on) the parameter space of string cosmology models.

Appendix 7A
Higher-derivative corrections to the tensor perturbation equations

When the curvature approaches the string scale, string theory predicts the appearance of higher-derivative (α') corrections modifying the effective action and the low-energy equations for the background fields. The equation describing the propagation of tensor metric fluctuations, on the other hand, is obtained by perturbing to linear order the background equations deduced from the string effective action, as illustrated in Section 7.1: we should expect, therefore, that the perturbation equation may also receive α' corrections for modes entering the high-curvature regime [12]. In this appendix we illustrate this effect, discussing an explicit example based on the following four-dimensional, first-order effective action,

$$S = S_{\text{tree}} + S_{\alpha'},$$

$$S_{\text{tree}} = -\frac{1}{2\lambda_s^2} \int d^4x \sqrt{-g}\, e^{-\phi} \left[R + (\nabla \phi)^2 \right],$$

$$S_{\alpha'} = \frac{1}{2\lambda_s^2} \frac{\alpha'}{4} \int d^4x \sqrt{-g}\, e^{-\phi} \left[R_{\text{GB}}^2 - (\nabla \phi)^4 \right],$$

(7A.1)

where $R_{\text{GB}}^2 = R^2 - 4R_{\mu\nu}^2 + R_{\mu\nu\alpha\beta}^2$ is the quadratic Gauss–Bonnet invariant. This action was introduced in Chapter 2, and used in Chapter 6 to describe the evolution of an inflationary string background in the high-curvature regime (see Eq. (6.12)).

In order to obtain the modified perturbation equation, and to define the canonical variable appropriate to the quantum normalization of the transverse and traceless part $\delta g_{\mu\nu} = h_{\mu\nu}$ of the metric fluctuations,

$$\nabla_\nu h^\nu_\mu = 0, \qquad g^{\mu\nu} h_{\mu\nu} = 0,$$

(7A.2)

we need the perturbed form of the action (7A.1) up to terms quadratic in $h_{\mu\nu}$:

$$\delta^{(2)} S = -\frac{1}{2\lambda_s^2} \int d^4x\, e^{-\phi} \left[\delta^{(2)} (\sqrt{-g} R) + \delta^{(2)} \left(\sqrt{-g} g^{\mu\nu} \partial_\mu \phi\, \partial_\nu \phi \right) \right.$$

$$\left. - \frac{\alpha'}{4} \delta^{(2)} \left(\sqrt{-g} R_{\text{GB}}^2 \right) + \frac{\alpha'}{4} \delta^{(2)} \left(\sqrt{-g} g^{\mu\nu} \partial_\mu \phi\, \partial_\nu \phi\, g^{\alpha\beta} \partial_\alpha \phi\, \partial_\beta \phi \right) \right].$$

(7A.3)

We are using the notations of Section 7.1 for $\delta^{(2)}$, and we are not perturbing the dilaton field ($\delta\phi = 0$), as we are only interested in the pure tensor part of the metric fluctuations (scalar, vector and tensor perturbations are decoupled to linear order; see [94] for a similar computation applied to the case of scalar perturbations).

The perturbative approach is exactly the same as that used in Section 7.1 so that, for the tree-level part of the action, we can directly apply the result obtained in that section. Choosing the comoving gauge for both the background metric $g_{\mu\nu}$ and the fluctuations $h_{\mu\nu}$ (see Eqs. (7.11) and (7.13), respectively), we obtain, from Eq. (7.28),

$$\delta^{(2)} S_{\text{tree}} = -\frac{1}{2\lambda_s^2} \int d^4x \, a^3 e^{-\phi} \text{Tr} \left[h\ddot{h} - \frac{1}{4} h \frac{\nabla^2}{a^2} h + \frac{3}{4} \dot{h}^2 \right.$$

$$\left. + 4\dot{h}hH + h^2 \left(\frac{3}{2}\dot{H} + 3H^2 - \frac{\dot{\phi}^2}{4} \right) \right], \tag{7A.4}$$

where we have set to zero the contribution of the extra dimensions ($n = 0$, $F = \dot{b}/b = 0$) and of the potential ($V = 0$).

This result must be added to the perturbation of the higher-derivative action, $\delta^{(2)} S_{\alpha'}$. Let us separately compute the various contributions, starting from the simplest one, the quartic dilaton term: applying Eq. (7.5) we immediately obtain

$$\delta^{(2)} \left[\sqrt{-g}(\nabla\phi)^4 \right] = (\nabla\phi)^4 \delta^{(2)} \sqrt{-g} = -\frac{1}{4} a^3 \dot{\phi}^4 \, \text{Tr} \, h^2. \tag{7A.5}$$

In the quadratic Gauss–Bonnet term we have contributions from the perturbations of the scalar curvature, of the Ricci tensor and of the Riemann tensor. The scalar curvature term gives

$$\delta^{(2)} (\sqrt{-g} R^2) = R^2 \delta^{(2)} \sqrt{-g} + 2\sqrt{-g} R \delta^{(2)} R + \sqrt{-g} \left(\delta^{(1)} R \right)^2. \tag{7A.6}$$

The last term, however, is vanishing because of the traceless condition $g^{ij} h_{ij} = 0$ (see Eq. (7.12) for R_{ij} and Eq. (7.23) for $\delta^{(1)} R_i{}^j$). We are left with

$$\delta^{(2)} (\sqrt{-g} R^2) = R^2 \delta^{(2)} \sqrt{-g} + 2\sqrt{-g} R \left(g^{\mu\nu} \delta^{(2)} R_{\mu\nu} + R_{\mu\nu} \delta^{(2)} g^{\mu\nu} + \delta^{(1)} g^{\mu\nu} \delta^{(1)} R_{\mu\nu} \right)$$

$$= -6a^3 (\dot{H} + 2H^2) \, \text{Tr} \left[2h\ddot{h} - \frac{h}{2} \frac{\nabla^2}{a^2} h + 8h\dot{h}H + \frac{3}{2}\dot{h}^2 + h^2 \left(\frac{3}{2}\dot{H} + 3H^2 \right) \right] \tag{7A.7}$$

(we have used the background of Eq. (7.12), and the results (7.4), (7.5), (7.22) and (7.26) for the metric perturbations to first and second order). Applying the same equations, with the addition of Eq. (7.23), and noting that $\delta^{(1)} R_i{}^0 = 0 = \delta^{(1)} R_0{}^0$, we can also immediately obtain the contribution of the Ricci squared term:

$$\delta^{(2)} (\sqrt{-g} R_\mu{}^\nu R_\nu{}^\mu) = R_{\mu\nu} R^{\mu\nu} \delta^{(2)} \sqrt{-g} + \sqrt{-g} \left(\delta^{(1)} R_\mu{}^\nu \delta^{(1)} R_\nu{}^\mu + 2R^{\mu\nu} \delta^{(2)} R_{\mu\nu} \right)$$

$$= a^3 \, \text{Tr} \left[\frac{\ddot{h}^2}{4} + \frac{3}{2} h\ddot{h}H - h\ddot{h}(4\dot{H} + 6H^2) + \frac{1}{4} \left(\frac{\nabla^2}{a^2} h \right)^2 \right.$$

$$- \frac{h^2}{2} \left(5\dot{H} + \frac{18}{4} H^2 \right) - \frac{1}{2} (\ddot{h} + 3H\dot{h} - \dot{H}h - 3H^2 h) \frac{\nabla^2}{a^2} h$$

$$\left. - h\dot{h}(12\dot{H}H + 24H^3) - h^2 (3\dot{H}^2 + 9\dot{H}H^2 + 9H^4) \right]. \tag{7A.8}$$

To complete the perturbation of R^2_{GB} we now need the contribution of the Riemann squared term:

$$\delta^{(2)}(\sqrt{-g}R_{\mu\nu}{}^{\alpha\beta}R_{\alpha\beta}{}^{\mu\nu}) = R^2_{\mu\nu\alpha\beta}\delta^{(2)}\sqrt{-g}$$

$$+ \sqrt{-g}\left(\delta^{(1)}R_{\mu\nu}{}^{\alpha\beta}\delta^{(1)}R_{\alpha\beta}{}^{\mu\nu} + 2R_{\mu\nu}{}^{\alpha\beta}\delta^{(2)}R_{\alpha\beta}{}^{\mu\nu}\right). \quad (7A.9)$$

We recall that, in our background, the non-vanishing components of the unperturbed Riemann tensor are the following:

$$R_{0i}{}^{0j} = \delta^j_i(\dot{H} + H^2) = -R_{0i}{}^{j0} = -R_{i0}{}^{0j},$$

$$R_{ik}{}^{jl} = H^2(\delta^j_i\delta^l_k - \delta^j_k\delta^l_i) = -R_{ik}{}^{lj} = -R_{ki}{}^{jl}. \quad (7A.10)$$

The first-order perturbation of the Riemann tensor,

$$\delta^{(1)}R_{\mu\nu\alpha}{}^{\beta} = \partial_\mu\delta^{(1)}\Gamma_{\nu\alpha}{}^{\beta} + \Gamma_{\nu\alpha}{}^{\rho}\delta^{(1)}\Gamma_{\mu\rho}{}^{\beta} + \Gamma_{\mu\rho}{}^{\beta}\delta^{(1)}\Gamma_{\nu\alpha}{}^{\rho} - (\mu \leftrightarrow \nu), \quad (7A.11)$$

using Eq. (7.15) for $\delta^{(1)}\Gamma$, then gives the following non-zero components:

$$\delta^{(1)}R_{0i}{}^{0j} = \frac{1}{2}\left(\ddot{h}_i{}^j + 2H\dot{h}_i{}^j\right),$$

$$\delta^{(1)}R_{0i}{}^{jk} = \frac{1}{2}\left(\partial^j\dot{h}_i{}^k - \partial^k\dot{h}_i{}^j\right),$$

$$\delta^{(1)}R_{ik}{}^{0j} = \frac{1}{2}\left(\partial_i\dot{h}_k{}^j - \partial_k\dot{h}_i{}^j\right), \quad (7A.12)$$

$$\delta^{(1)}R_{ik}{}^{jl} = \frac{1}{2}\left(\partial_i\partial^j h_k{}^l - \partial_k\partial^j h_i{}^l + \partial_k\partial^l h_i{}^j - \partial_i\partial^l h_k{}^j\right)$$

$$+ \frac{H}{2}\left(\delta^j_i\dot{h}_k{}^l - \delta^j_k\dot{h}_i{}^l + \delta^l_k\dot{h}_i{}^j - \delta^l_i\dot{h}_k{}^j\right).$$

At second order the perturbed expression is, in general,

$$\delta^{(2)}R_{\mu\nu\alpha}{}^{\beta} = \partial_\mu\delta^{(2)}\Gamma_{\nu\alpha}{}^{\beta} + \delta^{(1)}\Gamma_{\nu\alpha}{}^{\rho}\delta^{(1)}\Gamma_{\mu\rho}{}^{\beta}$$

$$+ \Gamma_{\mu\rho}{}^{\beta}\delta^{(2)}\Gamma_{\nu\alpha}{}^{\rho} + \Gamma_{\nu\alpha}{}^{\rho}\delta^{(2)}\Gamma_{\mu\rho}{}^{\beta} - (\mu \leftrightarrow \nu). \quad (7A.13)$$

Given the unperturbed tensor (7A.10), however, the only components we need for the computation (7A.9) are $\delta^{(2)}R_{0i}{}^{0j}$ and $\delta^{(2)}R_{ik}{}^{jl}$. Using the previous results for $\delta^{(1)}\Gamma$ and $\delta^{(2)}\Gamma$ we obtain

$$\delta^{(2)}R_{0i}{}^{0j} = -\frac{1}{2}\left(\ddot{h}_i{}^k h_k{}^j + \frac{1}{2}\dot{h}_i{}^k\dot{h}_k{}^j + 2H\dot{h}_i{}^k h_k{}^j\right),$$

$$R_{jl}{}^{ik}\delta^{(2)}R_{ik}{}^{jl} = -\frac{H^2}{2}\left(\dot{h}_i{}^j\dot{h}_j{}^i + 8H h_i{}^j\dot{h}_j{}^i - h_i{}^j\frac{\nabla^2}{a^2}h_j{}^i\right) \quad (7A.14)$$

(we have neglected all terms that, after integration by parts, do not contribute to the perturbed action because of the transversality condition $\partial_j h_i^j = 0$). Equation (7A.9) then provides

$$\delta^{(2)}(\sqrt{-g}R_{\mu\nu}{}^{\alpha\beta}R_{\alpha\beta}{}^{\mu\nu}) = a^3\text{Tr}\left[\ddot{h}^2 + 4H\dot{h}\ddot{h} - h\ddot{h}(4H^2 + 4\dot{H}) + \left(\frac{\nabla^2}{a^2}h\right)^2\right.$$

$$+2\dot{h}\frac{\nabla^2}{a^2}h+(H^2h-2H\dot{h})\frac{\nabla^2}{a^2}h+\dot{h}^2(2H^2-2\dot{H})$$

$$-h\dot{h}(8H\dot{H}+16H^3)-h^2\left(3\dot{H}^2+6\dot{H}H^2+6H^4\right)\Bigg]. \quad (7A.15)$$

We are now in a position to present the contribution of the first-order α' corrections to the perturbed action (7A.3). Collecting all terms (7A.5), (7A.7), (7A.8) and (7A.15) we are led to

$$\delta^{(2)}S_{\alpha'}=\frac{1}{2\lambda_s^2}\frac{\alpha'}{4}\int d^4x\,a^3\,e^{-\phi}\,\mathrm{Tr}\Bigg[-2H\dot{h}\ddot{h}-4H^2h\ddot{h}$$

$$+(2\ddot{h}+\dot{H}h+4H\dot{h}+H^2h)\frac{\nabla^2}{a^2}h+2\dot{h}\frac{\nabla^2}{a^2}\dot{h}-\dot{h}^2(\dot{H}+7H^2)$$

$$-h\dot{h}(8H\dot{H}+16H^3)+h^2\left(\frac{1}{4}\dot{\phi}^2-6\dot{H}H^2-6H^4\right)\Bigg]. \quad (7A.16)$$

Summing up this result to the tree-level result (7A.4) we can eventually integrate by parts all terms with more than two partial derivatives acting on h, as well as the terms containing $h\dot{h}$ and $h\ddot{h}$. The result (modulo a total derivative) is

$$\delta^{(2)}S=\frac{1}{2\lambda_s^2}\int d^4x\,a^3\,e^{-\phi}\mathrm{Tr}\left\{\frac{1}{4}\dot{h}^2(1-\alpha'H\dot{\phi})+\frac{1}{4}h\frac{\nabla^2}{a^2}h\left[1+\alpha'(\dot{\phi}^2-\ddot{\phi})\right]\right.$$

$$+h^2\left[\frac{1}{2}\ddot{\phi}+H\dot{\phi}-\frac{1}{4}\dot{\phi}^2-\dot{H}-\frac{3}{2}H^2\right.$$

$$\left.\left.+\frac{\alpha'}{4}\left(\frac{1}{4}\dot{\phi}^4+2H^2\ddot{\phi}-2H^2\dot{\phi}^2+4\dot{\phi}\dot{H}H+4\dot{\phi}H^3\right)\right]\right\}. \quad (7A.17)$$

The absence of terms with more than two derivatives of h follows from the Euler form of the higher-curvature corrections to the metric (the Gauss–Bonnet invariant R_{GB}^2). We may note, also, that all α' corrections disappear in the limit $\phi=$ const, since in that limit (and for $d=3$) the R_{GB}^2 part of the action (7A.1) reduces to a total derivative with no dynamical contributions.

A last simplification of the perturbed action (7A.17) is due to the vanishing of the coefficient of the h^2 term, thanks to the equations of motion of the unperturbed background (see, in particular, Eq. (6.20) for $d=3$). This leads us to the final result [12]

$$\delta^{(2)}S=\frac{1}{2\lambda_s^2}\frac{1}{4}\int d^4x\,a^3\,e^{-\phi}\left[\dot{h}_i^j\dot{h}_j^i\left(1-\alpha'H\dot{\phi}\right)+h_i^j\frac{\nabla^2}{a^2}h_j^i\left(1+\alpha'\dot{\phi}^2-\alpha'\ddot{\phi}\right)\right], \quad (7A.18)$$

and (by varying with respect to h) to the modified tensor perturbation equation,

$$\ddot{h}_i^j\left(1-\alpha'H\dot{\phi}\right)+\left[3H-\dot{\phi}-\alpha'\left(3H^2\dot{\phi}-H\dot{\phi}^2+\dot{H}\dot{\phi}+H\ddot{\phi}\right)\right]\dot{h}_i^j$$

$$-\frac{\nabla^2}{a^2}h_i^j\left[1+\alpha'\left(\dot{\phi}^2-\ddot{\phi}\right)\right]=0, \quad (7A.19)$$

generalizing (in $d = 3$) Eqs. (7.31) and (7.34) obtained by perturbing the tree-level action.

It is convenient, at this point, to decompose the matrix h_i^j into the two physical polarization modes h_A, $A = 1, 2$, associated with the propagation of a tensor (spin-two) wave in four dimensions. Using conformal time ($\mathrm{d}\eta = \mathrm{d}t/a$) the action (7A.18) can then be written, for each polarization mode h_A, as [12]

$$\delta^{(2)}S = \frac{1}{2} \int \mathrm{d}^3 x \, \mathrm{d}\eta \left[z^2(\eta) h'^2 + y^2(\eta) h \nabla^2 h \right], \qquad (7A.20)$$

where we have omitted the polarization index, and where

$$
\begin{aligned}
z^2(\eta) &= \frac{M_s^2}{2} e^{-\phi} \left(a^2 - \alpha' \frac{a'}{a} \phi' \right), \\
y^2(\eta) &= \frac{M_s^2}{2} e^{-\phi} \left[a^2 + \alpha' \left(\phi'^2 - \phi'' + \frac{a'}{a} \phi' \right) \right]
\end{aligned}
\qquad (7A.21)
$$

(the prime denotes differentiation with respect to η). By setting $\psi = zh$ the action becomes

$$\delta^{(2)}S = \frac{1}{2} \int \mathrm{d}^3 x \, \mathrm{d}\eta \left(\psi'^2 + \frac{y^2}{z^2} \psi \nabla^2 \psi + \frac{z''}{z} \psi^2 \right) \qquad (7A.22)$$

(modulo a total derivative). For each Fourier mode h_k we can thus define the variable $\psi_k = zh_k$, which has the correct canonical dimensions, diagonalizes the kinetic part of the action, and satisfies a Schrödinger-like evolution equation,

$$\psi_k'' + \left[k^2 - V_k(\eta) \right] \psi_k = 0, \qquad (7A.23)$$

with effective potential

$$V_k(\eta) = \frac{z''}{z} - \frac{k^2}{z^2} (y^2 - z^2). \qquad (7A.24)$$

The equations (7A.20)–(7A.24) generalize the canonical formulation of the dynamics of tensor fluctuations, described at the tree-level by Eqs. (7.46)–(7.50). The higher-curvature corrections are encoded into the new effective potential (7A.24), which is, in general, k-dependent because of the two different pump fields (z and y) appearing in the perturbed action. The difference $z \neq y$ also breaks the effective "duality" relating the evolution of the tensor fluctuation variable and its conjugate momentum (see Section 7.2), and is not peculiar to the considered model of α' corrections: similar results are indeed obtained by perturbing different, extended models, which include both α' and loop corrections, as shown in [56].

It seems appropriate to devote the last part of this appendix to a brief discussion of the possible effects of the α' corrections on the dynamics and the spectrum of the tensor perturbations. Such effects are more conveniently illustrated by rewriting the canonical equation (7A.23) as follows:

$$\psi_k'' + k^2 \left[1 + c(\eta) \right] \psi_k - \frac{z''}{z} \psi_k = 0, \qquad (7A.25)$$

where, for our model,

$$c(\eta) = \frac{y^2 - z^2}{z^2} = \alpha' \frac{\phi'^2 - \phi'' + 2 \frac{a'}{a} \phi'}{a^2 - \alpha' \frac{a'}{a} \phi'} = \alpha' \frac{\dot{\phi}^2 + H \dot{\phi} - \ddot{\phi}}{1 - \alpha' H \dot{\phi}}. \qquad (7A.26)$$

In the low-curvature regime $\alpha'H^2 \sim \alpha'\dot{\phi}^2 \ll 1$ one has $c \to 0$, the corrections to the pump fields become negligible, $z \sim y \sim a\exp(-\phi/2)$, and Eq. (7A.25) reduces to the tree-level perturbation equation (7.56). In the high-curvature regime $\alpha'H^2 \sim 1$ one obtains, instead, significant corrections to the canonical evolution equation, but such corrections are strongly model dependent, relying on the specific form of the pump field $z(\eta)$ and the frequency shift $c(\eta)$.

If, however, the high-curvature phase corresponds to a fixed point of the string cosmology equations, characterized by $\dot{\phi} = x_0 = \text{const}$, $H = x_1 = \text{const}$ (see Section 6.1), then all the α' corrections tend to stabilize to a constant value,

$$z^2 \to \xi^2(1 - \alpha'x_0x_1), \qquad y^2 \to \xi^2(1 + \alpha'x_0^2), \tag{7A.27}$$

where $\xi = (M_s/\sqrt{2})a\exp(-\phi/2)$ is the tree-level pump field. In this limit the effective potential z''/z reduces to the tree-level form ξ''/ξ, and the frequency shift becomes a constant,

$$c = \alpha'\,\frac{x_0x_1 + x_0^2}{1 - \alpha'x_0x_1}. \tag{7A.28}$$

Thus, in this case, there is no modification of the evolution of tensor perturbations outside the horizon, hence no modification of the perturbation spectrum with respect to that computed without α' corrections in the perturbation equation. The only effect of the high-curvature terms, in this case, is a constant shift of the asymptotic amplitude of the fluctuations, due to a shift of the horizon-crossing scale.

Let us consider, in fact, a phase of high and constant curvature, described for $\eta \to 0_-$ by the pump field $\xi(\eta) \sim (-\eta)^\alpha$, with $\alpha \le 1/2$, as typical of the string phase discussed in Section 7.3. Imposing the canonical normalization of the quantum fluctuations at horizon crossing, $|\psi_k|_{hc} = 1/\sqrt{2k}$ (see Section 7.2), and using the modified perturbation equation (7A.25), one finds, asymptotically, the power spectrum

$$k^{3/2}|\psi_k(\eta)| \sim k\frac{\xi(\eta)}{\xi_{hc}} = k|k\eta\sqrt{1+c}|^\alpha \sim k^{1+\alpha}. \tag{7A.29}$$

The k-dependence is the same as that obtained from the low-energy perturbation equation. The constant shift of the amplitude (by the factor $\sqrt{1+c}$) is *the same for all modes*, and is typically of order one – unless the denominator of Eq. (7A.28) goes to zero, which would imply a divergence of the perturbation amplitude, signaling a quantum gravitational instability of the type discussed in [22, 23].

These results are illustrated in Fig. 7.11, where we have plotted a numerical integration of the system formed by the high-curvature background equations (6.18)–(6.20) (for $d = 3$) and by the high-curvature perturbation equations (7A.19). We have compared, for a mode crossing the horizon in the high-curvature regime, the evolution of the amplitude $|h_k(t)|$ obtained from Eq. (7A.19) with the amplitude one would obtain (for the *same mode*, in the *same background*) neglecting the α' corrections in the perturbation equation, and using Eq. (7.34) instead of Eq. (7A.19). In both cases the amplitude oscillates inside the horizon, and the oscillations are damped outside the horizon, as expected. The effect of the α' corrections, when they become important near the fixed-point regime (see Fig. 6.2 for the background evolution), is to induce a shift of the comoving frequency, with a resulting shift of the final asymptotic amplitude. As long as this shift is of order one, the computations based on the low-energy perturbation equation provide a valid estimate of the spectrum.

This conclusion should not hide the fact that the α' corrections, modifying the background solutions, do also modify the tensor perturbation spectrum with respect to the results obtained in the small-curvature limit, in which the background is obtained by solving the tree-level

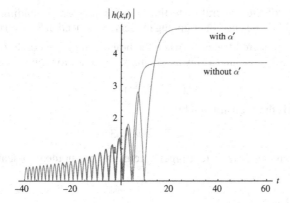

Figure 7.11 Numerical integration of the tensor perturbation equation (7A.19) in the high-curvature background of Fig. 6.2, for a tensor mode of comoving frequency $k = 1$ (in string units $\alpha' = 1$). The initial phase has been normalized in such a way that $h_k(\eta_i) = 0$, $\dot{h}_k(\eta_i) = -1$ at $\eta_i = -40$.

equations. One can easily evaluate such a spectral distortion, in particular, when the α' corrections are associated with a phase of constant curvature.

Let us consider, for instance, the high-curvature string phase possibly associated with a fixed point of the effective action, with a metric and dilaton evolution parametrized as in Eq. (7.196). In the constant-curvature regime the graviton spectrum is determined by the power α of the tree-level pump field,

$$\Omega_g \sim \omega^{3-2|\nu|}, \qquad \nu = \alpha - \frac{1}{2}, \tag{7A.30}$$

and the slope is flatter, in general, than the cubic slope $\Omega_g \sim \omega^3$ characterizing the low-frequency branch of the spectrum (see the discussion of Section 7.3). The modified slope, due to the modified background kinematics, is fully controlled by the constant background parameters $\dot{\phi} = x_0$, $H = x_1$: a simple integration gives for the fixed-point regime,

$$\phi \sim x_0 t, \qquad a \sim e^{x_1 t}, \tag{7A.31}$$

from which, in conformal time,

$$a(\eta) \sim \frac{1}{(-x_1\eta)}, \qquad \phi(\eta) \sim -\frac{x_0}{x_1} \ln(-\eta), \tag{7A.32}$$

with a corresponding pump field

$$z \sim a e^{-\phi/2} \sim (-\eta)^{-1+\frac{x_0}{2x_1}} \equiv (-\eta)^\alpha, \tag{7A.33}$$

and a corresponding spectral index

$$3 - 2|\nu| = 3 - |2\alpha - 1| = 3 - \left| \frac{x_0 - 3x_1}{x_1} \right|. \tag{7A.34}$$

This value may range, in principle, from the maximum slope $3 - 2|\nu| = 3$ (i.e. $x_0 = 3x_1$), corresponding to the saturation of the branch-changing condition (6.11), to the minimum $3 - 2|\nu| = 0$ (i.e. $x_0 = 0$), corresponding to a frozen-dilaton configuration. In practice, however, the range of the spectral distortion may be smaller, depending on the allowed ranges of x_0 and x_1. For the class of bouncing models introduced in Section 6.2, for instance, all allowed fixed points are located between the branch-changing condition $\ddot{\phi} < 0$ and the bounce condition (6.28), in such a way as to satisfy the condition [56]

$$-x_1 < x_0 - 3x_1 < 0. \tag{7A.35}$$

The slope (7A.34) is thus constrained by

$$2 < 3 - 2|\nu| < 3, \tag{7A.36}$$

and seems to remain too steep to trigger interesting phenomenological effects at low frequencies.

References

[1] L. P. Grishchuk and M. Solokhin, *Phys. Rev.* **D43** (1991) 2566.

[2] S. Weinberg, *Gravitation and Cosmology* (New York: Wiley, 1971).

[3] V. F. Mukhanov, H. A. Feldman and R. H. Brandenberger, *Phys. Rep.* **215** (1992) 203.

[4] M. Gasperini and M. Giovannini, *Class. Quantum Grav.* **9** (1992) L137.

[5] M. Demianski, A. G. Polnarev and P. Naselski, *Phys. Rev.* **D47**, 5275 (1993).

[6] M. Gasperini and M. Giovannini, *Class. Quantum Grav.* **14** (1997) 735.

[7] M. Gasperini and M. Giovannini, *Phys. Rev.* **D47** (1993) 1519.

[8] M. Giovannini, *Phys. Rev.* **D55** (1997) 595.

[9] L. P. Grishchuk, *Sov. Phys. JETP* **40** (1975) 409.

[10] L. P. Grishchuk and A. Polnarev, in *General Relativity and Gravitation*, ed. A. Held (New York: Plenum, 1980), p. 393.

[11] A. A. Starobinski, *JETP Lett.* **30** (1979) 682.

[12] M. Gasperini, *Phys. Rev.* **D56** (1997) 4815.

[13] C. W. Misner, K. S. Thorne and J. A. Wheeler, *Gravitation* (San Francisco: W. H. Freeman and Co., 1973).

[14] N. D. Birrel and P. C. W. Davies, *Quantum Fields in Curved Spaces* (Cambridge: Cambridge University Press, 1982).

[15] R. Brustein, M. Gasperini and G. Veneziano, *Phys. Lett.* **B431** (1998) 277.

[16] V. A. Rubakov, M. Sazhin and A. Veryaskin, *Phys. Lett.* **B115** (1982) 189.

[17] R. Fabbri and M. Pollock, *Phys. Lett.* **B125** (1983) 45.

[18] L. F. Abbot and M. B. Wise, *Nucl. Phys.* **B244** (1984) 541.

[19] B. Allen, *Phys. Rev.* **D37** (1988) 2078.

[20] V. Sahni, *Phys. Rev.* **D42** (1990) 453.

[21] M. Gasperini and G. Veneziano, *Phys. Rev.* **D50** (1994) 2519.

[22] S. Kawai, M. Sakagami and J. Soda, *Phys. Lett.* **B437** (1998) 284.

[23] S. Kawai and J. Soda, *Phys. Lett.* **B460** (1999) 41.

[24] B. Allen and J. D. Romano, *Phys. Rev.* **D59** (1999) 102001.

[25] R. Brustein, M. Gasperini, M. Giovannini, V. Mukhanov and G. Veneziano, *Phys. Rev.* **D51** (1995) 6744.

[26] M. Gasperini and M. Giovannini, *Phys. Lett.* **B282** (1992) 36.

[27] M. Gasperini, in *String Theory in Curved Space Times*, ed. N. Sanchez (Singapore: World Scientific, 1998), p. 333.

[28] L. P. Grishchuk and Y. V. Sidorov, *Class. Quantum Grav.* **6** (1989) L161.

[29] L. P. Grishchuk and Y. V. Sidorov, *Phys. Rev.* **D42** (1990) 3413.

[30] L. P. Grishchuk, *Phys. Rev. Lett.* **70** (1993) 2371.

[31] B. L. Schumaker, *Phys. Rep.* **135** (1986) 317.
[32] A. Messiah, *Quantum Mechanics*, Vol. 1 (Amsterdam: North Holland, 1972).
[33] M. Gasperini, M. Giovannini and G. Veneziano, *Phys. Rev.* **D48** (1993) R439.
[34] M. Gasperini and M. Giovannini, *Class. Quantum Grav.* **10** (1993) L133.
[35] R. Brandenberger, V. F. Mukhanov and T. Prokopec, *Phys. Rev. Lett.* **69** (1992) 3606.
[36] M. Gasperini and M. Giovannini, *Phys. Lett.* **B301** (1993) 334.
[37] G. De Risi and M. Gasperini, *Phys. Lett.* **B503** (2001) 140.
[38] J. Garriga and E. Verdaguer, *Phys. Rev.* **D39** (1989) 1072.
[39] M. Abramowitz and I. A. Stegun, *Handbook of Mathematical Functions* (New York: Dover, 1972).
[40] A. Buonanno, K. A. Meissner, C. Ungarelli and G. Veneziano, *JHEP* **9801** (1998) 004.
[41] R. Durrer and F. Vernizzi, *Phys. Rev.* **D66** (2002) 083503.
[42] R. Brustein, M. Gasperini, M. Giovannini and G. Veneziano, *Phys. Lett.* **B361** (1995) 45.
[43] A. Penzias and R. Wilson, *Ap. J.* **142** (1965) 419.
[44] G. F. Smooth *et al.*, *Ap. J.* **396** (1992) 1.
[45] C. L. Bennet *et al.*, *Ap. J.* **430** (1994) 423.
[46] M. Kamionkowski, A. Kosowsky and A. Stebbins, *Phys. Rev. Lett.* **78** (1997) 2058.
[47] M. Kamionkowski and A. Jaffe, *Int. J. Mod. Phys.* **A16** (2001) 116.
[48] M. Gasperini and G. Veneziano, *Astropart. Phys.* **1** (1993) 317.
[49] J. Khouri, B. A. Ovrut, P. J. Steinhardt and N. Turok, *Phys. Rev.* **D64** (2001) 123522.
[50] M. Giovannini, *Phys. Rev.* **D60** (1999) 123511.
[51] V. Kaspi, J. Taylor and M. Ryba, *Ap. J.* **428** (1994) 713.
[52] V. F. Schwartzmann, *JETP Lett.* **9** (1969) 184.
[53] R. Brustein, M. Gasperini and G. Veneziano, *Phys. Rev.* **D55** (1997) 3882.
[54] J. Khouri, B. A. Ovrut, N. Seiberg, P. J. Steinhardt and N. Turok, *Phys. Rev.* **D65** (2002) 086007.
[55] L. A. Boyle, P. J. Steinhardt and N. Turok, *Phys. Rev.* **D69** (2004) 127302.
[56] C. Cartier, E. J. Copeland and M. Gasperini, *Nucl. Phys.* **B607** (2001) 406.
[57] A. Buonanno, M. Maggiore and C. Ungarelli, *Phys. Rev.* **D55** (1997) 3330.
[58] V. Kaplunovsky, *Phys. Rev. Lett.* **55** (1985) 1036.
[59] M. Galluccio, F. Litterio and F. Occhionero, *Phys. Rev. Lett.* **79** (1997) 970.
[60] D. Babusci and M. Giovannini, *Int. J. Mod. Phys.* **D10** (2001) 477.
[61] L. E. Mendes and A. R. Liddle, *Phys. Rev.* **D60** (1999) 063508.
[62] T. Damour and A. Vilenkin, *Phys. Rev.* **D64** (2001) 064008.
[63] E. J. Copeland, R. C. Myers and J. Polchinski, *JHEP* **06** (2004) 013.
[64] S. Kachru, R. Kallosh, A. Linde *et al.*, *JCAP* **0310** (2003) 013.
[65] A. Vilenkin and E. P. S. Shellard, *Cosmic Strings and Other Topological Defects* (Cambridge: Cambridge University Press, 2000).
[66] M. S. Turner and F. Wilczek, *Phys. Rev. Lett.* **65** (1990) 3080.
[67] S. Y. Khlebnikov and I. I. Tkachev, *Phys. Rev.* **D56** (1997) 653.
[68] M. Gasperini and G. Veneziano, *Phys. Rep.* **373** (2003) 1.
[69] C. Ungarelli and A. Vecchio, *Phys. Rev.* **D63** (2001) 064030.
[70] M. Maggiore, *Phys. Rep.* **331** (2000) 283.
[71] G. Pizzella *et al.*, in *Gravitational Waves*, eds. I. Ciufolini *et al.* (Bristol: IOP Publishing, 2001), p. 89.
[72] A. Abramovici, W. E. Althouse, R. W. P. Drever *et al.*, *Science* **256** (1992) 325.
[73] B. Caron *et al.*, *Class. Quantum Grav.* **14** (1997) 1461.
[74] LIGO MIT/Caltech Groups, *Research and Development Program for Advanced LIGO Detectors*, LIGO-M970107-00-M (September 1997).

[75] B. J. Owen and B. S. Sathyaprakash, *Phys. Rev.* **D60** (1999) 022002.
[76] J. Hough, in *Proc. Second Edoardo Amaldi Conference on Gravitational Waves*, eds. E. Coccia *et al.* (Singapore: World Scientific, 1998), p. 97.
[77] N. Seto, S. Kawamura and T. Nakamura, *Phys. Rev. Lett.* **87** (2001) 221103.
[78] S. Phinney *et al.*, *NASA Mission Concept Study* (2003).
[79] K. Hotta, K. Kikkawa and H. Kunitomo, *Prog. Theor. Phys.* **98** (1997) 687.
[80] P. Astone *et al.*, in *Proc. Second Edoardo Amaldi Conference on Gravitational Waves*, eds. E. Coccia *et al.* (Singapore: World Scientific, 1998), p. 551.
[81] E. Coccia, V. Fafone, G. Frossati, J. Lobo and J. Ortega, *Phys. Rev.* **D57** (1998) 2051.
[82] A. Chincarini, G. Gemme, R. Parodi, P. Bernard and E. Picasso, gr-qc/0203024.
[83] E. Flanagan, *Phys. Rev.* **D48** (1993) 2389.
[84] B. Allen, in *Astrophysical Sources of Gravitational Waves*, eds. J. A. Marck and J. P. Lasota (Cambridge: Cambridge University Press, 1997), p. 373.
[85] D. Babusci, S. Foffa, G. Losurdo *et al.*, in *Gravitational Waves*, eds. I. Ciufolini *et al.* (Bristol: IOP Publishing, 2001), p. 179.
[86] P. Astone, J. Lobo and B. Schutz, *Class. Quantum Grav.* **11** (1994) 2093.
[87] P. Astone, G. Pallottino and G. Pizzella, *Class. Quantum Grav.* **14** (1997) 2019.
[88] S. Vitale, M. Cerdonio, E. Coccia and A. Ortolan, *Phys. Rev.* **D55** (1997) 17414.
[89] E. Coccia, V. Fafone, G. Frossati, J. Lobo and J. Ortega, *Phys. Rev.* **D57** (1998) 2051.
[90] B. Allen and R. Brustein, *Phys. Rev.* **D55** (1997) 3260.
[91] C. Ungarelli and A. Vecchio, in *Proc. Third Edoardo Amaldi Conference on Gravitational Waves*, ed. S. Meshkov, AIP Conference Proceedings **523** (2000) 90.
[92] D. Babusci and M. Giovannini, *Class. Quantum Grav.* **17**, 2621 (2000).
[93] B. Abbot *et al.*, *Phys. Rev. Lett.* **95** (2005) 221101.
[94] C. Cartier, J. C. Hwang and E. J. Copeland, *Phys. Rev.* **D64** (2001) 103504.

8

Scalar perturbations and the anisotropy of the CMB radiation

It has been shown in the previous chapter that the amplification of tensor fluctuations, and the corresponding formation of a cosmic background of relic gravitational waves, represent important sources of information on the dynamics of the inflationary Universe. This chapter is devoted to another important aspect of the inflationary kinematics, related to effects which have already been observed [1, 2], and whose measurements are becoming more and more accurate [3, 4, 5].

We are referring to the inflationary amplification of the scalar part of the metric fluctuations, and to the observed temperature anisotropies $\Delta T/T$ of the Cosmic Microwave Background (CMB) of electromagnetic radiation. Tensor perturbations can also contribute, in principle, to such an anisotropy. If the tensor spectrum is growing, however, its contribution turns out to be largely suppressed at the (very small) frequency scales relevant to the observed anisotropy. If the spectrum is flat or decreasing, on the contrary, then an observable tensor contribution is possibly allowed at frequencies near to the present Hubble scale; however, the tensor contribution tends rapidly to become negligible with respect to the scalar one as one considers higher-frequency modes, i.e. higher multipoles in the spherical-harmonic expansion of the temperature anisotropy (see e.g. [6]). We may thus regard scalar perturbations as in the main responsible for the peak structure (and the related oscillations) that we are presently observing in the power spectrum of $\Delta T/T$.

This chapter will start with a detailed study of the perturbations of the system of cosmological equations. After introducing in Section 8.1 a gauge-invariant formulation of the scalar perturbation equations, and discussing the choice of appropriate gauges, we will compute in Section 8.2 the angular power spectrum of the temperature anisotropies obtained from a generic initial spectrum of scalar metric perturbations. It will be shown, in particular, that an adiabatic and scale-invariant distribution of metric perturbations – like that obtained in the context of the standard (slow-roll) inflationary scenario – provides the appropriate initial

334

conditions for the post-inflationary evolution of the anisotropies, and for the reproduction of their currently observed structures.

In models of string cosmology inflation the natural implementation of a nearly flat and adiabatic spectrum of primordial scalar perturbations represents in general a hard, highly non-trivial problem. There are recent, possibly successful, attempts [7] based on type IIB superstrings and on the idea of D-brane–anti-D-brane inflation, which seem to be able to satisfy the conditions of moduli stabilization [8] and slow-roll inflaton potential (see Section 10.5). Another recent positive claim concerns the production of a flat, adiabatic spectrum of perturbations during an early Hagedorn phase in the context of the string-gas cosmology scenario [9], even without a period of inflation; this approach, however, seems generically to produce a strongly tilted spectrum, unless in the string phase the dilaton is fixed at a constant value by some unspecified strong coupling effect [10, 11]. The situation is also confused for the ekpyrotic scenario (see e.g. [12–15], to mention only a few among the many contributions to the discussion that have appeared in the literature); not to mention the pre-big bang scenario, where the produced spectrum of scalar metric perturbations is naturally too steep to be in direct agreement with the observed anisotropies [16, 17].

In all string theory models there are, however, other fundamental fields that can be easily amplified with a flat (or nearly flat) primordial spectrum: this may occur, for instance, to the axion field [18, 19] associated with the fluctuations of the NS–NS two-form $B_{\mu\nu}$. Section 8.3 will be devoted to illustrating the effect, interesting in itself, by which these exotic, "isocurvature", primordial fluctuations can induce a final spectrum of adiabatic and scale-invariant metric perturbations – the so-called "curvaton mechanism" [20, 21, 22]. This (more indirect) method of generating the observed CMB anisotropies has been investigated largely in the context of pre-big bang models [20, 23, 24], and has been shown to differ, in principle, from the (direct) amplification mechanism of the standard inflationary scenario by the possible presence of non-Gaussian signatures in the anisotropy spectrum [25, 26].

8.1 Scalar perturbations in a cosmological background

To discuss the evolution of scalar perturbations we shall consider, throughout this chapter, a particularly simple (but complete) geometric configuration represented by a four-dimensional and spatially flat isotropic background, with perfect-fluid matter and a self-interacting scalar ϕ as sources, described by the following E-frame action:

$$S = \int d^4x \sqrt{-g} \left[-\frac{R}{16\pi G} + \frac{1}{2}(\nabla\phi)^2 - V(\phi) \right] + S_m(g, \phi, \text{matter}). \qquad (8.1)$$

Here S_m describes the matter (fluid) sources, possibly coupled to the scalar field both intrinsically and as a consequence of the conformal rescaling required to transform the S-frame action to the diagonal form (8.1) (see Section 2.2). Separating the contributions of these two couplings, the variation with respect to the metric and to the scalar field leads, respectively, to the field equations,

$$G_\mu{}^\nu = \lambda_P^2 \left[T_\mu^\nu + \partial_\mu \phi \partial^\nu \phi - \frac{1}{2} \delta_\mu^\nu (\nabla \phi)^2 + \delta_\mu^\nu V \right], \tag{8.2}$$

$$\nabla^2 \phi + \frac{\partial V}{\partial \phi} + \frac{1}{2\mu} (\sigma + T) = 0 \tag{8.3}$$

(recall that $\lambda_P^2 = 8\pi G$). Here $T_{\mu\nu}$ is the (E-frame) stress tensor obtained from the matter action, T its trace, and σ is the possible E-frame transformation of the intrinsic scalar charge of the matter fields, already present in the S-frame (see Eqs. (2.54)–(2.59), where we have redefined σ to make explicit the dimensionful factor $2\mu = \sqrt{2} M_P$). The combination of the above equations leads to the generalized conservation law,

$$\nabla_\nu T_\mu{}^\nu = \frac{1}{2\mu} (\sigma + T) \nabla_\mu \phi. \tag{8.4}$$

For matter minimally coupled to gravity, evolving along the E-frame geodesic network, the coupling generated by $\sigma + T$ is absent, and one is left with the standard equations of general relativity. Such a case can be obtained as the limit $1/2\mu \to 0$ of the more general system of equations that we are considering.

Considering homogeneous, isotropic and spatially flat backgrounds it is convenient to work in the conformal-time gauge, where

$$ds^2 = a^2(\eta)(d\eta^2 - dx_i^2), \qquad \phi = \phi(\eta), \qquad \sigma = \sigma(\eta)$$
$$T_\mu{}^\nu = (\rho, -p\delta_i^j), \qquad \rho = \rho(\eta), \qquad p = p(\eta). \tag{8.5}$$

Then

$$\Gamma_{00}{}^0 = \mathcal{H}, \qquad \Gamma_{0i}{}^j = \mathcal{H}\delta_i^j, \qquad \Gamma_{ij}{}^0 = \mathcal{H}\delta_{ij},$$
$$R_{00} = -3\mathcal{H}', \qquad R_{ij} = (\mathcal{H}' + 2\mathcal{H}^2)\delta_{ij}, \qquad R = -\frac{6}{a^2}(\mathcal{H}' + \mathcal{H}^2), \tag{8.6}$$

where $\mathcal{H} = a'/a$, and the prime denotes differentiation w.r.t. η. The (00) and (ij) components of the Einstein equation (8.2) can be written, respectively, in the form

$$6\mathcal{H}^2 = 2\lambda_P^2 \left(\rho a^2 + \frac{1}{2} \phi'^2 + V a^2 \right), \tag{8.7}$$

$$4\mathcal{H}' + 2\mathcal{H}^2 = -2\lambda_P^2 \left(p a^2 + \frac{1}{2} \phi'^2 - V a^2 \right). \tag{8.8}$$

The dilaton equation (8.3) gives

$$\phi'' + 2\mathcal{H}\phi' + \frac{1}{2\mu}(\sigma + \rho - 3p)a^2 + a^2\frac{\partial V}{\partial \phi} = 0, \qquad (8.9)$$

and the conservation equation becomes

$$\rho' + 3\mathcal{H}(\rho + p) = \frac{1}{2\mu}(\sigma + \rho - 3p)\phi'. \qquad (8.10)$$

Finally, the subtraction of Eq. (8.7) from Eq. (8.8) leads to the useful identity

$$4(\mathcal{H}' - \mathcal{H}^2) + 2\lambda_P^2\left[\phi'^2 + a^2(\rho + p)\right] = 0. \qquad (8.11)$$

It can be easily checked that the above equations correspond to Eqs. (2.60)–(2.63), written in conformal time, for the case $d = 3$ and $\widetilde{V} = V$. It is also important to stress that the fluid is assumed to be perfect, but not necessarily barotropic, so that p/ρ may be time dependent, in general.

We now introduce a linear perturbation of the metric tensor, $g_{\mu\nu} \to g_{\mu\nu} + \delta g_{\mu\nu}$, where $g(\eta)$ is the unperturbed metric (8.5) and $\delta g(\eta, x_i)$ is a small, inhomogeneous correction. Also, we decompose δg into components transforming as irreducible representations of an appropriate isometry group of the unperturbed background: we use, in particular, the $SO(3)$ symmetry of the flat spatial sections, classifying perturbations as scalar, vector and tensor objects with respect to spatial rotations on a constant-time hypersurface. In a $(3+1)$-dimensional manifold, in particular, the 10 independent components of the symmetric tensor $\delta g_{\mu\nu}$ will be decomposed as follows: 2 degrees of freedom (or independent polarization states) associated with one massless tensor field (represented by a traceless, transverse tensor); 4 degrees of freedom associated with two massless vector fields (represented by two divergenceless vectors); and 4 degrees of freedom associated with four massless scalar fields [27, 28].

With respect to the spatial rotations the component $\delta g_{00} = A$ transforms as a scalar; the components $\delta g_{i0} = B_i$ transform as a $d = 3$ vector, which can be decomposed into a divergenceless (pure-vector) part $\overline{B}_i$, and the gradient of a scalar, $B_i = \overline{B}_i + \partial_i B$, $\partial_i \overline{B}_i = 0$; finally, the components $\delta g_{ij} = C_{ij}$ transform as a symmetric tensor, and may be decomposed into a transverse and traceless (pure-tensor) part, $\overline{\overline{C}}_{ij}$, plus vectors and scalar contributions: $C_{ij} = C_1\delta_{ij} + \partial_i\partial_j C_2 + \partial_{(i}\overline{C}_{j)} + \overline{\overline{C}}_{ij}$, where $\partial_i\overline{C}_i = 0$. Following the classical notations of [27], we can thus parametrize the perturbed line-element ds^2, expanded to first order, as follows:

$$
\begin{aligned}
ds^2 &= (g_{\mu\nu} + \delta g_{\mu\nu})dx^\mu dx^\nu \\
&= a^2(1 + 2\varphi)d\eta^2 - 2a^2(\partial_i B + S_i)dx^i d\eta \\
&\quad - a^2\left[(1 - 2\psi)\delta_{ij} + 2\partial_i\partial_j E + 2\partial_{(i}F_{j)} - h_{ij}\right]dx^i dx^j, \qquad (8.12)
\end{aligned}
$$

where φ, ψ, E, B are scalars, S_i, F_i are divergenceless vectors, and h_{ij} is a symmetric, transverse, traceless tensor:

$$\partial_i S_i = 0 = \partial_i F_i, \qquad \partial_i h^i_j = 0 = h_i{}^i. \tag{8.13}$$

It is important to note that, for a manifold with flat spatial sections, this decomposition of the metric perturbations is valid quite independently of the number of spatial dimensions: for any value of d the number of independent fields is the same, and what is varying is only the number of their components. In d spatial dimensions, in fact, a transverse, traceless tensor field has $(d^2 - d - 2)/2$ components, while a divergenceless vector has $d - 1$ components. Adding the components of one tensor, two vectors and four scalars one obtains $(d^2 + 3d + 2)/2$ degrees of freedom, which precisely corresponds to the number of independent components of the symmetric tensor $\delta g_{\mu\nu}$ in a $(d+1)$-dimensional manifold.

The decomposition (8.12), with the addition of other fields associated with the new (internal) degrees of freedom, can be directly extended also to higher-dimensional manifolds with *anisotropic* but *factorized* geometric structure, typical of the standard Kaluza–Klein scenario. In that case, the classification of perturbations as different irreducible representations of the rotation group is conveniently referred to the coordinate transformations of the "external" spatial submanifold [29, 30]. Such a procedure, however, cannot be directly applied to the case of *non-factorized* (or "warped") geometrical structures, like those appearing in the context of the Randall–Sundrum scenario [31] (see Chaper 10). In that case, typical of the embedding of a brane in the curved bulk manifold, the metric is associated with a different isometric structure, and requires a different classification of the independent components of its perturbations [32].

In the linear approximation, the components of $\delta g_{\mu\nu}$ with different transformation properties satisfy decoupled equations [27], and can be studied separately. In the presence of an accelerated (inflationary) background evolution, in particular, tensor metric perturbations are automatically amplified even without any direct coupling to the sources, as illustrated in the previous chapter; vector metric perturbations, on the contrary, tend to decay rapidly in the absence of a specific vector source [27], unless one considers particular models of multi-dimensional bouncing evolution [33]. In this chapter we concentrate our study on the pure scalar part of metric perturbations, described in general by the four functions $\{\varphi, \psi, E, B\}$. We set, therefore, $\delta g_{\mu\nu} = h_{\mu\nu}$, where

$$\delta g_{00} = h_{00} = 2a^2\varphi,$$

$$\delta g_{i0} = h_{i0} = -a^2\partial_i B, \tag{8.14}$$

$$\delta g_{ij} = h_{ij} = 2a^2(\psi\delta_{ij} - \partial_i\partial_j E).$$

The contravariant components of the perturbed metric are defined by the condition of matrix inversion,

$$(g_{\mu\nu} + \delta g_{\mu\nu})(g^{\alpha\nu} + \delta g^{\alpha\nu}) = \delta_\mu^\alpha, \tag{8.15}$$

which gives, to first order in h, $\delta g^{\mu\nu} = -h^{\mu\nu} \equiv -g^{\mu\alpha}g^{\nu\beta}h_{\alpha\beta}$. Thus,

$$\delta g^{00} = -h^{00} = -\frac{2\varphi}{a^2},$$

$$\delta g^{i0} = -h^{i0} = -\frac{\partial^i B}{a^2}, \tag{8.16}$$

$$\delta g^{ij} = -h^{ij} = -\frac{2}{a^2}(\psi\delta^{ij} - \partial^i\partial^j E).$$

Finally, the following remark concerning notations is in order: throughout this chapter the symbol of spatial gradient will simply denote a partial derivative, quite irrespective of the index position, i.e. $\partial^i \equiv \partial_i = \partial/\partial x^i$; all geometric factors related to the raising and lowering of indices in non-Euclidean manifolds will always be explicitly displayed.

Together with the metric pertubations we must introduce the scalar perturbations of the matter sources. For the scalar field we set $\phi \to \phi + \delta\phi$, where $\delta\phi(\eta, x_i)$ represents a small, inhomogeneous perturbation ($|\delta\phi| \ll |\phi|$) of the unperturbed scalar background $\phi(\eta)$; for simplicity, we also adopt the notation

$$\delta\phi \equiv \chi. \tag{8.17}$$

Other possible gravitational sources will be described hydrodynamically, as perfect fluid matter represented by the symmetric, rank-two tensor $T_{\mu\nu}$. As in the case of the metric, the perturbations of $T_{\mu\nu}$, classified as scalars with respect to spatial rotations, can be generally expressed in terms of four independent functions. A first scalar function is associated with δT_{00}; a second one with the gradient part of the mixed components, $\delta T_{i0} \sim \partial_i w$; two additional scalar functions are finally contained inside the spatial components, $T_{ij} \sim A\delta_{ij} + \partial_i\partial_j\sigma$. The last contribution associated with σ, called the "anisotropic stress perturbation" (or "scalar shear"), introduces however non-diagonal spatial components in the perturbed energy-momentum tensor, and vanishes in the perfect fluid approximation [27].

For a perfect fluid we have

$$T_\mu{}^\nu = (\rho + p)u_\mu u^\nu - p\delta_\mu^\nu, \tag{8.18}$$

where $u^\mu = (a^{-1}, \vec{0})$ (in the conformal gauge), if the fluid is at rest in the comoving system of coordinates. The perturbations $\delta T_\mu{}^\nu$ are thus fixed by $\delta\rho$, δp and δu^μ.

On the other hand, perturbing the (constant) squared modulus of the four-velocity vector, we obtain the conditions

$$\delta \left(g_{\mu\nu} u^\mu u^\nu \right) = \frac{\delta g_{00}}{a^2} + 2a\,\delta u^0 = 0,$$

$$\delta \left(g^{\mu\nu} u_\mu u_\nu \right) = \delta g^{00} a^2 + \frac{2}{a}\delta u_0 = 0,$$

(8.19)

from which, using Eqs. (8.14) and (8.16), we find that the velocity perturbation δu_0 is fully determined by the metric perturbation as

$$\delta u^0 = -\frac{\varphi}{a}, \qquad \delta u_0 = a\varphi. \tag{8.20}$$

We are left with δu_i as the only independent variables, and we can thus parametrize the scalar component of the velocity perturbations as

$$\delta u_i = a\,\partial_i w, \tag{8.21}$$

where w is the so-called "velocity potential" (we are following the conventions of [27]; see [34] for a possible alternative definition of the velocity potential). A direct perturbation of the stress tensor,

$$\delta T_\mu{}^\nu = (\delta\rho + \delta p)u_\mu u^\nu + (\rho + p)(\delta u_\mu u^\nu + u_\mu \delta u^\nu) - \delta p\,\delta_\mu^\nu, \tag{8.22}$$

then provides the relations

$$\delta T_0^0 = \delta\rho, \qquad \delta T_i^j = -\delta p\,\delta_i^j,$$

$$\delta T_i{}^0 = a^{-1}(\rho + p)\delta u_i = (\rho + p)\partial_i w,$$

(8.23)

which, together with Eqs. (8.14)–(8.17), complete the set of scalar variables describing the perturbations of the homogeneous background (8.5).

8.1.1 Gauge-invariant variables

It is important to stress, at this point, that none of these scalar variables is in general invariant under the local "gauge" transformation represented by the infinitesimal coordinate reparametrization,

$$x^\mu \to \widetilde{x}^\mu(x) = x^\mu + \xi^\mu(x), \tag{8.24}$$

and by its inverse, which, to first order in ξ, can be written as

$$x^\mu(\widetilde{x}) = \widetilde{x}^\mu - \xi^\mu(\widetilde{x}). \tag{8.25}$$

Transformations like (8.24) can in general introduce new vector components in the perturbed metric, even starting with metric perturbations of pure scalar type.

Let us compute the *local*, infinitesimal variations induced by such a transformation on the perturbed metric and matter fields. Starting from the metric, and from the exact tensor-transformation law

$$\tilde{g}_{\mu\nu}(\tilde{x}) = \frac{\partial x^\alpha}{\partial \tilde{x}^\mu} \frac{\partial x^\beta}{\partial \tilde{x}^\nu} g_{\alpha\beta}(x), \tag{8.26}$$

we compute the Jacobian matrix $(\partial x/\partial \tilde{x})$ from Eq. (8.25) (to first order in ξ), and we expand $g(x) \equiv g(\tilde{x} - \xi)$ in Taylor series around $x = \tilde{x}$. We obtain, to first order,

$$\tilde{g}_{\mu\nu}(\tilde{x}) = \left(\delta^\alpha_\mu - \partial_\mu \xi^\alpha\right)_{\tilde{x}} \left(\delta^\beta_\nu - \partial_\nu \xi^\beta\right)_{\tilde{x}} \left(g_{\alpha\beta} - \xi^\lambda \partial_\lambda g_{\alpha\beta}\right)_{\tilde{x}}$$
$$= g_{\mu\nu}(\tilde{x}) - g_{\mu\alpha}\partial_\nu \xi^\alpha - g_{\nu\alpha}\partial_\mu \xi^\alpha - \xi^\lambda \partial_\lambda g_{\mu\nu}, \tag{8.27}$$

where all terms, on the left- and right-hand sides, are computed *at the same space-time position* $\tilde{x}$. For a perturbed metric we may replace g by $g + h$ in the above equation, and we can separate the homogeneous, zeroth-order terms from the first-order terms (which are in general inhomogeneous), to obtain the local, infinitesimal transformation of the metric fluctuations:

$$h_{\mu\nu} \to \tilde{h}_{\mu\nu} = h_{\mu\nu} - g_{\mu\alpha}\partial_\nu \xi^\alpha - g_{\nu\alpha}\partial_\mu \xi^\alpha - \xi^\lambda \partial_\lambda g_{\mu\nu}. \tag{8.28}$$

One can easily check – using the definition of the Christoffel connection – that the last three terms of this equation can also be written in the (more compact) covariant form as

$$\tilde{h}_{\mu\nu} = h_{\mu\nu} - \nabla_\mu \xi_\nu - \nabla_\nu \xi_\mu. \tag{8.29}$$

Following the same procedure for the scalar field, we expand to first order the exact scalar-transformation law,

$$\tilde{\phi}(\tilde{x}) = \phi(x) = \phi(\tilde{x} - \xi) \simeq \phi(\tilde{x}) - \xi^\lambda(\tilde{x})\partial_\lambda \phi(\tilde{x}). \tag{8.30}$$

Replacing ϕ by $\phi + \chi$, and separating the homogeneous and inhomogeneous parts, we are led to the local transformation of the scalar fluctuations:

$$\chi \to \tilde{\chi} = \chi - \xi^\lambda \partial_\lambda \phi. \tag{8.31}$$

Finally, for the fluid stress tensor we start again from the exact transformation,

$$\tilde{T}_\mu{}^\nu(\tilde{x}) = \frac{\partial x^\alpha}{\partial \tilde{x}^\mu} \frac{\partial \tilde{x}^\nu}{\partial x^\beta} T_\alpha{}^\beta(x). \tag{8.32}$$

Expanding to first order, and separating the inhomogeneous contributions, we arrive at the local transformations of the perturbed matter sources:

$$\delta T_\mu{}^\nu \to \delta \tilde{T}_\mu{}^\nu = \delta T_\mu{}^\nu - \xi^\lambda \partial_\lambda T_\mu{}^\nu + T_\mu{}^\alpha \partial_\alpha \xi^\nu - T_\alpha{}^\nu \partial_\mu \xi^\alpha. \tag{8.33}$$

Armed with such results, we are now in a position to show that if the spatial part of the infinitesimal generator ξ^μ contains a pure traceless vector part, $\overline{\xi}^i$, then the coordinate transformation (8.25) does not preserve the scalar nature of the initial (metric or matter) perturbation. Suppose in fact that

$$\xi^\mu = \left(\xi^0, \xi^i\right), \qquad \xi^i = \overline{\xi}^i + \partial^i \xi, \qquad \partial_i \overline{\xi}^i = 0, \tag{8.34}$$

and compute the variation of the spatial part of the metric fluctuations, $h_{ij} \to \tilde{h}_{ij}$. Using the definition (8.14) of h_{ij}, and applying the transformation (8.28), we obtain

$$\tilde{h}_{ij} = 2a^2 \delta_{ij}(\psi + \mathcal{H}\xi^0) - 2a^2 \partial_i \partial_j (E - \xi) + a^2 (\partial_i \overline{\xi}_j + \partial_j \overline{\xi}_i). \tag{8.35}$$

Besides the transformed scalar variables, $\psi \to \psi + \mathcal{H}\xi^0$ and $E \to E - \xi$, the new perturbation $\tilde{h}$ contains a new irreducible vector component generated by $\overline{\xi}_i$. The same happens for the transformation of the mixed components, $h_{i0} \to \tilde{h}_{i0}$. Note that, in the absence of physical vector sources, this effect can be used to eliminate some vector components of the perturbations through an appropriate coordinate transformation (or choice of the gauge).

Another important consequence of the considered transformations – clearly illustrated by the result (8.35) – is that the variables $\{\varphi, \psi, E, B\}$, used to parametrize the scalar part of the metric fluctuations, *are not* gauge-invariant, not even with respect to the restricted class of transformations defined by $\overline{\xi}_i = 0$ and specified by two scalar parameters only, ξ^0 and ξ, such that

$$\eta \to \tilde{\eta} = \eta + \xi^0(\eta, x_i),$$
$$x^i \to \tilde{x}^i = x^i + \partial^i \xi(\eta, x_i). \tag{8.36}$$

However, for such a class of transformations which preserve the scalar nature of the initial perturbations, it is always possible to define an appropriate set of variables which are gauge invariant, and which are obtained through a linear combination of the variables introduced in Eq. (8.14).

Let us apply the general infinitesimal transformations of the metric and matter variables, Eqs. (8.28), (8.31) and (8.33), to the particular transformation (8.36). One easily obtains, for the various components,

$$\tilde{\varphi} = \varphi - \mathcal{H}\xi^0 - \xi'^0, \qquad \tilde{\psi} = \psi + \mathcal{H}\xi^0,$$
$$\tilde{E} = E - \xi, \qquad \tilde{B} = B + \xi^0 - \xi',$$
$$\tilde{\chi} = \chi - \phi'\xi^0, \qquad \delta\tilde{\rho} = \delta\rho - \rho'\xi^0, \tag{8.37}$$
$$\delta\tilde{p} = \delta p - p'\xi^0, \qquad \tilde{w} = w - \xi^0.$$

The combination $B - E'$, in particular, transforms as

$$\widetilde{B} - \widetilde{E}' = B - E' + \xi^0. \tag{8.38}$$

By exploiting this result, one can immediately introduce a set of new scalar variables $\{\Phi, \Psi, X, \mathcal{E}, \Pi, W\}$ which are "gauge-invariant" (i.e. which are preserved by the infinitesimal transformation (8.36)), by defining

$$\Phi = \varphi + \mathcal{H}(B - E') + (B - E')', \qquad \Psi = \psi - \mathcal{H}(B - E'),$$

$$X = \chi + \phi'(B - E'), \qquad \mathcal{E} = \delta\rho + \rho'(B - E'), \tag{8.39}$$

$$\Pi = \delta p + p'(B - E'), \qquad W = w + B - E'.$$

Notice that we have used only two gauge-invariant variables (Ψ and Φ) to represent the metric fluctuations since, with an appropriate choice of the arbitrary parameters ξ and ξ^0, it is always possible to impose two conditions on the transformed variables, and to eliminate two of the four components $\{\varphi, \psi, E, B\}$ which were initially present (see below). There is, of course, an infinite number of different choices of gauge-invariant variables, but the combination presented here has a particular phenomenological relevance, because the two metric variables Ψ and Φ (also called "Bardeen potentials" [35]) are the variables appearing as direct sources of the anisotropy of the CMB radiation, as we shall see in the next section. The definitions (8.39) will be used very frequently in the rest of the chapter, and will play a central role in the discussion of the scalar perturbation dynamics developed in this section.

It seems appropriate, at this point, to present a list of the most popular (and useful) gauge choices, often adopted in the specialized literature on scalar perturbations.

Longitudinal gauge (or conformally Newtonian gauge) fixed by the conditions $E = 0 = B$. In this case the two non-zero metric fluctuations automatically coincide with the Bardeen potentials, $\varphi = \Phi$ and $\psi = \Psi$. The system of coordinates turns out to be completely specified, in this gauge, because the choice $E \to 0$, $B \to 0$ uniquely fixes the parameters ξ^0 and ξ of Eq. (8.36), without any residual degrees of freedom.

Synchronous gauge fixed by the conditions $\varphi = 0 = B$. This choice does not completely specify the system of coordinates, as it determines ξ^0 and ξ only up to arbitrary functions of the spatial coordinates x_i [27]. One is left with residual degrees of freedom which may render the physical interpretation of the perturbative results difficult in this gauge.

Uniform-curvature gauge (or spatially flat, or flat-slicing gauge) fixed by the conditions $\psi = 0 = E$. In this case the coordinates are completely fixed, as in the longitudinal gauge. The alternative denominations, listed in parentheses, are due to the vanishing (in this gauge) of the perturbation of the *extrinsic* scalar curvature of the space-like hypersurfaces, $\delta(g^{ij}R_{ij}) = 0$. For some models of pre-big bang inflation, such a gauge (also called the "off-diagonal" gauge, in that context) has proved useful to get rid of a too fast growth of the metric fluctuations, and to restore the validity of the linear approximation even when the background is approaching the high-curvature regime [17].

Comoving gauge, fixed by the conditions $B = 0 = \chi$, if the gravitational model includes a scalar field which is left unperturbed. This choice is also called the "uniform-dilaton" gauge, with reference to a particular model of scalar source. If the comoving condition is instead referred to a fluid source, then the gauge is specified by the conditions $B = 0 = w$. In any case, these conditions do not completely fix the system of coordinates (see Eq. (8.37)), and one is left with a residual gauge freedom as in the synchronous case.

For any given choice of gauge, which reduces from eight to six the number of independent scalar variables, Eq. (8.39) defines the corresponding gauge-invariant fluctuations. It is always possible, therefore, to perform the computations in the more appropriate gauge, and then re-express the final result in gauge-invariant form using the relations obtained by inverting Eq. (8.39).

As an example of this technique we now show that the Bardeen variable Ψ is also the "potential" associated with the perturbation of the *intrinsic* scalar curvature of the space-like hypersurfaces, both in the longitudinal and in the uniform-curvature gauge. In the comoving gauge, on the contrary, the corresponding curvature perturbation is associated with another gauge-invariant variable [36], defined by

$$\mathcal{R}_\chi = \Psi + \frac{\mathcal{H}}{\phi'} X \tag{8.40}$$

if the comoving gauge is referred to the scalar field ($\chi = 0$), and by

$$\mathcal{R}_w = \Psi + \mathcal{H} W \tag{8.41}$$

if the comoving gauge is referred to a fluid ($w = 0$). The reader should be warned that other authors denote the curvature perturbation in the comoving gauge by the ζ symbol, while in this book we reserve such a symbol to denote the curvature perturbation in the uniform-density gauge $\delta\rho = 0$ (see below).

To compute the curvature perturbations let us start from the general expression of the perturbed Christoffel connection, to first order in $h_{\mu\nu}$, already presented in

Eq. (7.7). By applying such an equation to our scalar variables (8.14) and (8.16) we obtain, to first order:

$$\delta\Gamma_{00}{}^0 = \varphi',$$

$$\delta\Gamma_{i0}{}^0 = \partial_i(\varphi + \mathcal{H}B),$$

$$\delta\Gamma_{ij}{}^0 = -\partial_i\partial_j B - 2\mathcal{H}\delta_{ij}(\varphi + \psi) - \delta_{ij}\psi' + \partial_i\partial_j(2\mathcal{H}E + E'),$$

$$\delta\Gamma_{00}{}^i = \partial^i(\varphi + \mathcal{H}B + B'),\tag{8.42}$$

$$\delta\Gamma_{0i}{}^j = \partial_i\partial^j E' - \delta_i^j\psi',$$

$$\delta\Gamma_{ij}{}^k = \partial_i\partial_j\partial^k E + \delta_{ij}\partial^k\psi - \delta_i^k\partial_j\psi - \delta_j^k\partial_i\psi - \delta_{ij}\mathcal{H}\partial^k B.$$

In the longitudinal gauge we set $E = 0 = B$, and we note that, for the conformally flat metric that we are considering, the unperturbed spatial hypersurfaces $\eta = $ const are characterized by a Euclidean metric δ_{ij}, and thus by a vanishing intrinsic curvature, $\Gamma_{ij}{}^k = 0$, $R^{(3)} = 0$. The first-order perturbation of the intrinsic Ricci tensor then gives

$$\delta R_{ij}^{(3)} = \partial_k\delta\Gamma_{ij}{}^k - \partial_i\delta\Gamma_{kj}{}^k = \left(\partial_i\partial_j + \delta_{ij}\nabla^2\right)\psi,\tag{8.43}$$

and the perturbed (intrinsic) scalar curvature is

$$\delta R^{(3)} = g^{ij}\delta R_{ij}^{(3)} = -\frac{4}{a^2}\nabla^2\psi.\tag{8.44}$$

In the longitudinal gauge, on the other hand, $\psi = \Psi$ (see Eq. (8.39)), and we can rewrite the above result in gauge-invariant form as

$$\delta R^{(3)} = -\frac{4}{a^2}\nabla^2\Psi.\tag{8.45}$$

The Bardeen variable thus completely specifies the curvature perturbation in this gauge. The same is true for the uniform-curvature gauge: by setting $E = 0 = \psi$ in Eq. (8.42), we obtain the following perturbed components of the intrinsic Ricci tensor and Ricci scalar:

$$\delta R_{ij}^{(3)} = \partial_k\delta\Gamma_{ij}{}^k - \partial_i\delta\Gamma_{kj}{}^k = \mathcal{H}\left(\partial_i\partial_j - \delta_{ij}\nabla^2\right)B,$$

$$\delta R^{(3)} = g^{ij}\delta R_{ij}^{(3)} = \frac{2\mathcal{H}}{a^2}\nabla^2 B.\tag{8.46}$$

In this gauge, on the other hand, $\mathcal{H}B = -\Psi$ (see Eq. (8.39)), so that $\delta R^{(3)} \sim \nabla^2\Psi$, as before.

Let us now compute the curvature perturbation using the comoving gauge, setting $B = 0$ in the perturbed connection (8.42). It turns out that the contributions of E drop from the intrinsic curvature because of mutual cancelations, and one finds for $\delta R^{(3)}$ exactly the result (8.44) obtained in the longitudinal case. For the comoving gauge, however, the general definitions (8.39) imply

$$\psi = \Psi - \mathcal{H} E'. \tag{8.47}$$

So, if the comoving gauge is referred to an unperturbed scalar field, then the conditions $\chi = 0 = B$ give $E' = -X/\phi'$, and the curvature perturbation can be written explicitly in gauge-invariant form as

$$\delta R^{(3)}_\chi = -\frac{4}{a^2} \nabla^2 \left(\Psi + \frac{\mathcal{H}}{\phi'} X \right) \equiv -\frac{4}{a^2} \nabla^2 \mathcal{R}_\chi, \tag{8.48}$$

which defines the gauge-invariant potential $\mathcal{R}_\chi$ of Eq. (8.40). If, on the contrary, the comoving gauge is referred to fluid matter, then we have the conditions $w = 0 = B$ which, inserted into Eq. (8.39), give $E' = -W$. Using Eq. (8.47) we finally obtain

$$\delta R^{(3)}_w = -\frac{4}{a^2} \nabla^2 \left(\Psi + \mathcal{H} W \right) \equiv -\frac{4}{a^2} \nabla^2 \mathcal{R}_w, \tag{8.49}$$

which defines the gauge-invariant potential $\mathcal{R}_w$ of Eq. (8.41).

It may be important to note that, in the case of fluid sources, there is another useful gauge corresponding to the choice $B = 0 = \delta\rho$ (called the "uniform-density" gauge). In this case the definitions (8.39) give $E' = -\mathcal{E}/\rho'$ which, inserted into (8.47), leads to $\psi = \Psi + \mathcal{H}\mathcal{E}/\rho'$. This choice defines another gauge-invariant potential [37], that we shall call here ζ, related to the curvature perturbation as follows:

$$\delta R^{(3)}_\rho = -\frac{4}{a^2} \nabla^2 \left(\Psi + \frac{\mathcal{H}\mathcal{E}}{\rho'} \right) \equiv -\frac{4}{a^2} \nabla^2 \zeta. \tag{8.50}$$

The variable ζ controls the perturbation of the intrinsic scalar curvature on three-dimensional, space-like hypersurfaces characterized by a uniform distribution of the energy density ($\delta\rho = 0$); the variables $\mathcal{R}_\chi$ and $\mathcal{R}_w$ define the intrinsic curvature perturbation on hypersurfaces of uniform dilaton ($\chi = 0$) and uniform fluid velocity ($w = 0$) distributions, respectively.

Given the differential equations satisfied by ζ and $\mathcal{R}$ (see below), their Fourier components, in a typical cosmological background, only differ by terms of order $|k\eta|^2$: thus, the two variables tend to coincide as long as we are considering long-wavelength perturbations – in particular, when discussing the evolution of perturbations outside the horizon. From a more substantial point of view, however, there is an important difference between ζ and $\mathcal{R}$, due to the fact that $\mathcal{R}$ is closely connected to the canonical variable which diagonalizes the quadratic action for the scalar perturbations of the system {gravity + matter sources}. This connection can be easily illustrated in the simple case in which the matter sources are represented by a single scalar field.

Consider the Einstein action (8.1) with $S_m = 0$, perturb the metric and the scalar field according to Eqs. (8.14)–(8.17), and expand the perturbed action up to orders h^2 and χ^2, following the same procedure used in Section 7.1 for tensor perturbations (see [27] for a detailed computation in the scalar perturbation case). Using the unperturbed background equations (8.7)–(8.11), using the constraint equations following from the variation of the variables appearing as Lagrange multipliers in the perturbed action, and neglecting various total-derivative terms, the quadratic action can be written finally in canonical form as

$$\delta^{(2)}S = \frac{1}{2}\int d\eta\, d^3x \left(v'^2 + v\nabla^2 v + \frac{z''}{z}v^2 \right), \tag{8.51}$$

where [38]

$$v = a\chi + \frac{a\phi'}{\mathcal{H}}\psi = aX + \frac{a\phi'}{\mathcal{H}}\Psi, \qquad z = \frac{a\phi'}{\mathcal{H}} \tag{8.52}$$

(see [30] for a higher-dimensional generalization of the variables v and z to the case of a perturbed Kaluza–Klein background). The gauge-invariant variable v thus satisfies the canonical evolution equation,

$$v'' - \left(\nabla^2 + \frac{z''}{z} \right) v = 0. \tag{8.53}$$

Recalling the definition (8.40), we can write

$$v = z\left(\Psi + \frac{\mathcal{H}}{\phi'}X \right) \equiv z\mathcal{R}_\chi, \tag{8.54}$$

and we see that z plays the role of the "pump field" for the canonical evolution of the curvature perturbation $\mathcal{R}_\chi$, in close analogy with the relation $u = zh$ defining the canonical variable for tensor perturbations in Eq. (7.48). Note that in this chapter we are not assuming $M_P^2 = 2$, and that we are using standard units in which canonical scalar fields have dimensions of mass: thus, the curvature perturbation variable $\mathcal{R}$ is dimensionless, while z and v have the dimensions of mass appropriate to the canonical normalization, and to the quantization of the scalar perturbation spectrum.

We should mention, finally, that with a similar procedure one can show that the same linear combination of Ψ and $\mathcal{H}W$ defining $\mathcal{R}_w$ also leads to the canonical variable for the scalar perturbations of the gravitating fluid system [39, 40]. In the fluid case, however, one obtains a pump field z_w different from the variable z of Eq. (8.52) (see below, Eq. (8.118)).

8.1.2 Scalar perturbation equations

We now need the explicit set of differential equations governing the evolution of the various scalar variables that we have introduced. Such equations can be obtained by perturbing, to first order, the background equations (8.2)–(8.4). From the Einstein equations we obtain

$$
\delta G_\mu{}^\nu = \delta \left(g^{\nu\alpha} R_{\mu\alpha} \right) - \frac{1}{2} \delta^\nu_\mu \delta \left(g^{\alpha\beta} R_{\alpha\beta} \right)
$$

$$
\equiv -h^{\nu\alpha} R_{\mu\nu} + g^{\nu\alpha} \delta R_{\mu\alpha} - \frac{1}{2} \delta^\nu_\mu \left(-h^{\alpha\beta} R_{\alpha\beta} + g^{\alpha\beta} \delta R_{\alpha\beta} \right)
$$

$$
= \lambda_{\rm P}^2 \left[\delta T_\mu{}^\nu - h^{\nu\alpha} \partial_\mu \phi \, \partial_\alpha \phi + g^{\nu\alpha} \left(\partial_\mu \phi \, \partial_\alpha \chi + \partial_\mu \chi \, \partial_\alpha \phi \right) \right.
$$

$$
\left. - \frac{1}{2} \delta^\nu_\mu \left(-h^{\alpha\beta} \partial_\alpha \phi \, \partial_\beta \phi + 2 g^{\alpha\beta} \partial_\alpha \phi \, \partial_\beta \chi \right) + \delta^\nu_\mu \frac{\partial V}{\partial \phi} \chi \right]. \tag{8.55}
$$

The perturbation of the dilaton equation provides

$$
-h^{\mu\nu} \left(\partial_\mu \partial_\nu \phi - \Gamma_{\mu\nu}{}^a \partial_a \phi \right) + g^{\mu\nu} \left(\partial_\mu \partial_\nu \chi - \Gamma_{\mu\nu}{}^a \partial_a \chi - \delta\Gamma_{\mu\nu}{}^a \partial_a \phi \right)
$$

$$
+ \frac{\partial^2 V}{\partial \phi^2} \chi + \frac{1}{2\mu} \left(\delta\sigma + \delta T \right) = 0, \tag{8.56}
$$

where $\delta\sigma$ is a known function of the dilaton perturbation χ. Finally, it may be useful to consider the perturbation of the conservation equation, which gives

$$
\partial_\nu \delta T_\mu{}^\nu - \Gamma_{\nu\mu}{}^\alpha \delta T_\alpha{}^\nu + \Gamma_{\nu\alpha}{}^\nu \delta T_\mu{}^\alpha - \delta\Gamma_{\nu\mu}{}^\alpha T_\alpha{}^\nu + \delta\Gamma_{\nu\alpha}{}^\nu T_\mu{}^\alpha
$$

$$
= \frac{1}{2\mu} \left(\delta\sigma + \delta T \right) \partial_\mu \phi + \frac{1}{2\mu} \left(\sigma + T \right) \partial_\mu \chi. \tag{8.57}
$$

The information contained in this last equation can also be retrieved by combining the perturbed Einstein and dilaton equations.

For extracting the explicit evolution of the various scalar variables we need to compute, first of all, the perturbed components of the curvature tensor. We are free to perform the computations in the preferred gauge, with the understanding that the final result can be re-arranged in gauge-invariant form using the variables (8.39): we will thus assume – also in view of later applications – that $B = 0$. The perturbation of the Ricci tensor, according to our conventions,

$$
\delta R_{\nu\alpha} = \partial_\mu \delta\Gamma_{\nu\alpha}{}^\mu + \delta\Gamma_{\mu\rho}{}^\mu \Gamma_{\nu\alpha}{}^\rho + \Gamma_{\mu\rho}{}^\mu \delta\Gamma_{\nu\alpha}{}^\rho - (\mu \leftrightarrow \nu), \tag{8.58}
$$

then provides

$$\delta R_{00} = 3\left(\psi'' + \mathcal{H}\psi' + \mathcal{H}\varphi'\right) + \nabla^2\left(\varphi - E'' - \mathcal{H}E'\right),$$

$$\delta R_{i0} = 2\partial_i\left(\psi' + \mathcal{H}\varphi\right),$$

$$\delta R_{ij} = \partial_i\partial_j\left[\psi - \varphi + E'' + 2\mathcal{H}E' + (2\mathcal{H}' + 4\mathcal{H}^2)E\right] \qquad (8.59)$$

$$\qquad - \delta_{ij}\left[\psi'' + 5\mathcal{H}\psi' + \mathcal{H}\varphi' - \nabla^2\psi\right.$$

$$\qquad \left. + (2\mathcal{H}' + 4\mathcal{H}^2)(\varphi + \psi) - \mathcal{H}\nabla^2 E'\right],$$

from which we obtain the perturbed scalar curvature,

$$\delta R = \delta g^{\mu\nu}R_{\mu\nu} + g^{\mu\nu}\delta R_{\mu\nu} = \frac{2}{a^2}\nabla^2\left[-E'' - 3\mathcal{H}E' + \varphi - 2\psi\right]$$

$$\qquad + \frac{6}{a^2}\left[\psi'' + 3\mathcal{H}\psi' + \mathcal{H}\varphi' + 2(\mathcal{H}' + \mathcal{H}^2)\varphi\right]. \qquad (8.60)$$

The perturbed Einstein tensor is finally obtained, in mixed form, as follows:

$$\delta G_0^0 = 2\frac{\nabla^2}{a^2}(\psi + \mathcal{H}E') - 6\frac{\mathcal{H}}{a^2}(\psi' + \mathcal{H}\varphi),$$

$$\delta G_i{}^0 = \frac{2}{a^2}\partial_i(\psi' + \mathcal{H}\varphi), \qquad (8.61)$$

$$\delta G_i{}^j = \frac{1}{a^2}\partial_i\partial^j(\varphi - \psi - E'' - 2\mathcal{H}E')$$

$$\qquad + \frac{1}{a^2}\delta_i^j\left[\nabla^2(\psi - \varphi + E'' + 2\mathcal{H}E')\right.$$

$$\qquad \left. - 2\psi'' - 4\mathcal{H}\psi' - 2\mathcal{H}\varphi' - (4\mathcal{H}' + 2\mathcal{H}^2)\varphi\right].$$

We now have all the required ingredients to write down the various components of the perturbed system of equations. We can conveniently adopt the longitudinal gauge, also putting $E = 0$ in the above equations, since in this gauge all the remaining scalar variables exactly coincide with their gauge-invariant counterparts, i.e. $\varphi = \Phi$, $\psi = \Psi$, $\chi = X$, $\delta\rho = \mathcal{E}$, $\delta p = \Pi$, $w = W$ (see Eq. (8.39) with $E = B = 0$). This means that, with the choice $E = 0 = B$, the perturbed equations automatically appear in gauge-invariant form.

Let us start with the Einstein equations (8.55). The spatial, off-diagonal components $(i \neq j)$ give

$$\partial_i\partial_j(\Phi - \Psi) = 0, \qquad (8.62)$$

since we have no off-diagonal sources (the so-called "anisotropic stress") in the total perturbed energy-momentum tensor, at linear order. Going to momentum space, and noting that the equation $k_i k_j(\Phi - \Psi) = 0$ must hold for each mode separately, we must conclude that all Fourier coefficients Φ_k and Ψ_k coincide, so that [27]

$$\Phi = \Psi. \tag{8.63}$$

In the six following equations we keep the distinction between the two Bardeen potentials, to point out explicitly their different contributions to the system of perturbed equations, for possible applications to models with anisotropic stresses (where $\Phi \neq \Psi$). In all subsequent discussions and applications, however, we will use the constraint (8.63) in order to eliminate Φ by Ψ everywhere.

The mixed components $(i0)$ of the Einstein equations give the so-called "momentum constraint":

$$\Psi' + \mathcal{H}\Phi = \frac{\lambda_P^2}{2}\left[\phi'X + a^2(\rho+p)W\right]. \tag{8.64}$$

The time component (00) gives the so-called "Hamiltonian constraint":

$$\nabla^2\Psi - 3\mathcal{H}\Psi' - \left(3\mathcal{H}^2 - \frac{\lambda_P^2}{2}\phi'^2\right)\Phi = \frac{\lambda_P^2}{2}\left[a^2\mathcal{E} + \phi'X' + a^2\frac{\partial V}{\partial\phi}X\right]. \tag{8.65}$$

The spatial components (ij), tracing and dividing by 3, give

$$\Psi'' + 2\mathcal{H}\Psi' + \mathcal{H}\Phi' - \frac{1}{3}\nabla^2(\Psi - \Phi) + \left(2\mathcal{H}' + \mathcal{H}^2 + \frac{\lambda_P^2}{2}\phi'^2\right)\Phi$$

$$= \frac{\lambda_P^2}{2}\left[a^2\Pi + \phi'X' - a^2\frac{\partial V}{\partial\phi}X\right]. \tag{8.66}$$

The perturbed dilaton equation gives

$$X'' + 2\mathcal{H}X' - \nabla^2 X + a^2\frac{\partial^2 V}{\partial\phi^2}X = 2(\phi'' + 2\mathcal{H}\phi')\Phi + \phi'(\Phi' + 3\Psi')$$

$$-\frac{a^2}{2\mu}(\delta\sigma + \mathcal{E} - 3\Pi). \tag{8.67}$$

Finally, the time component $(\mu = 0)$ of the perturbed conservation equation (8.57) leads to

$$\mathcal{E}' + 3\mathcal{H}(\mathcal{E} + \Pi) - (\rho+p)\nabla^2 W - 3(\rho+p)\Psi' = \frac{1}{2\mu}(\delta\sigma + \mathcal{E} - 3\Pi)\phi'$$

$$+ \frac{1}{2\mu}(\sigma + \rho - 3p)X', \tag{8.68}$$

while the spatial component $(\mu = i)$ gives

$$W' + 4\mathcal{H}W - \Phi + \frac{\rho' + p'}{\rho+p}W - \frac{\Pi}{\rho+p} = \frac{1}{2\mu}\frac{\sigma+\rho-3p}{\rho+p}X. \tag{8.69}$$

The set of equations (8.63)–(8.69) describes in a gauge-invariant way the evolution of the scalar perturbations of the system {gravity + scalar field + fluid matter}, around the given homogeneous background (we recall that $\delta\sigma$ is assumed to be a known function of the dilaton perturbation, $\delta\sigma = \delta\sigma(X)$). Of these *seven* equations for the *six* variables {$\Phi, \Psi, X, \mathcal{E}, \Pi, W$} only *five* equations are independent, so that the system has to be closed by the addition of one further condition: for instance, by an appropriate "equation of state" $\Pi = \Pi(\mathcal{E})$ for the fluctuations, which specifies the model of matter we are considering. The above perturbation equations, moreover, are complemented by the set of background equations (8.7)–(8.10) and, in particular, by the useful relation (8.11) by which we can recast the equations in other forms, more convenient for phenomenological applications. In view of such applications we will now separately discuss the two simplest (and most frequently used) configurations, i.e. the case in which the gravitational source consists of either a scalar field, or a perfect fluid, only.

8.1.3 Scalar field source

In this subsection we restrict our attention to the case in which the only other field appearing in the action, besides the metric, is a scalar field (for instance the dilaton, minimally coupled to gravity in the low-energy Einstein action). We set to zero all fluid terms (sources ρ, p, σ, and fluctuations $\mathcal{E}$, Π, W, $\delta\sigma$) in the background and perturbation equations. We also use everywhere the constraint $\Phi = \Psi$. This situation can appropriately describe, for instance, a phase of primordial inflationary evolution in which the cosmological dynamics is dominated by the potential (or kinetic) energy of some cosmic scalar field.

In such a case it is possible to obtain a second-order, decoupled differential equation for the Bardeen variable $\Psi = \Phi$, which completely describes (in a gauge-invariant way) the time evolution of scalar metric perturbations. By subtracting Eq. (8.65) from (8.66), by eliminating X through the constraint (8.64), and the potential terms through the background equation (8.9), one is led to the decoupled equation

$$\Psi'' + 2\left(\mathcal{H} - \frac{\phi''}{\phi'}\right)\Psi' + 2\left(\mathcal{H}' - \mathcal{H}\frac{\phi''}{\phi'}\right)\Psi - \nabla^2\Psi = 0, \qquad (8.70)$$

which can also be written in the useful form

$$\Psi'' + 2\frac{\xi'}{\xi}\Psi' - \nabla^2\Psi + 2\left(\mathcal{H}' - \mathcal{H}^2 + \mathcal{H}\frac{\xi'}{\xi}\right)\Psi = 0, \qquad (8.71)$$

where $\xi = a/\phi'$. Introducing the new variable $V = \xi\Psi$ we can also obtain the "pseudo-canonical" form

$$V'' - \left(\nabla^2 + \frac{Z''}{Z}\right)V = 0, \tag{8.72}$$

where

$$V = \xi\Psi, \qquad \xi = \frac{a}{\phi'}, \qquad Z = \frac{\mathcal{H}}{a\phi'} = \frac{1}{z} \tag{8.73}$$

(note that this effective "pump field" Z is just the inverse of the field z introduced in Eq. (8.52)). The equivalence of Eqs. (8.70) and (8.72) can be easily checked by using the definitions of Z, ξ, and the background condition (8.11), which implies

$$2(\mathcal{H}^2 - \mathcal{H}') = \lambda_P^2\phi'^2, \qquad 2\mathcal{H}\mathcal{H}' - \mathcal{H}'' = \lambda_P^2\phi'\phi''. \tag{8.74}$$

It is important to note that Eq. (8.72) describes the classical evolution of the scalar perturbations, but cannot be used to impose the initial normalization to a quantum spectrum of vacuum fluctuations since neither Ψ nor V can be identified with the variable which diagonalizes the perturbed action. For the quantum normalization of the spectrum we must refer to the canonical variable v and/or to the curvature perturbation $\mathcal{R}_\chi$, related to v by Eq. (8.54). We should thus present, at this point, the evolution equation for $\mathcal{R}_\chi$, and the relation between Ψ and $\mathcal{R}_\chi$.

Supposing that we have not yet diagonalized the action, and that we do not know the relation $v = z\mathcal{R}_\chi$, where v satisfies the canonical equation (8.53), we could obtain the equation for $\mathcal{R}_\chi$ directly from its definition (8.40), combining the gauge-invariant equations for Ψ and X. There is, however, a simpler way to reach the same result, working in the *comoving, uniform-dilaton* gauge where $B = 0 = \chi$, and where the curvature perturbation simply coincides with the variable ψ:

$$\mathcal{R}_\chi = \Psi + \frac{\mathcal{H}}{\phi'}X = \psi + \mathcal{H}E' + \frac{\mathcal{H}}{\phi'}(-\phi'E') \equiv \psi, \tag{8.75}$$

according to Eq. (8.39). Let us follow this second approach, as an instructive exercise of gauge-invariant perturbation theory.

In the comoving, uniform-dilaton gauge we can use the result (8.61) for the perturbed form of the Einstein tensor, and we set $\chi = 0$ in the perturbed Einstein equation (8.55). The off-diagonal components $(i \neq j)$ then give

$$E'' + 2\mathcal{H}E' + \psi - \varphi = 0. \tag{8.76}$$

The mixed components $(i0)$ give

$$\psi' + \mathcal{H}\varphi = 0. \tag{8.77}$$

The time component (00), using the above equations and the background condition (8.74), gives

$$2\nabla^2(\psi + \mathcal{H} E') + 2(\mathcal{H}^2 - \mathcal{H}')\varphi = 0. \tag{8.78}$$

Finally, the diagonal $(i = j)$ components, again using the background, give

$$\psi'' + \mathcal{H}(2\psi' + \varphi') + (\mathcal{H}' + 2\mathcal{H}^2)\varphi = 0. \tag{8.79}$$

Note that only three equations of the above four are independent: the last equation, for instance, is an identity following from Eq. (8.77).

The decoupled equation for ψ is now obtained by subtracting (8.78) from (8.79), by differentiating (8.78) to eliminate E'', by using (8.76), (8.79), (8.78) and (8.77) to eliminate $\psi + 2\mathcal{H} E'$, φ', $\mathcal{H} E'$ and φ, respectively, and by exploiting the background conditions (8.74). We obtain

$$\psi'' + 2\left(\mathcal{H} + \frac{\phi''}{\phi'} - \frac{\mathcal{H}'}{\mathcal{H}}\right)\psi' - \nabla^2\psi = 0. \tag{8.80}$$

Recalling the definition $z = a\phi'/\mathcal{H}$, and the (gauge-dependent) relation $\psi = \mathcal{R}_\chi$, the above equation can be rewritten in explicit gauge-invariant form as

$$\mathcal{R}_\chi'' + 2\frac{z'}{z}\mathcal{R}_\chi' - \nabla^2\mathcal{R}_\chi = 0, \tag{8.81}$$

in full agreement with the evolution equation (8.53) for the canonical variable $v = z\mathcal{R}_\chi$.

As in the case of the tensor fluctuations, the evolution of the quantum fluctuations of the vacuum is now determined by the canonical normalization of $\mathcal{R}_\chi = v/z$, and the corresponding spectrum of the Bardeen variable is fixed by the relation between Ψ and $\mathcal{R}_\chi$. Such a relation is most easily obtained, again, in the comoving gauge where $\mathcal{R}_\chi = \psi$ and $\Psi = \psi + \mathcal{H} E'$. Combining Eqs. (8.77) and (8.78) we have

$$\psi' = -\mathcal{H}\varphi = \frac{\mathcal{H}}{\mathcal{H}^2 - \mathcal{H}'}\nabla^2(\psi + \mathcal{H} E'), \tag{8.82}$$

from which

$$\mathcal{R}_\chi' = \frac{\mathcal{H}}{\mathcal{H}^2 - \mathcal{H}'}\nabla^2\Psi = \frac{2\mathcal{H}}{\lambda_P^2\phi'^2}\nabla^2\Psi \tag{8.83}$$

(for the last equality we have used the background condition (8.74)).

This important equation, besides providing the precise relation between the Fourier spectrum of Ψ and the normalized spectrum of $\mathcal{R}_\chi$, also suggests the existence of an (approximate) "conservation law", $\mathcal{R}_\chi' \simeq 0$, in the limit in which $(\mathcal{H}/\phi'^2)\nabla^2\Psi \to 0$. Shifting to the Fourier components one can easily check that such a conservation becomes effective on large scales, for long wavelength modes satisfying $(k\eta)^2 \ll 1$, i.e. for modes well "outside the horizon" [27].

It may be noticed, finally, that in the limit $\nabla^2 \Psi = 0$ the equation $\mathcal{R}'_\chi = 0$ becomes exactly equivalent to the equation of motion of Ψ (this means, in other words, that the result $\mathcal{R}_\chi \simeq$ const outside the horizon is closely related to the dynamics of the Bardeen potential). Consider the gauge-invariant definition of $\mathcal{R}_\chi$ which, using the constraint (8.64), can be rewritten as

$$\mathcal{R}_\chi = \Psi + \frac{\mathcal{H}}{\phi'} X = \Psi + \frac{2}{\lambda_P^2 \phi'^2} \left(\mathcal{H}\Psi' + \mathcal{H}^2\Psi \right) = \Psi + \frac{\mathcal{H}\Psi' + \mathcal{H}^2\Psi}{\mathcal{H}^2 - \mathcal{H}'}. \quad (8.84)$$

Differentiating, and imposing $\mathcal{R}'_\chi = 0$, one obtains

$$\Psi'' + 2\left(\mathcal{H} - \frac{\phi''}{\phi'} \right)\Psi' + 2\left(\mathcal{H}' - \mathcal{H}\frac{\phi''}{\phi'} \right)\Psi = 0, \quad (8.85)$$

which exactly coincides with the Bardeen equation (8.70) with $\nabla^2 \Psi = 0$.

8.1.4 Perfect fluid source

Suppose now that the matter sources are represented by a perfect fluid, minimally coupled to gravity. This situation is possibly appropriate to describe the phase of standard cosmological evolution, where the scalar dilaton/inflaton field is frozen at a constant value, and its potential energy is negligible. We set to zero all scalar-field terms, in both the background and perturbation equations, and we look for a decoupled equation governing the evolution of the Bardeen potential $\Psi = \Phi$, as before. By subtracting Eq. (8.65) from (8.66) we obtain

$$\Psi'' + 6\mathcal{H}\Psi' - \nabla^2\Psi + (2\mathcal{H}' + 4\mathcal{H}^2)\Psi = \frac{\lambda_P^2}{2} a^2 (\Pi - \mathcal{E}). \quad (8.86)$$

For a more explicit interpretation of the source-term appearing on the right-hand side of this equation we need, at this point, a better identification of the fluid model we are considering. For a barotropic fluid, characterized by the equation of state $p/\rho = \gamma = $ const, the pressure perturbation δp is only a function of $\delta\rho$. However, for more general equations of state $p = p(\rho)$ (and, in particular, for the case in which the ratio γ is not a constant), the pressure perturbation may also depend on the perturbation of the fluid entropy density, δS. We can put, in general,

$$\delta p = c_s^2 \delta\rho + \tau \delta S, \quad (8.87)$$

where

$$c_s^2 = \left(\frac{\partial p}{\partial \rho} \right)_{S=\text{const}} = \frac{p'}{\rho'}, \qquad \tau = \left(\frac{\partial p}{\partial S} \right)_{\rho=\text{const}} \quad (8.88)$$

(we are following the notation of [27]). The first coefficient c_s^2 can be interpreted, in the context of relativistic hydrodynamics, as the velocity of sound perturbations.

For a barotropic fluid $c_s^2 = p/\rho = \gamma$, and there are no entropy perturbations. However, such perturbations appear (and may be important) if we consider a mixture of non-interacting barotropic fluids. Suppose that the fluid source contains n non-interacting barotropic components, satisfying the equation of state $p_i = \gamma \rho_i$, $\gamma_i = \text{const}$, $i = 1, \ldots, n$, and call $p = \sum_i p_i$ and $\rho = \sum_i \rho_i$ the total pressure and energy density of the fluid, respectively. The equation of state for this model is then determined by the (time-dependent) coefficient

$$\gamma = \frac{p}{\rho} = \frac{\sum_i \gamma_i \rho_i}{\sum_i \rho_i}. \tag{8.89}$$

The total perturbed pressure is

$$\delta p = \sum_i \gamma_i \delta \rho_i. \tag{8.90}$$

The sound-velocity coefficient, using the conservation equation of the various fluid components, $\rho_i' = -3\mathcal{H}(\rho_i + p_i)$, can be written as

$$c_s^2 = \frac{\partial p}{\partial \rho} = \frac{\sum_i p_i'}{\sum_i \rho_i'} = \frac{\sum_i \gamma_i \rho_i (1 + \gamma_i)}{\sum_i \rho_i (1 + \gamma_i)}. \tag{8.91}$$

Comparing with (8.87) we obtain

$$\tau \delta S = \sum_i \gamma_i \delta \rho_i - \frac{\sum_i \gamma_i (\rho_i + p_i)}{\sum_i (\rho_i + p_i)} \sum_i \delta \rho_i. \tag{8.92}$$

If we have only one component, then $\delta S = 0$. Consider, instead, the realistic example of a two-component fluid containing a mixture of matter and radiation, i.e. $\rho_1 = \rho_r$ and $\rho_2 = \rho_m$. Putting $\gamma_1 = 1/3$ and $\gamma_2 = 0$ in the previous equations one obtains

$$\gamma = \frac{1}{3} \left(1 + \frac{\rho_m}{\rho_r} \right)^{-1}, \qquad c_s^2 = \frac{1}{3} \left(1 + \frac{3}{4} \frac{\rho_m}{\rho_r} \right)^{-1} \neq \gamma. \tag{8.93}$$

The corresponding non-adiabatic contribution to the scalar-perturbation equations, according to Eq. (8.92), is represented by [27]

$$\tau \delta S = \frac{1}{3} \left(1 + \frac{4}{3} \frac{\rho_r}{\rho_m} \right)^{-1} \delta \rho_r - \frac{1}{3} \left(1 + \frac{3}{4} \frac{\rho_m}{\rho_r} \right)^{-1} \delta \rho_m, \tag{8.94}$$

which can be rewritten in the useful form

$$\tau \delta S = c_s^2 \rho_m \left(\frac{3}{4} \frac{\delta \rho_r}{\rho_r} - \frac{\delta \rho_m}{\rho_m} \right). \tag{8.95}$$

During a phase of standard cosmological evolution, on the other hand, the radiation is in thermal equilibrium at a proper temperature T, and $\rho_r \sim T^4$, so that the entropy density per non-relativistic particle of the matter fluid is proportional to $T^3/\rho_m \sim \rho_r^{3/4}/\rho_m$ (see Chapter 1). This gives a fractional entropy-density perturbation

$$\frac{\delta s}{s} = \frac{3}{4}\frac{\delta\rho_r}{\rho_r} - \frac{\delta\rho_m}{\rho_m}, \tag{8.96}$$

which allows rewriting Eq. (8.95) as $\tau\delta S = c_s^2\rho_m(\delta s/s)$, and explains the name "entropy perturbation" for such a contribution to the scalar fluctuation equations.

In the subsequent discussion we consider a fluid with possible non-adiabatic sources of perturbations, $\delta S \neq 0$, and we use the gauge-invariant counterpart of Eq. (8.87),

$$\Pi = c_s^2 \mathcal{E} + \Sigma, \tag{8.97}$$

where Σ represents the total entropy perturbations. Inserting into the Bardeen equation (8.86), and using Eq. (8.65) to eliminate $\mathcal{E}$, we arrive at the final equation

$$\Psi'' + 3\mathcal{H}(1+c_s^2)\Psi' + 2\mathcal{H}'\Psi + \mathcal{H}^2(1+3c_s^2)\Psi - c_s^2\nabla^2\Psi = \frac{\lambda_P^2}{2}a^2\Sigma. \tag{8.98}$$

Once this equation is solved, the time evolution of W, $\mathcal{E}$ and Π turns out to be uniquely fixed by the other perturbation equations (8.64)–(8.66). Note that a pure scalar field, with vanishing potential, can also be represented in hydrodynamical form as a perfect fluid with equation of state $p = \rho$. In that case $c_s^2 = \gamma = 1$, the perturbations are adiabatic ($\Sigma = 0$), and the above equation exactly reduces to the Bardeen equation (8.70) previously obtained for the scalar field (as can be checked by using the background equation (8.9)).

As in the case of the scalar field, it is possible to recast the Bardeen equation in a "pseudo-canonical" form by introducing the variable U defined by

$$U = \frac{2}{\lambda_P^2(\rho+p)^{1/2}}\Psi = \frac{\sqrt{2}a}{\lambda_P(\mathcal{H}^2 - \mathcal{H}')^{1/2}}\Psi, \tag{8.99}$$

where the second equality follows from the background relation

$$2(\mathcal{H}^2 - \mathcal{H}') = \lambda_P^2 a^2(\rho+p) \tag{8.100}$$

(see Eq. (8.11)). The Bardeen equation can then be rewritten as

$$U'' - \left(c_s^2\nabla^2 + \frac{\theta''}{\theta}\right)U = \mathcal{N}, \tag{8.101}$$

where

$$\theta = \frac{\lambda_P \mathcal{H}}{a\left[2(\mathcal{H}^2 - \mathcal{H}')\right]^{1/2}} = \frac{\lambda_P}{a}\left[\frac{\rho}{3(\rho+p)}\right]^{1/2}, \qquad \mathcal{N} = \frac{\lambda_P a^3 \Sigma}{\left[2(\mathcal{H}^2 - \mathcal{H}')\right]^{1/2}}$$

(8.102)

(the equivalence of Eqs. (8.98) and (8.101) can be verified by using the definition $c_s^2 = p'/\rho'$, the fluid conservation equation, and the condition (8.100)). It
must be stressed, however, that U is not the variable representing the canonical
perturbations of the gravitating fluid system, even in the particular case $\mathcal{N} = 0$.

For the canonical normalization of the spectrum one must refer, once again,
to the curvature perturbation, which for a fluid is defined by the gauge-invariant
equation (8.41). Using the momentum constraint (8.64) the curvature perturbation
can be written in various forms:

$$\mathcal{R}_w \equiv \Psi + \mathcal{H} W = \Psi + \frac{2}{\lambda_P^2}\frac{\mathcal{H}\Psi' + \mathcal{H}^2\Psi}{a^2(\rho+p)}$$

(8.103)

$$= \Psi + \frac{2}{3}\frac{\mathcal{H}^{-1}\Psi' + \Psi}{1+\gamma}$$

(8.104)

$$= \Psi + \frac{\mathcal{H}\Psi' + \mathcal{H}^2\Psi}{\mathcal{H}^2 - \mathcal{H}'}.$$

(8.105)

The second form (often used in the literature) follows from the first one by putting
$p = \gamma\rho$, and using the background equation (8.7). The third one has been obtained
using Eq. (8.100), and is formally identical to the result obtained for a scalar-
field source (see Eq. (8.84)). In any case, differentiating Eq. (8.103), using the
definition $c_s^2 = p'/\rho'$, the fluid conservation equation, and Eq. (8.100), we obtain

$$\mathcal{R}_w' = \frac{2\mathcal{H}}{\lambda_P^2 a^2(\rho+p)}\left[\Psi'' + 2\mathcal{H}\Psi' + 3(1+c_s^2)\mathcal{H}\Psi' + (1+3c_s^2)\mathcal{H}^2\Psi\right].$$

(8.106)

A quick comparison with the Bardeen equation (8.98) then leads to

$$\mathcal{R}_w' = \frac{2\mathcal{H} c_s^2}{\lambda_P^2 a^2(\rho+p)}\nabla^2\Psi + \frac{\mathcal{H}}{\rho+p}\Sigma$$

$$\equiv \frac{\mathcal{H} c_s^2}{\mathcal{H}^2 - \mathcal{H}'}\nabla^2\Psi + \frac{a^2\lambda_P^2\mathcal{H}}{2(\mathcal{H}^2 - \mathcal{H}')}\Sigma.$$

(8.107)

We thus recover the approximate conservation of the curvature perturbation
($\mathcal{R}_w' \simeq 0$), in the limit in which the sources are adiabatic ($\Sigma = 0$) and the spatial
gradients of the metric fluctuations are negligible (namely, in the case of long
wavelength modes outside the horizon). It is important to note that, in the fluid
case, the presence of entropy perturbations breaks in general the $\mathcal{R}_w$ conservation,

and may generate additional curvature perturbations: this effect is the basis of the "curvaton mechanism" that we will discuss in Section 8.3. In the adiabatic case the above equation determines the relative normalization of the Bardeen variable with respect to $\mathcal{R}_w$, just like Eq. (8.83) valid for the scalar-field source. The only difference is represented by the coefficient c_s^2 which depends on the fluid type, but which does not affect the k-dependence of the spectrum.

Let us complete our discussion, and the analogy with the scalar-field dominated scenario, by deriving the evolution equation for the curvature perturbation variable. We work in the gauge comoving with the fluid source, where $B = 0 = w$ and where, according to the definitions (8.39),

$$\mathcal{R}_w = \Psi + \mathcal{H}W = \psi + \mathcal{H}E' + \mathcal{H}(-E') \equiv \psi. \tag{8.108}$$

Using this gauge in the perturbed Einstein equations we obtain, from the $(i \neq j)$ and $(i0)$ components, exactly the same equations, (8.76) and (8.77), as before. The (00) and $(i = j)$ components lead, respectively, to the following new equations:

$$2\nabla^2(\psi + \mathcal{H}E') = \lambda_P^2 a^2 \delta\rho, \tag{8.109}$$

$$\psi'' + \mathcal{H}(\varphi' + 2\psi') + (2\mathcal{H}' + \mathcal{H}^2)\varphi = \frac{\lambda_P^2}{2} a^2 \delta p. \tag{8.110}$$

It is also convenient to take into account the conservation equation (8.57). In the comoving gauge, the $\mu = 0$ component gives

$$\delta\rho' + 3\mathcal{H}(\delta\rho + \delta p) + (\nabla^2 E' - 3\psi')(\rho + p) = 0, \tag{8.111}$$

while the $\mu = i$ component gives

$$\delta p = -(\rho + p)\varphi. \tag{8.112}$$

Using this last equation, together with the background condition (8.100), we may recast the spatial equations (8.110) in the form

$$\psi'' + \mathcal{H}(2\psi' + \varphi') + (\mathcal{H}' + 2\mathcal{H}^2)\varphi = 0, \tag{8.113}$$

which is formally identical to Eq. (8.79) obtained for the scalar-field source. Let us finally consider the combination of $\delta\rho$ and δp which, using Eqs. (8.87) and (8.112), provides

$$\lambda_P^2 a^2 \delta\rho = \lambda_P^2 a^2 c_s^{-2}(\delta p - \tau\delta S) = -2c_s^{-2}(\mathcal{H}^2 - \mathcal{H}')\varphi - \lambda_P^2 a^2 c_s^{-2}\tau\delta S. \tag{8.114}$$

This allows us to rewrite the Hamiltonian constraint (8.109) in the form

$$2c_s^2\nabla^2(\psi + \mathcal{H}E') + 2(\mathcal{H}^2 - \mathcal{H}')\varphi = -\lambda_P^2 a^2 \tau\delta S, \tag{8.115}$$

again reproducing the corresponding relation obtained for the scalar-field source, Eq. (8.78) (modulo the presence of the sound-velocity coefficient, and of possible contributions from entropy perturbations).

The decoupled equation for $\psi = \mathcal{R}_w$ is now easily obtained, in this gauge, by the same appropriate combinations of Eqs. (8.113), (8.115), (8.76) and (8.77), as in the scalar-field case. One is led to

$$\psi'' + 2\left(\mathcal{H} - \frac{c_s'}{c_s}\right)\psi' + \left(\frac{2\mathcal{H}'^2}{\mathcal{H}} - \mathcal{H}''\right)\frac{\psi'}{\mathcal{H}^2 - \mathcal{H}'} - c_s^2 \nabla^2 \psi = \Omega, \qquad (8.116)$$

where

$$\Omega = \frac{\lambda_{\rm P}^2}{2(\mathcal{H}^2 - \mathcal{H}')}\left[a^2 \tau \delta S\left(\mathcal{H}' + 2\mathcal{H}\frac{c_s'}{c_s} - 2\mathcal{H}^2\right) - (a^2 \tau \delta S)'\right] \qquad (8.117)$$

represents the total entropy contribution. Recalling that $\psi = \mathcal{R}_w$, and introducing the pump field

$$z_w = a\frac{[2(\mathcal{H}^2 - \mathcal{H}')]^{1/2}}{\lambda_{\rm P}\mathcal{H}c_s} \equiv \frac{1}{c_s \theta} \qquad (8.118)$$

(see the definition (8.102) for θ), the decoupled equation can be finally written, in fully gauge-invariant form, as

$$\mathcal{R}_w'' + 2\frac{z_w'}{z_w}\mathcal{R}_w' - c_s^2 \nabla^2 \mathcal{R}_w = \Omega. \qquad (8.119)$$

In the absence of entropy perturbations ($\Omega = 0$) the variable $u = z_w \mathcal{R}_w$ diagonalizes the quadratic action for the scalar perturbations of the gravitating fluid system [27], and satisfies the canonical equation

$$u'' - \left(c_s^2 \nabla^2 + \frac{z_w''}{z_w}\right)u = 0. \qquad (8.120)$$

The presence of the sound coefficient c_s^2 breaks the symmetry of "dual" inversion relating the two pump fields z_w and θ (for a scalar field, on the contrary, $z \equiv Z^{-1}$), but does not affect the canonical normalization of the spectrum.

8.1.5 Generalized comoving gauges

After the separate discussion of the perturbations for the scalar field and the fluid sources, minimally coupled to gravity, we conclude the section with two important observations.

The first observation concerns the possibility of describing in a unified manner, through the hydrodynamical formalism, both the fluid and the scalar-field sources. A homogeneous scalar field behaves as a fluid, characterized by an effective energy density and pressure given by

$$\rho_\phi = \frac{\phi'^2}{2a^2} + V, \qquad p_\phi = \frac{\phi'^2}{2a^2} - V, \qquad (8.121)$$

and the background equations (8.7)–(8.10) can be rewritten in terms of a unique fluid source, with total stress tensor defined by $\rho_T = \rho + \rho_\phi$, $p_T = p + p_\phi$. In such a case, however, the comoving gauge can be defined in three different ways, by referring to the evolution (1) of the total gravitational source, (2) of the fluid only, (3) of the scalar field only.

(1) In the first case one must consider the total velocity potential w_T, defined by

$$\delta T_i{}^0(\text{Tot}) = (\rho_T + p_T)\partial_i w_T = (\rho + p)\partial_i w + (\rho_\phi + p_\phi)\partial_i w_\phi \tag{8.122}$$

(see for instance [23]). Here w_ϕ is referred to the perturbed energy-momentum tensor of the scalar field,

$$\delta T_i{}^0(\phi) = g^{00}\phi'\partial_i\chi = \frac{1}{a^2}\phi'\partial_i\chi, \tag{8.123}$$

so that

$$\partial_i w_\phi = \frac{a^2}{\phi'^2}\delta T_i{}^0(\phi) = \frac{\partial_i\chi}{\phi'} = \frac{\partial_i\chi}{a\sqrt{\rho_\phi + p_\phi}}. \tag{8.124}$$

In the comoving gauge fixed by $w_T = 0 = B$ the curvature perturbation, $\mathcal{R} = \psi$, takes the explicit form $\mathcal{R} = \Psi + \mathcal{H}W_T$. On the other hand, the momentum constraint (8.64), and the background condition (8.11), can be written, respectively, as

$$\Psi' + \mathcal{H}\Psi = \frac{\lambda_P^2}{2}a^2(\rho_T + p_T)W_T,$$

$$\frac{\lambda_P^2}{2}a^2(\rho_T + p_T) = \mathcal{H}^2 - \mathcal{H}'. \tag{8.125}$$

We then recover the usual form of the curvature perturbation,

$$\mathcal{R} = \Psi + \frac{\mathcal{H}\Psi' + \mathcal{H}^2\Psi}{\mathcal{H}^2 - \mathcal{H}'}, \tag{8.126}$$

already introduced for the scalar case, Eq. (8.84), and the pure-fluid case, Eq. (8.105).

(2) If we choose to be "comoving" only with respect to the fluid source, $w = 0$, then the curvature perturbation is $\mathcal{R} = \psi = \Psi + \mathcal{H}W \equiv \mathcal{R}_w$. Using the momentum constraint and the background equations we obtain

$$\mathcal{R}_w = \Psi + \frac{\mathcal{H}\Psi' + \mathcal{H}^2\Psi - \frac{\lambda_P^2}{2}\mathcal{H}\chi\phi'}{\mathcal{H}^2 - \mathcal{H}' - \frac{\lambda_P^2}{2}\phi'^2}. \tag{8.127}$$

(3) Finally, if we choose to be "comoving" with respect to the scalar-field source, then $\chi = 0$, and we get $\mathcal{R} = \psi = \Psi + \mathcal{H}X/\phi' \equiv \mathcal{R}_\chi$. The momentum constraint and the background equations then lead to

$$\mathcal{R}_\chi = \Psi + \frac{\mathcal{H}\Psi' + \mathcal{H}^2\Psi - \frac{\lambda_{\mathrm{P}}^2}{2}a^2(\rho+p)\mathcal{H}W}{\mathcal{H}^2 - \mathcal{H}' - \frac{\lambda_{\mathrm{P}}^2}{2}a^2(\rho+p)}. \tag{8.128}$$

Note that the two variables $\mathcal{R}_w$ and $\mathcal{R}_\chi$, defined in the last two equations, are different from the corresponding objects obtained in the case of pure-fluid sources, Eq. (8.103), and pure scalar-field sources, Eq. (8.84).

8.1.6 Frame transformations

The second, final observation concerns the transformation properties of the gauge-invariant variables defined in this section. They are independent, to linear order, of the choice of the coordinate system on the given space-time manifold; however, they *are not* frame independent, in general, even to linear order. It may be useful, therefore, to compute how such variables are transformed when we move from the Einstein frame, with metric $g_{\mu\nu}$ (used in this chapter), to the string frame, with metric $\widetilde{g}_{\mu\nu}$. The transformations between the two frames (see Section 2.2)) are

$$\widetilde{g}_{\mu\nu} = g_{\mu\nu}\,\mathrm{e}^\phi, \qquad \widetilde{a} = a\,\mathrm{e}^{\phi/2}, \qquad \widetilde{\phi} = \phi,$$
$$\widetilde{\mathcal{H}} = \mathcal{H} + \phi'/2, \qquad \widetilde{T}_\mu{}^\nu = \mathrm{e}^{-2\phi}T_\mu{}^\nu. \tag{8.129}$$

We have used Eqs. (2.64) putting $d = 3$, and absorbing the factor $(\lambda_{\mathrm{s}}/\lambda_{\mathrm{P}})$ into the "tilded" variables, with the understanding that units are to be rescaled from the Planck to the string system. Also, from here to the end of this section, the tilded symbols will denote S-frame variables (variables without the tilde are instead referred to the E-frame). The prime will always denote derivation with respect to the conformal time, which is the same in both frames (see Section 5.2)).

The required transformations of the perturbed variables can be found by noting that an infinitesimal coordinate transformation acts in the same way in both frames: as a consequence, gauge-invariant variables can be defined formally in the same way in each frame. This means that the E-frame definitions presented in Eqs. (8.39), (8.40), (8.41), ... can also be interpreted as the definitions of S-frame gauge-invariant variables, provided both the background ($\mathcal{H}, \phi'$, ...) and perturbation ($\varphi, \psi, \delta\rho$, ...) variables are referred to the S-frame metric and fluid sources. The relations connecting the fluctuations of the two frames are then obtained by perturbing the transformations (8.129):

$$\delta \tilde{g}_{\mu\nu} = e^{\phi}(\delta g_{\mu\nu} + \chi g_{\mu\nu}) = e^{\phi} \delta g_{\mu\nu} + \chi \tilde{g}_{\mu\nu}, \qquad \delta \tilde{\phi} = \delta \phi = \chi,$$

$$\delta \tilde{T}_{\mu}{}^{\nu} = e^{-2\phi} \delta T_{\mu}{}^{\nu} - 2\chi \tilde{T}_{\mu}{}^{\nu}. \tag{8.130}$$

Using the definitions (8.14) and (8.23), valid in both frames, we obtain

$$\tilde{\varphi} = \varphi + \frac{\chi}{2}, \qquad\qquad \tilde{\psi} = \psi - \frac{\chi}{2},$$

$$\tilde{E} = E, \qquad\qquad \tilde{B} = B,$$

$$\tilde{\chi} = \chi, \qquad\qquad \tilde{w} = w \tag{8.131}$$

$$\delta \tilde{\rho} = e^{-2\phi} \delta \rho - 2\chi \tilde{\rho}, \qquad \delta \tilde{p} = e^{-2\phi} \delta p - 2\chi \tilde{p}.$$

We may note, in particular, that in the uniform-dilaton gauge ($\chi = 0$) all metric perturbations coincide in the two frames, while the fluid perturbations only differ by an overall rescaling factor. For the gauge-invariant variables we then obtain, from their definitions,

$$\tilde{\Phi} = \Phi + \frac{X}{2}, \qquad \tilde{\Psi} = \Psi - \frac{X}{2},$$

$$\tilde{X} = X, \qquad \tilde{W} = W, \qquad\qquad \tilde{\mathcal{R}}_{\chi} = \mathcal{R}_{\chi},$$

$$\tilde{\mathcal{E}} = e^{-2\phi}(\mathcal{E} - 2X\rho),$$

$$\tilde{\Pi} = e^{-2\phi}(\Pi - 2Xp), \tag{8.132}$$

$$\tilde{\mathcal{R}}_{w} = \mathcal{R}_{w} + \frac{1}{2}(\phi' W - X),$$

$$\tilde{\zeta} = \Psi - \frac{X}{2} + \left(\mathcal{H} + \frac{\phi'}{2}\right) \frac{\mathcal{E} - 2X\rho}{\rho' - 2\phi'\rho}.$$

The gauge-invariant variables X, W, $\mathcal{R}_{\chi}$, together with the combination $\Psi + \Phi$, are thus frame independent. The two Bardeen potentials, on the contrary, are not separately frame independent.

8.2 The anisotropy spectrum of the CMB radiation

The analysis of the temperature anisotropy of the CMB radiation is, at present, one of the most efficient tools to extract direct information on the primordial cosmological dynamics and, more specifically, to formulate precision tests of the various inflationary models. In this section it will be shown in detail how to connect the observed anisotropy spectrum with the spectrum of scalar metric perturbations amplified by inflation.

It will be shown, in particular, that the primordial perturbation spectrum produced by inflation provides the *initial conditions* for the evolution of the scalar fluctuations during the subsequent standard regime. Such initial conditions, as we shall see, have a crucial influence on the currently observed structure of our Universe on large scales. The discussion of this section may thus be divided into two parts: a first part, concerning the computation of the primordial spectrum of scalar perturbations produced in a generic inflationary context; and a second part where, starting from such a spectrum, we evaluate the induced inhomogeneities and anisotropies that we can observe today in the temperature of the cosmic radiation background.

For the first computation we consider a model of inflation generated by a cosmic (inflaton/dilaton/modulus) scalar field. We can thus apply the formalism and the results presented in the previous section, using the (gauge-invariant) canonical variable v of Eq. (8.52) to normalize the spectrum of scalar perturbations to the quantum fluctuations of the vacuum. The evolution of v, for each Fourier mode k, is governed by the canonical equation

$$v_k'' + \left(k^2 - \frac{z''}{z} \right) v_k = 0, \tag{8.133}$$

which is formally the same equation as the tensor-perturbation equation (7.56), modulo a different definition of the pump field z, given now by $z = a\phi'/\mathcal{H} = a\dot{\phi}/H$ (the dot denotes, as usual, the cosmic-time derivative). In the tensor case one has, instead, $z = a$.

We assume that the pump field, during inflation, can be parametrized by a power-law behavior in an appropriate, negative range of conformal-time values,

$$z = \frac{M_P}{\sqrt{2}} \left(-\frac{\eta}{\eta_1} \right)^\alpha, \qquad \eta < 0, \tag{8.134}$$

as typical of accelerated backgrounds (see Section 5.1). The equation for the canonical variable then becomes a Bessel equation,

$$v_k'' + k^2 v_k - \frac{\nu^2 - 1/4}{\eta^2} v_k = 0, \tag{8.135}$$

where $\nu^2 = \alpha(\alpha - 1) + 1/4$, and the general solution can be given as a linear combination of two Hankel functions of index $\nu = 1/2 - \alpha$ as in the case of tensor perturbations (see Chapter 7). Imposing on v_k the canonical normalization (7.59), and the same initial conditions (7.62) used for the tensor variable u_k, we are led to the exact scalar solution

$$v_k = e^{i\theta_k} \left(-\frac{\eta\pi}{4} \right)^{1/2} H_\nu^{(2)}(k\eta), \tag{8.136}$$

normalized to a vacuum fluctuation spectrum (θ_k is an arbitrary initial phase). By using the relation $v = z\mathcal{R}_\chi$ we can also immediately write down the solution for the (normalized) modes of curvature perturbations:

$$\mathcal{R}_k = \frac{v_k}{z} = \frac{e^{i\theta_k}}{M_P} \left(\frac{\pi\eta_1}{2}\right)^{1/2} \left(-\frac{\eta}{\eta_1}\right)^\nu H_\nu^{(2)}(k\eta), \tag{8.137}$$

(we have omitted the subscript χ, for simplicity). The above expression is formally the same as that obtained for the Fourier modes h_k of tensor perturbations (see Eq. (7.140)), with the important difference that the Bessel index ν is now determined by the time evolution of $a\phi'/\mathcal{H}$.

Like tensor modes, scalar modes also tend to be amplified as they are stretched outside the horizon, namely in the limit $\eta \to 0_-$ where $|k\eta| \ll 1$ (see the discussion of Section 7.2). In that regime, the exact solutions can be expanded using the small argument limit of the Hankel functions, Eqs. (7.141) and (7.150), and we obtain the following asymptotic form of v_k and $\mathcal{R}_k$,

$$v_k = e^{i\theta_k} \left(-\frac{\eta\pi}{4}\right)^{1/2} [p_\nu^*(k\eta)^\nu - iq_\nu(k\eta)^{-\nu}], \tag{8.138}$$

$$\mathcal{R}_k = \frac{v_k}{z} = \frac{e^{i\theta_k}}{M_P} \left(\frac{\pi\eta_1}{2}\right)^{1/2} \left[-iq_\nu(k\eta_1)^{-\nu} + p_\nu^* \left(\frac{\eta}{\eta_1}\right)^{2\nu} (k\eta_1)^\nu\right]. \tag{8.139}$$

Notice that the curvature perturbation contains a constant and a time-dependent part, just like the solution for the tensor metric perturbations h_k.

Suppose now that the kinematical power satisfies the condition $\alpha < 1/2$, namely $\nu > 0$ (indeed, as we shall see, α is expected to take values very near to -1, at least in conventional models of inflation). In such a case the time-dependent part of the curvature perturbation tends to die off as $\eta \to 0_-$, and $\mathcal{R}_k$ tends to stay frozen outside the horizon, with a constant spectral distribution which can be directly computed from Eq. (8.139),

$$|\delta_\mathcal{R}(k)|^2 \equiv k^3 |\mathcal{R}_k|^2 \sim k^{2+2\alpha} \sim k^{3-2\nu}. \tag{8.140}$$

In the frozen case, however, it may be convenient (and it has become customary) to express the spectrum in terms of the background variables evaluated when a given comoving scale k crosses the Hubble horizon, i.e. at $|k\eta| = 1$. To this purpose we may simply rewrite the asymptotic amplitude (8.139), for $\eta \to 0_-$, as

$$\mathcal{R}_k = -iq_\nu \frac{e^{i\theta_k}}{M_P} \left(\frac{\pi}{2k}\right)^{1/2} (k\eta_1)^\alpha = -iq_\nu e^{i\theta_k} \left(\frac{\pi}{4k}\right)^{1/2} (z^{-1})_{\text{hc}}, \tag{8.141}$$

where the subscript "hc" denotes that the quantity has to be evaluated at $|\eta| = k^{-1}$ (we have neglected the time-dependent part of $\mathcal{R}_k$ as $\eta \to 0_-$). The spectral distribution thus becomes

$$
|\delta_{\mathcal{R}}(k)| \equiv k^{3/2} |\mathcal{R}_k| = q_\nu \left(\frac{\pi}{4}\right)^{1/2} \left(\frac{k}{z}\right)_{\text{hc}} = q_\nu \left(\frac{\pi}{4}\right)^{1/2} (z\eta)_{\text{hc}}^{-1}
$$

$$
= q_\nu \left(\frac{\pi}{4}\right)^{1/2} \left(\frac{H}{a\dot\phi\eta}\right)_{\text{hc}} \sim \left(\frac{H^2}{\dot\phi}\right)_{\text{hc}} \tag{8.142}
$$

(we have used the assumption of asymptotic power-law evolution, leading to $a\eta \sim H^{-1}$). This well-known result (see for instance [28]) explicitly shows that the amplitude of the curvature perturbations tends to be extremely enhanced in the limit $\dot\phi \to 0$.

In order to make the k dependence explicit we can now introduce a reference time scale, η_1, with $(z\eta)_{\eta_1} = \dot\phi_1/H_1^2$ (we may conveniently refer, for instance, to the end of the inflationary regime). Using the pump-field evolution (8.134) we obtain

$$
k^3 |\mathcal{R}_k|^2 = \frac{\pi}{4} |q_\nu|^2 \left(\frac{H_1^2}{\dot\phi_1}\right)^2 \left(\frac{z_1\eta_1}{z\eta}\right)^2_{\text{hc}}
$$

$$
= \frac{\pi}{4} |q_\nu|^2 \left(\frac{H_1^2}{\dot\phi_1}\right)^2 \left(\frac{\eta_1}{\eta}\right)^{2+2\alpha}_{\text{hc}} = \frac{\pi}{4} |q_\nu|^2 \left(\frac{H_1^2}{\dot\phi_1}\right)^2 \left(\frac{k}{k_1}\right)^{3-2\nu}, \tag{8.143}
$$

where we have defined $k_1 = \eta_1^{-1}$. We have thus a spectrum of curvature perturbations whose frequency behavior is completely controlled by the pump-field kinematics, as in the tensor case. The main difference concerns the amplitude, controlled not only by the background curvature H_1, but also by the dilaton kinetic energy (in Planck units), $\dot\phi_1$. Another difference is due to the power α, which depends on both the metric and the dilaton kinematics. For $\alpha \to -1$, in particular, the spectrum tends to be scale-invariant, and it can be easily checked that this is what happens when the background approaches the regime of slow-roll inflation described in Section 1.2.

8.2.1 *Primordial spectrum and slow-roll inflation*

Consider a model where the cosmological dynamics is dominated by a scalar field which is slow-rolling along the potential $V(\phi)$, and in which the parameters ϵ_H, η_H, defined by Eqs. (1.109), (1.110), are very small and constant, to leading order. Let us compute, in this regime, the effective potential z''/z governing the evolution of scalar perturbations. Using the cosmic-time coordinate we have

$$\frac{z''}{z} = \frac{H}{\dot\phi}(a\ddot{z} + \dot{z}\dot{a}), \tag{8.144}$$

where

$$\dot{z} = a\dot\phi(1 + \epsilon_H - \eta_H), \qquad \ddot{z} = \dot{a}\dot\phi(1 - 2\eta_H + \epsilon_H), \tag{8.145}$$

to leading order in the slow-roll parameters. Thus

$$\frac{z''}{z} = 2a^2 H^2 \left(1 + \epsilon_H - \frac{3}{2}\eta_H\right) = \frac{2}{\eta^2}\left(1 + 3\epsilon_H - \frac{3}{2}\eta_H\right), \tag{8.146}$$

where we have used Eq. (1.112), which implies $aH = -(1 + \epsilon_H)\eta^{-1}$. The comparison with the Bessel equation (8.135) immediately gives us the Bessel index associated with a phase of slow-roll inflation:

$$\nu = \frac{3}{2} + 2\epsilon_H - \eta_H, \tag{8.147}$$

so that the perturbation spectrum (8.143) becomes

$$|\delta_{\mathcal{R}}(k)|^2 = \frac{\pi}{4}|q_\nu|^2 \left(\frac{H_1^4}{\dot\phi_1^2}\right)\left(\frac{k}{k_1}\right)^{-4\epsilon_H + 2\eta_H}. \tag{8.148}$$

The corresponding spectral index n_s, following the definition conventionally adopted for scalar perturbations (see for instance [28]), is then

$$n_s = 1 + \frac{d\ln|\delta_{\mathcal{R}}|^2}{d\ln k} = 1 - 4\epsilon_H + 2\eta_H \tag{8.149}$$

(notice the different convention with respect to the tensor index, Eq. (7.100)). A flat, Harrison–Zeldovich spectrum, corresponding to $n_s = 1$, is well approximated by the curvature spectrum produced by a phase of slow-roll inflation, characterized by $\epsilon_H \ll 1$ and $\eta_H \ll 1$.

For a specific model of slow-roll potential (for instance, $V \sim \phi^\beta$) the spectral index at a given scale k can be conveniently parametrized also in terms of the power β and of the number of e-folds of inflation occurring after that scale has left the horizon. Let us recall, in fact, that the slow-roll parameters can be expressed in terms of the slope of the inflaton potential, as illustrated in Chapter 1. Using in particular Eq. (1.115) one can then rewrite the spectral index as

$$n_s = 1 - 6\epsilon + 2\eta = 1 - \frac{6}{2\lambda_P^2}\left(\frac{V'}{V}\right)^2_{hc} + \frac{2}{\lambda_P^2}\left(\frac{V''}{V}\right)_{hc} = 1 - \frac{\beta(\beta+2)}{\lambda_P^2 \phi_{hc}^2}, \tag{8.150}$$

where we have used $V \sim \phi^\beta$, and where the subscript "hc" denotes the epoch of horizon exit of the scale we are considering. The value of ϕ_{hc}, on the other hand, can be eliminated by the above equation in terms of the number of e-folds N_{hc} between the epoch of horizon exit and the end of inflation, obtained from Eq. (1.119) as

$$\lambda_P^2 \phi_{hc}^2 \simeq 2\beta N_{hc} \tag{8.151}$$

(we have assumed ϕ_{hc}^2 to be significantly larger than the final value ϕ_f^2). Thus [41]

$$n_s \simeq 1 - \frac{2+\beta}{2N_{hc}}. \tag{8.152}$$

It is interesting to note that N_{hc} cannot be arbitrarily large, otherwise the given scale would not be inside the horizon at the present epoch. As a consequence, models of slow-roll inflation (with $\beta > 0$) naturally predict a scalar index (*near to but*) significantly *smaller than* one, in remarkable agreement with the three-year WMAP data [5], which imply

$$n_s = 0.951^{+0.015}_{-0.019} \tag{8.153}$$

(averaged over all scales).

In the context of models of slow-roll inflation, the same parameters β, N_{hc} can be used to express the spectral index of tensor perturbations and the relative amplitude of the scalar-to-tensor spectrum, for any given fixed scale. In fact, as already stressed, the evolution of scalar and tensor perturbations is described by a canonical equation which has the same form but different pump fields: $z = z_s = a\dot{\phi}/H$ for the scalar, and $z = z_g = M_P a/\sqrt{2}$ for the tensor canonical variables (see e.g. Eqs. (8.51) and (8.52) for the scalar case, and Eqs. (7.49)–(7.51) for the tensor case). By repeating the same computations leading to the spectrum (8.140) one then finds that the Bessel index ν controlling the tensor spectrum, $|\delta_h(k)|^2 \sim k^{3-2\nu}$, is related to the time evolution of the scale factor ($a \sim (-\eta)^\alpha$) by Eq. (8.135), which implies $\nu = 1/2 - \alpha$. During the slow-roll phase, on the other hand, we know that the scale factor has a power $\alpha = -1 - \epsilon_H$ (see Eq. (1.112)), and we can then express the tensor index as [41]

$$n_T = \frac{d \ln |\delta_h(k)|^2}{d \ln k} = 3 - 2\nu = 2 + 2\alpha = -2\epsilon_H \simeq -\frac{1}{\lambda_P^2} \left(\frac{V'}{V}\right)^2_{hc} \simeq -\frac{\beta}{2N_{hc}} \tag{8.154}$$

(again we have used a power-law potential, $V \sim \phi^\beta$, and the e-folds number N_{hc} defined in Eq. (8.151)).

Also, at fixed background, the differences between scalar and tensor spectra are only due to a different pump field: since the spectrum is determined by the pump field at horizon crossing, $|\delta(k)|^2 \sim (z\eta)^{-2}_{\text{hc}}$ (see Eq. (8.142)), we obtain that the relative amplitude of the two spectra is controlled by the ratio

$$r(k) = \frac{|\delta_h(k)|^2}{|\delta_{\mathcal{R}}(k)|^2} = \left(\frac{2z_s\sqrt{2}}{aM_{\text{P}}}\right)^2_{\text{hc}} = 8\left(\frac{\dot{\phi}}{HM_{\text{P}}}\right)^2_{\text{hc}} \tag{8.155}$$

(the factor 2 in front of the tensor spectrum is due to the contribution of the two graviton polarization modes, for unpolarized fluctuations). Using the slow-roll equations (1.113) for H and $\dot{\phi}$, the definition (1.115) of ϵ_H, and the parameter N_{hc}, we are finally led to the result [41]

$$r(k) = \frac{8}{\lambda_{\text{P}}^2}\left(\frac{V'}{V}\right)^2_{\text{hc}} \simeq 16\epsilon_H \simeq \frac{4\beta}{N_{\text{hc}}}. \tag{8.156}$$

The important consequence of the above discussion is that, in the context of slow-roll inflation, the measure of any two of the three parameters n_{s}, n_{T}, r (at the same given scale k) would allow a determination of β and N_{hc}, hence providing direct information on the primordial inflationary dynamics. Conversely, for any given (model-dependent) value of β, the present results for the spectral index, Eq. (8.153), can be used to predict the level of tensor contribution to the primordial spectrum of metric perturbations:

$$r(k) \simeq \frac{4\beta}{N_{\text{hc}}} = \frac{8\beta}{2+\beta}(1 - n_{\text{s}}). \tag{8.157}$$

In particular, for $\beta = 2$ (the simplest model of chaotic inflation [42]) one finds $r \simeq 0.1$, which should be detectable by near-future CMB polarization experiments.

Coming back to the dynamical evolution of scalar perturbations in a generic inflationary background, it must be observed that the knowledge of the super-horizon spectrum of curvature perturbations also automatically fixes the spectrum of the Bardeen potential, outside the horizon. Consider in fact Eq. (8.72) for the pseudo-canonical variable $V = \xi\Psi$. Expanding the solution in the regime $|k\eta| \ll 1$ (see e.g. Eq. (7.75)), one obtains, to leading order,

$$\Psi_k = \frac{V_k}{\xi} = \frac{Z}{\xi}\left[A_k + B_k \int_{\eta_{\text{ex}}}^{\eta} d\eta' Z^{-2}(\eta')\right], \tag{8.158}$$

where A_k, B_k are integration constants, and where the variables Z, ξ have been defined in Eq. (8.73). We can take, for illustrative purposes, a background in which the scale factor follows a power-law evolution, $a \sim (-\eta)^\alpha$, and in which the scalar field evolves logarithmically with a power β, namely $\lambda_{\text{P}}\phi' \sim \beta(-\eta)^{-1}$ (the slow-roll regime corresponds to the limits $\alpha \to -1$ and $\beta \to 0$). In such a case $Z/\xi \sim (-\eta)^{-(1+2\alpha)}$, while the integral over Z^{-2} evolves in time as $(-\eta)^{1+2\alpha}$.

It follows that the asymptotic amplitude (8.158) contains a constant and a time-dependent term, as usual.

We assume that the inflationary geometry is not too different from the de Sitter ($\alpha = -1$) solution, so that $\alpha < -1/2$. As a consequence, the modes Ψ_k tend to be frozen outside the horizon, and the Bardeen spectrum is determined by fixing its amplitude at horizon crossing. Considering the amplification of the quantum vacuum fluctuations we can apply Eq. (8.83) to relate Ψ_k and the normalized curvature perturbation $\mathcal{R}_k$. We then obtain, at the horizon-crossing scale $|k\eta| = 1$, that $(\Psi_k)_{hc} \simeq (\mathcal{R}_k)_{hc}$, so that, outside the horizon,

$$|\delta_\Psi(k)|^2 = k^3 |\Psi_k|^2 \simeq |\delta_\mathcal{R}(k)|^2, \qquad |k\eta| \ll 1 \qquad (8.159)$$

(the numerical coefficients relating Ψ_k and $\mathcal{R}_k$ can be computed exactly for any given model of inflation). We then arrive at the important result that the spectrum of super-horizon fluctuations, amplified by a phase of quasi-de Sitter inflation, is the same for both the curvature perturbations and the Bardeen potential. Models of slow-roll inflation thus predict a nearly flat spectrum for both $\mathcal{R}_k$ and Ψ_k.

It may be useful to compare these results with a drastically different situation arising for some models in a string cosmology context. For models of pre-big bang inflation, in particular, the spectrum of scalar perturbation tends to be strongly tilted towards the blue, following the trend of tensor perturbations.

Let us consider, for instance, the class of minimal models already used for the computation of the tensor perturbation spectrum in Section 7.3. The kinematics, for an isotropic $d = 3$ background, is simply parametrized (in the E-frame) by

$$a(\eta) \sim (-\eta)^{1/2}, \qquad \phi(\eta) \sim -2\sqrt{3} \ln a, \qquad z(\eta) = \frac{a\phi'}{\mathcal{H}} \sim a(\eta) \qquad (8.160)$$

(see Eqs. (4.110), (4.111)). The pump field has a power $\alpha = 1/2$ and this leads, according to Eq. (8.143), to a cubic slope of the spectrum of curvature perturbations [16],

$$|\delta_\mathcal{R}(k)|^2 = k^3 |\mathcal{R}_k|^2 \sim \left(\frac{H_1^4}{\dot{\phi}_1^2} \right) \left(\frac{k}{k_1} \right)^3 \qquad (8.161)$$

(modulo logarithmic corrections [17], due to the small argument expansion of the solutions with Bessel index $\nu = 1/2 - \alpha = 0$). The behavior in frequency is exactly the same as that of the tensor perturbation spectrum, Eq. (7.98): as a consequence, the amplitude turns out to be highly depressed at low frequencies, and thus cannot contribute to the observed anisotropy (a possible solution of this difficulty, for this class of models, is provided by the curvaton mechanism which will be illustrated in the following section).

8.2.2 "Conservation" of the Bardeen spectrum

To conclude the discussion about the inflationary amplification of scalar perturbations it is important to show that the amplitude of the Bardeen potential Ψ_k, *outside the horizon*, is transferred *almost unchanged* down to the matter-dominated phase: the final distribution of the various modes in momentum space, in particular, exactly reproduces their initial distribution. This implies, as we will see in the second part of this section, that the primordial spectrum of scalar fluctuations is directly reflected in the anisotropy spectrum that we are presently observing in the CMB radiation.

For a simplified illustration of this point we assume that the inflationary epoch is immediately followed by the regime of standard cosmological evolution, during which the inflaton/dilaton scalar field is frozen (or, in any case, has a negligible influence on the background dynamics). During the standard evolution the Universe is dominated by a perfect fluid source, and we can discuss the evolution of the Bardeen potential by applying the formalism developed in the previous section, using in particular Eq. (8.98), or the corresponding pseudo-canonical version, Eq. (8.101). We assume that the standard regime includes a radiation-dominated stage for $\eta_1 \le \eta \le \eta_{eq}$, and a matter-dominated stage for $\eta_{eq} \le \eta \le \eta_0$. We also assume that the evolution of perturbations is adiabatic, i.e. that $\Sigma = 0$ in Eqs. (8.98) and (8.101).

In the radiation-dominated epoch we have $c_s^2 = 1/3$, and the Bardeen equation (8.98) can be solved exactly (see for instance [27])). However, if we are only interested in the evolution of super-horizon modes, it may be instructive to consider the asymptotic expansion of the solution in the regime $c_s^2 k^2 \eta^2 \ll 1$. Starting from Eq. (8.101) one obtains (see also Eqs. (7.75) and (7.76)):

$$\Psi_k(\eta) = F(\eta)\left[A_k \left(1 - k^2 \int_{\eta_1}^{\eta} \theta^{-2}\, d\eta' \int_{\eta_1}^{\eta'} \theta^2\, d\eta'' \right) + kB_k \int_{\eta_1}^{\eta} \theta^{-2}\, d\eta' + \cdots \right],$$

$$(8.162)$$

$$\Psi_k'(\eta) = \frac{F(\eta)}{\theta^2}\left[kB_k - k^2 A_k \int_{\eta_1}^{\eta} \theta^2\, d\eta' + \cdots \right]$$
$$+ F'(\eta)\left[A_k + kB_k \int_{\eta_1}^{\eta} \theta^{-2}\, d\eta' + \cdots \right],$$

$$(8.163)$$

where

$$F(\eta) = \frac{\lambda_{\mathrm{P}}^2 \mathcal{H}}{2a^2}, \qquad \theta = \frac{\lambda_{\mathrm{P}} \mathcal{H}}{\sqrt{2}a(\mathcal{H}^2 - \mathcal{H}')^{1/2}}.$$

$$(8.164)$$

The expansion is valid for $\eta_1 < \eta < \eta_{eq}$ and for $|k\eta| \ll 1$, and the integration constants are fixed by the initial conditions at $\eta = \eta_1$:

$$A_k = \Psi_k(\eta_1)F_1^{-1}, \qquad kB_k = \frac{\theta_1^2}{F_1}\left[\Psi_k'(\eta_1) - \frac{F_1'}{F_1}\Psi_k(\eta_1)\right]. \qquad (8.165)$$

Let us insert, as initial conditions at the beginning of the standard evolution, the frozen amplitude of Ψ_k given by the primordial inflationary spectrum, i.e. $\Psi_k(\eta_1) = \Psi_{kr} = \text{const}$, $\Psi_k'(\eta_1) \simeq 0$. Also, let us use for the background geometry the standard radiation-dominated solutions, with $a \sim \eta/\eta_1$, $\mathcal{H} = \eta^{-1}$. The computation of A_k, B_k, and the integration of the first terms of the above expansion then gives, to leading order,

$$\Psi_k(\eta) = \Psi_{kr}, \qquad \Psi_k'(\eta) = \Psi_{kr}\left(\frac{\eta_1}{\eta^2} - \frac{1}{\eta}\right)(k\eta_1)^2, \qquad \eta_1 < \eta < \eta_{eq}. \quad (8.166)$$

The same result can be obtained from the exact solution, considering the super-horizon regime $k\eta \ll 1$.

In the matter-dominated phase ($\eta \geq \eta_{eq}$) the adiabatic evolution of the scalar fluctuations is characterized by $c_s^2 = 0$, and the background satisfies $a = (\eta/\eta_1)^2$, $\mathcal{H} = 2/\eta$, $2\mathcal{H}' + \mathcal{H}^2 = 0$. The Bardeen equation (8.98) reduces to the exact equation

$$\Psi'' + 3\mathcal{H}\Psi' = 0, \qquad (8.167)$$

whose solution can be simply written as

$$\Psi_k(\eta) = \Psi_{km} + \frac{C_k}{\eta^5}. \qquad (8.168)$$

The constants Ψ_{km} and C_k can be obtained by matching this solution to the radiation-dominated solution at η_{eq}. Even without performing the explicit matching, however, the asymptotic amplitude Ψ_{km} can be easily determined also using the conservation of the curvature perturbations in the super-horizon regime, $\mathcal{R}_k' = 0$. Starting from the definition (8.104), and imposing $\mathcal{R}_w(\text{rad}) = \mathcal{R}_w(\text{mat})$, we immediately obtain (neglecting the Ψ' terms)

$$\frac{3}{2}\Psi_{kr} = \frac{5}{3}\Psi_{km}. \qquad (8.169)$$

Deep enough in the matter-dominated era ($\eta \gg \eta_{eq}$) we thus recover, outside the horizon, the same spectral distribution produced by inflation, with a spectral amplitude $\Psi_k(\eta) = \Psi_{km} \simeq (9/10)\Psi_{kr} \simeq \text{const}$ [27, 34]. This is the primordial spectrum to be used for the computation of the anisotropy of the CMB temperature.

A final comment is in order. The computation of the post-inflationary amplitude of Ψ_k, as a function of its initial inflationary amplitude, has been performed by imposing the continuity of Ψ and Ψ' on some given $\eta = \text{const}$ hypersurface. In principle, however, there are other possible prescriptions for the computation of the final spectrum, based on the continuity of different perturbation variables (for instance, the continuity of $\mathcal{R}$ and $\mathcal{R}'$ [43]), and on different spatial hypersurfaces [44]. In the context of the standard inflationary scenario, where the transition to the post-inflationary regime corresponds to a transition from accelerated to decelerated expansion, the choice of the matching prescription has no crucial influence on the final spectrum of scalar perturbations. The choice may become crucial if the inflationary kinematics describes accelerated contraction, and the transition corresponds to a "bounce" of the curvature and of the scale factor, as in the case of pre-big bang models when they are represented in the Einstein frame [45], or in the case of ekpyrotic models [46]. In those cases, the final spectrum of scalar perturbations may strongly depend on the adopted model of transition, and on the adopted matching prescriptions [47].

For a discussion of these problems we refer the interested reader to the existing literature (see also Section 10.4). It should be remarked, however, that various explicit models of smooth bouncing transitions, studied up to now both analytically and numerically, seem to indicate that the curvature perturbation $\mathcal{R}_k$ goes smoothly through the bounce [48, 49, 50], while this is not the case, in general, for the Bardeen potential Ψ_k. If the detailed background evolution near the bounce is unknown, and the final spectrum has to be computed by imposing some matching condition across the bounce, the correct prescription seems to assume the continuity of $\mathcal{R}$ and $\mathcal{R}'$, eventually obtaining the Bardeen spectrum after the bounce through its general connection to the curvature perturbations.

8.2.3 Sachs–Wolfe effect

We are now ready to start the second part of the discussion, as anticipated at the beginning of this section. The scalar perturbations of the geometry and of the matter sources, which exist at the fundamental quantum level and are amplified by inflation, necessarily destroy the perfect homogeneity and isotropy typical of the unperturbed cosmological background. We may thus expect that the induced inhomogeneities may distort, at some level, the thermal spectrum of the CMB photons, and that such a distortion may be directly computed in terms of the primordial spectrum of scalar perturbations.

Let us first recall that in the standard cosmological scenario the history of the cosmic electromagnetic radiation is characterized by two important epochs, relatively close in time to each other, but significantly different. One is the epoch of matter–radiation equality, which occurs at a photon temperature $T_{eq} \simeq 2 \times 10^4\,\mathrm{K}(\Omega_m/0.3)h^2$ (see Eq. (1.73)), and corresponds to a redshift parameter $z_{eq} \simeq 0.7 \times 10^4(\Omega_m/0.3)h^2$; after equality the energy density of the radiation drops below the energy density of non-relativistic particles (baryons and electrons), and the Universe becomes matter dominated. The other is the epoch of decoupling (or "last scattering"), which occurs at a photon temperature

$$T_{\text{dec}} \simeq 3000\,\text{K}, \tag{8.170}$$

corresponding to $z_{\text{dec}} \simeq 1100$ (remember that, at large z, $T \sim a^{-1} \sim z$); after decoupling the mean-free-path of photons becomes larger than the Hubble radius H^{-1}, and the Universe becomes transparent to the electromagnetic radiation (see for instance [51, 52]).

For $\eta > \eta_{\text{dec}}$ the CMB photons are freely falling along the geodesic paths of a matter-dominated geometry, and can then transmit down to our epoch the faithful imprint of all inhomogeneities and anisotropies present on the surface of last scattering. Here we shall assume that such perturbations are due only to the primordial fluctuations of the geometry and of the matter sources produced by inflation, neglecting additional (non-gravitational) contributions, such as those possibly due to photon interactions with ionized plasma at later epochs. The distortion of the CMB temperature produced by such primordial perturbations is described by the so-called Sachs–Wolfe effect [53], which can be derived by perturbing the trajectory of a photon propagating from the decoupling to the present epoch, in a matter-dominated, spatially flat background.

Let us start by recalling that the energy (or the frequency) of a photon emitted at decoupling is redshifted, today, by the factor $\omega_0/\omega_{\text{dec}} = a_{\text{dec}}/a_0 = (z_{\text{dec}}+1)^{-1}$. The proper temperature of the CMB radiation, on the other hand, is redshifted in time (as the Universe expands) exactly like the photon frequency (see Eqs. (1.12), (1.70)). If the redshift suffered by photons is not the same in all directions and in all space positions, then we may expect similar fluctuations in the cosmic-temperature field that we are presently observing. In particular, one can characterize these temperature fluctuations $\Delta T/T$, along a direction $\widehat{n}$, at the position $\vec{x}_0$, as fractional perturbations of the redshift parameter, namely,

$$\frac{\Delta T}{T} \equiv \delta \ln(z_{\text{dec}}+1) = \frac{\delta(\omega_{\text{dec}}/\omega_0)}{\omega_{\text{dec}}/\omega_0} = \left(\frac{\delta\omega}{\omega}\right)_{\text{dec}} - \left(\frac{\delta\omega}{\omega}\right)_0 \equiv \left[\frac{\delta\omega}{\omega}\right]_0^{\text{dec}}. \tag{8.171}$$

For the computation of $\delta\omega/\omega$ we can use the standard results for the spectral shift of periodic signals in a conformally flat Friedman–Robertson–Walker (FRW) background. A photon of four-momentum p^μ and proper frequency ω, with respect to a physical observer associated with the velocity field u^μ, is characterized by the observable frequency

$$\omega(t) = p^\mu u_\mu = \frac{\overline{\omega}}{a(t)}\,\widehat{n}^\mu \widehat{u}_\mu. \tag{8.172}$$

Here $\widehat{n}_\mu$ and $\widehat{u}_\mu$ are the vectors associated with the photon momentum and with the observer velocity in the flat Minkowski space-time conformally related to the FRW geometry by $g_{\mu\nu} = a^2 \eta_{\mu\nu}$. Obviously $\widehat{u}_\mu \widehat{u}^\mu = 1$ and $\widehat{n}_\mu \widehat{n}^\mu = 0$, where $\widehat{n}^\mu = (1, \widehat{n})$, and $\widehat{n}$ is the unit tangent vector to the unperturbed (light-like) geodesic path of the photon in Minkowski space. The previous equation was obtained by setting $p^\mu = p^0 \widehat{n}^\mu$, and using the fact that the photon energy is parallelly transported along the null photon path $dx^i = \widehat{n}^i d\eta$. In conformal time we have thus the condition

$$dp^0 = -\Gamma_{\alpha\beta}{}^0 p^\alpha dx^\beta = -(\mathcal{H} p^0 d\eta + \mathcal{H} p^0 \widehat{n}_i dx^i) = -2\mathcal{H} p^0 d\eta, \qquad (8.173)$$

which gives $p^0 = \bar{\omega}/a^2$, or $p_0 = \bar{\omega}$ (see also Eq. (1.11)). If we consider, in particular, a photon emitted at the time $\eta_{\rm dec}$ and received at the time η_0 then, according to Eq. (8.172), its energy (or frequency) will be characterized by the unperturbed redshift factor

$$\frac{\omega_{\rm dec}}{\omega_0} = \frac{a_0}{a_{\rm dec}} \frac{(\widehat{n} \cdot \widehat{u})_{\rm dec}}{(\widehat{n} \cdot \widehat{u})_0} = \frac{T_{\rm dec}}{T_0} \frac{(\widehat{n} \cdot \widehat{u})_{\rm dec}}{(\widehat{n} \cdot \widehat{u})_0}. \qquad (8.174)$$

If u_0 and $u_{\rm dec}$ are both comoving observers of the FRW geometry, then $\widehat{u}_0^\mu = \widehat{u}_{\rm dec}^\mu = (1, \vec{0})$, and one recovers the standard (unperturbed) result $\omega_0/\omega_{\rm dec} = T_0/T_{\rm dec}$ (see Eq. (1.12)).

Let us now perturb the above relation by performing the logarithmic differentiation of both sides of the equation, and using the relation $T \sim \rho^{1/4}$ valid for radiation in thermal equilibrium. We obtain

$$\left[\frac{\delta\omega}{\omega}\right]_0^{\rm dec} = \left[\frac{\delta T}{T}\right]_0^{\rm dec} + \delta(\widehat{n} \cdot \widehat{u})_0^{\rm dec} = \frac{1}{4}\left[\frac{\delta\rho_r}{\rho_r}\right]_0^{\rm dec} + \delta(\widehat{n} \cdot \widehat{u})_0^{\rm dec}, \qquad (8.175)$$

where we have also taken into account a possible primordial perturbation of the photon energy density, $\delta\rho_r \neq 0$. The other contribution is due to the perturbation of the world-line of the comoving observer and of the photon geodesic, and is given by

$$\delta(\widehat{n}_\mu \widehat{u}^\mu) = \delta\widehat{n}_0 + \delta\widehat{u}^0 + \widehat{n}_i \delta\widehat{u}^i \qquad (8.176)$$

(we have used $\widehat{n}_0 = 1 = \widehat{u}^0$, $\widehat{u}^i = 0$). For the computation of $\delta\widehat{u}^0$ we can use the normalization of the four-velocity vector, which gives

$$\delta(\eta_{\mu\nu}\widehat{u}^\mu\widehat{u}^\nu) = 0 = \widehat{h}_{00} + 2\delta\widehat{u}^0, \qquad (8.177)$$

namely $\delta\widehat{u}^0 = -\widehat{h}_{00}/2$. Here we have denoted with $\widehat{h}_{\mu\nu}$ the fluctuations of the Minkowski metric, $\eta_{\mu\nu} \to \eta_{\mu\nu} + \widehat{h}_{\mu\nu}$, related to the FRW perturbations (8.12) and (8.14) by the conformal rescaling $h_{\mu\nu} = a^2 \widehat{h}_{\mu\nu}$.

For the computation of $\delta \widehat{n}_0$ we should recall that the vector $\widehat{n}^\mu = dx^\mu/d\tau$ is parallelly transported along a null Minkowski geodesic, i.e. $d\widehat{n}^\mu/d\tau = d^2 x^\mu/d\tau^2 = 0$, where τ is an appropriate affine parameter along the geodesics. The perturbed vector $\delta \widehat{n}^\mu$ thus satisfies the perturbed Minkowski geodesics, i.e.

$$\frac{d}{d\tau}\delta\widehat{n}^\mu + \delta\widehat{\Gamma}_{\alpha\beta}{}^\mu \widehat{n}^\alpha \widehat{n}^\beta = 0, \tag{8.178}$$

where

$$\delta\widehat{\Gamma}_{\alpha\beta}{}^\mu = \frac{1}{2}\eta^{\mu\nu}\left(\partial_\alpha \widehat{h}_{\beta\nu} + \partial_\beta \widehat{h}_{\alpha\nu} - \partial_\nu \widehat{h}_{\alpha\beta}\right). \tag{8.179}$$

Thus, to first order,

$$\frac{d}{d\tau}\delta\widehat{n}^\mu = -\eta^{\mu\nu}(\partial_\alpha \widehat{h}_{\beta\nu})\widehat{n}^\alpha \widehat{n}^\beta + \frac{1}{2}(\partial^\mu \widehat{h}_{\alpha\beta})\widehat{n}^\alpha \widehat{n}^\beta$$

$$= -\eta^{\mu\nu}\frac{d}{d\tau}(\widehat{h}_{\beta\nu}\widehat{n}^\beta) + \frac{1}{2}(\partial^\mu \widehat{h}_{\alpha\beta})\widehat{n}^\alpha \widehat{n}^\beta. \tag{8.180}$$

Integration over $d\tau$ from the decoupling epoch to t_0 gives, for the $\mu = 0$ component,

$$[\delta\widehat{n}^0]_0^{\text{dec}} = -\left[\widehat{h}_{00} + \widehat{h}_{0i}\widehat{n}^i\right]_0^{\text{dec}} + \frac{1}{2}\int_0^{\text{dec}} d\tau\, \widehat{h}'_{\alpha\beta}\widehat{n}^\alpha \widehat{n}^\beta, \tag{8.181}$$

where the prime denotes $d/d\eta$. Finally, perturbing the conditions $\widehat{n}_\mu \widehat{n}^\mu = 0$ and $\widehat{n}_i \widehat{n}^i = 1$, we obtain $\delta \widehat{n}^0 + \delta \widehat{n}_0 + \delta(\widehat{n}_i \widehat{n}^i) = 0$, from which $\delta \widehat{n}_0 = -\delta \widehat{n}^0$. Summing up all contributions, we can then rewrite Eq. (8.175) in the form

$$\left[\frac{\delta\omega}{\omega}\right]_0^{\text{dec}} = \left[\frac{1}{4}\frac{\delta\rho_r}{\rho_r} + \widehat{n}^i \partial_i \widehat{w} + \frac{1}{2}\widehat{h}_{00} + \widehat{h}_{0i}\widehat{n}^i\right]_0^{\text{dec}} - \frac{1}{2}\int_0^{\text{dec}} d\tau\, \widehat{h}'_{\alpha\beta}\widehat{n}^\alpha \widehat{n}^\beta, \tag{8.182}$$

where we have expressed the velocity perturbation $\delta \widehat{u}_i$ through its velocity potential $\widehat{w}$.

It should be observed, at this point, that the metric fluctuations perturbing the photon trajectory may contain, in general, scalar, vector and tensor components. The result of Eq. (8.182) can thus be used to evaluate the CMB anisotropies induced by any type of metric perturbation $\widehat{h}$. However, as already anticipated, we mainly concentrate the present discussion on the contribution of pure scalar perturbations: we can then directly evaluate $\delta\omega/\omega$ in the longitudinal gauge, $E = B = 0$, where the fluctuation variables coincide with their gauge-invariant counterpart, i.e. $\widehat{h}_{00} = 2\Phi$, $\widehat{h}_{ij} = 2\Psi\delta_{ij}$, $\widehat{h}_{0i} = 0$, $\widehat{w} = W$ (see Section 8.1). In this case we obtain

$$\left[\frac{\delta\omega}{\omega}\right]_0^{\text{dec}} = \left[\frac{1}{4}\frac{\mathcal{E}_r}{\rho_r} + \widehat{n}\cdot\nabla W + \Phi\right]_0^{\text{dec}} - \int_0^{\text{dec}} d\tau(\Phi' + \Psi'). \tag{8.183}$$

We also notice, in view of the multipole expansion to be performed later, that the perturbation terms evaluated at the present epoch, such as $\mathcal{E}_r(0)$ and $\Psi(0)$, give a contribution of monopole type, while the present velocity perturbation, $\partial_i W(0)$, gives a dipolar contribution. The dipole anisotropy, on the other hand, is dominated by the relative motion between the present local observer and the cosmic background, an effect which has been observed with good precision, determining a velocity of 369 ± 2 km/s for the motion of our Solar System with respect to the CMB rest frame (see for instance [54]). In the subsequent discussion we will thus neglect, as usual, both the monopole and dipole contributions, with the understanding that the relevant terms of the multipole expansion will start from the quadrupole contribution.

The overall temperature anisotropy induced by a primordial background of scalar perturbations, according to Eq. (8.183), can then be written in final form as

$$\frac{\Delta T}{T}(\widehat{n}, \vec{x}_0, \eta_0) = \left(\frac{1}{4}\frac{\mathcal{E}_r}{\rho_r} + \widehat{n} \cdot \nabla W + \Phi\right)(\eta_{\text{dec}}, \vec{x}_{\text{dec}}) + \int_{\eta_{\text{dec}}}^{\eta_0} d\eta (\Phi' + \Psi')(\eta, \vec{x}(\eta)),$$

(8.184)

where $\vec{x}_{\text{dec}} = \vec{x}(\eta_{\text{dec}})$. The perturbations are evaluated along the unperturbed photon trajectory, which for an observer at $\vec{x}_0$ is given by $\vec{x}(\eta) = \vec{x}_0 + \widehat{n}(\eta_0 - \eta)$ (we have used the conformal time as affine parameter along the geodesic). Thus, $\Delta T/T$ depends on the spatial directions $\widehat{n}$ along which the radiation is observed, on the observation time η_0, and on the observer position $\vec{x}_0$. The first three terms in round brackets represent the so-called "ordinary" Sachs–Wolfe (SW) effect, while the last term represents the "integrated" Sachs–Wolfe (ISW) effect. For a better comparison with the specialized literature it is useful to note, finally, that the anisotropy $\Delta T/T$ may be expressed also in terms of a different gauge-invariant variable for the radiation energy density, which is called $D_g^{(r)}$ [55], and which is related to our density parameter by $D_g^{(r)} = \mathcal{E}_r/\rho_r - 4\Psi$.

In the case of scalar metric perturbations one has to consider both the ordinary and the integrated SW effect. For the (transverse and traceless) part of tensor metric perturbations one obtains from Eq. (8.182) only the integrated contribution, leading to

$$\frac{\Delta T}{T}(\widehat{n}, \vec{x}_0, \eta_0) = \frac{1}{2}\int_{\eta_{\text{dec}}}^{\eta_0} h'_{ij}\widehat{n}^i\widehat{n}^j(\eta, \vec{x}(\eta)) \, d\eta,$$

(8.185)

where $\partial_i h^i_j = 0 = h^i_i$. Note, however, that scalar and tensor perturbations can both (simultaneously) contribute to the observed anisotropy, as we will discuss in the final part of this section.

Let us first discuss the scalar contribution (8.184), whose Fourier transform is

$$\frac{\Delta T}{T}(\hat{n}, \vec{k}, \eta_0) = \left(\frac{1}{4}\delta_r - i\hat{n}\cdot\vec{k}\,W + \Phi\right)(\eta_{\text{dec}}, \vec{k})\,e^{i\vec{k}\cdot\hat{n}(\eta_0-\eta_{\text{dec}})}$$

$$+ \int_{\eta_{\text{dec}}}^{\eta_0} d\eta(\Phi' + \Psi')(\eta, \vec{k})\,e^{i\vec{k}\cdot\hat{n}(\eta_0-\eta)} \tag{8.186}$$

(we have simplified the notation by calling $\delta_r \equiv \mathcal{E}_r/\rho_r$ the gauge-invariant density contrast of the radiation fluid). For the computation of $\Delta T/T$ one thus needs the Fourier components of the perturbations at the decoupling epoch, during the matter-dominated era. We are interested, in particular, in the large-scale anisotropy associated with the modes k that are still outside the horizon at decoupling, $k \leq k_{\text{dec}} = \eta_{\text{dec}}^{-1}$. For such super-horizon modes the Bardeen potential generated by inflation has been computed in Eq. (8.169). The velocity and density perturbations, on the other hand, can be computed using the gauge-invariant equations (8.68), (8.69), which, for an adiabatic and barotropic fluid with $\gamma = p/\rho = \Pi/\mathcal{E} = c_s^2 = $ const, can be written in Fourier space, respectively, as

$$\delta_k' + (1+\gamma)k^2 W_k - 3(1+\gamma)\Psi_k' = 0, \tag{8.187}$$

$$W_k' + (1-3\gamma)\mathcal{H}W_k - \Psi_k - \frac{\gamma}{1+\gamma}\delta_k = 0. \tag{8.188}$$

We have used the definition $\delta_k = \mathcal{E}_k/\rho$, the unperturbed conservation equation $\rho' = -3\mathcal{H}\rho(1+\gamma)$, and the constraint $\Phi = \Psi$. The solution of the above system of coupled differential equations requires initial conditions, and we recall here the two choices of initial conditions that are mainly studied in the current literature.

8.2.4 Adiabatic initial conditions

If the post-inflationary evolution of scalar perturbations is adiabatic, $\Sigma = 0$, it has been shown that the spectrum of the Bardeen potential outside the horizon, in the matter-dominated era, is exactly the spectrum inherited from inflation, characterized by a nearly constant amplitude,

$$\Psi_k(\eta) \simeq \Psi_{k0} = \text{const}, \qquad \Psi_k' \simeq 0 \tag{8.189}$$

(see Eq. (8.169)). Using this choice for the Bardeen potential, the system of equations (8.187) and (8.188) for the matter fluid with $\gamma = 0$ reduces to

$$\delta_m' = -k^2 W_m, \qquad W_m' + \mathcal{H}W_m = \Psi_{k0} \tag{8.190}$$

(we have omitted the Fourier index k, for simplicity, and we have inserted the subscript m to keep explicit track of the fact that we are considering matter perturbations). In the matter-dominated epoch $a \sim \eta^2$, $\mathcal{H} = 2/\eta$, and the above equations can be integrated exactly, with solution

$$W_m = \frac{1}{3}\eta\Psi_{k0}, \qquad \delta_m = -\frac{1}{6}(k\eta)^2\Psi_{k0} + c_m. \tag{8.191}$$

The integration constant c_m can be determined using the Hamiltonian constraint (8.65), relating Ψ and δ_m:

$$-(2k^2 + 6\mathcal{H}^2)\Psi_{k0} = \lambda_P^2 a^2 \rho_m \delta_m. \tag{8.192}$$

During the matter-dominated phase $\lambda_P^2 a^2 \rho_m = 3\mathcal{H}^2 = 12/\eta^2$, so that

$$\delta_m = -\frac{1}{6}(k\eta)^2\Psi_{k0} - 2\Psi_{k0}. \tag{8.193}$$

Once W_m and δ_m have been determined, one can also compute the corresponding perturbations of the radiation fluid. By setting $\gamma = 1/3$ in Eqs. (8.187) and (8.188), with $\Psi' = 0$, one obtains the equations

$$\delta_r' = -\frac{4}{3}k^2 W_r, \qquad W_r' = \Psi_{k0} + \frac{1}{4}\delta_r. \tag{8.194}$$

Their differentiation and combination leads to the decoupled equations

$$\delta_r'' + \frac{k^2}{3}\delta_r = -\frac{4}{3}k^2\Psi_{k0}, \qquad W_r'' + \frac{k^2}{3}W_r = 0, \tag{8.195}$$

with general solution

$$\delta_r = -4\Psi_{k0} + c_1 \cos\frac{k}{\sqrt{3}}\eta + c_2 \sin\frac{k}{\sqrt{3}}\eta,$$

$$W_r = \frac{\sqrt{3}}{4k}\left(c_1 \sin\frac{k}{\sqrt{3}}\eta - c_2 \cos\frac{k}{\sqrt{3}}\eta\right) \tag{8.196}$$

(we have used Eq. (8.194) for W'). The assumption of adiabatic evolution now plays a crucial role in the determination of the constants c_1 and c_2. In fact, for a perfect fluid composed of matter and radiation, the adiabatic condition implies $\delta_r = (4/3)\delta_m$ (see Eqs. (8.95) and (8.96)), which implies, in its turn, $W_r = W_m$, for long-wavelength modes with $k\eta \ll 1$ (see for instance [55]). Using the previous solutions for W_m and δ_m, and imposing the adiabatic conditions (for $k\eta \ll 1$), namely

$$\delta_r = \frac{4}{3}\delta_m = -\frac{8}{3}\Psi_{k0}, \qquad W_r = W_m = \frac{1}{3}\eta\Psi_{k0}, \tag{8.197}$$

one easily obtains $c_1 = 4\Psi_{k0}/3$, $c_2 = 0$. Thus

$$\delta_r = 4\Psi_{k0}\left(\frac{1}{3}\cos\frac{k}{\sqrt{3}}\eta - 1\right) \tag{8.198}$$

is the final expression for the density contrast of the radiation fluid during the matter-dominated era. These oscillations of δ_r (also called "acoustic oscillations", since $k\eta/\sqrt{3} = kc_s\eta$, where c_s represents the sound velocity during the matter-dominated era) will play a crucial role in the production of the typical peak structure observed at smaller scales in the CMB anisotropy.

We are now in a position to evaluate the CMB anisotropy induced by adiabatic scalar perturbations, by inserting our solutions in the general expression (8.186). We can neglect the ISW effect as $\Phi_k = \Psi_k \simeq$ const, and we can also neglect (on large scales) the velocity perturbations kW_k, as they are suppressed by the factor $k\eta$ with respect to the constant Bardeen mode Ψ_{k0} (see Eq. (8.197)). Using the solution (8.198), and the coincidence of Ψ and Φ in the absence of anisotropic stresses, we finally obtain

$$\frac{\Delta T}{T}(\hat{n}, \vec{k}, \eta_0) = \frac{1}{3}\Psi_{k0}\cos{(kc_s\eta_{dec})}\,e^{i\vec{k}\cdot\hat{n}(\eta_0 - \eta_{dec})}, \qquad (8.199)$$

which is the usual SW result [53], often concisely expressed as [27, 34, 55]

$$\frac{\Delta T}{T}(\hat{n}) = \frac{1}{3}\Psi(\eta_{dec}, \vec{x}_{dec}). \qquad (8.200)$$

Notice that we have introduced the generic notation c_s for the sound-velocity coefficient at the decoupling epoch, to take into account possible deviations (due to photon–baryon interactions) from the value $1/\sqrt{3}$ appearing in the simplified solution (8.198) (see below, in particular the discussion following Fig. 8.1.)

8.2.5 Isocurvature initial conditions

A second, alternative possibility is to assume that the Bardeen potential, at the beginning of the matter-dominated era, has a negligible amplitude but a non-negligible first derivative of the amplitude, namely

$$\Psi_k \simeq 0, \qquad \Psi_k' \simeq \text{const} \neq 0. \qquad (8.201)$$

In such a case the density and velocity perturbations are also negligible outside the horizon, and the leading contribution to the temperature anisotropy comes from the ISW effect. Inside the horizon, however, the integration of Eqs. (8.187) and (8.188) leads to a density contrast which oscillates in a sinusoidal way, with an opposite phase with respect to the case of adiabatic oscillations (see Eq. (8.199)). This behavior, as we shall see in a moment, is not consistent with the peak structure that we are presently observing in the CMB anisotropy (which seems to be instead in very good agreement with the adiabatic predictions). In the rest of this chapter, therefore, we will mainly concentrate our attention on the case of adiabatic initial conditions (see for instance [55, 34] for more detailed discussions of the isocurvature case). It should be stressed, however, that current observations do not yet exclude the possibility that the total temperature anisotropy is produced by a mixture of primordial perturbations which are dominated by the adiabatic component, but which also contain a (small enough) contamination of isocurvature perturbations [56, 57].

8.2.6 The angular power spectrum

For a direct comparison with observations it is convenient to expand the $\hat{n}$-dependence of $\Delta T/T$ in series of spherical harmonic functions, by setting

$$\frac{\Delta T}{T}(\hat{n}, \vec{x}_0, \eta_0) = \sum_{\ell,m} a_{\ell m}(\vec{x}_0, \eta_0) Y_{\ell m}(\hat{n}), \tag{8.202}$$

where $-\ell \leq m \leq \ell$, and ℓ ranges from 0 to ∞. As in the case of tensor perturbations (see Eq. (7.84)) the spectrum is characterized by the two-point correlation function, which in this case compares the temperature fluctuations along two different directions, $\hat{n}$ and $\hat{n}'$:

$$\xi(\hat{n}, \hat{n}') = \left\langle \frac{\Delta T}{T}(\hat{n}) \frac{\Delta T}{T}(\hat{n}') \right\rangle = \sum_{\ell,m} \sum_{\ell',m'} \langle a_{\ell m} a^*_{\ell' m'} \rangle Y_{\ell m}(\hat{n}) Y^*_{\ell' m'}(\hat{n}'). \tag{8.203}$$

The brackets denote, as usual, quantum expectation values if scalar perturbations are quantized, and the coefficients $a_{\ell m}$ are expressed in terms of annihilation and creation operators acting on the final "squeezed vacuum" state produced by inflation [58, 59]. In a classical context the brackets denote a statistical ensemble average, which can be expressed as a spatial average assuming the validity of the ergodic hypothesis.

In any case, for an isotropic and stochastic background of scalar fluctuations we may expect that the product $\langle a_{\ell m} a^*_{\ell' m'} \rangle$ depends neither on x_0, η_0 nor on m, and that the product vanishes for $\ell \neq \ell'$ and $m \neq m'$ (see also the discussion at the beginning of Section 7.4 for a stochastic background of tensor fluctuations). We thus define

$$\langle a_{\ell m} a^*_{\ell' m'} \rangle = \delta_{mm'} \delta_{\ell\ell'} C_\ell, \tag{8.204}$$

where C_ℓ are real positive coefficients, determining the so-called "angular power spectrum". For these stochastic fluctuations the correlation function (8.203) becomes

$$\xi(\hat{n}, \hat{n}') = \sum_{\ell,m} C_\ell Y_{\ell m}(\hat{n}) Y^*_{\ell m}(\hat{n}') = \frac{1}{4\pi} \sum_\ell (2\ell+1) C_\ell P_\ell(\hat{n} \cdot \hat{n}'), \tag{8.205}$$

where P_ℓ are Legendre polynomials, and we have used the addition theorem of spherical harmonics,

$$P_\ell(\widehat{n} \cdot \widehat{n}') = \frac{4\pi}{2\ell+1} \sum_{m=-\ell}^{\ell} Y_{\ell m}(\widehat{n}) Y_{\ell m}^*(\widehat{n}') \tag{8.206}$$

(see for instance [60]). It is now important to derive definite theoretical predictions for the angular coefficients C_ℓ, since the current measurements of the CMB anisotropy [1–5] directly provide the numerical values of such coefficients.

Let us first consider the simple case of the adiabatic spectrum (8.199), produced in the context of the standard inflationary scenario. We can apply the well-known expansion of a plane wave in polar coordinates [60]:

$$e^{i\vec{k} \cdot \widehat{n}(\eta_0 - \eta_{\rm dec})} = \sum_\ell (2\ell+1) i^\ell j_\ell(k\eta_0 - k\eta_{\rm dec}) P_\ell(\widehat{k} \cdot \widehat{n}), \tag{8.207}$$

where $\widehat{k} = \vec{k}/|\vec{k}|$, and j_ℓ are spherical Bessel functions of argument $(k\eta_0 - k\eta_{\rm dec})$. The spectrum (8.199) can then be rewritten as

$$\frac{\Delta T}{T}(\widehat{n}, \vec{k}, \eta_0) = \sum_\ell i^\ell \Delta_\ell(\vec{k}, \eta_0) P_\ell(\widehat{k} \cdot \widehat{n}), \tag{8.208}$$

where

$$\Delta_\ell = \frac{1}{3} \Psi_{k0} \cos(kc_s \eta_{\rm dec}) (2\ell+1) j_\ell(k\eta_0 - k\eta_{\rm dec}). \tag{8.209}$$

Performing the spatial average, and using the definition (7.64) of the Fourier transform, the two-point correlation function (8.203) becomes

$$\xi(\widehat{n}, \widehat{n}') = \frac{1}{V} \int d^3x \frac{\Delta T}{T}(\widehat{n}, \vec{x}) \frac{\Delta T}{T}(\widehat{n}', \vec{x})$$

$$= \int \frac{d^3k}{(2\pi)^3} \frac{\Delta T}{T}(\widehat{n}, \vec{k}) \frac{\Delta T}{T}^*(\widehat{n}', \vec{k})$$

$$= \int \frac{d^3k}{(2\pi)^3} \sum_{\ell, \ell'} i^{\ell-\ell'} \Delta_\ell(k) \Delta_{\ell'}^*(k) P_\ell(\widehat{k} \cdot \widehat{n}) P_{\ell'}^*(\widehat{k} \cdot \widehat{n}'). \tag{8.210}$$

The Legendre polynomials, on the other hand, can be expanded according to the addition theorem (8.206). Putting $d^3k = k^2\, dk\, d\Omega_{\widehat{k}}$, integrating over the solid angle $d\Omega_{\widehat{k}}$ and using the orthonormality of the spherical harmonic functions [60],

$$\int d\Omega_{\widehat{k}} Y_{\ell m}(\widehat{k}) Y_{\ell' m'}^*(\widehat{k}) = \delta_{\ell\ell'} \delta_{mm'}, \tag{8.211}$$

one can rewrite the correlation function as

$$\xi(\widehat{n}, \widehat{n}') = \frac{2}{\pi} \int k^2\, dk \sum_{\ell m} \frac{|\Delta_\ell(k)|^2}{(2\ell+1)^2} Y_{\ell m}(\widehat{n}) Y_{\ell m}^*(\widehat{n}'). \tag{8.212}$$

Comparing with Eq. (8.205) one finally obtains

$$C_\ell = \frac{2}{\pi} \int k^2 \, dk \, \frac{|\Delta_\ell(k)|^2}{(2\ell+1)^2}.$$

(8.213)

This result for C_ℓ is generally valid for any model in which the temperature anisotropy can be expanded as in (8.208). For the adiabatic fluctuations of the standard inflationary scenario we can use, in particular, Eq. (8.209), and we obtain

$$C_\ell^{\text{adia}} = \frac{2}{9\pi} \int \frac{dk}{k} \, k^3 |\Psi_{k0}|^2 \cos^2 (kc_s \eta_{\text{dec}}) \, j_\ell^2 (k\eta_0 - k\eta_{\text{dec}}).$$

(8.214)

We can also assume for the Bardeen potential a power-law spectrum which, according to Eqs. (8.148), (8.149) and (8.159), can be parametrized as follows:

$$k^3 |\Psi_{k0}|^2 = A_0^2 \left(\frac{k}{k_0} \right)^{n_s - 1}$$

(8.215)

(for later convenience, we have referred the spectrum to the present horizon scale $k_0 = \eta_0^{-1}$). In a realistic scenario, which agrees with the observed angular spectrum, the index n_s has to be very close to one, as we shall see in a moment.

Let us first consider the contribution of the comoving scales k for which we can neglect the $\cos^2$ modulation of the Bardeen potential, $k \ll k_{\text{dec}}/c_s = (c_s \eta_{\text{dec}})^{-1}$, namely the contribution of those scales which are outside the horizon at the decoupling epoch. In order to estimate their present range of angular values we may note that the comoving scale associated with the horizon decreases in time like η^{-1}. The angular separation corresponding to the horizon at decoupling, $\theta_{\text{dec}}(\eta_{\text{dec}}) = \pi$, is today associated with an angular separation which can be estimated as $\theta_{\text{dec}}(\eta_0) = \pi(\eta_{\text{dec}}/\eta_0) = \pi(T_0/T_{\text{dec}})^{1/2}$. We are thus considering scales corresponding to a present angular separation

$$\theta \gtrsim \frac{\theta_{\text{dec}}}{\sqrt{3}} = \pi \left(\frac{T_0}{3T_{\text{dec}}} \right)^{1/2} \sim 3°$$

(8.216)

(we have used Eq. (8.170), and $c_s = 1/\sqrt{3}$), i.e. to multipole moments with $\ell \sim \pi/\theta \lesssim 60$.

For such multipole moments, neglecting the $\cos^2$ factor, and using the approximation $k\eta_0 - k\eta_{\text{dec}} \simeq k\eta_0 \equiv x_0$, we can analytically integrate Eq. (8.214) to obtain [61]

$$C_\ell = \frac{2A_0^2}{9\pi} \int_0^\infty dx_0 \, x_0^{n_s - 2} \, j_\ell^2(x_0)$$

$$= \frac{A_0^2}{9} \frac{\Gamma(3-n_s)\Gamma(\ell+n_s/2-1/2)}{2^{3-n_s}\Gamma^2(2-n_s/2)\Gamma(\ell+5/2-n_s/2)}, \qquad \ell \lesssim 60$$

(8.217)

(a result valid for $-3 < n_s < 3$). The scale-invariant Harrison–Zeldovich spectrum [62, 63] corresponds to $n_s = 1$ and gives, in particular,

$$\ell(\ell+1)C_\ell = \left(\frac{A_0^2}{9\pi}\right) = \text{const}, \qquad \ell \lesssim 60, \qquad \theta \gtrsim 3°, \qquad (8.218)$$

in agreement with the observations of the COBE-DMR experiment [1, 2] on very large angular scales. The COBE measurements of the quadrupole ($\ell = 2$) anisotropy [64],

$$C_2 = (1.09 \pm 0.23) \times 10^{-10}, \qquad (8.219)$$

can then be used to normalize the amplitude of the primordial scalar spectrum, $A_0^2 = 54\pi C_2$, and to obtain (indirect) experimental information on the inflation scale and on the primordial dynamics.

On smaller angular scales, $\theta \leq 3°$, $\ell \gtrsim 60$, $k \gtrsim k_{\text{dec}}/c_s$, the $\cos^2$ modulation of the Bardeen spectrum – the so-called "acoustic oscillations" – becomes important, and the peaks of the $\cos^2$ function are mapped into peaks of the C_ℓ spectrum. Consider the integral (8.214), where the $\cos^2$ function peaks at values of $k = k_n$ such that $k_n c_s \eta_{\text{dec}} = n\pi$, with $n = 1, 2, \ldots$ Because of the general behavior of the spherical Bessel factor j_ℓ^2 present in the integral, the dominant contribution to C_ℓ comes from values of k such that the argument of j_ℓ satisfies $k(\eta_0 - \eta_{\text{dec}}) \simeq \ell$ (otherwise j_ℓ^2 is strongly suppressed). The peaks of the C_ℓ spectrum are thus determined by the condition

$$\ell_n \simeq k_n(\eta_0 - \eta_{\text{dec}}) = n\pi \frac{\eta_0 - \eta_{\text{dec}}}{c_s \eta_{\text{dec}}} \simeq n\pi \frac{\eta_0}{c_s \eta_{\text{dec}}}, \qquad n = 1, 2, \ldots \qquad (8.220)$$

Recalling that $\eta_0/\eta_{\text{dec}} = (T_{\text{dec}}/T_0)^{1/2}$ one can thus predict a first peak at $\ell_1 \sim 180$, and a constant separation of the various peaks, $\Delta\ell \sim \ell_1$. Such an oscillating behavior of the angular power spectrum is illustrated in Fig. 8.1, where we have numerically integrated C_ℓ from Eq. (8.214) for $n_s = 1$, and we have plotted $\ell(\ell+1)C_\ell$ in units of $2A_0^2/9\pi$.

The figure clearly displays the plateau observed by COBE at large scales, and the subsequent spectral oscillations. It must be stressed, however, that the integral (8.214) evaluates the anisotropy contribution of scalar perturbations with $k\eta_{\text{dec}} \lesssim 1$, and thus *cannot* be extended to values of ℓ which are too large with respect to the decoupling scale. For a complete and realistic description of the full anisotropy spectrum one must include smaller angular scales, and take into account two important physical effects which have a significant influence on the evolution of the scalar perturbations inside the horizon: the so-called "radiation driving" [65] and the "Silk damping" [66].

A detailed presentation of such effects is outside the purpose of this introductory discussion, and the interested reader is referred to more specialized publications

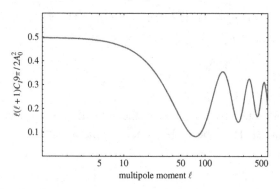

Figure 8.1 The angular power spectrum $\ell(\ell+1)C_\ell$ obtained through a numerical integration of Eq. (8.214) (courtesy of Antonio Marrone).

(see for instance [67, 68]). We only note that radiation driving has the effect of enhancing the height of the first peaks with respect to the large-scale normalization fixed by C_2, while Silk damping has the effect of suppressing the height of the subsequent peaks.

For a short illustration of the first effect we may consider the scalar modes that were already inside the horizon at decoupling, in particular those with $k\eta_{eq} > 1$, re-entering the horizon in the radiation-dominated epoch. By solving exactly Eq. (8.98) for the Bardeen potential in the radiation era (with adiabatic initial conditions), one obtains (modulo oscillations) for the sub-horizon modes the asymptotic behavior $\Psi_k(\eta) \simeq -3\Psi_{ki}/(k\eta/\sqrt{3})^2$ (see [27], and also Eq. (8.287) of the next section). Here Ψ_{ki} is the amplitude of the Bardeen potential at the beginning of the radiation era. The combination of the Hamiltonian constraint (8.65) for $\mathcal{E}_r$, and of the background equation (8.7) for ρ_r, then leads to

$$\delta_r \equiv \frac{\mathcal{E}_r}{\rho_r} \simeq -\frac{2}{3}(k\eta)^2 \Psi_k(\eta) \simeq 6\Psi_{ki}. \qquad (8.221)$$

These sub-horizon modes thus contribute to the SW effect (8.186) with an amplitude $\delta_r/4 = (3/2)\Psi_{ki}$, to be compared with the amplitude $\Psi_{k0}/3$ of super-horizon modes (see Eq. (8.199). Remembering the small shift in the asymptotic amplitude of the Bardeen potential due to the radiation–matter transition, leading from Ψ_{ki} to $\Psi_{k0} = (9/10)\Psi_{ki}$ (see Eq. (8.169)), we can eventually estimate the enhancement factor of the small-scale anisotropies, with respect to the large-scale ones, as $(3/2)\Psi_{ki}/(\Psi_{k0}/3) \simeq 5$.

A more precise calculation of such enhancement, which is in principle k-dependent, can be performed numerically [69] taking into account the velocity perturbations of the matter fluid, and the perturbations of the Boltzmann equation describing the photon–baryon interactions (see also [34, 55] for analytical discussions). The processes of photon diffusion during decoupling (and also earlier,

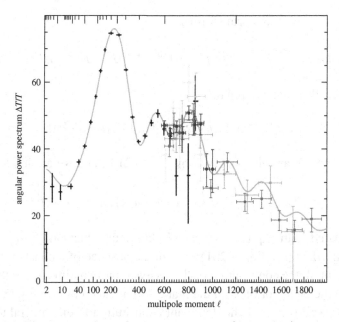

Figure 8.2 The measured angular power spectrum (data points), compared with the theoretical predictions (full curve) of a model based on adiabatic scalar perturbations (adapted from [76]).

during the so-called "recombination era") have indeed a crucial influence on the final height of the anisotropy peaks. These processes, together with the finite thickness of the last-scattering surface, are also sources of an exponential damping of the anisotropy oscillations at very high values of ℓ [66]. The inclusion of all these effects into a numerical computation of C_ℓ, always based on the assumption of adiabatic initial conditions, eventually leads to a precise prediction for the height and the position of the peaks of the angular power spectrum which is in good agreement with all present observations, as illustrated in Fig. 8.2.

The figure shows the data points of WMAP [4] (first year), BOOMERANG [70], MAXIMA [71], DASI [72], VSA [73], CBI [74] and ACIBAR [75]. The curve superimposed on the data [76] has been obtained through the numerical code CMBfast [69], and represents the predictions of an adiabatic model based on Eq. (8.214), but which takes into account the enhancement due to radiation driving and small-scale damping. In particular, the plotted curve corresponds to the following typical choice of the standard cosmological parameters: $\Omega_\Lambda = 0.72$, $\Omega_K = 0$, $\Omega_m h^2 = 0.12$, $\Omega_b h^2 = 0.024$ and $n_s = 1$.

In the case of isocurvature initial conditions the predicted anisotropy seems to be unable to match the observed peak structure. In the isocurvature case the dominant contribution to $\Delta T/T$ comes from the ISW effect. Expanding in spherical Bessel functions the Fourier transform (8.186),

$$\frac{\Delta T}{T}(\hat{n}, \vec{k}, \eta_0) = 2 \sum_{\ell} i^{\ell}(2\ell+1) \int_{\eta_{dec}}^{\eta_0} d\eta \, \Psi'(\eta, \vec{k}) j_{\ell}(k\eta_0 - k\eta) P_{\ell}(\widehat{k \cdot n}) \quad (8.222)$$

(we have used $\Phi = \Psi$), we obtain

$$\Delta_{\ell}(\vec{k}, \eta_0) = 2(2\ell+1) \int_{\eta_{dec}}^{\eta_0} d\eta \, \Psi'(\eta, \vec{k}) j_{\ell}(k\eta_0 - k\eta). \quad (8.223)$$

The coefficients C_{ℓ} are always given by Eq. (8.213), and we are led to the result

$$C_{\ell}^{iso} = \frac{8}{\pi} \int \frac{dk}{k} k^3 \left| \int_{\eta_{dec}}^{\eta_0} d\eta \, \Psi'(\eta, \vec{k}) j_{\ell}(k\eta_0 - k\eta) \right|^2 \quad (8.224)$$

(to be compared with Eq. (8.214)). For isocurvature fluctuations Ψ' is constant outside the horizon (see Eq. (8.201)) but, in the absence of a constant term in the Bardeen potential, the contribution of Ψ' decays rapidly in time inside the horizon, according to the behavior of Ψ in the matter-dominated era (see Eq. (8.168)). For super-horizon scales $k\eta > 1$ the dominant contribution to the integral then comes from the horizon-crossing time $\eta = k^{-1}$, and we can use the approximation

$$\int d\eta \, \Psi'(\eta, \vec{k}) j_{\ell}(k\eta_0 - k\eta) \simeq [\Psi_k]_{\eta=k^{-1}} j_{\ell}(k\eta_0). \quad (8.225)$$

Assuming that the spectrum of Ψ at horizon crossing is the same as in the adiabatic case (given in Eq. (8.217)) one then recovers for C_{ℓ} the same integral that controls the large-scale adiabatic fluctuations, Eq. (8.217), but with the factor $(2/9)A_0^2$ replaced by $8A_0^2$. It follows that the dependence of the angular spectrum on n_s is identical to that of the adiabatic case, but the contribution to the temperature anisotropy $\Delta T/T$ is enhanced by a factor 6 (i.e. the coefficients C_{ℓ} are enhanced by a factor 36).

On smaller angular scales the contribution of the isocurvature perturbations is dominated by the oscillations of the density contrast δ_r. As already mentioned, such oscillations are of sinusoidal type, controlled by the integral

$$\int \frac{dk}{k} k^3 \sin^2 (kc_s \eta_{dec}) j_{\ell}^2(k\eta_0 - k\eta_{dec}). \quad (8.226)$$

By applying the same arguments as before one then finds that the position and separation of the various peaks are determined by the condition

$$\ell_n \simeq k_n(\eta_0 - \eta_{dec}) = \left(n + \frac{1}{2}\right) \pi \frac{\eta_0 - \eta_{dec}}{c_s \eta_{dec}}, \qquad n = 0, 1, 2, \dots \quad (8.227)$$

The separation $\Delta\ell$ is the same as that for the adiabatic case, but the position of the first peak corresponds to a value of ℓ which is one-half of the adiabatic result. Such a different prediction is clearly not favored by the observational data of Fig. 8.2. In addition, the COBE normalization (8.218), (8.219), imposed on the isocurvature spectrum (8.224), determines a value of A_0 which is smaller (by a factor 6) with respect to the adiabatic case, with a corresponding suppression of the height of the first peak, a result which is also not favored by present observations.

Let us conclude the section with a short discussion of the tensor contribution to $\Delta T/T$, determined by the ISW effect according to Eq. (8.185). Projecting h_{ij} onto the two polarization states h_+, $h_\times$, and expanding their spatial dependence into spherical Bessel functions, one obtains the angular power spectrum [77, 6]

$$C_\ell^{\mathrm{T}} = \frac{2}{\pi}\frac{(\ell+2)!}{(\ell-2)!}\int \frac{dk}{k}k^3 \left|\int_{\eta_{\mathrm{dec}}}^{\eta_0} d\eta\, h'(\vec{k},\eta)\,\frac{j_\ell(k\eta_0-k\eta)}{(k\eta_0-k\eta)^2}\right|^2, \qquad (8.228)$$

where h is any one of the two polarization modes, and the tensor background is assumed to be unpolarized. This result is similar to that of the isocurvature integral, Eq. (8.224), with the only difference due to the factor $(k\eta_0-k\eta)^{-2}$ originating from the spin-two nature of the gravitational field h_{ij}, which induces $j_{\ell+2}$ and $j_{\ell-2}$ contributions for any j_ℓ term of the multipole expansion (see also [55])

The total CMB anisotropy is given, in principle, by the sum of the scalar and tensor contributions. It can be easily checked, however, that at small angular scales the tensor contribution is negligible with respect to the scalar one, *quite independently* of the spectral distribution of h_k in Fourier space.

During the matter-dominated era the general solution for $h(\vec{k},\eta)$ can be written as the linear combination of a term containing $\eta^{-3/2}J_{3/2}(k\eta)$, and another term containing $\eta^{-3/2}J_{-3/2}(k\eta)$, where J_ν are the ordinary Bessel functions (see for instance the solution of Eq. (7.168), with $\nu_3 = -3/2$). The leading contribution to the first derivative h' is proportional to $\eta^{-3/2}J_{5/2}$ so that, using the definitions of the spherical Bessel functions, $j_\ell(x) = (\pi/2x)^{1/2}J_{\ell+1/2}(x)$, we can approximate h' as follows:

$$h'(\vec{k},\eta) = \frac{\beta_k}{k\eta}\, j_2(k\eta), \qquad (8.229)$$

where β_k is a constant coefficient. On the other hand, using the properties of j_ℓ, it follows that the dominant contributions to the Fourier integral in Eq. (8.228) come from the values of k for which the argument of j_ℓ approximately coincides with the Bessel index ℓ. This imposes the two conditions $k\eta \simeq 2$ and $k\eta_0 - k\eta \simeq \ell$, from which $\ell \simeq 2(\eta_0-\eta)/\eta$. The allowed range of η is bounded by the lower limit $\eta = \eta_{\mathrm{dec}}$ of the time integral: thus, there are non-negligible contributions to C_ℓ^{T} only for

$$\ell \lesssim 2\frac{\eta_0-\eta_{\mathrm{dec}}}{\eta_{\mathrm{dec}}} \simeq 2\left(\frac{T_{\mathrm{dec}}}{T_0}\right)^{1/2} \sim 63. \qquad (8.230)$$

In this regime of large angular scales the tensor contribution can be estimated by noting that the dominant contribution of the mode k to the time integral of Eq. (8.228) comes from the horizon-crossing epoch, $\eta = k^{-1}$. The integral in $d\eta$ can thus be approximated by $[h(k, \eta)]_{\eta=k^{-1}} j_\ell(x_0) x_0^{-2}$, in a way similar to the case of isocurvature scalar perturbations. Using for h_k the generic spectrum

$$k^3 |h_k|^2 = A_T^2 \left(\frac{k}{k_0}\right)^n ,$$ (8.231)

we obtain the result

$$
\begin{aligned}
C_\ell^T &= \frac{2A_T^2}{\pi} \frac{(\ell+2)!}{(\ell-2)!} \int dx_0 \, x_0^{n-5} j_\ell^2(x_0) \\
&= A_T^2 \frac{(\ell+2)!}{(\ell-2)!} \frac{\Gamma(6-n)\Gamma(\ell-2+n/2)}{2^{6-n}\Gamma^2(7/2-n)\Gamma(\ell+4-n/2)}.
\end{aligned}
$$ (8.232)

In particular, for a scale-invariant $(n = 0)$ spectrum,

$$\ell(\ell+1)C_\ell^T = \frac{8A_T^2}{15\pi} \frac{\ell(\ell+1)}{(\ell-2)(\ell+3)}, \qquad \ell < 60,$$ (8.233)

to be compared with the scalar contribution, Eq. (8.218). Apart from the apparent singularity at $\ell = 2$ (which is only an artifact of our crude approximation), the resulting angular spectrum is rather flat in the considered range of scales (as also confirmed by more accurate numerical computations [6]).

In conclusion, a stochastic background of tensor metric perturbations cannot be responsible for the peak structure of the CMB anisotropy at small angular scales, but can contribute to the large-scale "plateau" observed by COBE-DMR, provided the primordial distribution of tensor perturbations is sufficiently flat. This means that, in principle, it is possible to obtain interesting constraints on the amplitude of the tensor spectrum from the large-scale measurements of the CMB anisotropy, as already stressed in Section 7.3. Comparing the ansatz (8.231) with Eq. (7.188) one obtains $A_T^2 = 6\pi^2 \Omega_g(\omega_0, t_0)$, and the exact numerical computation of C_2^T in terms of A_T^2 – together with the experimental result (8.219) – leads to the precise formulation of the upper limit presented in Eq. (7.189).

If the slope of the primordial tensor spectrum is too steep, as in the case of the minimal pre-big bang models illustrated in Section 7.3, then the result (8.232) cannot be matched to the observed anisotropy distribution. For such models the peak amplitude of the spectrum is normalized at the string scale, so that the tensor contribution to the large-scale anisotropy is certainly negligible (see Fig. 7.7). The same is true for the graviton spectrum of the ekpyrotic scenario, Eq. (7.216). String cosmology models, with the possible exception of models of D-brane–antibrane inflation [7], tend to differ from standard models of slow-roll inflation for the complete absence of tensor contributions to $\Delta T/T$.

The presence of tensor fluctuations on large scales cannot be easily distinguished from the presence of scalar fluctuations through their direct contribution to the C_ℓ spectrum; however, the presence of a tensor contribution could be detected through its specific influence on the polarization state of low-ℓ multipoles [78, 79]. Such a possibility will not be discussed here, but we recall that tensor perturbations should produce a characteristic "curl" component in the polarization of the CMB radiation, which is absent in the case of a purely scalar background of primordial perturbations. A missing detection of this polarization state (or upper limits on its presence), in future measurements of the large-scale anisotropy, could thus represent an important signal for discriminating among different models of inflation.

8.3 Adiabatic metric perturbations from the string theory axion

Scalar metric perturbations present at the decoupling epoch, characterized by a nearly constant amplitude ($\Psi'_k \simeq 0$) and a flat enough spectrum ($n_s \simeq 1$), can be the source – through the SW effect – of the observed structure of CMB anisotropies. It has been shown, in the previous section, that models of slow-roll inflation can easily generate a primordial spectrum of scalar metric perturbations satisfying the required (adiabatic) properties.

Models of pre-big bang inflation, on the contrary, tend to produce a spectrum of metric perturbations with a slope too steep to agree with the observed structures (see Eq. (8.161)). It is true that the end-point normalization of the spectrum, controlled by the string scale, implies that the amplitude of the large-scale fluctuations is far too small to provide significant contributions to the measured temperature fluctuations: this avoids embarassing conflict with observational data, but leaves open the problem of explaining the large-scale anisotropies.

However, a direct inflationary amplification of the metric fluctuations is not the only mechanism for efficient production of a nearly constant Bardeen potential at decoupling, characterized by the appropriate large-scale amplitude and spectral distribution. Another possibility is provided by the so-called "curvaton" mechanism [20–24, 80], based on the presence of a scalar (or pseudo-scalar) field (different from the inflaton) which, during inflation, is amplified with a flat spectrum, quite independently of the associated spectrum of metric fluctutations. After inflation such a field becomes massive and eventually decays, leaving a flat spectrum of curvature perturbations which, in their turn, are associated with a flat spectrum of metric perturbations described by a Bardeen potential with the required adiabatic properties. The metric perturbations produced in this way are to be added to those directly produced by inflation, and may represent the dominant contribution to the temperature anisotropies if the inflationary amplification of the Bardeen potential is absent, or negligible.

Such a mechanism of producing adiabatic scalar metric perturbations is active, in principle, in all inflationary models containing a self-interacting scalar field (dubbed "curvaton"), which evolves according to the above description. In this section it will be shown that, in a string cosmology context, the role of the curvaton may be played by the Kalb–Ramond axion σ, associated by space-time duality with the four-dimensional component of the NS–NS two-form $B_{\mu\nu}$ present in the low-energy string effective action [20, 23, 24].

The starting point of our discussion is the fact that the fluctuations of this axion field, amplified by inflation, may be characterized by a scale-invariant distribution of super-horizon modes, even in models where the corresponding spectrum of metric perturbations is very steep, as in the context of the pre-big bang scenario [18, 19]. In that context, the curvaton mechanism thus becomes a crucial ingredient for the construction of realistic inflationary models able to include a satisfactory explanation of the large-scale anisotropies.

The four-dimensional components of the NS–NS two-form are described by the following (dimensionally reduced, S-frame) action

$$S = \frac{1}{2\lambda_s^2} \int d^4x \sqrt{-g_s}\, \frac{e^{-\phi}}{12} H_{\mu\nu\alpha} H^{\mu\nu\alpha}, \qquad H_{\mu\nu\alpha} = \partial_\mu B_{\nu\alpha} + \partial_\nu B_{\alpha\mu} + \partial_\alpha B_{\mu\nu}$$

$$(8.234)$$

(throughout this section, we use the subscript "s" to denote the S-frame metric, related to E-frame metric by Eqs. (2.39) and (2.43)). In the absence of specific sources, the equations of motion for $H_{\mu\nu\alpha}$ are automatically satisfied by introducing the "dual" axion field σ, such that

$$H^{\mu\nu\alpha} = \frac{e^{\phi}}{\sqrt{-g_s}} \epsilon^{\mu\nu\alpha\beta} \partial_\beta \sigma. \qquad (8.235)$$

The S-frame action for σ is

$$S = \frac{1}{4\lambda_s^2} \int d^4x \sqrt{-g_s}\, e^{\phi}\, \partial_\mu \sigma \partial^\mu \sigma, \qquad (8.236)$$

with corresponding E-frame action:

$$S = \frac{1}{4\lambda_P^2} \int d^4x \sqrt{-g}\, e^{2\phi}\, \partial_\mu \sigma \partial^\mu \sigma. \qquad (8.237)$$

If the unperturbed axion background is vanishing, this is also the action for the axion fluctuations $\delta\sigma$. In conformal time, and for a conformally flat metric, we thus recover for $\delta\sigma$ the typical action of linear perturbations (see e.g. Eq. (7.46)),

$$S = \frac{1}{2} \int d\eta\, z^2(\eta) \left(\delta\sigma'^2 + \delta\sigma \nabla^2 \delta\sigma\right),$$

$$z = \frac{a}{\sqrt{2}} e^{\phi} = \frac{a_s}{\sqrt{2}} e^{\phi/2}, \qquad (8.238)$$

with pump field z, and Schrödinger-like equation for the canonical variable $v = z\delta\sigma$. Notice that the above pump field is dimensionless, since we have absorbed into the axion field the Planck length present in the E-frame action, $\sigma \to \sigma/\lambda_P$. Throughout this section we always use a canonically normalized axion field, with dimension one in Planck units.

We assume, as in the case of metric perturbations, that the accelerated evolution of the pump field during inflation can be parametrized as follows:

$$z \sim (-\eta)^\alpha, \qquad \alpha < 1/2, \qquad \eta \to 0_-. \qquad (8.239)$$

Solving the canonical equation for $\delta\sigma$, normalizing to an initial spectrum of vacuum fluctuations, using the small argument limit, and repeating exactly the procedure applied to the case of metric perturbations (see e.g. Section 8.2), one can easily find that the axion fluctuations tend to be frozen outside the horizon. Their spectral distribution,

$$|\delta_\sigma(k)|^2 = k^3 |\delta\sigma_k|^2 \sim k^{2+2\alpha}, \qquad (8.240)$$

is formally the same as that obtained for tensor metric perturbations, Eq. (7.91), and curvature perturbations, Eq. (8.140). The axion pump field (8.238) is different, however, from the E-frame scale factor (which represents the pump field of metric perturbations). We may thus expect for the axion a different spectrum, possibly flatter than the metric spectrum, even in models of pre-big bang inflation.

For a simple illustration of this possibility we consider here an example of low-energy, dilaton-driven anisotropic background, described by the Kasner-like (S-frame) solution (4.31)–(4.34). We may assume, in particular, that the inflationary regime is characterized by three accelerated expanding dimensions, with scale factor a_s, and by n "internal" contracting dimensions, with scale factors b_i^s, $i = 1, \ldots, n$. In conformal time such a solution can be parametrized, for $\eta \to 0_-$, as

$$a_s = \left(-\frac{\eta}{\eta_1}\right)^{\frac{\beta_0}{1-\beta_0}}, \qquad b_i^s = \left(-\frac{\eta}{\eta_1}\right)^{\frac{\beta_i}{1-\beta_0}},$$

$$\phi_d = \frac{\sum_i \beta_i + 3\beta_0 - 1}{1-\beta_0} \ln\left(-\frac{\eta}{\eta_1}\right), \qquad (8.241)$$

where the powers β_0, β_i satisfy the Kasner condition

$$\sum_i \beta_i^2 + 3\beta_0^2 = 1. \qquad (8.242)$$

The scalar ϕ_d is the higher-dimensional dilaton field appearing in the effective action before dimensional reduction, and is related to the four-dimensional dilaton ϕ by

$$\int d^{d+1}x \sqrt{-g_{d+1}^s}\, e^{-\phi_d} = \int d^4x \sqrt{-g_4^s}\, V_n e^{-\phi_d} = \int d^4x \sqrt{-g_4^s}\, e^{-\phi}, \qquad (8.243)$$

namely by

$$\phi = \phi_d - \ln V_n = \phi_d - \sum_i \ln b_i^s = \frac{3\beta_0 - 1}{1 - \beta_0} \ln\left(-\frac{\eta}{\eta_1}\right). \qquad (8.244)$$

The power-law evolution of the axion pump field, for this background, is given by

$$z \sim a_s e^{\phi/2} = a_s \left(\prod_i b_i^s\right)^{-1/2} e^{\phi_d/2} \sim (-\eta)^\alpha,$$

$$\qquad (8.245)$$

$$\alpha = \frac{5\beta_0 - 1}{2(1 - \beta_0)}.$$

It follows, from Eq. (8.240), that the axion spectral index,

$$n_\sigma = 1 + \frac{d \ln |\delta_\sigma|^2}{d \ln k} = 3 + 2\alpha = 2\left(\frac{1 + \beta_0}{1 - \beta_0}\right), \qquad (8.246)$$

is controlled by the Kasner power β_0 of the three-dimensional expanding space. In particular, a scale-invariant spectrum with $n_\sigma = 1$ can be obtained for $\beta_0 = -1/3$.

In the special case in which all $d = 3 + n$ dimensions are isotropically expanding, with $\beta_0 = \beta_i$, the Kasner condition (8.242) implies $\beta_0 = -1/\sqrt{d}$: interestingly enough, a fully scale-invariant spectrum thus corresponds to $d = 9$, i.e. just to the number of spatial dimensions in which critical superstrings consistently propagate. In the less special case in which the spatial background geometry can be factorized as the product of two isotropic, maximally symmetric spaces (three-dimensional and n-dimensional), one has instead $\beta_i = \beta \neq \beta_0$, with $3\beta_0^2 + n\beta^2 = 1$. In this case the spectral index can be expressed in terms of the parameter r,

$$r = \frac{1}{2}\left(\frac{\dot{V}_n}{V_n}\right)\left(\frac{\dot{V}_3}{V_3}\right)^{-1} = \frac{n\beta}{6\beta_0}, \qquad (8.247)$$

measuring the relative time evolution of the internal and external volumes [81] (the dot denotes differentiation with respect to the S-frame cosmic time t_s). Eliminating β in terms of β_0 through the Kasner condition, and replacing β_0 with r in Eq. (8.246), one can then parametrize the deviations of n_σ from one as the relative shrinking or expansion of the two spaces.

It should be stressed, at this point, that such a close correspondence between the kinematics of the background (8.241) and the resulting spectral index is lost in the case of tensor perturbations and curvature perturbations, which are instead characterized by a "universal" spectral index, fully independent of the particular values β_0 and β_i. For the solution (8.241) the pump field of metric perturbations coincides with the four-dimensional E-frame scale factor $a = a_s \exp(-\phi/2)$ – see e.g. Eq. (7.47) for tensor perturbations, and Eq. (8.160) for curvature perturbations. Therefore,

$$a_s e^{-\phi/2} = a_s \left(\prod_i b_i^s \right)^{1/2} e^{-\phi_d/2} \sim (-\eta)^\alpha,$$

(8.248)

$$\alpha = \frac{1}{1-\beta_0} \left[\beta_0 + \frac{1}{2} \sum_i \beta_i - \frac{1}{2} \left(3\beta_0 + \sum_i \beta_i - 1 \right) \right] \equiv \frac{1}{2},$$

which always leads to a cubic power spectrum $|\delta|^2 \sim k^{2+2\alpha} = k^3$ (modulo logarithmic corrections, see Section 7.2). In this sense, metric fluctuations are "insensitive" to the kinematical details of this low-energy class of pre-big bang backgrounds. The dependence of the spectrum on the kinematics reappears, however, for other types of fluctuations (not only for the axion but also, for instance, for the fluctuations of a background vector field, as will be discussed in Appendix 8A).

Let us now suppose that the phase of pre-big bang inflation has a kinematics suitable for the production of a flat (or nearly flat) spectrum of large-scale axion fluctuations. At the end of inflation, outside the horizon, we are thus left with a primordial "sea" of scalar perturbations of "isocurvature" type, since the metric and curvature perturbations – which are also necessarily present as components of this scalar background – have been amplified with a spectral slope much steeper than the axion component, and are certainly negligible with respect to the axion perturbations on super-horizon scales. Such a primordial background of scalar perturbations may contribute to the CMB anisotropy in two ways.

If the axion is massless, or light enough not only to have "survived" (without decaying) up to the present time, but also to determine an amplitude of the quadrupole anisotropy small enough to satisfy the COBE normalization of the spectrum [82, 83, 84], then the axion perturbations may play the role of "seeds" for the temperature anisotropies, with a mechanism that will be briefly illustrated in Appendix 8B. In that case, however, one would obtain an isocurvature contribution to the anisotropy which, as discussed in Section 8.2, seems to be excluded as the main source of the observed $\Delta T/T$.

If, on the contrary, the axion becomes massive after inflation, dominates the background evolution, and then decays (early enough to avoid disturbing the standard processes of baryogenesis and/or nucleosynthesis), then the axion fluctuations may produce a spectrum of adiabatic metric perturbations according to the curvaton mechanism. The conversion of the initial, inflationary spectrum of massless axion fluctuations into a post-inflationary spectrum of massive fluctuations, and the generation of a final spectrum of adiabatic metric perturbations (outside the horizon, in the radiation era), is an interesting cosmological effect which deserves detailed discussion. In this section we will sketch the main steps of such a process.

8.3.1 The curvaton mechanism

The crucial ingredient of the curvaton mechanism is the generation of a (non-perturbative) axion potential $V(\sigma)$ during the post-inflationary phase of radiation-dominated evolution, and the assumption that the axion emerges from this process, at some given time η_i, in an initial configuration characterized by a non-trivial background value $\sigma(\eta_i) \equiv \sigma_i \neq 0$, displaced from the minimum of the potential. In the present discussion we also assume that the potential can be approximated by the quadratic form $V = m^2\sigma^2/2$. It should be stressed, in fact, that the Kalb–Ramond axion we are considering, even if not necessarily identified with the "invisible" action invoked to solve the strong CP problem [85, 86, 87], is expected to be gravitationally coupled to photons and to the QCD topological current. The associated potential is periodic, with a periodicity related to the breaking of the Peccei–Quinn symmetry down to a discrete symmetry, and to shifting the QCD vacuum angle by multiples of 2π [88]. The quadratic approximation for $V(\sigma)$ is valid, therefore, for values of σ that are small compared to the periodicity: this is certainly the case for $\sigma_i \ll M_P$, but we assume it to be valid also for all values of σ_i/M_P not much larger than one – which is also the appropriate range of values for a consistent implementation of the curvaton mechanism, as we shall see later.

We start the discussion in the phase of standard, post-inflationary evolution where, for simplicity, the dilaton is assumed to be already frozen at its present value, with no effect on the cosmological dynamics. The only important gravitational sources are the self-interacting axion and the radiation fluid, and the unperturbed dynamics is described by the background equations (8.7)–(8.9), with the dilaton ϕ replaced by the minimally coupled axion σ. In particular,

$$3\mathcal{H}^2 = \lambda_P^2 a^2 (\rho_r + \rho_\sigma), \tag{8.249}$$

$$2\mathcal{H}' + \mathcal{H}^2 = -\lambda_P^2 a^2 (p_r + p_\sigma), \tag{8.250}$$

$$\sigma'' + 2\mathcal{H}\sigma' + m^2 a^2 \sigma = 0, \tag{8.251}$$

where $p_r = \rho_r/3$ and where

$$\rho_\sigma = \frac{1}{2}\left(\frac{\sigma'^2}{a^2} + m^2\sigma^2\right), \qquad p_\sigma = \frac{1}{2}\left(\frac{\sigma'^2}{a^2} - m^2\sigma^2\right). \tag{8.252}$$

The initial configuration, at a given curvature scale $H_i \lesssim H_1$ (where H_1 marks the beginning of the post-inflationary epoch), is assumed to describe a radiation-dominated Universe and a potential-dominated axion energy, with $\rho_\sigma \simeq V(\sigma_i) = m^2\sigma_i^2/2 \ll H_i^2$.

During this initial phase the axion energy density is sub-leading, and the axion field is slowly rolling along the potential towards the minimum, conventionally fixed at $\sigma = 0$. Such a configuration tends to change, however, as the curvature scale evolves in time towards smaller and smaller values. In particular, the axion background enters an oscillatory regime as soon as the curvature drops below the scale $\sim H_m = m$ (see Eq. (8.251)), and starts dominating the cosmological evolution at the scale $\sim H_\sigma(\eta) = m\lambda_P \sigma(\eta)$ (see Eq. (8.249)), unless it has already decayed. Because of its gravitational coupling to photons, in fact, the axion tends to decay into electromagnetic radiation at a rate $\Gamma = \tau^{-1} \sim m^3/M_P^2$, thus disappearing from the cosmological scene when the curvature reaches the decay scale $\sim H_d = \Gamma$. The detailed history of such an axion–radiation Universe and, in particular, the temporal hierarchy of the scales H_m, H_σ, H_d, strongly depend on the axion mass and on the initial value σ_i. We may consider, in general, three distinct possibilities.

(1) **Late dominance of the axion**, i.e. $\lambda_P \sigma_i \ll 1$. During the initial, radiation-dominated phase the time variation of the slow-roll axion can be neglected with respect to the time variation of the curvature ($\sigma \simeq \sigma_i = \text{const}$, $\sigma' \simeq 0$), so that the axion starts oscillating at $\eta = \eta_m$ when the Universe is still radiation dominated, since $H_m = m > H_\sigma$. During the oscillating phase we can neglect the friction term $H\sigma'$ in Eq. (8.251), as the time variation of $a(\eta)$ is much slower than the variation of $\sigma(\eta)$, and the axion evolution can be approximately described by the free equation $\ddot{\sigma} + m^2\sigma = 0$. The kinetic and potential energy terms are thus equal, on the average, during this phase: $\langle \sigma'^2/a^2\rangle = m^2\langle \sigma^2\rangle$. It follows that $\langle p_\sigma\rangle = 0$, and the axion behaves like a "dust" fluid, with $\langle \rho_\sigma\rangle \sim a^{-3}$. The radiation energy, on the other hand, is diluted faster ($\rho_r \sim a^{-4}$), so that ρ_σ/ρ_r grows in time, and the axion becomes dominant at $\eta = \eta_\sigma$ when the curvature reaches the scale $H_\sigma = m\lambda_P\sigma_{\text{dom}}$. The value of σ_{dom} can be computed by using the kinematics of the radiation-dominated regime, and considering the following ratio of axion energy densities:

$$\frac{\rho_\sigma(\eta_m)}{\rho_\sigma(\eta_\sigma)} = \frac{\sigma_i^2}{\sigma_{\text{dom}}^2} = \left(\frac{a_\sigma}{a_m}\right)^3 = \left(\frac{H_m}{H_\sigma}\right)^{3/2} = (\lambda_P\sigma_{\text{dom}})^{-3/2}. \tag{8.253}$$

This gives $\lambda_P\sigma_{\text{dom}} = (\lambda_P\sigma_i)^4$, or

$$H_\sigma = m\,(\lambda_P\sigma_i)^4. \tag{8.254}$$

The background then remains axion dominated until the axion decays, at a scale $H_d = \lambda_P^2 m^3$. The efficient production of adiabatic metric fluctuations requires, as we shall see, that the decay occurs after the beginning of the axion-dominated epoch, namely for $H_\sigma > H_d$. This is possible, for the case $\lambda_P \sigma_i < 1$, if we restrict consideration to the class of backgrounds satisfying

$$(m\lambda_P)^{1/2} < \lambda_P \sigma_i < 1. \tag{8.255}$$

Note that such a constraint is not so demanding, given the generous lower bounds on m following from the decay of a gravitationally coupled scalar field [89, 90] – typically, $m \gtrsim 10$ TeV, or $m\lambda_P \gtrsim 10^{-14}$, to avoid disturbing the standard nucleosynthesis, i.e. by requiring $H_d > H_N \sim \lambda_P (1\,\text{MeV})^2$. This leads to a class of backgrounds with the following temporal ordering of scales: $H_1 > H_i > H_m > H_\sigma > H_d$.

(2) **Planck-scale axion**, i.e. $\lambda_P \sigma_i \sim 1$. The discussion and the results of the previous case are valid also for this case, with the only difference that now $H_m = m \sim m\lambda_P \sigma_i = H_\sigma$, so that the beginning of the axion oscillations and of the axion-dominated phases is nearly simultaneous.

(3) **Early dominance of the axion**, i.e. $\lambda_P \sigma_i > 1$. In this case $H_\sigma > H_m$, namely the Universe becomes axion dominated before the axion starts oscillating. Thus, when $H = H_\sigma$, the Universe enters a phase of slow-roll "axionic" inflation, lasting until the final scale $H_f \sim H_m = m$ is reached, corresponding to the final value $\lambda_P \sigma_f \sim 1$ of the axion background. During the slow-roll phase we have $H^2 = m^2 \lambda_P^2 \sigma^2 /6$ from Eq. (8.249), and $\dot{\sigma} = \sigma'/a = -m^2 \sigma/(3H)$ from Eq. (8.251) (since $\ddot{\sigma} = 0$). Their combination gives

$$\frac{\dot{H}}{H^2} = -\frac{2}{(\lambda_P \sigma)^2}, \tag{8.256}$$

which relates the variation of H and σ (in cosmic time). After the inflationary phase, $H \lesssim m$, the background is dominated by the coherent oscillations of the axion, which then eventually decays at $H = H_d$. This scenario corresponds to $H_1 > H_i > H_\sigma = m\lambda_P \sigma_i$, namely to

$$1 < \lambda_P \sigma_i < \frac{H_i}{m} < \frac{H_1}{m}, \tag{8.257}$$

which is compatible with the above-mentioned limits on the axion mass.

After the above analysis we are now in a position to discuss the evolution of the coupled system of axion-metric perturbations, in the various types of background. We first show that the inflation-generated, super-horizon distribution of isocurvature axion fluctuations, $\delta\sigma_k$, produces super-horizon metric perturbations, Ψ_k, with the same spectral distribution and with an amplitude which, at the end of the axion-dominated phase, is always not smaller than the initial amplitude of the axion fluctuations.

The evolution in time of the perturbations is controlled by the coupled system of equations (8.63)–(8.69), where the fluid sources describe the cosmic radiation and its adiabatic perturbations, $p/\rho = \Pi/\mathcal{E} = 1/3$, and where the dilaton variables ϕ, X are replaced by the corresponding axion variables σ, $\delta\sigma$ (with $\mu \to \infty$ as we assume that the axion is minimally coupled to the metric). Also, $\partial V/\partial\phi$ is replaced by $m^2\sigma$, and $\partial^2 V/\partial\phi^2$ by m^2. The initial conditions at $\eta = \eta_i$ are

$$\delta\sigma(\eta_i) \equiv \delta\sigma_i \neq 0, \qquad \Psi(\eta_i) = 0, \qquad W(\eta_i) = 0, \qquad \delta_r(\eta_i) \equiv \frac{\mathcal{E}_r}{\rho_r}(\eta_i) = 0,$$

$$(8.258)$$

representing *isocurvature* axion fluctuations. The initial values of the first derivatives are determined by the momentum and Hamiltonian constraints (8.64), (8.65). In the radiation-dominated phase, in particular, we can set $\lambda_P^2 a^2 \mathcal{E} = 3\mathcal{H}^2\delta_r$, and the conservation equation (8.68) becomes

$$\delta_r' = 4\Psi' + \frac{4}{3}\nabla^2 W.$$

$$(8.259)$$

Thus, in the super-horizon regime where the spatial gradients can be neglected, $\delta_r = 4\Psi$, and the Hamiltonian constraint can be rewritten

$$\Psi' + 3\mathcal{H}\Psi = -\frac{\lambda_P^2}{6\mathcal{H}^2}\left(-\sigma'^2\Psi + \sigma'\delta\sigma' + m^2 a^2 \sigma\delta\sigma\right).$$

$$(8.260)$$

At the beginning of the radiation-dominated regime the axion field is slow-rolling, and we can set $\sigma \simeq \sigma_i = $ const, neglecting the derivatives σ'. In addition, for all super-horizon modes $\delta\sigma_k$, the equation (8.67) for the axion perturbations is identical to the equation (8.251) for the axion background, since the Bardeen potential is vanishing (or negligible): as a consequence, the super-horizon axion perturbations are also slow-rolling, and we can approximate $\delta\sigma_k' \simeq 0$, $\delta\sigma_k \simeq \delta\sigma_k(\eta_i) = $ const. Equation (8.260) reduces to

$$\Psi' + 3\mathcal{H}\Psi = -\frac{\lambda_P^2}{6\mathcal{H}^2}m^2 a^2 \sigma_i\delta\sigma_i,$$

$$(8.261)$$

and its direct integration gives

$$\Psi_k(\eta) = \frac{\lambda_P^2 Q^2}{42}\sigma_i\delta\sigma_k(\eta_i)\left[1 - (\eta/\eta_i)^4\right], \qquad \eta > \eta_i, \quad k\eta \ll 1, \quad (8.262)$$

where we have defined $Q = ma_i\eta_i$. The amplitude of the metric perturbations is thus monotonically growing (in modulus) during the slow-roll regime. The duration of this regime, on the other hand, is controlled by the initial value σ_i. Let us then separately consider the three cases listed above.

(1) In the case $\lambda_P\sigma_i < 1$ the phase of slow-roll ends at the oscillation scale $H = m$, namely at the time scale η_m such that

$$\left(\frac{a_i\eta_i}{a_m\eta_m}\right) = \left(\frac{\eta_i}{\eta_m}\right)^2 = ma_i\eta_i = Q.$$

$$(8.263)$$

The spectral amplitude of the (super-horizon) Bardeen potential, at this time scale, is given by

$$\Psi_k(\eta_m) = -\frac{\lambda_P^2}{42}\sigma_i \delta\sigma_k(\eta_i) \qquad \lambda_P \sigma_i < 1, \qquad k\eta < 1. \qquad (8.264)$$

The amplitude of Ψ_k is no longer vanishing, but still small with respect to the axion perturbations $\delta\sigma_k$.

In the case we are considering the axion starts oscillating when the Universe is still radiation dominated: we can thus continue to apply Eq. (8.260) to compute Ψ_k for $\eta > \eta_m$ (neglecting Ψ with respect to $\delta\sigma$ in the right-hand side). The evolution equations for σ and $\delta\sigma$ are still the same, to leading order, outside the horizon (see Eq. (8.251)), and can be solved exactly [24] in terms of the first-kind Bessel function, i.e. $\sigma \sim \delta\sigma \sim (\eta/\eta_i)^{-1/2}J_{1/4}(Q\eta^2/2\eta_i^2)$. In the oscillating regime (corresponding to the large argument limit of the Bessel function) one finds, averaging over many oscillations,

$$\langle \sigma'\delta\sigma' \rangle \sim m^2 a^2 \langle \sigma\delta\sigma \rangle \sim \frac{m^2}{a} \qquad (8.265)$$

and the integration of Eq. (8.260) gives $\Psi \sim a$, so that

$$\Psi_k(\eta) = \Psi_k(\eta_m)\left(\frac{a}{a_m}\right), \qquad k\eta < 1, \qquad \eta_m < \eta < \eta_\sigma \qquad (8.266)$$

(we have neglected, for simplicity, the average symbol).

This behavior of Ψ is valid until the axion-dominance scale H_σ, determined by Eq. (8.254). In the oscillating regime, on the other hand, $\sigma \sim a^{-3/2}$, so that $a_\sigma/a_m = (\sigma_i/\sigma_{\text{dom}})^{2/3} = (\lambda_P\sigma_i)^{-2}$. At the beginning of the axion-dominated phase the spectral amplitude of the Bardeen potential is then given by

$$\Psi_k(\eta_\sigma) = \frac{\epsilon_1}{\sigma_i}\delta\sigma_k(\eta_i), \qquad \lambda_P\sigma_i < 1, \qquad k\eta < 1, \qquad (8.267)$$

where ϵ_1 is a dimensionless numerical coefficient with modulus of order one. This (approximate) analytical result is in agreement with numerical integrations of the exact perturbation equations [24].

(2) In the case in which $\lambda_P\sigma_i \sim 1$ the time scales η_m and η_σ are nearly coincident, but the previous arguments are still valid, and lead to the result

$$\Psi_k(\eta_\sigma) = \epsilon_2\lambda_P\,\delta\sigma_k(\eta_i), \qquad \lambda_P\sigma_i \sim 1, \qquad k\eta < 1, \qquad (8.268)$$

where ϵ_2 is another numerical coefficient with modulus of order one.

(3) Finally, we have the case $\lambda_P\sigma_i > 1$. In this case the amplification of the Bardeen potential during the initial, radiation-dominated phase, is still described by Eq. (8.262). The radiation phase, however, ends before the beginning of

the oscillating regime, since the axion becomes dominant at the scale $H_\sigma \equiv (a_\sigma \eta_\sigma)^{-1} = m \lambda_P \sigma_i$. The corresponding time scale is determined by the condition

$$\left(\frac{a_i \eta_i}{a_\sigma \eta_\sigma}\right) = \left(\frac{\eta_i}{\eta_\sigma}\right)^2 = m a_i \eta_i \lambda_P \sigma_i = Q \lambda_P \sigma_i, \tag{8.269}$$

and the associated amplitude of the Bardeen potential is, according to Eq. (8.262),

$$\Psi_k(\eta_\sigma) = -\frac{1}{42} \frac{\delta \sigma_k(\eta_i)}{\sigma_i}, \qquad \lambda_P \sigma_i > 1, \qquad k\eta < 1. \tag{8.270}$$

This amplitude is still small with respect to the axion perturbations $\delta \sigma_k$, but the growth of the Bardeen potential continues during the subsequent phase of slow-roll inflation. This effect can be evaluated by using Eq. (8.85), which describes the super-horizon evolution of Ψ in a phase dominated by a scalar field. Neglecting second time derivatives (because of the slow-rolling regime) we get, in cosmic time,

$$H\dot{\Psi} + 2\dot{H}\Psi = 0, \tag{8.271}$$

namely

$$\Psi_k(\eta) \sim H^{-2} \sim \sigma^{-2} \tag{8.272}$$

(we have used the background equation (8.249), which implies $3H^2 = \lambda_P^2 \rho_\sigma \simeq \lambda_P^2 m^2 \sigma^2/2$). Taking into account this further growth, and noticing that inflation starts with $\lambda_P \sigma(\eta_\sigma) = \lambda_P \sigma_i$ and ends with $\lambda_P \sigma(\eta_m) \sim 1$, the resulting amplitude of the Bardeen potential, at the beginning of the oscillating phase ($\eta = \eta_m$), is given by

$$\Psi_k(\eta_m) = \frac{\sigma^2(\eta_\sigma)}{\sigma^2(\eta_m)} \Psi_k(\eta_\sigma) = \epsilon_3 \lambda_P^2 \sigma_i \delta \sigma_k(\eta_i), \qquad \lambda_P \sigma_i > 1, \qquad k\eta < 1, \tag{8.273}$$

where ϵ_3 is a dimensionless coefficient with modulus of order one.

In the subsequent phase dominated by the oscillating axion, and common to the three classes of backgrounds that we are considering, the amplitude of the Bardeen potential simply oscillates around the value determined by the preceding evolution, without further amplification (as shown by analytical and numerical integrations of the perturbation equations [23, 24]). The amplitude Ψ_k, which we have computed in the various cases, may thus be transferred (practically unchanged) down to the axion-decay scale η_d. The final result can be written in compact form, inclusive of all the three classes of backgrounds, by setting

$$|\Psi_k(\eta_d)| = \lambda_P |\delta \sigma_k(\eta_i)| f(\sigma_i), \qquad f(\sigma_i) = \frac{c_1}{\lambda_P \sigma_i} + c_2 + c_3 \lambda_P \sigma_i. \tag{8.274}$$

One may note that $f(\sigma_i)$ is approximately invariant under the transformation $\lambda_P \sigma_i \to (\lambda_P \sigma_i)^{-1}$ and, as a consequence, has a minimum of order one around $\lambda_P \sigma_i = 1$. A numerical integration (from η_i to η_d) of the coupled perturbation equations for different values of σ_i, and a fit of the final value $|\Psi_k|$ with the above form of $f(\sigma_i)$, leads, in particular, to the following numerical values for the c_i coefficients [24]:

$$c_1 \simeq 0.25, \qquad c_2 \simeq -0.01, \qquad c_3 \simeq 0.13. \qquad (8.275)$$

The obtained spectrum of scalar metric perturbations is thus fully determined by the primordial spectrum of axion perturbations $\delta \sigma_k(\eta_i)$, present outside the horizon during the initial radiation-dominated phase. At that time the axion fluctuations are still relativistic (the mass can be neglected for $\eta < \eta_m$), and the axion spectrum can be computed from the action (8.238), and from the corresponding canonical equation

$$(z\delta\sigma_k)'' + \left(k^2 - \frac{z''}{z}\right)(z\delta\sigma_k) = 0. \qquad (8.276)$$

We consider the usual transition at $\eta = -\eta_1$ between inflation and the standard radiation phase with frozen dilaton, so that we can exploit the results of the analysis already performed for tensor fluctuations in Section 7.2.

In particular, we can parametrize the axion pump field as in Eq. (7.133), where now we set $d = 3$ and $\alpha_2 = 1$ (after the transition, the pump field (8.238) simply coincides with the E-frame scale factor). The exact solution for the massless (canonically normalized) axion fluctuations in the radiation era is then obtained from Eq. (7.140) as

$$\lambda_P \delta\sigma_k(\eta) = \frac{1}{aM_P\sqrt{k}}\left[c_+(k)e^{-i(k\eta+2k\eta_1)} + c_-(k)e^{i(k\eta+2k\eta_1)}\right], \qquad \eta > \eta_1. \qquad (8.277)$$

The Bogoliubov coefficients $c_\pm(k)$ are fixed by the matching conditions at $\eta = -\eta_1$, and are given by Eq. (7.147) with $\nu_2 = -1/2$. Considering the regime $k\eta_1 \ll k\eta \ll 1$ we can use the small argument limit of the Hankel functions to obtain

$$|c_+| \simeq |c_-|, \qquad e^{-ik\eta_1}c_+ \simeq -c_-e^{ik\eta_1}. \qquad (8.278)$$

Also, from Eq. (7.153), one finds $|c_-| \sim (k/k_1)^{\alpha-1}$ (we are assuming, as usual, $\alpha < 1/2$). This leads to the following frozen spectrum of super-horizon axion fluctuations

$$\lambda_P k^{3/2}|\delta\sigma_k| \simeq \frac{2k\eta_1}{a_1 M_P\eta}|c_-(k)\sin(k\eta+k\eta_1)| \simeq \frac{H_1}{M_P}\left(\frac{k}{k_1}\right)^{\alpha+1},$$

$$k\eta < 1, \qquad k < k_1, \qquad (8.279)$$

where we have absorbed all numerical factors of order one into the transition scale $H_1 = k_1/a_1$, determined by the high-frequency cut-off $k_1 = \eta_i^{-1}$. This axion spectrum is valid at the initial epoch η_i, and is the source of the Bardeen spectrum according to Eq. (8.274):

$$k^3 |\Psi_k(\eta_d)|^2 = f^2(\sigma_i) \left(\frac{H_1}{M_P}\right)^2 \left(\frac{k}{k_1}\right)^{n_\sigma - 1}, \qquad k < k_1 \tag{8.280}$$

(we have reintroduced the spectral index $n_\sigma = 3 + 2\alpha$, determined by the kinematics of the solution Eq. (8.246)).

An important comment is in order, at this point, concerning the possible generalization of the previous expression to include the case in which the axion is still sub-dominant at the decay epoch η_d. This is possible for $\lambda_P \sigma_i < 1$, while for $\lambda_P \sigma_i > 1$ this option is impossible, at least for realistic values of the decay rate satisfying $\Gamma/m \sim (m/M_P)^2 < 1$. Let us thus consider the case $H_d > H_\sigma$, corresponding to the condition

$$\lambda_P \sigma_i < (m\lambda_P)^{1/2} < 1, \tag{8.281}$$

complementary to the condition (8.255) used in the previous discussion. In such a case the axion background, at the decay epoch, has an amplitude given by

$$\sigma_d = \sigma_i \left(\frac{a_m}{a_d}\right)^{3/2}_{\text{rad}} = \sigma_i \left(\frac{H_d}{H_m}\right)^{3/4} \simeq \sigma_i (m\lambda_P)^{3/2}, \tag{8.282}$$

and the corresponding energy density, in critical units, can be estimated as

$$\Omega_\sigma = \frac{\lambda_P^2}{3H_d^2} \rho_\sigma(\eta_d) = \frac{\lambda_P^2}{3H_d^2} m^2 \sigma_d^2 \simeq \frac{\lambda_P}{m} \sigma_i^2. \tag{8.283}$$

Note that this density always satisfies the bound $\Omega_\sigma < 1$, thanks to the condition (8.281). The amplitude of the Bardeen potential at the decay scale can now be determined by applying Eqs. (8.264) and (8.266) as before, but taking into account that the amplification of Ψ_k ceases at the scale η_d. Thus

$$\Psi_k(\eta_d) = \Psi_k(\eta_m) \left(\frac{a_d}{a_m}\right)_{\text{rad}} = \epsilon_1 \frac{\lambda_P \sigma_i}{m} \delta\sigma_k(\eta_i) \simeq \frac{\epsilon_1 \Omega_\sigma}{\sigma_i} \delta\sigma_k(\eta_i) \tag{8.284}$$

(when $\Omega_\sigma \sim 1$ one recovers the result (8.267)). Taking into account the case $\Omega_\sigma < 1$ we can thus rewrite the final amplitude of the Bardeen potential at η_d as in Eq. (8.274), but with $f(\sigma_i)$ replaced by $\overline{f}(\sigma_i)$, where

$$\overline{f}(\sigma_i) = c_1 \frac{\Omega_\sigma}{\lambda_P \sigma_i} + c_2 + c_3 \lambda_P \sigma_i. \tag{8.285}$$

The important consequence of this modification is a possible suppression of the metric perturbation spectrum (8.280) (at fixed H_1): the spectrum could have a minimum not at $f(\sigma_i) \sim 1$, but at

$$\overline{f}(\sigma_i) \sim \frac{\Omega_\sigma}{\lambda_P \sigma_i} = \frac{\sigma_i}{m} \ll 1. \qquad (8.286)$$

Axion models satisfying this property can match the large-scale normalization of the spectrum even with values of H_1/M_P much larger than in the case $f(\sigma_i) \sim 1$.

Too small values of σ_i/m (i.e. values of Ω_σ too far from one), however, tend to enhance the possible "non-Gaussian" properties of the scalar perturbation spectrum [25, 26]: in particular, such properties become dominant when the amplitude of the axion background and of its fluctuations are of comparable magnitude, $\sigma \sim \delta\sigma$. On the other hand, no significant deviation from Gaussianity has been definitely detected (up to now) in the observed CMB anisotropy, so that the possible non-Gaussianity of the primordial scalar background has to be small enough to be compatible with present observations [91]. We thus restrict the following discussion to the case $\Omega_\sigma = 1$, using the Bardeen spectrum (8.280) with the function $f(\sigma_i)$ of Eq. (8.274).

Such a spectrum has to be transferred from the end of the axion-dominated phase, $\eta = \eta_d$, down to the subsequent radiation- and matter-dominated epochs. To this purpose, we have to match the constant Bardeen potential prior to decay, Eq. (8.274), to the solution of Eq. (8.98) describing the evolution of Ψ in the radiation era. The general exact solution (for $c_s^2 = 1/3$ and $\Sigma = 0$) in the radiation era can be written as [27]

$$\Psi_k(\eta) = \frac{1}{\eta^3} \left[B_1(x \cos x - \sin x) + B_2(x \sin x + \cos x) \right],$$

$$x = \frac{k}{\sqrt{3}} \eta, \qquad \eta_s \leq \eta \leq \eta_{eq}, \qquad (8.287)$$

where B_1 and B_2 are integration constants to be determined by the matching of Ψ and Ψ' at η_d:

$$\Psi_k(\eta_d) = \Psi_{k0}, \qquad \Psi_k'(\eta_d) = 0, \qquad (8.288)$$

and where Ψ_{k0} is the constant amplitude of Eq. (8.274) (we have assumed a sudden transition). For super-horizon modes this gives

$$B_1 = -3 \left(\sqrt{3}/k \right)^3 \Psi_{k0}, \qquad B_2 \simeq 0, \qquad (8.289)$$

so that

$$\Psi_k(\eta) = 3\Psi_k(\eta_d) \left(\frac{\sin x}{x^3} - \frac{\cos x}{x^2} \right), \qquad k\eta \ll 1, \qquad \eta_s \leq \eta \leq \eta_{eq}, \quad (8.290)$$

is the final result for the axion-induced Bardeen potential in the radiation era. This estimate is confirmed by accurate numerical integrations [23, 24], taking into account the damping of the oscillations of the axion background for a more realistic description of its decay.

The scalar metric perturbations that we have obtained provide the starting point for the subsequent evolution of the CMB temperature fluctuations, and for the formation of their oscillatory pattern. It is thus important to note that, according to Eq. (8.290), super-horizon modes with $x \ll 1$ satisfy the conditions $\Psi_k \simeq \Psi_{k0} =$ const and $\Psi'_k \simeq 0$ not only at η_d, but also during the whole radiation phase (and during the subsequent matter-dominated evolution). This means that we are given "adiabatic" initial conditions (see Section 8.2) just as in the case of the standard inflationary scenario, in spite of the fact that Ψ has not been directly amplified from the vacuum during inflation, but has been generated by the axion during the post-inflationary evolution.

We can then repeat the computation of the CMB anisotropy exactly as done in the previous section for the case of adiabatic perturbations: in particular, Eq. (8.280) gives the primordial Bardeen spectrum to be inserted into the SW effect, and the observations may be faithfully reproduced provided the axion index n_σ is sufficiently near to one. The only possible difference is a (small) non-Gaussian component, present if $\Omega_\sigma < 1$ (i.e. if σ_i is small enough, as pointed out in the preceding discussion). In this context, the parameters H_1 and n_σ of the Bardeen spectrum depend on the details of the inflationary regime, and the large-scale normalization of the spectrum may impose important phenomenological constraints on the class of models that we are considering.

8.3.2 Normalization of the Bardeen spectrum

Let us consider the quadrupole coefficient C_2 obtained from the large-scale expression (8.217) of the angular power spectrum. Using the Bardeen potential (8.280) as the source of the ordinary SW effect we obtain

$$C_2 = \alpha_{n_\sigma}^2 f^2(\sigma_i) \left(\frac{H_1}{M_P}\right)^2 \left(\frac{\omega_0}{\omega_1}\right)^{n_\sigma - 1},$$

$$\alpha_{n_\sigma}^2 = \frac{2^{n_\sigma}}{72} \frac{\Gamma(3 - n_\sigma)\Gamma(3/2 + n_\sigma/2)}{\Gamma^2(2 - n_\sigma/2)\Gamma(9/2 - n_\sigma/2)},$$

(8.291)

where $\omega_1 = k_1/a_0$ is, as usual, the present value of the (proper) cut-off frequency, and $\omega_0 = k_0/a_0$ is the proper frequency crossing today the Hubble radius H_0. The value of C_2 depends explicitly on H_1, n_σ, σ_i and also, implicitly, on the axion mass m.

In fact, the computation of the cut-off parameter ω_1 requires the rescaling of the transition scale H_1 down to the present epoch, and thus depends on the kinematics (as well as on the duration) of the axion-dominated phase. In particular,

$$
\omega_1(t_0) = \begin{cases} H_1 \left(\frac{a_1}{a_\sigma}\right)_{\text{rad}} \left(\frac{a_\sigma}{a_d}\right)_{\text{mat}} \left(\frac{a_d}{a_{\text{eq}}}\right)_{\text{rad}} \left(\frac{a_{\text{eq}}}{a_0}\right)_{\text{mat}}, & \lambda_{\text{P}}\sigma_{\text{i}} \leq 1, \\ H_1 \left(\frac{a_1}{a_\sigma}\right)_{\text{rad}} \left(\frac{a_\sigma}{a_m}\right)_{\text{inf}} \left(\frac{a_m}{a_d}\right)_{\text{mat}} \left(\frac{a_d}{a_{\text{eq}}}\right)_{\text{rad}} \left(\frac{a_{\text{eq}}}{a_0}\right)_{\text{mat}}, & \lambda_{\text{P}}\sigma_{\text{i}} \geq 1, \end{cases}
$$

$$(8.292)$$

where we have considered the two possible types of post-big bang history. Replacing the ratios of scale factors with ratios of Hubble scales, and using Eq. (1.79) for H_0 and H_{eq}, one obtains (for $\Omega_m = 0.3$ and $h = 0.7$)

$$
\frac{\omega_1}{\omega_0} = \frac{H_1 a_1}{H_0 a_0} \simeq \begin{cases} 10^{29} \left(\frac{H_1}{M_{\text{P}}}\right)^{1/2} \left(\frac{mM_{\text{P}}}{\sigma_{\text{i}}^2}\right)^{1/3}, & \lambda_{\text{P}}\sigma_{\text{i}} \leq 1, & (8.293) \\ 10^{29} \left(\frac{\sigma_{\text{i}}H_1}{M_{\text{P}}^2}\right)^{1/2} \left(\frac{m}{M_{\text{P}}}\right)^{1/3} Z_\sigma^{-1}, & \lambda_{\text{P}}\sigma_{\text{i}} \geq 1, & (8.294) \end{cases}
$$

where $Z_\sigma = a_m/a_\sigma$ denotes the expansion factor associated with the axion-dominated phase of slow-roll inflation. The result of the COBE measurements for C_2, Eq. (8.219), thus imposes the bounds

$$
10^{-29(n_\sigma-1)} \alpha_{n_\sigma}^2 f^2(\sigma_{\text{i}}) \left(\frac{H_1}{M_{\text{P}}}\right)^{\frac{5-n_\sigma}{2}} \left(\frac{\sigma_{\text{i}}}{M_{\text{P}}}\right)^{\frac{2}{3}(n_\sigma-1)} \left(\frac{m}{M_{\text{P}}}\right)^{\frac{1-n_\sigma}{3}} \simeq 10^{-10},
$$

$$\lambda_{\text{P}}\sigma_{\text{i}} \leq 1, \qquad (8.295)$$

$$
10^{-29(n_\sigma-1)} \alpha_{n_\sigma}^2 f^2(\sigma_{\text{i}}) Z_\sigma^{n_\sigma-1} \left(\frac{H_1}{M_{\text{P}}}\right)^{\frac{5-n_\sigma}{2}} \left(\frac{\sigma_{\text{i}}}{M_{\text{P}}}\right)^{\frac{1-n_\sigma}{2}} \left(\frac{m}{M_{\text{P}}}\right)^{\frac{1-n_\sigma}{3}} \simeq 10^{-10},
$$

$$\lambda_{\text{P}}\sigma_{\text{i}} \geq 1. \qquad (8.296)$$

More precise constraints can be obtained if we do not fix the values of Ω_m and h, and include them among the arbitrary parameters.

In order to discuss the allowed region in the four-dimensional parameter space, spanned by $\{H_1, n_\sigma, m, \sigma_{\text{i}}\}$, we first note that the above two constraints have to be supplemented, respectively, by the conditions (8.255) and (8.257), required for the consistency of the corresponding classes of backgrounds. Both constraints are to be intersected with the experimentally allowed range of the spectral index,

$$
0.932 \lesssim n_\sigma \lesssim 0.966 \qquad (8.297)
$$

(see Eq. (8.153)). In addition, we must take into account the (conservative) nucleosynthesis bound on the axion mass, $m\lambda_{\text{P}} \geq 10^{-14}$, required to avoid a too-late axion decay which could destroy the light nuclei already formed [89, 90];

however, it may be noted that for $n_\sigma \simeq 1$ the coefficient C_2 is almost insensitive to the value of m, according to Eqs. (8.295) and (8.296).

Finally, there is a further constraint to be imposed on Z_σ in the case $\lambda_P \sigma_i > 1$, since we are implicitly assuming that there is no contribution to C_2 arising from scalar metric perturbations directly amplified from the vacuum, during the phase of axion-dominated inflation. This means, roughly, that the proper frequency of a mode crossing the horizon at the beginning of inflation, ω_σ, has to be larger than the frequency scale re-entering the horizon at decoupling, i.e. $\omega_\sigma(t_0) = H_\sigma a_\sigma/a_0 > \omega_{\text{dec}}(t_0) = H_{\text{dec}} a_{\text{dec}}/a_0$. This gives the constraint

$$Z_\sigma \lesssim 10^{28} \lambda_P \sigma_i \left(\frac{m}{M_P} \right)^{5/6}, \qquad (8.298)$$

to be added to Eq. (8.257) for $\lambda_P \sigma_i > 1$.

We refer to the literature for a detailed study of the region of parameter space determined by the intersection of all constraints [24]. Here we note that a fully scale-invariant spectrum ($n_\sigma = 1$) is consistent with the COBE normalization provided that

$$\alpha_1 f(\sigma_i) H_1 \simeq 10^{-5} M_P. \qquad (8.299)$$

More generally, for n_σ varying in the restricted range (8.297), the allowed values of H_1 and σ_i are illustrated in Fig. 8.3. In the figure we have plotted the conditions (8.295) and (8.296) for the two limiting values of n_σ, using Eq. (8.275) for $f(\sigma_i)$, and using the exponential parametrization $Z_\sigma = \exp\left[(\lambda_P^2 \sigma_i^2 - 1)/4 \right]$ for the expansion factor of a phase of slow-roll inflation with quadratic potential (see Eq. (1.123)). Also, we have plotted the two cases $m = 10^{-3} M_P$ (bold solid curves) and $m = 10^{-14} M_P$ (dashed curves), for a concrete illustration of the very weak – practically unappreciable – mass dependence of the results in the given range of

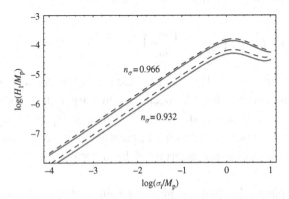

Figure 8.3 Allowed values of H_1 as a function of σ_i, for $n_\sigma = 0.932$ and $n_\sigma = 0.966$.

the spectral index. Finally, we have truncated the plot at $\lambda_P \sigma_i \simeq 10$, since higher values of σ_i would imply a violation of the constraint (8.298).

The curves plotted in Fig. 8.3 might suggest that, given some reasonable assumption about the axion amplitude σ_i (for instance, $\lambda_P \sigma_i \sim 1$), one could directly interpret a measure of n_σ as a measure of the inflation scale H_1. In the class of string cosmology models that we are considering, on the other hand, we may expect, generally, $H_1 \sim M_s$: thus, one might think of "*weighing the string mass with the CMB data*" [92]. However, as a warning against a too enthusiastic application of CMB observations – more generally, against a too naive extrapolation of low-energy data to determine high-energy parameters – let us conclude this section by showing that even the knowledge of all the three parameters m, n_σ and σ_i might be not enough for a complete determination of H_1.

Consider, for instance, a non-minimal inflationary scenario in which the primordial spectrum of relativistic axion fluctuations has two branches: a low-frequency branch, which is flat enough to match large-scale observations, and a high-frequency branch which is steeper, and which matches the string scale normalization at the end-point of the spectrum. In such a context, the axion spectrum (8.279) has to be replaced by

$$\lambda_P^2 k^3 |\delta\sigma_k|^2 = \begin{cases} \left(\dfrac{H_1}{M_P}\right)^2 \left(\dfrac{k}{k_1}\right)^{n_\sigma - 1 + \delta}, & k_s < k < k_1, \\[4mm] \left(\dfrac{H_1}{M_P}\right)^2 \left(\dfrac{k_s}{k_1}\right)^{n_\sigma - 1 + \delta} \left(\dfrac{k}{k_s}\right)^{n_\sigma - 1}, & k < k_s, \end{cases} \tag{8.300}$$

where k_s is the break scale, and the power $\delta > 0$ parametrizes the deviations of the high-frequency branch with respect to the spectral behavior at lower frequencies. Examples of backgrounds producing such a spectrum in the context of the pre-big bang scenario have been presented, for instance, in [84].

The computation of C_2, in this case, leads to the result

$$C_2 = \alpha_{n_\sigma}^2 f^2(\sigma_i) \left(\frac{H_1}{M_P}\right)^2 \left(\frac{\omega_0}{\omega_1}\right)^{n_\sigma - 1} \left(\frac{\omega_s}{\omega_1}\right)^\delta \simeq 10^{-10}, \tag{8.301}$$

to be compared with Eq. (8.291). It is then clear that the steeper and/or the longer the high-frequency branch of the spectrum, the larger the suppression at low-frequency scales, and the wider the range of values of H_1 matching the measured anisotropies. We have new dimensions in parameter space, spanned by k_s and δ, and the scale H_1 is no longer determined by a measure of n_σ, contrary to the indications of Fig. 8.3.

It follows, in particular, that the allowed values of the inflation scale H_1 can be higher than those illustrated in Fig. 8.3 – and thus more consonant with the

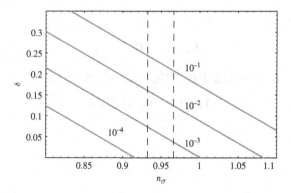

Figure 8.4 Relation between δ and n_σ for different values of H_1/M_P, ranging from 10^{-1} to 10^{-4} (bold curves). The allowed range of n_σ is within the vertical dashed lines.

usual expectation $H_1 \sim M_s \sim 0.1 M_P$ – but still in agreement with the experimental results on n_σ. For a flat spectrum, Eq. (8.299) is indeed replaced by

$$\alpha_1 f(\sigma_i) H_1 \simeq 10^{-5} M_P \left(\frac{\omega_1}{\omega_s}\right)^{\delta/2}. \tag{8.302}$$

For a more general spectrum, with $n_\sigma \neq 1$, the relation between n_σ and δ imposed by the normalization condition (8.301) is illustrated in Fig. 8.4, for various values of H_1/M_P. We have used the "natural" value $\lambda_P \sigma_i = 1$, and we have set $m\lambda_P = 10^{-9}$ (but the curves remain stable even if we change m by various orders of magnitude). Finally, we have identified ω_s with the equality scale $\omega_{eq} = H_{eq} a_{eq}/a_0$, in such a way that the flat branch of the spectrum contains all modes relevant to the CMB anisotropies. With these assumptions, the new factor ω_1/ω_s appearing in Eq. (8.301) can be estimated as follows:

$$\frac{\omega_1}{\omega_s} = \frac{\omega_1}{\omega_{eq}} = \frac{H_1}{H_{eq}} \left(\frac{a_1}{a_\sigma}\right)_{rad} \left(\frac{a_\sigma}{a_d}\right)_{mat} \left(\frac{a_d}{a_{eq}}\right)_{rad} \simeq 10^{27} \left(\frac{H_1}{M_P}\right)^{1/2} \left(\frac{m}{M_P}\right)^{1/3}. \tag{8.303}$$

As clearly shown in Fig. 8.4, even a small departure from the minimal spectrum (8.279) may be enough to make the scale $H_1 = 0.1 M_P$ compatible with the measured values of C_2 and n_σ. On the other hand, no deviation at all is needed if, for some dynamical mechanism (see e.g. [93]), the scale of string cosmology inflation is lowered down to the GUT level $H_1 \simeq 10^{-3} M_P$.

Appendix 8A
Photon–dilaton interactions and cosmic magnetic fields

A long and fast enough period of primordial inflation can amplify not only the scalar and tensor components of the metric fluctuations, but also the fluctuations of any other field coupled to the accelerated evolution of the background. Interesting examples are the quantum fluctuations of the axion background, analyzed in Section 8.3, and the fluctuations of the electromagnetic (e.m.) field, which will be discussed here. Just as metric fluctuations may act as "seeds" for the CMB anisotropy, it will be shown here that the e.m. fluctuations of the vacuum, appropriately amplified in a string cosmology context, can provide the required seeds for the cosmic magnetic fields observed on galactic (and intergalactic) scales.

The origin of such cosmic fields (with coherence scale $\gtrsim 10$ kpc, and typical strength $\sim 10^{-6}$ gauss) is still, to a large extent, an open astrophysical problem (see for instance [94, 95]). Almost all mechanisms able to generate large-scale fields, such as the galactic "dynamo" [96, 97], require the presence of primordial seed fields, large enough to trigger the subsequent e.m. amplification. The inflationary amplification of the quantum fluctuations could represent, in principle, the most natural origin of these primordial seed fields, as first pointed out in [98]. The minimal coupling of the e.m. field to the geometry of a four-dimensional manifold, described by the Maxwell Lagrangian $\sqrt{-g}g^{\mu\alpha}g^{\nu\beta}F_{\mu\nu}F_{\alpha\beta}$, is, however, conformally invariant. As a consequence, there is no amplification of e.m. fluctuations propagating in a conformally flat metric background, which is typically the case for the standard inflationary scenario.

There are various possibilities to avoid this conclusion, at least in principle. One may assume, for instance, that the geometry is not exactly conformally flat [99] (a typical example is the case of a higher-dimensional, factorized geometry [100]), or that the conformal invariance of the photon–graviton interaction is broken (for instance by non-minimal couplings to the curvature [98], or by quantum trace-anomaly effects [101], or by more exotic "trans-Planckian" effects [102]). Alternatively, one may consider additional, non-conformally invariant couplings of the photon to other background fields such as the inflaton [103], the axion [104], the dilaton [105, 106], charged scalar fields [107, 108], supersymmetric vector fields [109] or graviphotons [110].

The analysis of this appendix concentrates on the possibility offered by the direct coupling of the photon to the dilaton, typical of the string cosmology scenario. In such a case the e.m. fluctuations may remain minimally and conformally coupled to the four-dimensional geometry, and the amplification is driven by the time evolution of the dilaton background. This mechanism can be efficient enough to produce the required magnetic seeds directly from the vacuum, as will be explicitly illustrated in this appendix with an example based on a class of models of pre-big bang inflation [105, 106].

We start from the general form of the S-frame, tree-level action for the effective interactions of the dilaton with the four-dimensional e.m. field $F_{\mu\nu}$:

$$S = -\frac{1}{4} \int d^4x \sqrt{-g} \, f(b_i) \, e^{-\epsilon\phi_d} \, F_{\mu\nu} F^{\mu\nu} \tag{8A.1}$$

(we are using string units $2\lambda_s^2 = 1$). Here ϕ_d is the dilaton appearing in the higher-dimensional action, and the coupling function $f(b_i)$ explicitly represents the possible dynamical contribution of the internal moduli fields b_i, $i = 1, \ldots, n$, after reduction from $D = 4 + n$ to four dimensions. Finally, ϵ is a model-dependent constant parametrizing the strength of the photon–dilaton coupling in the higher-dimensional action. For instance, $\epsilon = 1, 1/2$ for the heterotic and type I superstring models, respectively, if the e.m. $U(1)$ symmetry is identified as a component of the non-Abelian gauge symmetry appearing in the 10-dimensional effective action (see Eqs. (3B.110) and (3B.87)). The value of ϵ could be different if the e.m. field were identified, after dimensional reduction, with a one-form present in the R–R or NS–NS sector of other superstring models (see Appendix 3B); in that case, the coupling to the internal moduli would also be different [111].

The example we discuss refers to the simple case in which $f(b_i)$ always corresponds to the volume factor of the n-dimensional internal manifold, $f(b_i) = V_n \equiv \prod_{i=1}^{n} b_i$, so that different string models will be characterized by different values of ϵ only. After all, when the internal moduli are stabilized, the coupling function $f(b_i)$ becomes trivial, and the only relevant parameter is ϵ. In any case, different values of ϵ amount to different rescaling of the photon–dilaton coupling, and may have significant impact on cosmological processes where large variations of the dilaton field, for long periods of time, may occur.

If the unperturbed e.m. background is vanishing ($A_\mu = 0$) the action (8A.1) may be directly interpreted as the quadratic action for the first-order e.m. perturbations δA_μ, with $F_{\mu\nu} = \partial_\mu \delta A_\nu - \partial_\nu \delta A_\mu$. Since we are considering free radiation in a spatially flat geometry it is convenient to adopt for the metric the conformal-time gauge, $g_{\mu\nu} = a^2(\eta)\eta_{\mu\nu}$, and to impose on the fluctuations the radiation gauge, defined by $\delta A_0 = 0 = \partial_i \delta A_i$. After partial integration the action (8A.1) becomes, in these gauges,

$$S = \frac{1}{2} \int d\eta \, z_\gamma^2(\eta) \left[(\delta A_i')^2 + \delta A_i \nabla^2 \delta A_i \right],$$

$$z_\gamma = \left(\prod_{i=1}^{n} b_i \right)^{1/2} e^{-\epsilon\phi_d/2}, \tag{8A.2}$$

and can be recast in canonical form by using the variable $\psi_i = z_\gamma \delta A_i$. For each polarization mode ψ_i we then recover the usual, Schrödinger-like equation

$$\psi_i'' - \left[\nabla^2 + U(\eta) \right] \psi_i = 0,$$

$$\psi_i = z_\gamma \delta A_i, \qquad U = z_\gamma''/z_\gamma, \tag{8A.3}$$

describing the evolution of the canonically normalized e.m. fluctuations. This equation is formally the same as the canonical equation (7.50) for tensor perturbations, with the only difference that now the pump field z_γ is fully determined by the dilaton and by the internal moduli fields, according to Eq. (8A.2). The complete absence of coupling to the four-dimensional geometry follows from the conformal invariance of the action (8A.1), and implies that in a frozen ($\phi = $ const) or decoupled ($\epsilon = 0$) dilaton background the amplification of the e.m. fluctuations is possibly induced only by the internal moduli, if they are time dependent.

For a simple example of magnetic seed production we consider here the "minimal" model of pre-big bang inflation already used in Section 7.3 for the discussion of graviton production (see Fig. 7.6). The initial, low-energy dilaton-driven phase is described by the exact solution (8.241)–(8.244) and we obtain, for the photon pump field,

$$z_\gamma \sim (-\eta)^{\epsilon'\sqrt{3}/2}, \qquad \epsilon' = \frac{\sum_i \beta_i - \epsilon(\sum_i \beta_i + 3\beta_0 - 1)}{\sqrt{3}(1-\beta_0)}, \qquad \eta \le -\eta_s \qquad (8A.4)$$

(note that $\epsilon' = \epsilon$ when the internal dimensions are frozen, i.e. when $\beta_i = 0$ and $\beta_0 = -1/\sqrt{3}$). In the subsequent discussion we consider, for simplicity, an isotropic internal geometry, so that β_i can be eliminated everywhere in terms of β_0 through the Kasner condition (8.242). Also, for consistency with a realistic scenario in which the Universe evolves from the higher-dimensional perturbative vacuum towards the four-dimensional strong coupling regime, we should take into account that the internal dimensions are shrinking ($\beta_i > 0$), and that the four-dimensional coupling, controlled by $\phi = \phi_d - \ln V_n = \phi - \sum_i \ln b_i$, is growing ($\beta_0 < 1/3$). The intersection with the Kasner condition then defines the following allowed ranges of values for β_0 and β_i:

$$-1/\sqrt{3} \le \beta_0 < 1/3, \qquad 0 \le \beta_i \le 1/\sqrt{n}. \qquad (8A.5)$$

For a direct application to superstrings we also assume $n = 6$ everywhere in the subsequent discussion.

In the second, high-curvature phase we assume that the internal dimensions are frozen and we obtain, according to Eq. (7.196),

$$z_\gamma = e^{-\epsilon\phi/2} \sim (-\eta)^{\epsilon\beta}, \qquad -\eta_s \le \eta \le -\eta_1. \qquad (8A.6)$$

Finally, $z_\gamma = $ const during the radiation-dominated, post-big bang phase $\eta \ge -\eta_1$, where the dilaton is also frozen, the effective potential U is vanishing, and the Fourier components of the canonical field oscillate with constant amplitude determined by the Bogoliubov coefficients $c_\pm(k)$:

$$\psi^i(k) = \frac{1}{\sqrt{2k}} \left[c_+^i(k)e^{-ik\eta} + c_-^i(k)e^{+ik\eta} \right]. \qquad (8A.7)$$

The computation of $c_\pm$ can be performed exactly as in Chapter 7, solving the canonical equation in the dilaton and string phases in terms of Hankel functions, with Bessel indices

$$\nu = (1 - \epsilon'\sqrt{3})/2, \qquad \mu = (1 - 2\epsilon\beta)/2, \qquad (8A.8)$$

respectively, and matching the solutions at η_s and η_1. The presence of two transition scales produces two branches in the spectrum, as illustrated in Fig. 7.6. Higher-frequency modes ($k > k_s = \eta_s^{-1}$) are affected only by the transition at η_1: by imposing the canonical normalization at $\eta \to -\infty$ we can then apply the results of Eq. (7.151) with $\alpha_1 = \epsilon\beta$ and $\alpha_2 = 0$. Taking into account the leading terms, for both $\epsilon\beta > 1/2$ and $\epsilon\beta < 1/2$, we obtain that the production of high-frequency photons from the vacuum, for each polarization mode, is controlled by the following Bogoliubov coefficient:

$$|c_-(\omega)|^2 \sim \left(\frac{\omega}{\omega_1} \right)^{-2|\mu|-1}, \qquad \omega_s < \omega < \omega_1 \qquad (8A.9)$$

(to be compared with the corresponding result (7.198) for the graviton spectrum). We omit, for simplicity, the polarization index.

Lower-frequency modes, with $k < k_s$, are affected by both background transitions. The first matching at η_s provides the intermediate transition coefficients

$$
c_+^2 e^{i\theta} = i\frac{\pi}{4} x_s \left[-H_\nu^{(1)} H_{\mu+1}^{(2)} + H_{\nu+1}^{(1)} H_\mu^{(2)} + \frac{2}{x_s}(\mu-\nu) H_\nu^{(1)} H_\mu^{(2)} \right]_{x_s},
$$

$$
c_-^2 e^{-i\theta} = i\frac{\pi}{4} x_s \left[-H_{\nu+1}^{(1)} H_\mu^{(1)} + H_\nu^{(1)} H_{\mu+1}^{(1)} - \frac{2}{x_s}(\mu-\nu) H_\nu^{(1)} H_\mu^{(1)} \right]_{x_s},
$$

$$(8A.10)$$

where all the Hankel functions are evaluated at $x_s = k\eta_s$ (the phase θ is a real parameter depending on the initial normalization). The above coefficients can be directly obtained from Eq. (7.201) by replacing the Bessel indices 0 and ν with ν and μ, respectively. The second matching at η_1 leads to the final Bogoliubov coefficient

$$
c_-^3 = \frac{\pi}{4} x_1 \left[\left(c_+^2 e^{i\theta} H_\mu^{(1)} + c_-^2 e^{-i\theta} H_\mu^{(2)} \right) \left(2\frac{\mu+1}{x_1} H_{1/2}^{(2)} - H_{3/2}^{(2)} \right) \right.
$$

$$
\left. - H_{1/2}^{(2)} \left(c_+^2 e^{i\theta} H_{\mu+1}^{(1)} + c_-^2 e^{-i\theta} H_{\mu+1}^{(2)} \right) \right]_{x_1}
$$

$$(8A.11)$$

(all the Hankel functions are evaluated at $x_1 = k\eta_1$, except those contained inside $c_\pm^2$). This coefficient can also be obtained from the corresponding result for the graviton spectrum, Eq. (7.200), with the obvious replacement $\nu \to \mu$ and $-1/2 \to 1/2$ (the second replacement is due to the fact that, in the radiation era, the graviton pump field leads to a Bessel index $1/2 - 1 = -1/2$; the photon pump field, on the contrary, is constant, and the Bessel index is $1/2 - 0 = 1/2$). The approximated estimate of the mean number of photons produced is then obtained by considering the limit $k \ll k_s \ll k_1$, and using the small-argument expansion of the Hankel functions. This gives, to leading order in $x_1 \ll x_s \ll 1$,

$$
|c_-(\omega)|^2 \sim \left(\frac{\omega}{\omega_1}\right)^{-1} \left(\frac{\omega}{\omega_s}\right)^{-2|\nu|} \left(\frac{\omega_1}{\omega_s}\right)^{2|\mu|}, \qquad \omega < \omega_s \qquad (8A.12)
$$

(modulo a real coefficient of order one). This branch of the spectrum matches continuously to the high-frequency branch (8A.9) at $\omega = \omega_s$.

We now insert the two results (8A.9) and (8A.12) into the spectral energy density (7.160), using the definition (7.163) for the radiation density $\Omega_r(t)$, and absorbing into H_1 all numerical factors of order one arising from the computation of c_- (as in the case of the graviton spectrum). The full energy density spectrum of the e.m. fluctuations can be written in final form as follows:

$$
\Omega_\gamma(\omega, t_0) = \begin{cases} \Omega_r(t_0) \left(\dfrac{H_1}{M_P}\right)^2 \left(\dfrac{\omega}{\omega_1}\right)^{3-|2\epsilon\beta-1|}, & \omega_s < \omega < \omega_1, \\[4mm] \Omega_r(t_0) \left(\dfrac{H_1}{M_P}\right)^2 \left(\dfrac{\omega_s}{\omega_1}\right)^{3-|2\epsilon\beta-1|} \left(\dfrac{\omega}{\omega_s}\right)^{3-|\epsilon'\sqrt{3}-1|}, & \omega < \omega_s. \end{cases}
$$

$$(8A.13)$$

This result is valid for $\epsilon > 0$, since for $\epsilon = 0$ there is no amplification of the fluctuations during the string phase, when the internal dimensions are frozen. Comparing with the similar expression (7.205) for the graviton spectrum, and remembering that $\beta = 1 + \alpha$

for the given class of inflationary backgrounds, one can obtain (for any value of ϵ) a close correspondence between the spectral distribution of the produced photons and the produced gravitons (see [112] for the case $\epsilon = 1$, and [113] for the case $\epsilon = 1/2$).

Let us now impose the condition that the radiation produced contains a magnetic component strong enough to seed the cosmic magnetic fields B_G, of microgauss strength, currently observed on a galactic scale. We must require that the amplified e.m. fluctuations be coherent and large enough, over a proper length scale that today roughly corresponds to the megaparsec scale, as first pointed out in [98]. For a conservative estimate of the required field strength [98] we can then assume the existence of the standard galactic dynamo mechanism, operating since the epoch of structure formation, and characterized by an amplification factor $\sim 10^{13}$; also, we can take into account the additional amplification ($\sim 10^4$) due to magnetic flux conservation in the collapse of the galactic structure from the Mpc to the 10 kpc scale. We obtain, in this way, the lower bound $B_s/B_G \gtrsim 10^{-17}$, i.e. $B_s \gtrsim 10^{-23}$ gauss, on the present amplitude of the magnetic seeds at the Mpc scale; the identification of the seeds with the inflationary spectrum of e.m. fluctuations eventually leads to the condition [98]

$$\frac{B_s^2(\omega_G, t_0)}{B_G^2(t_0)} \simeq \frac{\rho_\gamma(\omega_G, t_0)}{\rho_r(t_0)} = \frac{\Omega_\gamma(\omega_G, t_0)}{\Omega_r(t_0)} \gtrsim 10^{-34}, \qquad \omega_G = (1\,\mathrm{Mpc})^{-1} \sim 10^{-14}\,\mathrm{Hz}.$$

$$(8A.14)$$

We have used the approximate equality of the present energy density associated with the galactic magnetic field ($B_G \sim 10^{-6}$ gauss) and the energy density of the CMB radiation.

This lower bound on Ω_γ has to be complemented by a competing upper bound, since the energy density stored in the amplified fluctuations cannot be too large – to avoid destroying the large-scale homogeneity of the cosmological background, and to be consistent with the linearized treatment of the fluctuations as small perturbations, with negligible backreaction. This imposes the stringent, model-independent constraint [105, 112]

$$\Omega_\gamma(\omega, t) \leq \Omega_r(t),$$
$$(8A.15)$$

to be satisfied at all times, for all frequency scales of the amplified spectrum. Remarkably enough, both conditions (8A.14) and (8A.15) can be satisfied, without fine-tuning, in a wide region of the parameter space spanned by the variables $\{\omega_1, \omega_s, \beta, H_1, \beta_0, \epsilon\}$.

In order to illustrate this possibility we can first express the parameter β in terms of the ratio between the value of the string coupling at the beginning and at the end of the high-curvature phase – following Eq. (7.196) – as $g_s/g_1 = (\eta_1/\eta_s)^\beta$; next, we can relate the break-point frequency ω_s to the time duration of the string phase, using the redshift factor $z_s = \eta_s/\eta_1 = \omega_1/\omega_s$. In this way we can express the photon spectrum (and the related constraints) through a new, but equivalent, set of parameters, $\{z_s, g_s, g_1, H_1, \beta_0, \epsilon\}$. These new variables are more convenient from a phenomenological point of view because, in the minimal string cosmology scenario that we are considering, the transition scale is controlled by the string mass scale, $H_1 \simeq M_s$, and the final value of the string coupling is also fixed [114], at a value quite close to the presently expected value $g_1 \simeq M_s/M_P$ (remember that the dilaton is assumed to be frozen during the subsequent standard evolution).

In the following discussion we keep g_1 and H_1 fixed at these typical values, assuming, in particular, $M_s = 0.1 M_P$, consistent with string models of unified gravitational and gauge interactions [115]. The number of free parameters thus reduces to four, with the two variables $\{z_s, g_s\}$ parametrizing the details of the background evolution, and the other two variables $\{\beta_0, \epsilon\}$ parametrizing the strength of the photon coupling to the inflationary background. We plot the allowed regions in the plane $\{z_s, g_s\}$ at different, constant values of β_0 and ϵ, chosen appropriately for the e.m. spectrum. For practical purposes we work with a decimal logarithmic scale, introducing the variables

$$x = \log z_s = \log \left(\frac{\omega_1}{\omega_s} \right) > 0,$$

$$y = \log \left(\frac{g_s}{g_1} \right) = -\beta \log z_s = -\beta x < 0. \tag{8A.16}$$

In terms of these variables we can rewrite the photon spectrum as

$$\frac{\Omega_\gamma}{\Omega_r}(\omega, t) = \begin{cases} g_1^2 \left(\dfrac{\omega}{\omega_1} \right)^{3-|1+2\epsilon y/x|}, & 10^{-x} < \dfrac{\omega}{\omega_1} < 1, \\[3mm] g_1^2 \left(\dfrac{\omega}{\omega_1} \right)^{3-|1-\epsilon'\sqrt{3}|} 10^{-x|1-\epsilon'\sqrt{3}|+|x+2\epsilon y|}, & \dfrac{\omega}{\omega_1} < 10^{-x}, \end{cases} \tag{8A.17}$$

where g_1 and ω_1 are both determined by M_s as

$$g_1 = \frac{M_s}{M_P}, \qquad \omega_1 = 4 \times g_1^{1/2} 10^{11} \text{Hz} \tag{8A.18}$$

(see Eq. (7.206)).

In order to impose the homogeneity condition (8A.15) let us first note that the low-frequency branch of the spectrum (8A.13), controlled by ϵ', is always growing for the values of β_0, β_i included in the range (8A.5), and for all ϵ varying from 0 to 1. The high-frequency (string) branch of the spectrum, on the contrary, is growing for $y > -(2x/\epsilon)$, and decreasing for $y < -(2x/\epsilon)$.

In the first case the peak of the spectrum is localized at $\omega = \omega_1$, and the homogeneity condition requires $g_1^2 < 1$ (which is always satisfied by our choice $g_1 = 0.1$). In the second case the peak is at $\omega = \omega_s = 10^{-x}\omega_1$, and the spectrum is marginally compatible with the homogeneity condition, which requires $y > -(2x/\epsilon) + \epsilon^{-1} \log g_1$. In that case, however, one should take into account also the slightly more stringent (but model-dependent) bound $\Omega_\gamma \lesssim 0.1\,\Omega_r$, – following from the presence of strong magnetic fields at the nucleosynthesis epoch [116] – which implies $y > -(2x/\epsilon) + \epsilon^{-1}(\log g_1 + 1)$. For $g_1 = 0.1$ such a condition pratically rules out a decreasing photon spectrum. We thus restrict our subsequent discussion to the case of a monotonically growing spectrum, satisfying the constraint

$$-\frac{2x}{\epsilon} < y < 0. \tag{8A.19}$$

We are left with the seed condition (8A.14), for which we must separately consider the two cases $\omega_G < \omega_s$ and $\omega_s < \omega_G < \omega_1$. We impose the seed condition on the spectrum (8A.17), plotting in the $\{x, y\}$ plane the allowed region determined by the intersection with Eq. (8A.19). We consider the particular cases of the heterotic ($\epsilon = 1$) and type I ($\epsilon = 1/2$) superstring models, and four different pairs of values of the parameters (β_0, β_i), to take into explicit account the effects of the internal dynamics on the low-energy branch of the spectrum. The resulting allowed regions, labeled by "photons", are reported in Fig. 8.5, where we have marked with a bold border the simplest case in which the extra dimensions are frozen, $\beta_0 = -1/\sqrt{3}$, $\beta_i = 0$, and with a dashed border the cases corresponding to the other pairs of values of (β_0, β_i). We have chosen, respectively, the pairs $(-1/3, 1/3)$,

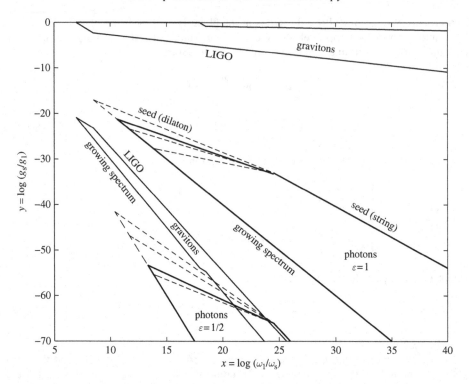

Figure 8.5 Allowed regions determined by the conditions (8A.14) and (8A.19) imposed on the photon spectrum (8A.17), and by the conditions (8A.20), (8A.21) and (8A.22) imposed on the graviton spectrum (7.205). We have fixed $H_1 = M_s = 0.1 M_P$, $g_1 = 0.1$ and $n = 6$. The photon regions are plotted for the two cases $\epsilon = 1$ and $\epsilon = 1/2$, and for appropriate values of β_0 and β_i, illustrating the maximal and minimal extension of the allowed region in the case of a non-trivial evolution of the moduli fields (dashed lines).

$(0, 1/\sqrt{6})$, $(1/3, 1/3)$ for $\epsilon = 1$, and $(-0.568, 0.072)$, $(-1/3, 1/3)$, $(0.063, 0.406)$ for $\epsilon = 1/2$, where the numerical values of β_0 and β_i have been selected so as to span the allowed region from its maximal to its minimal extension, compatible with the range of variations of the parameters β_0 and β_i given in Eq. (8A.5) (with $n = 6$).

Also plotted in Fig. 8.5, for comparison, are the regions of parameter space compatible with a detectable production of cosmic gravitons, in the same class of inflationary backgrounds. The graviton spectrum $\Omega_g(\omega, t)$ of Eq. (7.205), produced in the context of the "minimal" pre-big bang scenario, depends in fact only on the four parameters $\{\omega_s, \omega_1, \beta, H_1\}$, and with the choice $g_1 = M_s/M_P = 0.1$ it can be rewritten in terms of the two independent variables x and y, exactly like the photon spectrum (8A.17). As already stressed in Section 7.3, such a spectrum is monotonically growing even at high frequency, because of the condition $0 < \beta \leq 3$, which implies, in the $\{x, y\}$ plane,

$$-3x < y < 0. \tag{8A.20}$$

As a consequence, it is thus automatically consistent with the homogeneity and nucleosynthesis conditions imposed on the peak value of the spectrum. If we require, in addition, that the graviton background be strong enough to be detectable by Advanced LIGO,

$$h^2 \Omega_g(\omega_L, t_0) \gtrsim 10^{-11}, \qquad \omega_L = 10^2 \, \text{Hz} \qquad (8\text{A}.21)$$

(see Section 7.4), and weak enough to be compatible with the pulsar limit

$$h^2 \Omega_g(\omega_P, t_0) \lesssim 10^{-8}, \qquad \omega_P = 10^{-8} \, \text{Hz} \qquad (8\text{A}.22)$$

(see Section 7.3), we obtain the allowed regions marked by the thin lines, and labeled "gravitons", in Fig. 8.5. There are two regions because of the absolute value present in the slope of the spectrum (7.205): the upper region corresponds to $\alpha = \beta - 1 < 1/2$, the lower one to $\alpha = \beta - 1 > 1/2$.

Two comments are now in order. The first concerns the possibility of obtaining an efficient seed production, even in the context of the minimal scenario considered here, for both $\epsilon = 1$ and $\epsilon = 1/2$. In both cases, however, the efficiency of the mechanism requires a small enough value of the dilaton at the beginning of the string phase, and a long enough duration of the string phase [105]. Otherwise the amplified fluctuations are too weak to seed the galactic dynamo.

The second comment concerns the overlap of the graviton and photon regions [113]. As clearly illustrated in Fig. 8.5, efficient production of seeds for the cosmic magnetic fields is in principle compatible with the associated production of a cosmic graviton background detectable by Advanced LIGO if $\epsilon = 1/2$ (as for type I superstrings), and incompatible if $\epsilon = 1$ (as for heterotic superstrings). This may give us direct experimental information on the possible primordial strength of the photon–dilaton coupling, and on the choice of superstring model most appropriate to the early cosmological evolution.

Let us suppose, for instance, that further studies and observations of cosmic magnetic fields may give us direct independent confirmation of the expected seed production, exactly as predicted in a string cosmology context: a future detection of cosmic gravitons by the next generation of gravitational antennas will thus provide support in favour of type I models, while the absence of detection (at the same sensitivity level) should be interpreted as more in favor of the heterotic model of coupling. Such an argument cannot be applied, however, to cosmological models where photons are decoupled from the dilaton, since the amplification due to the internal moduli is in general too small to be effective (see however [100]).

It should be noted, also, that the extension of the low-energy (dilaton) part of the e.m. allowed regions is rather strongly dependent on the values of β_0 and β_i. A precise experimental determination of the spectrum of the e.m. fluctuations may thus, in principle, open a direct window on the primordial dynamics of the extra dimensions (see also [111] for previous studies on higher-dimensional modifications of the photon spectrum).

A final remark concerns the fact that the photon spectrum is in general flatter than the corresponding spectrum of metric perturbations, suggesting the possible role of e.m. fluctuations as seeds for an isocurvature component of the observed CMB anisotropy [83]. A more appropriate candidate for this purpose is, however, a massless axion field, as will be discussed in Appendix 8B.

Appendix 8B
Seeds for the CMB anisotropy

The metric fluctuations, sources of the CMB anisotropy through the SW effect (see Section 8.2), can be produced by inflation in two ways: either *directly*, through the parametric amplification of the quantum fluctuations of the metric tensor, or *indirectly*, through an auxiliary field whose fluctuations are amplified with an appropriate (nearly flat) spectrum. In this second case there are two possible mechanisms of indirect production: (*i*) the so-called "curvaton" mechanism (illustrated in Section 8.3), in which the background component $\langle \sigma \rangle$ of the auxiliary field is non-zero, becomes dominant and then decays, leaving a flat spectrum of *adiabatic* metric perturbations; (*ii*) the so-called "seed" mechanism [117], in which the background component of the auxiliary field is vanishing, $\langle \sigma \rangle = 0$, but its fluctuations contribute as quadratic sources to the metric perturbation equations, generating a final spectrum of *isocurvature* metric fluctuations.

In view of the constraints obtained from current observations, the seed mechanism may at most provide a sub-dominant contribution to the total CMB anisotropy, which seems to be largely dominated by the contributions of adiabatic metric perturbations [56, 57]. The seed mechanism, however, is interesting in itself, and typical of the string cosmology scenario where there are, in principle, many background fields in the ground state of the effective action. In the example presented in this appendix, the role of seeds is played by the axion fluctuations $\delta\sigma$, amplified with a flat spectrum (as in the curvaton scenario), but with a vanishing background even in the post-inflationary era.

In the curvaton mechanism, generation of the metric perturbations is triggered by the axion fluctuations present at the beginning of the radiation era, but the subsequent amplification of the Bardeen potential is mainly due to the presence of the axion background. As a consequence, the final Bardeen spectrum (outside the horizon, in the matter era), only depends on the initial post-inflationary spectrum of relativistic axion fluctuations. In the seed mechanism the amplification of the metric perturbations is continuously controlled by the evolution of the axion perturbations, and the final Bardeen spectrum at decoupling is determined by the final spectrum of axion fluctuations in the matter-dominated era.

The present discussion concentrates on the case of massless axion fluctuations [82, 83] (but the seed mechanism can be implemented also in the massive case [84], provided the seeds are light enough to survive down to the present epoch). In such a case there is no need for non-relativistic corrections to the axion spectrum, and the evolution of the axion fluctuations is always governed by Eq. (8.276); we must take into account, however, that for low-frequency modes ($k < k_{eq} = \eta_{eq}^{-1}$) there is a further amplification due to the background transition at η_{eq}, as discussed in Section 7.3 for the graviton case.

The final axion spectrum thus has two branches, and can be obtained from the corresponding spectrum of tensor perturbations, Eq. (7.178), simply by replacing the graviton spectral index $2 + 2\alpha = 3 - 2|\nu|$ with the axion spectral index $3 - 2|\mu| = n_\sigma - 1$, where the values of μ and n_σ are those given in Eqs. (8.240) and (8.246). For the high-frequency branch of the axion spectrum it is possible that $\mu \simeq 3/2$ (i.e. $n_\sigma \simeq 1$), with an appropriate model of background evolution (see Section 8.3). For the low-frequency sector, which is the one relevant to our discussion, the spectrum is given by

$$\Omega_\sigma(\omega, \eta) = g_1^2 \Omega_r(\eta) \left(\frac{\omega}{\omega_1}\right)^{3-2|\mu|} \left(\frac{\omega}{\omega_{eq}}\right)^{-2}, \qquad \omega < \omega_{eq}, \quad \eta_{eq} < \eta. \tag{8B.1}$$

We have defined $g_1 = H_1/M_P$, and we have used $\Omega_r = (H_1/H)^2 (a_1/a)^4$ to denote, as usual, the critical fraction of radiation energy density.

We now have to compute the Bardeen spectrum induced by these axionic seeds, and the corresponding (indirect) contribution to the CMB anisotropy. We start by assuming that there is no (appreciable) direct amplification of the metric perturbations during inflation, so that $\Phi = 0 = \Psi$ outside the horizon, at the beginning of the matter-dominated era. The amplification of super-horizon metric perturbations is fully "seeded" by the axion sources which generate $\Psi' \neq 0$, and thus contribute to the temperature anisotropy with isocurvature initial conditions. As a consequence, the large-scale amplitude of $\Delta T/T$ will be controlled by the spectrum of the Bardeen potential at horizon crossing, $\Psi_k(\eta = k^{-1})$, according to Eqs. (8.224) and (8.225).

A quick estimate of the Bardeen spectrum can be obtained by including the seed source into the general, gauge-invariant perturbation equations (8.62)–(8.69), and neglecting the adiabatic contributions of the perturbed matter sources. From the $(i \neq j)$, $(i0)$ and (00) components of the scalar perturbation equations one obtains, respectively, the conditions

$$\partial_i \partial^j (\Phi - \Psi) = \lambda_P^2 a^2 \tau_i^j, \tag{8B.2}$$

$$\partial_i (\Psi' + \mathcal{H}\Phi) = \frac{\lambda_P^2}{2} a^2 \tau_i^0, \tag{8B.3}$$

$$\nabla^2 \Psi - 3\mathcal{H}(\Psi' + \mathcal{H}\Phi) = \frac{\lambda_P^2}{2} a^2 \tau_0^0, \tag{8B.4}$$

where τ_μ^ν is the stress tensor associated with the axion fluctuations:

$$\tau_\mu{}^\nu = \partial_\mu(\delta\sigma)\partial^\nu(\delta\sigma) - \frac{1}{2}\delta_\mu^\nu \partial_\alpha(\delta\sigma)\partial^\alpha(\delta\sigma). \tag{8B.5}$$

Moving to momentum space, it is convenient to define the set of seed variables f_π, f_v, f_ρ such that

$$\begin{aligned} a^2 \tau_i^j &= k_i k^j f_\pi, \qquad i \neq j, \\ a^2 \tau_i^0 &= ik_i f_v, \qquad a^2 \tau_0^0 = f_\rho. \end{aligned} \tag{8B.6}$$

From Eq. (8B.2), and from the combination of Eqs. (8B.3) and (8B.4), we then obtain the relations

$$\Psi - \Phi = \lambda_P^2 f_\pi, \tag{8B.7}$$

$$-k^2 \Psi = \frac{\lambda_P^2}{2} (f_\rho + 3\mathcal{H} f_v) \tag{8B.8}$$

(we have omitted, for simplicity, the Fourier index k). The combination $\Psi + \Phi$, appearing in the ISW effect (see Eq. (8.186)), is then given by

$$\Psi + \Phi = -\frac{\lambda_P^2}{k^2} \left(k^2 f_\pi + f_\rho + 3\mathcal{H} f_v \right) . \tag{8B.9}$$

In the case of massless axions, the application of Eqs. (8B.5) and (8B.6) at horizon crossing ($k\eta = 1$) leads to

$$\left(k^2 f_\pi \right)_{\text{hc}} \sim \left(f_\rho \right)_{\text{hc}} \sim (kf_v)_{\text{hc}} \tag{8B.10}$$

(as confirmed by an explicit computation of $\tau_{\mu\nu}$ in terms of the exact solution for $\delta\sigma$ in the matter-dominated era [82, 83]). Thus,

$$\left| \Phi_k + \Psi_k \right|_{\text{hc}} \sim \frac{\lambda_P^2}{k^2} \left| f_\rho \right|_{\text{hc}} = \lambda_P^2 \left(\frac{a^2 \rho_\sigma}{k^2} \right)_{\text{hc}} , \tag{8B.11}$$

where $\rho_\sigma = \tau_0^0$. At horizon crossing, on the other hand, $(k/a)_{\text{hc}} = \omega_{\text{hc}} = H$, so that we can explicitly insert the critical density in the previous equation, since $(\lambda_P/\omega)_{\text{hc}}^2 \sim G/H^2 \sim \rho_c^{-1}$. One is then naturally led to the conclusion that the spectrum of the Bardeen potentials, at horizon crossing, is directly controlled by the spectral energy density of the axion fluctuations (in critical units) as follows:

$$k^{3/2} \left| \Phi_k + \Psi_k \right|_{\text{hc}} \sim (\Omega_\sigma)_{\text{hc}} . \tag{8B.12}$$

The precise relation between the Bardeen spectrum and the energy spectrum of the axion seeds is to be determined, of course, by computing the Fourier transform of the two-point correlation function of both sides of Eq. (8B.9). The two-point correlation of the left-hand side is given by

$$\xi_{\Phi+\Psi}(x, x') \equiv \langle (\Phi_x + \Psi_x)(\Phi_{x'} + \Psi_{x'}) \rangle = \int \frac{d^3 k}{(2\pi k)^3} e^{i\vec{k}\cdot(\vec{x}-\vec{x}')} \left| \delta_{\Psi+\Psi}(k) \right|^2 , \tag{8B.13}$$

and the usual interpretation of the correlation brackets as a spatial average (see Eqs. (7.84)–(7.86)) leads to the spectral amplitude $\left| \delta_{\Psi+\Psi} \right| = k^{3/2} \left| \Phi_k + \Psi_k \right|$. The correlation of the right-hand side requires instead the computation of the two-point function for the components of the seed stress tensor, $\langle \tau_\mu^\nu(x) \tau_\mu^\nu(x') \rangle$, which becomes a four-point function when explicitly expressed through the fluctuating field $\delta\sigma$.

Let us consider, for instance, the correlation associated with the energy density term, $f_\rho = a^2 \rho_\sigma$. The two-point function is

$$\xi_\rho(x, x') = \langle (\rho_\sigma(x) - \langle \rho_\sigma \rangle)(\rho_\sigma(x') - \langle \rho_\sigma \rangle) \rangle = \langle \rho_\sigma(x)\rho_\sigma(x') \rangle - (\langle \rho_\sigma \rangle)^2 , \tag{8B.14}$$

where, from Eq. (8B.5),

$$\rho_\sigma = \frac{1}{2a^2} \left(\delta\sigma'^2 + \delta\sigma_i^2 \right) \tag{8B.15}$$

(for simplicity, we denote spatial gradients with a spatial index, i.e. $\delta\sigma_i \equiv \partial_i(\delta\sigma)$). We have subtracted the term $\langle \rho_\sigma \rangle$, since we are correlating the fluctuations of a quadratic variable with non-zero average value $\langle \rho_\sigma \rangle \neq 0$ (unlike the linear fluctuations Ψ, $\delta\sigma$, which are represented by stochastic variables with zero expectation value). The various terms to be computed are then the following:

$$\Delta(x, x') = \langle \delta\sigma'^2(x)\delta\sigma'^2(x') \rangle - (\langle \delta\sigma'^2 \rangle)^2, \tag{8B.16}$$

$$\Delta_{ij}(x, x') = \langle \delta\sigma_i^2(x)\delta\sigma_j^2(x') \rangle - \langle \delta\sigma_i^2 \rangle\langle \delta\sigma_j^2 \rangle, \tag{8B.17}$$

$$\Delta_i(x, x') = \langle \delta\sigma'^2(x)\delta\sigma_i^2(x') \rangle - \langle \delta\sigma'^2 \rangle\langle \delta\sigma_i^2 \rangle. \tag{8B.18}$$

For the explicit computation it is convenient to observe that the spatial correlation of a generic variable $A(x)$,

$$\langle A(x)A(x+r) \rangle = \int d^3x \int \frac{d^3k}{(2\pi)^3} \frac{d^3k'}{(2\pi)^3} e^{i\vec{k}\cdot\vec{x}} e^{i\vec{k'}\cdot(\vec{x}+\vec{r})} A_k A_{k'}$$

$$= \int \frac{d^3k}{(2\pi)^3} d^3k'\, \delta^3(k+k')e^{i\vec{k'}\cdot\vec{r}} A_k A_{k'} = \int \frac{d^3k}{(2\pi)^3} e^{-i\vec{k}\cdot\vec{r}} |A_k|^2 \tag{8B.19}$$

(we have used the reality condition $A_{-k} = A_k^*$, and the definition (7.64) of Fourier transform), can also be obtained by replacing the spatial average procedure, $V^{-1}\int d^3x$, with an appropriate average prescription for the Fourier components:

$$\langle A_k A_{k'}^* \rangle = \frac{(2\pi)^3}{V} \delta^3(k-k')|A_k|^2, \tag{8B.20}$$

namely

$$\langle A(x)A(x+r) \rangle \equiv V \int \frac{d^3k}{(2\pi)^3} \frac{d^3k'}{(2\pi)^3} e^{i\vec{k}\cdot\vec{x}} e^{i\vec{k'}\cdot(\vec{x}+\vec{r})} \langle A_k A_{k'} \rangle. \tag{8B.21}$$

Thus, the spatial averages required by the computations of the correlation functions (8B.16)–(8B.18) can be automatically accounted for by imposing the so-called "stochastic" conditions on the fluctuation variables,

$$\langle \delta\sigma'_k \delta\sigma'^*_{k'} \rangle = \frac{(2\pi)^3}{V} \delta^3(k-k')|\delta\sigma'_k|^2, \tag{8B.22}$$

$$\langle \delta\sigma_i(k)\delta\sigma_j^*(k') \rangle = k_i k_j \frac{(2\pi)^3}{V} \delta^3(k-k')|\delta\sigma_k|^2, \tag{8B.23}$$

$$\langle \delta\sigma'(k)\delta\sigma_i^*(k') \rangle = -\langle \delta\sigma_i(k)\delta\sigma'^*(k') \rangle = -ik_i \frac{(2\pi)^3}{V} \delta^3(k-k')|\delta\sigma'_k \delta\sigma_k|. \tag{8B.24}$$

We present here an explicit example by performing a detailed computation of the correlation function $\Delta(x, x')$ of Eq. (8B.16) (see [83] for the other contributions to ρ_σ, and for the correlation of the other components of $\tau_{\mu\nu}$). Let us start with the mean value $\langle \delta\sigma'^2 \rangle$. Using Eq. (8B.22) we obtain

$$\langle \delta\sigma'^2 \rangle = V \int \frac{d^3k}{(2\pi)^3} \frac{d^3k'}{(2\pi)^3} \langle \delta\sigma'_k \delta\sigma'_{k'} \rangle e^{i(k+k')\cdot x} = \int \frac{d^3k}{(2\pi)^3} |\delta\sigma'_k|^2, \tag{8B.25}$$

where the integrand is a convolution of Fourier transforms, since

$$|\delta\sigma'_k|^2 = \sqrt{V} \int d^3x \int \frac{d^3p}{(2\pi)^3} \frac{d^3q}{(2\pi)^3} e^{-ik\cdot x} e^{i(p+q)\cdot x} \delta\sigma'_p \delta\sigma'_q$$

$$= \sqrt{V} \int \frac{d^3 p}{(2\pi)^3} \delta\sigma'_p \delta\sigma'_{k-p}. \tag{8B.26}$$

The four-point part of Eq. (8B.16), on the other hand, can be written as follows:

$$\langle \delta\sigma'^2(x) \delta\sigma'^2(x') \rangle = V \int \frac{d^3 k}{(2\pi)^3} \frac{d^3 k'}{(2\pi)^3} \langle \delta\sigma'^2_k \delta\sigma'^2_{k'} \rangle e^{i(k\cdot x + k'\cdot x')}$$

$$= V^2 \int \frac{d^3 k}{(2\pi)^3} \frac{d^3 k'}{(2\pi)^3} \frac{d^3 p}{(2\pi)^3} \frac{d^3 q}{(2\pi)^3}$$

$$\langle \delta\sigma'_p \delta\sigma'_{k-p} \delta\sigma'_q \delta\sigma'_{k'-q} \rangle e^{i(k\cdot x + k'\cdot x')}. \tag{8B.27}$$

Decomposing the four-point bracket as

$$\langle \delta\sigma'_p \delta\sigma'_{k-p} \rangle \langle \delta\sigma'_q \delta\sigma'_{k'-q} \rangle + \langle \delta\sigma'_p \delta\sigma'_q \rangle \langle \delta\sigma'_{k-p} \delta\sigma'_{k'-q} \rangle + \langle \delta\sigma'_p \delta\sigma'_{k'-q} \rangle \langle \delta\sigma'_{k-p} \delta\sigma'_q \rangle, \tag{8B.28}$$

and using the stochastic condition (8B.22), one finds that the first term exactly reproduces (and cancels) the quadratic average $(\langle \delta\sigma'^2 \rangle)^2$, while the contribution of the other two terms is identical, so that

$$\Delta(x, x') = 2 \int \frac{d^3 k}{(2\pi)^3} e^{ik\cdot(x-x')} \int \frac{d^3 p}{(2\pi)^3} |\delta\sigma'_p|^2 |\delta\sigma'_{k-p}|^2. \tag{8B.29}$$

With similar computations we can also express the other correlation functions as convolutions of the Fourier transforms $\delta\sigma_k$ and $\delta\sigma'_k$:

$$\Delta_{ij}(x, x') = 2 \int \frac{d^3 k}{(2\pi)^3} e^{ik\cdot(x-x')} \int \frac{d^3 p}{(2\pi)^3} p_i p_j |\delta\sigma_p|^2 (k_i - p_i)(k_j - p_j) |\delta\sigma_{k-p}|^2,$$

$$\Delta_i(x, x') = -2 \int \frac{d^3 k}{(2\pi)^3} e^{ik\cdot(x-x')} \int \frac{d^3 p}{(2\pi)^3} p_i(k_i - p_i) |\delta\sigma_p \delta\sigma'_p| |\delta\sigma'_{k-p} \delta\sigma_{k-p}| \tag{8B.30}$$

(no sum over i, j). Summing all contributions one then obtains, for the correlation of ρ_σ,

$$\xi_\rho(x, x') = \frac{1}{a^4} \xi_{f_\rho}(x, x') = \frac{1}{4a^4} \left[\Delta(x, x') + 2 \sum_i \Delta_i(x, x') + \sum_{ij} \Delta_{ij}(x, x') \right]$$

$$= \frac{1}{2a^4} \int \frac{d^3 k}{(2\pi k)^3} e^{ik\cdot(x-x')} k^3 \int \frac{d^3 p}{(2\pi)^3} F_\rho(k, p), \tag{8B.31}$$

where

$$F_\rho(k, p) = |\delta\sigma'_p|^2 |\delta\sigma'_{k-p}|^2 + |\vec{p} \cdot (\vec{k} - \vec{p})|^2 |\delta\sigma_p|^2 |\delta\sigma_{k-p}|^2$$

$$- 2\vec{p} \cdot (\vec{k} - \vec{p}) |\delta\sigma_p \delta\sigma'_p| |\delta\sigma_{k-p} \delta\sigma'_{k-p}|. \tag{8B.32}$$

The comparison of this result with the left-hand side correlation (8B.13) immediately gives the contribution of f_ρ to the Bardeen spectrum:

$$k^3 |\Phi_k + \Psi_k|^2 = \frac{\lambda_P^4}{k^4} \frac{k^3}{2} \int \frac{d^3 p}{(2\pi)^3} F_\rho(k, p). \tag{8B.33}$$

To complete our computation we now need the explicit solution for the axion fluctuations $\delta\sigma_k(\eta)$ in the matter-dominated era. We need, in particular, the whole spectrum (not only the super-horizon sector $k\eta < 1$, and not only the low-frequency branch $k < k_{eq}$), since the convolutions are performed by integrating over all modes, from zero to the high-frequency cut-off k_1.

We solve the canonical equation (8.276) by assuming the simplest model of background evolution, characterized by a first transition from the inflation phase to the radiation phase at $\eta = -\eta_1$, and by a second transition from the radiation phase to the matter phase at $\eta = \eta_{eq}$ (the same background as used in Section 7.3 to discuss graviton production). The axion pump field (8.238) can be parametrized as

$$z^1 = \frac{1}{\sqrt{2}}\left(-\frac{\eta}{\eta_1}\right)^\alpha, \qquad\qquad \eta \le -\eta_1,$$

$$z^2 = \frac{1}{\sqrt{2}}\left(\frac{\eta+2\eta_1}{\eta_1}\right), \qquad -\eta_1 \le \eta \le \eta_{eq}, \qquad (8B.34)$$

$$z^3 = \frac{1}{\sqrt{2}}\left(\frac{\eta+2\eta_1}{\eta_1}\right)^2\left(\frac{\eta_{eq}+2\eta_1}{\eta_1}\right)^{-1}, \qquad \eta_{eq} \le \eta.$$

The exact solution for the axion perturbations, normalized to an initial spectrum of quantum fluctuations of the vacuum, can then be obtained directly from the corresponding solution for tensor perturbations, Eqs. (7.167) and (7.168), with the obvious replacement $\alpha_1 = \alpha$, $\alpha_2 = 1$, $\alpha_3 = 2$, namely $\nu_1 = \mu = 1/2 - \alpha$, $\nu_2 = -1/2$, $\nu_3 = -3/2$:

$$\delta\sigma_k^1(\eta) = e^{i\theta}\left(\frac{\pi\eta_1}{2}\right)^{1/2}\left(\frac{x}{x_1}\right)^\mu H_\mu^{(2)}(x), \qquad\qquad \eta \le -\eta_1,$$

$$\delta\sigma_k^2(\eta) = \left(\frac{\pi\eta_1}{2}\right)^{1/2}\left(\frac{y}{x_1}\right)^{-1/2}\left[c_+^2 H_{-1/2}^{(2)}(y) + c_-^2 H_{-1/2}^{(1)}(y)\right], \qquad -\eta_1 \le \eta \le \eta_{eq},$$

$$\delta\sigma_k^3(\eta) = \left(\frac{\pi\eta_1}{2}\right)^{1/2}\left(\frac{y}{x_1}\right)^{-3/2}\left(\frac{y_{eq}}{x_1}\right)\left[c_+^3 H_{-3/2}^{(2)}(y) + c_-^3 H_{-3/2}^{(1)}(y)\right], \qquad \eta_{eq} \le \eta.$$

$$(8B.35)$$

We recall that $x = k\eta$, $y = x + 2x_1$, $x_1 = k\eta_1$, $y_{eq} = x_{eq} + 2x_1$, and that the Bogoliubov coefficients $c_\pm^2$, $c_\pm^3$ are to be determined by matching the solutions at $\eta = -\eta_1$ and $\eta = \eta_{eq}$.

We can separately evaluate the two branches of the axion spectrum, starting from the high-frequency modes $k \gg k_{eq} = \eta_{eq}^{-1}$ which are only affected by the first background transition. For such modes $c_\pm^2 = c_\pm^3$, and for $c_\pm^2$ we can apply the general result (7.169). Using the explicit form (7.171) of $H_{\pm 1/2}^{(1,2)}$, and the small argument limit $x_1 \ll 1$, we obtain the relation

$$c_+^2(k)e^{-ix_1} = -c_-^2(k)e^{ix_1} = \beta_\mu x_1^{-|\mu|-1/2}, \qquad (8B.36)$$

where β_μ is a complex number (with modulus of order one) determined by the coefficients of the Hankel expansion (see also Eq. (8.278)). Inserting the coefficients $c_\pm^2 = c_\pm^3$ into the general solution for $\delta\sigma_k^3$, and using for $H_{-3/2}^{(1,2)}(y)$ the large argument approximation of Eq. (7.137), we eventually obtain the high-frequency fluctuations in the form

$$\delta\sigma_k(\eta) \simeq \frac{c(k)}{a\sqrt{k}}\sin k\eta, \qquad \eta_{\mathrm{eq}} < \eta, \qquad k_{\mathrm{eq}} < k < k_1, \tag{8B.37}$$

where

$$|c(k)| \sim (k/k_1)^{-|\mu|-1/2}, \tag{8B.38}$$

modulo a numerical factor of order one. We have used the scale factor of the matter-dominated era, normalized as in Eq. (8B.34).

For the low-frequency solutions, $k < k_{\mathrm{eq}}$, we must take into account also the second background transitions, so that the coefficients $c_{\pm}^3$ are determined by the matching of $\delta\sigma_k^2$ and $\delta\sigma_k^3$ at $\eta = \eta_{\mathrm{eq}}$. The result of such a computation, for a generic background kinematics, has already been presented in Eq. (7.170). The application to our particular background leads to the exact result

$$c_-^3 = -\mathrm{i}\left(\frac{\pi y_{\mathrm{eq}}}{8}\right)^{1/2}\left[-\mathrm{i}H_{-3/2}^{(2)}(y_{\mathrm{eq}})\left(c_+^2 e^{-\mathrm{i}y_{\mathrm{eq}}} - c_-^2 e^{\mathrm{i}y_{\mathrm{eq}}}\right)\right.$$

$$\left. + \left(c_+^2 e^{-\mathrm{i}y_{\mathrm{eq}}} + c_-^2 e^{\mathrm{i}y_{\mathrm{eq}}}\right)\left(\sqrt{\frac{2}{\pi y_{\mathrm{eq}}}}e^{-\mathrm{i}y_{\mathrm{eq}}} + \frac{2}{y_{\mathrm{eq}}}H_{-3/2}^{(2)}(y_{\mathrm{eq}})\right)\right],$$

$$\tag{8B.39}$$

$$c_+^3 = -\mathrm{i}\left(\frac{\pi y_{\mathrm{eq}}}{8}\right)^{1/2}\left[\mathrm{i}H_{-3/2}^{(1)}(y_{\mathrm{eq}})\left(c_+^2 e^{-\mathrm{i}y_{\mathrm{eq}}} - c_-^2 e^{\mathrm{i}y_{\mathrm{eq}}}\right)\right.$$

$$\left. - \left(c_+^2 e^{-\mathrm{i}y_{\mathrm{eq}}} + c_-^2 e^{\mathrm{i}y_{\mathrm{eq}}}\right)\left(\sqrt{\frac{2}{\pi y_{\mathrm{eq}}}}e^{\mathrm{i}y_{\mathrm{eq}}} + \frac{2}{y_{\mathrm{eq}}}H_{-3/2}^{(1)}(y_{\mathrm{eq}})\right)\right].$$

Let us now use the relation (8B.36), and keep only the leading terms for $x_1 \ll x_{\mathrm{eq}} \ll 1$. Taking into account the explicit coefficients (7.150) of the small argument expansion of $H_{-3/2}$ (in particular, using the relation $p_{-3/2} = p_{-3/2}^*$), we are led to

$$c_-^3(k) = -c_+^3(k) = \gamma_\mu x_1^{-|\mu|-1/2} x_{\mathrm{eq}}^{-1}, \tag{8B.40}$$

where γ_μ is a numerical factor determined by the Hankel expansion.

By inserting these coefficients into Eq. (8B.35) we can finally distinguish, in the low-frequency branch $k\eta_{\mathrm{eq}} < 1$ of the solution, the modes which are already inside the horizon ($k\eta > 1$), from the modes which are still outside ($k\eta < 1$). In the first case the large argument limit of $H_{-3/2}^{(1,2)}$ leads to

$$\delta\sigma_k(\eta) \simeq \frac{c(k)}{a\sqrt{k}}\left(\frac{k}{k_{\mathrm{eq}}}\right)^{-1}\sin k\eta, \qquad \eta_{\mathrm{eq}} < \eta, \quad k < k_{\mathrm{eq}}, \quad k\eta > 1. \tag{8B.41}$$

In the second case, the small argument limit of $H_{-3/2}^{(1,2)}$ leads to

$$\delta\sigma_k(\eta) \simeq \frac{c(k)}{a\sqrt{k}}\left(\frac{k}{k_{\mathrm{eq}}}\right)^{-1}(k\eta)^2, \qquad \eta_{\mathrm{eq}} < \eta, \quad k < k_{\mathrm{eq}}, \quad k\eta < 1. \tag{8B.42}$$

The three branches (8B.37), (8B.41) and (8B.42) completely specify the axion spectrum, and can be used to compute the convolution (8B.31), and its contribution to the Bardeen potential (8B.33).

Following the same procedure for the other axion contributions appearing on the right-hand side of Eq. (8B.9), and evaluating the final spectrum at horizon crossing, one finally obtains the result anticipated by Eq. (8B.12) (see [82, 83] for a detailed computation). The axion energy spectrum, on the other hand, is given in Eq. (8B.1) where, in the matter-dominated era, $\Omega_r(\eta) = (a_{eq}/a) = (\eta_{eq}/\eta)$. At horizon crossing we have $\Omega_r(\text{hc}) = (k\eta_{eq})^2 = (\omega/\omega_{eq})^2$, so that the resulting (axion-induced) Bardeen spectrum can be written as

$$k^3 |\Phi_k + \Psi_k|^2_{\text{hc}} \sim (\Omega_\sigma)^2_{\text{hc}} = g_1^4 \left(\frac{\omega}{\omega_1}\right)^{6-4|\mu|}. \tag{8B.43}$$

The corresponding isocurvature contribution to the angular power spectrum, according to Eqs. (8.224) and (8.225), is then

$$
\begin{aligned}
C_\ell^{\text{iso}} &\simeq \frac{2}{\pi} \int \frac{dk}{k} k^3 |\Phi_k + \Psi_k|^2_{\text{hc}} j_\ell^2(k\eta_0) \\
&\simeq \frac{2}{\pi} g_1^4 \left(\frac{\omega_0}{\omega_1}\right)^{n_s-1} \int dx_0 \, x_0^{n_s-2} j_\ell^2(x_0) \\
&\simeq g_1^4 \left(\frac{\omega_0}{\omega_1}\right)^{n_s-1} \frac{\Gamma(3-n_s)\Gamma(\ell+n_s/2-1/2)}{2^{3-n_s}\Gamma^2(2-n_s/2)\Gamma(\ell+5/2-n_s/2)},
\end{aligned} \tag{8B.44}
$$

where we have set $x_0 = k\eta_0 = \omega/\omega_0$, $n_s - 1 = 6 - 4|\mu|$, and we have used the result (8.217).

At very large scales, the COBE normalization (8.219) imposes the important constraint

$$C_2 = 8\alpha_{n_s}^2 \left(\frac{H_1}{M_P}\right)^4 \left(\frac{\omega_0}{\omega_1}\right)^{n_s-1} \simeq 10^{-10}, \tag{8B.45}$$

where α_{n_s} is the same n-dependent coefficient of Eq. (8.291) appearing in the context of the curvaton mechanism (note, however, that in the curvaton case C_2 is proportional to the second power of the ratio H_1/M_P). Given the very large value of the ratio $\omega_1/\omega_0 \sim (H_1/M_P)^{1/2}10^{29}$ (see Eqs. (7.180), (7.181)), it is clear that a very small, positive tilt $n_s > 1$ of the relativistic axion spectrum may be enough to make compatible the previous constraint even with minimal models of string cosmology inflation [82, 83, 92], where $H_1 \sim M_s \sim (10^{-1} - 10^{-2})M_P$.

Unfortunately, however, the seed mechanism has difficulties in reproducing, by itself, the observed peak structure of the CMB anisotropy at smaller angular scales, even taking into account non-minimal models with a break in the axion spectrum [119], and even adjusting other cosmological parameters [120]. Such a mechanism may thus contribute, at most, to a small fraction of the total observed temperature anisotropy. Nevertheless, it may be important to stress that the metric fluctuations Φ_k induced by the seeds are quadratic functions of a Gaussian stochastic variable $\delta\sigma_k$, and are thus intrinsically non-Gaussian. The seed mechanism may then represent a possible interesting source of non-Gaussianity, to be confronted with present [91] and future analyses of the statistical properties of the anisotropy spectrum.

References

[1] G. F. Smooth *et al.*, *Ap. J.* **396** (1992) 1.
[2] C. L. Bennet *et al.*, *Ap. J.* **430** (1994) 423.
[3] P. de Bernardis *et al.*, *Nature* **404** (2000) 955.
[4] D. N. Spergel *et al.*, *Ap. J. Suppl.* **148** (2003) 175.
[5] D. N. Spergel *et al.*, astro-ph/0603449.
[6] R. Crittenden *et al.*, *Phys. Rev. Lett.* **71** (1993) 324.
[7] S. Kachru, R. Kallosh, A. Linde *et al.*, *JCAP* **0310** (2003) 013.
[8] S. Kachru, R. Kallosh, A. Linde and S. P. Trivedi, *Phys. Rev.* **D68** (2003) 046005.
[9] A. Nayeri, R. H. Brandenberger and C. Vafa, *Phys. Rev. Lett.* **97** (2006) 021302.
[10] N. Kaloper, L. Kofman, A. Linde and V. Mukhanov, *JCAP* **0610** (2006) 006.
[11] R. Brandenberger, A. Nayeri, S. P. Patil and C. Vafa, hep-th/0608121.
[12] J. Khouri, B. A. Ovrut, P. J. Steinhardt and N. Turok, *Phys. Rev.* **D66** (2002) 046005.
[13] R. Brandenberger and F. Finelli, *JHEP* **11** (2001) 056.
[14] D. H. Lyth, *Phys. Lett.* **B526** (2002) 173.
[15] R. Durrer and F. Vernizzi, *Phys. Rev.* **D66** (2002) 083503.
[16] M. Gasperini and G. Veneziano, *Phys. Rev.* **D50** (1994) 2519.
[17] R. Brustein, M. Gasperini, M. Giovannini, V. Mukhanov and G. Veneziano, *Phys. Rev.* **D51** (1995) 6744.
[18] E. J. Copeland, R. Easther and D. Wands, *Phys. Rev.* **D56** (1997) 874.
[19] E. J. Copeland, J. E. Lidsey and D. Wands, *Nucl. Phys.* **B506** (1997) 407.
[20] K. Enqvist and M. Sloth, *Nucl. Phys.* **B626** (2002) 395.
[21] D. H. Lyth and D. Wands, *Phys. Lett.* **B524** (2002) 5.
[22] T. Moroi and T. Takahashi, *Phys. Lett.* **B522** (2001) 215.
[23] V. Bozza, M. Gasperini, M. Giovannini and G. Veneziano, *Phys. Lett.* **B543** (2002) 14.
[24] V. Bozza, M. Gasperini, M. Giovannini and G. Veneziano, *Phys. Rev.* **D67** (2003) 063514.
[25] D. Lyth, C. Ungarelli and D. Wands, *Phys. Rev.* **D67** (2003) 023503.
[26] N. Bartolo, S. Matarrese and A. Riotto, *Phys. Rev.* **D69** (2004) 043503.
[27] V. F. Mukhanov, H. A. Feldman and R. H. Brandenberger, *Phys. Rep.* **215** (1992) 203.
[28] A. R. Liddle and D. H. Lyth, *Cosmological Inflation and Large Scale Structure* (Cambridge: Cambridge University Press, 2000).
[29] R. B. Abbot, B. Bednarz and S. D. Ellis, *Phys. Rev.* **D33** (1986) 2147.
[30] M. Gasperini and M. Giovannini, *Class. Quantum Grav.* **14** (1997) 735.

[31] L. Randall and R. Sundrum, *Phys. Rev. Lett.* **83** (1999) 4690.

[32] C. van de Bruck, M. Dorca, R. H. Brandenberger and A. Lukas, *Phys. Rev.* **D62** (2000) 123515.

[33] M. Giovannini, *Phys. Rev.* **D70** (2004) 103509.

[34] M. Giovannini, *Int. J. Mod. Phys.* **D14** (2005) 363.

[35] J. Bardeen, *Phys. Rev.* **D22** (1980) 1822.

[36] D. H. Lyth, *Phys. Rev.* **D31** (1985) 1792.

[37] R. Brandenberger, R. Kahn and W. Press, *Phys. Rev.* **D28** (1983) 1809.

[38] V. F. Mukhanov, *Sov. Phys. JETP* **67** (1988) 1297.

[39] V. N. Lukash, *Sov. Phys. JETP* **52** (1980) 807.

[40] G. V. Chibisov and V. N. Mukhanov, *Mon. Not. R. Astron. Soc.* **200** (1982) 535.

[41] A. D. Liddle and D. Lyth, *Phys. Lett.* **B291** (1992) 391.

[42] A. D. Linde, *Phys. Lett.* **B129** (1983) 177.

[43] N. Deruelle and V. F. Mukhanov, *Phys. Rev.* **D52** (1995) 5549.

[44] R. Durrer and F. Vernizzi, *Phys. Rev.* **D66** (2002) 083503.

[45] M. Gasperini and G. Veneziano, *Mod. Phys. Lett.* **A8** (1993) 3701.

[46] S. Gratton, J. Khoury, P. J. Steinhardt and N. Turok, *Phys. Rev.* **D69** (2004) 103505.

[47] C. Cartier, R. Durrer and E. J. Copeland, *Phys. Rev.* **D67** (2003) 103517.

[48] M. Gasperini, M. Giovannini and G. Veneziano, *Phys. Lett.* **B569** (2003) 113.

[49] M. Gasperini, M. Giovannini and G. Veneziano, *Nucl. Phys.* **B694** (2004) 206.

[50] L. E. Allen and D. Wands, *Phys. Rev.* **D70** (2004) 063515.

[51] S. Weinberg, *Gravitation and Cosmology* (New York: Wiley, 1971).

[52] E. W. Kolb and M. S. Turner, *The Early Universe* (Redwood City, CA: Addison-Wesley, 1990).

[53] R. K. Sachs and M. Wolfe, *Ap. J.* **147** (1967) 73.

[54] Particle Data Group webpage, at www.pdg.lbl.gov.

[55] R. Durrer, *J. Phys. Stud.* **5** (2001) 177.

[56] K. Moodley, M. Bucher, J. Dunkley, P. J. Ferreira and C. Skordis, *Phys. Rev.* **D70** (2004) 103520.

[57] K. Moodley, M. Bucher, J. Dunkley, P. J. Ferreira and C. Skordis, *Phys. Rev. Lett.* **93** (2004) 081301.

[58] L. P. Grishchuk, *Phys. Rev. Lett.* **70** (1993) 2371.

[59] L. P. Grishchuk, *Phys. Rev.* **D48** (1993) 3513.

[60] A. Messiah, *Quantum Mechanics* (Amsterdam: North Holland, 1972).

[61] I. S. Gradshteyn and I. M. Ryzhik, *Tables of Integrals, Series and Products* (New York: Academic Press, 1965).

[62] E. R. Harrison, *Phys. Rev.* **D1** (1970) 2726.

[63] Y. B. Zel'dovich, *Mon. Not. R. Astron. Soc.* **160** (1972) 1.

[64] A. J. Banday *et al.*, *Ap. J.* **475** (1997) 393.

[65] W. Hu and N. Sugiyama, *Ap. J.* **444** (1995) 489.

[66] J. Silk, *Ap. J.* **151** (1968) 459.

[67] W. Hu, *Ann. Phys.* **303** (2003) 203.

[68] T. Padmanabhan, *Theoretical Astrophysics*, Vol. 3 (Cambridge: Cambridge University Press, 2002).

[69] U. Seljak and M. Zaldarriaga, *Ap. J.* **469** (1996) 437.

[70] C. B. Netterfield *et al.*, *Ap. J.* **571** (2002) 604.

[71] A. T. Lee *et al.*, *Ap. J.* **561** (2001) L1.

[72] N. W. Alverson *et al.*, *Ap. J.* **568** (2002) 38.

[73] A. C. Taylor *et al.*, *Mon. Not. R. Astron. Soc.* **341** (2003) 1066.

[74] J. L. Sievers *et al.*, *Ap. J.* **591** (2003) 599.

[75] C. L. Kuo *et al.*, *Ap. J* **600** (2004) 32.

[76] M. Tegmark *et al.*, *Phys. Rev.* **D69** (2004) 103501.

[77] L. F. Abbot and M. B. Wise, *Nucl. Phys.* **B244** (1984) 541.

[78] U. Seljac and M. Zaldarriaga, *Phys. Rev. Lett.* **78** (1997) 2054.

[79] M. Kamionkowski, A. Kosowsky and A. Stebbins, *Phys. Rev. Lett.* **78** (1997) 2058.

[80] S. Mollerach, *Phys. Rev.* **D42** (1990) 313.

[81] A. Melchiorri, F. Vernizzi, R. Durrer and G. Veneziano, *Phys. Rev. Lett.* **83** (1999) 4464.

[82] R. Durrer, M. Gasperini, M. Sakellariadou and G. Veneziano, *Phys. Lett.* **B436** (1998) 66.

[83] R. Durrer, M. Gasperini, M. Sakellariadou and G. Veneziano, *Phys. Rev.* **D59** (1999) 43511.

[84] M. Gasperini and G. Veneziano, *Phys. Rev.* **D59** (1999) 43503.

[85] K. Kim, *Phys. Rev. Lett.* **43** (1979) 103.

[86] M. Dine, W. Fischler and M. Srednicki, *Phys. Lett.* **B104** (1981) 199.

[87] J. Preskill, M. Wise and F. Wilczek, *Phys. Lett.* **B120** (1983) 127.

[88] P. Di Vecchia and G. Veneziano, *Nucl. Phys.* **B171** (1980) 253.

[89] J. Ellis, D. V. Nanopoulos and M. Quiros, *Phys. Lett.* **B174** (1986) 176.

[90] J. Ellis, C. Tsamish and M. Voloshin, *Phys. Lett.* **B194** (1987) 291.

[91] E. Komatsu *et al.*, *Ap. J. Suppl.* **148** (2003) 119.

[92] M. Gasperini, in *Proc. Fifth Paris Cosmology Colloquium*, eds. H. J. de Vega and N. Sanchez (Paris: Publications Observatoire de Paris, 1999), p. 317.

[93] G. Veneziano, *JHEP* **0206** (2002) 051.

[94] D. Grasso and H. R. Rubinstein, *Phys. Rep.* **348** (2001) 163.

[95] M. Giovannini, *Int. J. Mod. Phys.* **D13** (2004) 391.

[96] E. N. Parker, *Cosmical Magnetic Fields* (Oxford: Clarendon Press, 1979).

[97] Y. B. Zeldovich, A. A. Ruzmaikin and D. D. Sokoloff, *Magnetic Fields in Astrophysics* (New York: Gordon and Breach, 1983).

[98] M. S. Turner and L. M. Widrow, *Phys. Rev.* **D37** (1988) 2473.

[99] K. H. Lotze, *Class. Quantum Grav.* **7** (1990) 2145.

[100] M. Giovannini, *Phys. Rev.* **D62** (2000) 123505.

[101] A. Dolgov, *Phys. Rev.* **D48** (1993) 2499.

[102] A. Ashoorion and R. B. Mann, *Phys. Rev.* **D71** (2005) 103509.

[103] B. Ratra, *Ap. J. Lett.* **391** (1992) L1.

[104] W. D. Garretson, G. B. Field and S. M. Carroll, *Phys. Rev.* **D57** (1992) 5346.

[105] M. Gasperini, M. Giovannini and G. Veneziano, *Phys. Rev. Lett.* **75** (1995) 3796.

[106] D. Lemoine and M. Lemoine, *Phys. Rev.* **D52** (1995) 1955.

[107] M. Giovannini and M. Shaposhnikov, *Phys. Rev.* **D62** (2000) 103512.

[108] A. Davis, K. Dimopoulos and T. Prokopec, *Phys. Lett.* **B501** (2001) 165.

[109] O. Bertolami and D. F. Mota, *Phys. Lett.* **B455** (1999) 96.

[110] M. Gasperini, *Phys. Rev.* **D63** (2001) 047301.

[111] A. Buonanno, K. A. Meissner, C. Ungarelli and G. Veneziano, *JHEP* **9801** (1998) 004.

[112] M. Gasperini, in *String Gravity and Physics at the Planck Energy Scale*, eds. N. Sanchez and A. Zichichi (Dordrecht: Kluwer Academic Publishers, 1996), p. 305.

[113] M. Gasperini and S. Nicotri, *Phys. Lett.* **B633** (2006) 155.

[114] R. Brustein, M. Gasperini and G. Veneziano, *Phys. Rev.* **D55** (1997) 3882.

[115] V. Kaplunowski, *Phys. Rev. Lett.* **55** (1985) 11036.

[116] D. Grasso and H. R. Rubinstein, *Astropart. Phys.* **3** (1995) 95.
[117] R. Durrer, *Phys. Rev.* **D42** (1990) 2533.
[118] R. Durrer, *Fund. Cosm. Phys.* **15** (1994) 209.
[119] A. Melchiorri, R. Durrer and G. Veneziano, *Phys. Rev. Lett.* **83** (1999) 4464.
[120] F. Vernizzi, A. Melchiorri and R. Durrer, *Phys. Rev.* **D63** (2001) 063501.

9

Dilaton phenomenology

The dilaton field is an essential component of all superstring models, and thus of the cosmological scenarios based on the string effective action. In particular, as shown in the previous chapters, the dilaton may control the inflationary dynamics and play a fundamental role in the generation of the primordial spectra of quantum fluctuations amplified by inflation. Also, it is the dilaton which controls the intensity of the various coupling strengths, and which may drive the Universe towards a phase of strong coupling possibly preceding the standard decelerated evolution. Thus, we can say that the dilaton (together with the other moduli fields, associated with the dynamics of the extra dimensions) is one of the most typical ingredients of models of string cosmology inflation, and is the basis of the main differences between the string models and the standard inflationary models based on the general relativistic equations.

In the post-inflationary epochs we may expect that the dilaton, like the other moduli fields, tends to approach a stabilized configuration either under the action of an appropriate potential (attracting it to a local minimum), or simply as a consequence of the standard, radiation-dominated dynamics: the low-energy string cosmology equations admit in fact asymptotic, radiation-dominated solutions at constant dilaton (see Eq. (4.58), and the discussion of Section 6.2). The dilaton, at this point, would seem to disappear from the cosmological scene (or, at least, to reject the primary role played at earlier epochs). In this chapter it will be shown, instead, that the phase of standard cosmological evolution (and, in particular, our present epoch) might also be characterized by an interesting dilaton phenomenology. We will discuss, in particular, two possible effects.

The first effect concerns the amplification of the quantum fluctuations of the dilaton background – in other words, the quantum production of dilaton pairs from the vacuum – under the action of the inflationary Universe. Depending on the mass acquired by the dilaton after inflation, the cosmic dilaton background produced in this way could survive until the present epoch, and could be accessible

428

to direct detection by gravitational antennas of appropriate sensitivity, as will be discussed in Sections 9.1 and 9.2.

The second effect concerns the dilaton background itself, quite irrespective of its possible fluctuations. The freezing (and/or the sub-dominant contribution to the cosmological dynamics) of the dilaton during the radiation era might correspond only to a temporary "hibernation" of this field: the dilaton could "wake up" after the equilibrium epoch, possibly driving our Universe towards a phase of late-time accelerated expansion. Various aspects of such a dilatonic dark energy scenario, including the possible strong coupling to some exotic dark matter component, and the corresponding impact on the cosmic coincidence problem, will be discussed in Section 9.3.

9.1 Spectral intensity of a massive dilaton background

As already stressed in the previous chapters, the low-energy string effective action contains (even to lowest order) at least two fundamental bosonic fields: a tensor field, the metric $g_{\mu\nu}$, and a scalar field, the dilaton ϕ. The dilaton controls the effective strength of all gauge couplings [1] in the context of "grand-unified" models of all fundamental interactions, and can also be geometrically interpreted as the effective "radius" of the eleventh dimension [2] in the M-theory context (see Appendix 3B). Quite independently of its possible interpretation, the dilaton is present in all (super)string models as the scalar partner of the graviton associated with the conformal invariance of the world-sheet action (see Chapter 3 for a detailed discussion).

During inflation, the accelerated evolution of the cosmological background induces a parametric amplification of the transverse and traceless tensor part $h_{\mu\nu} = \delta g_{\mu\nu}$ of the metric perturbations. As discussed in Chapter 7 such an effect can be described as a process of graviton production from the vacuum, with the consequent formation of a cosmic background of relic gravitons. In the same way, as we shall see in this chapter, a phase of inflation also parametrically amplifies the fluctuations $\chi = \delta\phi$ of the scalar dilaton background, thus producing pairs of dilatons, and eventually forming a cosmic background of relic dilatons [3, 4].

The amplifications of tensor and scalar waves are very similar processes, conceptually: from a technical point of view, however, there are two important differences. The first difference is due to the fact that the dilaton fluctuations χ are tightly coupled to the scalar perturbations of the metric and the other gravitational sources (unlike the tensor fluctuations $h_{\mu\nu}$, which are completely decoupled, to linear order). This might, in principle, differentiate the graviton and dilaton spectra even at a primordial level.

The second difference is due to the fact that dilatons, unlike gravitons, could become massive in the course of the standard post-inflationary evolution. Indeed, the dilaton *should* acquire a mass according to standard models of supersymmetry breaking [5]. Also, the dilaton *must* acquire a mass (in order to avoid long-range violations of the equivalence principle) if it is non-universally coupled with gravitational strength to standard macroscopic matter [6, 7]. The induced mass may substantially modify the primordial spectrum in the non-relativistic regime.

For a general computation of the dilaton spectrum we should then start from the full coupled system of scalar perturbation equations, which includes the two metric variables (for instance φ, ψ in the longitudinal gauge), the dilaton variable χ, and the fluid variables $\delta\rho, \delta p, w$. Such a system has been already written in gauge-invariant form in Eqs. (8.62)–(8.67). A complete computation requires, however, not only the exact solution of such coupled equations, but also the diagonalization of the perturbed (quadratic) action for the full system of perturbation variables: this is a necessary step to obtain the correct canonical variable, and to normalize the initial amplitude of the perturbations to the quantum fluctuations of the vacuum. The analytical approach to such a program is in general quite complicated, and only approximate solutions are presently available (see for instance [8]) for the full, general system of equations.

An exact solution can be easily obtained, however, if we limit ourselves to a simplified picture in which there are no matter sources in the initial, dilaton-dominated inflationary phase ($T_{\mu\nu} = 0 = \delta T_{\mu\nu}$), and if we assume that the subsequent radiation-dominated phase, for $\eta > -\eta_1$, is characterized by adiabatic fluid perturbations (with $\mathcal{E} = 3\Pi$, $\delta\sigma = 0$), with the dilaton frozen at the minimum of its potential ($\phi' = 0 = \phi''$, $\partial V/\partial\phi = 0$). In that case the canonical variable for the initial gravi-dilaton system is known, and is given by the gauge-invariant variable $v = z\mathcal{R}$ already introduced in Eqs. (8.51)–(8.54). Also, in the subsequent radiation-dominated era, the gauge-invariant dilaton perturbation X turns out to be decoupled from the other (metric and matter) scalar perturbations (see Eqs. (8.64)–(8.67)), and the canonical variable $\overline{X} = aX$ satisfies the free evolution equation

$$\overline{X}_k'' + (k^2 + m^2 a^2)\overline{X}_k = 0. \tag{9.1}$$

We have denoted with $m^2 = \partial^2 V/\partial\phi^2$ the possible mass term induced by the potential in Eq. (8.67).

We assume that the induced mass is small enough in string units ($m \ll M_s$) so that, soon after the transition to the radiation era, the high-frequency sector of the spectrum contains relativistic modes with proper momentum $p = k/a \gg m$. For such modes we can neglect the mass term, and we can match the normalized solution for the variable v_k in the initial inflationary phase (given in Eq. (8.136), and valid for $\eta < -\eta_1$) to the free oscillating solutions of Eq. (9.1),

$$X_k = \frac{1}{\sqrt{2k}}\left[c_+(k)e^{-ik\eta} + c_-(k)e^{ik\eta}\right], \qquad \eta \geq -\eta_1. \qquad (9.2)$$

A direct computation then leads to the Bogoliubov coefficients

$$c_\pm(k) = \pm c(k)e^{\mp ix_1}, \qquad |c(k)| \simeq x_1^{-|\nu|-1/2}, \qquad (9.3)$$

where $x_1 \equiv k\eta_1 = k/k_1$, so that

$$X_k = \frac{\overline{X}_k}{a} \simeq \frac{c(k)}{a\sqrt{k}}\sin k\eta \qquad (9.4)$$

(see also the analogous computation of Appendix 8B for the relativistic axions, Eqs. (8B.36)–(8B.38)). The corresponding energy density distribution, in critical units, for the relativistic fluctuations re-entering the horizon during the radiation era, is then given by

$$\Omega_\chi = \frac{1}{\rho_c}\frac{d\rho_\chi(k)}{d\ln k} = \frac{k^3}{2a^2\rho_c}\left(|X_k'|^2 + k^2|X_k|^2\right)$$

$$\sim \left(\frac{k}{a}\right)^4|c(k)|^2 \sim p^{3-2|\nu|}, \qquad p_{eq} < p < p_1, \qquad m < p, \qquad (9.5)$$

where $p = k/a$ is the proper momentum of the mode k.

It is important to stress that the Bessel index $\nu = 1/2 - \alpha$, determining the power of the dilaton spectrum, is fixed by the pump field $z = a\phi'/\mathcal{H} \sim (-\eta)^\alpha$ associated with the evolution of the canonical scalar variable during inflation. We have seen in Chapter 8 that during the dilaton-dominated phase z evolves in time exactly as the E-frame scale factor, i.e. like the pump field of tensor perturbations. As a consequence, the parameters α and ν are the same for gravitons and dilatons, and we recover for the relativistic dilaton fluctuations exactly the same spectrum as that of tensor metric perturbations [4].

For the class of minimal pre-big bang models discussed in Section 7.3 one finds $\Omega_\chi \sim p^3 \ln^2 p$. More generally, if we want to take into account the possible modifications due to the high-curvature and strong coupling regime, we may para-metrize the high-frequency branch of the dilaton spectrum with a slope parameter δ as follows:

$$\Omega_\chi(p, t) = g_1^2\left(\frac{H_1}{H}\right)^2\left(\frac{a_1}{a}\right)^4\left(\frac{p}{p_1}\right)^\delta, \qquad m < p < p_1 \qquad (9.6)$$

(we have set $g_1 = H_1/M_P$, and we have absorbed all numerical factors of order one inside the parameter H_1). For typical string cosmology models the slope is growing, $0 \leq \delta \leq 3$, but the precise value of δ is strongly model dependent. Finally, the cut-off scale $p_1 = k_1/a \simeq H_1 a_1/a$ may be expected to be controlled by the string scale ($H_1 \sim M_s$), as in the case of scalar and tensor metric perturbations.

The slope of the spectrum is the same as that one would obtain for gravitons, in the same class of inflationary backgrounds. There is, however, a difference in the dilaton spectrum due to the fact that we cannot identify proper momentum p and proper energy ω in Eq. (9.6) since, in general, we are dealing with massive particles with $\omega^2 = p^2 + m^2$. It should be noted, also, that the spectrum (9.6) cannot be extrapolated down to momentum scales re-entering the horizon after equality (i.e. for $p < p_{eq} = k_{eq}/a$), since in the matter-dominated era (where $3\Pi \neq \mathcal{E}$, in general) the dilaton fluctuations are no longer decoupled from scalar metric and matter perturbations, even if the dilaton background remains frozen at the minimum of the potential (see Eq. (8.67)).

Let us now consider the spectrum of the non-relativistic modes with $p = k/a \ll m$. Even if the mass is very small, and negligible at the beginning of the radiation era ($m < k/a_1$), the proper momentum $p = k/a(t)$ is continuously redshifted with respect to the mass in the course of the subsequent evolution, so that all modes tend to approach the non-relativistic regime. For the exact computation of the massive-mode spectrum, on the other hand, we should obtain exact solutions of the canonical equation (9.1) with the mass term included, and impose their matching to the corresponding (massless) inflationary solution (see below).

A quick estimate of the non-relativistic spectrum can be obtained, however, by noting that the number density of the dilatons produced (parametrized by the Bogoliubov coefficient $c(k)$) is determined by the asymptotic amplitude of the oscillating solutions inside the horizon. For modes becoming non-relativistic well inside the horizon, at a time scale t_{nr} such that $p(t_{nr}) \sim m \gg H(t_{nr})$, the number of produced dilatons remains the same as in the relativistic case, and the only effect of the mass is a simple rescaling of the energy density,

$$\Omega_\chi^{rel} \rightarrow \Omega_\chi^{nr} = \left(\frac{m}{p}\right)\Omega_\chi^{rel}, \tag{9.7}$$

according to the definition of Ω_χ. Using the relativistic dilaton spectrum (9.6), and the definition $p_1 = H_1 a_1/a$, one immediately obtains

$$\Omega_\chi(p, t) = g_1^2 \left(\frac{m}{H_1}\right)\left(\frac{H_1}{H}\right)^2 \left(\frac{a_1}{a}\right)^3 \left(\frac{p}{p_1}\right)^{\delta-1}, \qquad p_m < p < m. \tag{9.8}$$

The validity range of this spectrum is limited by the momentum scale p_m of a mode that becomes non-relativistic just at the time it re-enters the horizon, such that $p_m(t_{nr}) = m = H(t_{nr})$. Such a limiting scale is thus related to the cut-off scale p_1 by [3, 4]

$$\frac{p_m}{p_1} = \frac{m \, a_{nr}}{H_1 a_1} = \frac{m}{H_1} \sqrt{\frac{H_1}{H_{nr}}} = \sqrt{\frac{m}{H_1}}. \tag{9.9}$$

Lower momentum scales, $p < p_m$, become non-relativistic when they are still outside the horizon, i.e. at a time t_{nr} such that $p(t_{nr}) \sim m \ll H(t_{nr})$. For such modes the final energy distribution turns out to be determined by the background kinematics at the time of their horizon exit (because of the freezing of the fluctuations and of their conjugate momentum, see Section 7.2). The effective number density of non-relativistic dilatons, in this case, is determined by the continuity of the spectrum at $p = p_m$ [9], in such a way that Ω_χ has the same momentum dependence as in the absence of mass ($\sim (p/p_1)^\delta$), and a time dependence of non-relativistic type ($\rho_\chi \sim a^{-3}$). One then obtains the spectrum

$$\Omega_\chi(p, t) = g_1^2 \left(\frac{m}{H_1}\right)^{1/2} \left(\frac{H_1}{H}\right)^2 \left(\frac{a_1}{a}\right)^3 \left(\frac{p}{p_1}\right)^\delta, \qquad p < p_m. \tag{9.10}$$

The three branches (9.6), (9.8) and (9.10) parametrize the full spectrum of massive dilatons re-entering the horizon during the radiation era, for the simple model of inflationary background that we are considering. It may be useful to note, for further applications, that the non-relativistic sector of the spectrum stays constant in time during the matter-dominated era. For $t > t_{eq}$ we have

$$\left(\frac{H_1}{H}\right)^2 \left(\frac{a_1}{a}\right)^3 = \left(\frac{H_1}{H_{eq}}\right)^2 \left(\frac{a_1}{a_{eq}}\right)^3 = \left(\frac{H_1}{H_{eq}}\right)^{1/2}, \tag{9.11}$$

and we can then express the above spectrum at the present time t_0 as follows:

$$\Omega_\chi(p, t_0) = g_1^2 \, \Omega_r(t_0) \left(\frac{p}{p_1}\right)^\delta, \qquad m < p < p_1,$$

$$= g_1^2 \left(\frac{m^2}{H_1 H_{eq}}\right)^{1/2} \left(\frac{p}{p_1}\right)^{\delta-1}, \qquad p_m < p < m,$$

$$= g_1^2 \left(\frac{m}{H_{eq}}\right)^{1/2} \left(\frac{p}{p_1}\right)^\delta, \qquad p_{eq} < p < p_m. \tag{9.12}$$

At this point, two important comments are in order. The first is that we have applied the non-relativistic corrections to a single relativistic branch, with slope parameter δ. The procedure can be easily extended, however, to a more general case in which there are two (or more) branches in the relativistic sector of the spectrum: a possible example of this scenario is provided by pre-big bang models with a long enough high-curvature phase, modifying the slope of the spectrum above some given scale $p = p_s$. In such a case, the non-relativistic corrections will affect *only the lower*, or *also the higher*, frequency bands of the spectrum depending on whether $m < p_s$ or $m > p_s$, respectively.

The second comment concerns the validity of the dilaton spectrum (9.12), obtained by a simple kinematic rescaling of the spectral distribution computed in the radiation era. This procedure is justified provided the high-frequency modes $p > p_{eq}$ are consistently described by the free equation (9.1) also after the equality, i.e. when the Universe enters the phase of matter domination. During this phase, however, the dilaton perturbation equation might be modified because of two possible effects (see Eq. (8.67)): (*i*) a running dilaton ($\phi' \neq 0 \neq \phi''$), shifted away from the minimum of its potential, and (*ii*) a coupling to the matter perturbations $\delta\rho$ and $\delta\sigma$. Both effects depend on the quantum loop corrections which become operative in the strong coupling regime, as will be discussed in detail later. We only anticipate here that the coupling to the matter energy density tends to be diluted in time, but the direct coupling to the dilaton charge density of the dark matter could remain strong, and even grow in time, as we shall see in Section 9.3. These effects could modify the dilaton spectrum (9.12), distorting it from a faithful picture of the primordial inflationary dynamics. For such reasons, in the following discussion, we will always treat the spectral index δ as a pure phenomenological parameter with arbitrary (model-dependent) values.

The results obtained for the non-relativistic sector of the spectrum can be confirmed by solving exactly the equation for the massive dilaton fluctuations in the radiation era. Using the explicit behavior of the scale factor ($a/a_1 = \eta/\eta_1 = a_1 H_1 \eta$), we rewrite Eq. (9.1) in the form

$$\overline{X}_k'' + (k^2 + \alpha^2 \eta^2)\overline{X}_k = 0, \qquad \alpha = mH_1 a_1^2, \tag{9.13}$$

and we note that the general solution, with $m \neq 0$, may be given in terms of the parabolic cylinder functions [10]. For our applications it is convenient to distinguish, as before, modes becoming non-relativistic inside ($k > k_m$) and outside ($k < k_m$) the horizon, where $k_m = k_1(m/H_1)^{1/2}$ is the limiting scale of Eq. (9.9). We consider the two cases separately [11].

(1) In the case $k > k_m$ we first rewrite Eq. (9.13) as

$$\frac{d^2\overline{X}_k}{dx^2} + \left(\frac{x^2}{4} - b\right)\overline{X}_k = 0, \tag{9.14}$$

where

$$x = (2\alpha)^{1/2}\eta, \qquad b = -\frac{k^2}{2\alpha}, \tag{9.15}$$

and give its general solution in terms of the Weber cylinder functions [10] $W(b, x)$ as

$$\overline{X}_k = AW(b, x) + BW(b, -x). \tag{9.16}$$

The integration constants A and B can be fixed by imposing that this solution reduces to the normalized massless solution (9.4) in the relativistic limit in which

$$\frac{p^2}{m^2} = \frac{k^2}{m^2 a^2} = \frac{k^2}{\alpha^2 \eta^2} = -\frac{4b}{x^2} \gg 1. \tag{9.17}$$

In this limit, since we are considering modes becoming non-relativistic inside the horizon, we also have

$$\frac{k^2}{k_m^2} = \frac{k^2 H_1}{k_1^2 m} = \frac{k^2}{H_1 m a_1^2} = \frac{k^2}{\alpha} \sim (-b) \gg 1, \tag{9.18}$$

so that we can approximate the W functions for large values of $|b|$ and moderate values of x. Using the corresponding W expansion [10] one then recovers the relativistic solution (9.4) provided $A = 0$ and $B = c(k)\alpha^{-1/4}$, corresponding to the particular exact solution

$$\overline{X}_k = \frac{c(k)}{\alpha^{1/4}} W(b, -x). \tag{9.19}$$

Expanding this normalized solution in the opposite limit $x^2 \gg 4|b|$ we are led to the explicit form of the non-relativistic fluctuations inside the horizon:

$$X_k = \frac{\overline{X}_k}{a} \simeq \frac{c(k)}{a\sqrt{ma}} \sin\left(\frac{ma\eta}{2}\right), \qquad k > k_m \tag{9.20}$$

(we have used the definition (9.13) of α, to set $x^2/4 = ma\eta/2 = m/(2H)$). Comparison with the relativistic solution (9.4), and with the associated energy distribution (9.5), finally gives the rescaling $\rho_\chi^{\text{nr}}(k) = (ma/k)\rho_\chi^{\text{rel}}(k)$, which leads to the non-relativistic branch (9.8) of the dilaton spectrum.

(2) The other case $k < k_m$ corresponds to modes that become non-relativistic when they are still outside the horizon. For such modes $|b| < 1$, so that we cannot use the large $|b|$ expansion of the Weber functions: it is more convenient, in this case, to rewrite the general solution of Eq. (9.14) in the form

$$\overline{X}_k = Ay_1(b, x) + By_2(b, x), \tag{9.21}$$

where y_1 and y_2 are the even and odd parts of the parabolic cylinder functions [10]. The matching to the massless solution (9.4), in the relativistic limit $x \to 0$, now gives $A = 0$ and

$$\overline{X}_k = \left(\frac{k}{2\alpha}\right)^{1/2} c(k) y_2(b, x). \tag{9.22}$$

In the non-relativistic limit $x^2 \gg |b|$, on the contrary, it is possible to use the Weber expansion [10],

$$y_2(b, x) \sim W(b, x) - W(b, -x) \sim \frac{1}{\sqrt{x}} \sin \left(\frac{x^2}{4} \right), \tag{9.23}$$

to obtain

$$X_k = \frac{\overline{X}_k}{a} \simeq \frac{c(k)}{a\sqrt{ma}} \left(\frac{k}{k_1} \right)^{1/2} \left(\frac{H_1}{m} \right)^{1/4} \sin \left(\frac{ma\eta}{2} \right), \qquad k < k_m. \tag{9.24}$$

A comparison with the complementary branch (9.20) of the non-relativistic solution then leads to the relation

$$\rho_\chi(k < k_m) \simeq \left(\frac{H_1}{m} \right)^{1/2} \left(\frac{p}{p_1} \right) \rho_\chi(k > k_m), \tag{9.25}$$

from which one easily recovers the lowest energy branch (9.10) of the dilaton spectrum.

It should be mentioned, as a final comment, that this last result for the non-relativistic spectrum is only valid if the dilaton becomes massive *before* the limiting scale p_m crosses the horizon. More precisely, calling T_m the temperature scale at which the mass turns on, and calling p_T the momentum mode re-entering the horizon just at the same epoch, we can say that the spectrum (9.10) is valid for $p_m < p_T$. In the opposite case ($p_m > p_T$) the role of transition-scale separating modes that become non-relativistic inside and outside the horizon is played by p_T, and the lowest frequency branch (9.10) of the spectrum has to be replaced by [9]

$$\Omega_\chi(p, t) = g_1^2 \left(\frac{m^2}{H_1 H_T} \right)^{1/2} \left(\frac{H_1}{H} \right)^2 \left(\frac{a_1}{a} \right)^3 \left(\frac{p}{p_1} \right)^\delta, \qquad p < p_T, \qquad p_T < p_m \tag{9.26}$$

(obtained by a continuous match to the branch (9.8), which now extends down to the lower limit $p = p_T$). Here H_T is the curvature scale at the time the mode p_T crosses the horizon, i.e. $H_T = k_T/a_T = a p_T/a_T$. For $H_T = m$ one is led back to the result (9.10).

In any case, the overall effect of the non-relativistic corrections is to enhance the amplitude of the low-energy part of the dilaton spectrum, as clearly illustrated in Fig. 9.1 where we have plotted the spectrum (9.12) for the two cases $\delta = 2$ and $\delta = 1/2$. It is important to note that, even if the relativistic part of the spectrum is growing ($\delta > 0$), the non-relativistic part may contain a flat (or decreasing) branch if $\delta \leq 1$. Also, if the slope is flat enough (i.e. $\delta < 1$), and the branch $p_m < p < m$ is sufficiently extended, then the spectrum may be peaked *not* at the cut-off frequency p_1 but at the intermediate frequency p_m (we will come back to

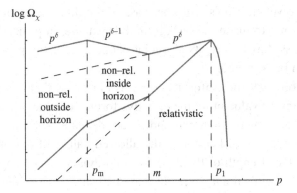

Figure 9.1 Examples of dilaton spectra with non-relativistic corrections. The spectrum (9.12) is plotted for the two cases $\delta = 2$ (lower curve) and $\delta = 1/2$ (upper curve).

this point at the end of this section). Finally, if the mass is larger than the present value of the cut-off scale p_1, namely if

$$m > p_1(t_0) = \frac{k_1}{a_0} = \frac{H_1 a_1}{a_0} \sim g_1^{1/2} 10^{11}\,\text{Hz} \sim \left(\frac{H_1}{M_\text{P}}\right)^{1/2} 10^{-4}\,\text{eV} \qquad (9.27)$$

(we have used Eq. (7.180)), then all modes are today non-relativistic, so that the massless branch (9.6) disappears from the dilaton spectrum (in other words, the branch (9.8) extends from p_m to p_1).

9.1.1 Dilaton mass and couplings

For the simple example of inflationary scenario considered in this section, the final amplitude of the dilaton spectum Ω_χ turns out to be controlled by two parameters: the inflation scale (in Planck units) $g_1 = H_1/M_\text{P}$, and the dilaton mass m. Assuming $H_1 \sim M_\text{s}$, and taking into account also the non-relativistic branch of the spectrum, the question: *How strong is the relic background today?* may thus be rephrased as: *How large is the dilaton mass?*

It would be interesting, from a phenomenological point of view, to have a mass small enough to avoid dilaton decay before the present epoch, so that the primordial background would be still around us, and accessible, in principle, to direct detection. Given the dilaton decay rate into two photons, $\Gamma_d \sim m^3/M_\text{P}^2$, associated with a decay scale $H_d = \Gamma_d$, the present existence of a relic dilaton background is guaranteed by the condition $H_d < H_0$, which imposes on the mass a first upper bound $m \lesssim 10^2$ MeV. Also, as we shall see in Section 9.2, a cosmic dilaton background may resonantly interact with the present gravitational antennas only in the case of very small masses. The amplitude of the non-relativistic sector of the spectrum, however, grows with the mass of the dilaton, so that large enough masses are required to make the background detectable, in principle. The precise value of the mass thus represents a crucial parameter for the phenomenology of the relic dilaton background.

From the theoretical side, unfortunately, it is fair to say that there is no compelling prediction on the value of m: the effective form and value of $V(\phi)$, even at the present epoch, are largely unknown and model dependent. From the phenomenological side, however, we know that the allowed values of the mass are strictly correlated with the strength of the coupling to ordinary macroscopic matter.

With reference to this point we recall that in all string theory models the matter fields are coupled not only to the metric but also to the dilaton field. This implies that a matter distribution is associated, in general, with a dilaton charge density σ (see Eq. (2.12)), and that the matter stress tensor is not separately conserved in the presence of an external gravi-dilaton background. There are macroscopic forces generated by the gradients of the metric *and* of the dilaton field, according to the generalized conservation equation (2.19),

$$\nabla_\nu T^{\mu\nu} = \frac{\sigma}{2}\nabla^\mu \phi,\tag{9.28}$$

which can also be rewritten as

$$\partial_\nu(\sqrt{-g}T^{\mu\nu}) + \sqrt{-g}\,\Gamma_{\alpha\nu}{}^\mu T^{\alpha\nu} = \frac{1}{2}\sqrt{-g}\,\sigma\nabla^\mu\phi.\tag{9.29}$$

It follows that the equation of motion of a test particle is no longer a geodesic, as can be checked by integrating Eq. (9.29) over the (space-like) $t = $ const hypersurface Σ:

$$\int_\Sigma d^3x'\,\partial_i(\sqrt{-g}\,T^{\mu i}) + \frac{d}{dt}\int_\Sigma d^3x'\,\sqrt{-g}\,T^{\mu 0}(x') + \int_\Sigma d^3x'\,\sqrt{-g}\,\Gamma_{\alpha\nu}{}^\mu(x')T^{\alpha\nu}(x')$$

$$= \frac{1}{2}\int_\Sigma d^3x'\,\sqrt{-g}\,\sigma(x')\nabla^\mu\phi(x').\tag{9.30}$$

Assuming that the energy-momentum and charge distributions, $T_{\mu\nu}$ and σ, are non-zero only inside a narrow "world-tube", centered around the world-line $x^\mu(\tau)$ of the center of mass of the test body, we can use the Gauss theorem for the first integral of Eq. (9.30), and we can expand the external fields Γ and $\nabla\phi$ around $x^\mu(\tau)$, applying the so-called "multipole expansion",

$$\Gamma_{\alpha\beta}{}^\mu(x') = \Gamma_{\alpha\beta}{}^\mu(x) + (x'^\nu - x^\nu)\partial_\nu\Gamma_{\alpha\beta}{}^\mu(x) + \cdots,$$
$$\nabla_\mu\phi(x') = \nabla_\mu\phi(x) + (x'^\nu - x^\nu)\partial_\nu\nabla_\mu\phi(x) + \cdots.\tag{9.31}$$

In the point-particle (or "pole-particle" [12]) approximation, we can neglect all possible internal "momenta" of the test body, limiting ourselves to the zeroth-order terms:

$$\frac{d}{dt}\int_{\Sigma} d^3x' \sqrt{-g}\, T^{\mu 0}(x') + \Gamma_{\alpha\nu}{}^{\mu}(x(\tau))\int_{\Sigma} d^3x' \sqrt{-g}\, T^{\alpha\nu}(x')$$

$$= \frac{1}{2}\nabla^{\mu}\phi(x(\tau))\int_{\Sigma} d^3x' \sqrt{-g}\, \sigma(x'). \qquad (9.32)$$

In this approximation, and in the limit in which the radius of the world-tube shrinks to zero, $x'^{\mu} \to x^{\mu}(\tau)$, we can define the usual gravitational "current density", $T_{\mu\nu}$, and the dilaton charge density, σ, as follows:

$$T^{\mu\nu}(x') = \frac{p^{\mu}p^{\nu}}{\sqrt{-g}\,p^0}\, \delta^3(x' - x(\tau)),$$

$$\frac{1}{2}\sigma(x') = q\frac{m^2}{\sqrt{-g}\,p^0}\, \delta^3(x' - x(\tau)). \qquad (9.33)$$

Here $p^{\mu} = mu^{\mu} = m\,dx^{\mu}/d\tau$, while q represents the dilaton charge per unit of gravitational mass [13], i.e. the relative intensity of scalar to tensor forces. The integration of Eq. (9.32), multiplied by $p^0/m^2 = m^{-1}(dt/d\tau)$, eventually leads to the generalized (non-geodesic) equation of motion

$$\frac{du^{\mu}}{d\tau} + \Gamma_{\alpha\beta}{}^{\mu}u^{\alpha}u^{\beta} = q\nabla^{\mu}\phi. \qquad (9.34)$$

This equation, in combination with the dilaton equation of motion obtained from the string effective action, clearly implies that strong dilaton couplings (i.e. scalar charges $q \geq 1$) must be associated with a sufficiently large value of the dilaton mass: in that case the effective range of the scalar macroscopic forces has to be sufficiently small, to avoid contradictions with the standard gravitational phenomenology [14] (see also [15] for a recent compilation of bounds on deviations from Newtonian gravity at small distances). Small enough couplings with $q \ll 1$, on the contrary, may be compatible also with long-range forces, i.e. with small dilaton masses. In the discussion of the phenomenological bounds on the mass we are thus led to the related question: *How strong is the dilaton coupling?*

For a more precise formulation of this question we should note, first of all, that the coupling of the dilaton to the fundamental quark and lepton fields (building up macroscopic matter) is described by an effective action which should also include all the dilaton loop corrections, at least in the regime of moderately strong coupling ($g_s \sim 1$), which seems typical of our present cosmological state. The effective coupling may thus depend also on the level of approximation adopted, i.e. on the order of the truncated perturbative expansion. In such a context, however, it is always possible to give a convenient, frame-independent definition of the scalar charge by expressing the effective interaction Lagrangian in terms of the canonical fields diagonalizing the kinetic part of the action [4]. Such an action can be written, in general, as follows (see also Section 3.2):

$$S = \frac{1}{2\lambda_s^2} \int d^4x \sqrt{-g} \left[-Z_R(\phi)R - Z_\phi(\phi)(\nabla\phi)^2 - V(\phi) \right.$$

$$\left. + Z_k^i(\phi)(\nabla\psi_i)^2 - M_i^2 Z_m^i(\phi)\psi_i^2 \right]. \tag{9.35}$$

We have used, for simplicity, a scalar model of fundamental matter fields ψ_i, and we have called Z the dilaton "form factors" due to the loop corrections. All fields ϕ, ψ_i are dimensionless, and the loop corrections are referred to the fundamental S-frame metric appearing in the sigma model action (as clearly stressed by the presence of the string length parameter λ_s, controlling all dimensional factors of the above Lagrangian).

In the given frame, the effective masses (m_i) and dilaton couplings (g_i) of the fields ψ_i can be computed by introducing the rescaled variables $\widehat{\psi}_i$ which restore the canonical form of the kinetic energy terms in the matter action, and have canonical dimensions

$$\widehat{\psi}_i = M_s (Z_k^i)^{1/2} \psi_i. \tag{9.36}$$

The matter action becomes

$$S_m = \frac{1}{2} \int d^4x \sqrt{-g} \left[(\nabla\widehat{\psi}_i)^2 - \mu_i^2(\phi)\widehat{\psi}_i^2 \right],$$

$$\mu_i^2(\phi) = M_i^2 Z_m^i (Z_k^i)^{-1}, \tag{9.37}$$

modulo higher-order, derivative interactions between ϕ and ψ_i. Assuming that ϕ is stabilized by its potential V, we expand the interaction Lagrangian $-\mu_i^2 \widehat{\psi}_i^2/2$ around the value of ϕ which extremizes the potential (and which can always be assumed to coincide with $\phi = 0$, after a trivial shift):

$$L_i(\phi, \widehat{\psi}_i) \equiv -\frac{1}{2}\mu_i^2(\phi)\widehat{\psi}_i^2 = -\frac{1}{2}m_i^2 \widehat{\psi}_i^2 - g_i \phi \widehat{\psi}_i^2 + \mathcal{O}(\phi^2) + \cdots. \tag{9.38}$$

This expansion defines the low-energy masses and couplings, respectively, as

$$m_i^2 = \left[\mu_i^2(\phi)\right]_{\phi=0} = M_i^2 \left[Z_m^i (Z_k^i)^{-1}\right]_{\phi=0}, \tag{9.39}$$

$$g_i = \frac{1}{2}\left(\frac{\partial\mu_i^2}{\partial\phi}\right)_{\phi=0} = \frac{m_i^2}{2}\left[\frac{\partial}{\partial\phi}\ln\mu_i^2\right]_{\phi=0}, \tag{9.40}$$

with corresponding dimensionless dilaton charge

$$q_i = \frac{g_i}{m_i^2} = \frac{1}{2}\left[\frac{\partial}{\partial\phi}\ln\left(\frac{Z_m^i}{Z_k^i}\right)\right]_{\phi=0}.$$ (9.41)

The coupling defined in this way, however, turns out to be strongly frame dependent (like the coupling appearing in Eq. (9.33)). Consider, for instance, a pure Brans–Dicke model of scalar-tensor gravity, corresponding to an S-frame action (9.35) with no-dilaton couplings to the $\widehat{\psi}_i$ fields, i.e. $\partial_\phi Z_k^i = 0 = \partial_\phi Z_m^i$. The corresponding dilaton charge (9.41) is vanishing in the S-frame, but it is non-vanishing, in general, in other frames, where $\widetilde{Z}_m^i \neq \widetilde{Z}_k^i$, and $\partial_\phi \widetilde{Z}_k^i \neq 0 \neq \partial_\phi \widetilde{Z}_m^i$. Such a frame dependence of the coupling is due to the fact that, in a generic frame, the field ϕ and the metric are non-trivially mixed through the Z_ϕ and Z_R coupling functions, in such a way that the charge (9.41) actually controls the matter couplings *not to the pure scalar* part, but to a *mixture of scalar and tensor* parts of the gravi-dilaton field.

A frame-independent definition of the coupling can thus be given only when the full kinetic part of the action (9.35) – including the gravi-dilaton sector – is diagonalized [4, 6], and the action is given in terms of the canonical scalar fields $\widehat{\phi}$, $\widehat{\psi}_i$, and of the canonical (E-frame) metric $\widehat{g}_{\mu\nu}$, defined by the rescaling

$$\widehat{g}_{\mu\nu} = \left(\frac{\lambda_P}{\lambda_s}\right)^2 g_{\mu\nu} Z_R.$$ (9.42)

Applying to the action (9.35) the transformation rules (2.39)–(2.42) (with $d = 3$ and $\psi = -\ln Z_R$), one easily finds that the full diagonalized action takes the form

$$S = \int d^4x\sqrt{-g}\left[-\frac{\widehat{R}}{2\lambda_P^2} + \frac{1}{2}(\widehat{\nabla}\widehat{\phi})^2 - \widehat{V} + \frac{1}{2}(\widehat{\nabla}\widehat{\psi}_i)^2 - \frac{1}{2}\widehat{\mu}_i(\widehat{\phi})\widehat{\psi}_i^2\right].$$ (9.43)

Here $\widehat{R} = R(\widehat{g})$, $\widehat{\nabla} = \nabla(\widehat{g})$, and

$$\frac{d\widehat{\phi}}{d\phi} = M_P\left[\frac{3}{2}\left(\frac{d}{d\phi}\ln Z_R\right)^2 - \frac{Z_\phi}{Z_R}\right]^{1/2},$$

$$\widehat{V} = \frac{1}{2}\frac{\lambda_s^2}{\lambda_P^4}Z_R^{-2}V,$$ (9.44)

$$\widehat{\psi}_i = M_P\left(\frac{Z_k^i}{Z_R}\right)\psi_i,$$

$$\widehat{\mu}_i^2 = M_i^2\left(\frac{\lambda_s}{\lambda_P}\right)^2\frac{Z_m^i}{Z_k^i Z_R}.$$

We can now expand, as before, the interaction term $\widehat{\mu}_i^2$ around the value $\widehat{\phi} = 0$ to obtain the effective masses,

$$\widehat{m}_i^2 = \widehat{\mu}_i^2(0) = M_i^2 \left(\frac{\lambda_s}{\lambda_P}\right)^2 \left[\frac{Z_m^i}{Z_k^i Z_R}\right]_{\widehat{\phi}=0}, \tag{9.45}$$

and couplings,

$$\widehat{g}_i = \frac{1}{2}\left(\frac{\partial \widehat{\mu}_i^2}{\partial \widehat{\phi}}\right)_{\widehat{\phi}=0} = \frac{\widehat{m}_i^2}{2}\left(\frac{d\phi}{d\widehat{\phi}}\right)_{\widehat{\phi}=0}\left(\frac{\partial}{\partial \phi}\ln \widehat{\mu}_i^2\right)_{\widehat{\phi}=0}, \tag{9.46}$$

with corresponding (frame-independent) dilaton charge per unit mass,

$$\widehat{q}_i = \frac{\widehat{g}_i}{\widehat{m}_i} = \frac{\widehat{m}_i}{2}\left(\frac{d\phi}{d\widehat{\phi}}\right)_{\widehat{\phi}=0}\left[\frac{\partial}{\partial \phi}\ln\left(\frac{Z_m^i}{Z_k^i Z_R}\right)\right]_{\widehat{\phi}=0}. \tag{9.47}$$

In the weak coupling limit we have $Z_R \simeq Z_\phi \simeq \exp(-\phi)$, so that the factor

$$\frac{\widehat{m}_i}{2}\frac{d\phi}{d\widehat{\phi}} \simeq \frac{1}{\sqrt{2}}\frac{\widehat{m}_i}{M_P} = \sqrt{4\pi G}\,\widehat{m}_i \tag{9.48}$$

reduces (in our units) to the standard, dimensionless "gravitational charge" (appearing, for instance, in the Poisson equation for the Newton potential φ_N: $\nabla^2\varphi_N = 4\pi G\rho$). One then finds, in the weak coupling limit, that the frame-independent dilaton charge $\widehat{q}$ deviates from the standard gravitational charge by the dimensionless factor

$$\overline{q}_i \equiv \frac{\widehat{q}_i}{\widehat{m}_i\sqrt{4\pi G}} \simeq 1 + \left[\frac{\partial}{\partial \phi}\ln\left(\frac{Z_m^i}{Z_k^i}\right)\right]_{\phi=0}. \tag{9.49}$$

For a pure Brans–Dicke model, in particular, $\overline{q}_i = 1$. For a generic string theory model, however, the coupling factors $\overline{q}_i$ tend to deviate from 1 in a non-universal way, as the value of $\overline{q}_i$ depends on the form factors Z^i, which are in principle different for different fields. The previous question about the strength of the dilaton coupling can then be formulated, more precisely, as follows: *How large is $\overline{q}$?* There are two possible (alternative) theoretical scenarios.

The more conventional scenario, based on the fact that the loop corrections determining the dilaton coupling are the same as those determining the effective mass of the given particle, suggests a rather large dilaton charge ($\overline{q}_i \sim 40–50$) for the confinement-generated components of the hadronic masses [6, 7], and a smaller charge, of gravitational intensity ($\overline{q}_i \sim 1$) for the leptonic masses (see also [16]). If this is the case, the total dilaton charge of a macroscopic body tends to be large (in gravitational units), and composition dependent.

Consider a body of total mass M, composed of B baryons of mass m_b and dilaton charge $\widehat{q}_b$, and of Z electrons of mass m_e and dilaton charge $\widehat{q}_e$. Assuming

$Z \sim B$, and using $m_e \ll m_b$, $\widehat{q}_e \ll \widehat{q}_b$, one obtains that the total charge per unit mass is given by

$$\widehat{q} = \frac{\sum_i m_i \widehat{q}_i}{\sum_i m_i} \simeq \frac{B}{M} m_b \widehat{q}_b = \left(\frac{B}{\mathcal{M}}\right) \widehat{q}_b, \qquad (9.50)$$

where $\mathcal{M} = M/m_b$ is the mass of the body in units of baryonic masses. Since $B/\mathcal{M} \sim 1$, the total dilaton charge of the macroscopic body, in gravitational units, turns out to be controlled by the large coupling of the baryons, $\overline{q} \sim \overline{q}_b \gg 1$. Also, since the factor $B/\mathcal{M}$ depends on the internal (nuclear) structure of the body, the effective coupling $\widehat{q}$ turns out to be composition dependent, with variations which are typically of order

$$\frac{\Delta \widehat{q}}{\widehat{q}} = \Delta \left(\frac{B}{\mathcal{M}}\right) \sim 10^{-3}, \qquad (9.51)$$

across different types of ordinary matter. We know, on the other hand, that a large and composition-dependent component of the gravitational force is excluded by present experimental tests, down to the millimeter scale [14, 15]. It follows that the range of the dilaton force has to be smaller than this scale of distance, i.e. that the dilaton mass has to satisfy the phenomenological bound $m \gtrsim 10^{-4}$ eV.

This bound can be relaxed, however, according to a second, alternative scenario in which the dilaton coupling to the fundamental matter fields is universal to a very high degree of accuracy (namely, it is the same for all fields, to higher orders in the loop expansion), and in which the combined loop corrections are fine-tuned to produce a highly suppressed ($\overline{q} \ll 1$) dilaton charge [17, 18]. In that case the dilaton force may be a long-range one, and the associated dilaton mass may be arbitrarily small, or even zero (actually, if the coupling is small but non-zero, a small but finite mass is expected to be generated by radiative corrections, as will be discussed in the next section).

In any case, the precise measurements of the effective gravitational forces over a wide range of distances provide exclusion plots in the $\{m, \overline{q}^2\}$ plane, and thus define the presently allowed values of the dilaton mass, for any given theoretical prediction of the coupling strength (see [14] for a comprehensive compilation of bounds on possible "Yukawa" deviations from Newtonian gravity). Thus, we can discuss the phenomenology of the relic dilaton background by taking into account two alternative possibilities: (*i*) (heavy) massive dilatons, gravitationally (or more strongly) coupled to macroscopic matter; (*ii*) very light (or massless) dilatons, universally and weakly coupled to matter. In the rest of this chapter we will focus our attention on this second (phenomenologically more interesting) possibility, assuming that the dilaton is arbitrarily light and correspondingly weakly coupled (relaxing, however, the assumption of universal dilaton interactions, which seems to be rather unnatural in a string theory context). A detailed discussion of the strongly coupled, heavy dilaton case may be found in [4, 19, 20].

9.1.2 Light and weakly coupled dilatons

We assume, first of all, that the dilatons are light enough to have not yet decayed (i.e. $m \lesssim 10^2$ MeV), and that the relic background is still available, at least in principle, to a direct experimental observation. The amplitude of this background depends on m, and a larger mass (compatible with the phenomenological bounds) corresponds to a stronger background of non-relativistic dilatons. For the range of parameters that we are considering, the most constraining upper bound on the mass comes from the observed critical density [4, 19, 20]: the total energy density of the background, integrated over all modes, has to be sub-critical,

$$h^2 \int_0^{p_1} \mathrm{d}(\ln p)\, \Omega_\chi(p, t) < 1, \tag{9.52}$$

at all times, to avoid a Universe over-dominated by dilatons. Now, the interesting question becomes: *Which value of mass does correspond to the strongest, non-relativistic dilaton background, with almost critical intensity?* The answer depends on the shape of the dilaton spectrum. There are two alternative possibilities for a cosmic background dominated by the non-relativistic branch of the spectrum (see Fig. 9.2).

(1) The first possibility corresponds to the case $m > p_1(t_0)$ and $\delta \geq 1$, in which all modes are presently non-relativistic (see Eq. (9.27)), and the spectrum (9.12) is peaked at $p = p_1$. The critical density bound (9.52) can then be approximated by the condition $\Omega_\chi(p_1) < 1$, which implies

$$m < \left(\frac{H_{\mathrm{eq}} M_{\mathrm{P}}^4}{H_1^3}\right)^{1/2}. \tag{9.53}$$

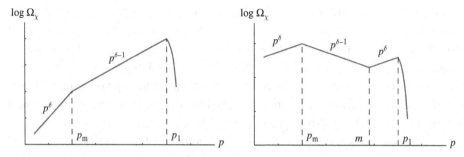

Figure 9.2 Two possible examples of dilaton spectra dominated by non-relativistic modes: the case $m > p_1$ and $\delta \geq 1$ (left panel), and the case $m < p_1$, $0 < \delta < 1$ (right panel).

This bound, for $H_1 \sim M_s \sim 0.1 M_P$, is saturated by a dilaton mass $m \sim 10^2$ eV. If $\delta < 1$ the bound is relaxed, but the mass is still constrained to be in the range $m > p_1(t_0) \sim g_1^{1/2} 10^{-4}$ eV, which is still too large for a resonant interaction with present gravitational detectors (as we shall see in Section 9.2).

(2) The second, phenomenologically more interesting, possibility, is the case in which $m < p_1(t_0)$ and the spectrum (9.12) is flat enough ($\delta < 1$) to be dominated by the non-relativistic branch peaked at $p = p_m = p_1(m/H_1)^{1/2}$ (see Fig. 9.2). The critical density can now be approximated by the condition $\Omega_\chi(p_m) < 1$, which implies

$$m < \left(H_{eq} M_P^4 H_1^{\delta - 4} \right)^{\frac{1}{\delta + 1}}. \tag{9.54}$$

For $H_1 \sim M_s$ and $\delta \to 0$, this bound can be saturated by masses as small as

$$m \sim H_{eq} \left(\frac{M_P}{M_s} \right)^4 \sim 10^{-23} \quad \text{eV}. \tag{9.55}$$

In this second case the dilaton mass could be in the appropriate range for a resonant interaction with the present gravitational antennas (see the next section), and a new, interesting possibility emerges: the weakness of the coupling, required for the phenomenological consistency of long-range interactions, could be compensated by a large (i.e. near-to-critical) intensity of the non-relativistic background.

This is to be contrasted with the case of the (relativistic) graviton background, which is coupled to matter (and then to the detectors) with gravitational strength, but which has an amplitude constrained by the nucleosynthesis bound (7.195). Since this bound only applies to the massless (or relativistic) part of the primordial cosmic backgrounds, it is possible, in principle, to envisage a situation in which the peak values of the graviton and of the dilaton backgrounds satisfy the condition

$$h^2 \Omega_g^{rel}(t_0) \lesssim 10^{-6} \ll h^2 \Omega_\chi^{nr}(t_0) \lesssim 1. \tag{9.56}$$

If this is the case, it may be appropriate to discuss the possible response of the gravitational antennas not only to a cosmic background of relic gravitons, but also to a relic background of massive, non-relativistic dilatons. Such a discussion will be the object of the next section.

9.2 Interaction with gravitational antennas

The interaction of a cosmic dilaton background with a gravitational antenna is governed by the so-called equation of "geodesic deviation", which is the basis of the operation mechanism of all detectors of gravitational radiation. Such an equation is obtained (see for instance [21]) starting from the world-lines of two

Dilaton phenomenology

neighbouring test particles, $x^\mu(\tau)$ and $x'^\mu(\tau)$, with identical scalar charges q, and with a space-like separation parametrized by the infinitesimal vector η^μ, i.e.

$$x'^\mu(\tau) = x^\mu(\tau) + \eta^\mu(\tau), \qquad \eta^\mu \dot{x}_\mu = 0 \tag{9.57}$$

(the dot denotes derivation with respect to the proper time τ). In the presence of gravi-dilaton interactions each world-line satisfies the equation of motion (9.34) so that, to first order in η^μ,

$$\ddot{x}'^\mu = \ddot{x}^\mu + \ddot{\eta}^\mu = -\Gamma_{\alpha\beta}{}^\mu(x')(\dot{x}^\alpha \dot{x}^\beta + 2\dot{x}^\alpha \dot{\eta}^\beta) + q \nabla^\mu \phi(x'). \tag{9.58}$$

Expanding the external fields as in Eq. (9.31), and using the equation of motion for x^μ, we obtain, to first order,

$$\ddot{\eta}^\mu = -2\Gamma_{\alpha\beta}{}^\mu(x)\dot{x}^\alpha \dot{\eta}^\beta - \eta^\nu \partial_\nu \Gamma_{\alpha\beta}{}^\mu(x)\dot{x}^\alpha \dot{x}^\beta + q\eta^\nu \partial_\nu \nabla^\mu \phi(x). \tag{9.59}$$

Shifting to covariant derivatives, on the other hand, we have

$$\frac{D\eta^\mu}{D\tau} = \dot{\eta}^\mu + \Gamma_{\lambda\rho}{}^\mu \dot{x}^\lambda \eta^\rho,$$

$$\frac{D^2\eta^\mu}{D\tau^2} = \frac{d}{d\tau}\left(\dot{\eta}^\mu + \Gamma_{\lambda\rho}{}^\mu \dot{x}^\lambda \eta^\rho\right) + \Gamma_{\alpha\beta}{}^\mu\left(\dot{\eta}^\beta + \Gamma_{\lambda\rho}{}^\beta \dot{x}^\lambda \eta^\rho\right)\dot{x}^\alpha. \tag{9.60}$$

In this last equation we can eliminate $\ddot{\eta}^\mu$ through Eq. (9.59), and $\ddot{x}^\lambda$ through the equation of motion for $x^\mu(\tau)$. We can thus rewrite the accelerated variation of the separation vector η^μ in the compact, covariant form,

$$\frac{D^2\eta^\mu}{D\tau^2} + R_{\nu\alpha\beta}{}^\mu \eta^\nu \dot{x}^\alpha \dot{x}^\beta = q\eta^\nu \nabla_\nu \nabla^\mu \phi, \tag{9.61}$$

which represents the scalar–tensor generalization [13] of the standard equation of geodesic deviation obtained in the theory of general relativity.

The Riemann term appearing in this equation describes the usual coupling of a point-like test particle to the second derivatives of the metric background, and is obtained from the geodesic part of the equation of motion. The term on the right-hand side describes a new coupling to the second derivatives of the dilaton background, induced by the scalar charge of the test body. A gravitational detector can thus interact with the cosmic background of dilaton radiation in two ways:

(1) *indirectly*, through the *geodesic* coupling of the gravitational charge of the detector to the scalar part of the metric fluctuations induced by the dilaton, and contained inside the Riemann tensor [22, 23, 24];
(2) *directly*, through the *non-geodesic* coupling of the gravitational charge of the detector to the gradients of the dilaton background itself [13, 25].

The indirect (or geodesic) coupling is characterized by a gravitational strength ($q = 1$), but the corresponding amplitude of the scalar background is expected to be strongly suppressed (in critical units $\Omega \ll 1$), at least in the case of primordial relic radiation. The direct (or non-geodesic) coupling may be characterized by a much higher background amplitude ($\Omega \lesssim 1$), but the corresponding coupling strength has to be weaker ($q \ll 1$), since we are considering long-range dilaton interactions (see the discussion of the previous section).

The generalized equation of geodesic deviation (9.61) provides the basic starting point for the computation of the so-called "pattern functions" $F(\hat{n})$, describing the response of a detector as a function of the direction $\hat{n}$ of the incident radiation. For a background of massive scalar waves, like the dilaton background we are considering, the pattern functions are in general different from those of the tensor graviton background studied in Section 7.4. The differences, as we shall see in this section, are due not only to the presence of the new non-geodesic coupling, but also to the possible longitudinal polarizations present in the massive radiation background.

The ideal response of a gravitational detector can be schematically represented by considering a mechanical system of two test masses, which at rest are separated by a proper (space-like) distance $L^\mu = \text{const}$. By setting $\eta^\mu = L^\mu + \xi^\mu(\tau)$, where ξ^μ represents the infinitesimal displacements induced by the incident radiation, the relative acceleration of the two masses, in the non-relativistic and weak field limit, is given by Eq. (9.61) as

$$\ddot{\xi}^i = -L^k R_{k00}{}^i + qL^k \partial_k \partial^i \phi \equiv -L^k M_k{}^i. \tag{9.62}$$

Here M_{ij} is the total (scalar-tensor) stress tensor describing the "tidal" forces associated with the equation of geodesic deviation. For its computation we have to take into account that the "electric" components of the Riemann tensor ($R_{k00}{}^i$) may include both scalar and tensor contributions from the fluctuations of the metric background. Working in the weak field approximation we can neglect the gradients of the local, static gravi-dilaton field at the detector position, considering only the perturbations induced by the incident radiation: we can then compute M_{ij} by expanding $\phi = \phi_0 + \delta\phi$, $\phi_0 = \text{const}$, and by considering the linear perturbations of the Minkowski metric, using e.g. Eq. (8.12) with $a^2 = 1$. The perturbed components of the Christoffel connection have been given in Eq. (7.15) for the tensor part, and in Eq. (8.42) for the scalar part of the fluctuations. The final result can be easily expressed in terms of the gauge-invariant Bardeen potentials Φ, Ψ, and of the dilaton fluctuation X, defined in Eq. (8.39), as follows:

$$M_{ij} = \delta R_{i00j} - q\partial_i\partial_j X = -\frac{1}{2}\ddot{h}_{ij} + \partial_i\partial_j\Phi - \delta_{ij}\ddot{\Psi} - q\partial_i\partial_j X. \tag{9.63}$$

This result, inserted into Eq. (9.62), directly provides the relative acceleration of the two test masses responding to the given scalar-tensor fluctuations.

A realistic gravitational antenna, however, may be characterized by a more general, tensor-like geometrical structure, different from the simple vector-like structure associated with a pair of test masses. In the general case, the physical strain $h(t)$ induced by external radiation is to be computed by projecting M_{ij} onto the tensor D^{ij} specifying the geometrical configuration and the spatial orientation of the arms of the detector (see Section 7.4). It is convenient, for this purpose, to adopt for M_{ij} a Fourier expansion, taking into account that tensor fluctuations are massless, and can be expanded into frequency modes $h_{ij}(\nu, \widehat{n})$, with $\nu = E(p) = p$, as in Eq. (7.234); scalar fluctuations, on the contrary, are possibly massive, and can be expanded into momentum modes as $X(p, \widehat{n})$, with $\nu = E(p) = (p^2 + m^2)^{1/2}$ (here, as in Section 7.4, we are using "unconventional" units in which $h = 1$, i.e. $\hbar = 1/2\pi$, for an easier comparison with experimental variables). We then expand the dilaton field as

$$X = \frac{1}{2} \int_{-\infty}^{\infty} dp \int_{\Omega_2} d^2\widehat{n} \left[X(p, \widehat{n}) e^{2\pi i(p\widehat{n} \cdot \vec{x} - Et)} + X^*(p, \widehat{n}) e^{-2\pi i(p\widehat{n} \cdot \vec{x} - Et)} \right], \quad (9.64)$$

and the same for Φ and Ψ. The unit vector $\widehat{n}$ specifies the propagation direction on the two-sphere Ω_2, so that for $m \to 0$, $p \to E = \nu$, we exactly recover the gravity-wave expansion (7.234). The tidal stress tensor (9.63) thus becomes

$$M_{ij} = \frac{1}{2} \int_{-\infty}^{\infty} dp \int_{\Omega_2} d^2\widehat{n} \, (2\pi E)^2 \left[\frac{1}{2} \epsilon_{ij}^A h_A + \delta_{ij} \Psi - n_i n_j \Phi \right.$$

$$\left. + \frac{m^2}{E^2} n_i n_j \Phi + q \frac{p^2}{E^2} n_i n_j X \right] e^{2\pi i(p\widehat{n} \cdot \vec{x} - Et)} + \text{h.c.}, \quad (9.65)$$

where ϵ_{ij}^A is the transverse, traceless polarization tensor of the (spin-two) gravitational radiation (we have used the relation $p^2 = E^2 - m^2$ for the Φ component of the scalar fluctuations).

We now assume that the dilaton is the only source of scalar metric perturbations so that, in the absence of anisotropic stresses, $\Phi = \Psi$ (see Section 8.1). Introducing the transverse and longitudinal projections of the scalar stresses, defined respectively by

$$T_{ij} = \delta_{ij} - n_i n_j, \qquad L_{ij} = n_i n_j, \quad (9.66)$$

the tidal stress tensor can then be rewritten as

$$M_{ij} = \frac{1}{2} \int_{-\infty}^{\infty} dp \int_{\Omega_2} d^2\widehat{n} \, (2\pi E)^2 \left[\frac{1}{2} \epsilon_{ij}^A h_A + \left(T_{ij} + \frac{m^2}{E^2} L_{ij} \right) \Psi + q \frac{p^2}{E^2} L_{ij} X \right]$$

$$\times e^{2\pi i(p\widehat{n} \cdot \vec{x} - Et)} + \text{h.c.} \quad (9.67)$$

Finally, we define $M_{ij} = -\ddot{F}_{ij}$, so as to make explicit the second derivatives appearing in the components of the tidal tensor. Projecting onto D^{ij}, and separating the physical strain $h = D^{ij}F_{ij}$ into tensor, scalar geodesic and scalar non-geodesic parts, we eventually obtain

$$h(t) = \frac{1}{2}\int_{-\infty}^{\infty} dp \int_{\Omega_2} d^2\hat{n} \left[F^A(\hat{n})h_A(p,\hat{n}) + F^{\text{geo}}(\hat{n})\Psi(p,\hat{n}) + F^{\text{ng}}(\hat{n})X(p,\hat{n}) \right]$$
$$\times e^{2\pi i(p\hat{n}\cdot\vec{x}-Et)} + \text{h.c.}, \tag{9.68}$$

with the following antenna-pattern functions,

$$F^A = \frac{1}{2}D^{ij}\epsilon^A_{ij}, \tag{9.69}$$

$$F^{\text{geo}} = D^{ij}\left(T_{ij} + \frac{m^2}{E^2}L_{ij}\right), \tag{9.70}$$

$$F^{\text{ng}} = q\frac{p^2}{E^2}D^{ij}L_{ij}. \tag{9.71}$$

Note that F^A exactly coincides with the tensor pattern function already introduced in Section 7.4.

An important difference from the case of pure tensor radiation is that the scalar component of the radiation background also contributes to the response of the detector with its longitudinal polarization states. Such a longitudinal contribution is present even in the ultrarelativistic limit $m \to 0$, $p \to E$, because of the direct coupling (9.71). In the opposite, non-relativistic limit $p \to 0$, $E \to m$, the geodesic strain tends to become isotropic, $T_{ij} + (m/E)^2L_{ij} \to \delta_{ij}$, while the non-geodesic one becomes sub-leading.

9.2.1 Pattern functions for interferometric detectors

For a concrete illustration of the important differences between scalar and tensor pattern functions we consider here an interferometric model of a gravitational antenna, whose arms are aligned along the orthogonal unit vectors $\hat{u}$ and $\hat{v}$. As already mentioned in Section 7.4, the response of an interferometric detector can be described in terms of two tensors: the so-called "differential mode", D^{ij}_-, and the "common mode", D^{ij}_+, defined by [27]

$$D^{ij}_{\pm} = u^i v^j \pm v^i u^j. \tag{9.72}$$

The theoretical and experimental analyses of the response to tensor gravitational radiation are usually concentrated on the differential mode, not only for practical motivations (the common mode is, in general, much more "noisy"), but also because the common mode is "blind" to one of the two tensor polarization states, as we shall see below. In the case of massive scalar radiation and longitudinal polarization states both modes may provide, in principle, an efficient response (modulo technical difficulties possibly associated with the higher level of noise present in D_+^{ij}).

To illustrate the response of the different modes of the interferometer to the different polarization states we may consider a convenient reference frame in which $\widehat{u}$ and $\widehat{v}$ are coaligned with the x_1 and x_2 axes, respectively, and in which the propagation direction of a generic incident wave, associated with the unit vector $\widehat{n}$, is specified by the polar and azimuthal angles φ and θ, respectively (see Fig. 9.3). Thus, in this frame

$$\widehat{u} = (1, 0, 0), \qquad \widehat{v} = (0, 1, 0), \qquad \widehat{n} = (\sin\theta\cos\varphi, \sin\theta\sin\varphi, \cos\theta). \quad (9.73)$$

For an explicit representation of the tensor polarizations, ϵ_{ij}^A, we also introduce two unit vectors $\widehat{x}$ and $\widehat{y}$ orthogonal to $\widehat{n}$ and to each other (see Fig. 9.3), with components

$$\widehat{x} = (\sin\varphi, -\cos\varphi, 0), \qquad \widehat{y} = (\cos\theta\cos\varphi, \cos\theta\sin\varphi, -\sin\theta). \quad (9.74)$$

In terms of these transverse vectors, the two independent (traceless, symmetric) "*plus*" and "*cross*" tensor polarization states (see Section 7.1) can then be parametrized as [27]

$$\epsilon_{ij}^{(+)} = x_i x_j - y_i y_j, \qquad \epsilon_{ij}^{(\times)} = x_i y_j + y_i x_j, \quad (9.75)$$

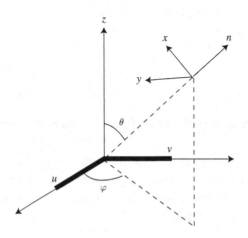

Figure 9.3 Relative orientation of the axes $\widehat{u}, \widehat{v}$ of the detector and the propagation direction $\widehat{n}$ of the incident radiation. The plane orthogonal to $\widehat{n}$ is spanned by the unit vectors $\widehat{x}$ and $\widehat{y}$.

where

$$x_i^2 = y_i^2 = 1, \qquad\qquad x_i y_i = x_i n_i = y_i n_i = 0. \qquad (9.76)$$

We are now in a position to compute the antenna pattern functions for the different polarizations of the incident radiation, and for the two modes of the interferometer. We start with the differential mode D_-^{ij}, considering separately scalar and tensor polarization states. For the scalar part we find the same pattern function (modulo a sign) for both transverse and longitudinal modes,

$$F_-^T(\hat{n}) \equiv D_-^{ij} T_{ij} = (u^i u^j - v^i v^j)(\delta_{ij} - n_i n_j) = -(u^i u^j - v^i v^j) n_i n_j$$

$$= -F_-^L(\hat{n}) \equiv -D_-^{ij} L_{ij} = -\sin^2\theta \cos 2\varphi, \qquad (9.77)$$

as a consequence of the fact that the response tensor is traceless, $D_-^{ij}\delta_{ij} = 0$. For the tensor part we obtain

$$F_-^{(+)}(\hat{n}) \equiv D_-^{ij}\epsilon_{ij}^{(+)} = (u^i u^j - v^i v^j)(x_i x_j - y_i y_j) = -(1+\cos^2\theta)\cos 2\phi,$$

$$(9.78)$$

$$F_-^{(\times)}(\hat{n}) \equiv D_-^{ij}\epsilon_{ij}^{(\times)} = (u^i u^j - v^i v^j)(x_i y_j + y_i x_j) = 2\cos\theta \sin 2\phi. \qquad (9.79)$$

The different responses of the differential mode to the scalar and tensor parts of the incident radiation are illustrated in Fig. 9.4.

The pattern functions for the common mode D_+^{ij} can be computed exactly with the same procedure. For the scalar component we separate transverse and longitudinal polarizations and we obtain, respectively,

$$F_+^T(\hat{n}) \equiv D_+^{ij} T_{ij} = (u^i u^j + v^i v^j)(\delta_{ij} - n_i n_j) = 1+\cos^2\theta, \qquad (9.80)$$

$$F_+^L(\hat{n}) \equiv D_+^{ij} L_{ij} = (u^i u^j + v^i v^j) n_i n_j = \sin^2\theta. \qquad (9.81)$$

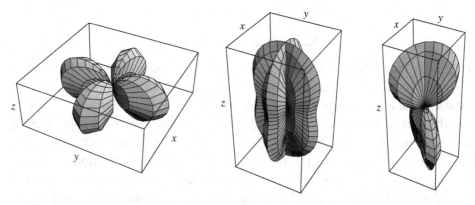

Figure 9.4 Parametric plots of the antenna pattern functions for the differential mode of an interferometer. The left panel illustrates the response to scalar radiation, according to Eq. (9.77); the central and right panels to tensor radiation, according to Eqs. (9.78) and (9.79), respectively.

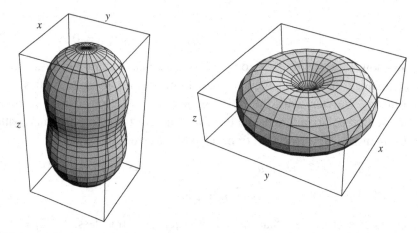

Figure 9.5 Parametric plots of the antenna pattern functions for the common mode of an interferometer. The left panel illustrates the response to the transverse scalar polarization, according to Eq. (9.80); the right panel to the longitudinal scalar polarization and to the "plus" tensor polarization, according to Eqs. (9.81) and (9.82), respectively.

For the tensor part we find that the response to the "plus" polarization state, $\epsilon_{ij}^{(+)}$, is the same as the response to scalar longitudinal radiation, while there is no response to the "cross" polarization:

$$F_+^{(+)}(\widehat{n}) \equiv D_+^{ij}\epsilon_{ij}^{(+)} = (u^i u^j + v^i v^j)(x_i x_j - y_i y_j) = \sin^2\theta, \qquad (9.82)$$

$$F_+^{(\times)}(\widehat{n}) \equiv D_+^{ij}\epsilon_{ij}^{(\times)} = (u^i u^j + v^i v^j)(x_i y_j + y_i x_j) = 0. \qquad (9.83)$$

The angular response of the common mode is illustrated in Fig. 9.5.

9.2.2 Signal-to-noise ratio

The rest of this chapter will be devoted to discussing the response of the interferometric antennas to a stochastic background of scalar radiation (see Section 7.4 for the discussion of the tensor component). Once the physical strain, Eq. (9.68), and the pattern functions, Eqs. (9.70), (9.71), are defined, the computation of the signal-to-noise ratio (SNR) proceeds as in the tensor case. One correlates the outputs $s_i(t) = h_i(t) + n_i(t)$, $i = 1, 2$, of two independent detectors, assuming that the noises n_i are statistically independent and much larger than the physical strains h_i (see Eqs. (7.239)–(7.241)). The relic dilaton background is assumed to be a stochastic collection of massive scalar waves, isotropic and stationary [27], satisfying the average conditions

$$\langle X(p, \widehat{n}) \rangle = 0,$$

$$\langle X(p, \widehat{n}) X(p', \widehat{n}') \rangle = 0 = \langle X^*(p, \widehat{n}) X^*(p', \widehat{n}') \rangle, \tag{9.84}$$

$$\langle X(p, \widehat{n}) X^*(p', \widehat{n}') \rangle = \delta(p - p') \frac{1}{4\pi} \delta^2(\widehat{n}, \widehat{n}') \frac{1}{2} S_\chi(p),$$

(equivalent to quantum expectation values on an n-particle state if the scalar field is quantized). Similar conditions apply also to the metric scalar background $\Psi(p, \widehat{n})$.

The dilaton strain density $S_\chi(p)$ has been normalized as in the case of tensor fluctuations (see Eq. (7.242)), and can be computed in terms of the spectral energy density Ω_χ, starting from the stress tensor of the free (canonically normalized) dilaton fluctuations,

$$T_{\mu\nu} = \frac{M_P^2}{2} \left[\partial_\mu X \partial_\nu X - \frac{1}{2} g_{\mu\nu} (\partial X)^2 + \frac{m^2}{2} X^2 \right]. \tag{9.85}$$

The energy density of a stochastic background of massive standing waves is given by the stochastic average (or quantum expectation value) of the T_0^0 component,

$$\rho_\chi = \langle T_0^0 \rangle = \frac{M_P^2}{4} \langle \dot{X}^2 + (\partial_i X)^2 + m^2 X^2 \rangle. \tag{9.86}$$

Expanding the dilaton fluctuations as in Eq. (9.64), using the conditions (9.84), and the overall background isotropy, we obtain

$$m^2 \langle X^2 \rangle = \frac{m^2}{2} \int_0^\infty dp \, S_\chi(p),$$

$$\langle (\partial_i X)^2 \rangle = \frac{(2\pi)^2}{2} \int_0^\infty dp \, S_\chi(p) p^2 \tag{9.87}$$

$$\langle \dot{X}^2 \rangle = \frac{(2\pi)^2}{2} \int_0^\infty dp \, S_\chi(p) E^2,$$

namely

$$\rho_\chi = \frac{M_P^2}{4} 4\pi^2 \int_0^\infty dp \, S_\chi(p) E^2(p), \tag{9.88}$$

from which, in units of critical energy density $\rho_c = 3H_0^2 M_P^2$,

$$S_\chi(p) = \frac{1}{\pi^2 M_P^2 E^2} \frac{d\rho_\chi}{dp} = \frac{3H_0^2}{\pi^2 |p| E^2} \Omega_\chi(p). \tag{9.89}$$

A similar equation relates the strain density of scalar metric perturbations, S_ψ, to the associated energy density spectrum, Ω_ψ. In the massless case $p = E = \nu$, one recovers the corresponding relation for a stochastic graviton background, Eq. (7.224), modulo the numerical factor $1/4$. This factor is due to the fact that the graviton spectrum $\Omega_g(\nu)$ contains, for a given $S_g(\nu)$, the contribution of two polarization states, each of them associated with two possible helicity configurations, ± 2.

We can now correlate the outputs s_i of the two detectors over a given observation time T, using an appropriate "filter function" $Q(t)$ (to be chosen in such a way as to optimize SNR). One first obtains

$$\langle S \rangle = \int_{-T/2}^{T/2} dt\, dt'\, \langle h_1(t) h_2(t') \rangle Q(t-t'), \tag{9.90}$$

exactly as in the case of a tensor background, Eq. (7.246). The scalar strains h_i may contain in general both the geodesic and non-geodesic contributions, according to Eq. (9.68): in order to correlate the signals, however, the two contributions have to be discussed separately. In the subsequent computations we will consider, in particular, the direct (non-geodesic) coupling of the dilaton fluctuations, producing the following strains at the detector positions $\vec{x}_i$ ($i = 1, 2$):

$$h_i(t) = \frac{1}{2} \int_{-\infty}^{\infty} dp \int_{\Omega_2} d^2\hat{n} \left[F_i^{\mathrm{ng}}(\hat{n}) X(p,\hat{n})\, e^{2\pi i(p\hat{n}\cdot\vec{x}_i - Et)} + \mathrm{h.c.} \right]. \tag{9.91}$$

The computation of the geodesic contribution proceeds in the same way, with the obvious replacements $F^{\mathrm{ng}} \to F^{\mathrm{geo}}$ and $X \to \Psi$.

Inserting the above expansion into Eq. (9.90), using the stochastic average conditions, and the reality of the $F_i(\hat{n})$ functions, one is led to

$$\langle S \rangle = \frac{1}{4} \int_{-T/2}^{T/2} dt\, dt' \int_{-\infty}^{+\infty} dE'\, Q(E') \int_{-\infty}^{+\infty} dp \int_{\Omega_2} d^2\hat{n}\, \frac{S_X(p)}{8\pi} F_1^{\mathrm{ng}}(\hat{n})\, F_2^{\mathrm{ng}}(\hat{n})$$
$$\times e^{2\pi i p\hat{n}\cdot\Delta\vec{x}} \left[e^{-2\pi i(E+E')(t-t')} + e^{2\pi i(E-E')(t-t')} \right],$$
$$\tag{9.92}$$

where $\Delta\vec{x} = \vec{x}_1 - \vec{x}_2$ is the spatial separation of the two detectors, and $Q(E)$ is the Fourier transform of the filter function,

$$Q(t-t') = \int_{-\infty}^{+\infty} dE\, Q(E) e^{-2\pi i E(t-t')}. \tag{9.93}$$

Finally, F_i^{ng} are generic non-geodesic pattern functions referred to the response tensor D_{ij} of a generic antenna. Their angular integration defines, as in the tensor case, the "overlap reduction function",

$$\gamma(p) = \frac{1}{N} \int_{\Omega_2} d^2\hat{n}\, F_1^{\mathrm{ng}}(\hat{n})\, F_2^{\mathrm{ng}}(\hat{n})\, e^{2\pi i p\hat{n}\cdot\Delta\vec{x}}, \tag{9.94}$$

which modulates the correlated signal $\langle S \rangle$ according to the relative orientation and separation of the two detectors (the normalization factor N can be chosen in such a way that $\gamma = 1$ for coincident and coaligned detectors).

In order to compute the time integrals we assume that the observation time T is much larger than the typical time intervals over which $Q \neq 0$. We can thus approximate the integral over dt' by extending the limits to $\pm\infty$, as in the case of tensor radiation: this leads to the delta functions $\delta(E - E')$ and $\delta(E + E')$. The integrations over dE' and dt then become trivial, and we obtain

$$\langle S \rangle = \frac{NT}{16\pi} \int_{-\infty}^{+\infty} dp\, \gamma(p) Q(E(p)) S_\chi(p)$$

$$= NT \frac{3H_0^2}{16\pi^3} \int_{-\infty}^{+\infty} \frac{dp}{|p|E^2} \gamma(p) Q(E(p)) \Omega_\chi(p), \qquad (9.95)$$

where we have assumed $Q(E) = Q(-E)$ (i.e. $Q(t - t') = Q(t' - t)$). Switching to the frequency domain, and setting $E = \nu = (p^2 + m^2)^{1/2}$, $dp = (\nu/p)d\nu$, the momentum integral becomes an integral over all frequencies $|\nu| \geq m$, and can be rewritten as

$$\langle S \rangle = NT \frac{3H_0^2}{16\pi^3} \int_{-\infty}^{+\infty} \frac{d\nu}{\nu(\nu^2 - m^2)} \gamma\left(\sqrt{\nu^2 - m^2}\right) Q(\nu) \Omega_\chi\left(\sqrt{\nu^2 - m^2}\right)$$

$$\times [\theta(\nu - m) + \theta(-\nu - m)], \qquad (9.96)$$

where θ is the Heaviside step function. In the limit $m \to 0$ one recovers the result (7.250) relative to the case of tensor gravitational radiation (modulo a numerical factor due to polarization differences). Besides this formal analogy, however, there are important differences due to the spectral energy density of the cosmic background, $\Omega_\chi(p)$, and to the different pattern functions.

The computation of the variance, $\Delta S^2 = \langle S^2 \rangle - \langle S \rangle^2$, is identical to the computation already performed for the tensor radiation background (see Eq. (7.255)), and will not be repeated here. Using the inner product defined in Eq. (7.257) we can then write the SNR as

$$(\text{SNR})^2 = \frac{\langle S \rangle^2}{(\Delta S)^2} = T \left(\frac{3NH_0^2}{8\pi^3}\right)^2 \frac{(Q, A)^2}{(Q, Q)}, \qquad (9.97)$$

where

$$A = \frac{\gamma\left(\sqrt{\nu^2 - m^2}\right) \Omega_\chi\left(\sqrt{\nu^2 - m^2}\right)}{|\nu| (\nu^2 - m^2) P_1(|\nu|) P_2(|\nu|)} [\theta(\nu - m) + \theta(-\nu - m)]. \qquad (9.98)$$

As in the graviton case the value of SNR is maximized by the optimal filtering choice, corresponding to $Q = \lambda A$. Using this choice, and switching to an integral over the positive momentum domain, we are led to the final result [26]

$$\text{SNR} = \frac{3N\sqrt{T}H_0^2}{8\pi^3}(A, A)^{1/2}$$

$$= \frac{3NH_0^2}{8\pi^3}\left[2T\int_0^\infty \frac{dp}{p^3(p^2+m^2)^{3/2}}\frac{\gamma^2(p)\,\Omega_\chi^2(p)}{P_1(\sqrt{p^2+m^2})\,P_2(\sqrt{p^2+m^2})}\right]^{1/2}.$$

$$(9.99)$$

By replacing F_i^{ng} with F_i^{geo} in the overlap function (9.94), and $\Omega_\chi(p)$ with $\Omega_\psi(p)$, one can also immediately obtain the corresponding SNR for the geodesic strain associated with scalar metric perturbations. These results are valid for all types of detectors, i.e. for any given form of the response tensor D_{ij} appearing in the definition of the pattern functions.

9.2.3 Non-relativistic backgrounds

Apart from the different spectrum of the relic radiation background, the obtained signal-to-noise ratio differs from the previous result (7.260) in two main respects: (*i*) the overlap function $\gamma(p)$, defined in terms of the scalar pattern functions (9.70) and (9.71), and (*ii*) the presence of the mass in the argument of the noise power spectra P_i. As already stressed, the scalar strains h_i are different from those induced by the tensor (spin-two) radiation because of the different polarization states; in addition, as we shall now discuss, the mass dependence of the noise has an important impact on the resonant response of the detector.

In a typical power spectrum, $P_i(\nu)$, the minimum level of noise is reached around a (rather narrow) frequency band ν_0 (see for instance Figs. 7.9 and 7.10). Outside this band the noise diverges, $P_i \to \infty$, and the signal becomes negligible, SNR $\to 0$. As $\nu = (p^2+m^2)^{1/2}$ there are, in principle, three possibilities.

(1) If $m \gg \nu_0$ then the noise is always outside the sensitivity band (i.e. far from the minimum), since $P_i(\sqrt{p^2+m^2}) \gg P_i(\nu_0)$ for all modes p. The induced signal is expected to be negligible for both the relativistic and non-relativistic branches of the spectrum.

(2) If $m \ll \nu_0$ then the sensitivity band of the detector may overlap with the relativistic branch of the scalar spectrum, for those modes with $p \sim \nu \sim \nu_0$. The non-relativistic branch $p < m$ always corresponds to a very high noise, $P_i(m) \gg P_i(\nu_0)$, and to a negligible signal.

(3) If $m \sim \nu_0$ the noise stays at a minimum for the whole non-relativistic branch of the scalar spectrum, since $P_i(\nu) \simeq P_i(m) \sim P_i(\nu_0)$ for all modes with $0 \le p \lesssim m$. The relativistic sector $p \gg m$, on the contrary, corresponds to a high noise, $P_i \gg P(\nu_0)$, and to a negligible signal.

Therefore, it is possible to obtain a resonant response even to a *massive, non-relativistic* background of scalar particles, provided their mass lies within the frequency range of maximal sensitivity of the two detectors [25, 26]. For the presently operating, Earth-based, gravitational antennas the resonant band may vary from the 1 Hz to the 1 kHz range: the maximal sensitivity to a non-relativistic background is thus in the mass range

$$10^{-15}\,\text{eV} \lesssim m \lesssim 10^{-12}\,\text{eV}. \tag{9.100}$$

The rest of this section will be devoted to discussing the detector response to a non-relativistic scalar background; in the relativistic case we can neglect the mass everywhere in Eq. (9.99), and the subsequent analysis is exactly the same as that performed for a background of tensor radiation in Section 7.4, except for the different polarization states of the scalar particles.

Present studies on the detection of cosmic scalar backgrounds concern both interferometric [25, 26, 28] and spherical (resonant-mass) [29] detectors. Here we start by considering the interaction of a scalar background with the differential mode of an interferometer, described by the response tensor D^{ij}_- (the interaction with the common mode D^{ij}_+ will be discussed in Appendix 9A). For such a mode the detector tensor is traceless, $D^{ij}_- \delta_{ij} = 0$, so that the geodesic and non-geodesic pattern functions (9.70) and (9.71) turn out to be proportional, namely

$$F^{\text{geo}}_- = -\left(\frac{p}{E}\right)^2 D^{ij}_- L_{ij},$$

$$F^{\text{ng}}_- = -qF^{\text{geo}}_- = -q\left(\frac{p}{E}\right)^2 F^{\text{rel}}_-, \tag{9.101}$$

where we have called F^{rel}_- the pattern function of a relativistic scalar mode with $p = E$. This relation also shows that the response to non-relativistic modes is strongly suppressed with respect to the relativistic response, because of the factor $p/E \ll 1$. The overlap function, in particular, is quadratic in F, so that

$$\gamma^{\text{nr}}(p) = \left(\frac{p}{E}\right)^4 \gamma^{\text{rel}}(p) \simeq \left(\frac{p}{m}\right)^4 \gamma^{\text{rel}}(p), \qquad p \lesssim m. \tag{9.102}$$

Such a suppression, however, may become ineffective if the scalar spectrum is peaked at $p = m$. If, in addition, the peak intensity approaches the limiting value saturating the critical density bound, $\Omega_\chi \sim 1$, then the experimental detection may become compatible with both the expected sensitivities of present (and advanced) interferometers, and the present experimental limits on the dilaton coupling strength [25, 26].

In order to illustrate this interesting possibility, let us consider the simplified situation in which the non-relativistic dilaton spectrum is peaked at $p = m$, and interacts with two identical interferometers with spectral noises $P_1 = P_2 = P$, dilaton charges $q_1 = q_2 = q$, ideally arranged in the setup of maximal overlap in which $\gamma = 1$ for the relativistic branch of the spectrum (the associated normalization factor is then $N = 4\pi/15$, as computed in [24, 26]). The non-geodesic response to a non-relativistic background is thus governed by the overlap factor $\gamma^{ng} = q^2(p/m)^4$, while in the geodesic case we have the same expression with $q = 1$ (see Eq. (9.101)). Assuming that the SNR integral (9.99) is dominated by the peak value of the dilaton spectrum, $\Omega_\chi = \Omega_m$ at $p = m$, the condition of detectable signal, SNR > 1, can be expressed as

$$\sqrt{2T} \frac{q^2 H_0^2}{10\pi^2} \frac{\Omega_m}{m^{5/2} P(m)} \gtrsim 1, \qquad (9.103)$$

namely as

$$m^{5/2} P(m) \lesssim 10^{-33} \, \text{Hz}^{3/2} \left(\frac{T}{4 \times 10^7 \, \text{s}}\right)^{1/2} q^2 h^2 \Omega_m. \qquad (9.104)$$

We can eventually consider, as a possible realistic example of available interferometric sensitivity, the noise power spectrum of LIGO, illustrated in Fig. 7.10 for the present and advanced generations. The intersection of that power spectrum with the condition (9.104) provides a rough estimate of the mass values possibly compatible with detection, for the given class of relic dilaton backgrounds.

The allowed mass window depends on the relic intensity Ω_m and on the strength q of the dilaton coupling. Large couplings ($q^2 \sim 1$), associated with the range of very small masses that we are considering, can only refer to the indirect (geodesic) response of the interferometer to the spectrum of scalar metric perturbations induced by the dilaton. In that case, however, the background intensity is expected to be much smaller than the limiting value $\Omega_m \sim 1$. In the case of direct (non-geodesic) coupling to the detectors, the background intensity can be higher, but the scalar charge has to be appropriately suppressed to avoid contradictions with known gravitational phenomenology. In particular, in the relevant mass range of Eq. (9.100), the existing phenomenological bounds [14] can be parametrized as follows:

$$\log q^2 \lesssim \begin{cases} -7, & 1 \, \text{Hz} \lesssim m \lesssim 10 \, \text{Hz}, \\ -7 + \log(m/10 \, \text{Hz}), & 10 \, \text{Hz} \lesssim m \lesssim 1 \, \text{kHz}, \end{cases} \qquad (9.105)$$

for universal dilaton interactions, and

$$\log q^2 \lesssim \begin{cases} -8, & 1 \, \text{Hz} \lesssim m \lesssim 10 \, \text{Hz}, \\ -8 + \log(m/10 \, \text{Hz}), & 10 \, \text{Hz} \lesssim m \lesssim 1 \, \text{kHz}, \end{cases} \qquad (9.106)$$

for composition-dependent dilaton interactions.

For a common discussion of the geodesic and non-geodesic case we can use as free parameter the factor $q^2 h^2 \Omega_m$ appearing in Eq. (9.104), taking into account for q^2 the limits (9.105) and (9.106). The condition (9.104) of detectable background,

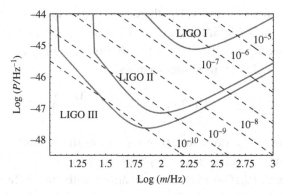

Figure 9.6 Noise power spectrum of the three LIGO generations (bold curves), compared with the condition (9.104) of detectable background (dashed lines), plotted at various fixed values of the parameter $q^2 h^2 \Omega_m$ ranging from 10^{-10} to 10^{-5}.

at various fixed values of the parameter $q^2 h^2 \Omega_m$, is then compared in Fig. 9.6 with the spectral noises of the three LIGO generations, Eqs. (7.230)–(7.232), for $T = 4 \times 10^7$ s. The allowed region of the plane $\{m, P(m)\}$ corresponding to a detectable background is placed *above* the bold noise curves and *below* the dashed lines, representing the upper limit (9.104) for different values of $q^2 h^2 \Omega_m$. This limit may be interpreted either as a constraint on the intensity Ω_m of a background geodesically coupled ($q^2 = 1$) to the detector, or as a limit on the non-geodesic coupling strength q^2 of a scalar background of given energy density Ω_m.

Quite independently of the possible interpretation, the main message of Fig. 9.6 is that the sensitivity of the next-generation interferometers is (in principle) already sufficient to detect a cosmic background of non-relativistic scalar particles, even in the case of very weak coupling to macroscopic matter, provided their density is sufficiently close to the critical one, and the mass is within the resonant frequency band of the antennas. The approximate results of this qualitative discussion have been confirmed by a more accurate numerical analysis [26] which takes into account the real parameters of the LIGO pair of interferometers: the analysis has been performed by parametrizing the non-relativistic dilaton spectrum as

$$\Omega_\chi(p) = \Omega_m \left(\frac{p}{m}\right)^\delta, \qquad p \leq m, \qquad 0 \leq \delta \leq 3, \qquad (9.107)$$

and integrating numerically Eq. (9.99), using for q^2 the values saturating the bounds (9.105) and (9.106).

Concerning the value of the dilaton mass, it is perhaps worth stressing that the very small value required for a resonant response of the detectors might be not so unrealistic if we consider a mass generated by the perturbative mechanism of radiative corrections. For a scalar field, gravitationally coupled (with dimensionless strength q) to fermions of mass M_f, there are in fact quantum loop corrections to the mass of order $qM_f(\Lambda/M_P)$, where Λ/M_P is the cut-off scale (typically 1 TeV) in Planck units (see for instance [30]). The dilaton coupling to ordinary baryonic matter (with $M_f \sim 1$ GeV) thus induces a mass

$$m \sim q\,10^{-6}\,\mathrm{eV} \left(\frac{\Lambda}{1\,\mathrm{Tev}}\right)\left(\frac{M_f}{1\,\mathrm{Gev}}\right). \tag{9.108}$$

Assuming that this is the dominant contribution to the dilaton mass, it turns out (perhaps surprisingly) that a value of q smaller than, but not very far from, the present upper limits (9.106) may be compatible with the preferred mass range (9.100).

It should be noted, finally, that the assumption of a non-relativistic peak at $p = m$ (used to obtain the result (9.104)) cannot be applied to the "minimal" spectrum (9.12) when it is peaked at $p = p_m = p_1(m/H_1)^{1/2}$: the condition $p_m = m$ would imply, in fact, $m \sim 10^{-8}$ (eV$^2/M_P$), well below the resonant frequency band (9.100). If, on the contrary, the mass does lie in the required range, then the amplitude of the spectrum (9.12) at $p = m$ is always well below its possible maximal intensity (see Fig. 9.1), and this strongly disfavors its possible detection [26].

9.2.4 Signal enhancement for flat spectra

The presence of a peak at $p = m$ was previously assumed to compensate the strong suppression factor p/m which characterizes the response of the differential mode to non-relativistic radiation. Such a suppression is closely related to the fact that D_-^{ij} is traceless, and may be absent for other response modes and/or for different gravitational detectors: there is no such suppression, for instance, for the common mode of an interferometer [28] (see Appendix 9A), and for the monopole mode of a resonant spherical detector [29].

This last case is characterized by the trivial response tensor $D_0^{ij} = \delta^{ij}$, so that $D_0^{ij}T_{ij} = 2$, $D_0^{ij}L_{ij} = 1$. As a consequence, both the geodesic and non-geodesic pattern functions are isotropic:

$$F_0^{\mathrm{geo}} = \frac{2p^2 + 3m^2}{p^2 + m^2}, \qquad F_0^{\mathrm{ng}} = q\,\frac{p^2}{p^2 + m^2}. \tag{9.109}$$

The non-relativistic response is still suppressed for non-geodesic interactions, while the suppression disappears in the geodesic case (actually, there is an enhancement factor $3/2$ with respect to the relativistic modes with $m = 0$). The corresponding geodesic overlap function, for two spheres with spatial separation $|\vec{x}_1 - \vec{x}_2| = d$, is then given by Eq. (9.94) as

$$\gamma(p) = \frac{15}{4\pi} \int_{\Omega_s} d^2\hat{n} \left(F_0^{geo}\right)^2 e^{2i\pi p\hat{n}\cdot(\vec{x}_1 - \vec{x}_2)}$$

$$= \frac{15}{2\pi} \left(\frac{2p^2 + 3m^2}{p^2 + m^2}\right)^2 \frac{\sin(2\pi pd)}{pd} \tag{9.110}$$

(we have chosen the same normalization factor as in the interferometric case, for an easier comparison). In the limit $p \to 0$ we thus obtain $\gamma \to 135$ and we find, as a first result, that the value of SNR is greatly enhanced (by a fixed amount) with respect to the correlation of the differential modes of two interferometers, in particular if the spectrum is not peaked at $p = m$.

The fact that γ tends to a constant in the infrared limit $p \to 0$ has, however, a second important consequence if the non-relativistic scalar spectrum is sufficiently flat. According to Eq. (9.99), the SNR is proportional to the integral

$$(SNR)^2 \sim T \int_0^{p_1} dp \frac{\gamma^2(p)\Omega^2(p)}{p^3 E^3 P_1(E) P_2(E)}, \tag{9.111}$$

where we can assume that the spectral distribution $\Omega(p)$ is a power-law function of p, with an ultraviolet cut-off at $p = p_1$. In the massless case ($p = E$) this integral is always convergent, even in the infrared limit because, when $p = E \to 0$, the physical strains are produced outside the sensitivity band of the detectors, where the noises shoot up to infinity, $P_i(E) \to \infty$. For $m \neq 0$, on the contrary, one finds that in the infrared limit the noises stay frozen at the frequency scale fixed by the mass of the scalar background,

$$P_i(E) \to P_i(m) = \text{const}, \qquad p \to 0, \tag{9.112}$$

and the behavior of the integral depends on $\gamma(p)$ and $\Omega(p)$.

For the differential mode of two interferometers the overlap function contains the suppression factor $\gamma^2 \sim (p/m)^8$, which may be reasonably expected to force the integral to be convergent in the limit $p \to 0$ (modulo spectra with dramatically strong infrared divergences). For the monopole modes of two spherical detectors, however, $\gamma(p) \to \gamma_0 = \text{const}$, when $p \to 0$. For the non-relativistic sector ($p < m$), and for a generic power-law spectrum, $\Omega \sim p^\delta$, we then find that the SNR integral is dominated by the infrared limit, and we obtain

$$(SNR)^2 \sim \frac{T\gamma_0^2}{m^3 P_1 P_2} \left[p^{2(\delta-1)}\right]_0^m, \tag{9.113}$$

i.e. the integral diverges for all spectra (even blue, $\delta > 0$) with $\delta < 1$! This would seem to imply an infinite signal for the geodesic response of two spherical detectors interacting with a flat enough massive scalar spectrum, quite independently of the level of instrumental noise and of the absolute intensity of the background.

This infrared divergence is unphysical, of course, and can be removed by taking into account that the observation time T is *not* infinite, and is therefore associated with a minimum (non-zero) resolvable frequency interval, defined by the condition $\Delta\nu = \Delta E \gtrsim T^{-1}$. For $p \to 0$, on the other hand, $E \simeq m + p^2/2m$, so that the uncertainty condition defines a *minimum* momentum scale [29]

$$p \gtrsim p_{\min} = \left(\frac{2m}{T}\right)^{1/2},\tag{9.114}$$

acting as infrared cut-off and regularizing the momentum integral. Indeed, modes with $p < p_{\min}$ cannot be resolved by instruments working during a finite time interval T, and must be included in the constant background over which scalar perturbations propagate, without contributing to the signal. The lower limit $p = 0$ in Eq. (9.113) has thus to be replaced by $p = p_{\min}$, and this implies a modified dependence of SNR on the integration time T in the case of flat enough spectra:

$$\text{SNR} \sim T^{1/2}\left[p^{\delta-1}\right]_{p_{\min}}^{m} \sim \begin{cases} T^{1/2}, & \delta > 1, \\ T^{1-\delta/2}, & \delta < 1 \end{cases}\tag{9.115}$$

(for $\delta = 1$ there is only an unimportant logarithmic correction to the standard time dependence $T^{1/2}$). The modification depends on whether the integral is dominated (or not) by its lower limit, and disappears in the case of a massless background.

Such an anomalous (and, in particular, faster) growth of SNR with T, for $\delta < 1$, may produce an important enhancement of the sensitivity of resonant spherical detectors to a cosmic background of non-relativistic scalar particles. For a quantitative illustration of this effect we compute here the *minimum* detectable non-relativistic background geodesically interacting with the monopole mode of two cross-correlated spheres.

We consider, for simplicity, a background characterized by the spectrum (9.107), using the geodesic overlap function (9.110) for two identical ($P_1 = P_2 = P$) and coincident ($d = 0$) detectors [29]. By imposing on Eq. (9.99) the detectability condition SNR $\gtrsim 5$, we obtain, for the two cases $\delta = 3/2$ and $\delta = 1/2$, respectively,

$$h^2\Omega_m \gtrsim \begin{cases} 10^{-5}\left(\dfrac{\text{SNR}}{5}\right)\left(\dfrac{T}{10^7\,\text{s}}\right)^{-1/2}\left(\dfrac{P}{10^{-46}\,\text{Hz}^{-1}}\right)\left(\dfrac{m}{3\times10^3\,\text{Hz}}\right)^{5/2}, & \delta = 3/2, \\[3mm] 10^{-8}\left(\dfrac{\text{SNR}}{5}\right)\left(\dfrac{T}{10^7\,\text{s}}\right)^{-3/4}\left(\dfrac{P}{10^{-46}\,\text{Hz}^{-1}}\right)\left(\dfrac{m}{3\times10^3\,\text{Hz}}\right)^{9/4}, & \delta = 1/2 \end{cases}\tag{9.116}$$

(the result for $\delta = 3/2$ also applies to all spectra with $\delta > 1$). We have used, as reference values, a typical observation time ~ 1 year, and the minimum (expected) noise density of a resonant sphere [31, 32] in the frequency band $\nu_0 = 3 \times 10^3$ Hz. We have also assumed that the scalar mass is in the same frequency range, otherwise, for $m \ll \nu_0$, the infrared cut-off (9.114) becomes ineffective, since the low-energy part of the integral is suppressed by the very high instrumental noise. What should be noted in the above result (besides the remarkable sensitivity to scalar backgrounds with energy density well below the critical limit) is the greatly enhanced sensitivity to backgrounds with spectral index $\delta < 1$, *at fixed observation time* T, fixed spectral noise P, and fixed scalar mass m.

It should be mentioned, as a final remark, that spherical-mass detectors (unlike interferometers) may offer the interesting possibility of tuning the resonant frequency over a rather wide range [31, 32, 33], thus opening up the sensitivity bandwidth. Scanning the corresponding mass window may lead to important information on the presence of massive, ultra-light scalar backgrounds of primordial origin, and provide unique constraints on minimal and non-minimal string cosmology models of inflation.

9.3 Dilaton dark energy and late-time acceleration

According to the standard scenario, the cosmological phase subsequent to the radiation-dominated era (including also the present epoch) should be dominated by a non-relativistic and incoherent ($p = 0$) distribution of cosmic matter, and thus characterized by a decelerated expansion ($\dot{a} > 0$, $\ddot{a} < 0$), according to Eqs. (1.25). Assuming a negligible spatial curvature, as predicted by inflation (and as confirmed by recent observations, see e.g. [34]), one can then deduce from the Einstein equation (1.26) a present matter density of critical order, $\rho_m(t_0) = \rho_c(t_0) = 3M_P^2 H_0^2 \sim 10^{-29}$ g/cm^3. The difficulty that only a tiny fraction (about 1%) of this energy density turns out to be optically visible – the so-called "missing mass" problem (see e.g. [35]) – can be easily solved by assuming that the present Universe is dominated on large scales by some non-baryonic (and possibly exotic) "dark-matter" component. The possible composition (WIMPS, axions, neutralinos, ...) and properties of such dark-matter fluid have been under active study for more than two decades (see for instance [36]).

As already pointed out in Chapter 1, this standard picture has been challenged by a series of recent observations, primarily by the study of the Hubble diagram of Type Ia Supernovae (SNIa) [37, 38]: the optimal fit of their luminosity distance–redshift distribution seems to require a model in which our present Universe undergoes a phase of accelerated expansion, $\ddot{a} > 0$, and is thus dominated by a cosmic component with negative pressure (with $p < -\rho/3$, according to the

Einstein equations), dubbed "dark energy". Recent data of SNIa at $z > 1$ [39], as well as more extended analyses including data from CMB anisotropies, large-scale structure and Hubble space telescope [40, 41], seem to confirm that the pressureless dark-matter component is only a minority fraction of the present critical density: $0.22 \lesssim \Omega_m \lesssim 0.35$. The remaining, dominating contribution $1 - \Omega_m$ should come from the dark-energy component which, if represented as a barotropic perfect fluid, may be characterized by an equation of state $p/\rho = w$ with $-1.38 \lesssim w \lesssim -0.82$ (see also Section 1.1 for more recent measurements).

The present large-scale observations are thus perfectly compatible, in principle, with a Universe dominated by a cosmological constant Λ, which, as stressed in Section 1.1, is dynamically equivalent to a perfect fluid with equation of state $p/\rho = -1$. This simplest explanation of the observational data is affected, however, by two important conceptual difficulties. The first one concerns a "naturalness" (or fine-tuning) problem: why is $\Lambda \sim \rho_c(t_0) \ll M_{\rm P}^4$ so small with respect to the (apparently) most natural particle physics prediction $\Lambda \sim M_{\rm P}^4$? The second difficulty concerns a problem of "cosmic coincidence": why are the dark-energy density Λ and the dark-matter density $\rho_m(t)$ of the same order just at the present epoch $t = t_0$? After all, the value of Λ is frozen at a constant, while $\rho(t)$ is running as a^{-3}, and may intersect the value of Λ at one given epoch only.

The problems just mentioned could be solved, or at least alleviated, in the context of a less-trivial scenario in which the dark-energy density is not a constant, and its variation in time is connected, in some way, to the variation of the dark-matter density. This idea is the basis of the scenario of *quintessence* [42–45], in which the role of the cosmic dark energy is played by a scalar field ϕ, slow-rolling along an appropriate self-interaction potential. Choosing, for instance, $V(\phi) \sim \phi^{-\alpha}$, $\alpha > 0$ one can obtain the so-called "tracking" solutions [46, 47] in which the effective equation of state of the scalar field changes in time, following the background evolution: the effective pressure, in particular, may become negative as the Universe transforms from radiation- to matter-dominated, so that the scalar potential energy is doomed to become critical at late enough times, $V(\phi) \sim \rho_c$, quite irrespective of the given initial conditions.

In such a context, however, the dark-energy density $\rho_\phi \sim V(\phi) \sim H^2$, and the dark-matter density $\rho_m \sim a^{-3}$, have asymptotically different time dependence, and the cosmic coincidence can hardly be explained. A possible solution to this problem can be obtained either by including an appropriate bulk viscosity term into the dark-matter stress tensor [48, 49], or by introducing a direct (and strong enough) non-minimal gravitational coupling between ρ_ϕ and ρ_m, as proposed in models of "coupled quintessence" [50, 51, 52]. The aim of this section is to show that the string theory dilaton may provide a natural implementation of the coupled quintessence scenario, provided the cosmological run of the dilaton does not stop after entering the strong coupling regime $\exp\phi \gtrsim 1$. We consider, in particular, the scenario in which the dilaton keeps rolling to plus infinity along an exponentially suppressed (non-perturbative) potential [53], and in which the limit $\phi \to \infty$ is characterized by the asymptotic saturation of all the dilaton loop corrections [54], in such a way that they approach a finite limit as $t \to \infty$.

It should be noted that in this case a unique fundamental field, the dilaton, might be responsible for both the primordial (high-energy) inflationary phase and the present cosmic acceleration, occurring at an exceedingly lower curvature scale: the situation is similar to what happens in the context of the cyclic/ekpyrotic scenario [55], where a single modulus field, the interbrane distance, controls the primordial phase of brane collision and the present large-scale acceleration (see Section 10.4). The dilaton model that we will consider here is not cyclic, unlike the scenario illustrated in Section 10.4; the model, however, could be easily extended to become cyclic, through a suitable modification of the effective potential producing a bounce of the dilaton motion in the strong coupling regime, and preparing the Universe for a new, future, pre-big bang phase (see the discussion at the end of Section 10.4).

9.3.1 Saturation of the loop corrections

In order to illustrate such a dilaton dark-energy scenario let us consider again the S-frame string effective action (9.35): we are working to lowest order in α' (as we are interested in the late-time, small-curvature regime), but we are including the dilaton-dependent quantum loop corrections (possibly to all orders), together with a non-perturbative potential $\widetilde{V}(\phi)$. We rewrite Eq. (9.35), for later convenience, as follows:

$$S = -\frac{1}{2\lambda_s^2} \int d^4x \sqrt{-\widetilde{g}} \left[e^{-\psi(\phi)} \widetilde{R} + Z(\phi)(\widetilde{\nabla}\phi)^2 + 2\lambda_s^2 \widetilde{V}(\phi) \right] + S_m(\widetilde{g}, \phi, \text{matter}),$$

(9.117)

where the tilde is to remind us that we are using the S-frame variables, and $\psi(\phi)$, $Z(\phi)$ are the dilaton "form factors" due to the loop corrections (other corrections are included inside the potential and the matter action S_m). In the weak coupling limit $\phi \to -\infty$ we have $\exp(-\psi) = Z = \exp(-\phi)$, and we recover the lowest-order string effective action with the possible addition of an instantonically suppressed potential, $\widetilde{V} \sim \exp[-\exp(-\phi)]$.

In the opposite, strong "bare coupling" limit $\phi \to \infty$ we assume the validity of an asymptotic Taylor expansion in inverse powers of the bare coupling constant $g_s^2 = \exp(\phi)$, following the spirit of the "induced gravity" models where the gravitational and gauge couplings saturate at small (finite) values because of the very large number N of fundamental gauge bosons entering the loop corrections

[53, 54]. Applying this assumption to the loop form factors, to the potential and to the scalar dilaton charge $q(\phi)$ (controlling the ratio of the charge density $\tilde{\sigma}$ to the energy density $\tilde{\rho}$ of a homogeneous gravtational source), we set, for $\phi \to \infty$,

$$
\begin{aligned}
e^{-\psi(\phi)} &= c_1^2 + b_1 e^{-\phi} + \mathcal{O}(e^{-2\phi}), \\
Z(\phi) &= -c_2^2 + b_2 e^{-\phi} + \mathcal{O}(e^{-2\phi}), \\
\tilde{V}(\phi) &= V_0 e^{-\phi} + \mathcal{O}(e^{-2\phi}), \\
q(\phi) &= q_0 + \mathcal{O}(e^{-2\phi})
\end{aligned}
\tag{9.118}
$$

(see also Eqs. (3.55)).

A few comments on the coefficients of the above expansion are in order. The dimensionless coefficients c_1^2 and c_2^2 are typically of order $N \sim 10^2$, because of their quantum loop origin, and because of the large dimensions of the GUT gauge groups. We note, in particular, that c_1^2 asymptotically controls the fundamental ratio between the string and the Planck scale, $c_1^2 = (\lambda_s/\lambda_P)^2$ (see Eq. (3.56)), and which is expected, indeed, to be a number of the above order. The coefficients b_1 and b_2, on the contrary, are dimensionless numbers of order one. The mass scale V_0, being of non-perturbative origin, should be related to the string scale in a typically instantonic way, namely,

$$
V_0 = M_s^4 e^{-4/\beta \alpha_{\text{GUT}}}, \tag{9.119}
$$

where $\alpha_{\text{GUT}} \simeq 1/25$ is the asymptotic value of the GUT gauge coupling, and β is some model-dependent loop coefficient. Finally, the asymptotic value q_0 of the dilaton charge is strongly dependent on the considered type of matter field: we may expect, for a small dilaton mass, that $q_0 \simeq 0$ (corresponding to an exponential suppression of the dilaton coupling) for electromagnetic radiation and ordinary macroscopic matter (such as baryons), in order to avoid unacceptably large deviations from the standard gravitational phenomenology. For the (possibly more exotic) dark-matter components, however, there is no phenomenological need for such suppression, and the asymptotic charge q_0 could be non-zero, and even large, in principle. If this is the case we are led to interesting, late-time deviations from the standard cosmological scenario.

For a simple discussion of such deviations let us include in the matter sources a radiation fluid, with $\tilde{\rho} = 3\tilde{p}$, a pressureless baryon-matter component, $\tilde{\rho}_b$, and the usual dark-matter component $\tilde{\rho}_m$. Considering a homogeneous, conformally flat metric background we can then write the cosmological equations for the action (9.117) using, for instance, the cosmic-time gauge. For the purposes of this section it will be convenient to concentrate our discussion on the transformed E-frame equations, defined by the conformal rescaling $\tilde{g}_{\mu\nu} = c_1^2 g_{\mu\nu} \exp\psi(\phi)$. The various cosmological variables are transformed as follows (see also Section 2.2):

$$\tilde{a} = c_1 a\, e^{\psi/2}, \qquad d\tilde{t} = c_1 dt\, e^{\psi/2}, \qquad V = c_1^4 \tilde{V} e^{2\psi},$$

$$\rho = c_1^2 \tilde{\rho}\, e^{2\psi}, \qquad p = c_1^2 \tilde{p}\, e^{2\psi} \qquad \sigma = c_1^2 \tilde{\sigma}\, e^{2\psi}. \tag{9.120}$$

In units $2\lambda_P^2 = 1$, and in the cosmic-time gauge, we can then write the E-frame gravitational equations in the form

$$6H^2 = \rho_r + \rho_b + \rho_m + \rho_\phi,$$

$$4\dot{H} + 6H^2 = -\frac{\rho_r}{3} - p_\phi, \tag{9.121}$$

where

$$\rho_\phi = \frac{k^2(\phi)}{2}\dot{\phi}^2 + V, \qquad p_\phi = \frac{k^2(\phi)}{2}\dot{\phi}^2 - V,$$

$$k^2(\phi) = 3\psi'^2 - 2e^{\psi}Z \tag{9.122}$$

(the prime denotes differentiation with respect to ϕ, the dot with respect to the E-frame cosmic time t). We assume that the dilaton charge densities of the baryon and radiation fluids can be safely neglected, $\sigma_r = 0 = \sigma_b$, and we call $q(\phi) = \sigma_m/\rho_m$ the dilaton charge of the (homogeneous) dark-matter component. The transformed (E-frame) equation for the dilaton field then takes the form

$$k^2(\ddot{\phi} + 3H\dot{\phi}) + kk'\dot{\phi}^2 + V' + \frac{1}{2}\left[\psi'\rho_b + (\psi' + q)\rho_m\right] = 0. \tag{9.123}$$

The combination of Eqs. (9.121) and (9.123) leads to the separate energy-conservation equation of the various fluid components:

$$\dot{\rho}_r + 4H\rho_r = 0, \tag{9.124}$$

$$\dot{\rho}_b + 3H\rho_b - \frac{1}{2}\dot{\phi}\psi'\rho_b = 0, \tag{9.125}$$

$$\dot{\rho}_m + 3H\rho_m - \frac{1}{2}\dot{\phi}\left(\psi' + q\right)\rho_m = 0. \tag{9.126}$$

Finally, using the definitions of ρ_ϕ and p_ϕ, we can also rewrite the dilaton equation in fluidodynamical form,

$$\dot{\rho}_\phi + 3H(\rho_\phi + p_\phi) + \frac{1}{2}\dot{\phi}\left[\psi'\rho_b + (\psi' + q)\rho_m\right] = 0. \tag{9.127}$$

It is important to note, at this point, that this system of coupled equations contains two types of dilaton coupling to the dust matter sources ρ_b and ρ_m. A first coupling is controlled by the loop form factor ψ', and is doomed to an exponential decay since, according to the asymptotic expansion (9.118), $\psi' \simeq c_1^{-2} \exp(-\phi)$ for $\phi \to \infty$. The other type of coupling is induced by the dilaton charge, and is only active for the dark-matter component ρ_m: this coupling, on the contrary, tends to grow with ϕ, approaching the constant asymptotic value q_0. A detailed analysis of the above coupled equations [53] shows that in a realistic cosmological scenario the effects of the dilaton charge (as well as those of the dilaton potential) may become important only at late enough epochs, when the Universe has already entered the matter-dominated regime. This means, in our context, that an efficient scenario of coupled quintessence must be a "two-phases" scenario, in which the dark-matter component evolves from an initial regime of weak and growing dilaton coupling, to a final regime in which the coupling is strong and asymptotically constant. For a qualitative illustration of such a background evolution let us now consider in some detail the behavior of the various cosmological components.

First of all we note that in the initial radiation-dominated phase, soon after inflation, no modification of the standard cosmological evolution is to be expected. Indeed, ρ_r is decoupled from the dilaton (see Eq. (9.124)), and since ρ_m, ρ_b and V are initially negligible in the standard scenario, then $\rho_\phi \sim a^{-6}$ (according to Eq. (9.127)), so that the dilaton components also rapidly become negligible (even starting from an initial configuration with $\rho_\phi \sim \rho_r$). Such a fast dilution of ρ_ϕ cannot continue, however, for the whole radiation-dominated phase because at late enough time, as soon as ρ_ϕ drops below ρ_m, the dilaton coupling $\rho_m \dot{\phi} \psi'$ becomes important and tends to "damp" the evolution of ϕ: the dilaton then enters a "focusing" regime in which its energy density evolves as $\rho_\phi \sim a^{-2}$, tending to approach the higher values of ρ_r and ρ_m (see [53] for a detailed discussion).

This effect stops at the equality epoch so that, at the beginning of the phase dominated by the dark-matter component, we can assume that ρ_ϕ is still sub-dominant (as well as ρ_b) with respect to ρ_m, and that $q(\phi)$ and $V(\phi)$ are also negligible. However, the (weak) coupling between ρ_ϕ and ρ_m, due to ψ', is responsible for a "dragging" effect producing the same time evolution for the dark-matter and the dark-energy densities, $\rho_\phi \sim \rho_m$, together with a slight modification of the standard behavior $\rho_m \sim a^{-3}$.

9.3.2 The dragging and freezing regimes

For an explicit, analytic description of the dragging regime we can note, according to the asymptotic expansion (9.118), that the loop coefficient $k(\phi)$ tends to a constant for $\phi \to \infty$, i.e. $k \to \sqrt{2}c_2/c_1$; in this limit we can thus rewrite the dilaton and dark-matter equations (9.123) and (9.126) in terms of the rescaled variable $\widehat{\phi} = k\phi$ (associated with a canonical kinetic term $\dot{\widehat{\phi}}^2/2$), respectively as follows:

$$\ddot{\widehat{\phi}} + 3H\dot{\widehat{\phi}} + \frac{\epsilon}{2}\rho_m = 0, \tag{9.128}$$

$$\dot{\rho}_m + 3H\rho_m - \frac{\epsilon}{2}\rho_m\dot{\widehat{\phi}} = 0. \tag{9.129}$$

Here $\epsilon = \psi'/k \simeq e^{-\phi}/(\sqrt{2}c_1 c_2) \ll 1$ is the small (loop-induced) coupling parameter. Neglecting its time dependence with respect to H and $\widehat{\phi}$ (for small enough time intervals), one easily finds that Eqs. (9.128) and (9.121) are solved by

$$\dot{\widehat{\phi}} = -2\epsilon H, \tag{9.130}$$

so that, from Eq. (9.129),

$$\rho_m \sim a^{-(3+\epsilon^2)} \sim H^2 \sim \dot{\widehat{\phi}}^2 \sim \rho_\phi,$$
$$a \sim t^{2/(3+\epsilon^2)}. \tag{9.131}$$

The (fully kinetic) dilaton dark energy ρ_ϕ is thus "dragged" along by the dark matter density. Because of this dragging (and even for a sub-dominant ρ_ϕ) the background evolution slightly deviates from the typical behavior of a dust-dominated Universe, in which $\rho \sim a^{-3}$, $a \sim t^{2/3}$. The cosmological expansion, however, remains decelerated.

We must now take into account that, as time goes on, the coupling parameter $\epsilon = \psi'/k$ tends to zero, while the dilaton charge grows, and tends to be stabilized at the constant value q_0. The Universe approaches, asymptotically, a late-time regime in which baryons (as well as radiation, and any other component of ordinary macroscopic matter) are decoupled from the dilaton, while dark matter becomes strongly coupled with a charge q_0. (We are considering here, for simplicity, a model with a unique type of exotic dark-matter component; but the discussion can be easily generalized to the case of multicomponent dark matter, with different dilaton charges for the different components.) Eventually, when the dilaton potential (9.118) comes into play, the Universe enters an asymptotic "freezing" phase in which the ratio ρ_m/ρ_ϕ becomes frozen at a final value, controlled by q_0 and by the leading loop coefficients c_1 and c_2. Interestingly enough, such a final configuration is accelerated, provided $q_0 > 1$.

For a convenient description of this asymptotic regime we can again use the canonical dilaton variable $\widehat{\phi} = k\phi$, in such a way that the dark-matter and dilaton equations, (9.126) and (9.127), can be written in the form

$$\dot{\rho}_m + 3H\rho_m - \frac{q_0}{2k}\rho_m\dot{\widehat{\phi}} = 0, \tag{9.132}$$

$$\dot{\rho}_\phi + 6H\rho_k + \frac{q_0}{2k}\rho_m\dot{\hat{\phi}} = 0. \tag{9.133}$$

We have defined

$$\rho_k = \frac{\dot{\hat{\phi}}^2}{2}, \qquad \rho_\phi = \rho_k + \rho_V, \qquad \rho_V = V(\hat{\phi}) = \hat{V}_0 e^{-\hat{\phi}/k}, \tag{9.134}$$

according to Eqs. (9.118) and (9.120). The two equations for ρ_ϕ and ρ_m, together with the gravitational equations (9.121) (with $\rho_r = 0 = \rho_b$), can be satisfied by an asymptotic configuration in which ρ_m, ρ_ϕ, V and H^2 scale in time in the same way, so that the critical fractions of dark-matter energy density, $\Omega_m = \rho_m/6H^2$, potential energy density, $\Omega_V = V/6H^2$, and kinetic energy density, $\Omega_k = \dot{\hat{\phi}}^2/(12H^2)$, are also separately constant.

Let us look, in fact, for solutions with frozen dark-matter over dark-energy ratio, characterized by the conditions

$$\frac{\dot{\rho}_m}{\rho_m} = \frac{\dot{\rho}_\phi}{\rho_\phi}, \qquad \frac{\dot{\rho}_m}{\rho_m} = \frac{\dot{\rho}_V}{\rho_V}. \tag{9.135}$$

The first condition, using the conservation equations (9.132) and (9.133), and the Einstein equations (9.121) rewritten in the form

$$1 = \Omega_m + \Omega_\phi = \Omega_m + \Omega_k + \Omega_V, \tag{9.136}$$

leads to

$$\frac{\dot{\hat{\phi}}}{H} = \frac{6k}{q_0}(\Omega_V - \Omega_k). \tag{9.137}$$

The second condition, also using the explicit form of the potential energy (9.134), leads to

$$\frac{\dot{\hat{\phi}}}{H} = \frac{6k}{q_0 + 2}. \tag{9.138}$$

Their combination gives

$$\Omega_V = \Omega_k + \frac{q_0}{q_0 + 2}. \tag{9.139}$$

Expressing $\dot{\hat{\phi}}$ through Ω_k we can rewrite Eq. (9.138) as

$$\Omega_k^{1/2} = \frac{\dot{\hat{\phi}}}{\sqrt{12}H} = \frac{\sqrt{3}k}{q_0 + 2}. \tag{9.140}$$

This completely fixes the constant (asymptotic) critical fractions of dark-energy and dark-matter density,

$$\Omega_k = \frac{3k^2}{(q_0+2)^2}, \qquad \Omega_V = \frac{3k^2 + q_0(q_0+2)}{(q_0+2)^2},$$

$$\Omega_\phi = \Omega_k + \Omega_V, \qquad \Omega_m = 1 - \Omega_\phi, \tag{9.141}$$

and the constant, asymptotic dark-energy equation of state,

$$w = \frac{p_\phi}{\rho_\phi} = \frac{\Omega_k - \Omega_V}{\Omega_k + \Omega_V} = -\frac{q_0(q_0+2)}{6k^2 + q_0(q_0+2)}. \tag{9.142}$$

The behavior of Ω_ϕ and w, as a function of the charge q_0 and of the loop parameter $k = \sqrt{2}c_2/c_1$, is illustrated in Fig. 9.7 in a range of parameters corresponding to realistic values of the dark-energy density and of its equation of state. Assuming that the present Universe is already inside the asymptotic freezing regime, a fit of the observed values of $\{\Omega_\phi, w\}$ could then provide direct information on the leading asymptotic parameters of the string effective action (9.117).

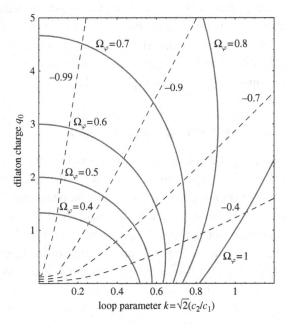

Figure 9.7 Plot of the curves $\Omega_\phi = $ const (bold curves) and $w = $ const (dashed curves), for the solutions (9.141) and (9.142). The dark-energy density Ω_ϕ ranges from 0.4 to 1, the equation of state from $w = -0.4$ to $w = -0.99$.

The explicit time dependence of the asymptotic solution can be determined by using the Einstein equation (9.121) for $\dot{H}$, rewritten in the form

$$1 + \frac{2\dot{H}}{3H^2} = \Omega_V - \Omega_k, \tag{9.143}$$

from which

$$\frac{\ddot{a}}{aH^2} = 1 + \frac{\dot{H}}{H^2} = \frac{3}{2}(\Omega_V - \Omega_k) - \frac{1}{2} = \frac{q_0 - 1}{q_0 + 2}. \tag{9.144}$$

A double integration of $\dot{H}$ then gives

$$a \sim t^{(q_0+2)/3}, \qquad H \sim a^{-3/(q_0+2)}, \tag{9.145}$$

from which

$$\rho_m \sim \frac{\dot{\phi}^2}{2} \sim \widehat{V}_0 e^{-\widehat{\phi}/k} \sim H^2 \sim a^{-6/(q_0+2)}. \tag{9.146}$$

Finally, the integration of $\dot{\widehat{\phi}}$ in Eq. (9.138) leads to

$$\phi = \phi_0 + 2k \ln t. \tag{9.147}$$

It is important to stress that, according to Eq. (9.144), the above solution describes accelerated expansion provided that $q_0 > 1$, as previously anticipated. The case $q_0 < -2$ would also correspond to a positive acceleration (of super-inflationary type, with $\dot{H} > 0$, see Chapter 5), describing a background evolving towards a "big rip" singularity [56]: however, such a possibility is to be excluded in our context, as it would imply $\Omega_m < 0$ (see Eqs. (9.141)).

9.3.3 A numerical example

A global representation of the modified cosmological evolution, illustrating the smooth transition from the initial radiation phase to the intermediate dragging phase and to the final, asymptotic freezing regime, can be obtained by performing numerical integrations of the string cosmology equations (9.121)–(9.127), using an appropriate parametrization of the loop form factors. For our illustrative purpose we adopt here the expansion (9.118) truncated to first order in $g_s^{-2} = \exp(-\phi)$, setting $b_1 = b_2 = 1$, $c_1^2 = 100$ and $c_2^2 = 30$. We also model the asymptotic rise of the dilaton coupling to dark matter with the function

$$q(\phi) = q_0 \frac{e^{q_0\phi}}{c^2 + e^{q_0\phi}}, \tag{9.148}$$

where we set $c^2 = 150$ and $q_0 = 2.5$.

Finally, we must specify the explicit form of the dilaton potential. In agreement with its non-perturbative origin, and with the assumption of asymptotic exponential suppression in the strong bare-coupling limit $\phi \to \infty$, the simplest choice for the S-frame potential is a difference of exponential terms like this,

$$\tilde{V}(\phi) = m_V^2 \left(e^{-\beta_1 e^{-\phi}} - e^{-\beta_2 e^{-\phi}} \right), \qquad 0 < \beta_1 < \beta_2, \qquad (9.149)$$

which leads to the large-ϕ behavior of Eq. (9.118), with $V_0 = m_V^2 (\beta_2 - \beta_1)$. In the E-frame, taking into account the rescaling (9.120), the potential becomes (in units $M_P^2 = 2$)

$$V(\phi) = c_1^4 m_V^2 \frac{e^{2\phi}}{(b_1 + c_1^2 e^{\phi})^2} \left(e^{-\beta_1 e^{-\phi}} - e^{-\beta_2 e^{-\phi}} \right). \qquad (9.150)$$

We fix $\beta_1 = 0.1$, $\beta_2 = 0.2$ and $c_1^2 m_V = 10^{-3} H_{eq}$. This last choice, which implies $m_V \sim H_0$, is crucial to obtain a realistic scenario in which the Universe starts accelerating at a phenomenologically acceptable epoch [53]. Note that, given the instantonic relation (9.119) between the amplitude of the potential and the fundamental string scale, a value of β slightly smaller than the coefficient of the QCD beta function is enough to move m_V from the QCD scale down to the Hubble scale, as already stressed in [53]. This should not hide, however, the fine-tuning required to adjust very precisely the amplitude of the potential, and which seems to be common to all quintessence scenarios characterized by a running scalar field. We will comment on this point further at the end of this section.

We now have all the ingredients for the numerical integration of Eqs. (9.124)–(9.127) determining the time evolution of ρ_r, ρ_b, ρ_m, ρ_ϕ. Using the first equation (9.121) as a constraint on the set of initial data, and imposing the initial conditions $\rho_{\phi i} = \rho_{ri}$, $\rho_{mi} = 10^{-20} \rho_{ri}$, $\rho_{bi} = 7 \times 10^{-21} \rho_{ri}$, $\phi_i = -2$, at the initial scale $H_i = 10^{40} H_{eq}$, we obtain the result illustrated in Fig. 9.8.

The figure clearly displays, at early times in the radiation era, the presence of the focusing effect by which the (till sub-dominant) dilaton energy density tends to approach the energy density of the other cosmological components. In the subsequent dragging phase, the dilaton (kinetic-dominated) energy closely follows the evolution of the dark-matter density. In the final freezing phase, the dilaton potential comes into play, and the dilaton and dark-matter energy densities become closely tied together, with a ratio fixed forever at a number of order one. For the particular values of q_0, c_1 and c_2 of our numerical example, the asymptotic configuration corresponds to the critical fractions $\Omega_\phi = 0.73$ and $\Omega_m = 0.27$, with a dark-energy equation of state $w = -0.76$.

It seems appropriate to conclude the present discussion by listing the main properties of the dilaton model presented in this section, in order to stress the

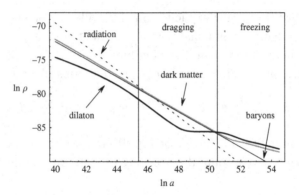

Figure 9.8 Time evolution (on a logarithmic scale) of the energy densities of
the various gravitational sources. Note that, at late enough times, baryons and
radiation become uncoupled to the dilaton, and thus obey the standard scaling
behavior $\rho_r \sim a^{-4}$, $\rho_b \sim a^{-3}$.

possible phenomenological differences with other models of quintessential dark
energy.

(1) A first point, which is in common with all models of coupled quintessence
[50, 51, 52], is that the coincidence problem – if not solved – is at least relaxed,
because the (dilaton) dark-energy density and the dark-matter density are of the
same order *not only today*, but also in the future (forever), and possibly (for a
significant amount of time) also in the past, depending on the beginning of the
freezing epoch (i.e. on the specific amplitude V_0 of the potential).

(2) A second, important point (also evident in Fig. 9.8) concerns the faster
dilution in time of baryons with respect to dark matter, because of their weaker
coupling to the dilaton. This effect is maximal in the final accelerated phase,
where the baryon-to-dark-matter ratio decreases in time as

$$\frac{\rho_b}{\rho_m} \sim a^{-3q_0/(2+q_0)}. \tag{9.151}$$

This effect, together with an early enough beginning of the accelerated regime,
could explain why the fraction of baryons today is so small ($\sim 10^{-2}$) in critical
units. Experimental information on the past value of the ratio ρ_b/ρ_m, compared
with its present value, would immediately provide a direct and unambiguous test
of this class of models.

(3) A third point concerns the beginning of the cosmic acceleration, which
is in principle allowed to start at earlier epochs than in models of uncoupled
quintessence with fixed equation of state.

Consider the Einstein equations for a two-component cosmological model with
dark matter, ρ_m, and (uncoupled) dark energy, ρ_Q, as the only relevant gravita-

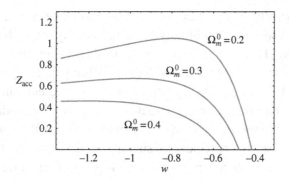

Figure 9.9 The redshift scale marking the beginning of the accelerated epoch, z_{acc}, in models of uncoupled dark energy and constant equation of state w, for various possible values of the present dark-matter density Ω_m^0.

tional sources:

$$6H^2 = \rho_m + \rho_Q,$$

$$4\dot{H} + 6H^2 = -p_Q = -w\rho_Q. \tag{9.152}$$

Here $w = p_Q/\rho_Q$ is a constant parameter ranging, say, from $-1/3$ to $-4/3$. In such a context $\rho_m \sim a^{-3}$ is diluted faster than $\rho_Q \sim a^{-3(1+w)}$. So, even if today ρ_Q dominates and the expansion is accelerated ($\ddot{a}/a = \dot{H} + H^2 > 0$), at early enough times the Universe was dominated by ρ_m, and the expansion was decelerated. In particular, the acceleration switches off at the scale a_{acc} such that

$$\left(\frac{\ddot{a}}{aH^2}\right)_{acc} = 1 + \left(\frac{\dot{H}}{H^2}\right)_{acc} = -\frac{1}{12H^2}\left[\rho_m + (1+3w)\rho_Q\right]_{acc} = 0. \tag{9.153}$$

Using the present critical fractions $\Omega_m^0 = \rho_m^0/6H_0^2$ and $\Omega_Q^0 = 1 - \Omega_m^0$, the above condition can be rewritten as

$$\Omega_m^0\left(\frac{a_{acc}}{a_0}\right)^{-3} = (1+3w)\left(\Omega_m^0 - 1\right)\left(\frac{a_{acc}}{a_0}\right)^{-3(1+w)}, \tag{9.154}$$

and fixes the beginning of the acceleration at the relative redshift scale

$$z_{acc} = \frac{a_0}{a_{acc}} - 1 = \left[(1+3w)\left(\frac{\Omega_m^0 - 1}{\Omega_m^0}\right)\right]^{-1/3w} - 1. \tag{9.155}$$

One can then easily check that $z_{acc} \lesssim 1$ for realistic (i.e. observationally compatible) values of Ω_m^0 and w, as clearly illustrated in Fig. 9.9.

For the model of dark energy illustrated in this section the dilaton-dominated, accelerated phase is characterized by an acceleration parameter $\ddot{a}/aH^2$, which is constant (see Eq. (9.144)), as ρ_ϕ and ρ_m scale in time in the same way. Therefore, the beginning of the acceleration regime is not determined by the present value of the dark-matter density. The past extension of the accelerated phase is constrained, instead, by the present fraction of baryon energy density, Ω_b, which grows with respect to Ω_m and Ω_ϕ as we go back in time, tending to become dominant. Even with this bound, however, we are left with the possibility of a very early beginning of the accelerated phase, $z_{\rm acc} \gg 1$, given the very tiny present value of Ω_b.

More stringent constraints on the initial scale $z_{\rm acc}$, for the model we are considering, can be obtained by combining present experimental information on the so-called density contrast σ_8 (characterizing the level of dark-matter fluctuations over the distance scale of 8 Mpc) with current SNIa observations. One obtains [57] that an early beginning of the dilaton-dominated epoch is allowed up to $z_{\rm acc} = 3.5$ (for the best fit of the data), and up to $z_{\rm acc} \lesssim 5$ within one standard sigma deviation. Such results provide important experimental information on the allowed scale of the non-perturbative dilaton potential, as well as on other parameters of the string effective action [57]. A value of $z_{\rm acc}$ significantly larger than one seems to be compatible even with the recently discovered, high-redshift supernovae [39], and with the possibility that our present cosmological state is approaching (but is not fully coincident with) the asymptotic freezing regime. In this last case, the present cosmological configuration may contain a significant (even if non-dominant) fraction of dark matter uncoupled to the dilaton [58]. Type Ia supernovae of the SNLS dataset [59] are also consistent with an early beginning of the acceleration of $z_{acc} \simeq 3$ and higher [60].

Another important property of this dilaton model is that, in contrast with other quintessential models, the dark-energy density is always characterized by a "conventional" equation of state satisfying $w = p_\phi/\rho_\phi \geq -1$ (see Eq. (9.142)). As a consequence, the accelerated expansion is always associated with a decreasing curvature state with $\dot{H} = -3q_0 H^2/(2+q_0) \leq 0$, according to Eq. (9.144). This excludes the possibility of "phantom" gravitational sources [61] with *supernegative* equation of state, $w < -1$, and the possible occurrence of future "big rip" singularities. Such a conclusion is valid also for a more general classe of models in which the leading asymptotic term of the dilaton potential is of the form $V_0 \exp(-\lambda\phi)$, with $\lambda \neq 1$ [62] (the condition $\dot{H} > 0$, in particular, turns out to be always incompatible with the requirement $\Omega_m > 0$).

Let us conclude with a further comment on the asymptotic amplitude of the nonperturbative dilaton potential. Consider, for instance, the explicit form (9.150) used in our numerical example, and recall that for a realistic scenario the amplitude of the potential must correspond to an effective dilaton mass roughly of the order of the present Hubble scale, $m \sim m_V \sim 10^{-5} H_{\rm eq} \sim H_0$. This is a fine-tuning problem which seems to affect all presently known models of quintessence, and which requires, for the dilaton potential, suitable protection against possible contributions to the mass induced by the radiative corrections.

For the dilaton, the coupling with charge q to ordinary baryonic matter generates the mass term given in Eq. (9.108). The condition $m \lesssim H_0$ thus requires fine-tuning of the baryon–dilaton coupling,

$$q \lesssim 10^{-27} \left(\frac{1\,\text{Tev}}{\Lambda} \right) \left(\frac{1\,\text{GeV}}{M_f} \right), \qquad (9.156)$$

which, by the way, seems to exclude the possible resonant interaction of a relic dilaton background with the gravitational detectors (according to the discussion of Section 9.2). Conversely, the coupling with strength $q \sim 1$ to a possible non-baryonic (but fermionic) dark-matter component, as required by models of coupled quintessence, would seem to imply a fine-tuning to extremely low values of the fermionic mass M_f.

This conclusion may be avoided, however, if the dilaton is exponentially coupled to the fermions [30], and the dilaton charge is related to the potential slope of the (canonically normalized) dilaton field $\widehat{\phi}$ by $q_0 = 1/k$. This possibility does not seem to be excluded, at least from a phenomenological point of view, according to the results illustrated in Fig. 9.7. Also, this possibility seems to be supported by a recently proposed argument based on the AdS/CFT correspondence [63], suggesting that scalar masses comparable to the cosmological curvature scale ($m \sim H_0$) could be safe from radiative corrections.

Appendix 9A
The common mode of interferometric detectors

The expression (9.99) for the signal-to-noise ratio is valid, in general, for any given overlap functions $\gamma(p)$, computed in terms of two arbitrary pattern functions according to the definition (9.94). In Section 9.2 we have considered the overlap between the differential mode D^{ij}_- of two interferometers, and the overlap between the monopole mode D^{ij}_0 of two resonant spheres. We have also stressed, however, that the interferometric antennas can efficiently respond to a stochastic background of scalar radiation through their common mode D^{ij}_+ (see Fig. 9.5). In this appendix we will discuss the possible advantages of the common mode (with respect to the differential one) for the detection of a non-relativistic background of scalar particles.

We start by considering the "mixed" configuration in which one correlates the common mode of one interferometer with the differential mode of another (or even the same) interferometer. In such a case, the results for the overlap function are strongly dependent on the relative geometric arrangement of the two detectors. In particular, if the arms of the two interferometers have exactly the same angular separation, the mixed overlap is identically vanishing in the case of co-planar detectors, quite independently of the relative distance and arm orientation [28]. The overlap is vanishing also for detectors lying on two parallel planes, separated by an arbitrary distance. The overlap may be non-zero, however, if the angular separation of the arms is different for the two interferometers, even for co-planar detectors and for vanishing spatial separation, $\Delta \vec{x} = 0$.

In order to illustrate this possibility we consider here an experimental setup in which the arms of the first interferometer $\widehat{u}_1$ and $\widehat{v}_1$ are orthogonal in the polar plane of Fig. 9.3, while the arms $\widehat{u}_2$ and $\widehat{v}_2$ of the second interferometer have angular separation α in the same plane, with the arm $\widehat{u}_2$ coincident and coaligned with $\widehat{u}_1$:

$$\widehat{u}_1 = \widehat{u}_2 = (1, 0, 0), \qquad \widehat{v}_2 = (\cos\alpha, \sin\alpha, 0), \qquad \widehat{v}_1 = (0, 1, 0) \qquad (9A.1)$$

(see Fig. 9.10). Note that such an experimental configuration could be simply realized, in principle, by adding a third, non-orthogonal arm to existing interferometers.

We now have to compute the scalar pattern functions for the differential and common modes of the two interferometers. Let us first concentrate on their geodesic response, described by Eq. (9.70). For the first, orthogonal interferometer we can directly exploit the results of Eqs. (9.77), (9.80) and (9.81), to obtain

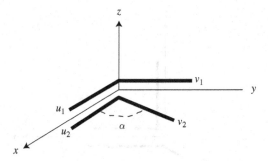

Figure 9.10 Example of two co-planar interferometers with vanishing spatial separation and different angular separation of their arms.

$$F_{1-}^{\text{geo}} = -\left(\frac{p}{E}\right)^2 \sin^2\theta \cos 2\varphi,$$

$$F_{1+}^{\text{geo}} = 2 - \left(\frac{p}{E}\right)^2 \sin^2\theta.$$

(9A.2)

For the second, non-orthogonal interferometer, we obtain instead from the definition (9.70) an α-dependent response:

$$F_{2-}^{\text{geo}}(\alpha) = -\left(\frac{p}{E}\right)^2 \sin^2\theta \left(\cos 2\varphi \sin^2\alpha - \sin 2\varphi \sin\alpha \cos\alpha\right),$$

$$F_{2+}^{\text{geo}}(\alpha) = 2 - \left(\frac{p}{E}\right)^2 \sin^2\theta \left(1 + \cos 2\varphi \cos^2\alpha + \sin 2\varphi \sin\alpha \cos\alpha\right).$$

(9A.3)

Let us compute the mixed overlap function between differential and common modes, according to Eq. (9.94) at $\Delta\vec{x} = 0$. The result is identically zero, independently of α, if we overlap the common mode of the orthogonal interferometer, F_{1+}, with the differential mode of the non-orthogonal one, F_{2-}. In the opposite case (i.e. overlapping F_{1-} with F_{2+}) we obtain, instead, a non-zero result:

$$\gamma_{-+}^{\text{geo}}(p,\alpha)_{\Delta x=0} = \frac{1}{N}\int_{\Omega_2} d^2\hat{n}\, F_{1-}^{\text{geo}}(p,\hat{n})\, F_{2+}^{\text{geo}}(p,\hat{n},\alpha) = \frac{16\pi}{15N}\left(\frac{p}{E}\right)^4 \cos^2\alpha. \quad (9A.4)$$

However, in this case we also find a strong suppression for the response to non-relativistic radiation, since the function $\gamma(p)$ is controlled by the same factor $(p/E)^4$ which characterizes the overlap of two differential modes (see Eq. (9.102)). Similar results can be obtained by considering the mixed overlap of the non-geodesic response of the two detectors [28].

Another possible configuration with a non-zero mixed-overlap function is the configuration in which the arms of the interferometers do not lie on parallel planes, even if the angular separation of the arms is the same, and even if the spatial separation of the two central stations tends to zero.

For a simple illustration of this possibility we may consider the limiting case in which both interferometers are centered at the origin of the same reference frame: the arms of the first interferometer, $\hat{u}_1$ and $\hat{v}_1$, are aligned along the x and y axes, respectively, while the arms of the second interferometer, $\hat{u}_2$ and $\hat{v}_2$, are aligned along y and z, respectively. The two arms $\hat{v}_1$ and $\hat{u}_2$ are coaligned and coincident (see Fig. 9.11). The computation

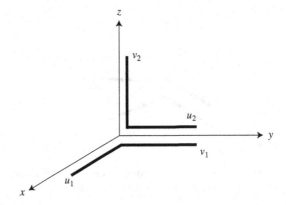

Figure 9.11 Example of two interferometers with vanishing spatial separation, orthogonal arms and non-zero mixed overlap function.

of the mixed, geodesic overlap function, for such a configuration, leads to the result

$$\gamma_{+-}^{\rm geo}(p)_{\Delta x=0} = \frac{1}{N}\int_{\Omega_2} d^2\hat{n}\, F_{1+}^{\rm geo} F_{2-}^{\rm geo} = \frac{\pi}{N}\left(\frac{p}{E}\right)^2\left(2-\frac{7}{15}\frac{p^2}{E^2}\right). \qquad (9A.5)$$

The response to non-relativistic radiation is still suppressed, but the suppression is lower than in previous cases for the infrared sector of the scalar spectrum: indeed, for $p \to 0$, the overlap goes to zero as $(p/E)^2$, instead of $(p/E)^4$. For the non-geodesic response one finds a non-zero mixed overlap with the standard (i.e. quartic) non-relativistic suppression [28].

Another substantial improvement of the non-relativistic signal can be obtained if we compute the overlap function taking into account the common mode of *both* interferometers. In that case, one finds that the overlap is non-zero even if the interferometers are co-planar, and their arms have the same angular separation. Considering such a simplified configuration, computing the pattern functions $F_+^{\rm geo}$, $F_+^{\rm ng}$ from the definitions (9.70) and (9.71) for a generic spatial separation $|\Delta\vec{x}| = d$, and using the integral representation of the spherical Bessel functions, one obtains the following overlap functions:

$$\gamma_+^{\rm ng}(p) = \frac{1}{N}\int_{\Omega_2} d^2\hat{n}\, F_{1+}^{\rm ng} F_{2+}^{\rm ng}\, e^{2i\pi\hat{n}\cdot\Delta\vec{x}}$$

$$= q_1 q_2 \frac{4\pi}{N}\left(\frac{p}{E}\right)^4\left[j_0(x)-\frac{2}{x}j_1(x)+\frac{3}{x^2}j_2(x)\right], \qquad (9A.6)$$

$$\gamma_+^{\rm geo}(p) = \frac{1}{N}\int_{\Omega_2} d^2\hat{n}\, F_{1+}^{\rm geo} F_{2+}^{\rm geo}\, e^{2i\pi\hat{n}\cdot\Delta\vec{x}}$$

$$= \frac{4\pi}{N}\left[\left(4-4\frac{p^2}{E^2}+\frac{p^4}{E^4}\right)j_0(x)+\frac{1}{x}\left(4\frac{p^2}{E^2}-2\frac{p^4}{E^4}\right)j_1(x)+\frac{3}{x^2}\left(\frac{p}{E}\right)^4 j_2(x)\right].$$

$$(9A.7)$$

Here $x = 2\pi p d$, and j_0, j_1, j_2 are spherical Bessel functions, satisfying

$$j_0(x) = \frac{\sin x}{x},$$

$$j_1(x) = \frac{j_0(x)}{x} - \cos x, \qquad (9A.8)$$

$$j_{l+1(x)} = \frac{2l+1}{x} j_l(x) - j_{l-1}(x), \qquad l \geq 1.$$

We may note that, for co-planar interferometers, the overlap γ_+ is independent of the relative orientation of their axes, thanks to the rotational symmetry of F_+ with respect to the polar angle φ.

If we now consider the non-relativistic limit $p \to 0$ we can easily check that the non-geodesic overlap γ_+^{ng} goes to zero as $(p/E)^4$, exactly as in the case of two differential modes, Eq. (9.102). The geodesic overlap function, however, goes to a constant,

$$p \to 0 \qquad \Longrightarrow \qquad \gamma_+^{geo}(p) \to \frac{16\pi}{N} = \text{const}, \qquad (9A.9)$$

as in the case of two resonant spheres (see Eq. (9.110)). We may thus repeat exactly the discussion presented in Section 9.2 for the monopole response tensor of a sphere: not only is the suppression of the non-relativistic modes avoided, but also, and primarily, the dependence of SNR on the observation time T is modified, according to Eq. (9.115), if the scalar background has a flat enough spectrum.

The cross-correlated response of the common mode of two interferometers, geodesically interacting with a (flat enough) cosmic background of non-relativistic scalar particles, may produce a signal-to-noise ratio growing with T faster than at the standard rate $T^{1/2}$. By extending the observation time it becomes possible, in principle, to enhance the signal associated with the scalar background with respect to the corresponding one produced by the differential mode, possibly compensating the higher level of noise expected to affect the experimental analyses of the common mode data.

References

[1] E. Witten, *Phys. Lett.* **B149** (1984) 351.
[2] E. Witten, *Nucl. Phys.* **B443** (1995) 85.
[3] M. Gasperini, *Phys. Lett.* **B327** (1994) 314.
[4] M. Gasperini and G. Veneziano, *Phys. Rev.* **D50** (1994) 2519.
[5] T. Banks, D. B. Kaplan and A. E. Nelson, *Phys. Rev.* **D49** (1994) 779.
[6] T. Taylor and G. Veneziano, *Phys. Lett.* **B213** (1988) 459.
[7] J. Ellis, S. Kalara, K. A. Olive and C. Wetterich, *Phys. Lett.* **B228** (1989) 264.
[8] N. Deruelle, G. Gundlach and D. Langlois, *Phys. Rev.* **D46** (1992) 5337.
[9] M. Gasperini and G. Veneziano, *Phys. Rev.* **D59** (1999) 43503.
[10] M. Abramowitz and I. Stegun, *Handbook of Mathematical Functions* (New York: Dover, 1972).
[11] R. Durrer, M. Gasperini, M. Sakellariadou and G. Veneziano, *Phys. Rev.* **D59** (1999) 43511.
[12] A. Papapetrou, *Proc. R. Soc. London* **A209** (1951) 284.
[13] M. Gasperini, *Phys. Lett.* **B470** (1999) 67.
[14] E. Fischbach and C. Talmadge, *Nature* **356** (1992) 207.
[15] C. D. Hoyle, *et al.*, *Phys. Rev.* **D70** (2004) 042004.
[16] I. Antoniadis, in *Gravitational Waves and Experimental Gravity*, eds. J. Tran Than Van *et al.* T. Domour and A.M. Polyetron, (Singapore: World Scientific, 2000).
[17] T. Damour and A. M. Polyakov, *Nucl. Phys.* **B423** (1994) 352.
[18] T. Damour and A. M. Polyakov, *Gen. Rel. Grav.* **26** (1994) 1171.
[19] M. Gasperini, in *General Relativity and Gravitational Physics*, eds. M. Bassan *et al.* (Singapore: World Scientific, 1997), p. 181.
[20] M. Gasperini and G. Veneziano, *Phys. Rep.* **373** (203) 1.
[21] C. Misner, K. Thorne and J. A. Wheeler, *Gravitation* (San Francisco: Freeman, 1973).
[22] M. Bianchi *et al.*, *Phys. Rev.* **D57** (1998) 4525.
[23] M. Brunetti *et al.*, *Phys. Rev.* **D59** (1999) 44027.
[24] M. Maggiore and A. Nicolis, *Phys. Rev.* **D62** (2000) 024004.
[25] M. Gasperini, *Phys. Lett.* **B477** (2000) 242.
[26] M. Gasperini and C. Ungarelli, *Phys. Rev.* **D64** (2001) 064009.
[27] B. Allen and J. D. Romano, *Phys. Rev.* **D59** (1999) 102001.
[28] N. Bonasia and M. Gasperini, *Phys. Rev.* **D71** (2005) 104020.
[29] E. Coccia, M. Gasperini and C. Ungarelli, *Phys. Rev.* **D65** (2002) 067101.
[30] M. Doran and J. Jaeckel, *Phys. Rev.* **D66** (2002) 043519.
[31] W. W. Johnson and S. M. Merkowitz, *Phys. Rev. Lett.* **70** (1993) 2367.

[32] E. Coccia *et al.*, *Phys. Rev.* **D57** (1998) 2051.

[33] M. Cerdonio *et al.*, *Phys. Rev. Lett.* **87** (2001) 031101.

[34] D. N. Spergel *et al.*, astro-ph/0603449.

[35] S. Weinberg, *Gravitation and Cosmology* (New York: Wiley, 1971).

[36] G. Bertone, D. Hooper and J. Silk, *Phys. Rep.* **405** (2005) 279.

[37] S. Perlmutter *et al.*, *Nature* **391** (1998) 51.

[38] A. G. Riess *et al.*, *Astron. J.* **116** (1998) 1009.

[39] A. G. Riess *et al.*, *Astrophys. J.* **607** (2004) 665.

[40] S. Hamerstad and E. Morstel, *Phys. Rev.* **D66** (2002) 063508.

[41] A. Melchiorri *et al.*, *Phys. Rev.* **D68** (2003) 043509.

[42] B. Ratra and P. J. E. Peebles, *Phys. Rev.* **D37** (1988) 3406.

[43] C. Wetterich, *Nucl. Phys.* **B302** (1988) 668.

[44] M. S. Turner and C. White, *Phys. Rev.* **D56** (1997) 4439.

[45] R. R. Caldwell, R. Dave and P. J. Steinhardt, *Phys. Rev. Lett.* **80** (1998) 1582.

[46] I. Zlatev, L. Wang and P. J. Steinhardt, *Phys. Rev. Lett.* **82** (1999) 896.

[47] I. Zlatev, L. Wang and P. J. Steinhardt, *Phys. Rev.* **D59** (1999) 123504.

[48] L. P. Chimento, A. S. Jacubi and D. Pavon, *Phys. Rev.* **D62** (2000) 063508.

[49] S. Sen and A. Sen, *Phys. Rev.* **D63** (2001) 124006.

[50] L. Amendola, *Phys. Rev.* **D62** (2000) 043511.

[51] L. Amendola and D. Tocchini-Valentini, *Phys. Rev.* **D64** (2001) 04359.

[52] L. Amendola and D. Tocchini-Valentini, *Phys. Rev.* **D66** (2002) 043528.

[53] M. Gasperini, F. Piazza and G. Veneziano, *Phys. Rev.* **D65** (2002) 023508.

[54] G. Veneziano, *JHEP* **0206** (2002) 051.

[55] P. J. Steinhardt and N. Turok, *Phys. Rev.* **D65** (2002) 126003.

[56] R. R. Caldwell, M. Kamionkowski and N. N. Weinberg, *Phys. Rev. Lett.* **91** (2003) 07130.

[57] L. Amendola, M. Gasperini, D. Tocchini-Valentini and C. Ungarelli, *Phys. Rev.* **D67** (2003) 043512.

[58] L. Amendola, M. Gasperini and F. Piazza, *JCAP* **09** (2004) 014.

[59] P. Astier *et al.*, *Astron. Astrophys.* **447** (2006) 31.

[60] L. Amendola, M. Gasperini and F. Piazza, *Phys. Rev.* **D74** (2006) 127302.

[61] R. R. Caldwell, *Phys. Lett.* **B545** (2002) 23.

[62] M. Gasperini, *Int. J. Mod. Phys.* **D13** (2004) 2267.

[63] S. S. Gubser and P. J. E. Peebles, *Phys. Rev.* **D70** (2004) 123511.

10

Elements of brane cosmology

We have shown, in Chapter 3, that the quantum consistency of string theory requires a higher-dimensional target manifold, with $D = 9 + 1$ (superstrings) or $D = 10 + 1$ (M-theory) space-time dimensions (see in particular Appendix 3B). In our present Universe, however, only four space-time dimensions seem to be accessible to direct observations and measurements. For a possible identification of the target manifold of string theory with the space-time in which we live, we should thus explain why $D - 4$ spatial dimensions turn out to be invisible (at least to present technological investigation).

There are two possible approaches to this problem. The first one assumes that the extra $D - 4$ dimensions are compact, and that their proper volume is stabilized at an extremely small length scale L_c, which may become accessible to direct exploration only with processes of proper energy $E \gtrsim L_c^{-1}$. In the simplest (and older) version of the so-called Kaluza–Klein scenario [1, 2], for instance, the compactification scale of the extra dimensions is controlled by the coupling constant appearing in the gravitational action: one then naturally obtains a compactification volume of Planckian (or stringy) size $(L_c^{-1} \sim M_P)$, certainly inaccessible to present direct observation. What is required, however, from a phenomenological point of view, is a compactification scale not necessarily Planckian, but simply compatible with the present explorable energy range, say $L_c^{-1} \gtrsim 1$ TeV. This observation has recently led to the formulation of models with "large extra dimensions" [3, 4], which might alleviate the hierarchy problems arising from the huge difference between the gravitational scale, M_P, and the scale of electroweak interactions, M_W, and which are currently under active study (see e.g. [5]).

For a simple illustration of this possibility we may consider the gravi-dilaton string effective action in $D = 4 + n$ space-time dimensions:

$$S = -\frac{1}{2\lambda_s^{D-2}} \int d^D x \sqrt{|g_D|} e^{-\phi} \left[R_D + g^{AB} \partial_A \phi \partial_B \phi \right] \tag{10.1}$$

(throughout this chapter, capital roman indices will run from 0 to $D-1$). According to the Kaluza–Klein scenario we can factorize the higher-dimensional space-time as $\mathcal{M}_4 \times \mathcal{K}_n$, denoting with x^μ, $\mu = 0, \ldots, 3$ the coordinates on $\mathcal{M}_4$, and y^a, $a = 1, \ldots n$ the coordinates on the extra-dimensional manifold $\mathcal{K}_n$, supposed to be Ricci flat ($R_{ab} = 0$). For the purpose of our discussion it will be enough to suppose that all fields $(g, \phi, \ldots)$ are independent of the extra spatial coordinates y^a, and that $\mathcal{K}_n$ is a (globally flat) n-dimensional torus $\mathcal{T}^n$. The complete D-dimensional metric g_{AB} can then be conveniently parametrized as follows [6],

$$g_{AB} = \begin{pmatrix} g_{\mu\nu} + f_{ab}A_\mu^a A_\nu^b & A_\mu^b f_{ab} \\ f_{ab}A_\mu^b & f_{ab} \end{pmatrix}, \tag{10.2}$$

where the $n(n+1)/2$ scalar fields $f_{ab}(x)$ associated with the (symmetric) metric tensor of $\mathcal{K}_n$ are four-dimensional "moduli" fields describing the "shape" of the torus, and $A_\mu^a(x)$ are n Abelian vector fields associated with the process of dimensional reduction. One finds, with this parametrization, that $\det g_{AB} = \det g_{\mu\nu} \det f_{ab}$; also, computing the connection and the Ricci scalar R_D, and substituting into Eq. (10.1), one is led (after integration by parts) to the following factorized action:

$$S = -\frac{1}{2\lambda_s^{2+n}} \int d^n y \sqrt{|f|} \int d^4 x \sqrt{|g|} e^{-\phi} \left[R_4(g) - \frac{1}{4} g^{\mu\nu} \partial_\mu f^{ab} \partial_\nu f_{ab} \right.$$

$$\left. + g^{\mu\nu} \partial_\mu \left(\phi - \ln \sqrt{|f|} \right) \partial_\nu \left(\phi - \ln \sqrt{|f|} \right) + \frac{1}{4} f_{ab} G_{\mu\nu}^a G^{b\mu\nu} \right]. \tag{10.3}$$

Here $f = \det f_{ab}$, $g = \det g_{\mu\nu}$, $G_{\mu\nu}^a = \partial_\mu A_\nu^a - \partial_\nu A_\mu^a$ and $R_4(g)$ is the four-dimensional scalar curvature for the metric $g_{\mu\nu}$.

Apart from the additional scalar-vector terms induced by the process of dimensional reduction, the relevant term for the present discussion is the factor $V_n = \int d^n y \sqrt{|f|}$, which is nothing but the proper (finite) volume of the extra-dimensional space. From the comparison of the tensor part of the action (10.3) with the four-dimensional Einstein action,

$$-\frac{M_P^2}{2} \int d^4 x \sqrt{|g|} R_4(g), \tag{10.4}$$

we are thus led to relate the effective gravitational coupling and the extra-dimensional volume as follows:

$$M_s^{2+n} V_n = g_s^2 M_P^2, \tag{10.5}$$

where $g_s^2 = \exp \phi$. For a typical "stringy" size of the compactified dimensions, $V_n \sim M_s^{-n}$, one then finds the usual tree-level relation connecting string and Planck mass, $M_s \sim g_s M_P$.

However, as clearly shown by the above equation, even much larger compactification volumes ($V_n \gg M_s^{-n}$) are compatible with the standard phenomenology (i.e.
with $g_s \sim 0.1$ and $M_P \sim 10^{18}$ GeV), provided the mass scale M_s appearing in the
higher-dimensional action (10.1) is much smaller than Planckian. Let us consider,
for instance, an isotropic volume, and set $V_n = L_c^n$: one then obtains, from Eq. (10.5),

$$L_c = M_s^{-1} \left(\frac{g_s M_P}{M_s} \right)^{2/n} \simeq 2 \times 10^{-17} \text{cm } 10^{28/n} \left(\frac{1 \text{ Tev}}{M_s} \right)^{1+2/n} \left(\frac{g_s M_P}{10^{17} \text{ Gev}} \right).$$

$$(10.6)$$

Thus, in the presence of at least two extra dimensions ($n \geq 2$), a string scale M_s
as small as the TeV scale is already compatible with tests of Newtonian gravity at
short distances, which are presently limiting the possible size of extra dimensions
to $L_c \lesssim 0.1$ mm (see e.g. [7]). Conversely, any given upper bound on L_c implies
(through Eq. (10.5)) a phenomenological lower bound on the allowed values of M_s.

A second, and probably more drastic, explanation of the invisibility of the
extra spatial dimensions does not rely on their small-scale compactification, but
is based on the possibility that the standard fundamental interactions – which are
our tools to explore the surrounding space-time – may propagate only through
a four-dimensional "slice" (spanned by the propagation of a three-dimensional
extended object, or 3-brane) of the higher-dimensional space-time (also called
"bulk" manifold). This is the so-called "brane-world" scenario [8], which has
recently found strong support in a superstring theory context where the gauge
fields associated with the ends of open strings, satisfying Dirichlet boundary
conditions, are indeed localized on p-dimensional hypersurfaces, or D_p-branes
(see Appendix 3B). There is a difficulty in this approach due to the fact that closed
strings and gravitational interactions can propagate equally through all spatial
dimensions, without feeling the effects of preferred hypersurfaces. This problem,
however, could be resolved by the so-called Randall–Sundrum (RS) mechanism
[9], or by other related mechanisms [10], able to explain (at least for some special
backgrounds) why even gravity should be localized on the brane representing our
Universe (or in its nearest neighborhood).

This chapter will concentrate on the possible cosmological applications of the
brane-world scenario which, in view of the many and deep string theory motivations, are fully entitled to be included in a book devoted to the basic aspects of string
cosmology. After discussing in Section 10.1 the modifications to the Einstein equations required to describe the effective gravitational dynamics on the brane and, in
Section 10.2, the possible "confinement" of the long-range part of the gravitational
interaction, we will briefly illustrate three possible types of cosmological scenarios.

In Section 10.3 we will introduce the so-called "brane-world cosmology",
where inflation can be implemented through conventional sources localized on the

brane, but interacting through modified gravitational equations. Section 10.4 will be devoted to the so-called "ekpyrotic" scenario [11, 12], which tries to evade the necessity of a primordial inflationary expansion, solving the standard cosmological problems through the collision of two branes; a possible generalization of this scenario, leading to a cyclic model of the Universe [13, 14], will also be briefly discussed. Finally, in Section 10.5, we will present a model of inflation based on the interaction of two Dirichlet branes [15] (in particular, with opposite charges of the R–R fields [16, 17]), where the interbrane distance plays the role of the inflaton field, and is governed by an effective potential which may be flat enough to sustain a phase of slow-roll inflation [18].

10.1 Effective gravity on the brane

Let us suppose that our Universe can be represented as a $(p+1)$-dimensional hypersurface Σ spanned by the time evolution of a p-brane (i.e. a p-dimensional extended object), embedded in a D-dimensional bulk manifold with one extra spatial dimension (i.e. $D = p+2$). The effective equations determining the gravitational field on the brane then depend on both the bulk gravitational equations and the so-called Israel junction conditions [19, 20], controlling the matching of the metric and its first derivative across the brane. Such matching conditions can be obtained in two ways, either treating the codimension-one hypersurface Σ as a "domain wall" splitting the bulk hypervolume into two distinct regions [21], or inserting the brane as a dynamical source of the bulk gravity equations, and performing a covariant Gauss–Codacci projection [22].

In the first case the junction conditions are directly obtained from the action as field equations for the extrinsic curvature of the brane. In this case, in fact, the "world-hypersurface" Σ swept by the time evolution of the $(D-2)$-brane splits the bulk manifold into two regions $\Omega_\pm$, with boundaries $\partial\Omega_\pm$ at infinity and $\Sigma_\pm$ on the two sides of the brane (see Fig. 10.1). The gravitational action (including the Gibbons–Hawking boundary terms) can then be decomposed as follows:

$$S = \int_{\Omega_+} d^D x \sqrt{|g|} \left(-\frac{R}{2\lambda^{D-2}} + L_{\text{bulk}} \right) + \int_{\Omega_-} d^D x \sqrt{|g|} \left(-\frac{R}{2\lambda^{D-2}} + L_{\text{bulk}} \right)$$

$$+ \frac{1}{2\lambda^{D-2}} \int_{\partial\Omega_+} d^{D-1} x \sqrt{|h|}\, K + \frac{1}{2\lambda^{D-2}} \int_{\partial\Omega_-} d^{D-1} x \sqrt{|h|}\, K$$

$$+ \frac{1}{2\lambda^{D-2}} \int_{\Sigma_+} d^{D-1}\xi \sqrt{|h|}\, K + \frac{1}{2\lambda^{D-2}} \int_{\Sigma_-} d^{D-1}\xi \sqrt{|h|}\, K$$

$$- \int_{\Sigma} d^{D-1}\xi \sqrt{|h|}\, L_{\text{brane}}. \tag{10.7}$$

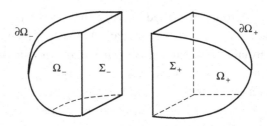

Figure 10.1 Splitting of the D-dimensional bulk manifold into the two domains $\Omega_{\pm}$, separated by the $(D-1)$-dimensional hypersurface Σ swept by the brane.

Here λ is some fundamental length scale governing the strength of the gravitational bulk interactions (possibly, but not necessarily, identified with the phenomenological Planck scale λ_{P}); L_{bulk} and L_{brane} are the Lagrangian densities for the matter fields present in the bulk and confined on the brane, respectively, including a possible bulk cosmological constant and the tension of the brane (and entering the action with the opposite sign, to follow standard conventions). The first line of the above equation gives the standard Einstein action (we are working, for simplicity, in the E-frame), the second line gives the boundary terms at spatial infinity, and the third line gives the boundary terms on the brane, where h and K are, respectively, the determinant of the induced metric h_{AB} on the boundary, and the trace of its intrinsic curvature K_{AB}. In terms of the space-like unit vector n^A, normal to the boundary, h and K are defined by

$$h_{AB} = g_{AB} + n_A n_B, \qquad K_{AB} = h_A{}^C h_B{}^D \nabla_C n_D, \tag{10.8}$$

where

$$g_{AB} n^A n^B = -1, \qquad h_{AB} n^B = 0, \qquad K_{AB} n^B = 0. \tag{10.9}$$

The last two conditions tell us that both the induced metric and the extrinsic curvature are tangential to the boundary, and in particular to Σ, as they describe geometric properties of the space-time manifold *at* the brane (space-time) position.

Let us vary the action (10.7) with respect to the bulk metric g_{AB}, by observing that the vector n^A also depends on the metric through the normalization (10.9), whose variation implies

$$\delta n_A = -\frac{1}{2} n_A n^M n^N \delta g_{MN}. \tag{10.10}$$

Using the standard result (2.6) for the variation of the Einstein action, and summing up all contributions, one finds that all boundary terms of the type $\nabla_A \delta g^{MN}$ are canceled by the variation of the Gibbons–Hawking action (see Chapter 2), on both the boundaries $\partial \Omega_{\pm}$ at infinity and the two sides $\Sigma_{\pm}$ of the brane. The variation of $\sqrt{|h|} K$, however, also provides a boundary contribution which is proportional to the extrinsic curvature, and which is vanishing on $\partial \Omega_{\pm}$, but not necessarily vanishing on $\Sigma_\pm$. Thus, the stationarity of the action (10.7), in general, imposes the condition [21]

$$
\begin{aligned}
\delta S = \frac{1}{2} \int_{\Omega_+} d^D x \sqrt{|g|} \left(-G_{AB} + \lambda^{D-2} T_{AB}^{\text{bulk}} \right) \delta g^{AB} \\
+ \frac{1}{2} \int_{\Omega_-} d^D x \sqrt{|g|} \left(-G_{AB} + \lambda^{D-2} T_{AB}^{\text{bulk}} \right) \delta g^{AB} \\
- \int_{\Sigma_+} d^{D-1} \xi \sqrt{|h|} \left(K_{AB} - h_{AB} K \right) \delta g^{AB} \\
+ \int_{\Sigma_-} d^{D-1} \xi \sqrt{|h|} \left(K_{AB} - h_{AB} K \right) \delta g^{AB} \\
+ \frac{\lambda^{D-2}}{2} \int_{\Sigma} d^{D-1} \xi \sqrt{|h|} \, \overline{T}_{AB}^{\text{brane}} \delta g^{AB} = 0.
\end{aligned}
\tag{10.11}
$$

We have multiplied by λ^{D-2}, and we have applied the usual definition (1.3) of the energy-momentum tensor associated with the matter Lagrangian. Note, however, that $\overline{T}_{AB}^{\text{brane}}$ has been defined on Σ, and has dimensions $[M^{D-1}]$, while T_{AB}^{bulk} has dimensions $[M^D]$ (we use an overline to distinguish this brane tensor from the stress tensor of the brane covariantly defined with respect to the whole bulk manifold, see below). The difference in sign between the two boundary terms on $\Sigma_\pm$ is due to the opposite orientation of the normal vector on the two sides of the hypersurface.

We have thus obtained two sets of equations valid, respectively, on the brane and on the bulk exterior to the brane. On the portion of bulk manifold defined by the union of the subspaces Ω_+ and Ω_- we have the standard Einstein equations,

$$
(G_{AB})_\Omega = \lambda^{D-2} \left(T_{AB}^{\text{bulk}} \right)_\Omega, \qquad \Omega = \Omega_+ \cup \Omega_-.
\tag{10.12}
$$

On the hypersurface Σ which separates the two domains we obtain an additional condition,

$$
(K_{AB} - h_{AB} K)_{\Sigma_+} - (K_{AB} - h_{AB} K)_{\Sigma_-} = \lambda^{D-2} \left(\overline{T}_{AB}^{\text{brane}} \right)_\Sigma,
\tag{10.13}
$$

controlling the possible discontinuity of the first derivative of the metric tensor across the brane, and determining the difference in the extrinsic curvature on the two sides of Σ in terms of the gravitational energy density located on the brane. Tracing with respect to the induced metric h_{AB}, using Eqs. (10.8) and (10.9), we obtain

$$
(2 - D)(K^+ - K^-) = \lambda^{D-2} \overline{T}^{\text{brane}},
\tag{10.14}
$$

and we can then rewrite Eq. (10.13) in the form usually adopted to express the Israel junction conditions:

$$
K_{AB}^+ - K_{AB}^- = \lambda^{D-2}\left(\overline{T}_{AB} - \frac{\overline{T}}{D-2}h_{AB}\right)^{\text{brane}}, \tag{10.15}
$$

where $K_{AB}^\pm = K_{AB}|_{\Sigma_\pm}$. If the model we are considering is Z_2 symmetric, with the position of the brane as the fixed point of the symmetry, then $K_{AB}^+ = -K_{AB}^- \equiv K_{AB}$, and the extrinsic curvature of the brane is fully determined by its energy momentum tensor as follows [22],

$$
K_{AB} = \frac{\lambda^{D-2}}{2}\left(\overline{T}_{AB} - \frac{\overline{T}}{D-2}h_{AB}\right)^{\text{brane}}. \tag{10.16}
$$

10.1.1 Covariant projection on the brane

We now present an alternative procedure to obtain the Israel junction conditions, based on a covariant approach in which the energy-momentum of the brane also acts directly as a source of the gravitational equations. This approach is convenient, in particular, to obtain the modified equations describing the total effective gravitational forces experienced by observers living on the brane, but which are also affected by the gravitational field present in the bulk, outside the brane.

We start with an action defined in the whole bulk manifold,

$$
S = \int_\Omega d^D x \sqrt{|g|}\left(-\frac{R}{2\lambda^{D-2}} + L_{\text{bulk}}\right) + S_{\text{brane}}, \tag{10.17}
$$

where S_{brane} is the action of the brane and all possible gravitational sources living on it. Such an action can be parametrized, as before, in Nambu–Goto form using the induced metric h_{AB} (the last term of Eq. (10.7)), or in Polyakov form,

$$
S_{\text{brane}} = -\int_\Omega d^D x' \int_\Sigma d^{D-1}\xi\, \delta^D(x' - X(\xi))\sqrt{|\gamma|}\, L_{\text{brane}}
$$
$$
\times \left[\gamma^{\mu\nu}\frac{\partial X^A}{\partial\xi^\mu}\frac{\partial X^B}{\partial\xi^\nu}g_{AB}(X) - (D-3)\right] \tag{10.18}
$$

where $X^A = X^A(\xi)$ are functions describing the parametric embedding of the brane into the bulk manifold, ξ^μ are the coordinates spanning the hypersurface Σ swept by the brane (obviously, Greek indices run from 0 to $D-2$) and $\gamma_{\mu\nu}$ is the intrinsic metric on the brane, acting as an auxiliary field (see for instance [23]). For the purpose of this section the explicit form of L_{brane} is not important: it is sufficient to observe that the variation with respect to the metric g_{AB}

provides a covariant contribution to the bulk gravitational equations, leading to the conventional Einstein form,

$$G_{AB} = \lambda^{D-2} \left(T_{AB}^{\text{bulk}} + T_{AB}^{\text{brane}} \right), \qquad (10.19)$$

where we have assumed that the boundary terms at infinity are canceled by the variation of an appropriate Gibbons–Hawking action. (See [24, 25] for explicit cosmological applications of the covariant Polyakov form of the brane action.)

The above equations are covariantly defined over the whole manifold spanned by the bulk coordinates x^A, but we should recall that the energy-momentum density of the brane has a distributional form, being strictly localized on a $(D-1)$-dimensional hypersurface through the Dirac delta function. Suppose, for instance, that the only gravitational contribution of the brane is generated by its tension, so that $L_{\text{brane}} = \mu^{D-1}/2 = \text{const}$, where μ is the typical mass scale of the brane. The functional differentiation of the Polyakov action with respect to $g^{AB}(x)$ then leads to the energy-momentum tensor

$$T_{AB}^{\text{brane}}(x) = \frac{2}{\sqrt{|g(x)|}} \frac{\delta S_{\text{brane}}}{\delta g^{AB}(x)}$$

$$= \frac{\mu^{D-1}}{\sqrt{|g(x)|}} \int_{\Sigma} d^{D-1}\xi \sqrt{|\gamma|} \, \gamma^{\mu\nu} \frac{\partial X_A}{\partial \xi^\mu} \frac{\partial X_B}{\partial \xi^\nu} \delta^D(x - X(\xi)). \qquad (10.20)$$

Note that T_{AB}^{brane} has dimensions $[M^D]$, and is thus dimensionally homogeneous with T_{AB}^{bulk}, different from the tensor $\overline{T}_{AB}^{\text{brane}}$ of Eq. (10.11) defined only on Σ.

Outside the brane $(x^A \neq X^A)$ Eq. (10.19) coincides with the previous equation (10.12). In order to obtain the junction conditions (10.15) we have to project the bulk equations (10.19) on the world-hypersurface Σ, matching all terms which are discontinuous across it. Here *projecting* means, more precisely, that all D-dimensional geometric objects (Riemann tensor, Ricci tensor, scalar curvature, …) appearing in the gravitational equations have to be expressed in terms of the corresponding $(D-1)$-dimensional objects defined intrinsically on Σ; in addition, the bulk indices of all covariant objects have to be contracted with the induced metric tensor h_A^B, in such a way as to result tangential to Σ [22].

It is convenient, for this purpose, to trace the Einstein equations in order to eliminate the scalar curvature: we obtain

$$R = -\frac{2\lambda^{D-2}}{D-2} \left(T^{\text{bulk}} + T^{\text{brane}} \right), \qquad (10.21)$$

and

$$R_{AB} = \lambda^{D-2} \left(T_{AB} - \frac{T}{D-2} g_{AB} \right)^{\text{bulk}} + \lambda^{D-2} \left(T_{AB} - \frac{T}{D-2} g_{AB} \right)^{\text{brane}}. \qquad (10.22)$$

The contraction with the induced metric then gives

$$R_{MN} h_A^M h_B^N = \lambda^{D-2} \left(T_{MN} h_A^M h_B^N - \frac{T}{D-2} h_{AB} \right)^{\text{bulk}}$$

$$+ \lambda^{D-2} \left(T_{AB} - \frac{T}{D-2} h_{AB} \right)^{\text{brane}}, \qquad (10.23)$$

where we have used the property $h_A^M h_{MB} = h_{AB}$, following from the definition (10.8); also, we have used the fact that T_{AB}^{brane} is confined on the brane, i.e. tangential to Σ, so that $T_{AB}^{\text{brane}} n^B = 0$ (like the induced metric and the extrinsic curvature, see Eq. (10.9)).

In order to express the bulk objects in terms of the intrinsic geometric objects of the brane we can use the Gauss equation [22], relating the bulk Riemann tensor R_{ABCD} to the $(D-1)$-dimensional *intrinsic* Riemann tensor defined on Σ, and constructed with the induced metric h_{AB} (let us call it $R_{ABCD}^\Sigma(h)$):

$$R_{ABCD}^\Sigma(h) = R_{MNPQ} h_A^M h_B^N h_C^P h_D^Q - K_{AD} K_{BC} + K_{AC} K_{BD}. \qquad (10.24)$$

The contraction with g^{AD} gives the corresponding relations between the Ricci tensors,

$$R_{BC}^\Sigma(h) = R_{NP} h_B^N h_C^P + R_{MNPQ} h_B^N h_C^P n^M n^Q - K K_{BC} + K_{AC} K_B^A. \qquad (10.25)$$

Finally, the application of a useful geometric identity (see e.g. [26]) allows us to rewrite the second term on the right-hand side of this equation in terms of the extrinsic curvature and of its covariant derivatives, and we are led to

$$R_{MN} h_A^M h_B^N = R_{AB}^\Sigma(h) + n^M \nabla_M K_{AB} + K_{MB} \nabla_A n^M + K_{AM} \nabla_B n^M$$

$$- 2K_{AM} K_B^M + K K_{AB}. \qquad (10.26)$$

Note that $R_{AB}^\Sigma(h)$ is symmetric and tangential to Σ, i.e. $R_{AB}^\Sigma(h) n^B = 0$.

We are now in a position to obtain the junction conditions by inserting the above relation into Eq. (10.23), integrating all terms of the equation along the normal direction $n_A dx^A$ from $-\epsilon$ to $+\epsilon$ across the position of Σ, and then performing the limit $\epsilon \to 0$ [26]. We observe that all *intrinsic* brane quantities (such as the induced metric h_{AB}, the Ricci tensor R_{AB}^Σ, the brane stress tensor, etc.) do have, of course, the same value on either side of Σ: thus, they give no contribution to this integration procedure, unless they are characterized by a singular distribution at $\epsilon = 0$ (i.e. on Σ), like T_{AB}^{brane} (see e.g. Eq. (10.20)). *Extrinsic* quantities (like K_{AB}) may have different values on the two sides of Σ, thus generating a (step-like) discontinuity across it: such a discontinuity does not contribute after performing the integration and the limiting procedure, but the *normal derivative* of such a discontinuity gives a divergent contribution on Σ which survives the integration procedure, and which has to be matched to the delta-function contribution of T_{AB}^{brane}.

Assuming that T_{AB}^{bulk} is non-singularly distributed over the whole bulk (and thus on Σ), and that the intrinsic curvature of Σ is also non-singular, the only singular contribution of Eq. (10.23) – beside that of T_{AB}^{brane} – may come from the normal derivative of the extrinsic curvature, $n^M \partial_M K_{AB}$, appearing on the right-hand side of Eq. (10.26). Integrating such a term along the normal direction from $-\epsilon$ to $+\epsilon$, and performing the limit $\epsilon \to 0$, we obtain the extrinsic curvature evaluated on the two opposite sides of Σ, and we are eventually led to the condition

$$K_{AB}^{+} - K_{AB}^{-} = \lambda^{D-2} \left(\overline{T}_{AB} - \frac{\overline{T}}{D-2} h_{AB} \right)^{\text{brane}}, \tag{10.27}$$

where $\overline{T}_{AB}^{\text{brane}}$ (with dimensions $[M^{D-1}]$) is the finite part of T_{AB}^{brane}, defined on Σ according to Eq. (10.11). This equation exactly coincides with the Israel junction condition (10.15).

In view of our subsequent cosmological applications it is now convenient to project on Σ the gravitational equations in the standard Einstein form (10.19). We again contract the Ricci tensor (10.25) with g^{BC}, to obtain a relation between the bulk scalar curvature R and the instrinsic curvature R^{Σ} of the brane,

$$R^{\Sigma} = R + 2R_{AB} n^A n^B - K^2 + K_A{}^B K_B{}^A. \tag{10.28}$$

Projecting the bulk Einstein tensor, $G_{MN} h_A^M h_B^N$, and combining Eqs. (10.25) and (10.28), we can then express the intrinsic Einstein tensor on Σ as follows.

$$G_{AB}^{\Sigma}(h) \equiv R_{AB}^{\Sigma} - \frac{1}{2} h_{AB} R^{\Sigma} = G_{MN} h_A^M h_B^N - h_{AB} R_{MN} n^M n^N$$

$$- K K_{AB} + K_{MB} K_A{}^M + \frac{1}{2}(K^2 - K_M{}^N K_N{}^M) h_{AB} + R_{MNPQ} h_A^N h_B^P n^M n^Q. \tag{10.29}$$

Before using the Einstein equations (to eliminate the bulk variables G_{MN}, R_{MN}) it is convenient to rewrite the last term of the above equation by decomposing the Riemann tensor into its Weyl, Ricci and scalar components, according to the well-known relation which, in D dimensions, takes the form [27]

$$R_{MNPQ} = C_{MNPQ} - \frac{2}{D-2} \left(R_{M[P} g_{Q]N} - R_{N[P} g_{Q]M} \right)$$

$$+ \frac{2R}{(D-1)(D-2)} g_{M[P} g_{Q]N} \tag{10.30}$$

(C_{ABCD} is the Weyl tensor). Using the property of the induced metric we obtain

$$R_{MNPQ}h_A^N h_B^P n^M n^Q = E_{AB} - \frac{1}{D-2}R_{MN}h_A^M h_B^N + \frac{1}{D-2}R_{MN}n^M n^N h_{AB}$$

$$+ \frac{R}{(D-1)(D-2)}h_{AB}, \tag{10.31}$$

where

$$E_{AB} = C_{MNPQ}\, h_A^N h_B^P n^M n^Q \tag{10.32}$$

is a symmetric and trace-free tensor, $E_A{}^A = 0$, by virtue of the Weyl tensor symmetries.

Let us now insert this last expression into Eq. (10.29), and eliminate everywhere the bulk Einstein tensor G_{MN} through the Einstein equations (10.19), the bulk Ricci tensor R_{MN} through the equivalent form (10.22), and the bulk scalar through Eq. (10.21). We should note, however, that Eqs. (10.29) and (10.31) are valid on the bulk (on the two sides of the world-hypersurface Σ), and are to be evaluated on Σ with a limiting procedure. Thus, when using the bulk equations for G_{MN}, R_{MN} and R we have to drop the contribution of T_{AB}^{brane} which is sharply localized on Σ, and vanishing outside it. Summing up all similar terms we finally obtain the equation

$$G_{AB}^\Sigma = \left(\frac{D-3}{D-2}\right)\lambda^{D-2}\left[T_{MN}h_A^M h_B^N - h_{AB}\left(T_{MN}n^M n^N + \frac{T}{D-1}\right)\right]^{\text{bulk}}$$

$$+ E_{AB} + S_{AB}, \tag{10.33}$$

where

$$S_{AB} = -KK_{AB} + K_{MB}K_A{}^M + \frac{1}{2}h_{AB}\left(K^2 - K_M{}^N K_N{}^M\right), \tag{10.34}$$

and which generalizes to the D-dimensional case previous results obtained in $D = 5$ [22, 26].

The gravitational contribution of the brane, in this equation, is fully contained inside the extrinsic curvature terms, related to $\overline{T}_{AB}^{\text{brane}}$ by the Israel junction conditions. In order to make this contribution explicit let us consider a simple brane-world scenario in which the hypersurface Σ is sitting at the fixed point of the Z_2 symmetry, and then its extrinsic curvature is determined by the energy-momentum tensor of the brane according to Eq. (10.16). The source tensor S_{AB} can thus be expressed as follows:

$$S_{AB} \equiv S_{AB}(\overline{T}^{\text{brane}}) = \left(\frac{\lambda^{D-2}}{2}\right)^2\left[\overline{T}_{MB}\overline{T}_A{}^M - \frac{\overline{T}}{D-2}\overline{T}_{AB}\right.$$

$$\left. - \frac{1}{2}h_{AB}\left(\overline{T}_M{}^N\overline{T}_N{}^M - \frac{\overline{T}^2}{D-2}\right)\right]^{\text{brane}}. \tag{10.35}$$

We should recall now that $\overline{T}_{AB}^{\text{brane}}$ may contain, in general, a contribution due to the tension μ^{D-1} (acting as a vacuum energy density, or cosmological constant, on the brane), and another contribution due to the energy-momentum density of other matter fields possibly present (and localized) on the brane [28, 29]. We can thus split the brane contribution as follows:

$$\overline{T}_{AB}^{\text{brane}} = \mu^{D-1} h_{AB} + \tau_{AB},\tag{10.36}$$

where τ_{AB} represents the contribution of all matter sources contained in L_{brane} and strictly confined on Σ (of course, τ_{AB} has to be tangential to the brane, i.e. $\tau_{AB} n^B = 0$). From Eq. (10.35) we then obtain

$$S_{AB}(\overline{T}^{\text{brane}}) = \left(\frac{D-3}{D-2}\right) \mu^{D-1} \left(\frac{\lambda^{D-2}}{2}\right)^2 \tau_{AB} + \frac{1}{2}\left(\frac{D-3}{D-2}\right) \left(\frac{\mu^{D-1}\lambda^{D-2}}{2}\right)^2 h_{AB}$$
$$+ S_{AB}(\tau),\tag{10.37}$$

where $S_{AB}(\tau)$ is given exactly by the expression (10.35) with $\overline{T}_{AB}^{\text{brane}}$ replaced by τ_{AB}. Note that all terms of this equation have dimensions $[L^{-2}]$, thus consistently matching the dimensions of the Einstein tensor in Eq. (10.33).

Inserting this result into Eq. (10.33), and noting that all terms are tangential to Σ (i.e. they have vanishing contraction with the normal vector n^A), we can rewrite the final equations using Greek indices, ranging from 0 to $D-2$. We want to stress, in this way, that such equations describe the effective gravitational dynamics localized on the brane, dimensionally reduced from the higher-dimensional bulk interactions. The result is

$$G_{\alpha\beta}^{\Sigma} = \left(\frac{D-3}{D-2}\right) \mu^{D-1} \left(\frac{\lambda^{D-2}}{2}\right)^2 \tau_{\alpha\beta} + \frac{1}{2}\left(\frac{D-3}{D-2}\right) \left(\frac{\mu^{D-1}\lambda^{D-2}}{2}\right)^2 h_{\alpha\beta}$$
$$+ \left(\frac{D-3}{D-2}\right) \lambda^{D-2} \left[T_{\mu\nu} h_\alpha^\mu h_\beta^\nu - h_{\alpha\beta}\left(T_{MN} n^M n^N + \frac{T}{D-1} \right) \right]^{\text{bulk}}$$
$$+ \left(\frac{\lambda^{D-2}}{2}\right)^2 \left[\tau_\alpha^{\ \mu} \tau_{\mu\beta} - \frac{\tau}{D-2} \tau_{\alpha\beta} - \frac{1}{2} h_{\alpha\beta}\left(\tau_\mu^{\ \nu} \tau_\nu^{\ \mu} - \frac{\tau^2}{D-2} \right) \right]$$
$$+ E_{\alpha\beta},\tag{10.38}$$

where $E_{\alpha\beta}$ is defined in Eq. (10.32). This result clearly shows that there are three types of corrections to the standard Einstein equations with cosmological terms (corresponding to the first line of the above equation): (*i*) corrections due to the possible presence of bulk matter (second line), (*ii*) corrections quadratic in the brane stress tensor (third line), possibly important in the high-density regime, and (*iii*) corrections due to the curvature of the bulk geometry (fourth line).

Two final remarks are in order. The first concerns the coefficient of the term linear in $\tau_{\alpha\beta}$, representing the effective gravitational coupling constant appearing in the low-energy Einstein equations (and thus determining the phenomenological Planck length of the brane):

$$8\pi G_{\text{brane}} \equiv \lambda_{\text{P}}^{D-3} = \left(\frac{D-3}{D-2}\right)\mu^{D-1}\left(\frac{\lambda^{D-2}}{2}\right)^2. \qquad (10.39)$$

This effective coupling is controlled by the bulk gravitational scale λ and by the brane-tension scale μ, but there is no explicit volume factor (compare with Eq. (10.5)) since, in this case, there are no compactified dimensions. Nevertheless, also in this case, it is possible to reproduce the standard value of the Planck length λ_{P} even starting from a very different (in particular, much larger) value of the higher-dimensional gravitational scale λ, by an appropriate tuning of the tension of the brane.

The second remark concerns the fact that Eq. (10.38) describes the gravitational field on the brane, but its complete solution needs information also about the bulk gravitational field, in order to determine the Weyl contribution $E_{\alpha\beta}$ appearing as a non-local source to an observer confined on the brane [26]. This means, in other words, that the evolution of the gravitational field (and, in particular, of its perturbations) cannot be completely determined from initial conditions set solely on the brane, but requires a fully higher-dimensional analysis [30].

The cosmological applications of the modified gravitational equations (10.38) will be discussed in the following sections.

10.2 Warped geometry and localization of gravity

This section is devoted to a simple but important example which illustrates how the bulk geometry may affect the gravitational interaction on the brane. It will be shown, in particular, that an appropriate bulk curvature can force the massless component of the gravitational fluctuations to be confined on the brane, thus generating long-range gravitational forces which are insensitive to the presence of extra spatial dimensions normal to the brane [9]. This effect is of crucial phenomenological importance, since in string theory the mechanism of localization of gauge fields on the brane cannot be directly applied to the gravitational field, which is free to propagate along all bulk directions.

Our discussion will be concentrated on a particular exact solution of the generalized gravitational equations introduced in the previous section, describing a flat Minkowski brane embedded in an anti-de Sitter (AdS) bulk, and will proceed with the study of tensor metric perturbations around this background solution. We start by considering a model described by the action (10.17) in which the only gravitational source in the bulk is a cosmological constant,

$$L_{\text{bulk}} = -\frac{\Lambda}{\lambda^{D-2}}, \qquad (10.40)$$

and the only gravitational source on the brane is its tension, so that S_{brane} coincides with Eq. (10.18) with

$$L_{\text{brane}} = \frac{\mu^{D-1}}{2}. \qquad (10.41)$$

The variation of the action with respect to g^{AB} provides the following bulk Einstein equations,

$$G_A{}^B = \Lambda \delta_A^B + \frac{\lambda^{D-2}\mu^{D-1}}{\sqrt{|g|}} \int_\Sigma d^{D-1}\xi \sqrt{|\gamma|}\, \gamma^{\mu\nu} \partial_\mu X_A \partial_\nu X^B \delta^D(x-X), \qquad (10.42)$$

where $\partial_\mu X^A \equiv \partial X^A / \partial \xi^\mu$ (see also Eq. (10.20)). The variation with respect to X^A gives the equations governing the evolution of the brane in the bulk space-time,

$$\partial_\mu \left[\sqrt{|\gamma|}\, \gamma^{\mu\nu} \partial_\nu X^B g_{AB}(x) \right]_{x=X(\xi)} = \frac{1}{2} \left[\sqrt{|\gamma|}\, \gamma^{\mu\nu} \partial_\mu X^M \partial_\nu X^N \partial_A g_{MN}(x) \right]_{x=X(\xi)} \qquad (10.43)$$

(analogous to the string equations of motion for the Polyakov action, see Eq. (3.7)). Finally, the variation with respect to $\gamma^{\mu\nu}$ gives

$$\partial_\mu X^A \partial_\nu X_A - \frac{1}{2} \left(\gamma^{\alpha\beta} \partial_\alpha X^A \partial_\beta X_A \right) \gamma_{\mu\nu} + \frac{1}{2}(D-3)\gamma_{\mu\nu} = 0, \qquad (10.44)$$

from which, tracing with respect to γ, we obtain $\partial_\mu X^A \partial^\mu X_A = D-1$, and we arrive at the identification of $\gamma_{\mu\nu}$ with the induced metric

$$\gamma_{\mu\nu} = \partial_\mu X^A \partial_\nu X^B g_{AB} \equiv h_{\mu\nu}. \qquad (10.45)$$

We may note, incidentally, that inserting this result into the Polyakov action (10.18) we exactly recover the Nambu–Goto form of the brane action,

$$S_{\text{brane}} = -\mu^{D-1} \int_\Sigma d^{D-1} \sqrt{|h|}\,. \qquad (10.46)$$

We now look for particular solutions of the above equations in which the bulk metric is conformally flat, $g_{AB} = f^2(z)\eta_{AB}$, with a conformal factor $f(z)$ which only depends on the coordinate z normal to the brane, and in which the (globally flat) brane is rigidly fixed at $z = 0$, and described by the trivial embedding

$$X^A(\xi) = \delta_\mu^A \xi^\mu, \qquad A = 0, \ldots, D-2,$$

$$X^{D-1} \equiv z = 0. \qquad (10.47)$$

We also impose that the solution be symmetric under $z \to -z$ reflections, so that $f = f(|z|)$. We thus consider a Z_2-even, higher-dimensional background characterized by the following "warped" geometrical structure,

$$ds^2 = f^2(|z|) \left(\eta_{\mu\nu} \, dx^\mu \, dx^\nu - dz^2 \right), \tag{10.48}$$

different from the "factorized" geometrical structure typical of the Kaluza–Klein scenario, in which the dimensionally reduced metric is translationally invariant along the extra spatial dimensions.

For the chosen type of background the induced metric (10.45) reduces to $\gamma_{\mu\nu} = f^2 \eta_{\mu\nu}$, and the brane equation (10.43) turns out to be identically satisfied. Also, for the metric (10.48), we find the following non-vanishing components of the connection,

$$\Gamma_{zz}{}^z = \frac{f'}{f} \equiv F, \qquad \Gamma_{\mu\nu}{}^z = F \, \eta_{\mu\nu}, \qquad \Gamma_{z\mu}{}^\nu = F \, \delta_\mu^\nu, \tag{10.49}$$

and of the Ricci tensor,

$$R_{\mu\nu} = \left[F' + (D-2)F^2 \right] \eta_{\mu\nu}, \qquad R_{zz} = -(D-1)F', \tag{10.50}$$

where the prime denotes differentiation with respect to z. Their combination leads to the Einstein tensor,

$$G_\mu{}^\nu = -\frac{1}{f^2} \left[(D-2)F' + \frac{1}{2}(D-2)(D-3)F^2 \right] \delta_\mu^\nu,$$

$$G_z{}^z = -\frac{1}{2f^2}(D-1)(D-2)F^2. \tag{10.51}$$

Inserting these results into the Einstein equation (10.42) we obtain, from the component normal and tangential to the brane, respectively, the equations

$$-\frac{1}{2}(D-1)(D-2)F^2 = \Lambda f^2, \tag{10.52}$$

$$-(D-2)F' - \frac{1}{2}(D-2)(D-3)F^2 = \Lambda f^2 + \lambda^{D-2} \mu^{D-1} f \delta(z). \tag{10.53}$$

To solve this system of equations we must recall that the warp function f depends on the modulus of z, so that the second derivatives of f contain the derivative of the sign function, which generates a delta-function contribution also on the left-hand side of Eq. (10.53): we have thus to match separately the finite part of the equation and the coefficients of the singular contributions at $z = 0$. We obtain, in this way, two conditions: one determining the functional form of the warp factor, the other imposing a consistency relation between the brane tension and the bulk cosmological constant.

Let us put $f = f(|z|)$, and define

$$y \equiv |z| = z \, \text{sign}\{z\} = z[\theta(z) - \theta(-z)] \equiv z\epsilon(z),$$

$$f' = \frac{df}{dy} y' = \frac{df}{dy} \epsilon(z),$$

(10.54)

where θ is the Heaviside step function. From Eq. (10.52) we obtain

$$\left(\frac{df}{dy}\right)^2 = -\frac{2\Lambda}{(D-1)(D-2)} f^4,$$

(10.55)

which admits a real solution provided $\Lambda < 0$. Assuming that the bulk cosmological constant is negative, and integrating, we are led to the particular exact solution

$$f(|z|) = (1 + k|z|)^{-1}, \qquad k = \left[\frac{-2\Lambda}{(D-1)(D-2)}\right]^{1/2},$$

(10.56)

which inserted into the metric (10.48) describes an anti-de Sitter bulk geometry, in the conformally flat parametrization.

We have still to satisfy the other Einstein equation (10.53), which contains the explicit contribution of the brane. We observe that using the following properties of the sign function,

$$y' = \epsilon, \qquad \epsilon^2 = 1, \qquad \epsilon' = 2\delta(z),$$

(10.57)

we can rewrite Eq. (10.53) in terms of y as

$$-\frac{(D-2)(D-5)}{2f^2}\left(\frac{df}{dy}\right)^2 - \frac{(D-2)}{f}\frac{d^2f}{dy^2} - 2\frac{(D-2)}{f}\frac{df}{dy}\delta(z)$$

$$= \Lambda f^2 + \lambda^{D-2}\mu^{D-1}f\delta(z).$$

(10.58)

Inserting the solution (10.56) one then finds that the finite part of this equation is identically satisfied, while the coefficients of the terms containing the delta-function are exactly matched provided we impose the condition

$$\frac{1}{2}\lambda^{D-2}\mu^{D-1} = (D-2)k \equiv \left[\frac{-2(D-2)\Lambda}{(D-1)}\right]^{1/2},$$

(10.59)

relating the tension of the brane to the curvature scale of the AdS bulk geometry [9].

It is important to note that precisely this condition is responsible for the flat geometry of the brane: this condition implies that the intrinsic vacuum energy of the brane (due to its tension) is canceled by an opposite contribution generated by the bulk cosmological constant (a mechanism of "off-loading" gravity from the brane to the bulk, see [31]).

The above condition can also be directly deduced from the effective equation (10.38) projected on the brane, presented in the previous section. For a conformally flat metric, in fact, the Weyl term is vanishing ($E_{AB} = 0$, see e.g. [26]); also, $\tau_{\alpha\beta} = 0$, as there are no matter sources on the brane. We are thus left with only the contribution of the brane tension and the bulk cosmological constant, which is associated with the bulk stress tensor

$$\lambda^{D-2} T_{AB}^{\text{bulk}} = \Lambda g_{AB} \tag{10.60}$$

(see Eq. (10.42)). Inserting this contribution into Eq. (10.38), and summing up all non-vanishing terms, we find that the brane geometry is compatible with a Ricci-flat metric (satisfying $G_{\alpha\beta}^{\Sigma} = 0$) provided the following condition holds,

$$\frac{1}{2}\left(\frac{D-3}{D-2}\right)\left(\frac{\mu^{D-1}\lambda^{D-2}}{2}\right)^2 + \frac{D-3}{D-1}\Lambda = 0, \tag{10.61}$$

exactly equivalent to the previous equation (10.59). Note that, besides the flat Minkowski metric, this condition is also compatible with less trivial (for instance, Kasner-like [25]) vacuum geometries on the brane.

It shoud be stressed, also, that the same condition relates the tension mass scale to the effective renormalized coupling strength of gravity on the brane. Such a coupling strength, in fact, can be obtained from the gravitational part of the bulk action (10.17), computing the bulk scalar curvature R for a perturbed metric $\tilde{g}_{\mu\nu}$ which includes the tensor fluctuations on the brane, namely for

$$ds^2 = f^2(|z|)\left(\tilde{g}_{\mu\nu}\, dx^\mu\, dx^\nu - dz^2\right). \tag{10.62}$$

As we shall see in more detail below, the part of the action quadratic in $\tilde{g}$ defines an effective action for the (weak-field) gravity on the brane which is of the general form

$$-\frac{1}{2\lambda^{D-2}}\int dz\, f^D f^{-2}\int d^{D-1}x\sqrt{|\tilde{g}|}\, R_{D-1}(\tilde{g}), \tag{10.63}$$

where the factor f^D comes from the determinant $\sqrt{|g|}$ of the D-dimensional metric tensor g_{AB}, and the factor f^{-2} comes from the inverse metric g^{AB} (required for the computation of the scalar curvature, see Eqs. (10.50) and (10.51)). For the solution (10.56), in particular, the z-dependence of the above action can be integrated exactly,

$$\int_{-\infty}^{+\infty} dz\, f^{D-2} = \int_{-\infty}^{+\infty} dz\, (1+k|z|)^{2-D} = \frac{2}{k(D-3)}, \tag{10.64}$$

and the comparison with the Einstein action on the brane,

$$-\frac{1}{2\lambda_P^{D-3}}\int d^{D-1}x\sqrt{|g_{D-1}|}\, R_{D-1}, \tag{10.65}$$

leads immediately to the effective coupling constant

$$\lambda_P^{D-3} = \frac{k}{2}(D-3)\lambda^{D-2}, \tag{10.66}$$

with k given in Eq. (10.59). For $D = 5$, for instance, one obtains the usual RS definition [9] of the Planck scale on the brane, $\lambda_P^2 = k\lambda^3$. Finally, using Eq. (10.59) relating k to the tension of the brane, one obtains

$$\lambda_P^{D-3} = \left(\frac{D-3}{D-2}\right)\mu^{D-1}\left(\frac{\lambda^{D-2}}{2}\right)^2, \tag{10.67}$$

which exactly reproduces the effective gravitational constant (10.39), following from the covariant projection of the Einstein equations on the brane. This confirms the equivalence of the two approaches, but also shows that the full, higher-dimensional approach is required for a complete analysis of the bulk geometry.

Let us now discuss in more detail the dynamics of tensor fluctuations on the brane. We start by perturbing the bulk metric at fixed brane position, introducing the expansion

$$g_{AB} \to g_{AB} + \delta g_{AB}, \qquad \delta g_{AB} \equiv h_{AB}, \qquad \delta X^A = 0, \tag{10.68}$$

and then compute the perturbations of the total action (10.17) up to terms quadratic in the first-order fluctuations h_{AB}. We are interested, in particular, in the transverse and traceless part of the metric fluctuations on the brane, $h_{\mu\nu}$, which in the linear approximations are decoupled from other (scalar and extra-dimensional) components of δg_{AB}. The perturbed configuration we study is thus characterized by

$$h_{zA} = 0, \qquad h_{\mu\nu} = h_{\mu\nu}(x^\mu, z), \qquad g^{\mu\nu}h_{\mu\nu} = 0 = \partial^\nu h_{\mu\nu}. \tag{10.69}$$

For the computation of the perturbed, quadratic action we may follow the standard procedure already introduced in Chapter 7, applying it to the unperturbed metric background (10.48). After using the unperturbed background equations we thus obtain (see in particular Eq. (7.23))

$$\delta^{(2)}S = -\frac{1}{8\lambda^{D-2}}\int d^D x \sqrt{|g|}\, h^\nu_\mu \nabla_A \nabla^A h^\mu_\nu$$

$$= -\frac{1}{8\lambda^{D-2}}\int d^D x\, f^{D-2}\left[h^\nu_\mu \Box h^\mu_\nu - h^\nu_\mu h''^\mu_\nu - (D-2)Fh^\nu_\mu h'^\mu_\nu\right], \tag{10.70}$$

where $\Box \equiv \partial_t^2 - \partial_i^2$ is the usual d'Alembert operator in flat Minkowski space. Integrating by parts, and tracing over the spin-two polarization tensor (see Section 7.1), the action for each polarization mode $h_A(t, x^i, z)$ can be written in the form

$$\delta^{(2)}S = \frac{1}{4\lambda^{D-2}} \int dz \int d^{D-1}x \, f^{D-2} \left(\dot{h}^2 + h\nabla^2 h - h'^2 \right), \qquad (10.71)$$

where the dot denotes differentiation with respect to t, and $\nabla^2 = \delta^{ij}\partial_i\partial_j$ is the usual Laplace operator on the $(D-2)$-dimensional Euclidean sections of the brane (we have omitted, for simplicity, the polarization index). The variation with respect to h finally provides the vacuum propagation equation for the weak-field gravitational perturbations:

$$\Box h - h'' - (D-2)Fh' = 0. \qquad (10.72)$$

This equation differs from the free wave equation in the Minkowski space of the brane, since the gravitational fluctuations are coupled to the AdS bulk geometry through the gradients of the warp factor.

To solve this equation we note that the bulk and the brane coordinates can be separated by setting

$$h(x^\mu, z) = \sum_m v_m(x^\mu)\psi_m(z), \qquad (10.73)$$

where the new variables v and ψ satisfy the eigenvalue equations

$$\Box v_m = -m^2 v_m,$$
$$\psi_m'' + (D-2)F\psi_m' \equiv f^{2-D} \left(f^{D-2}\psi_m' \right)' = -m^2\psi_m \qquad (10.74)$$

(for the continuous part of the eigenvalue spectrum the sum of Eq. (10.73) is clearly replaced by the integration over m). We note, also, that the term with the first derivative present in the second equation can be eliminated by introducing the variable

$$\widehat{\psi}_m = \left(\frac{f^{D-2}}{\lambda} \right)^{1/2} \psi_m, \qquad (10.75)$$

where the constant factor $\lambda^{-1/2}$ has been inserted for later convenience. In terms of $\widehat{\psi}$, the second equation then takes the form of a one-dimensional Schrödinger-like equation,

$$\widehat{\psi}_m'' + \left[m^2 - V(z) \right] \widehat{\psi}_m = 0, \qquad (10.76)$$

with effective potential

$$V(z) = \frac{D(D-2)}{4} \frac{k^2}{(1+k|z|)^2} - \frac{k(D-2)}{(1+k|z|)} \delta(z). \qquad (10.77)$$

This potential has the so-called "volcano-like" shape, since the first term of the potential has a peak at $z = 0$, but in correspondence with it there is a negative delta function, like the crater of a volcano.

As is well known from elementary quantum mechanics, the one-dimensional Schrödinger equation with an attractive delta-function potential $-\delta(z)$ admits one bound state only, with a square-integrable function which is localized around the position of the potential at $z = 0$. The bound state, in our context, is represented by the massless mode solution of Eq. (10.76). For $m = 0$ the equation has the Z_2-even solution

$$\widehat{\psi}_0 = c_0 f^{(D-2)/2},\qquad (10.78)$$

which is normalizable (with respect to inner products with measure dz, as in conventional one-dimensional quantum mechanics), even for an infinite extension of the coordinate z normal to the brane:

$$\int dz \left|\widehat{\psi}_0(z)\right|^2 = \int_{-\infty}^{+\infty} dz \frac{c_0^2}{(1+k|z|)^{D-2}} = \frac{2c_0^2}{k(D-3)} < \infty.\qquad (10.79)$$

Thus, the massless components of the metric fluctuations are localized on the brane at $z = 0$, not because the extra dimensions are compactified on a volume of very small size (as in the standard Kaluza–Klein scenario), but because the bulk curvature (due in this case to the AdS geometry) forces such fluctuations to be peaked around the brane position.

Concerning the normalization condition we note, finally, that $\widehat{\psi}_0(z)$ defined as in Eq. (10.75) has the correct canonical dimensions to belong to the space L_2 of square-integrable functions with measure dz. The orthonormality condition can be equivalently expressed in terms of ψ_m, however, using inner products with measure $f^{D-2}dz/\lambda$ (see Eq. (10.75)):

$$\int \frac{dz}{\lambda} f^{D-2} \psi_m \psi_n = \delta(m, n),\qquad (10.80)$$

where $\delta(m, n)$ denotes the Kronecker symbol for the discrete part of the spectrum, and the Dirac delta function for the continuous part. In any case, with this normalization, we obtain from Eq. (10.79) $c_0 = [k(D-3)/2]^{1/2}$, so that

$$\psi_0 = c_0 \lambda^{1/2} = \left[\frac{(D-3)k\lambda}{2}\right]^{1/2} = \text{const.}\qquad (10.81)$$

Besides the massless mode ψ_0, the metric fluctuations also contain massive components ψ_m – solutions of Eq. (10.76) with $m \neq 0$ – which are not localized on the brane, and are characterized by a continuous spectrum which extends over all positive values of m, up to infinity. To obtain these solutions we may follow that standard quantum-mechanical treatment of the delta-function potential: looking for Z_2-even solutions, $\widehat{\psi}_m = \widehat{\psi}_m(|z|)$, and using Eqs. (10.54) and (10.57), we first rewrite Eq. (10.76) as

$$\frac{d^2\widehat{\psi}_m}{dy^2} + 2\delta(z)\frac{d\widehat{\psi}_m}{dy} + \left[m^2 - V(z)\right]\widehat{\psi}_m = 0. \tag{10.82}$$

Outside the origin ($z \neq 0$) we have thus the equation

$$\frac{d^2\widehat{\psi}_m}{dy^2} + \left[m^2 - \frac{D(D-2)}{4}\frac{k^2}{(1+ky)^2}\right]\widehat{\psi}_m = 0, \tag{10.83}$$

whose general exact solution can be written as a combination of Bessel functions [32] of index $\nu = (D-1)/2$, and argument $m(1+ky)/k = m/fk$,

$$\widehat{\psi}_m = f^{-1/2}\left[A_m J_{\frac{D-1}{2}}\left(\frac{m}{kf}\right) + B_m Y_{\frac{D-1}{2}}\left(\frac{m}{kf}\right)\right]. \tag{10.84}$$

Imposing that this solution be valid also at $z = 0$, and equating the coefficients of the delta-function terms, we obtain the additional condition

$$2\frac{d\widehat{\psi}_m}{dy}(0) + k(D-2)\widehat{\psi}_m(0) = 0, \tag{10.85}$$

which provides a relation between the two integration constants A_m and B_m,

$$B_m = -A_m \frac{J_{(D-3)/2}(m/k)}{Y_{(D-3)/2}(m/k)}. \tag{10.86}$$

The general solution can thus be rewritten as follows:

$$\widehat{\psi}_m = c_m f^{-1/2}\left[Y_{\frac{D-3}{2}}\left(\frac{m}{k}\right)J_{\frac{D-1}{2}}\left(\frac{m}{kf}\right) - J_{\frac{D-3}{2}}\left(\frac{m}{k}\right)Y_{\frac{D-1}{2}}\left(\frac{m}{kf}\right)\right], \tag{10.87}$$

and the overall constant factor c_m is finally fixed by the orthonormality condition (10.80), which leads to

$$c_m = \left(\frac{m}{2k}\right)^{1/2}\left[J^2_{\frac{D-3}{2}}\left(\frac{m}{k}\right) + Y^2_{\frac{D-3}{2}}\left(\frac{m}{k}\right)\right]^{-1/2} \tag{10.88}$$

(see also [33]).

These massive modes form a continuous spectrum of solutions in the interval $0 < m \leq +\infty$, and are asymptotically oscillating for $z \to \pm\infty$, so that they cannot be localized on the brane as in the case of the massless fluctuations. We may thus expect that such massive components may induce short-range gravitational corrections which "feel" the presence of the extra spatial dimensions exterior to the brane, and which are thus directly affected by their geometrical properties.

For a quantitative estimation of such corrections we need the effective gravitational coupling of the massive fluctuations, which can be obtained from the canonical form of the action (10.71), and from the relative canonical normalization of ψ_0 and ψ_m. Inserting the expansion (10.73) in the action we note, first of all, that the first derivatives with respect to z can be eliminated (modulo a total derivative) by virtue of the second equation (10.74), which implies

$$\int dz f^{D-2} h'^2 = \sum_{m,n} v_m v_n \int dz\, f^{D-2} \psi'_m \psi'_n$$

$$= \sum_{m,n} v_m v_n \int dz \left[\frac{d}{dz} \left(f^{D-2} \psi_m \psi'_n \right) - \psi_m \left(f^{D-2} \psi'_n \right)' \right]$$

$$= \sum_{m,n} v_m v_n \int dz\, f^{D-2} m^2 \psi_m \psi_n. \tag{10.89}$$

Integrating over z, and using the orthormality condition (10.80), we are then led to a dimensionally reduced action which contains only the components $v_m(x)$ of the fluctuations,

$$\delta^{(2)} S = \sum_m \delta^{(2)} S_m = \sum_m \frac{1}{4\lambda^{D-3}} \int d^{D-1}x \left(\dot{v}_m^2 + v_m \nabla^2 v_m - m^2 v_m^2 \right), \tag{10.90}$$

where the symbol $\sum_m$ denotes that the contribution of the massless mode $m = 0$ has to be summed to the integral (from 0 to ∞) over the continuous spectrum of the massive mode contributions. Finally, introducing the fields $\overline{h}_m$ which represent the effective gravitational fluctuations on the brane,

$$\overline{h}_m(x^\mu) \equiv [h_m(x^\mu, z)]_{z=0} = v_m(x^\mu) \psi_m(0), \tag{10.91}$$

we obtain the effective action

$$\delta^{(2)} S = \sum_m \frac{1}{4\lambda^{D-3} \psi_m^2(0)} \int d^{D-1}x \left(\dot{\overline{h}}_m^2 + \overline{h}_m \nabla^2 \overline{h}_m - m^2 \overline{h}_m^2 \right), \tag{10.92}$$

which defines the canonical variables for the effective gravitational interaction on the brane,

$$u_m(x^\mu) = \frac{\overline{h}_m(x^\mu)}{(2\lambda^{D-3})^{1/2} \psi_m(0)}. \tag{10.93}$$

Comparison with the canonical form of the tensor fluctuation components (see for instance Chapter 7, Eqs. (7.49) and (7.51)) leads immediately to defining the effective coupling for a generic mode $\overline{h}_m$:

$$\lambda_P^{D-3}(m) \equiv \lambda^{D-3} \psi_m^2(0). \tag{10.94}$$

For the massless mode, using the solution (10.81), we thus recover the relation (10.66) defining the phenomenological Planck scale of the brane,

$$\lambda_P^{D-3} \equiv 8\pi G_{D-1} \equiv \lambda_P^{D-3}(0) = \frac{k}{2}(D-3)\lambda^{D-2}, \tag{10.95}$$

which is the typical scale of long-range gravitational interactions (see also Eqs. (10.67) and (10.39)). For the massive modes, instead, the effective coupling is mass dependent: using the relation $\psi_m(0) = \sqrt{\lambda}\,\widehat{\psi}(0)$, and the solution (10.87) and (10.88), we obtain

$$\lambda_P^{D-3}(m) \equiv 8\pi G_{D-1}(m) = \lambda^{D-2} F^2\left(\frac{m}{k}\right), \qquad (10.96)$$

where $(x \equiv m/k)$

$$F^2(x) = \left(\frac{x}{2}\right)\frac{\left[Y_{\frac{D-3}{2}}(x)\,J_{\frac{D-1}{2}}(x) - J_{\frac{D-3}{2}}(x)\,Y_{\frac{D-1}{2}}(x)\right]^2}{J^2_{\frac{D-3}{2}}(x) + Y^2_{\frac{D-3}{2}}(x)}. \qquad (10.97)$$

This gives the effective coupling parameter of the mode $\overline{h}_m$ in the infinitesimal mass interval between m and $m+dm$.

10.2.1 Short-range corrections

We are now in a position to provide a precise estimate of the contribution of the massive modes to the effective gravitational interaction on the brane. We consider a simple, but instructive, example corresponding to the static field produced by a point-like source of mass M confined on the brane, with energy density $\rho(x^\mu, z) = M\delta^{d-1}(x)\delta(z)$. For a more direct comparison with standard phenomenology we choose the realistic case of a 3-brane, spanning a $(3+1)$-dimensional world-hypervolume which can be identified with the four-dimensional space-time in which we live, and which is assumed to be embedded in a $D = 5$ bulk manifold.

We start by recalling that, in the flat Minkowski space of the brane, the quadratic action (10.92) is generated by the contribution of the perturbed Ricci tensor (see Eq. (7.23)), and its variation leads to the usual (apart from the mass term) linearized form of the Einstein equations in flat space: for a generic component μ, ν, and a generic mode m, of the weak-field fluctuations $\overline{h}$ on the brane, they read

$$-\frac{1}{2}\left(\Box + m^2\right)\overline{h}_m^{\mu\nu} = \lambda_P^2(m)\left(T^{\mu\nu} - \frac{1}{2}\eta^{\mu\nu}\tau\right), \qquad (10.98)$$

where we have used the definition (10.94) of the canonical coupling to the matter stress tensor. In the static limit $\Box \to -\nabla^2$, $\tau_i^{\ j} \to 0$, $\tau = \eta^{\mu\nu}T_{\mu\nu} \to T_0^{\ 0} = \rho$, and $\overline{h}_m^{00} \to 2\overline{\varphi}_m$, where $\overline{\varphi}_m$ is the effective gravitational potential generated by the mode m on the brane. The (00) component of Eq. (10.98) then gives the equation

$$(-\nabla^2 + m^2)\overline{\varphi}_m(x) = -\frac{1}{2}\lambda_P^2(m)\,\rho(x), \qquad (10.99)$$

which controls the contribution of the generic mode m to the static gravitational potential.

The solution to this equation can be easily obtained, for a generic source, using the static limit of the retarded Green function, i.e. by setting

$$\bar{\varphi}_m = -\frac{1}{4\pi} \int d^3 x' \mathcal{G}_m(x, x') \frac{1}{2} \lambda_P^2(m) \rho(x'), \qquad (10.100)$$

where $\mathcal{G}_m(x, x')$ satisfies

$$(-\nabla^2 + m^2) \mathcal{G}_m(x, x') = 4\pi \delta^3(x - x'). \qquad (10.101)$$

Thus, Fourier transforming,

$$\mathcal{G}_m(x, x') = 4\pi \int \frac{d^3 p}{(2\pi)^3} \frac{e^{i p \cdot (x - x')}}{p^2 + m^2}. \qquad (10.102)$$

For the massless mode one then obtains, using polar coordinates,

$$\mathcal{G}_0(x, x') = \frac{2}{\pi} \int_0^\infty dp \frac{\sin p |\vec{x} - \vec{x}'|}{p |\vec{x} - \vec{x}'|} = \frac{1}{|\vec{x} - \vec{x}'|}, \qquad (10.103)$$

so that, from Eq. (10.100), and for a point-like source $\rho(x) = M \delta^3(x)$,

$$\bar{\varphi}_0 = -\frac{\lambda_P^2}{8\pi} \int d^3 x' \frac{\rho(x')}{|\vec{x} - \vec{x}'|} = -\frac{GM}{r}, \qquad (10.104)$$

where $r = |\vec{x}|$. For the massive modes one obtains, in the same way,

$$\mathcal{G}_m(x, x') = \frac{2}{\pi} \int_0^\infty dp \frac{p^2}{p^2 + m^2} \frac{\sin p |\vec{x} - \vec{x}'|}{p |\vec{x} - \vec{x}'|} = \frac{e^{-m |\vec{x} - \vec{x}'|}}{|\vec{x} - \vec{x}'|}, \qquad (10.105)$$

and

$$\bar{\varphi}_m = -\frac{\lambda_P^2(m)}{8\pi} \int d^3 x' \frac{e^{-m|\vec{x} - \vec{x}'|}}{|\vec{x} - \vec{x}'|} \rho(x') = -\frac{\lambda_P^2(m)}{8\pi} \frac{M e^{-mr}}{r}. \qquad (10.106)$$

The total static field produced by the source is finally given by the sum of all (massless and massive) contributions, namely by

$$\bar{\varphi} = \sum_m \bar{\varphi}_m = \bar{\varphi}_0 + \int_0^\infty dm \, \bar{\varphi}_m$$

$$= -\frac{GM}{r} \left[1 + \frac{1}{8\pi G} \int_0^\infty dm \, \lambda_P^2(m) e^{-mr} \right], \qquad (10.107)$$

where we have to take into account the m-dependence of the effective gravitational coupling, specified by Eqs. (10.96) and (10.97). For an approximate analytical estimate, the above integral can be evaluated using for $\lambda_P(m)$ the expression obtained in the small-argument limit $m \to 0$ of the Bessel functions (indeed, at the long distances typical of the weak-field limit, the dominant contribution to the integral comes from the small-mass regime). From the definition (10.97) of $F(x)$ one has, in general,

$$\lim_{x \to 0} F^2(x) \simeq \left(\frac{x}{2}\right)^{D-4} \Gamma^{-2} \left(\frac{D-3}{2}\right),$$ (10.108)

where Γ is the Euler function. In this regime, and for $D = 5$,

$$\lambda_{\mathrm{P}}^2(m) = \frac{m \lambda^3}{2k} = \frac{m}{2k^2} 8\pi G,$$ (10.109)

where we have used the definitions (10.96) and (10.97). The effective potential thus becomes

$$\overline{\varphi} = -\frac{GM}{r} \left(1 + \frac{1}{2k^2} \int_0^\infty dm\, m\, \mathrm{e}^{-mr}\right)$$

$$= -\frac{GM}{r} \left(1 + \frac{1}{2k^2 r^2}\right).$$ (10.110)

It is important to note that the higher-dimensional corrections come into play only at a distance which is sufficiently small with respect to the bulk curvature scale [34], namely for $\lesssim k^{-1}$, where k^{-1} is the curvature radius of the bulk AdS geometry external to the brane. At large enough distances the gravitational interaction experienced on the brane is thus effectively four-dimensional, *quite irrespective* of the compactification of the extra dimensions. These results can be directly extended also to cases in which the unperturbed geometry of the brane is described by Ricci flat solutions different from the trivial Minkowski metric, such as the warped Kasner metric discussed in [25].

10.3 Brane-world cosmology

In the previous section we have analyzed a geometric configuration describing a p-brane embedded in a $(p+2)$-dimensional bulk manifold, with the tension of the brane and the bulk cosmological constant as the only gravitational sources. In this section we will study more realistic configurations in which the brane representing our Universe also contains other conventional sources, contributing to the geometry through linear and quadratic terms according to the generalized Einstein equations (10.38). We will also consider the so-called "induced gravity" terms, possibly produced (via quantum loop corrections [35, 36, 37]) by the matter localized on the brane, and we will briefly discuss the associated cosmological effects, possibly obtained even if the brane is embedded in a flat bulk space-time [10].

Even without induced-gravity terms the brane-world scenario requires important modifications of the standard cosmological equations [38, 39], as can be easily deduced from the effective equations (10.38). Let us consider, for simplicity, a cosmological constant Λ as the only source of the bulk stress tensor, defined as in Eq. (10.60). Inserting such a contribution into Eq. (10.38), the gravitational equations on the brane can be written as

$$G_{\alpha\beta} = \lambda_{\mathrm{P}}^{D-3} \tau_{\alpha\beta} + \Lambda_{D-1} h_{\alpha\beta} + E_{\alpha\beta}$$
$$+ \left(\frac{\lambda^{D-2}}{2}\right)^2 \left[\tau_\alpha{}^\mu \tau_{\mu\beta} - \frac{\tau}{D-2} \tau_{\alpha\beta} - \frac{1}{2} h_{\alpha\beta} \left(\tau_\mu{}^\nu \tau_\nu{}^\mu - \frac{\tau^2}{D-2} \right) \right].$$

(10.111)

Here $\lambda_{\mathrm{P}}^{D-3} \equiv 8\pi G_{D-1}$ is the effective gravitational coupling (related to the brane tension by Eq. (10.67)); λ^{D-2} is the coupling scale of D-dimensional bulk gravity (see Eq. (10.17)); finally, Λ_{D-1} is the total effective vacuum energy density of the brane,

$$\Lambda_{D-1} = \frac{1}{2}\left(\frac{D-3}{D-2}\right)\left(\frac{\mu^{D-1}\lambda^{D-2}}{2}\right)^2 + \frac{D-3}{D-1}\Lambda$$
$$\equiv \frac{1}{2}\mu^{D-1}\lambda_{\mathrm{P}}^{D-3} + \frac{D-3}{D-1}\Lambda \qquad (10.112)$$

(see also Eq. (10.61)), including the contributions of the tension and the bulk cosmological constant. Assuming the validity of the local conservation equation for the matter stress tensor,

$$\nabla_\beta \tau_\alpha{}^\beta = 0, \qquad (10.113)$$

(following from the absence of bulk matter contributions, see e.g. [26]), we then obtain from the Bianchi identity ($\nabla_\beta G_\alpha{}^\beta = 0$) and the metricity condition ($\nabla_\alpha h_{\alpha\beta} = 0$) a useful "conservation equation" relating the Weyl contribution $E_{\alpha\beta}$ to the quadratic source term [22]:

$$\nabla_\beta E_\alpha{}^\beta = \left(\frac{\lambda^{D-2}}{2}\right)^2 \left[\tau_\mu{}^\beta \left(\nabla_\beta \tau_\alpha{}^\mu - \nabla_\alpha \tau_\beta{}^\mu \right) - \frac{1}{D-2} \left(\tau_\alpha{}^\beta \nabla_\beta \tau - \tau \nabla_\alpha \tau \right) \right].$$

(10.114)

We look for homogeneous, isotropic and spatially flat metric solutions on the brane, assuming that the stress tensor $\tau_{\alpha\beta}$ represents a barotropic, perfect fluid source. We can thus set, in the cosmic-time gauge,

$$h_{\mu\nu} = \mathrm{diag}\left(1, -a^2(t)\,\delta_{ij}\right)$$
$$\tau_\mu{}^\nu = \mathrm{diag}\left(\rho, -p\,\delta_i^j\right), \qquad \frac{p}{\rho} = \gamma = \mathrm{const.} \qquad (10.115)$$

To obtain the explicit components of Eq. (10.111), however, we must also discuss the possible form of the term $E_{\alpha\beta}$, sourced by the bulk Weyl tensor, which in general is left undetermined by the local energy-momentum distribution of the sources localized on the brane. To the brane-bounded observer the tensor $E_{\alpha\beta}$, being traceless, may be interpreted as the effective stress tensor of some radiation fluid, also called "dark radiation" [40]. Such a tensor may in general contain an anisotropic part, whose evolution cannot be determined from initial conditions set solely on the brane [26]. To be consistent with a homogeneous and isotropic metric $h_{\alpha\beta}$, however, such an anisotropic part has to vanish, as well as the velocity of the effective Weyl fluid in the comoving frame of the metric (10.115): thus, in our particular case, we can write $E_{\alpha\beta}$ in the comoving form

$$E_\mu{}^\nu = \text{diag}\left(\rho_{\rm W}, -p_{\rm W}\delta_i^j\right), \qquad p_{\rm W} = \frac{\rho_{\rm W}}{D-2}. \qquad (10.116)$$

On the other hand, if $\tau_{\mu\nu}$ describes a homogeneous perfect fluid separately conserved according to Eq. (10.113), one can easily check that the right-hand side of Eq. (10.114) is identically vanishing in the background (10.115). It follows that $E_{\alpha\beta}$ is also separately conserved, $\nabla_\beta E_\alpha{}^\beta = 0$, and that the energy density of the dark radiation fluid can be given in the form

$$\rho_{\rm W} = \frac{\rho_0}{a^{D-1}}, \qquad (10.117)$$

where ρ_0 is an integration constant determined by the properties of the bulk geometry.

We can now provide the explicit form of all the independent components of Eq. (10.111). Let us first evaluate the quadratic contribution $S_{\alpha\beta}(\tau)$, whose time and space components are reduced, respectively, to

$$S_0{}^0 = \frac{1}{2}\left(\frac{D-3}{D-2}\right)\left(\frac{\lambda^{D-2}}{2}\right)^2 \rho^2,$$

$$\qquad (10.118)$$

$$S_i{}^j = -\left(\frac{D-3}{D-2}\right)\left(\frac{\lambda^{D-2}}{2}\right)^2 \rho\left(p+\frac{\rho}{2}\right)\delta_i^j.$$

The (00) component of Eq. (10.111) then gives the modified Friedman equation

$$\frac{1}{2}(D-2)(D-3)H^2 = \Lambda_{D-1} + \frac{C}{a^{D-1}} + \lambda_{\rm P}^{D-3}\rho\left(1+\frac{\rho}{2\mu^{D-1}}\right), \qquad (10.119)$$

where we have used Eq. (10.117) for the dark-radiation contribution (C is a constant fixed by the initial density of the Weyl fluid on the brane), and we have also used the definition (10.67) of the gravitational coupling λ_P^{D-3}. The spatial components $(i=j)$ of Eq. (10.111) give the second cosmological equation,

$$(D-3)\dot{H}+\frac{1}{2}(D-2)(D-3)H^2 = \Lambda_{D-1}-\frac{C}{(D-2)a^{D-1}}$$

$$-\lambda_P^{D-3}\left[p+\frac{\rho}{\mu^{D-1}}\left(p+\frac{\rho}{2}\right)\right]. \qquad (10.120)$$

The combination of these two equations, and the application of the dark-radiation conservation equation,

$$\dot{\rho}_W+(D-1)H\rho_W = 0, \qquad (10.121)$$

leads consistently to the local conservation equation of the matter fluid on the brane,

$$\dot{\rho}+(D-2)H(p+\rho) = 0, \qquad (10.122)$$

in agreement with the assumption (10.113).

The cosmological system we are considering is thus characterized by the usual conservation equations of the matter sources, but the dynamical evolution of the geometry is governed by equations which differ from the standard ones in two respects: (*i*) the contribution of the Weyl radiation, (*ii*) the corrections quadratic in the energy density of the matter sources. Such modified equations can also be interpreted as equations describing the motion of the brane through a static but curved bulk [41], whose geometry is of the Schwarzschild–anti-de Sitter (SAdS) type (this is no longer true, however, when the bulk contains a scalar dilaton field [42]). Indeed, in the absence of dilaton sources, the SAdS geometry is the most general static and vacuum bulk geometry compatible with an FRW metric on the brane [43], with the mass of the bulk black hole which is directly related to the integration constant C appearing in the cosmological equations.

In the case $C=0$ the black hole disappears, and we obtain a configuration describing an FRW brane embedded in a pure AdS geometry. Since the bulk cosmological constant Λ is negative, we can also assume that its contribution exactly cancels the energy density associated with the tension of the brane, as in the case of the RS background of Eqs. (10.56) and (10.59) – and as advocated by the proposed "self-tuning" mechanism [44, 45], by which the brane solutions might naturally seek out fixed points with small values of their vacuum energy density.

If we consider, in particular, an AdS bulk geometry satisfying the RS tuning condition (10.59) or (10.61), then $C=0$, $\Lambda_{D-1}=0$, and the simplified system of equations (10.119)–(10.122) can be solved analytically. Using the barotropic equation of state $p/\rho = \gamma$ one obtains, for the scale factor [38, 39],

$$a \sim t^{1/q} \left(1 + \frac{qt}{2t_0} \right)^{1/q} , \qquad (10.123)$$

where

$$q = (D-2)(1+\gamma), \qquad \frac{1}{t_0} = \left[\frac{\lambda_P^{D-3} \mu^{D-1}}{(D-2)(D-3)} \right]^{1/2} . \qquad (10.124)$$

Using Eq. (10.67) for λ_P^{d-3}, and Eq. (10.61) for Λ, one finds that the parameter t_0^{-1} exactly coincides with the curvature scale of the AdS background,

$$\frac{1}{t_0} = \left[\frac{-2\Lambda}{(D-1)(D-2)} \right]^{1/2} \equiv k, \qquad (10.125)$$

namely with the parameter k of the warped RS metric (see Eq. (10.56)), which separates the regime of higher-dimensional bulk gravity from the regime of standard Einstein gravity on the brane. The solution (10.123), in particular, provides a smooth interpolation between an early phase ($t \ll t_0$) at high energy density ($\rho \gg \mu^{D-1}$), in which the Friedman equation is dominated by the ρ^2 contribution and the evolution is unconventional ($a \sim t^{1/q}$), and a late-time phase ($t \gg t_0$) at low energy density and standard Friedman evolution, $a \sim t^{2/q}$.

The corrections to the standard cosmological evolution, induced by the modified gravitational equations on the brane, may be confronted by the existing phenomenological constraints, in order to extract information on the allowed values of the parameters μ, λ and Λ (see for instance [46, 47]). A modified evolution, in particular, may be acceptable provided it is not in contrast with the standard nucleosynthesis scenario: this means that the possible corrections to the standard cosmological equations may come into play only at earlier epochs (i.e. at higher densities) with respect to the typical scale of nucleosynthesis. This provides a constraint on the quadratic corrections of Eq. (10.119), which implies a lower limit on the tension of the brane,

$$\mu^{D-1} \gtrsim \rho_{\rm Nuc} \sim (1\,{\rm MeV})^{D-1} . \qquad (10.126)$$

The tension, on the other hand, is related to the Planck scale λ_P by Eq. (10.67). Using the phenomenological value $\lambda_P \sim (10^{18}\,{\rm GeV})^{-1}$ we can then obtain a bound on the effective coupling scale λ of bulk gravity,

$$\lambda^{2-D} \gtrsim (1\,{\rm MeV})^{(D-1)/2} \lambda_P^{(3-D)/2} , \qquad (10.127)$$

or, in terms of the mass scales $M_P = \lambda_P^{-1}$, $M = \lambda^{-1}$,

$$\frac{M}{M_P} \gtrsim \left(\frac{1\,\text{MeV}}{M_P}\right)^{\frac{D-1}{2(D-2)}} \sim 10^{-\frac{21(D-1)}{2(D-2)}}. \tag{10.128}$$

For the phenomenological case of a 3-brane embedded in a $D = 5$ bulk manifold we find, in particular, the lower limit

$$M \gtrsim 10^{-14} M_P \sim 10\,\text{Tev}, \tag{10.129}$$

compatible with present phenomenology.

Much more stringent constraints on M, for the case in which the fine-tuning relation $\Lambda_{D-1} = 0$ is valid, arise however from small-scale gravity experiments [7], which imply

$$k^{-1} \lesssim 0.1\,\text{mm} \tag{10.130}$$

(for the consistency of the experimental results with the model prediction (10.110)). Using Eq. (10.66), connecting the bulk and brane curvature scale in the RS scenario, one immediately obtains

$$\left(\frac{\lambda}{\lambda_P}\right)^{D-2} \lesssim \frac{2}{D-3}\left(\frac{0.1\,\text{mm}}{\lambda_P}\right), \tag{10.131}$$

which implies

$$\frac{M}{M_P} \gtrsim 10^{-\frac{30}{D-2}}. \tag{10.132}$$

Thus, $M \gtrsim 10^5$ TeV in $D = 5$.

Another firm observational constraint concerns the possible presence of the isotropic Weyl contribution (10.116): the energy density ρ_W, behaving as massless radiation, has to satisfy the nucleosynthesis constraint [47]. We should recall that any "non-standard" degree of freedom contributing to the gravitational sources at the nucleosynthesis epoch cannot exceed, roughly, about one-tenth of the total energy density dominating the Universe at that epoch (see Chapter 7). This gives $\rho_W/\rho_r \lesssim 0.1$ at $t = t_{\text{Nuc}}$ and implies, in critical units at the present epoch,

$$h^2 \Omega_W(t_0) = h^2 \frac{\rho_W}{\rho_c}(t_0) \lesssim 10^{-5} \tag{10.133}$$

(see Eq. (1.34)).

10.3.1 Inflation on the brane

If we relax the assumption $\Lambda_{D-1} = 0$, detuning the tension from the RS value (10.61) (or, equivalently, introducing on the brane a perfect fluid with equation of state $p = -\rho = $ const, $\rho > 0$), we can obtain inflationary solutions on the brane driven by a positive cosmological constant Λ_{D-1}.

We illustrate this possibility by assuming, for simplicity, that there are no matter sources either in the bulk ($T_{AB}^{\text{bulk}} = 0$) or in the brane ($\tau_{\mu\nu} = 0$), and that the only gravitational sources are the cosmological constant Λ and the tension μ^{D-1}, whose combination produces a vacuum energy density Λ_{D-1}, according to Eq. (10.112). We are then left with the same set of equations (10.40)–(10.47) already considered in Section 10.2 when discussing the RS background; in this case, however, we look for a more general solution representing a de Sitter brane, described by the metric

$$ds^2 = f^2(|z|) \left(g_{\mu\nu} \, dx^\mu \, dx^\nu - dz^2 \right),$$

(10.134)

where

$$g_{\mu\nu} = \text{diag} \left(1, -e^{2H_0 t} \delta_{ij} \right), \qquad H_0 = \text{const}$$

(10.135)

(for $H_0 \to 0$ one has $g_{\mu\nu} \to \eta_{\mu\nu}$, and one is led back to the RS background (10.48)).

For this metric background the induced metric (specified by the embedding equations (10.47)) is given by $\gamma_{\mu\nu} = \delta_\mu^A \delta_\nu^B g_{AB} = f^2 g_{\mu\nu}$, and the equations of motion of the brane are trivially satisfied. The connection for the metric (10.134) has non-vanishing components

$$\Gamma_{\mu\nu}{}^z = F g_{\mu\nu}, \qquad \Gamma_{zA}{}^B = F \delta_A^B,$$

$$\Gamma_{0i}{}^j = H_0 \, \delta_i^j, \qquad \Gamma_{ij}{}^0 = H_0 e^{2H_0 t} \, \delta_{ij},$$

(10.136)

where $F = f'/f$ and $f' = df/dz$. The Einstein tensor is then given by

$$G_z{}^z = -\frac{1}{2f^2}(D-1)(D-2)\left(F^2 - H_0^2\right),$$

$$G_\mu{}^\nu = -\frac{1}{f^2}\left[(D-2)F' + \frac{1}{2}(D-2)(D-3)\left(F^2 - H_0^2\right)\right]\delta_\mu^\nu,$$

(10.137)

and the components of the Einstein equations (10.42) normal and parallel to the brane reduce, respectively, to

$$-(D-1)(D-2)\left(F^2 - H_0^2\right) = 2\Lambda f^2,$$

(10.138)

$$-(D-2)F' - \frac{1}{2}(D-2)(D-3)\left(F^2 - H_0^2\right) = \Lambda f^2 + \lambda^{D-2} \mu^{D-1} f \delta(z).$$

(10.139)

They differ from the previous equations (10.52) and (10.53) only by the presence of the constant contribution H_0^2.

The integration of these equations can be performed following the same procedure used in the case of the Minkowski brane, setting $|z| = y = z\epsilon(z)$, and applying the relations (10.54). Equation (10.138) can then be rewritten as

$$\left(\frac{df}{dy}\right)^2 = f^2 \left(H_0^2 + k^2 f^2\right), \qquad k^2 = \frac{-2\Lambda}{(D-1)(D-2)} > 0, \qquad (10.140)$$

where the bulk curvature scale k is defined exactly as before in Eq. (10.56). This equation can be easily integrated, and we choose the particular exact solution

$$f(|z|) = \frac{H_0}{k \sinh \left[H_0 \left(y + k^{-1}\right)\right]}, \qquad (10.141)$$

which for $H_0 \to 0$ reduces to the previous solution (10.56).

We now have to satisfy the second Einstein equation (10.139), which we rewrite in terms of y as

$$-2\frac{(D-2)}{f}\frac{df}{dy}\delta(z) + \frac{(D-2)}{f^2}\left(\frac{df}{dy}\right)^2 - \frac{(D-2)}{f}\frac{d^2f}{dy^2}$$

$$-\frac{1}{2}(D-2)(D-3)\left[\frac{1}{f^2}\left(\frac{df}{dy}\right)^2 - H_0^2\right] = \Lambda f^2 + \lambda^{D-2}\mu^{D-1}f\delta(z). \qquad (10.142)$$

Inserting the solution (10.141) we find that the finite part of this equation ($z \neq 0$) is identically satisfied. Imposing the solution to be valid also at $z = 0$ we have to match the coefficients of the delta functions,

$$-2(D-2)\frac{df}{dy}(0) = \lambda^{D-2}\mu^{D-1}f^2(0), \qquad (10.143)$$

and we obtain the condition

$$\frac{1}{2}\lambda^{D-2}\mu^{D-1} = k(D-2)\cosh\left(\frac{H_0}{k}\right), \qquad (10.144)$$

which determines the de Sitter scale H_0 in terms of the background parameters λ, μ and Λ.

For the Minkowski brane with $H_0 = 0$ we recover the RS condition (10.59) which is equivalent to a vanishing cosmological constant on the brane, Eq. (10.61). For $H_0 \neq 0$ we can use in the above condition the definition (10.112) of the brane cosmological constant Λ_{D-1}. Using also the definition (10.140) of k^2 we obtain

$$\Lambda_{D-1} = \left(\frac{D-3}{D-2}\right)\Lambda\left[1 - \cosh^2\left(\frac{H_0}{k}\right)\right]$$

$$= -\left(\frac{D-3}{D-2}\right)\Lambda \sinh^2\left(\frac{H_0}{k}\right), \qquad (10.145)$$

which is positive for $\Lambda < 0$, and which quantifies the detuning of the vacuum energy density versus the de Sitter curvature scale H_0 of the brane.

The linear evolution of tensor metric perturbations on the de Sitter brane can be studied analytically, since the tensor perturbation equation can be separated in the variables x^μ and z, as in the case of the Minkowski brane. The effective potential for the evolution along z is still characterized by the typical volcano-like shape which localizes the massless mode on the brane, and guarantees the correct "confined" behavior of long-range gravitational interactions. However, the potential goes asymptotically to the constant non-zero value $9H_0^2/4$, and the tensor fluctuation spectrum is thus characterized by a "mass gap", $\Delta m = 3H_0/2$, separating the zero mode from the continuous sector of the massive fluctuations [48]. This modifies the spectrum of gravitational radiation produced during the phase of de Sitter inflation on the brane, enhancing the amplitude of the graviton background if inflation occurs at energies higher than the bulk curvature scale [49], i.e. if $H_0 \gg k$. A similar effect (but with a different origin) also occurs in models where the accelerated evolution of the brane is described by an anisotropic Kasner geometry [25].

More realistic models of brane inflation can be realized by including in the energy-momentum tensor $\tau_{\mu\nu}$ a scalar field, self-interacting through an appropriate slow-roll potential [50]. One then finds, interestingly enough, that the high-energy corrections to the gravitational equations tend to assist the inflationary process, with respect to the same process occurring in the absence of extra dimensions normal to the brane [51].

For a simple illustration of this effect we may consider the modified cosmological equations (10.119) and (10.120) with $C = 0$ and $\Lambda_{D-1} = 0$, restricted to the realistic case of a 3-brane embedded in a $D = 5$ bulk manifold. We also assume that the only matter source on the brane is a self-interacting scalar field, which in the homogeneous and isotropic limit may be treated as a perfect fluid with energy density $\rho = \dot{\phi}^2/2 + V$ and pressure $\dot{\phi}^2/2 - V$. Eliminating H^2 in Eq. (10.120) through Eq. (10.119) we obtain the following system of cosmological equations:

$$3H^2 = \lambda_P^2 \left(\frac{\dot{\phi}^2}{2} + V \right) \left[1 + \frac{1}{2\mu^4} \left(\frac{\dot{\phi}^2}{2} + V \right) \right],$$

$$2\dot{H} = -\lambda_P^2 \dot{\phi}^2 \left[1 + \frac{1}{\mu^4} \left(\frac{\dot{\phi}^2}{2} + V \right) \right],$$

(10.146)

to which we can add the conservation equation (10.122), which for a scalar field takes the form

$$\ddot{\phi} + 3H\dot{\phi} + V' = 0$$

(10.147)

(the prime denotes differentiation with respect to ϕ).

In the regime where the slow-roll approximation is valid, i.e. $\dot{\phi}^2 \ll V$, $\dot{H} \ll H^2$, $\ddot{\phi} \ll H\dot{\phi}$ (see Section 1.2), the previous three equations reduce, respectively, to

$$3H^2 = \lambda_{\rm P}^2 V \left[1 + \frac{V}{2\mu^4}\right],\tag{10.148}$$

$$2\dot{H} = -\lambda_{\rm P}^2 \dot{\phi}^2 \left[1 + \frac{V}{\mu^4}\right],\tag{10.149}$$

$$3H\dot{\phi} = -V'.\tag{10.150}$$

The terms in square brackets represent the brane-world corrections to the standard Einstein equations, which become negligible in the limit $V \ll \mu^4$. Differentiating Eq. (10.148) with respect to ϕ, and dividing by $6H^2$, we obtain the relation

$$\frac{H'}{H} = \frac{V'}{2V} \left[\frac{1 + V/\mu^4}{1 + V/2\mu^4}\right],\tag{10.151}$$

which will be useful for our subsequent discussion. We can also rewrite Eq. (10.149) using $\dot{H} = H'\dot{\phi}$, to obtain

$$2H' = -\lambda_{\rm P}^2 \dot{\phi} \left[1 + \frac{V}{\mu^4}\right].\tag{10.152}$$

The last two equations generalize, respectively, Eqs. (1.114) and (1.107) obtained in the context of the standard inflationary scenario.

We can now calculate the typical parameters of a model of slow-roll inflation, applying the standard definitions but using the above modified equations. Consider, for instance, the parameter ϵ_H defined in Eq. (1.109). Putting $\dot{H} = H'\dot{\phi}$, using Eq. (10.152) for $\dot{\phi}$ and Eq. (10.151) for H', we obtain

$$\epsilon_H = -\frac{\dot{H}}{H^2} = -\frac{H'\dot{\phi}}{H^2} = \frac{2}{\lambda_{\rm P}^2} \frac{H'^2}{H^2} \left[1 + \frac{V}{\mu^4}\right]^{-1}$$

$$= \frac{1}{2\lambda_{\rm P}^2} \left(\frac{V'}{V}\right)^2 \left[\frac{1 + V/\mu^4}{(1 + V/2\mu^4)^2}\right].\tag{10.153}$$

Let us now compute η_H, defined by Eq. (1.110). Differentiating Eq. (10.152) with respect to t to obtain $\ddot{\phi}$, dividing by $H\dot{\phi}$, differentiating Eq. (10.151) with respect to ϕ to obtain H'', and combining the two results, we obtain, to leading order,

$$\eta_H = -\frac{\ddot{\phi}}{H\dot{\phi}} = -\epsilon_H + \eta,$$

$$\eta = \frac{1}{\lambda_P^2} \left(\frac{V''}{V}\right) \left[\frac{1}{1 + V/2\mu^4}\right].$$

(10.154)

We may note that at inflation scales sufficiently low with respect to the tension of the brane, $V \ll \mu^4$, the parameters ϵ_H and η reduce to the form (1.115) obtained in the standard inflationary context. In the opposite limit $V \gg \mu^4$ we have, instead,

$$\epsilon_H \rightarrow \frac{1}{2\lambda_P^2} \left(\frac{V'}{V}\right)^2 \frac{4\mu^4}{V} \ll \frac{1}{2\lambda_P^2} \left(\frac{V'}{V}\right)^2,$$

$$\eta \rightarrow \frac{1}{\lambda_P^2} \left(\frac{V''}{V}\right) \frac{2\mu^4}{V} \ll \frac{1}{\lambda_P^2} \left(\frac{V''}{V}\right).$$

(10.155)

The slow-roll parameters are thus strongly suppressed, at high enough energy scales, by the brane-induced corrections, and this can make inflation possible on the brane even with potentials that would be too steep in a standard cosmological context. We can say, in this sense, that brane-world effects *ease* slow-roll inflation.

The modified gravitational equations of the brane also improve the efficiency of the inflationary expansion by modifying the relation between the inflaton potential and the number of e-folds $N(t)$ between a given time t and the end of inflation t_f,

$$N(t) = \ln \frac{a_f}{a(t)} = \int_t^{t_f} H \, dt.$$

(10.156)

In fact, dividing Eq. (10.148) by Eq. (10.150) we obtain, in the slow-roll approximation,

$$N(t) = \int_\phi^{\phi_f} \frac{H}{\dot{\phi}} \, d\phi = \lambda_P^2 \int_{\phi_f}^\phi \frac{V}{V'} \left[1 + \frac{V}{2\mu^4}\right] d\phi,$$

(10.157)

to be compared with the standard result (1.119). The brane-world corrections thus increase the effective Hubble factor, yielding more inflation for the same given initial values of the inflaton field – or, equivalently, the same inflation for a smaller initial value of the inflaton.

The predictions of models of brane-world inflation may be compared with present observations, and with the phenomenological predictions of other classes of inflationary models, by studying the perturbations of the coupled brane–bulk system. We will not perform such an analysis here, referring the interested reader to the existing literature (see for instance [26] for a detailed introduction to the various aspects of this problem). We wish to remark, however, that a complete study of the amplification and propagation of the background fluctuations necessarily requires a higher-dimensional approach based on the full bulk equations: the brane point of view, although instructive in some cases, is not sufficient in general, since brane and bulk perturbations are coupled. We may recall, for instance, that the anisotropic part of the Weyl stress tensor projected on the brane is vanishing when the induced metric is homogeneous and isotropic: these symmetry properties of the background are lost, however, when the metric is perturbed, and an associated fluctuation of the Weyl tensor is generated even in the bulk, outside the brane.

10.3.2 Induced gravity on the brane

The cosmological examples that we have discussed, up to this point, were based on the simplest model of brane-world gravity in which the gravitational dynamics of the bulk is simply projected down to the brane: in this case, the forces experienced by the brane-bound observer are simply the "shadows" of more fundamental interactions taking place in the bulk manifold.

This scenario, however, can be generalized in various ways, and an interesting (and possibly phenomenologically important) generalization concerns the introduction in the brane action of the scalar contribution of its *intrinsic* scalar curvature, $R_{D-1}(h)$. This contribution may be induced, in general, by the quantum corrections [35, 36, 37] to the coupling of bulk gravity to the matter fields living on the brane. Irrespective of the possible (quantum or geometric) origin of this new term, it implies that the Einstein tensor of the brane plays the role of a new effective source of bulk gravity, thus providing (as noted in [52]) an unexpected realization of the old proposal of Lorentz and Levi-Civita of using the Einstein tensor defined on three-dimensional spatial sections as the energy momentum tensor of the gravitational field in four dimensions. The important consequence of this generalization, for the cosmological applications of this chapter, is that it may lead the Universe to a late-time regime of "self-accelerated" expansion [53, 54], without introducing any cosmological constant or dark-energy field.

For a brief discussion of this model we can again start from the general action (10.17), working however in the complete absence of bulk sources, and thus setting $L_{\text{bulk}} = 0$ (we also eliminate the cosmological constant Λ which was a basic ingredient of the previous models characterized by bulk AdS geometry). The model still contains a codimension-one brane, described by the action (10.18). The brane is tensionless, $\mu = 0$, but contains matter sources and, in addition, an induced gravity term represented by the intrinsic Ricci scalar constructed with the induced metric, $R_{D-1}(h)$, and with an effective coupling scale that we call λ_P. Thus,

$$L_{\text{brane}} = \frac{1}{2}\left(\frac{R_{D-1}}{2\lambda_P^{D-3}} - L_{\text{matter}}\right). \tag{10.158}$$

Using the Nambu–Goto form of the brane action, and inserting this Lagrangian into Eq. (10.17), we are led to the action

$$S = -\frac{1}{2\lambda^{D-2}}\int_\Omega d^D x\sqrt{|g|}\,R + \int_\Sigma d^{D-1}\xi\sqrt{|h|}\left(L_{\text{matter}} - \frac{R_{D-1}}{2\lambda_P^{D-3}}\right). \tag{10.159}$$

The effective interaction on the brane is thus described by Eq. (10.38), where $\mu = 0$, $T_{\mu\nu}^{\text{bulk}} = 0$, and where the intrinsic stress tensor $T_{\mu\nu}$ – defined by varying L_{brane} with respect to the induced metric – contains the "true" energy-momentum tensor of the matter sources $(t_{\mu\nu})$, obtained from the variation of L_{matter}, and the Einstein tensor intrinsic to the brane $(G_{\mu\nu}^\Sigma)$, obtained from the variation of R_{D-1}:

$$T_{\mu\nu} = t_{\mu\nu} - \frac{1}{\lambda_P^{D-3}}G_{\mu\nu}^\Sigma. \tag{10.160}$$

We have assumed that the boundary terms arising from the variation are canceled by an appropriate Gibbons–Hawking action, defined in terms of the extrinsic curvature of the brane. Inserting these sources into Eq. (10.38) we thus obtain the dimensionally reduced field equations:

$$G_\alpha{}^\beta = E_\alpha{}^\beta + S_\alpha{}^\beta(t) + \left(\frac{1}{\lambda_P^{D-3}}\right)^2 S_\alpha{}^\beta(G)$$

$$+\frac{1}{\lambda_P^{D-3}}\left(\frac{\lambda^{D-2}}{2}\right)^2\left[-t_\alpha{}^\mu G_\mu{}^\beta - G_\alpha{}^\mu t_\mu{}^\beta + \frac{1}{D-2}\left(t\,G_\alpha{}^\beta + G\,t_\alpha{}^\beta\right)\right.$$

$$\left.+\delta_\alpha{}^\beta t_\mu{}^\nu G_\nu{}^\mu - \frac{t\,G}{D-2}\delta_\alpha^\beta\right], \tag{10.161}$$

where $E_{\mu\nu}$ is the Weyl stress tensor (10.32), $S_{\mu\nu}(t)$ and $S_{\mu\nu}(G)$ are given by the quadratic expressions (10.35) computed with $t_{\mu\nu}$ and $G_{\mu\nu}$, respectively, and $G = -(D-3)R/2$ is the trace of the Einstein tensor of the brane (we have omitted the superscript Σ on $G_{\mu\nu}$ everywhere). We note that the above equation can also be directly obtained from Eq. (10.111) omitting the cosmological term and the term linear in $T_{\mu\nu}$, and using for $T_{\mu\nu}$ the expression (10.160).

Let us now look for solutions characterized by a trivially flat bulk geometry, by a conformally flat cosmological metric on the brane (as in

Eq. (10.115)), and by a perfect fluid stress tensor $t_{\mu\nu}$. We can thus set, for this background,

$$E_\alpha{}^\beta = 0,$$

$$t_\alpha{}^\beta = \text{diag}\left(\rho, -p\,\delta_i^j\right),$$ (10.162)

$$G_\alpha{}^\beta = \text{diag}\left(\rho_G, -p_G\,\delta_i^j\right),$$

where

$$\rho_G = \frac{1}{2}(D-2)(D-3)H^2,$$

$$p_G = -(D-3)\dot{H} - \frac{1}{2}(D-2)(D-3)H^2,$$ (10.163)

are the effective energy density and pressure due to the intrinsic curvature of the brane. The total stress tensor of the effective gravitational sources localized on the brane can thus be written as

$$\tau_\alpha{}^\beta = \text{diag}\left[\rho - \frac{\rho_G}{\lambda_P^{D-3}}, -\left(p - \frac{p_G}{\lambda_P^{D-3}}\right)\delta_i^j\right].$$ (10.164)

For this homogeneous and isotropic regime we can then obtain the modified cosmological equations directly from Eqs. (10.119) and (10.120) by setting to zero C, Λ_{D-1}, and the terms linear in ρ and p, and by inserting into the quadratic source terms (10.118) the components of the generalized stress tensor (10.164). We obtain

$$\frac{1}{2}(D-2)(D-3)H^2 = \frac{1}{2}\left(\frac{D-3}{D-2}\right)\left(\frac{\lambda_P^{D-2}}{2}\right)^2\left(\rho - \frac{\rho_G}{\lambda_P^{D-3}}\right)^2,$$ (10.165)

$$(D-3)\dot{H} + \frac{1}{2}(D-2)(D-3)H^2$$

$$= -\left(\frac{D-3}{D-2}\right)\left(\frac{\lambda_P^{D-2}}{2}\right)^2\left(\rho - \frac{\rho_G}{\lambda_P^{D-3}}\right)\left(p - \frac{p_G}{\lambda_P^{D-3}} + \frac{\rho}{2} - \frac{\rho_G}{\lambda_P^{D-3}}\right).$$ (10.166)

We may add to this system the conservation equation of the matter fluid, $\nabla_\nu t_\mu{}^\nu = 0$, which takes the form (10.122), and which is still valid thanks to the Bianchi identity $\nabla_\nu G_\mu{}^\nu = 0$.

For the phenomenological applications the modified Friedman equation (10.165) can be conveniently rewritten as

$$\frac{1}{2}(D-3)H^2 = r_c^2\left[\frac{\lambda_P^{D-3}}{D-2}\rho - \frac{1}{2}(D-3)H^2\right]^2,$$ (10.167)

where

$$r_c^2 = \frac{1}{2}(D-3)\left(\frac{\lambda^{D-2}}{2\lambda_P^{D-3}}\right)^2 \tag{10.168}$$

is a crossover length scale, typical of this model. Solving for H^2, the above equation can be finally rewritten in the useful form [53]

$$\frac{1}{2}(D-3)H^2 = \left[\left(\frac{8\pi G_{D-1}}{D-2}\rho + \frac{1}{4r_c^2}\right)^{1/2} \pm \frac{1}{2r_c}\right]^2, \tag{10.169}$$

where $8\pi G_{D-1} \equiv \lambda_P^{D-3}$ is the effective gravitational coupling defining the Planck scale on the brane. This form of the modified Friedman equation clearly shows how the length scale r_c, controlled by the ratio of the bulk and brane gravitational couplings, determines the transition between the regime of higher-dimensional gravity and effective gravity confined on the brane.

In this model, in particular, the effects of the bulk corrections are the opposite of those produced in models of brane-world cosmology considered before, in the sense that such effects become important in the regime of sufficiently *low* energy scales: one recovers the standard form of the Friedman equation only if the density is large enough with respect to r_c^{-2}, namely for

$$8\pi G_{D-1}\rho\, r_c^2 \gg 1. \tag{10.170}$$

The possible application of this model to the most recent cosmological stages seems to require a parameter r_c not much smaller than the present Hubble radius, $r_c \gtrsim H_0^{-1}$, which imposes strong constraints on the bulk gravity scale. Nevertheless, it should be stressed that the model provides a remarkably simple and explicit example showing that brane gravity, in an appropriate regime, may become completely insensitive to the presence of extra spatial dimensions, even if such dimensions are *non-compact* (actually, infinitely extended), and even *flat* (different from the "warped" backgrounds discussed in the previous examples).

The higher-dimensional effects of bulk gravity, in this model, become important in the limit of small enough densities, $8\pi G_{D-1}\rho\, r_c^2 \ll 1$. In this regime, the modified cosmological solutions are characterized by two branches, depending on the choice of sign in Eq. (10.169). Choosing the $-$ sign, and expanding the equation for $\rho \to 0$, we recover the behavior $H^2 \sim \rho^2$ typical of the high-energy limit of previous models (see Eq. (10.119)). Choosing the $+$ sign we find, instead, that for $\rho \to 0$ the system evolves towards a phase of asymptotic accelerated expansion at constant curvature, $H^2 \sim r_c^{-2} = \text{const}$ [54]. This model thus naturally contains a phase of late-time cosmic acceleration, without adding exotic sources or a cosmological constant on the brane (thus implementing a "self-acceleration" scenario). The validity of this branch of the cosmological solution, however, has been challenged by the possible presence of ghost instability [55, 56].

10.4 Ekpyrotic and cyclic scenario

In the cosmological models illustrated in the previous sections our Universe is identified with a brane embedded in a higher-dimensional (possibly curved, and not necessarily compact) bulk space-time. The bulk, however, might contain two (or more) branes, and these branes could interact among themselves, move through the bulk, and eventually collide. The scattering of the brane representing our Universe with another brane might simulate the "big bang" marking the origin of the phase of standard cosmological evolution. The so-called "ekpyrotic scenario" [11, 12, 57] tries to explain in this way the presence of the cosmic radiation background and its temperature anisotropies, proposing a model heavily based on the process of brane collision without resorting to a phase of standard inflationary expansion.

The scenario is inspired by the M-theory model of Horawa and Witten [58, 59, 60], based on $E_8 \times E_8$ heterotic superstring theory, in which there are two 3-branes at the boundaries of an effectively five-dimensional bulk manifold [61, 62] (the remaining six spatial dimensions are compactified on a Calabi–Yau manifold, at a scale which is at least one order of magnitude smaller). The bulk possibly contains other floating branes, free to move along the fifth dimension, and eventually to collide with the boundary branes [11]. Alternatively, the two boundary branes may collide against each other [12, 57], in which case the size of the extra dimension orthogonal to the brane (controlled by the strength of the string coupling in the M-theory context) shrinks to zero before the collision, and then bounces back to increase again when the boundary branes separate after the collision. With an appropriate effective potential, controlling the dynamics of the interbrane distance modulus, the two branes could keep separating and colliding an (almost) infinite number of times, thus implementing the so-called "cyclic scenario" [13, 14, 63].

The action of the ekpyrotic scenario, if we neglect the presence of the six "internal" dimensions compactified on a smaller scale, can be written as the action of a model of brane-world cosmology in which there are two (or more) 3-branes embedded in a $D = 5$ bulk manifold, but with two main differences: (i) the bulk action is not the Einstein action but the action of M-theory, containing additional fields interacting with the branes besides gravity; (ii) a suitable potential energy term, $S_{\text{interaction}}$, has to be added to describe the interbrane interaction, to control their evolution and their possible scattering. Thus,

$$S = S_{\text{het}} + \sum_i S^i_{\text{brane}} + S_{\text{interaction}}. \tag{10.171}$$

Let us consider here, for simplicity, the simplest non-trivial background of five-dimensional heterotic M-theory in which we set to zero all fields except those which are directly coupled to the 3-branes. We are thus left with the graviton, the dilaton, and a four-form gauge potential A_{ABCD}, leading to the following E-frame action:

$$S_{\text{het}} = -\frac{1}{2\lambda^3} \int_{\mathcal{M}_5} d^5x \sqrt{|g|} \left(R - \frac{1}{2}\nabla_A\phi\nabla^A\phi - \frac{e^{2\phi}}{5!} F^2_{ABCDE} \right), \qquad (10.172)$$

where $F_5 = dA_4$ is the five-form field strength. The brane action can be written as

$$S^i_{\text{brane}} = -\int_{\Sigma_i} d^4\xi \left[\sqrt{|h_i|} \left(\frac{3\alpha_i}{\lambda^3} e^{-\phi} - L^i_{\text{matter}} \right) \right.$$

$$\left. -\frac{\epsilon^{\mu\nu\alpha\beta}}{4!} \partial_\mu X^A_i \partial_\nu X^B_i \partial_\alpha X^C_i \partial_\beta X^D_i A_{ABCD} \right], \qquad (10.173)$$

where α_i/λ^3 is the tension of the ith brane (we are following the conventions of [11]), $h^i_{\mu\nu}$ is the induced metric, and $X^A_i(\xi)$ are the functions describing its embedding. The tension coefficients satisfy the condition $\sum_i \alpha_i = 0$, as explained in [64]. Finally, L^i_{matter} describes the matter fields (and radiation) produced on the branes by the collision process (the so-called "ekpyrosis"), starting from an initial configuration in which all branes (except for quantum fluctuations) are empty.

The initial configuration of the ekpyrotic scenario is obtained by considering the limit in which the interaction among the branes is negligible, $S_{\text{interaction}} \simeq 0$, and the system approaches the highly symmetric (and supersymmetric) Bogolmon'y–Prasad–Sommerfeld (BPS) state in which the branes are flat, parallel and static, described by the trivial embedding

$$X^A_i(\xi) = (\xi^\mu, y_i). \qquad (10.174)$$

The coordinate y parametrizes the fifth dimension orthogonal to the brane, and y_i specifies the initial brane position. Here we consider the case in which there are only two boundary branes, initially located at $y_1 = 0$ and $y_2 = Y$, and with tension coefficients $\alpha_1 = -\alpha_2 = -\alpha$, with $\alpha > 0$: we thus concentrate on the second version of the ekpyrotic scenario [12, 57], where no role is played by the bulk branes. For the given initial configuration the equations of motion for the metric and the other background fields of the action (10.171) are satisfied by the following BPS solution [61]:

$$ds^2 = D(y) \left[N^2 dt^2 - A^2 dx^2_i - B^2 D^3(y) dy^2 \right],$$

$$F_{0123y} = -\alpha A^3 N B^{-1} D^{-2}(y), \qquad e^\phi = B D^3(y), \qquad (10.175)$$

$$D(y) = \alpha y + C,$$

where $0 \leq y \leq Y$, and N, A, B, C and Y are constants. Note that the proper volume of the bulk is growing with $D(y)$ as we move from the first boundary at $y = 0$ to the second one at $y = Y$, and that for $C > 0$ the singularity $D = 0$ does not fall between the boundary branes.

We now have to specify $S_{\text{interaction}}$, which is required to shift the system away from the initial static vacuum, and which may be expected to be generated by supersymmetry-breaking (non-perturbative) M-theory interactions between the 3-branes (possibly mediated by the exchange of wrapped 2-branes). Such an interaction should depend on the interbrane distance along y: in this model, however, the fifth dimension normal to the brane (and to the other six dimensions rolled up in the Calabu–Yau threefolds) corresponds to the eleventh dimension of the full M-theory framework. Thus, the modulus controlling its size corresponds to the tree-level coupling $g_s^2 = \exp \phi$, as evident from Eq. (10.175) (see also the discussion at the end of Appendix 3B). As a consequence, the interaction between the boundary branes should vanish as they approach each other and the fifth dimension shrinks to zero.

As the exact form of the brane interaction predicted by the full, higher-dimensional M-theory is unknown, the proposal of the ekpyrotic scenario is to specify such an interaction in the so-called "moduli space approximation". In this approximation, appropriate to the slow-motion regime, the evolution of the bulk+brane system is described as an evolution in the vacuum space spanned by the variation of the integration constants of the BPS solution (10.175). These constants (N, A, B, C, Y) are thus promoted to moduli functions depending on the space-time coordinates orthogonal to y (and thus only on time for a homogeneous, isotropic, spatially flat background). This generalized background is inserted into the action (10.171), where the explicit dependence on y (contained in D) can be factorized, and easily integrated away (see [11] for a detailed calculation). One thus arrives at an effective four-dimensional action where one finally adds a covariant potential V, a function of the interbrane distance Y, which should be eventually computable from heterotic M-theory.

The model is reduced, in this way, to the study of a four-dimensional cosmological system containing a (non-canonical) scalar field (the modulus Y), minimally coupled to gravity. It seems possible to apply the standard methods of scalar-field inflation, with the interbrane separation playing a role similar to that of the inflaton during slow-roll inflation. A possible effective potential for the attractive interbrane interaction has been parametrized in exponential form as follows [11]:

$$V(Y) = -f(Y) V_0 e^{-cY}, \qquad (10.176)$$

where V_0 and c are positive constants, and the function f modulates the amplitude of the potential imposing $f(Y) = 0$ for $Y = 0$, and $f(Y) \simeq 1$ everywhere else (but also power potentials, satisfying $VV'' \simeq V'^2$, may be acceptable for a successful scenario). With this potential, the solutions describing the phase in which the two branes approach each other, before the collision, are associated with the shrinking of the scale factor of the effective four-dimensional geometry (see below, Eq. (10.181)): one obtains, in this context, an epoch of pre-big bang evolution characterized (even in the S-frame [12]) by *isotropic contraction* and *decreasing dilaton* (as the fifth dimension shrinks to zero), to be contrasted with the duality-motivated phase of pre-big bang evolution characterized by expansion and growing dilaton (see Fig. 4.3 and the subsequent discussion).

The crucial aspect of this scenario is the collision and rebound of the two boundary branes, and the passage of the system through the singular point $y = 0$. In the context of a four-dimensional geometric description the model must contain a bounce not only of the curvature, but also of the scale factor (from big crunch to big bang [12]), for the production of a post-collision state approaching the standard cosmological configuration. From a phenomenological point of view one has then to face the important problem of matching perturbations across the bounce, to obtain the spectrum of primordial perturbations produced by the phase of ekpyrotic (pre-collision) evolution. From a more fundamental/theoretical point of view, however, there is also the problem of providing a correct description of the passage of the system through $y = 0$.

This second problem is similar, in various respects, to the "graceful exit" problem of the pre-big bang scenario (see Chapter 6), with the possible simplification that the ekpyrotic bounce occurs in the perturbative regime $g_s^2 \to 0$ (but not necessarily at small curvature), and a possible complication due to the fact that there is no fundamental frame where the evolution of the ekpyrotic scale factor is monotonic (like the S-frame of the pre-big bang scenario). We refer the reader to the literature for a deepened discussion [12, 65, 66, 67], but we stress here that it is possible, in principle, for the transverse dimension to collapse to a point and then re-expand without the curvature and the energy density (of the matter created on the brane) becoming singular [68]. However, the energy density created on one of the two branes has to be *negative* – a condition which is reminiscent of the anisotropic, regular solutions with $\rho < 0$ also found in the context of the low-energy equations of the pre-big bang scenario (see Appendix 4B).

The most relevant problem is probably the phenomenological one concerning the predictions for the primordial spectrum of scalar perturbations produced by the quantum fluctuations of the interbrane distance (which in the effective, four-dimensional version of the model are treated as the fluctuations of a scalar inflaton field). The main differences from the standard inflationary scenario are that the phase of fast accelerated expansion is replaced by a regime of slow contraction (preceding the collision), and that the transition to the standard radiation-dominated era is accompanied by a (possibly singular) bounce of the scale factor.

The first difference does not represent a real difficulty, as the system of cosmological perturbation equations can be treated in a standard way quite irrespective of the given background kinematics (see Chapter 8). The true problem comes from the second difference, since to date, and to the best of my knowledge, there is no unambiguous, strongly motivated and universally accepted prescription for matching the solutions of the perturbation equations across a singular hypersurface where the scale factor vanishes. On the other hand, the spectral distributions inherited by the various fluctuation variables in the final (post-bounce) expanding regime, and directly imprinting the present CMB fluctuations, do strongly depend on the chosen matching conditions (see e.g. [69, 70]).

In the context of the ekpyrotic scenario it has been shown that, with the choice of a sufficiently flat potential $V(Y)$, and the matching of suitable variables which remain finite across the bounce, it is possible to obtain a final spectrum of scalar metric perturbations which is nearly scale invariant [57, 71, 72]. Such a choice of matching prescriptions, however, is not supported by analytical and numerical computations of the perturbations in *smooth* models of bouncing [73, 74], where the background singularity is regularized by appropriate repulsive effects near the crossover point $a \to 0$, so that it becomes possible to follow the evolution of perturbations throughout their cosmological history. Such computations, actually, seem to disprove the ekpyrotic prescriptions, in the sense that they support a smooth crossover of the bounce of the curvature perturbations variable $\mathcal{R}$, and not of the Bardeen potential Ψ, as assumed in the ekpyrotic scenario (where the singularity $a \to 0$ is not smoothed out by regularizing high-energy corrections). This seems to denote that the singular nature of the bounce might play a fundamental role in obtaining the correct phenomenological predictions for the ekpyrotic scenario. (See, however, [75] for recent progress on a new mechanism of entropy-generated curvature perturbations.)

A generalization of the ekpyrotic model has recently led to the formulation of the so-called "cyclic" scenario [13, 14, 63], where the attraction, the collision and the subsequent separation of the Horawa–Witten boundary branes repeat cyclically to infinity (or at least for 10^{30} times, if the entropy produced in a cycle is not lost, and accumulates, eventually to saturate the entropy bound associated with the de Sitter horizon [76]). This cyclic picture completes the ekpyrotic model, in the sense that it includes, after the brane collision, a late-time phase of slow, accelerated expansion (similar to that we are presently experiencing), with the function of smoothing, flattening and emptying the branes, as well as diluting the produced entropy, thus preparing the highly symmetric and vacuum BPS state which represents the initial configuration of a new, forthcoming ekpyrotic stage.

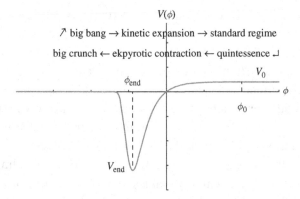

Figure 10.2 Qualitative behavior of the effective potential representing the inter-brane interaction of the ekpyrotic scenario, as a function of the distance modulus ϕ (appropriately shifted in such a way that the attraction begins at $\phi = 0$). Also shown in the picture are the main stages of the cyclic scenario – corresponding to the phases of $V(\phi)$ – for the post-bounce regime with ϕ growing (upper line), and the pre-bounce regime with ϕ decreasing (lower line).

A concrete realization of the cyclic scenario has been proposed, up to now, only in an effective four-dimensional context, "inspired" by heterotic M-theory, but not yet "derived" from it. Such a model can be simply represented by a scalar field ϕ, minimally coupled to gravity according to the standard equations of general relativity,

$$3H^2 = 8\pi G \left(\frac{\dot{\phi}^2}{2} + V \right), \tag{10.177}$$

$$3\frac{\ddot{a}}{a} = -8\pi G \left(\frac{\dot{\phi}^2}{2} - V \right) \tag{10.178}$$

(see Eqs. (1.25) and (1.26) with a scalar field as source), and self-interacting through the potential $V(\phi)$, whose qualitative behavior is illustrated in Fig. 10.2.

The potential approaches a small, positive constant value as $\phi \to +\infty$, has a steep, exponential fall down to a negative minimum $V_{\mathrm{end}} < 0$ for intermediate ϕ in the range $\phi_{\mathrm{end}} < \phi < 0$, and then increases again, approaching zero from negative values as $\phi \to -\infty$. The modulus field ϕ, which controls the inter-brane distance of the underlying ekpyrotic scenario, is rolling back and forth through this potential (which represents the energy associated with the ekpyrotic interactions between the branes); in particular, ϕ varies between a maximum positive value ϕ_0, corresponding to the maximum elongation of the ekpyrotic system along the y axis, and a minimum $\phi = -\infty$, corresponding to the collision point at $y = 0$.

As the modulus ranges from ϕ_0 to $-\infty$ and back to ϕ_0, the system undergoes various stages of cosmic evolution, which can be easily represented in analytical form assuming for the potential the following simple approximation [63]:

$$V(\phi) = V_0 \left(1 - e^{-c\phi/M_P}\right) \theta(\phi - \phi_{end}). \qquad (10.179)$$

Here c is a dimensionless coeffcient, large enough for the model to reproduce an acceptable (i.e. flat enough) spectrum of scalar perturbations [13, 14]; V_0 is a constant energy density, set roughly equal to the present vacuum energy density (no explanation is attempted of such a small number, but the fine-tuning of V_0 is a common problem of all quintessential models based on a running scalar field, see Section 9.3). Finally, the damping of the potential for $\phi < \phi_{end}$ has been approximated by the Heaviside step function, for computational convenience (more realistically, one might also assume that, for $\phi \to -\infty$, the potential is instantonically suppressed as $\exp(-\alpha/g_s^2) = \exp[-\alpha \exp(-\phi)]$, as done in the plot of Fig. 10.2).

With the above potential, the Einstein equations can be easily solved during the various stages of evolution, and one obtains the cosmological phases illustrated below, and summarized in Fig. 10.2.

(1) Let us start with the modulus nearly stationary around its maximum value $\phi \simeq \phi_0 > 0$, a configuration which should roughly correspond to the present epoch in which the Universe expands at an accelerated rate, dominated by the potential energy $V(\phi)$ acting as dark energy ($\dot{\phi}^2 \ll V_0$):

$$H^2 \sim V_0 \simeq \text{const}, \qquad a \sim e^{Ht}, \qquad 0 < \phi \lesssim \phi_0. \qquad (10.180)$$

After reaching the maximum ϕ starts rolling very slowly towards decreasing values (because of the slope of the potential), but its kinetic energy is still negligible with respect to V, and the Universe stays in a state of accelerated expansion identical to that of standard slow-roll inflation, except that the expansion rate is exceedingly slower ($\sim H_0 \sim 10^{-61} M_P$), and the corresponding time scale exceedingly longer. This phase has the virtue of producing an efficient dilution of all cosmic relics (matter, radiation, entropy) associated with the preceding big bang (i.e. brane collision). As ϕ approaches zero V decreases, while the kinetic energy grows and starts to become comparable to the potential energy, so that the acceleration stops (see Eq. (10.178)), but the Universe continues to expand, even if damped by the potential when it drops below zero.

(2) When V becomes sufficiently negative, the total (kinetic plus potential) energy of the modulus hits zero, and the Universe becomes momentarily static, $H = 0$ (see Eq. (10.177)). On the other hand, $\ddot{a} < 0$ from Eq. (10.178), so that the Universe enters a phase of slow contraction dominated by a combination of scalar kinetic and potential energy density. In this regime the constant part of

the potential (10.179) can be neglected, and one is led to a potential of the form (10.176), just typical of the (previously discussed) ekpyrotic picture. From the Einstein equations one then obtains the scaling solution

$$a \sim (-t)^{2/c^2}, \qquad \phi \sim \frac{2}{c} \ln(-t),$$

(10.181)

$$t < t_{end} < 0, \qquad \phi_{end} < \phi < 0,$$

describing contraction and decreasing dilaton, and valid for $\phi > \phi_{end}$.

(3) When ϕ drops below ϕ_{end} the potential energy is rapidly switched off, and the ekpyrotic phase enters a regime of fast, kinetic-dominated contraction described by the solution

$$a \sim (-t)^{1/3}, \qquad \phi \sim \sqrt{\frac{2}{3}} \ln(-t),$$

(10.182)

$$t_{end} < t < 0, \qquad -\infty < \phi < \phi_{end},$$

lasting until the brane collision at $t = 0$, $\phi = -\infty$. This solution describes the accelerated contraction also obtained in the E-frame representation of the vacuum, dilaton-dominated solution of the pre-big bang scenario (see Section 4.2), with the difference that here the dilaton is decreasing.

(4) At the big crunch/big bang collision, matter and radiation are generated on the brane representing our Universe, which starts expanding driven by the kinetic energy of the modulus field, which gets a kick at the bounce and reverses its direction of motion. The background evolution is still described by the kinetic-dominated solution (10.182), but t is now increasing in the positive range $0 < t < \infty$, so that a and ϕ are also growing.

(5) As the Universe expands, the kinetic energy of the modulus is redshifted as $\dot{\phi}^2 \sim a^{-6}$, so that the radiation ($\rho_r \sim a^{-4}$) and matter ($\rho_m \sim a^{-3}$) produced at the bounce are doomed to become dominant, and the Universe eventually enters the phase of standard cosmological evolution,

$$a \sim t^{1/2}, \qquad \rho \sim \rho_r, \qquad t_r < t < t_m,$$

(10.183)

$$a \sim t^{2/3}, \qquad \rho \sim \rho_m, \qquad t_m < t < t_0.$$

During this phase the motion of ϕ is rapidly damped, and the modulus tends to converge towards its maximum value, to stop around this value in a nearly stationary state with $V \simeq V_0$, and then to turn back, evolving very slowly towards negative values. As soon as ρ_m drops below V_0 the Universe undergoes the transition to the phase of late-time slow-roll/dark-energy inflation, and the cycle begins anew.

A detailed study of scalar perturbations in this scenario [57, 71, 72] suggests that the fluctuation spectrum of the modulus ϕ produced during the ekpyrotic phase may be transformed into a (nearly) scale-invariant and adiabatic spectrum of primordial metric fluctuations in the post-bounce, expanding phase (see also the alternative mechanism recently proposed in [74a]). Such a spectrum is expected to be virtually indistinguishable from that generated by a conventional inflationary mode, even if the physical mechanisms are very different in the two cases. Thus, one might wonder whether the cyclic scenario can be observationally discriminated by other models of inflation. The answer is yes, thanks to the differences arising in the spectrum of tensor metric perturbations.

In fact, as discussed in Chapter 7, conventional (e.g. slow-roll) models of inflation predict a flat (or slightly red) spectrum of primordial gravitational waves, which could substantially contribute to the low multipoles of the large-scale CMB anisotropy, and/or cause distinct signals in the CMB polarization (see the discussions of Sections 7.3 and 8.2). In the cyclic scenario, on the contrary, the produced background of cosmic gravitational waves [63] is strongly suppressed at the low-frequency scales relevant to the CMB anisotropy (see Eq. (7.216)), and unable to produce detectable signals. Thus, near-future astrophysical observations (to be performed, for instance, by the Planck satellite) might help in discriminating between the cyclic/ekpyrotic and the conventional inflationary scenarios.

It should be stressed, however, that the complete absence of detectable signals, induced by tensor perturbations on the large-scale CMB anisotropy, is not peculiar to the ekpyrotic picture but is also typical of other string cosmology models of inflation, such as the minimal pre-big bang models illustrated in Section 7.3. Thus, it is fair to say that the search for direct/indirect effects of the primordial gravitational radiation on the CMB radiation may be a crucial test for distinguishing *conventional* models of inflation from *string theory based* models of the early Universe, but not for distinguishing between different string cosmology models, in general.

For what concerns the ekpyrotic and pre-big bang scenarios, however, a clear observational discrimination could follow from *direct* measurements of the primordial gravitational radiation at the frequency scales typical of the present detectors. Indeed, as discussed in Sections 7.3 and 7.4, the relic background produced by pre-big bang models is, in principle, high enough to fall within the sensitivity range of (near-future) advanced detectors, while the ekpyrotic background is not.

Thus, a combined non-observation of tensor polarization effects on the CMB radiation and non-observation of relic gravitons by advanced detectors would be in favor of the ekpyrotic/cyclic scenario. Conversely, a combined non-observation of polarization effects and a direct observation of relic gravitons would be more in favor of the pre-big bang scenario. We recall, for comparison, that the expected signature of conventional inflationary models is a combined observation of tensor polarization effects and non-observation of the relic background. Near-future, cross-correlated observations of the electromagnetic and gravitational relic radiation backgrounds will thus give us important (and surprising?) information on our past cosmological history.

We conclude this section with a comment concerning the generalization of the ekpyrotic scenario to its extended, cyclic version. Such an extension is possible because in the ekpyrotic case, unlike in models of standard inflation, the big bang is not regarded as a very special, unique and singular event marking the beginning of everything (including space-time), but only as a transition (even if of dramatic importance, in various respects) between two different cosmological regimes, thus representing only one of the many stages of the full cosmological history. It is possible, therefore, that the same process is repeated whenever the appropriate initial conditions are reproduced.

The same conceptual approach – i.e. the big bang as a cosmological transition – also underlies the self-dual pre-big bang scenario, illustrated in Chapter 4. In that case, the pre-big bang phase marks the transition between two duality-related cosmological regimes, and is expected to be a (high-energy but) smooth and regular process, self-produced by the gravi-dilaton dynamics and, in principle, reproducible. Also in that case, therefore, we could speculate about a possible *cyclic* extension of the pre-big bang scenario: instead of a cyclic alternation of the phases of attraction and separation of two branes one would find a cyclic alternation of the phase of standard cosmological evolution and of its string theory dual.

For a more specific understanding of the possible cyclic extension of the pre-big bang scenario we should recall, at this point, that the rhythmic series of cycles, in the ekpyrotic case, is due to the fact that the distance modulus ϕ is traveling back and forth between the two ends $\phi = \phi_0$ and $\phi = -\infty$, across a suitable potential. In particular, ϕ is decreasing in the pre-collision phase, and growing in the post-collision phase. In the model of self-dual evolution the modulus is the dilaton, which evolves however in a monotonic way: it grows from $-\infty$ during the pre-big bang phase, and keeps growing (or becomes asymptotically frozen) even in the subsequent post-big bang regime. To implement a cyclic scenario (i.e. to reproduce the initial conditions, and prepare the system for a new big bang transition) the Universe should eventually exit from the late-time phase of standard (or quintessential, see Section 8.3) evolution. Thus, the dilaton should start decreasing, to lead the system back to the perturbative regime, at the beginning of the self-dual cycle.

An obvious possibility, in order to achieve the bounce back of the dilaton in the post-big bang regime, is the addition to the action of a suitable, duality-breaking, effective potential (as suggested by the example of the ekpyrotic scenario). Interestingly enough, however, a post-big bang configuration with an asymptotically decreasing dilaton can also be obtained by considering a higher-dimensional, anisotropic background in which the final decelerated expansion of the "external" dimensions is accompanied by the decelerated contraction of at least one "internal" dimension (see Eq. (4B.55)).

This suggests that a cyclic pre-big bang scenario should contain extra spatial dimensions undergoing a contracting phase, as in the ekpyrotic case. In that case, a cyclic alternation of duality-related anisotropic phases could even exchange, at each cycle, the expanding and contracting dimensions among themselves. In fact, this is what happens at the pre- to post-big bang transition in the example described by the analytical and numerical solutions presented in [77]: the simultaneous sign inversion of $\dot{a}_i$ for the various scale factors seems to be typical of anisotropic backgrounds characterized by a smooth bounce transition in the context of the low-energy string cosmology equations (see also the regular and bouncing boosted Milne background illustrated in Fig. 4.6).

10.5 Brane–antibrane inflation

In the ekpyrotic model of the previous section the branes are "domain walls" of a manifold with topology $\mathcal{M}_{10} \times S_1/Z_2$, and represent the space-time boundaries located at the fixed points of the orbifold S_1/Z_2. Our Universe (factorized as $\mathcal{M}_4 \times CY_6$, where CY is a Calabi–Yau manifold) is identified with one of these branes, and the standard cosmological phase is reproduced as a consequence of the collision of the two branes (without any previous epoch of standard inflationary expansion).

Besides the topological branes, generally present in higher-dimensional models of gravity, string theory contains another class of higher-dimensional extended objects, the so-called Dirichlet branes (or D_p-branes, where p is the dimensionality of their spatial extension). They are peculiar to string theory, being related to the choice of particular boundary conditions in the open string equations of motion (see Eq. (3A.27) and the subsequent discussion). From a geometric point of view they are associated with $(p+1)$-dimensional hypersurfaces Σ, spanned by the ends of open Dirichlet strings which are strictly confined on Σ and cannot move along the transverse $D - (p+1)$ directions. As the ends of open strings may carry non-Abelian charges, it follows that a D_p-brane might be a good candidate to represent a "world" in which the standard gauge interactions are localized in $p+1$ space-time dimensions, and are insensitive to any other "transverse" direction. In a superstring theory context these branes play a fundamental role in type I and type II models [78], and in establishing duality relations between different supersymmetric configurations.

An important property of D_p-branes is that they are coupled not only to gravity but also to the antisymmetric fields of the R–R sector, and the combined action of all their couplings generates an effective interaction among D_p-branes which can be easily computed at large distances. It turns out, in particular, that the net interaction is attractive if two identical branes have opposite R–R charges [16, 17], namely if they represent a configuration of a D_p-brane and an anti-D_p-brane (or $\overline{D}_p$-brane). Given such a pair of parallel branes, separated by a finite space-like distance, they will tend to approach each other moving through the bulk, and accelerating as they come closer and closer along the transverse direction (say, the fifth dimension).

On the other hand, as pointed out in Section 10.3, the motion in the bulk through a warped geometry may correspond to an expansion of the effective cosmological metric on the brane, and an accelerated motion may correspond to accelerated expansion, i.e. to inflation: the mechanism of brane–antibrane attraction has thus stimulated the study of models of $D_p - \overline{D}_p$-brane inflation [16, 17, 79] (see also [80] for a detailed introduction). In these models the accelerated evolution of the four-dimensional geometry is driven by the effective potential which controls the interbrane interaction, and which is a scalar function of the interbrane distance: one obtains, in this way, dimensionally reduced cosmological equations where the distance modulus plays the role of the inflaton. The qualitative picture is reminiscent of the ekpyrotic scenario (see Section 10.4), but the physics is very different, as the brane–antibrane interaction induces *inflationary expansion* (instead of ekpyrotic contraction).

For a brief discussion of this promising inflationary scenario we start by recalling that the interaction between two identical D_p-branes, characterized initially by a vacuum, static, supersymmetric BPS configuration, is identically vanishing: the gravi-dilaton attraction is compensated for by an opposite repulsion due to the exchange of R–R fields. More precisely, the computation of the interaction amplitude between the branes – due to the tree-level exchange of bulk closed strings or, equivalently, to the one-loop exchange of open strings (see for instance [78, 80]) – leads to an exact cancelation between the contributions of the R–R and NS–NS sectors, because of the highly supersymmetric state of the system. Thus, the system is and remains static (unless one introduces additional non-perturbative interactions, as in the case of the ekpyrotic scenario [11]).

One way to avoid this conclusion is to consider a less symmetric configuration in which supersymmetry is broken, and the axion, the dilaton and the R–R fields become massive, while gravity remains long-ranged. In that case the previous cancelation of forces no longer holds, and a non-zero interaction potential develops which is generally attractive and which, at large distances compared with the string length, should take the form [15]

$$V(Y) \simeq \Lambda + \frac{k}{Y^{n-2}} \left(1 + \sum_{NS} e^{-m_{NS} Y} - \sum_{R} e^{-m_R Y} \right). \tag{10.184}$$

Here Λ is a constant vacuum contribution due to the tension of the two branes, k is a model-dependent constant, Y is the separation of the branes, and n is the number of transverse dimensions (for instance, $n = 6$ for a D_3-brane embedded in the $D = 10$ bulk space-time of superstring theory).

In the massless (supersymmetric) limit the coefficients in round brackets (i.e. the total contribution of the R–R and NS–NS sectors) sum to zero, and the Y-dependent part of the potential vanishes. But when the fields are massive the form of the potential may be appropriate (i.e. sufficiently flat) to lead to inflation, and to satisfy naturally the slow-roll conditions [15]. However, the explicit computation of the potential requires a specific model of supersymmetry breaking (not supplied in [15]); also, the potential has to be completed by the addition of string corrections which appear when $Y \sim \lambda_s$, and which are expected to be important to describe the end of inflation and the regime of brane collision.

Another (probably more promising) possibility for developing a complete model of D_p-brane inflation is based on the interactions of a brane–antibrane pair. Their mutual interaction is non-zero because, for the $D_p - \bar{D}_p$ pair, the forces due to the exchange of NS–NS fields remain attractive, while those due to the R–R sector change sign, and the two contributions add to give naturally a non-zero attractive interaction [16, 17].

The effective potential for the brane–antibrane system can be computed perturbatively, starting from the action

$$S = S_{\text{bulk}} + S_{D_p} + S_{\bar{D}_p}, \tag{10.185}$$

where the 10-dimensional bulk action is specified by the choice of the superstring model, and the $(p+1)$-dimensional brane action, with $p \geq 3$, is obtained by expanding the DBI action (6.52):

$$S_{D_p} = - \int_{\Sigma} d^{p+1} \xi \sqrt{|\gamma|} \left(T_p + \cdots \right), \tag{10.186}$$

where the ellipsis denotes other possible fields interacting with the brane. Here $T_p \sim \lambda_s^{-(p+1)} \exp(-\phi/2)$ is the tension of the D_p-brane, and the metric $\gamma_{\mu\nu}^i = \partial_\mu X_i^A \partial_\nu X_i^B g_{AB}$ is specified by the embedding functions $X_i^A(\xi)$ (conventions: Greek indices run from 0 to p, and the subscript $i = 1, 2$ refers to D_p and $\bar{D}_p$, respectively).

Assuming the branes to be parallel we set

$$X_i^A = (\xi^\mu, X_i^m), \tag{10.187}$$

where X_i^m are the coordinates of the brane positions along the transverse directions ($m, n = p+1, p+2, \ldots, 9$); we can then separate the relative transverse motion of the branes, parametrized by $Y^m = (X_1^m - X_2^m)$, by the motion of their center of mass, parametrized by $\overline{X}^m = (X_1^m + X_2^m)$, by setting

$$X_1^m = \frac{1}{2}(\overline{X}^m + Y^m), \qquad X_2^m = \frac{1}{2}(\overline{X}^m - Y^m). \tag{10.188}$$

The computation of $\gamma_{\mu\nu}$ then gives

$$\gamma_{\mu\nu}^1 = h_{\mu\nu} + \frac{1}{4}\partial_\mu Y^m \partial_\nu Y^n g_{mn} + \frac{1}{4}\partial_\mu \overline{X}^m \partial_\nu \overline{X}^n g_{mn} + \frac{1}{2}\partial_\mu \overline{X}^m \partial_\nu Y^n g_{mn},$$

$$\gamma_{\mu\nu}^2 = h_{\mu\nu} + \frac{1}{4}\partial_\mu Y^m \partial_\nu Y^n g_{mn} + \frac{1}{4}\partial_\mu \overline{X}^m \partial_\nu \overline{X}^n g_{mn} - \frac{1}{2}\partial_\mu \overline{X}^m \partial_\nu Y^n g_{mn},$$

$$\tag{10.189}$$

where $h_{\mu\nu} = \delta_\mu^A \delta_\nu^B g_{AB}$, and we have assumed that the bulk metric has a block structure, choosing a frame where $g_{\mu n} = 0$.

We are interested, in particular, in the relative transverse motion of the branes, i.e. in the dynamics of the modulus Y^m. Inserting the above results into the action (10.186), summing the two brane actions, and expanding $\sqrt{|\gamma|}$ to the lowest non-trivial order in powers of $\partial_\mu Y^m$, we obtain

$$S_{D_p} + S_{\overline{D}_p} = -\int d^{p+1}\xi \sqrt{|h|}\, T_p \left(2 + \frac{1}{4}h^{\mu\nu}\partial_\mu Y^m \partial_\nu Y^n g_{mn} + \cdots \right), \tag{10.190}$$

where $h = \det h_{\mu\nu}$. The above action specifies the kinetic term for the moduli Y^m, and can be used to fix their canonical normalization (see below).

To compute the interaction potential we assume, for simplicity, that the brane–antibrane system is initially sitting in a highly symmetric BPS configuration, and we choose a frame with one spatial axis oriented along the direction normal to the brane, so that $Y^m = (Y, 0, 0, \ldots)$. Let us also assume, for the moment, that the geometry is flat along the transverse directions, so that $g_{mn} = -\delta_{mn}$, and

$$\partial_\mu Y^m \partial^\mu Y^n g_{mn} = -\partial_\mu Y \partial^\mu Y. \tag{10.191}$$

The calculation of the one-loop amplitude for the open string exchange between D_p and $\overline{D}_p$, in the limit of large separations $Y \gg \lambda_s$, then leads to the effective (Coulomb-like) potential [16, 17]:

$$V_{D-\overline{D}} = -e^\phi \lambda_s^8 T_p^2 \frac{\beta}{Y^{n-2}}, \tag{10.192}$$

where $e^\phi \lambda_s^8 \equiv 8\pi G_{10}$ is the effective gravitational coupling appearing in the 10-dimensional bulk action, $n = 10 - (p+1)$ is the number of spatial dimensions transverse to the brane, and β is a dimensionless number of order one (which can be computed exactly as a function of n). We may also note that, for a system of D_3-branes, $e^\phi \lambda_s^8 T_p^2 \sim 1$, since $T_3 \sim g_s^{-1}(\alpha')^{-2} \sim e^{-\phi/2}\lambda_s^{-4}$.

We are now in the position of completing the action (10.190) by adding the above interaction term, and integrating over the $p - 3$ extra spatial dimensions in order to obtain a four-dimensional effective action. We assume that the $p - 3$ "parallel" dimensions (along the brane) and the n "orthogonal" directions (transverse to the brane) are compact (even if not necessarily small), and that the moduli describing their shape and size (as well as the dilaton modulus) have been stabilized by some appropriate mechanism (see the discussion at the end of the section). In this case we can treat $T_p(\phi)$ and the volume of the space parallel, $V_\parallel$, and orthogonal, $V_\perp$, to the brane, as constant parameters: the volumes, in particular, can be expressed in terms of their typical length scales, $r_\parallel$ and $r_\perp$, respectively, as

$$V_\parallel = r_\parallel^{p-3}, \qquad V_\perp = r_\perp^n, \qquad n = 9 - p. \tag{10.193}$$

Note that these parameters are not independent, for consistency with the four-dimensional mass scale M_P to which the volumes are related by the process of dimensional reduction. They must satisfy the constraint (10.5) which, in this case, reduces to

$$e^\phi M_P^2 = M_s^8 V_\perp V_\parallel. \tag{10.194}$$

Inserting the interaction (10.192), integrating over the parallel dimensions, and normalizing the kinetic term in canonical form, we obtain from Eq. (10.190) the following dimensionally reduced effective action,

$$S = \int d^4\xi \sqrt{|g|} \left[\frac{1}{2}\partial_\mu \psi \, \partial^\mu \psi - V(\psi) \right], \tag{10.195}$$

where

$$\psi = \left(\frac{T_p V_\parallel}{2} \right)^{1/2} Y, \qquad V(\psi) = A - B\psi^{2-n}, \tag{10.196}$$

and

$$A = 2T_p V_\parallel,$$

$$B = \beta e^\phi \lambda_s^8 T_p^2 V_\parallel \left(\frac{T_p V_\parallel}{2} \right)^{(n-2)/2}. \tag{10.197}$$

We have thus introduced a canonical scalar field, self-interacting, and minimally coupled to gravity, and we can now study its possible ability to sustain a phase of long and efficient inflation. The situation is strongly reminiscent of a brane-world cosmological scenario with the inflaton scalar field localized on the brane (see Section 10.3), but with two important differences.

The first is that the scalar field has a precise physical/geometric identification as the distance modulus of the pair D_p–$\overline{D}_p$. As a consequence, its properties – in particular, the slope and the amplitude of its potential – cannot be arbitrarily prescribed, but are to be computed from the given string theory model. The second difference is that the present scenario, as also noted in [17], permits two possible interpretations and applications. The scalar field action (10.195) may be regarded either as an effective gravitational source for the dimensionally reduced bulk action (obtained by integrating over all six compact dimensions), or as a gravitational source localized on the brane which contains our four-dimensional Universe. In this second case, however, the gravitational equations on the brane should be obtained by projecting the bulk gravitational equations and imposing the appropriate junction conditions, as discussed in Section 10.1.

In the first case, one can instead directly apply the standard Einstein equations with the scalar field as an inflaton source. One then finds that is not easy for the obtained potential, in the model of background we have considered, to satisfy the conditions of slow-roll inflation. The computation of the slow-roll parameter η (see Section 1.2) gives, in fact,

$$\eta \equiv \frac{M_P^2}{V} \frac{\partial^2 V}{\partial \psi^2} \simeq -\beta(n-1)(n-2) \left(\frac{r_\perp}{Y}\right)^n, \qquad (10.198)$$

where we have used the definitions (10.196) and (10.197), and the constraint (10.194). The brane–antibrane potential (10.192), on the other hand, is valid for transverse separations which are larger than the string length λ_s, but which cannot be larger than the proper size of the compact transverse dimensions, i.e. $Y \lesssim r_\perp$. Thus, the slow-roll condition $|\eta| \ll 1$ would require $\beta \ll 1$, which cannot be satisfied by any realistic string theory model, as string theory implies $\beta \sim 1$ for a pair D_p–$\overline{D}_p$ of interacting BPS branes [80].

A possible solution to this problem has been suggested in [16] for the case in which the compact transverse manifold is topologically an n-dimensional torus, with uniform circumference of size $r_\perp$. When the antibrane is separated by a distance $Y \sim r_\perp$ one must include in the interbrane potential the contributions of all the topological "images" of the other brane, forming the n-dimensional lattice $(R/Z)^n$. Each contribution gives a term of the form (10.192), and the total effective interaction is obtained by summing over all the lattice sites occupied by the brane images. The resulting potential is quartic in the displacement z of the antibrane from the center of the hypercubic cell of the lattice, and can satisfy the required slow-roll conditions as long as the interbrane separation remains in the range $Y \sim r_\perp$.

When $Y \ll r_\perp$ inflation stops since there are no longer contributions from the lattice images, and the potential reduces to the form (10.196). The interaction, however, remains attractive, so that the system D_p–$\overline{D}_p$ keeps collapsing. When the limiting separation $Y \sim \lambda_s$ is reached, the large distance approximation breaks down, and the potential acquires corrections due to the exchange of an open string state, T, which has a tachyonic mass for $Y < \lambda_s$ [81]. Including such a tachyon condensate, the interaction then takes the approximate (short distance) form,

$$V(T, Y) \sim \frac{1}{\lambda_s^2} \left(\frac{Y^2}{\lambda_s^2} - 1 \right) T^2 + c\, T^4 + \cdots \tag{10.199}$$

(where c is a constant), which should be appropriate to describe the final process of brane–antibrane collision/annihilation. Remarkably, the above potential (as a function of T and Y) is precisely of the form required by the so-called models of "hybrid" inflation [82], which automatically include a mechanism to exit from the inflationary phase.

Using the physics of D_p–$\overline{D}_p$ interactions, string theory seems thus to be able to suggest a complete and satisfactory model of standard inflation, in which the inflaton potential naturally evolves from the initial slow-roll regime (associated with a period of weak brane attraction) to the final hybrid regime (associated with tachyonic instability and brane annihilation). The considered model, however, contains a major assumption concerning the stabilization of the dilaton and of the volume moduli of the compact dimensions. In addition, the slow-roll condition is satisfied for special initial conditions, and for a special choice of the topology of the transverse dimensions.

A different possibility to ensure the validity of the slow-roll approximation, in the context of D_p–$\overline{D}_p$ interactions, is based on the presence of a warped bulk geometry [18]. There are various valid motivations for choosing such a configuration: for instance, a warped geometry may also help in stabilizing moduli, thus avoiding the main difficulty mentioned above. Furthermore, a warped compactification of the extra dimensions naturally admits solutions with D_3-branes and $\overline{D}_3$-branes transverse to the six compact dimensions [83], thus providing a natural arena for the mechanism of brane–antibrane inflation.

The influence of a warped geometry on the effective interaction of the $D_p - \overline{D}_p$ pair has been studied in detail for the case of a five-dimensional AdS bulk manifold [18], where the $\overline{D}_3$-brane is held fixed in the asymptotic regime of small proper volumes of the AdS_5 background, while the D_3-brane is free to move through the curved bulk, driven by the attractive force towards the $\overline{D}_3$-brane. The resulting effective potential describing this interaction is much flatter than the potential between the same brane-antibrane pair in flat (compact) space.

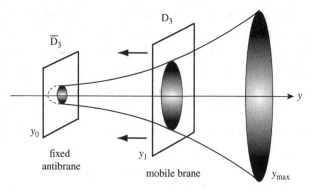

Figure 10.3 Schematic view of the warped background (10.200). The mobile D_p-brane is attracted towards the fixed $\overline{D}_p$-brane located at the infrared end y_0.

For an illustration of this possibility we start by assuming that no other moduli, or the effects which stabilize them, interfere with the relative motion of the two branes (unfortunately this assumption cannot be easily satisfied, as we shall discuss later). Also, we consider a 10-dimensional background solution of the type IIB superstring action which, in a suitable region of the target space manifold, can be approximated by the factorized background $AdS_5 \times \mathcal{M}_5$, where $\mathcal{M}_5$ is a five-dimensional Einstein space of constant curvature. We use for AdS_5 the convenient parametrization

$$ds^2 = \frac{y^2}{R^2} \left(dt^2 - dx_i^2 \right) - \frac{R^2}{y^2} dy^2, \tag{10.200}$$

where the coordinate y parametrizes the fifth dimension, and R^{-1} is the AdS curvature scale (assumed to be sufficiently small, i.e. $R \gg \lambda_s$, for the validity of a perturbative analysis of the background based on the low-energy effective action [18]).

Let us suppose, finally, that the solution (10.200) is valid for $y_0 \leq y \leq y_{max}$: for $y > y_{max}$ the background is expected to evolve smoothly into a different (Calabi–Yau) type of compactification, while for $y \sim y_0$ the AdS geometry is expected to have a smooth tip of finite size (as in the case of the Klebanov–Strasser solution [84]), which can be modeled by cutting off the coordinate y at some minimum value y_0. The "warp factor" $(y/R)^2$ – representing the effective gravitational redshift (due to the bulk curvature) of the four-dimensional space-time sections spanned by the coordinates (t, x_i) – thus decreases with y down to the minimum value $y_0 \ll R$ associated with the fixed position of the $\overline{D}_p$-brane, which is placed just at the "infrared" end of the background (see Fig. 10.3).

In order to specify the brane dynamics we call y_1 the position of the brane mobile along the fifth dimension, assuming $y_1 \gg y_0$. The interbrane distance is then $Y = y_1 - y_0 \simeq y_1$. The kinetic term is again of the type (10.190), with $p = 3$ and Y replaced by y_1. The attractive potential, taking into account that for $p = 3$ there are $n = 6$ dimensions transverse to the brane, will behave as $V(Y) \sim -Y^{-4} \sim -y_1^{-4}$, according to Eq. (10.192). The computation of $V(Y)$, however, must take into account that the interaction takes place in the curved background described by the metric (10.200): the energy density to be inserted into the action has to be written in a covariant form. This prescription has two consequences: (*i*) the coordinate distance has to be replaced by the proper distance, $y_1 \to y_1 \sqrt{-g_{yy}} = y_1 (R/y)$; (*ii*) the scalar potential must become a scalar density, $V \to V \sqrt{-g} = V(y/R)^4$. As a consequence, the effective interaction energy experienced by the $\overline{D}_p$-brane located at y_0 will take the form

$$V(y_1, y_0) \simeq - \left(\frac{y_0}{R} \right)^8 \frac{1}{y_1^4}. \tag{10.201}$$

The strong suppression factor $(y_0/R)^8 \ll 1$ is what makes this potential easily compatible with the slow-roll condition, as we shall see in a moment.

A more precise computation of the interbrane potential can be performedd by evaluating the perturbation of the background produced by the brane located at $y = y_1$, and then computing the resulting energy of the antibrane fixed at $y = y_0$ in this perturbed background. We sketch here the main steps of this procedure, referring to the original paper [18] for further details.

We start by recalling that, in the background we are considering, the curvature of the 10-dimensional space-time $AdS_5 \times \mathcal{M}_5$ is sourced by the stress tensor of a five-form field strength, $F_5 = dA_4$, and that the complete set of background fields can be parametrized in terms of a scalar function $h(y)$ as follows:

$$ds_{10}^2 = h^{-1/2}(y) \left(dt^2 - dx_i^2 \right) - h^{1/2}(y) \left(dy^2 + \frac{y^2}{R^2} g_{ab} \, dz^a \, dz^b \right), \tag{10.202}$$

$$F_{01234} = \partial_y h^{-1}, \qquad A_{0123} = h^{-1}.$$

Here g_{ab} is the metric on $\mathcal{M}_5$, spanned by the coordinates z^a, $a = 5, 6, \ldots, 9$, and h satisfies the equation $\nabla_6^2 h = 0$, where ∇_6^2 is the covariant Laplace operator of the six-dimensional space with metric

$$ds_6^2 = dy^2 + \frac{y^2}{R^2} g_{ab} \, dz^a \, dz^b. \tag{10.203}$$

The unperturbed background is described by the exact solution $h_0(y) = (R/y)^4$, which leads to the metric (10.200). The perturbation sourced by a D_3-brane transverse to the directions spanned by the coordinates $\{y, z^a\}$, and localized at the origin of the $\mathcal{M}_5$ coordinates, i.e. at $y = y_1$, $z^a = 0 = \text{const}$, can still be parametrized in the form (10.202), by setting $h = h_0 + h_1$. One then finds that h_1 must satisfy the source equation $\nabla_6 h_1 = c\delta(y - y_1)\delta^5(z)$, where c is a constant depending on the tension T_3. The solution of this equation [18] is given by $h_1 = (2\pi^2 T_3 y_1^4)^{-1}$, so that the perturbed background (including the D_3-brane contribution), is described by the total warp factor

$$h(y) = h_0 + h_1 = \frac{R^4}{y^4} + \frac{1}{2\pi^2 T_3 y_1^4}. \tag{10.204}$$

Let us now consider the interaction of the $\overline{D}_3$-brane with this perturbed background. The total antibrane action is given by the DBI action (10.186), which describes the interaction with the metric, plus the Chern–Simons action which describes the interaction with the four-form potential A_4 [78]. As the antibrane is fixed at $y = y_0$ there is no kinetic term and we have, for $p = 3$, that $\sqrt{-\gamma} = h^{-1}(y_0) = A_{0123}(y_0)$, so that

$$S_{\overline{D}_3} = S_{DBI} + S_{CS} = -T_3 \int d^4 x \sqrt{-\gamma} - T_3 \int A_{0123} \, dt \, d^3 x$$

$$= -2T_3 \int d^4 x \, h^{-1}(y_0). \tag{10.205}$$

The D_3-brane has the same tension but opposite five-form charge: thus, there is an exact cancelation between the two interaction terms above, and we are left with the kinetic term only (see Eq. (10.190)):

$$S_{D_3} = S_{DBI} + S_{CS} = T_3 \int d^4 x \, \frac{1}{4} \partial_\mu y_1 \, \partial^\mu y_1. \tag{10.206}$$

Expanding $h^{-1}(y_0)$ for $y_0 \ll y_1$ (and, of course, for $T_3 R^4 \gg 1$), and summing $S_{\overline{D}_3}$, S_{D_3}, we obtain for the brane system the following four-dimensional effective action:

$$S = \int d^4 x \left[\frac{T_3}{4} \partial_\mu y_1 \, \partial^\mu y_1 - 2T_3 \left(\frac{y_0}{R}\right)^4 + \frac{1}{\pi^2} \left(\frac{y_0}{R}\right)^8 \frac{1}{y_1^4} \right] \tag{10.207}$$

(note that the attractive interaction term is in agreement with the previous estimate (10.98)).

Defining, as before, the canonical field $\psi = y_1 (T_3/2)^{1/2}$, we are led eventually to the warped potential

$$V(\psi) = 2T_3 \left(\frac{y_0}{R}\right)^4 \left[1 - \frac{T_3}{8\pi^2} \left(\frac{y_0}{R}\right)^4 \psi^{-4}\right], \tag{10.208}$$

differing from the previous result obtained in flat transverse space for the large suppression factor $(y_0/R)^4$ of the second term with respect to the constant part of the potential (compare with Eq. (10.197)). The most restrictive slow-roll parameter, in this case, is given by

$$\eta = \frac{M_P^2}{V} \frac{\partial^2 V}{\partial \psi^2} \simeq -\frac{5}{2\pi^2} M_P^2 T_3 \left(\frac{y_0}{R}\right)^4 \psi^{-6}. \tag{10.209}$$

On the other hand, the number of e-folds between a given time t and the end of inflation t_f, according to Eq. (1.119), is given by

$$N = \int_t^{t_f} H \, dt = \frac{1}{M_P^2} \int_{\psi_f}^{\psi} \frac{V}{V'} \, d\psi \simeq \frac{\pi^2}{3M_P^2 T_3} \left(\frac{R}{y_0}\right)^4 \psi^6. \qquad (10.210)$$

It follows that $|\eta| \simeq N^{-1}$, and that a sufficiently large number of e-folds (say, $N \gtrsim 60$) automatically guarantees the condition $|\eta| \ll 1$ for the slow-roll approximation to be valid.

Brane–antibrane interactions in a warped background thus seem to provide a promising scenario for the formulation of efficient models of slow-roll inflation. The example we have discussed, however, was based on a crucial assumption: the stabilization of all the moduli (except y) of the compactification manifold. As discussed in [18], this is a highly non-trivial issue. For instance, if one applies the stabilization mechanism illustrated in [85], one finds that generic volume-stabilizing superpotentials also generate a large mass term for the inflaton field, making inflation impossible. In more general models, the (non-perturbative) stabilizing superpotential could depend on both the volume modulus and the inflaton (i.e. the interbrane distance modulus). If such a dependence is generic, however, inflation does not occur, as shown in [18].

In spite of these difficulties it is possible to find "non-generic" examples where inflation is possible, but then the inflationary predictions are strongly dependent on the details of the stabilization mechanism. However, the degree of fine-tuning required to implement slow-roll inflation, in these examples, seems to be low (of order 1%), and particularly acceptable in models of eternal inflation characterized by indefinitely large and ever growing volume of the inflationary domain [18]. Furthermore, in view of the many possible existing realizations of string theory at low energies (also called string theory "vacua" [86]), most of which are largely in contrast with standard phenomenology, one could even argue that all string theory realizations not leading to inflation should be discarded, because they are incompatible with the Universe in which we live (thus adapting to cosmology the proposed anthropic approach to the "landscape" problem of string theory [87]).

We can say, in summary, that the physics of branes has produced (and is still producing) novel approaches to the picture of a multi-dimensional Universe, new interesting ideas in the field of primordial cosmology and, in particular, new possible schemes for the realization of the inflationary scenario. It also seems fair to say, however, that brane cosmology, and brane inflation above all, are probably still in their infancy: many points need clarification, and many problems are still to be solved. Thus, let me exhort the reader not to be scared of the present difficulties, and to address the large existing literature for a deeper study of this field (which, after the initial "explosion" at the end of the last century, is now evolving at a constant and continuous rate).

After all, brane cosmology models – in various ways and at various levels – are all deeply rooted in string theory, which is at present our best candidate for a unified theory of all fundamental interactions. Thus, brane cosmology offers us two possibilities. On one hand, through a top-down approach, it gives string theory the opportunity to formulate phenomenological predictions, to be directly compared with present (or near-future) observations. On the other hand, through a bottom-up approach, and through the input of observational data more and more abundant and accurate, it stimulates a more and more detailed study of string theory, boosting the understanding of many obscure aspects still present in this theory.

References

[1] T. Kaluza, *Sitzungsber. Preuss. Akad. Wiss. Berlin* **1921** (1921) 966.

[2] O. Klein, *Z. Phys.* **37** (1926) 895.

[3] N. Arkani Hamed, S. Dimopoulos and G. R. Dvali, *Phys. Lett.* **B429** (1998) 263.

[4] I. Antoniadis, *Phys. Lett.* **B246** (1990) 377.

[5] I. Antoniadis, *J. Phys. Conf. Ser.* **33** (2006) 170.

[6] J. Maharana and J. H. Schwarz, *Nucl. Phys.* **B390** (1993) 3.

[7] E. G. Adelberg, B. R. Heckel and A. E. Nelson, *Ann. Rev. Nucl. Part. Sci.* **53** (2003) 77.

[8] V. A. Rubakov and M. Shaposhnikov, *Phys. Lett.* **B125** (1983) 136.

[9] L. Randall and R. Sundrum, *Phys. Rev. Lett.* **83** (1999) 4960.

[10] G. R. Dvali, G. Gabadadze and M. Porrati, *Phys. Lett.* **B485** (2000) 208.

[11] J. Khoury, B. A. Ovrut, P. J. Steinhardt and N. Turok, *Phys. Rev.* **D64** (2001) 123522.

[12] J. Khoury, B. A. Ovrut, N. Seiberg, P. J. Steinhardt and N. Turok, *Phys. Rev.* **D65** (2002) 086007.

[13] P. J. Steinhardt and N. Turok, *Phys. Rev.* **D65** (2002) 126003.

[14] P. J. Steinhardt and N. Turok, *Science* **296** (2002) 1436.

[15] G. Dvali and S. Tye, *Phys. Lett.* **B450** (1999) 72.

[16] C. Burgess, M. Majumdar, D. Nolte *et al.*, *JHEP* **0107** (2001) 047.

[17] G. Dvali, Q. Shafi and S. Solganik, hep-th/0105203.

[18] S. Kachru, R. Kallosh, A. Linde *et al.*, *JCAP* **0310** (2003) 013.

[19] W. Israel, *Nuovo Cimento* **B44** (1966) 1.

[20] W. Israel, *Nuovo Cimento* **B48** (1967) 463.

[21] H. A. Chamblin and H. S. Reall, *Nucl. Phys.* **B562** (1999) 133.

[22] T. Shiromizu, K. Maeda and M. Sasaki, *Phys. Rev.* **D62** (2000) 024012.

[23] M. J. Duff, *Class. Quantum Grav.* **5** (1988) 189.

[24] V. Bozza, M. Gasperini and G. Veneziano, *Nucl. Phys.* **B619** (2001) 191.

[25] M. Cavaglià, G. De Risi and M. Gasperini, *Phys. Lett.* **B610** (2005) 9.

[26] R. Maartens, *Living Rev. Rel.* **7** (2004) 1.

[27] R. M. Wald, *General Relativity* (Chicago: University of Chicago Press, 1984).

[28] C. Csaki, M. Graesser, C. F. Kolda and J. Terning, *Phys. Lett.* **B462** (1999) 34.

[29] J. M. Cline, C. Grojean and G. Servant, *Phys. Rev. Lett.* **83** (1999) 4245.

[30] R. Maartens, *Phys. Rev.* **D62** (2000) 084023.

[31] C. Csaki, in *Boulder 2002, TASI Lectures on Particle Physics and Cosmology*, eds. H. E Haber and A. E. Nelson (Singapore: World Scientific, 2004), p.605.

[32] M. Abramowicz and I. A. Stegun, *Handbook of Mathematical Functions* (New York: Dover, 1927).
[33] V. A. Rubakov, *Phys. Usp.* **44** (2001) 871.
[34] J. Garriga and T. Tanaka, *Phys. Rev. Lett.* **84** (2000) 2778.
[35] A. D. Sakharov, *Sov. Phys. Dokl.* **12** (1968) 1040.
[36] A. Zee, *Phys. Rev. Lett.* **42** (1979) 417.
[37] S. L. Adler, *Phys. Rev. Lett.* **44** (1980) 1567.
[38] P. Binetruy, C. Deffayet and D. Langlois, *Nucl. Phys.* **B565** (2000) 269.
[39] P. Binetruy, C. Deffayet, U. Ellwanger and D. Langlois, *Phys. Lett.* **B477** (2000) 285.
[40] P. Kraus, *JHEP* **9912** (1999) 011.
[41] D. Langlois, R. Maartens and D. Wands, *Phys. Rev. Lett.* **88** (2000) 259.
[42] C. Grojean, F. Quevedo, G. Tasinato and I. Zavala, *JHEP* **0108** (2001) 005.
[43] S. Mukohyama, T. Shiromizu and K. Maeda, *Phys. Rev.* **D62** (2000) 024028.
[44] N. Harkani-Hamed, S. Dimopoulos, N. Kaloper and R. Sundrum, *Phys. Lett.* **B480** (2000) 193.
[45] S. Kachru, M. B. Schulz and E. Silverstein, *Phys. Rev.* **D62** (2000) 045021.
[46] B. Brax, C. van der Bruck and A. C. Davis, *Rep. Prog. Phys.* **67** (2004) 2183.
[47] D. Langlois, in *Dublin 2004, General Relativity and Gravitation*, eds. P. Florides, B. Nolan and A. Ottewill (Singapore: World Scientific, 2005), p. 63.
[48] J. Garriga and M. Sasaki, *Phys. Rev.* **D62** (2000) 043523.
[49] D. Langlois, R. Maartens and D. Wands, *Phys. Lett.* **B489** (2000) 259.
[50] R. Maartens, D. Wands, B. A. Basset and I. Head, *Phys. Rev.* **D62** (2000) 041301.
[51] E. J. Copeland, A. R. Liddle and J. E. Lisey, *Phys. Rev.* **D64** (2001) 023509.
[52] R. Dick, *Class. Quantum Grav.* **18** (2001) R1.
[53] C. Deffayet, *Phys. Lett.* **B502** (2001) 199.
[54] C. Deffayet, G. R. Dvali and G. Gabadadze, *Phys. Rev.* **D65** (2002) 044023.
[55] M. A. Luty, M. Porrati and R. Rattazzi, *JHEP* **0309** (2003) 029.
[56] A. Nicolis and R. Rattazzi, *JHEP* **0406** (2004) 059.
[57] J. Khoury, B. A. Ovrut, P. J. Steinhardt and N. Turok, *Phys. Rev.* **D66** (2002) 046005.
[58] P. Horawa and E. Witten, *Nucl. Phys.* **B460** (1996) 506.
[59] P. Horawa and E. Witten, *Nucl. Phys.* **B475** (1996) 94.
[60] E. Witten, *Nucl. Phys.* **B471** (1996) 135.
[61] A. Lukas, B. A. Ovrut, K. S. Stelle and D. Waldram, *Phys. Rev.* **D59** (1999) 086001.
[62] A. Lukas, B. A. Ovrut, K. S. Stelle and D. Waldram, *Nucl. Phys.* **B552** (1999) 246.
[63] L. A. Boyle, P. J. Steinhardt and N. Turok, *Phys. Rev.* **D69** (2004) 127302.
[64] R. Y. Donagi, J. Khoury, B. A. Ovrut, P. J. Steinhardt and N. Turok, *JHEP* **0111** (2001) 041.
[65] H. Liu, G. Moore and N. Seiberg, *JHEP* **0206** (2002) 045.
[66] A. J. Tolley and N. Turok, *Phys. Rev.* **D66** (2002) 106005.
[67] G. T. Horowitz and J. Polchinski, *Phys. Rev.* **D66** (2002) 103512.
[68] S. Rasanen, *Nucl. Phys.* **B626** (2002) 183.
[69] R. Durrer and F. Vernizzi, *Phys. Rev.* **D66** (2002) 083503.
[70] C. Cartier, R. Durrer and E. J. Copeland, *Phys. Rev.* **D67** (2003) 103517.
[71] A. J. Tolley, N. Turok and P. J. Steinhardt, *Phys. Rev.* **D69** (2004) 106005.
[72] S. Gratton, J. Khoury, P. J. Steinhardt and N. Turok, *Phys. Rev.* **D69** (2004) 103505.
[73] M. Gasperini, M. Giovannini and G. Veneziano, *Nucl. Phys.* **B694** (2004) 206.

[74] L. E. Allen and D. Wands, *Phys. Rev.* **D70** (2004) 063515.

[75] J. Lehners, P. McFadden, N. Turok and P. J. Steinhardt, hep-th/0702153.

[76] N. Turok and J. P. Steinhardt, *Physica Scripta* **T117** (2005) 76.

[77] G. De Risi and M. Gasperini, *Phys. Lett.* **B521** (2001) 335.

[78] J. Polchinski, *String Theory* (Cambridge: Cambridge University Press, 1998).

[79] S. Alexander, *Phys. Rev.* **D65** (2002) 023507.

[80] F. Quevedo, *Class. Quantum Grav.* **19** (2002) 5721.

[81] A. Sen, *JHEP* **9808** (1998) 012.

[82] A. Linde, *Phys. Rev.* **D49** (1994) 748.

[83] S. Giddings, S. Kachru and J. Polchinski, *Phys. Rev.* **D66** (2002) 106006.

[84] I. Klebanov and M. J. Strasser, *JHEP* **0008** (2000) 052.

[85] S. Kachru, R. Kallosh, A. Linde and S. P. Trivedi, *Phys. Rev.* **D68** (2003) 046005.

[86] S. Ashrok and M. Douglas, *JHEP* **0401** (2004) 060.

[87] L. Susskind, hep-th/0302219.

Index